BENZOFURANS

This is the Twenty-Ninth Volume in the Series

THE CHEMISTRY OF HETEROCYCLIC COMPOUNDS

THE CHEMISTRY OF HETEROCYCLIC COMPOUNDS

A SERIES OF MONOGRAPHS

ARNOLD WEISSBERGER and EDWARD C. TAYLOR

Editors

BENZOFURANS

AHMED MUSTAFA

CHEMISTRY DEPARTMENT
CAIRO UNIVERSITY

AN INTERSCIENCE ® PUBLICATION

JOHN WILEY & SONS
NEW YORK • LONDON • SYDNEY • TORONTO

An Interscience® Publication

Library of Congress Cataloging in Publication Data:

Mustafa, Ahmed.
Benzofurans.

(The Chemistry of heterocyclic compounds, v. 29)
Includes bibliographical references.
1. Benzofuran. I. Title.

QD405.M92 547'.592 73-4780

ISBN O-471-38207-8

Printed in the United States of America

10 9 8 7 6 5 4 3 2 1

The Chemistry of Heterocyclic Compounds

The chemistry of heterocyclic compounds is one of the most complex branches of organic chemistry. It is equally interesting for its theoretical implications, for the diversity of its synthetic procedures, and for the physiological and industrial significance of heterocyclic compounds.

A field of such importance and intrinsic difficulty should be made as readily accessible as possible, and the lack of a modern detailed and comprehensive presentation of heterocyclic chemistry is therefore keenly felt. It is the intention of the present series to fill this gap by expert presentations of the various branches of heterocyclic chemistry. The subdivisions have been designed to cover the field in its entirety by monographs which reflect the importance and the interrelations of the various compounds, and accommodate the specific interests of the authors.

In order to continue to make heterocyclic chemistry as readily accessible as possible new editions are planned for those areas where the respective volumes in the first edition have become obsolete by overwhelming progress. If, however, the changes are not too great so that the first editions can be brought up-to-date by supplementary volumes, supplements to the respective volumes will be published in the first edition.

ARNOLD WEISSBERGER

Research Laboratories
Eastman Kodak Company
Rochester, New York

EDWARD C. TAYLOR

Princeton University
Princeton, New Jersey

Preface

The benzofurans occupy a prominent position among the plant phenols, a group of organic substances of extraordinary variety and interest. Their study reveals a close structural and chemical interrelationship that appears to reflect a similarly close relationship in the processes by which these compounds are formed in plants.

The study of the oxygen heterocyclic constituents of plants has developed as a branch of organic chemistry, and it makes important and distinctive contributions to other fields, such as agriculture, medicine, and so forth.

Somewhat related groups of natural products are discussed in the last chapters of this book. The first group consists of a range of compounds formed by fungi. The fungal metabolites include some of the relatively few natural organic products containing chlorine, among them the interesting spirocyclic compound, griseofulvin. A survey of the chemistry of usnic acid, a lichen constituent known for more than a century, is especially timely now that the seal has been set on a long series of degradation studies and clear biogenetic implication is revealed.

The varied nomenclature was most confusing. In this book alternative nomenclature and numbering have been indicated so that no confusion would result. The synonyms can usually be found in "Konstitution und Vorkommen der Organischen Pflanzenstoffe", by W. Karrer; "The Chemistry of Flavonoid Compounds", by T. A. Geissman; "Naturally Occurring Oxygen Ring Compounds", by F. M. Dean; "Compounds Containing Two Hetero-Oxygen Atoms in Different Rings", by W. B. Whalley in Elderfield "Heterocyclic Compounds", and in "Oxygen Heterocyclic Fungal Metabolites", by W. B. Whalley in "Progress in Organic Chemistry", edited by J. W. Cook. Some of these works discuss not only chemical matters but also a variety of related subjects of economic importance.

Every effort has been made to include papers indexed by *Chemical Abstracts* and papers in the more important journals up to and including December 1971. The Chemical Abstracts reference is listed in addition to the primary reference for any article not consulted in the original.

It is hoped that the arrangement and discussion of the closely related classes that are included here will arouse greater interest in and impart a new viewpoint to the chemistry of the respective classes and of the individual substances. The author has aimed to bring together the knowledge of the respective compounds that has so far been gained, and to present a systematic survey and bibliography of the present position from which further progress can be made. Present-day studies on the synthesis, stereochemistry, physiological activity, and biosynthesis of benzofurans continue to add new information to the body of knowledge that already exists.

I wish to acknowledge the understanding of my wife, Dr. W. Asker, who not only suffered patiently all the problems of writing a book but helped me to solve many of them by proofreading and indexing. The names of all those who have played important parts in the development of benzofuran chemistry cannot be mentioned in a brief introduction. Particular thanks are extended to those who assisted in the preparation of this volume including, among others, Drs. I. A. Hafez, M. Ali, M. H. El-Nagdy, N. Kassab, and S. M. Fahmy.

A. Mustafa

Cairo, Egypt
January 1973

Contents

BENZOFURANS

This is the Twenty-Ninth Volume in the Series

THE CHEMISTRY OF HETEROCYCLIC COMPOUNDS

CHAPTER I

Benzofurans

1. Introduction and Nomenclature

Benzofuran, as the name implies, contains a ring system obtained by the fusion of a benzene nucleus with a furan ring in the manner of formula **1**. Another name for this compound is coumarone, a somewhat unhappy choice because it is easily confused with coumarin. It was coined as a consequence of an early synthesis of benzofuran-2-carboxylic acid (coumarilic acid) from coumarin which, in turn, was decarboxylated to benzofuran:

This older name for benzofuran (coumarone) is nearly obsolete; in addition other equally confusing terminologies for derivatives of benzofuran have found their way into the literature. Among these are coumaran and coumaranone. In this book 2,3-benzofuran, benzo(b)furan, and coumarone are referred to as benzofuran, and its derivatives are related to the parent substance. The numbering system denoting the position of substituents is adopted after "Ring Index No. 1328".[1]

The benzofuran nucleus is a common one in natural products and appears in many guises. These products are usually complicated, and the furan ring especially may be present in a somewhat modified form, for example, in morphine, lignin (which is largely composed of benzofuran residues), and some alkaloids derived from isobenzofuran but not from benzofuran. These difficult and highly specialized matters exceed the scope of this book.

Benzofurans resemble naphthalenes much less than do thionaphthenes. Nevertheless they tend to behave as condensed aromatic systems and commonly afford picrates. Benzofuran (and its homologs) can be found in certain fractions of coal tar, lignite tar, and in the tar from beechwood. It is remarkably stable toward alkali, but polymerizes readily by action of concentrated sulfuric acid, and thus arises its technical importance as the basis for cheap, chemically relatively inert resins.[2-4]

2. Benzofuran and Its Alkyl Derivatives

A. Preparation

a. Catalytic Dehydrocyclization

Benzofuran is obtained by passing ethylene through phenol at 170 to 180°, and then over Fe_2O_3–Al_2O_3 catalyst at 650°.[5] The vapor-phase catalytic dehydrocyclization of 2-ethylphenol to yield benzofuran in the presence of cobalt sulfide or hydrogen sulfide and a catalyst, such as magnesium oxide and aluminum oxide, gives overall conversions of 85 and 75% with selectivities of 99 and 98%, respectively. It is suggested that dehydrogenation of the ethyl group to a vinyl group, followed by ring closure to dihydrobenzofuran takes place first, and then dehydrogenation of the latter to the benzofuran **(1)**.[6]

Catalytic dehydrocyclization of *o*-alkylphenols in presence of various catalysts has been extensively studied leading to the formation of alkyl-substituted benzofurans.[7-13]

b. Cyclization of Allylphenols

Cyclization of 2,4-dimethyl-6-allylphenol with pyridine hydrochloride has been successfully used for the preparation of 2,5,7-trimethylbenzofuran; similarly direct ring closure of *O*-α-methylallylphenol,[14] and of *O*-allyl-*m*-cresol yields 2,3-dimethyl-, and 2,6-dimethylbenzofuran, respectively.[15] Thermal treatment of phenol and 1,1-dichloro-2-butyl-, and/or 1,1-dichloro-2-cyclopropane furnishes 2-methyl-3-butyl-, and 2-phenyl-3-methylbenzofuran, respectively.[16]

c. Cyclodehydration of Aryloxyketones

Cyclization of aryloxyketones by sulfuric acid or polyphosphoric acid has established a promising route for the preparation of a large number of alkyl-substituted benzofurans.[17-22]

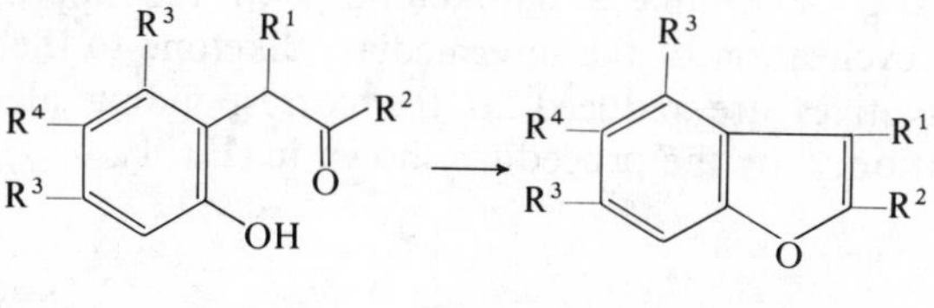

2

2a $R^1 = R^3 = CH_3$, $R^2 = R^4 = H$
2b $R^1 = R^4 = Me$, $R^2 = R^3 = H$
2c $R^1 = R^2 = Me$, $R^3 = H$, $R^4 = OMe$
2d $R^1 = R^2 = R^3 = Me$, $R^4 = H$
2e $R^1 = R^2 = -(CH_2)_4$, $R^3 = H$, $R^4 = Me$
2f $R^1 = C_6H_5$, $R^2 = R^3 = H$, $R^4 = OMe$

The scope of this reaction has been recently studied.[23,24] The 3-(*m*-substituted phenoxy)-2-butanones (**3**) undergo cyclization to yield benzofurans **4** or **5**; the latter benzofurans are formed when R is an electron donor.[24] The presence of a methoxyl group on the acyl ring of *ω*-aryloxyacetophenones lowers the reaction temperature of the cyclodehydration by polyphosphoric acid (PPA), and also facilitates the migration of the *p*-anisyl group from the 3 position.[25]

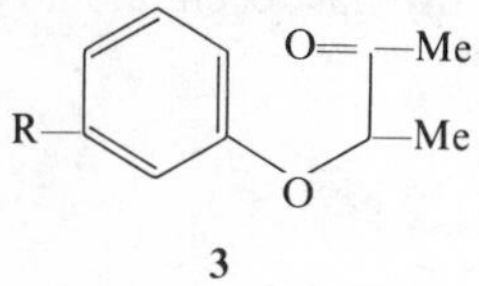

3 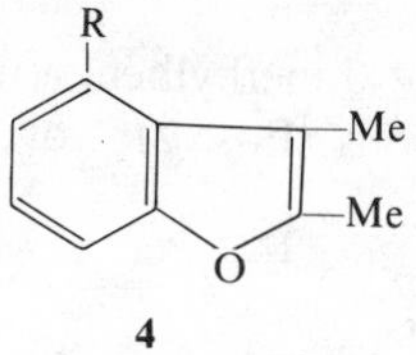 **4** 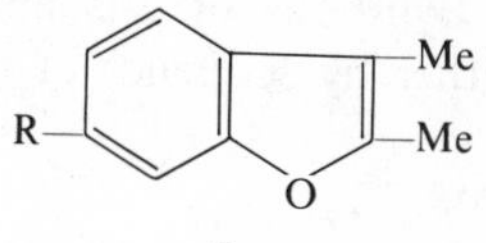 **5**

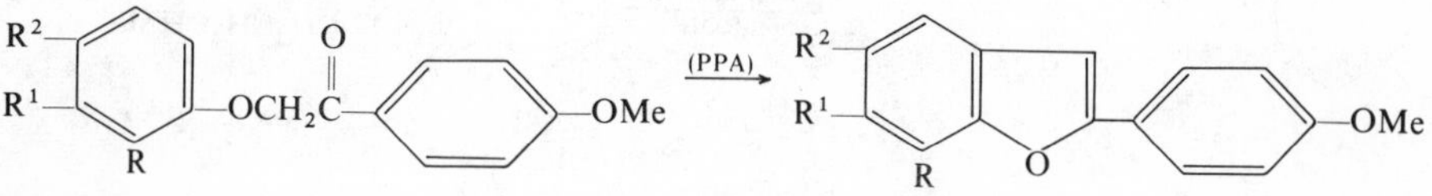

(PPA)

$R = R^1$ and R^2 = H or Me or OMe

Cyclodehydration of 4-(α-aryloxypropionyl)anisoles is carried out by action of phosphoric acid and phosphorus pentoxide; that with phosphorus oxychloride is brought out as with 3-aryloxy-2-butanones.[26]

The 5-, and 7-methylbenzofurans are obtained by cyclization of *p*-, and *o*-cresoxyacetic acids by action of hydrogen fluoride to the corresponding methyl-substituted benzofuran-3-ones, followed by reduction with lithium aluminum hydride, and dehydration by repeated distillation.[27]

Benzofurans substituted at C-5 with a butenolide ring (**7**) are prepared from *o*-hydroxy-acetophenones containing the same ring (**6**) via condensation of an aliphatic or aromatic haloketone with the appropriate phenol, followed by cyclization of the intermediate diketone to the furan ring. The keto derivatives are reduced to the corresponding alcohols, followed by dehydration,[28] by the procedure shown in (Eq. 1).

6 + R^1/Ar–C(=O)–CH_2X → R–C(=O)–aryl–O–CH_2–C(=O)–R^1/Ar $\xrightarrow{-H_2O}$ **7** (1)

R and R^1 = alkyl, Ar = aryl, X = Cl or Br

Synthesis of 2-isoamyl-3,4-dimethylbenzofuran (**8**) has been achieved either by sequence of reactions (Eqs. 2)[29] or (3).[30]

Me, Me OH, CH–CH, $CH_2CH_2CHMe_2$, OMe, OH $\xrightarrow[-MeOH]{TosCl,\ C_5H_5N,}$ Me, Me, $CH_2CH_2CHMe_2$, O, OTos $\xrightarrow{LiAlH_4}$

Me, Me, $CH_2CH_2CHMe_2$, O $\xrightarrow{Se}$

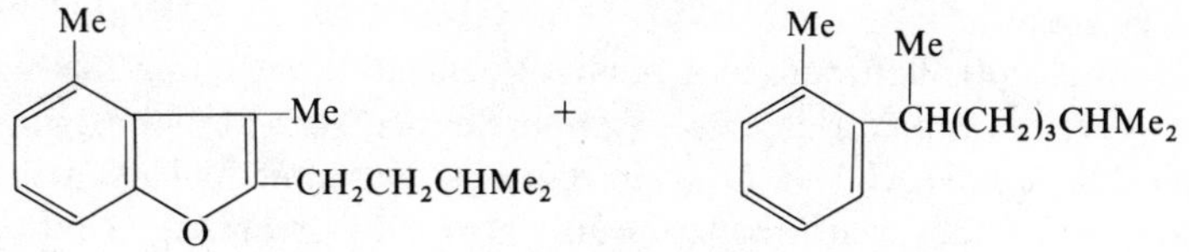

$$\text{8} + \text{2-Me-C}_6\text{H}_4\text{–CH(Me)(CH}_2)_3\text{CHMe}_2 \qquad (2)$$

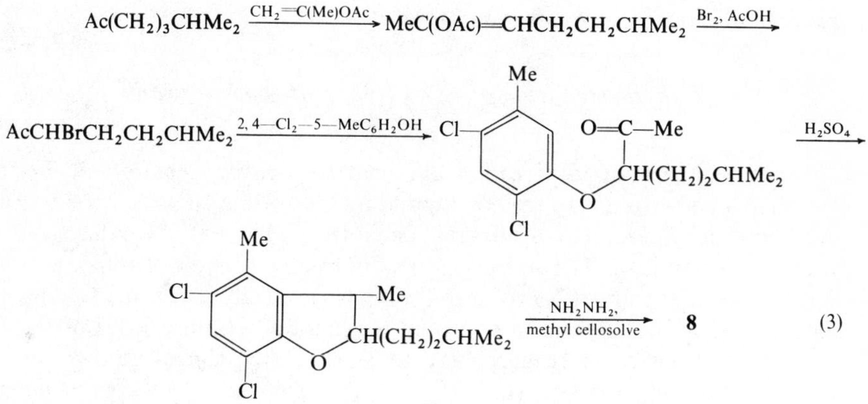

$$Ac(CH_2)_3CHMe_2 \xrightarrow{CH_2{=}C(Me)OAc} MeC(OAc){=}CHCH_2CH_2CHMe_2 \xrightarrow{Br_2,\ AcOH}$$

$$AcCHBrCH_2CH_2CHMe_2 \xrightarrow{2,4\text{—}Cl_2\text{—}5\text{—}MeC_6H_2OH} \text{ArO–CH(COMe)(CH}_2)_2\text{CHMe}_2 \xrightarrow{H_2SO_4}$$

$$\text{Cl}_2\text{Me-benzofuran (3-Me, 2-CH(CH}_2)_2\text{CHMe}_2) \xrightarrow[\text{methyl cellosolve}]{NH_2NH_2,} \mathbf{8} \qquad (3)$$

d. *Rearrangement of O-Aryloximes*

Rearrangement of *O*-aryloximes to *o*-hydroxyaryl ketimine salts, for example, **9** provides possible intermediates for benzofurans, for example, **10** (Eq. 4).[31]

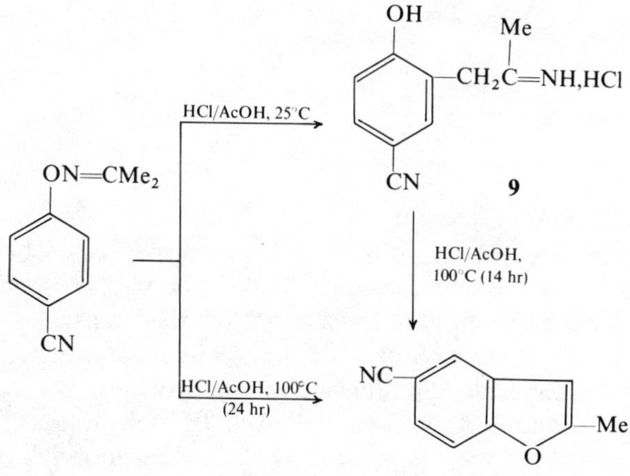

$$\text{4-NC-C}_6\text{H}_4\text{–ON=CMe}_2 \xrightarrow{HCl/AcOH,\ 25°C} \mathbf{9}\ (\text{2-OH-5-CN-C}_6\text{H}_3\text{–CH}_2\text{C(Me)=NH,HCl}) \xrightarrow{HCl/AcOH,\ 100°C\ (14\ hr)} \mathbf{10}$$

$$\text{4-NC-C}_6\text{H}_4\text{–ON=CMe}_2 \xrightarrow{HCl/AcOH,\ 100°C\ (24\ hr)} \mathbf{10}\ (\text{5-NC-2-Me-benzofuran}) \qquad (4)$$

Thus treatment of the metal salt of an oxime with an activated halide, such as *p*-halonitrobenzene, in a polar solvent and quenching the mixture in water gave the appropriate *O*-aryloxime. Rearrangement of the oximes, for example $Me_2C{=}NOC_6H_4NO_2$-*p*, in refluxing alcoholic hydrogen chloride gave the expected 5-nitrobenzofuran. However, rearrangement of 2-butanone oxime effected the formation of predominantly 2,3-dimethyl-, together with 3-ethyl-5-nitrobenzofuran. Cyclization is hindered, but not prevented by the presence of the nitro groups.[32]

e. Dehydrogenation of Bz-Alkyldihydrobenzofurans

Bz-alkylbenzofurans are readily obtained by dehydrogenation of Bz-2,3-dihydrobenzofurans to the corresponding benzofurans by the action of stoichemetric amount of sulfur,[33] palladized charcoal,[34] and/or *N*-bromosuccinimide.[35] The agents, which normally convert ketones into alcohols, usually convert benzofuran-3-ones (**11**) directly into benzofurans; dehydration of the intermediate alcohol is generally spontaneous. For the same reasons, Grignard reagents lead to 3-substituted benzofurans,[36] for example, from (**11**), in one step.

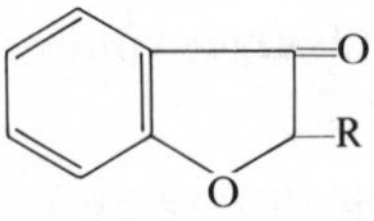

O R O **11**

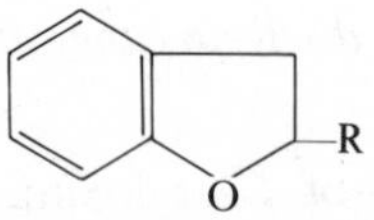

R O **12**

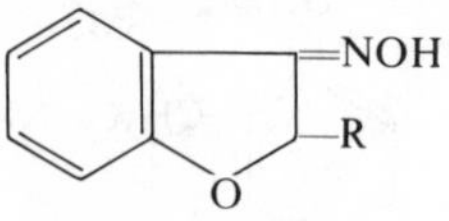

NOH R O **13**

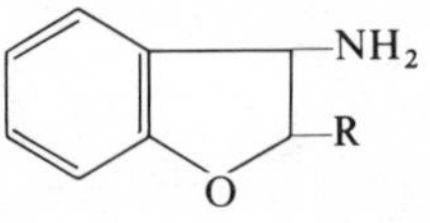

NH_2 R O **14**

Dehydrogenation of 2,3-dihydro-2-alkylbenzofurans (**12**),[36-38] and deamination of the reduction product (**14**) of 2-alkyl-3(2*H*)-benzofuranone oxime (**13**) seem to be entirely satisfactory for the preparation of 2-alkylbenzofurans.[39-42] Reduction of 2-alkyl-3(2*H*)-benzofuranones of type **15** with sodium borohydride or lithium aluminum hydride to the corresponding dihydrobenzofuranols of type **16**, followed by dehydration, furnishes a convenient route for the preparation of 2-alkylbenzofurans of type **17**[44] (Eq. 5).

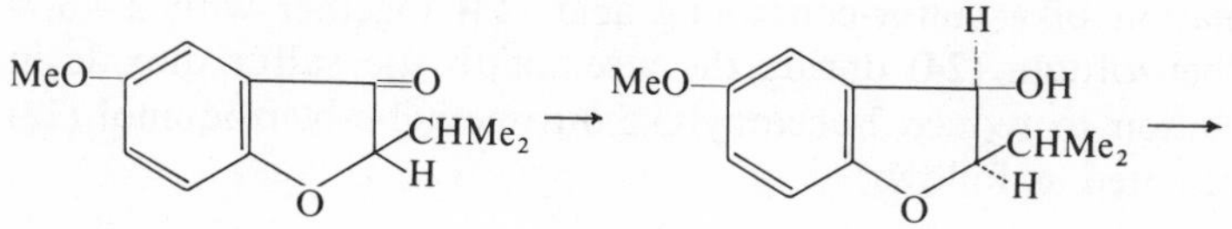

15 → **16** → **17** (5)

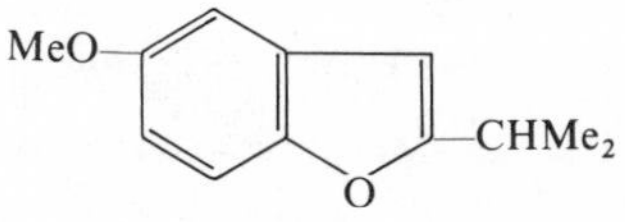

Reductive dehalogenation of 2,3-dibromobenzofuran with sodium hydrazide and hydrazine results in the formation of benzofuran in 80% yield.[43]

f. Reduction of 2-Acetonyl-o-benzoquinols

2-Acetonyl-*o*-benzoquinols (**18**) are readily reduced by the action of zinc and acetic acid to the expected phenols (**19**).[48a] The observed phenomenon that **19a–c** do not show the strong carbonyl band in the 1698–1718 cm^{-1} region in their infrared spectra, but show a carbonyl band at 1693 cm^{-1} as melt or as solution in carbon tetrachloride, is attributed to the fact that **19a–c** in the crystalline state are the cyclic hemiketals (**20a–c**). The alkyl substituent in the 6 position in **19a–c** seems to favor ketalization. Moreover, dehydration of cyclic ketals (**20a–c**) to the benzofurans (**21a–c**) seems to be catalyzed by an increased hydrogen ion concentration as it occurred only after adding mineral acid.

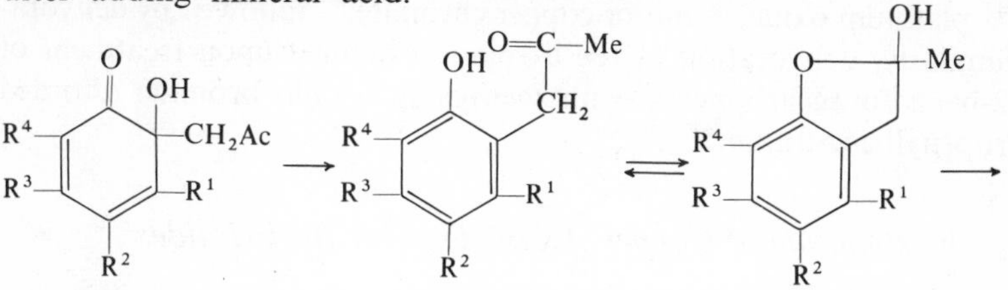

18 → **19** ⇌ **20** 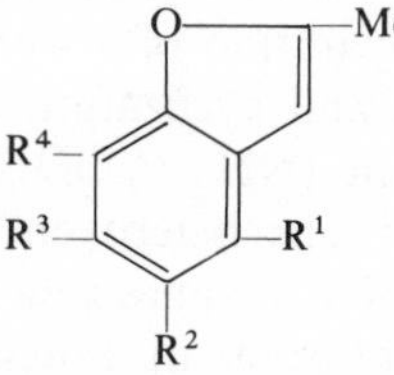 → **21**

a $R^1 = R^3 = R^4 = Me$, $R^2 = H$
b $R^1 = R^3 = H$, $R^2 = R^4 = CMe_3$
c $R^1 = R^3 = R^4 = Me$, $R^2 = OH$

The formation of a sulfur-containing acid (**23**) together with 2,4,6,7-tetramethylbenzofuran (**24**) during the attempt to use sulfur dioxide in aqueous solution to reduce 2-acetonyl-3,5,6-trimethyl-*o*-benzoquinol (**22**) can be represented as follows:

g. *Hydrogenation of* 2-*Acetylbenzofuran*

2-Vinylbenzofuran has been successfully obtained by hydrogenation of 2-acetylbenzofuran to the carbinol in the presence of copper–chromium oxide,[45] platinum oxide,[46] and/or copper chromite,[47] followed by dehydration. Similarly, dehydration of the carbinol, obtained upon treatment of ethyl 2-benzofurancarboxylate with methylmagnesium bromide afforded 2-isopropenylbenzofuran.[48]

h. *Reaction of Copper Acetylides with Aryl Halides*

The reaction of cuprous acetylides with aryl halides has been proved to be an effective pathway for the preparation of a wide variety of aromatic acetylenes. Moreover, substitution and cyclization of halides, bearing a neighboring nucleophilic substituent (NH_2, COOH, OH, etc.), are the basis for obtaining benzofurans.[49] The heterogeneous character of the cyclization in heterocyclic synthesis is emphasized by conversion of the acetylenic intermediates for benzofurans to other products when the reaction is run homogeneously. This striking reaction (Eq. 6) affords no benzofuran, although it is the exclusive product obtained with undissolved cuprous salt. With cuprous chloride, in addition to 2,3-dihydrobenzofuran, the iodo alcohol is converted to the corresponding chloride. The benzofuran is converted to the corresponding chloride. The benzofuran synthesis

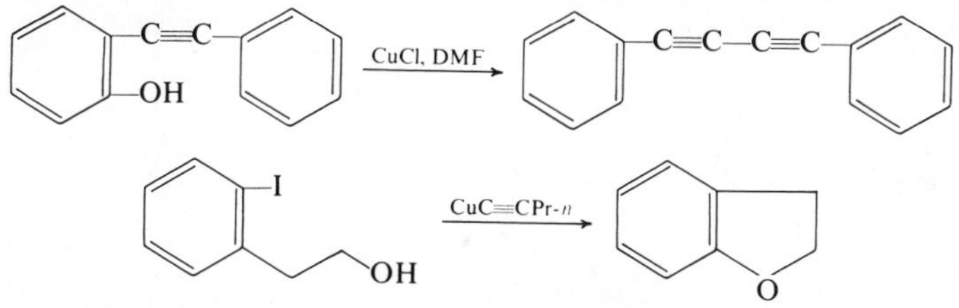

(6)

proceeds smoothly in pyridine, and in no case has *o*-hydroxytolan been isolated. The ease of the reaction and the heightened lability of the halogen adjacent to the phenolic hydroxy are stressed (Eq. 7).

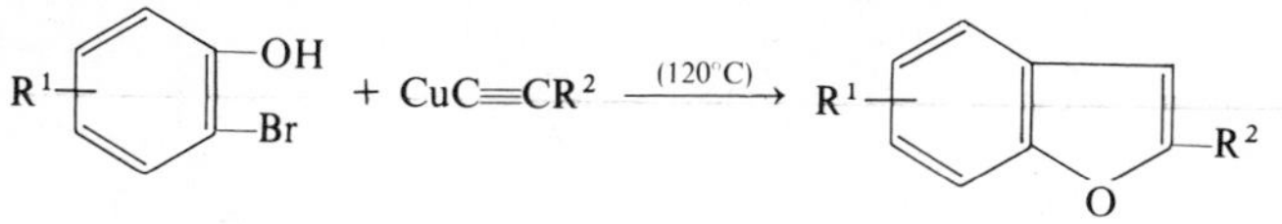

(7)

i. Decarboxylation of Benzofurancarboxylic Acids

Another approach to benzofuran synthesis[50-52] was made possible by Perkin's discovery that 3-bromocoumarins (**25**), when heated with alkali, yield benzofurancarboxylic acids (coumarilic acids) (**27**). The reaction appears to involve ring fission, followed by extrusion of halogen by the phenoxide ion as in **26**; it is very satisfactory when the acids are wanted but, as benzofurancarboxylic acids are not always decarboxylated, not necessarily convenient for preparing the parent benzofurans.

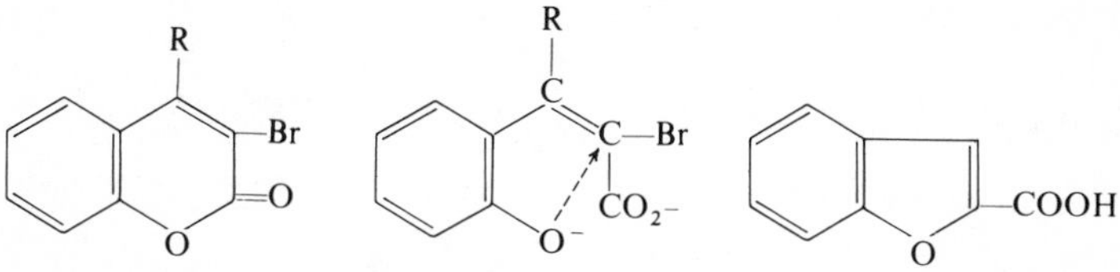

25 **26** **27**

j. Photochemical Formation of Benzofurans

Aryloxyacetones (**28**) and *m*-substituted aryloxyacetones (**29**), on irradiation in methanol and depending on the substitution in the benzene ring, yield the dimethylacetals **30** and **31**, and the benzofurans **32** and **33** together with the corresponding phenols that are formed by fission of the $ArOCH_2^-$ bond.[53]

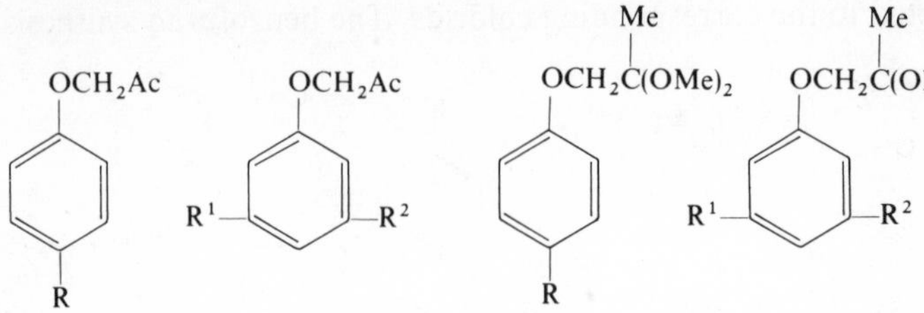

OCH_2Ac R **28**

OCH_2Ac R^1 R^2 **29**

Me $OCH_2C(OMe)_2$ R **30**

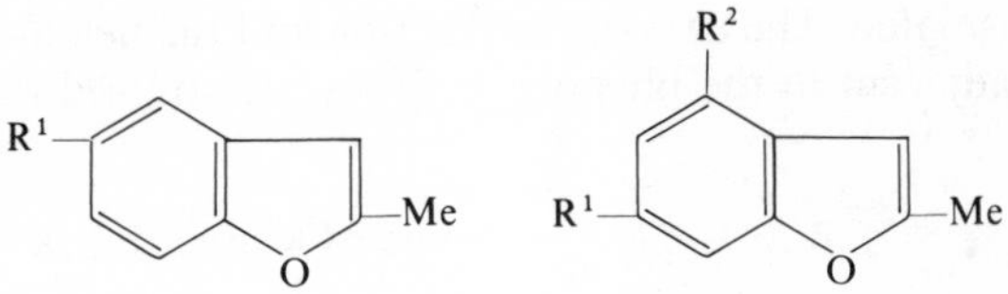

Me $OCH_2C(OMe)_2$ R^1 R^2 **31**

R^1 Me O **32**

R^2 R^1 Me O **33**

Chloro, bromo, and nitro substituents retard or prevent the photoreaction. The isolation of the intermediate **34**, leading to the formation of the corresponding benzofuran in the case of *p*-methoxyphenoxyacetone, has shown that such an intermediate is a product of photo *o*-rearrangement[54] (Eq. 8), which cyclized during working up.

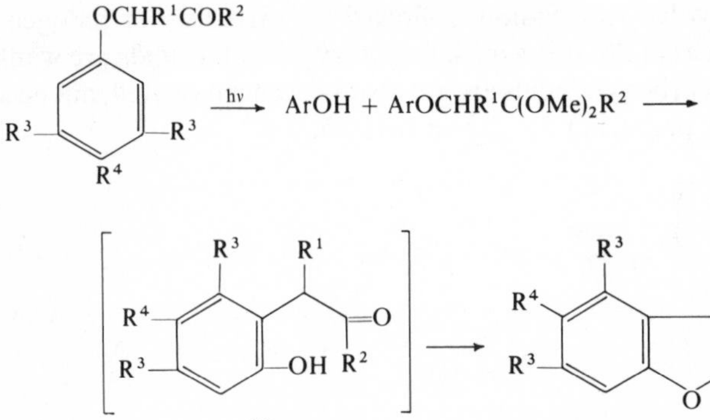

$OCHR^1COR^2$ R^3 R^3 R^4 $\xrightarrow{h\nu}$ $ArOH + ArOCHR^1C(OMe)_2R^2$ $\longrightarrow$

[R^3 R^1 R^4 =O R^3 OH R^2] **34** $\longrightarrow$ R^3 R^4 R^1 R^3 R^2 O (8)

Photolysis of 2-benzyloxy-4-methoxybenzophenone in benzene yields 2,3-diphenyl-6-methoxybenzofuran (**36**), presumably via internal hydrogen transfer in the excited state, followed by cyclization of the resulting diradical (**35**)[55] (Eq. 9).

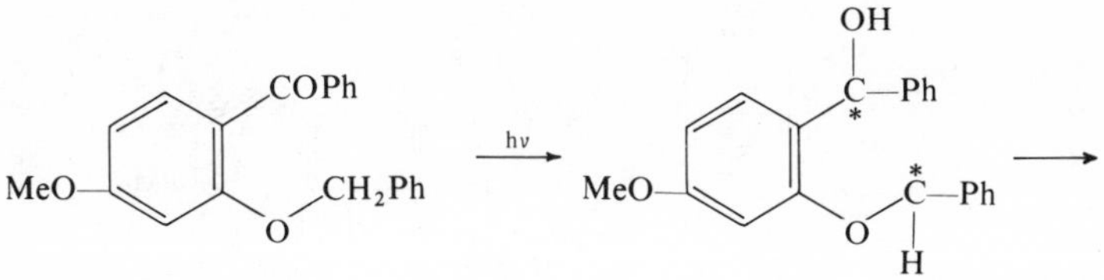

(9)

35

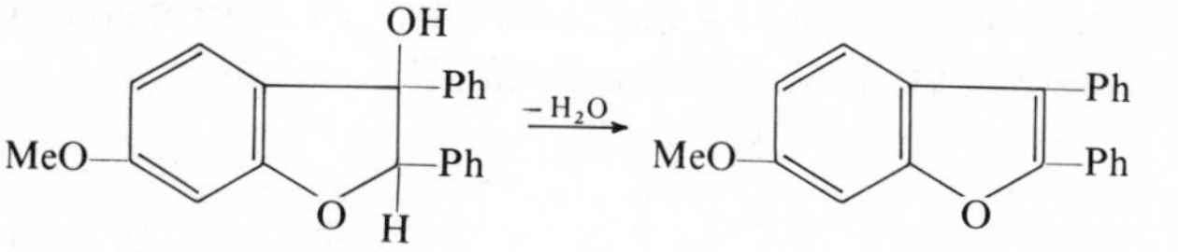

36

Irradiation of cyclooctatetraene oxide (**37**) in pentane affords the dihydrobenzofuran **38**.[56]

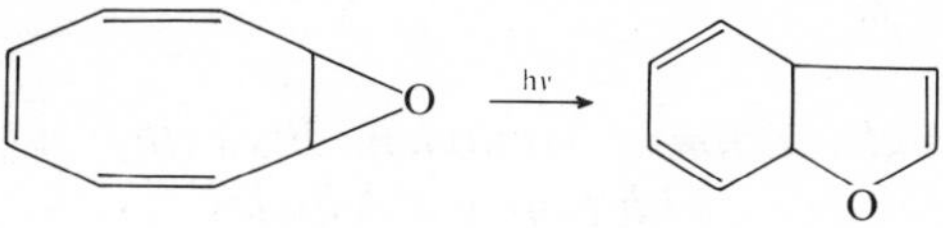

37

38

k. Adsorptive Cyclization

1,1,6,6-Tetraphenyl-2,4-hexadiyne-1,6-diol has been shown to undergo rearrangement under the influence of hydrobromic–acetic acid mixture to 1,1,6,6-tetraphenyl-3,4-dibromo-1,2,4,5-hexatriene.[57] However the dibenzofuranylacetylene (**41**) is unexpectedly obtained in the eluate from alumina column chromatography of the reaction product of hexadiynediol (**39**), which has an *o*-methoxy group at both C-1 and C-6. By isolation of benzofuranylbutatriene bromide (**40**) as the intermediate product, it has been proved that this unusual intramolecular cyclization reaction proceeds by two steps. The second step proceeds easily by absorption of **40** on adsorbent such as alumina and/or silica gel[58] (Eq. 10).

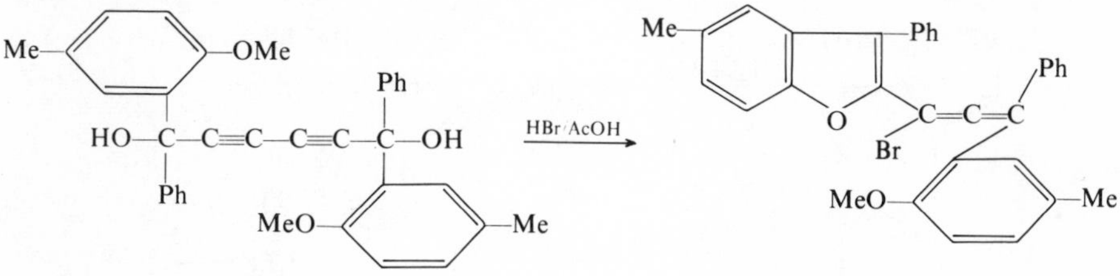

HBr/AcOH

39

40

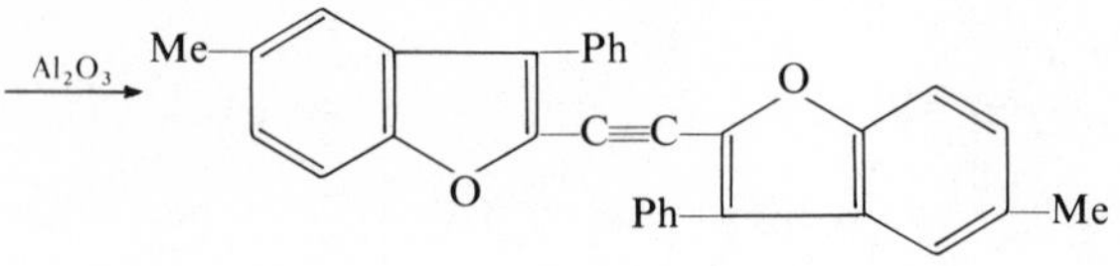

$\xrightarrow{Al_2O_3}$ (10)

41

Acid hydrolysis of 2-cyano-4-methylbenzo(*f*)(1,3)-oxazepine (**42**) results in the formation of 2-methylbenzofuran.[59]

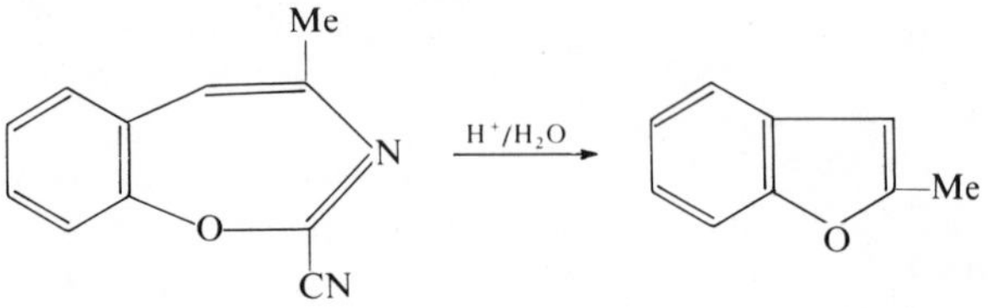

$\xrightarrow{H^+/H_2O}$

42

I. Condensation of Methylene Bis (ethyl sulfone) with Salicylaldehydes

Aldol condensation of methylene bis(ethyl sulfone) with salicylaldehydes in the presence of a basic catalyst effected the elimination of an ethyl sulfonyl group as ethyl sulfinic acid and the formation of 2-ethylsulfonylbenzofurans (**43**).[60a-b] Formation of 43 probably arises via the reaction sequence shown in (Eq. 11). Following the initial condensation–dehydration sequence, intramolecular attack of the phenolate anion upon the carbon bearing the sulfone group results in ring closure with the concomitant elimination of ethanesulfinic acid, yielding **43**. Any pathway to **43** involving attack of phenolate or piperidine at the olefinic position β to the sulfone group in **43**, and the fact that the phenolate anion can supply electrons to the sulfone olefin, effects the strong inductive effect of the sulfone group (**44a,b**) and materially reduces the tendency of **44** to add nucleophiles. Hydrogen peroxide oxidation of 2-methylthiomethylbenzofuran in acetic acid afforded 2-methylsulfonylbenzofuran.[60c]

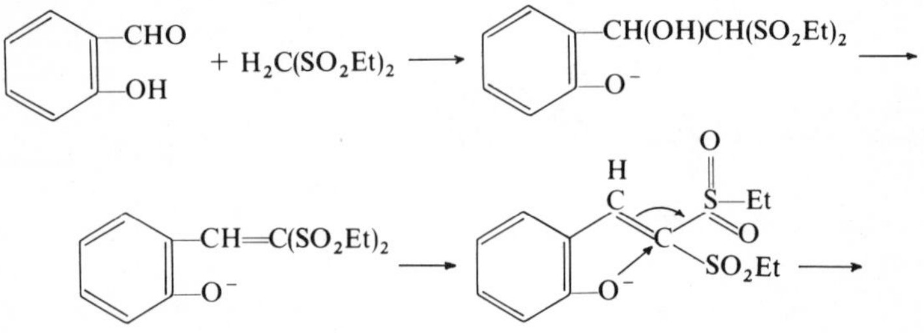

$+ H_2C(SO_2Et)_2 \longrightarrow$ $CH(OH)CH(SO_2Et)_2$ $\longrightarrow$

$CH{=}C(SO_2Et)_2$ $\longrightarrow$ $\longrightarrow$

$EtSO_2^-$ + [structure 43: 2-(ethylsulfonyl)benzofuran, SO_2Et]

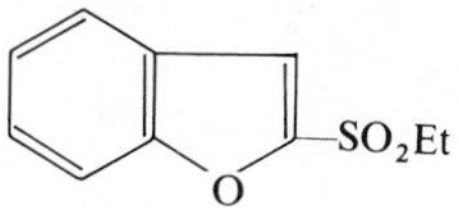

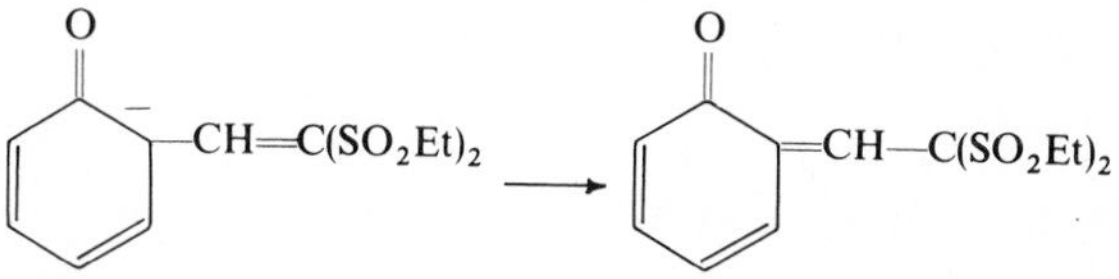

43

[structure 44a: $CH{=}C(SO_2Et)_2$] → [structure 44b: $=CH{-}C(SO_2Et)_2$] (11)

44a **44b**

Benzofuran, upon treatment with $C_5H_5N{-}SO_3$ at 100°C, is quantitatively converted to 2-benzofuransulfonic acid.[61a,b] The reaction of the complex between dimethylformamide with $S_2O_5Cl_2$ and nuceophiles with available nucleophilic properties yields the sulfonyl chloride in preference to aldehyde thus forming 2-benzofuransulfonyl chloride.[61c]

Rausch and Siegel[62] have reported the preparation of 2-ferrocenylbenzofuran (**45**), upon treatment of *o*-iodophenol and ferrocenylethynyl copper.

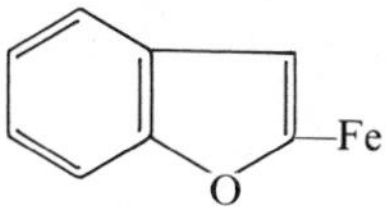

[structure 45: 2-ferrocenylbenzofuran, Fe]

45

3. Arylbenzofurans

A. Preparation

a. Cyclodehydration of ω-Aryloxyacetophenones

The literature records several methods for the preparation of 2-arylbenzofurans;[63-65] cyclodehydration of ω-aryloxyacetophenones, in presence of polyphosphoric acid as a cyclizing agent, seems to be a promising route. Cyclodehydration of a number of ω-phenoxyacetophenones of type **46a–c** has been studied; at 80°C the normal product 3-phenylbenzofuran is formed; at 132°C the rearranged product 2-phenylbenzofuran is produced; at intermediate temperatures a mixture of 2- and 3-phenylbenzofuran is obtained. The 2-phenylbenzofurans yield complex compounds with 2,4,7-trimethylfluorenone, whereas the 3-phenylbenzofurans do not.[21] The presence of a methoxy group on the acyl ring of **47a–c** lowers the reaction temperature for cyclodehydration of **47a–c**, and facilitates the migration of the *p*-anisyl group from position 3, thus affording **48**.[25]

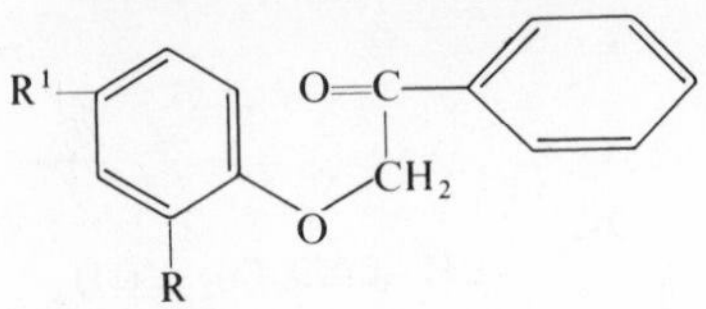
R¹
O=C
CH₂
O
R

46a R = OMe, R^1 = H
46b R = H, R^1 = OMe
46c R = H, R^1 = Me

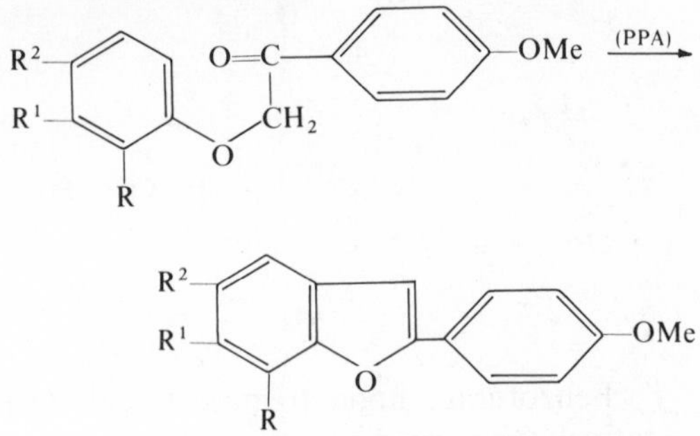
R²
R¹
O=C
CH₂
O
R
OMe
(PPA)
R²
R¹
O
R
OMe

48

47a R = R^1 = R^2 = H
47b R = R^1 = H, R^2 = Me
47c R = R^1 = H, R^2 = OMe

ω-Tolyloxyacetophenone, obtained either by condensation of substituted salicylaldehyde and α-bromophenylacetic acid (Eq. 12), or by treatment of tolyloxyacetyl chloride with aromatic hydrocarbon in presence of aluminum chloride (Eq. 13), undergoes cyclodehydration to give the corresponding benzofurans.[65-68]

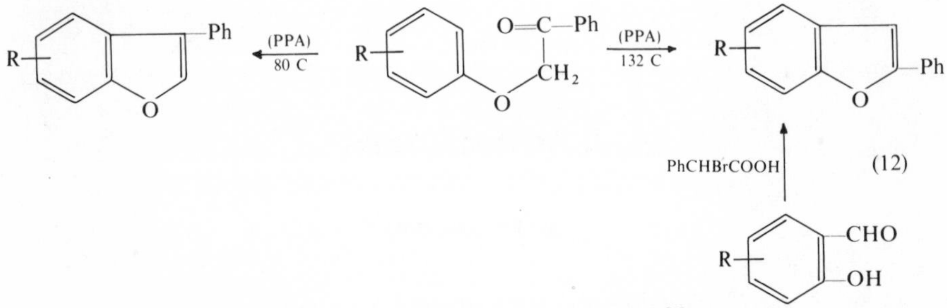
R
Ph
O
(PPA)
80 C
R
O=C—Ph
CH₂
O
(PPA)
132 C
R
O
Ph
PhCHBrCOOH
R
CHO
OH

(12)

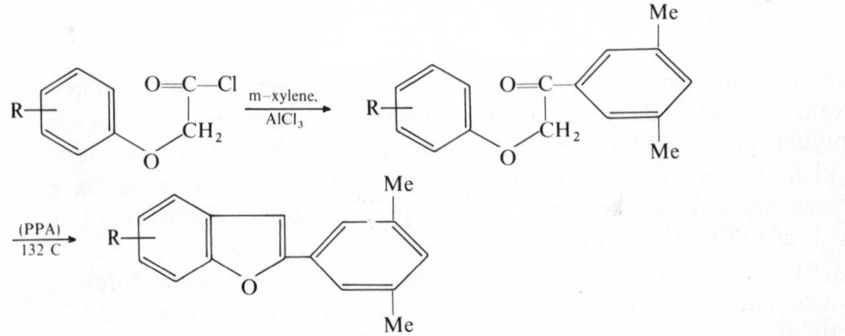
R
O=C—Cl
CH₂
O
m-xylene,
AlCl₃
R
O=C
CH₂
O
Me
Me
(PPA)
132 C
R
O
Me
Me

(13)

Davies and Middleton[69] have suggested mechanism "A" for the rearrangement of 3- to 2-phenylbenzofuran and the migration of a phenyl group during cyclization of ω-phenoxyacetophenone. However, an alternative mechanism "B" involves the fission between oxygen atom and the CH_2 group (which could occur if the oxonium salt, **49,** was initially formed).

49

50 **51**

52 **53**

Use of ^{14}C has been applied to distinguish between the two mechanisms. ω-Phenoxy(carbonyl-^{14}C)acetophenone (**50**) was prepared from (I-^{14}C)-acetyl bromide and treated with polyphosphoric acid at 132°C to give predominantly **51**. This was reduced with sodium and ethanol to *o*-hydroxydibenzyl which, with methyl iodide, yielded the *o*-methoxy compound (**52**). Bromination with *N*-bromosuccinimide followed by dehydrobromination with triethylamine resulted in the formation of *o*-methoxystilbene (**53**), which was oxidized with potassium permanganate to a mixture of benzoic acid and *o*-methoxybenzoic acid. In this sequence, mechanism "A" would give inactive benzoic acid and active *o*-methoxybenzoic acid. However, mechanism "B" would afford active benzoic acid and inactive *o*-methoxybenzoic acid. Treatment of the mixed acids with *p*-phenylazophenacyl bromide, and the azoesters separated by chromatography, resulted in the observation that benzoic acid ester contained only 0.3% of the combined activities of both esters. Thus, under the conditions used $\ngtr$ 0.3% of 2-phenylbenzofuran is formed by mechanism "B" or any other mechanism not involving migration of the phenyl group.

Rearrangement of 3- to 2-phenylbenzofuran is apparently acid catalyzed since 3-phenylbenzofuran is not rearranged when heated alone; the reaction may be initiated by proton attack on the benzofuran nucleus. Since benzofuran undergoes electrophilic attack, predominantly at the 2 position, it would be expected that proton attack on 3-phenylbenzofuran would afford the carbonium ion (**54**). This, however, would not be expected to rearrange, so the reaction should not proceed further.

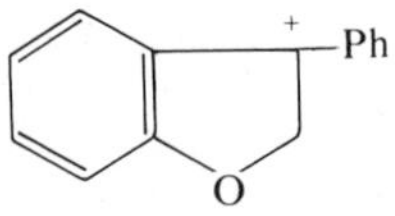

+ Ph
O

54

The formation of a positive charge on a carbon atom (carbonium center) results, in a number of cases, in skeletal isomerization, as occurs for example in Wagner-rearrangement. It has been shown that the rearrangement of 3-substituted benzofurans into their 2 isomers has an intramolecular character and passes through the following series of transformations. The primary act is the protonation of the heterocycle molecule. The proton attacks the 3 position, heterocyclic opening of the 2,3- double bond occurs, and the charge is localized to a considerable extent in the 2 position. It must be pointed out that the anion of the acid has a substantial influence on the course of the reaction. Thus, when rearrangement is conducted in presence of sulfuric acid and trifluoroacetic acid, yields of the two isomers are low (10–30%), and considerable amounts of water-soluble substances are formed.[70] After protonation, migration of the group occurs with accompanying hydride shift; direct elimination of a proton with reestablishment of the aromatic structure is also theoretically possible. However, route 1 is more likely because 2-methyl-3-phenylindole, for example, undergoes interchange of substituents in presence of aluminum chloride in keeping with the greater tendency of the phenyl group to migrate.

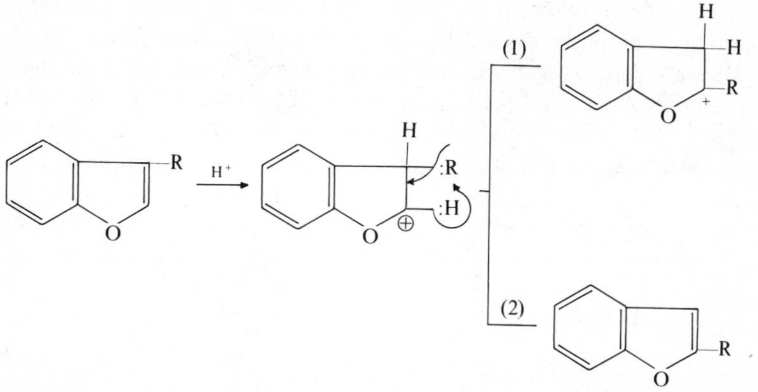

Cyclodehydration reagents vary, thus, whereas hydrofluoric acid has been used in the case of ω-phenoxyacetophenone,[71] polyphosphoric acid readily brings out the cyclization of ω-phenoxy-l-, and ω-phenoxy-2-acetylnaphthalene to 2-(1′-naphthyl)- (**55a**) and 2-(2′-naphthyl)benzofuran (**55b**).[72] In a similar manner, the benzofurans **55c**[72] and **55d**[73] have been obtained.

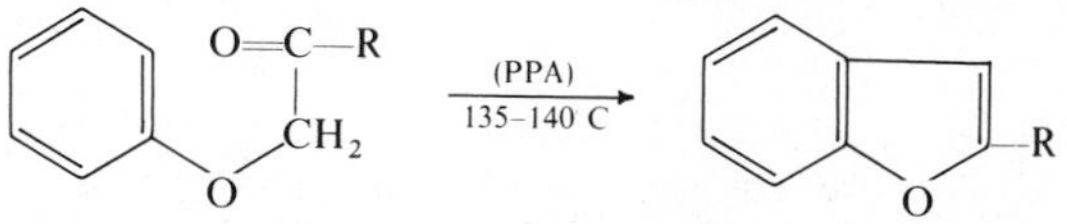

(PPA) 135–140 C

55a R = 1-$C_{10}H_7$
55b R = 2-$C_{10}H_7$
55c R = C_6H_4OMe-m
55d R = C_6H_5, 3-Me

Royer and others[74-78] have investigated thermal rearrangement of α-aryloxyaceto-(or propio-)phenones (**56**) (Chart 1) which include: (1) rupture of the linkage between aryloxy and the carbon chain carrying the carbonyl group (this leads to the phenol **57** and the ketone **58**); (2) rupture accompanied by rearrangement (this leads to the formation of benzaldehyde and benzoic acid); and (3) cyclodehydration occasionally reported[76,79,80] to afford arylbenzofurans of type **59** and water.

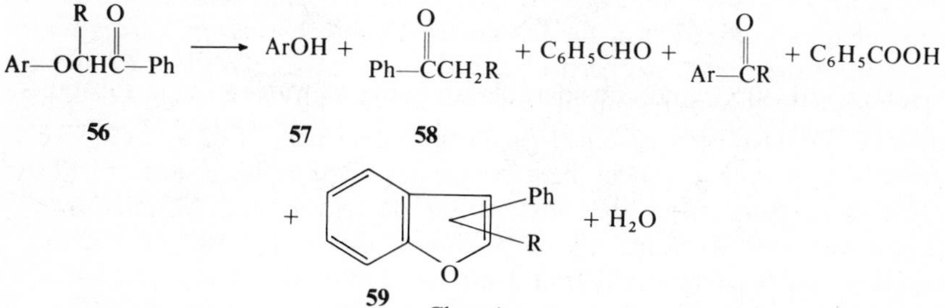

Ar—OCH(R)C(=O)—Ph (**56**) ⟶ ArOH (**57**) + Ph—C(=O)CH$_2$R (**58**) + C_6H_5CHO + Ar—C(=O)R + C_6H_5COOH + **59** + H_2O

Chart 1.

Thermal rearrangement of α-(4-chlorophenoxy)- (**60**), and α-(4-bromophenoxy)propiophenone (**61**) gave, with other degradation products, 9–16% of 4-chloro- and 4-bromoacetophenone [thus confirming the rearrangement reaction (ArOC— → Ar—C=O)], and 4% of 2-phenyl-3-methyl-5-chloro- (**64**) and 2-phenyl-3-methyl-5-bromobenzofuran (**65**), which were formed during the original rearrangement. Under similar conditions, 4′-chloro-1-phenoxy-1-phenylacetone (**62**) afforded *p*-chlorophenol and phenylacetone, together with 12% of 2-methyl-3-phenyl-5-chlorobenzofuran (**66**).

X— O=C—Ph, CH—Me, O
60 X = Cl
61 X = Br

(300°C) →

X— Me, Ph, O
64 X = Cl
65 X = Br

i) CrO_3 ii) KOH →

X— Ac, OH
+ C_6H_5COOH
67 X = Cl
68 X = Br

(PPA) (120°C) ↓

X— Ph, Me, O
66 X = Cl
70 X = Br

i) CrO_3 ii) KOH →

X— COPh, OH
69 X = Cl
71 X = Br

(PPA) or H_2SO_4 (0°C) ↑

X— O=CMe, CHPh, O
62 X = Cl
63 X = Br

Normal cyclodehydration in presence of polyphosphoric acid or sulfuric acid has been successfully used to bring out the formation of the benzofuran **66** from the ketone **60**; under similar conditions the benzofuran **70** was obtained on cyclization of the ketone **61**. Cyclodehydration, accompanied with rearrangement, has been observed in the case of ketones **62** and **63**. This leads to the formation of benzofurans **66** and **70**, respectively, which meanwhile have been obtained by thermal treatment of **62** and **63**. Cyclodehydration of **62** and **65** in the presence of polyphosphoric acid or sulfuric acid is in contrast to the observation of Davies and Middleton.[65,69] Moreover, this reaction cannot be announced as a "general reaction," since it has been reported that 1-aryloxy-1-phenylacetones and some α-aryloxypropiophenones[81] underwent cyclodehydration in presence of sulfuric acid without rearrangement of the substituents.[79,82] Pyrorearrangement of arloxypropiophenones, for example **72**, to give benzofurans of type **73** proceeds via an intramolecular cyclodehydration.

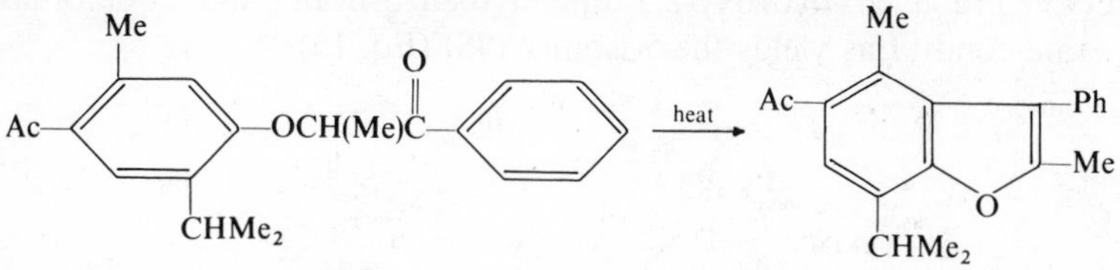

72 → (heat) → **73**

Pyrodecomposition of resorcinol phenacyl ethers leads to the formation of 3-phenyl-6-hydroxybenzofuran;[79] on the other hand, condensation of dihydroresorcinol with phenacyl bromide afforded 2-phenylbenzofuran (Eq. 14).[83]

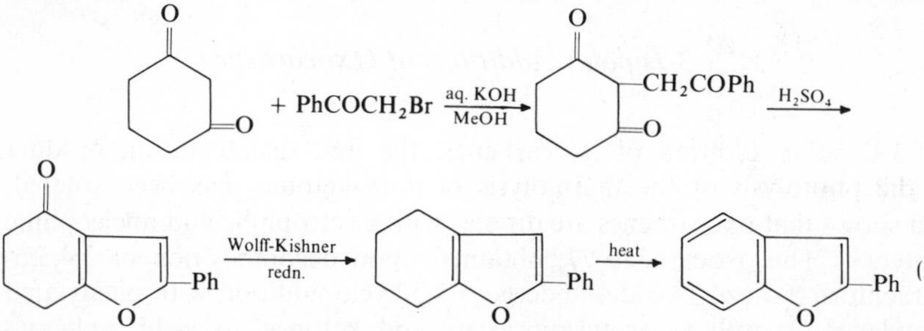

+ $PhCOCH_2Br$ (aq. KOH, MeOH) → CH_2COPh (H_2SO_4) → (Wolff-Kishner redn.) → (heat) → Ph (14)

Cyclodehydration of 2-hydroxy-α-phenylacetophenone[84] and of α-(*o*-hydroxyphenyl)acetophenone[85] leads to the formation of 2-phenylbenzofuran. A simple method for the preparation of 2-phenylbenzofuran includes treatment of α-cyano-α-(*o*-methoxyphenyl) acetophenone (**74**) with hydrobromic acid in acetic acid.[72b] This involves hydrolysis, decarboxylation, demethylation, and cyclization in one operation.

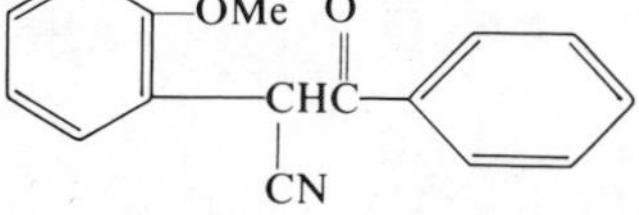

74

b. Condensation of Benzoins with Phenols

Phenol reacts with benzoin in presence of boron oxide at 150–160°C to yield 2,3-diphenylbenzofuran. When *o*-, *p*-cresol and/or thymol are allowed to condense with benzoin, 7-methyl-, 5-methyl-, and 4-methyl-7-isopropyl-2,3-diphenylbenzofurans are obtained, respectively.[86,87] Dihydric phenols, for example resorcinol, react with benzoin in presence of sulfuric acid (73%) to give 6-hydroxy-2,3-diphenylbenzofuran (**76**);[88] quinol under the same conditions yields the 5 isomer (**75**) (Eq. 15).[88,89]

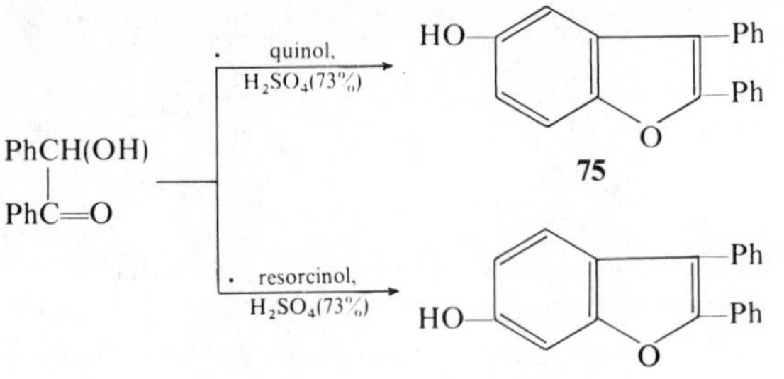

PhCH(OH)–PhC=O → (quinol, H_2SO_4(73%)) **75** (5-hydroxy-2,3-diphenylbenzofuran); → (resorcinol, H_2SO_4(73%)) **76** (6-hydroxy-2,3-diphenylbenzofuran) (15)

c. 1,3-*Dipolar Additions of Oxocarbenes*

1,3-Dipolar addition of oxocarbenes, the first disintegration product of the photolysis or the thermolysis of diazoketones, has been studied, and shows that oxocarbenes are dipoles with electrophilic and nucleophilic centers.[90] The oxocarbene **77**, obtained upon decomposition of 3,4,5,6-tetrachloro-2-diazo-1-oxide, undergoes 1,3-cyclo addition with phenylated acetylenes, as well as acetylenic esters and ketones, to yield 2-phenylbenzofuran derivatives which, when hydrogenated over Raney nickel catalyst, give 2-phenylbenzofuran and *o*-hydroxybibenzyl (Eq. 16).[91]

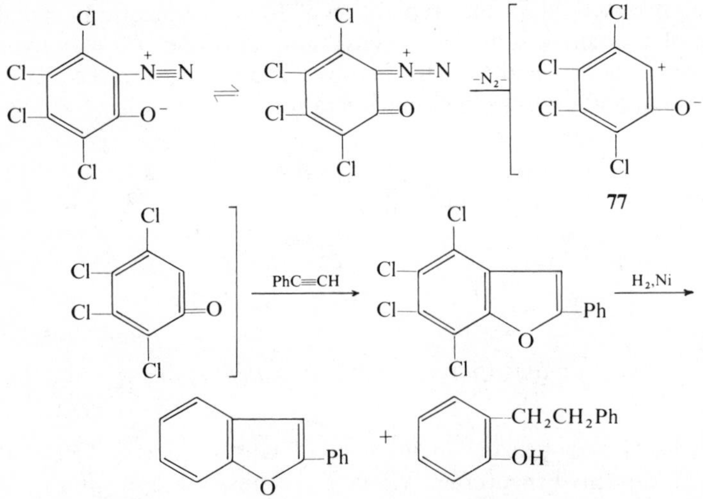

Tetrachloro-2-diazonio-phenoxide ⇌ tetrachloro-6-diazocyclohexadienone →($-N_2$) [**77** ⇌ tetrachloro oxocarbene] →(PhC≡CH) 4,5,6,7-tetrachloro-2-phenylbenzofuran →(H_2,Ni) 2-phenylbenzofuran + 2-(CH_2CH_2Ph)phenol (OH) (16)

The oxocarbene (**77**) undergoes 1,3,-cyclo-addition with styrene and *cis*-, and *trans*-stilbene to yield the corresponding 4,5,6,7-tetrachloro-2,3-dihydrobenzofuran (**78**), which is dehydrogenated to the benzofurans of type **79** (Eq. 17).[92]

$$\mathbf{77} \xrightarrow[\substack{C_6H_4(OH)_2\text{-}1,4,\\ (140^\circ C)}]{RCH{=}CHR^1} \mathbf{78} \xrightarrow[\text{heat}]{\text{Se}} \mathbf{79} \qquad (17)$$

d. *Copper-Catalyzed Decomposition of Diazoketones*

α-(*o*-Hydroxyphenyl)acetophenone, produced during copper-catalyzed decomposition of α-diazoacetophenone in benzene solution containing phenol, undergoes cyclodehydration during working up of the reaction mixture to give 2-phenylbenzofuran (Eq. 18).[93]

$$PhCOCHN_2 + PhOH \xrightarrow{Cu} PhCOCH_2OPh + N_2$$

$$\longrightarrow o\text{-}(CH_2COPh)C_6H_4OH \xrightarrow{-H_2O} \text{2-phenylbenzofuran} \qquad (18)$$

e. *Ethynation of p-Benzoquinone*

Addition of lithium salts of monosubstituted acetylenes (ethynaton) to *p*-benzoquinone in liquid ammonia or dry toluene yields, depending on the reaction conditions, the mono- or diquinol adduct (**80**) which, upon upon acidification, gives the acetylene derivatives **81** and **82**. In cold formic acid, **80** affords **82** which, when refluxed with sodium hydroxide solution, gave 2-phenyl-5-phenylethenylbenzofuran (**83**). However heating **82** slowly with formic acid yielded **84**; the latter is readily cyclized, upon reflux with sodium hydroxide, to give 2-phenyl-5-phenacetylbenzofuran (**85**) (Eq. 19).[94]

$$p\text{-benzoquinone} \xrightarrow[\text{Li}]{PhC{\equiv}CH,} \mathbf{80} \xrightarrow{H^+} \mathbf{81} + \mathbf{82} \xrightarrow{\text{NaOH}}$$

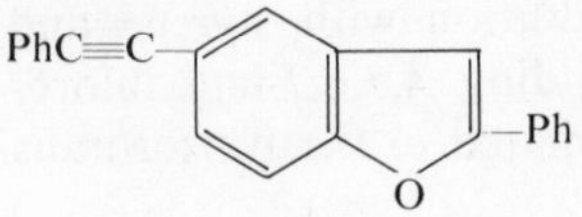

PhC≡C– –Ph O

83

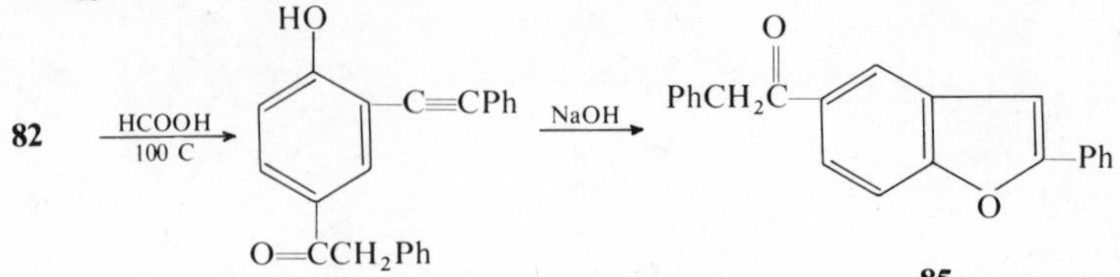

82 $\xrightarrow[100\text{ C}]{\text{HCOOH}}$ HO –C≡CPh O=CCH$_2$Ph **84** $\xrightarrow{\text{NaOH}}$ PhCH$_2$C(=O)– –Ph O **85** (19)

Wessely and Zbiral[95] have reported the unexpected formation of **87**, obtained upon treatment of **86** with PhC≡CMgBr, and its ready cyclization, in presence of sodium hydroxide, to 2-phenyl-5-methylbenzofuran (**88**) (Eq. 20).

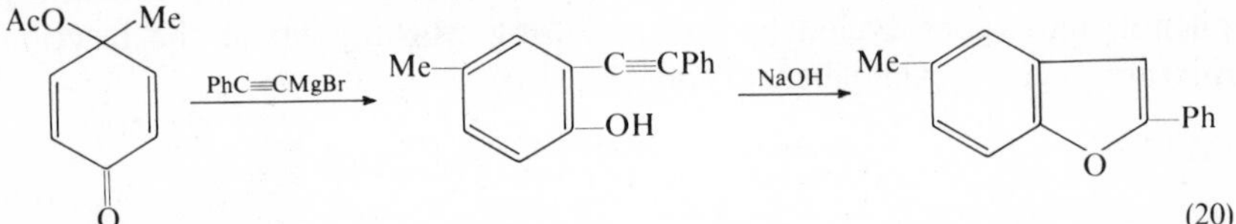

AcO Me O **86** $\xrightarrow{\text{PhC≡CMgBr}}$ Me– –C≡CPh –OH **87** $\xrightarrow{\text{NaOH}}$ Me– –Ph O **88** (20)

f. Oxidation of Flavylium and Pyrylium Salts

A favorable approach for the preparation of 2-arylbenzofurans has been reported.[85,96] Hydrogen peroxide oxidation of 3-alkyl-, alkoxy-, and/or aryloxyflavylium salts effected the formation of 3-acyl- and 3-carboxy-2-arylbenzofurans (Eq. 21), for example 3-acetyl-2-(4-hydroxyphenyl)benzofuran.

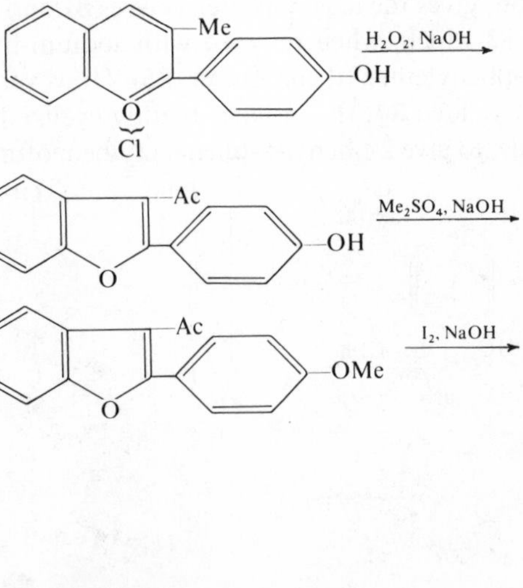

–Me –OH O Cl $\xrightarrow{\text{H}_2\text{O}_2\text{, NaOH}}$

–Ac –OH O $\xrightarrow{\text{Me}_2\text{SO}_4\text{, NaOH}}$

–Ac –OMe O $\xrightarrow{\text{I}_2\text{, NaOH}}$

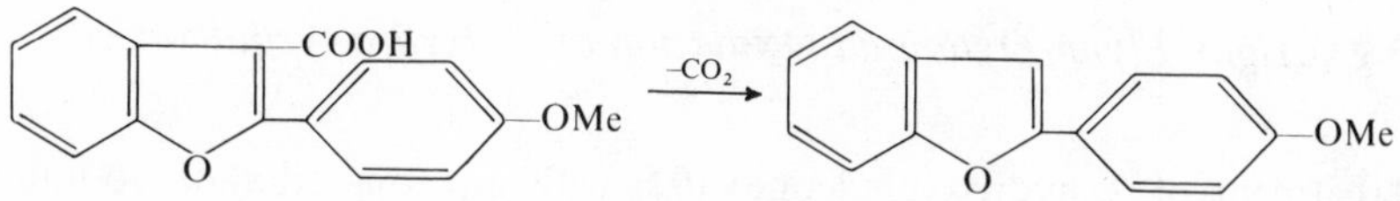

(21)

2-Phenylbenzofuran has been obtained upon treatment of benzopyrylium salts with hydrogen peroxide (Eq. 22).[85]

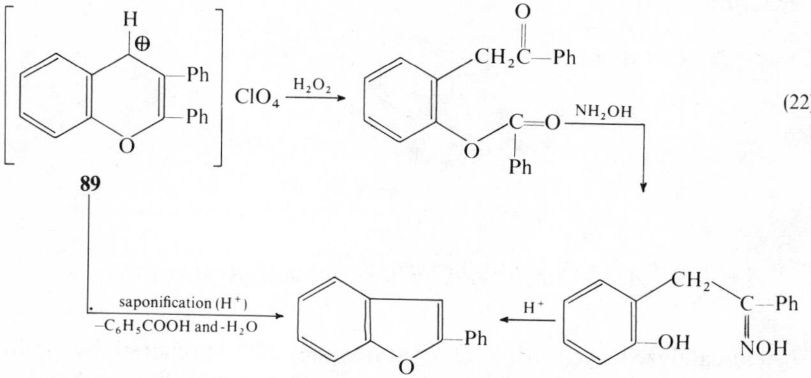

(22)

89

Flavan-3,4-diols (leucoanthocyanidins) are split by lead tetraacetate. Thus flavan-3,4-diol (**90**) yields small amounts of benzaldehyde and salicylaldehyde together with 70% of 2-formyl-2,3-dihydro-3-hydroxy-2-phenylbenzofuran (**91**). This is readily converted to 2-phenylbenzofuran either upon its treatment or treatment of its oxime derivative with dilute sulfuric acid, and/or upon sublimation, whereby elements of formic acid are readily lost (Eq. 23).[97] It is believed that the first formed dialdehyde "A" underwent an aldol condensation. Oxidation of **91** with silver oxide in methanol or with potassium permanganate in acetone gave benzoic acid and salicylic acid.

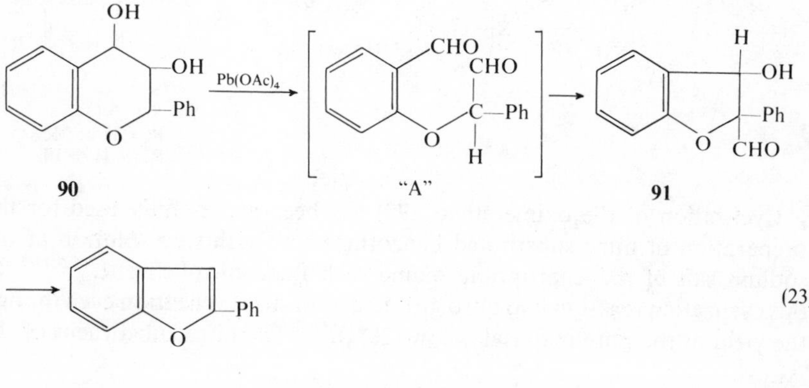

90 "A" **91**

(23)

g. *Algar-Flynn-Oyamada Oxidation of 2′-Hydroxychalcones*

Treatment of 2′-hydroxychalcones (**92**) with aqueous ethanol, sodium hydroxide, and hydrogen peroxide at room temperature for up to one week resulted in the formation of the corresponding flavonol (ca. 30%) and acid (**93**) (up to 30%) which, on decarboxylation, gave the 2-arylbenzofuran (**94**).[98]

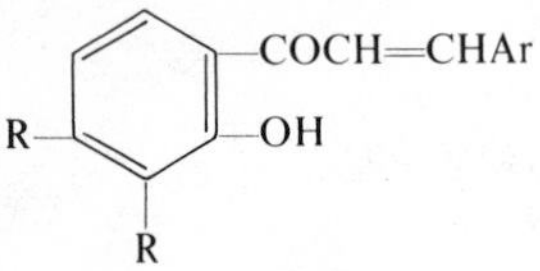

92 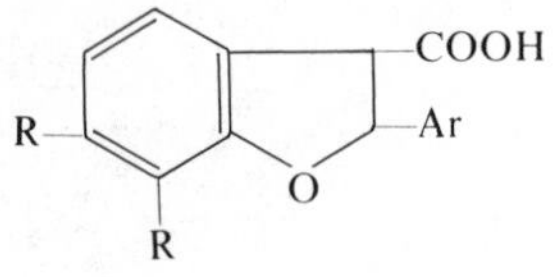 **93** 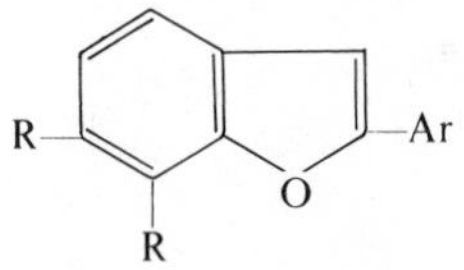 **94**

h. *Acid-Catalyzed Cyclization of O-Aryloximes*

Acid-catalyzed cyclization of *O*-aryloximes (**95**), prepared by boiling the ketones with *O*-phenylhydroxyamine with boron trifluoroetherate in acetic acid at 100°C, leads to the formation of 2-arylbenzofurans (**96**).[99] The reaction mechanism is probably similar to that suggested for the Fischer indole synthesis, and proceeds via intermediates A and B in (Eq. 24).

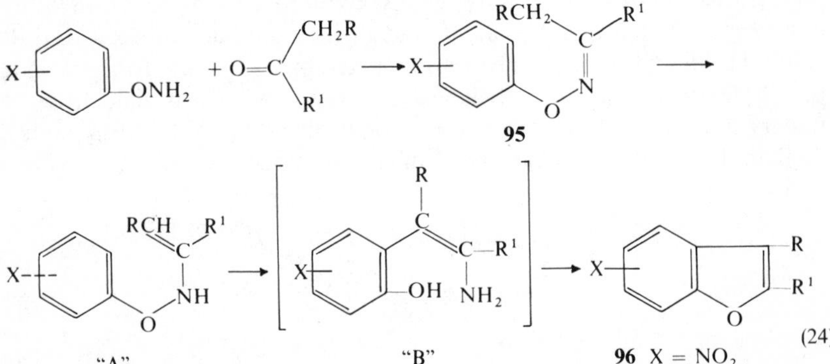

95

"A" "B"

96 $X = NO_2$
$R = H$ or Me
$R^1 = H$ or Ph

(24)

Cyclization of the oxime ethers (**95**) has been successfully used for the preparation of nitro-substituted benzofurans via refluxing solution of the sodium salt of the appropriate oxime with fluoronitrobenzene.[100,101] In this cyclization reaction two nitro groups do not stop cyclization even though the yield of the pure material is low (20%).[101] The nitro substituent of the oximes can be replaced by other electron-withdrawing groups, for example a trifluoromethyl group, without having any adverse effect upon the subsequent cyclizations.[101,102]

i. Photolytic Cyclizations

Benzoin acetate has been observed to undergo photolytic cyclization to 2-phenylbenzofuran.[103] Desyl chloride, benzoin tosylate, 4,4′-dimethoxybenzoin acetate, 3,3′-dimethoxybenzoin acetate, and desyldimethylamine hydrochloride also cyclize to the corresponding 2-arylbenzofurans (Table I).

TABLE I

Reactants (Solvent)	Products of irradiation
Benzoin acetate (benzene) Desyl chloride (benzene) Benzoin tosylate (benzene) Desyldimethylamine HCl (water)	2-Phenylbenzofuran
4,4′-Dimethoxybenzoin acetate (benzene)	2-(4′-Methoxyphenyl)-6-methoxybenzofuran
3,3′-Dimethoxybenzoin acetate (benzene)	2-(3′-Methoxyphenyl)-5-methoxybenzofuran + 2-(3′-methoxyphenyl)-7-methoxybenzofuran

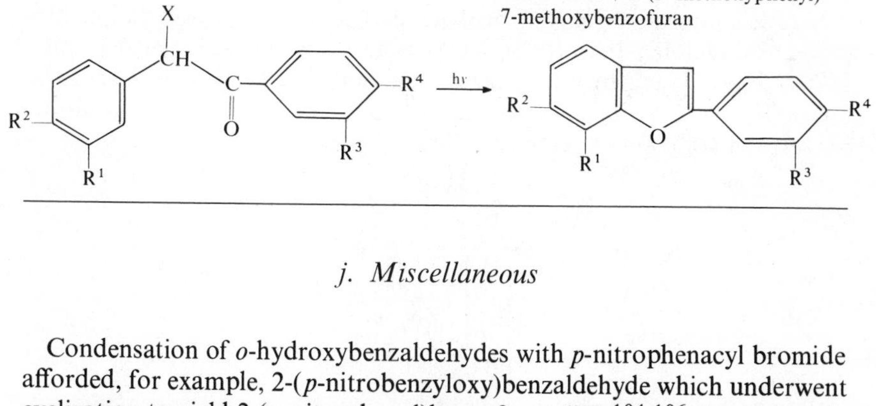

X, CH, C, O, R¹, R², R³, R⁴, hν, R², O, R¹, R³, R⁴

j. Miscellaneous

Condensation of *o*-hydroxybenzaldehydes with *p*-nitrophenacyl bromide afforded, for example, 2-(*p*-nitrobenzyloxy)benzaldehyde which underwent cyclization to yield 2-(*p*-nitrophenyl)benzofuran (**97**).[104-106] Consequently reduction of **97** furnished the corresponding aminobenzofurans (**98**).[107] Also **97** can be obtained by condensation of *p*-hydroxybenzaldehyde with *p*-nitrophenacyl bromide in presence of potassium carbonate.[106]

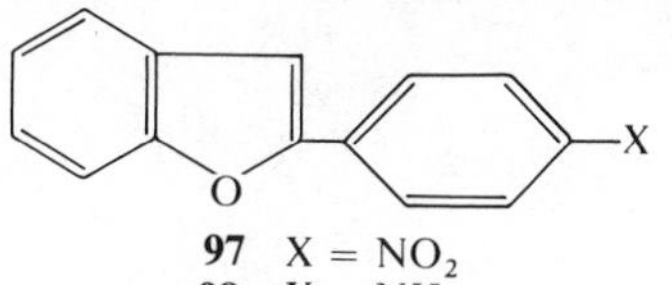

X, O

97 $X = NO_2$
98 $X = NH_2$

3-Halo-substituted arylbenzofurans have been found to be accessible via condensation of *p*-chloro-, 2,4-dichloro-, and/or 2,4,5-trichlorophenol with $ClCH_2CH(OMe)_2$ in presence of sulfuric acid (96%), thus yielding 3-(2-hydroxy-5-chlorophenyl)-5-chloro- (**99**), 3-(2-hydroxy-3,5-dichlorophenyl)-5,7-dichloro- (100), and 3-(2-hydroxy-3,5,6-trichlorophenyl)-4,5,7-trichlorobenzofuran (**101**),[108] respectively.

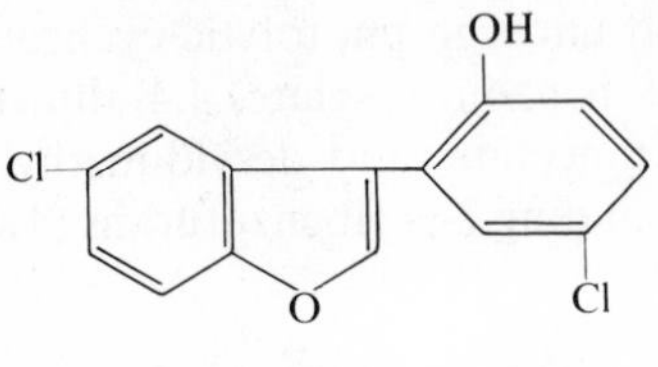

99

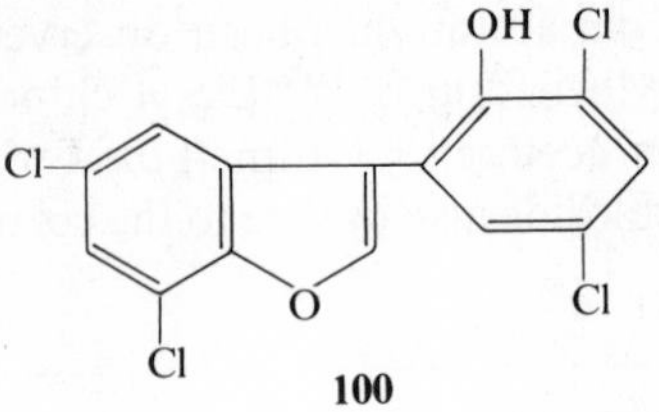

100

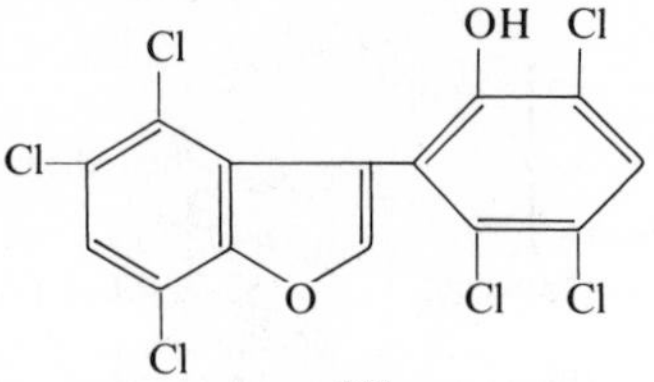

101

Pyrodecomposition of the benzoyl derivative of *o*-hydroxydiphenylacetic acid lactone, and of *o*-hydroxy-*p*-chlorodiphenylacetic acid lactone at 280–290°C liberated water, carbon dioxide, and benzoic acid with the formation of 2,3-diphenylbenzofuran (**102**) and 2,3-diphenyl-4-chlorobenzofuran (**103**),[109] respectively.

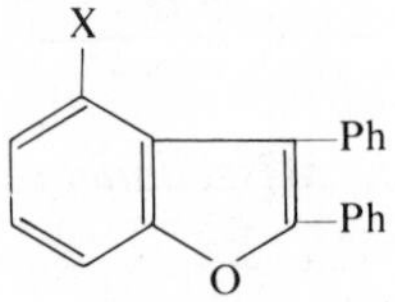

102 X = H
103 X = Cl

Aryl iodides bearing ortho nucleophilic substituents, when refluxed with copper acetylides in pyridine, underwent cyclization to the appropriate heterocycle extensively (Eq. 25).[110]

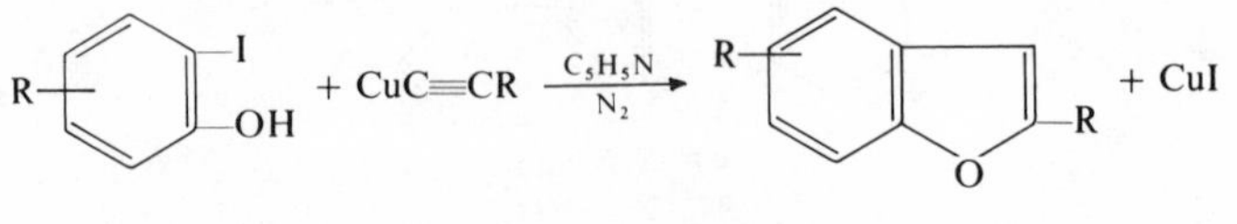

$$\text{R–}C_6H_3(\text{I})(\text{OH}) + \text{CuC}\equiv\text{CR} \xrightarrow[N_2]{C_5H_5N} \text{R-benzofuran-2-R} + \text{CuI} \qquad (25)$$

Treatment of the addition product between cyclohexanone and benzil (**104**) with acidic dehydrating agents resulted in the formation of the dihydrobenzofuran **105** which, upon distillation, underwent disproportionation yielding the tetrahydrobenzofuran **106** and the polymeric material (Eq. 26).[111]

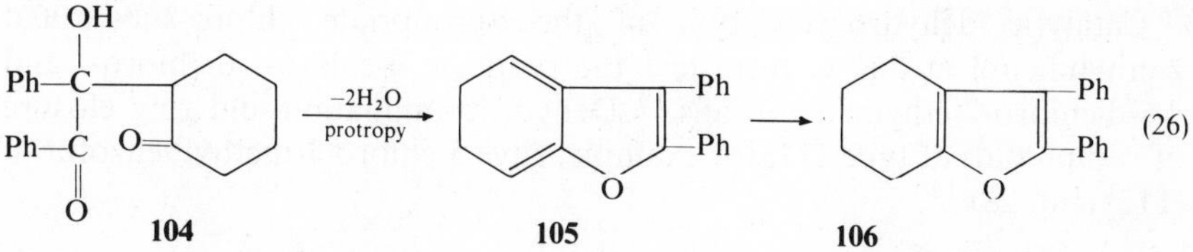

$-2H_2O$, protropy

104 **105** **106** + $(C_{20}H_{14}O)_x$ (26)

It should be noted that, whereas in these cyclohexanones the unsaturation is not a part of a stable aromatic system, dehydration and aromatization occur simultaneously with the addition reaction as was shown with 3,5-dimethylcyclohexenone. This substance gave rise to the phenolic ketone **107,** which was smoothly converted to 2,3-diphenyl-4,6-dimethylbenzofuran (**108**) by distillation under reduced pressure (Eq. 27).

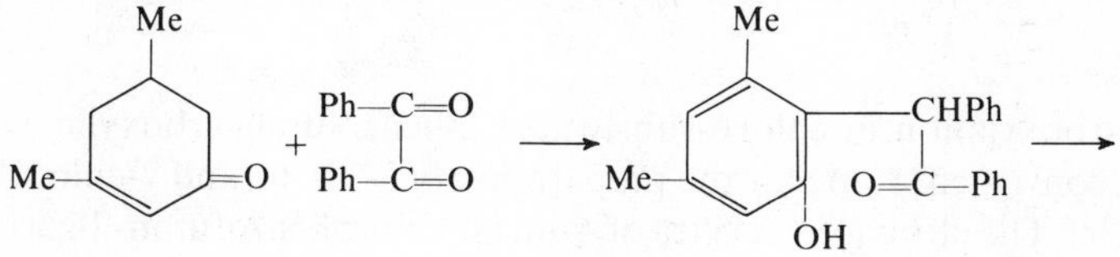

107

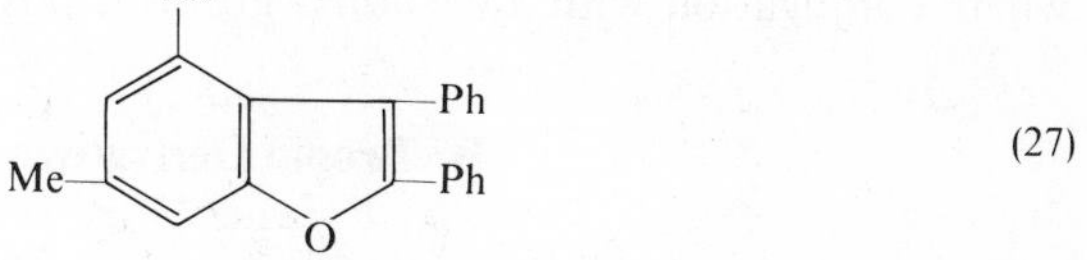

108 (27)

Treatment of compound **109,** obtained by condensation of ethyl α,β-dioxobutyrate with *p*-cresol in presence of anhydrous zinc chloride and acetic acid–hydrochloric acid mixture, gave high yield of **110.**[112] This reaction is of a wide scope, and has been successfully studied with other phenols, leading to analogs of **110.**

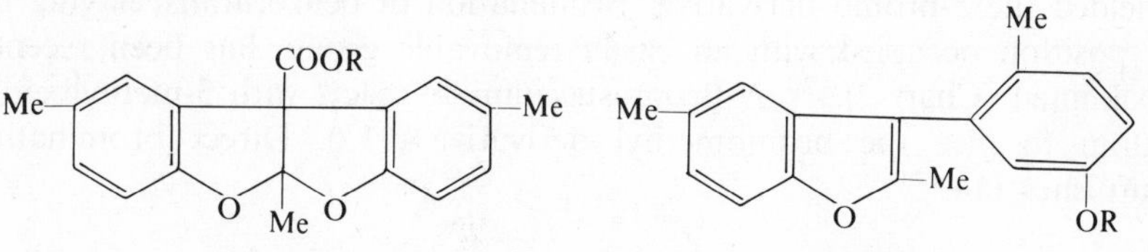

109 R=Et **110** R=H

4. Halobenzofurans

A. Chloro Derivatives

Catalytic dehydrocyclization of the appropriate chloro-substituted 2-ethylphenol at 250°C furnished the requisite 4-chloro-, 6-chloro-, and 4,6-dichloro-2-ethylbenzofurans.[12] Dehydrobromination and ring closure of compounds of type **111**, for example, gave 5-chloro-2-methylbenzofuran (**112**) (Eq. 28).[113]

Cl— —$CH_2CH{=}CH_2$ —OAc $\xrightarrow{Br_2}$ Cl— —$CH_2CHBrCH_2Br$ —OAc $\xrightarrow{NaOH,\ EtOH}$

111

Cl— —Me O (28)

112

Decarboxylation of chloro-substituted 2-benzofurancarboxylic acids affords a convenient route for the preparation of 3-, 5-, 6-, and 7-chlorobenzofurans.[114] The ultraviolet spectra of 4- and 6-chlorobenzofuran-2-carboxylic acid exhibit enhanced bathochromic effects which are caused by chlorine where conjugation with a carbonyl group is possible.

B. Bromo Derivatives

Bromination of benzofuran was reported to give the unstable 2,3-dibromobenzofuran.[115-117] However 2-methyl- and 2-phenylbenzofuran gave the corresponding 3-bromo derivatives. The potassium salt of 2-benzofurancarboxylic acid gave the 2-bromo derivative on bromination with elimination of the carboxyl group.[118] However Smith[119] found that 2-benzofurancarboxylic acid did not undergo bromination; its ethyl ester yielded the 5-bromo derivative. Bromination of benzofurans, having the 2 position occupied with an easily removable group, has been recently examined (Chart 2).[120] *N*-Bromosuccinimide reacts with 3-methylbenzofuran to give the bromomethyl derivative (**113**). Direct bromination furnishes **114**.[121]

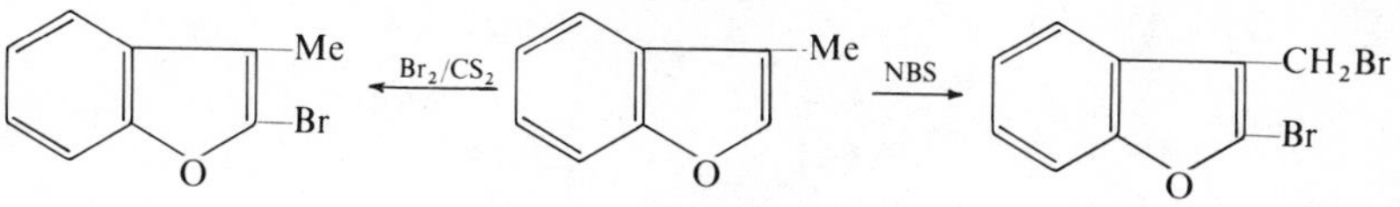

114 113

C. Iodo Derivatives

In the "iodocyclization reaction," iodine combines by means of a double bond when a nucleophilic group is present in the molecule to form a heterocyclic ring.[122] The iodocyclization reaction has led to the formation of almost quantitative yields of 2-iodomethyl-5,7-dichloro-2,3-dihydrobenzofuran (**115**) and of 2-iodomethyl-7-methyl-5-propyl 2, 3-dihydrobenzofuran (**116**).

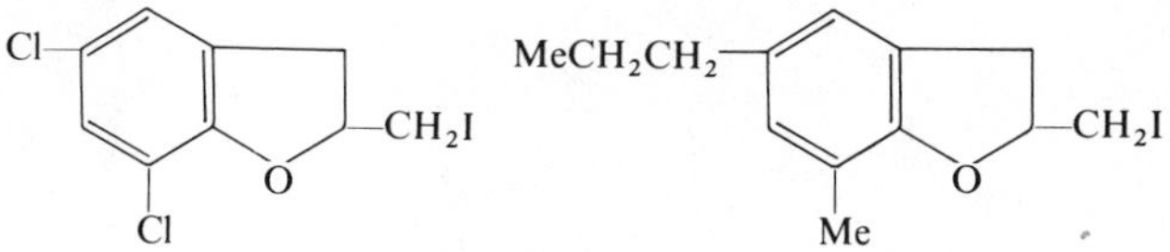

115 116

D. Fluoro Derivatives

Fluoro-substituted 2-ethylbenzofurans at C-4 and/or C-6 have been obtained upon cyclization of the requisite fluoro-2-ethylphenol.[12] 2,4,5-Trifluoro-3-methoxyphenoxyacetic acid, obtained by condensation of 2,4,5-trifluoro-3-methoxyphenol with bromoacetic acid in presence of potassium carbonate, is treated with butyllithium in tetrahydrofuran, followed by carbonation to yield 4,5,7-trifluoro-6-methoxy-3(2H)-benzofuranone (**117**). This, on reduction with sodium borohydride, is followed by dehydration with phosphorus pentoxide and affords 4,5,7-trifluoro-6-methoxybenzofuran (**118**) (Eq. 29).[123]

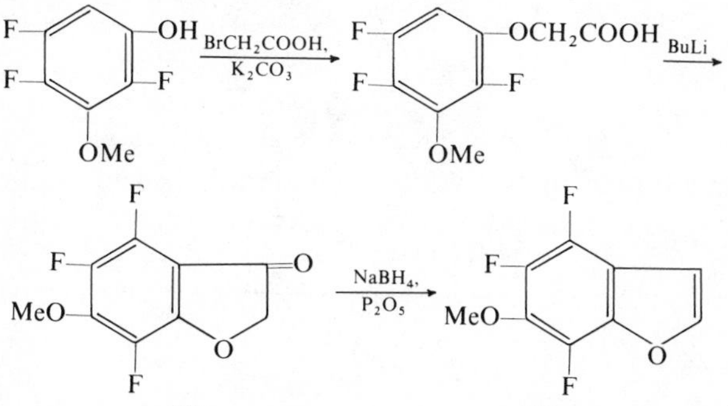

(29)

117 118

C_6H_6, $AlCl_3$

Br_2(1 mol.), AcOH

Br_2/AcOH ; heat

Me_2SO_4, NaOH

excess Br_2/AcOH

Br_2/AcOH ; room temp.

Me_2SO_4, NaOH

Cu, quinoline

liquid Br_2, overnight

Ac_2O, NaOAc

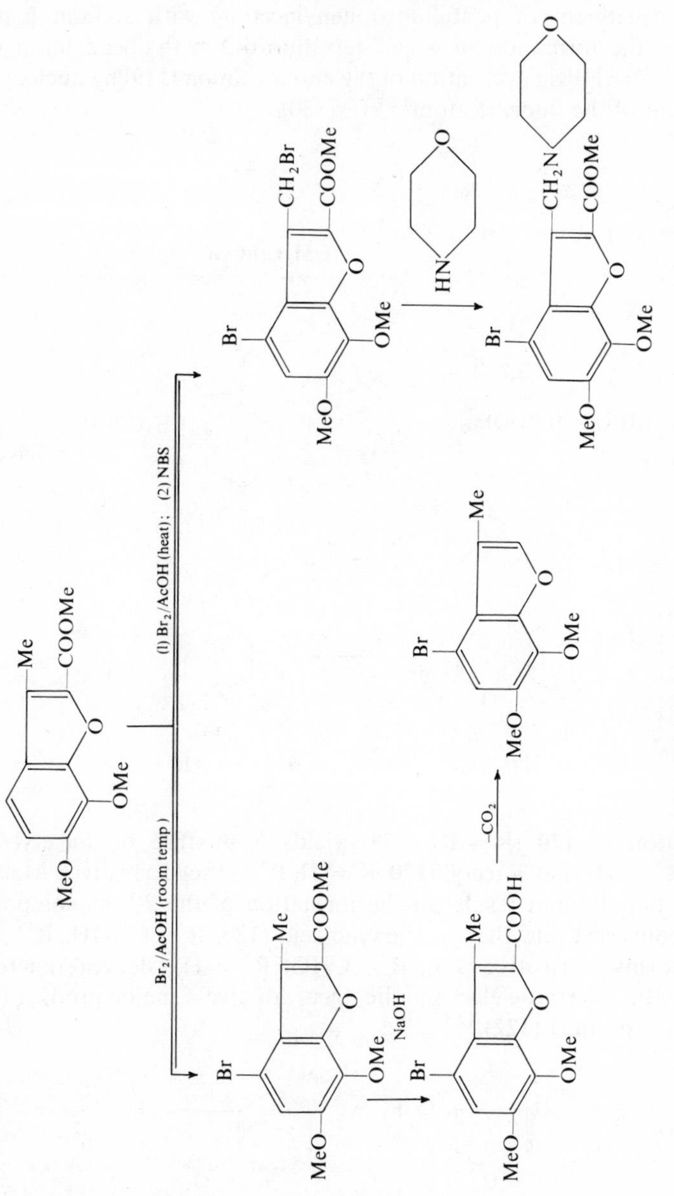

NBS = N—bromosuccinimide

Chart 2.

Similarly, treatment of pentafluorophenylacetone with sodium hydride resulted in the formation of 4,5,6,7-tetrafluoro-2-methylbenzofuran (**120**, R = Me, R^1 = H) via cyclization of the enolate anion (**119**) by nucleophilic replacement of the fluorine atom[124] (Eq. 30).

$C_6F_5CH_2COCl$ + $EtOMgCH(COOEt)_2$ ⟶

$C_6F_5CH_2COCH(COOEt)_2$ $\xrightarrow[\text{boil (1.5 hr)}]{H_2SO_4,}$ $C_6F_5CH_2COCH_3$ $\xrightarrow[\text{2) } HCONMe_2]{\text{1) NaH,}}$

[**119**] ⟶ **120** (30)

Acetylation of **120** (R = R^1 = H) yields a mixture of 2-acetyl-(**120**, R = Ac, R^1 = H), and 3-acetyl-(**120**, R = H, R^1 = Ac) derivatives. Metallation with butyllithium results in the formation of the 2-lithiocompound, which is converted into the 2-carboxylic acid (**120**, R = COOH, R^1 = H) and the formyl derivative (120, R = CHO, R^1 = H). Benzofuran reacts with CF_3OF, a versatile electrophilic agent, to give a major product (**121**) and a minor product (**122**).[125]

121 **122**

5. Nitrobenzofurans

Direct nitration of benzofuran with nitric acid in acetic acid medium gives 2-nitrobenzofuran, and 5-nitrobenzofuran.[18,116,126] When the 2 position of benzofuran is occupied by a group replaced only with difficulty, nitration occurs in the benzene ring (Eq. 31).[126] However, when a group capable of being replaced is present in the 2 position, displacement sometimes occurs on nitration.[116] 2-Nitrobenzofuran has been obtained via condensation of salicylaldehyde with bromonitromethane in the presence of sodium hydroxide (Eq. 32).[127]

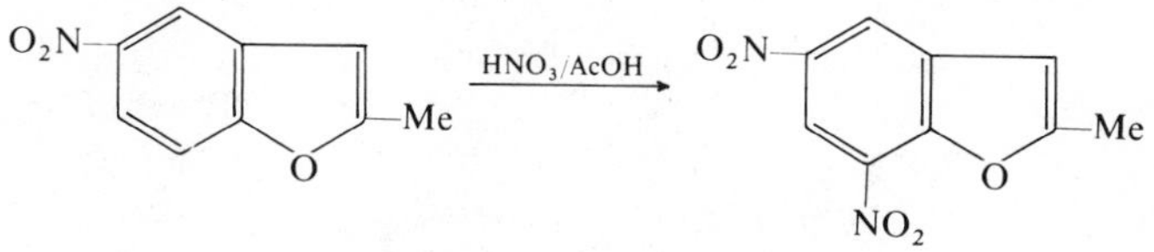

(31)

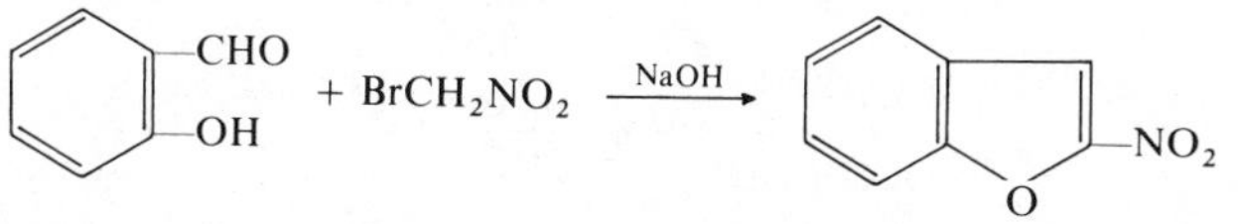

(32)

5-Nitro-2-methylbenzofuran was the earliest example which was obtained via cyclization of 2-hydroxy-5-nitrophenylacetone.[127] Cyclodehydration of 3-(*m*- and/or *p*-nitrophenoxy)butanones has been recently achieved, thus leading to the formation of the corresponding nitro-substituted 2,3-dimethylbenzofurans, for example, Eq. 33.[128]

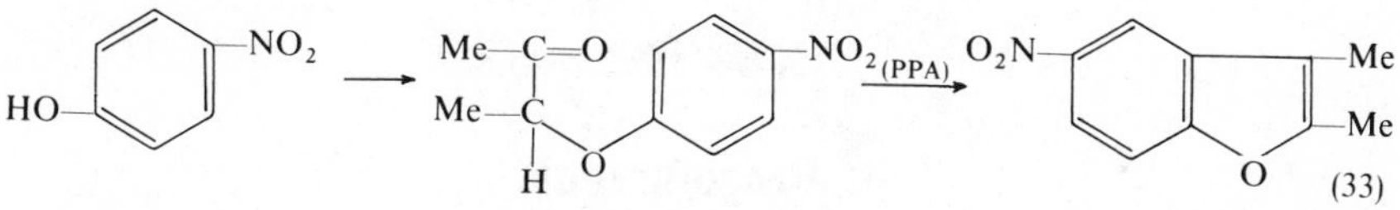

(33)

Severin et al.[129] have synthesized 2-methyl- (**123**), and 2-phenyl-4,6-dinitrobenzofuran (**124**) after the sequence of reactions shown in Eq. 34.

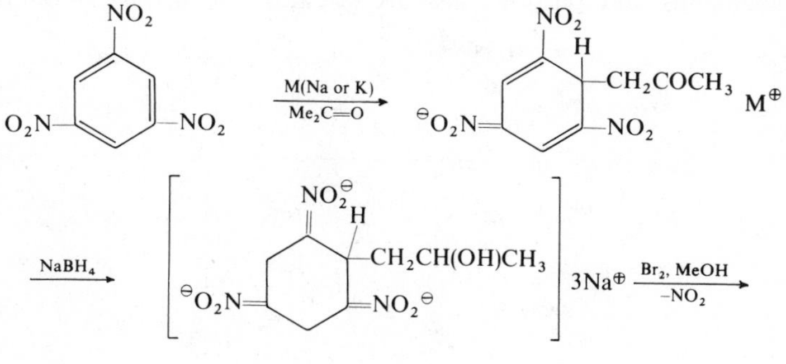

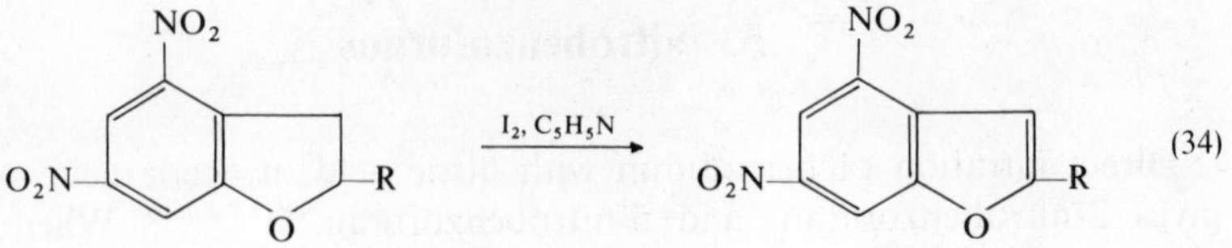

(34)

123 R = Me
124 R = Ph

Synthesis of 5-nitro-7-methoxybenzofuran (**125**) has recently been achieved (Eq. 35).[130]

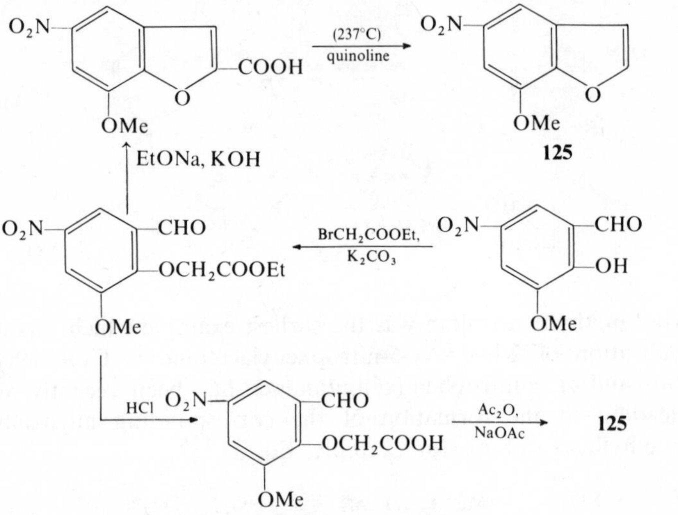

125

(35)

6. Benzofuranols

A general synthesis of benzofuranols,[131] including condensation of alkoxybenzoins and phenols, has been achieved (Eq. 36). With phenol

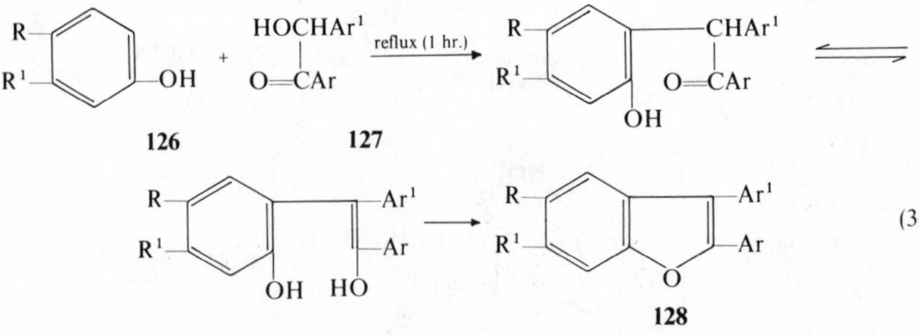

126 **127**

(36)

128

itself the furan synthesis occurs only to the extent of 4% to yield **128** (R = R^1 = H, Ar = Ar^1 = C_6H_4OMe-*p*), because benzyl alcohols condense preferably with a position para to a phenolic hydroxyl group.[132] When the para position is occupied with *p*-cresol for example, the yield of the benzofuran is increased. With resorcinol, it is possible to omit the methylation and to isolate 6-hydroxybenzofuran, for example (**128**, R = H, R^1 = OH, Ar = Ar^1 = C_6H_4OMe-*p*).

Benzoin does not react with phenol in presence of hydrogen chloride,[131,132] but introduction of a methoxyl group ortho or para to the alcoholic group leads to the formation of the benzofuran. 2,2′-Dihydroxy-5,5′-dibromobenzoin (**129**) gave 2-(2′-hydroxy-5′-bromophenyl)-3-methoxy-5-bromobenzofuran (**130**). However saturation of methanolic solution of **129** with hydrogen chloride at − 15°C led to the formation of 2-(2′-hydroxy-5′-bromophenyl)-3-hydroxy-5-bromobenzofuran (**131**) (Eq. 37).[133,134]

MeOH/HCl without cooling

129

MeOH/HCl (− 15°C)

131

130

(37)

Decarboxylation of hydroxy-substituted benzofuran-2-carboxylic acids, obtained by treatment of the appropriate hydroxy-substituted 3-bromocoumarin, afforded a route for the preparation of benzofuranols. Thus the acids, obtained by condensation of 4-methoxy-, 5-methoxy-, and 4,5-dimethoxysalicaldehyde with ethyl bromomalonate and/or methyl ethyl ketone in presence of potassium carbonate, with copper and quinoline gave 6-methoxy-, 5-methoxy-, and 5,6-dimethoxybenzofuran, respectively.[135,136] Similarly, 7-benzyl-6-hydroxy-3-methylbenzofuran is prepared from 8-benzyl-7-hydroxy-4-methylcoumarin.[137,138] Saponification of 5-methoxybenzofuran-3-carboxylic esters with concentrated sulfuric acid in acetic acid medium is accompanied by easy decarboxylation.[138]

Another approach is the condensation of resorcinol, pyrocatechol, hydroquinone, *o*-nitrophenol, 4-ethoxyphenol, and/or 2-methoxyphenol with the appropriate α-haloketone in presence of potassium carbonate, leading to the formation of 6-hydroxy- and 5-hydroxy-substituted benzofurans.[139,140]

5-Methoxybenzofurans are readily obtained by the interaction of quinones and metallic enolates. Smith and MacMullen[141,142] have reported the formation of **132** and **133** via reaction of trimethylquinone with enolate of acetoacetic ester. It is believed that such reaction proceeds through a primary 1,4-addition of the enolate to the quinone-conjugated system which carried no substituent in the β position to give **134**, followed by usual cleavage and ring closure. Similarly, **133** has been obtained (Eq. 38).

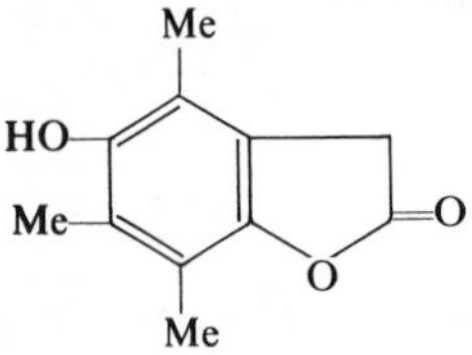

Me, HO, Me, Me, O, O

132

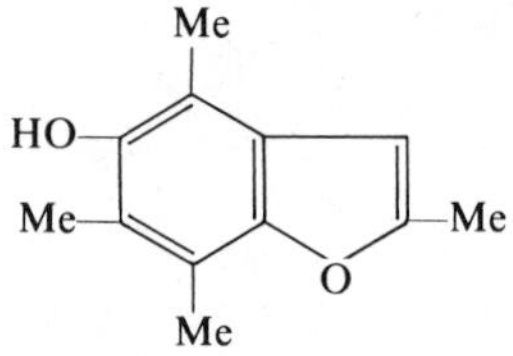

Me, HO, Me, Me, Me, O

133

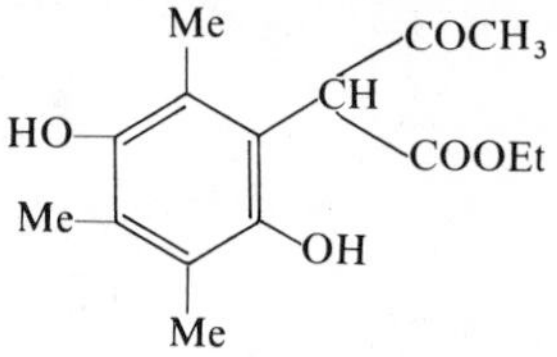

Me, HO, Me, Me, CH, $COCH_3$, COOEt, OH

134

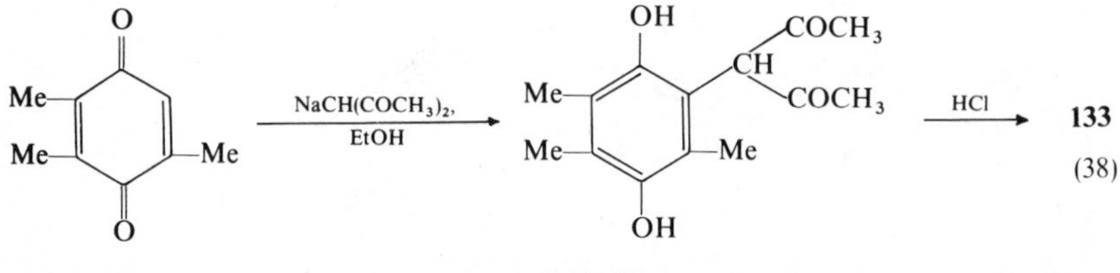

O, Me, Me, Me, O $\xrightarrow[\text{EtOH}]{\text{NaCH(COCH}_3)_2,}$ OH, Me, Me, Me, OH, CH, $COCH_3$, $COCH_3$ $\xrightarrow{\text{HCl}}$ **133** (38)

This reaction has been extensively investigated and establishes a convenient procedure for the preparation of polyalkylated benzofuranols.[143-145]

The following represents the synthesis of 4-benzofuranols, which has been achieved by Reichstein and Hirt[146] (Eq. 39), and Marathey et al. (Eq. 40).[147]

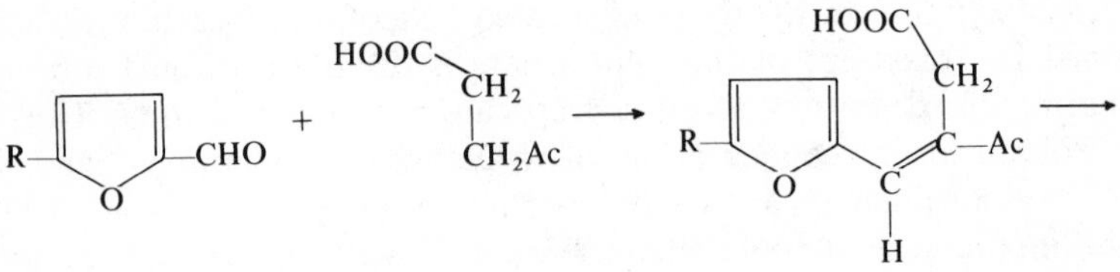

R, O, CHO + HOOC, CH_2, CH_2Ac ⟶ R, O, HOOC, CH_2, C—Ac, C, H ⟶

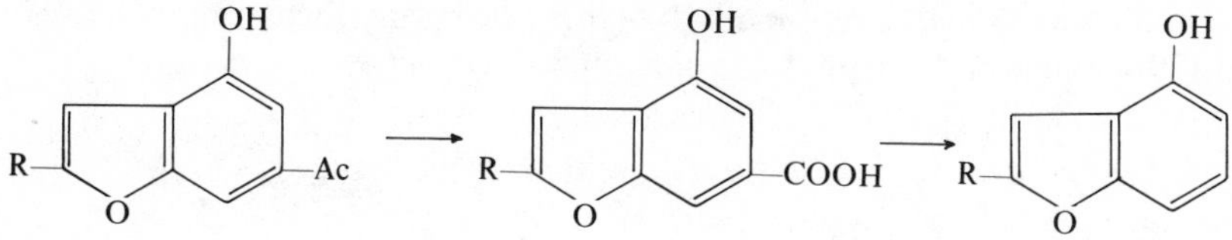

(39)

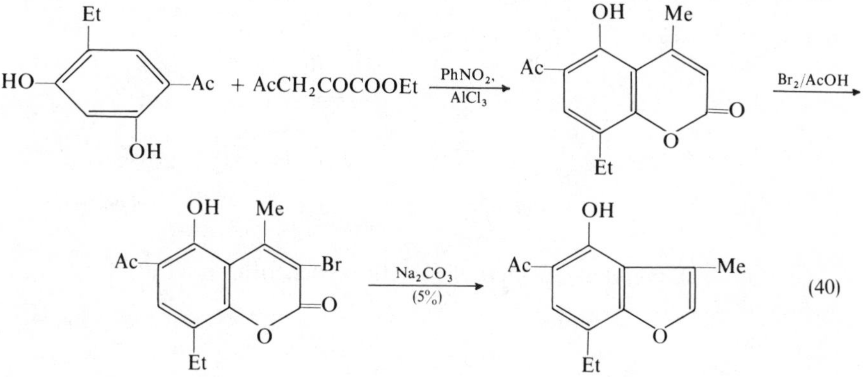

(40)

6-Methoxy-3-acetoxybenzofuran, upon catalytic hydrogenation, loses the elements of acetic acid readily to yield 6-methoxybenzofuran.[148]

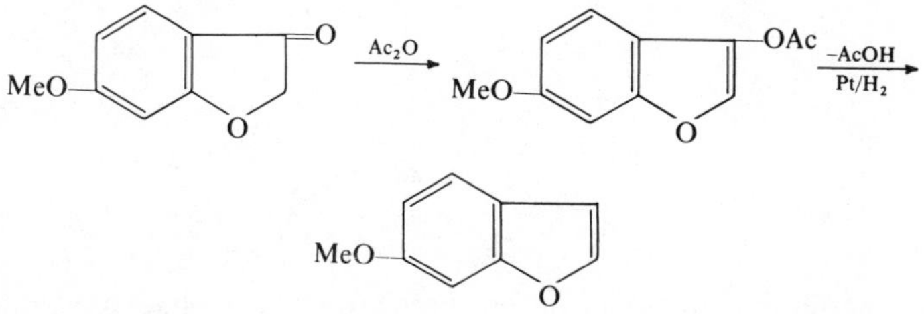

Acidic cleavage of anisopinacolone (*ω,ω,ω*-tris(*p*-methoxyphenyl)-*p*-methoxyacetophenone) (**135**) yielded 2,3-bis(*p*-methoxyphenyl)-6-methoxybenzofuran (**136**) and anisole (Eq. 41).[149]

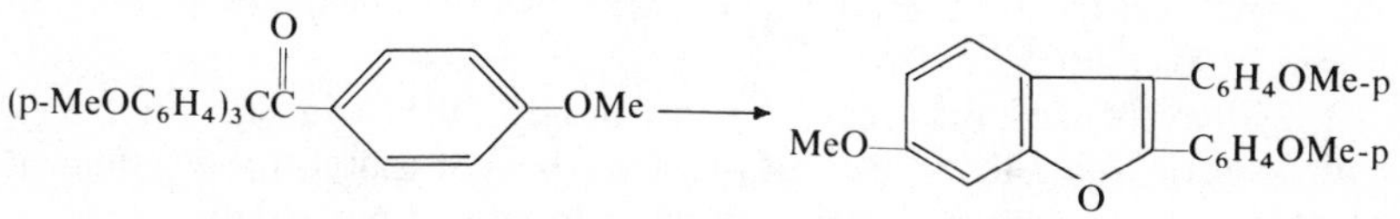

135 **136** (41)

Alkylation of evernininaldehyde (**137**) with bromoacetaldehyde in presence of potassium carbonate is accompanied by ring closure and dehydration to give 6-methoxy-4-methyl-2-formylbenzofuran (**138**). Treatment of **138** with methoxymethylenetriphenylphosphorane yields a 2:1 mixture of *cis*- and *trans*-6-methoxy-4-methyl-2-(β-methoxyvinyl)benzofuran (**139a** and **139b**), separated chromatographically.[150]

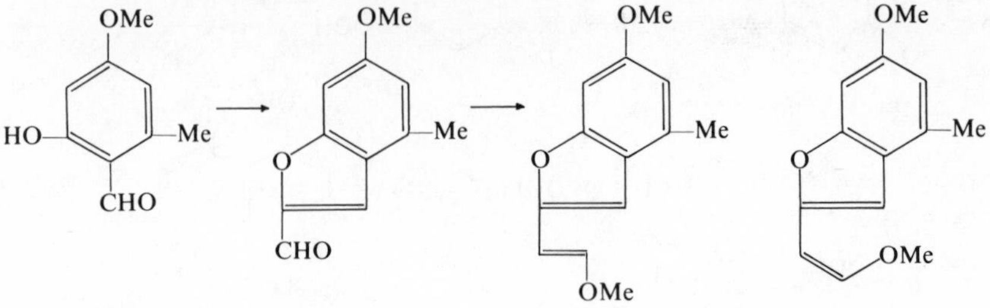

137 **138** **139a** **139b**

Synthesis of 2-phenyl-3-methyl-6-methoxybenzofuran (**140**) has been achieved (Eq. 42).[151]

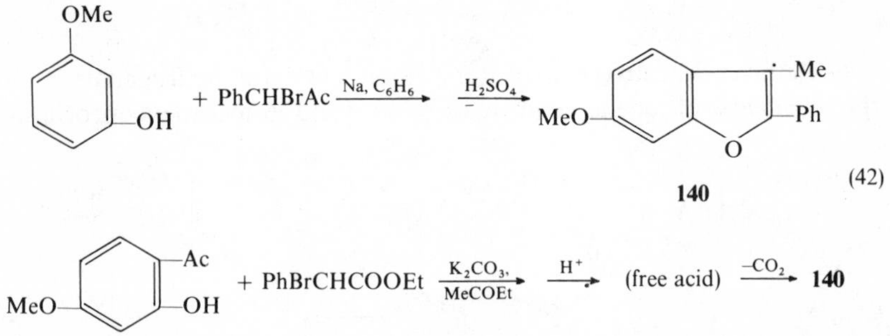

$$\text{+ PhCHBrAc} \xrightarrow{\text{Na, } C_6H_6} \xrightarrow{H_2SO_4} \mathbf{140} \qquad (42)$$

$$\text{+ PhBrCHCOOEt} \xrightarrow[\text{MeCOEt}]{K_2CO_3,} \xrightarrow{H^+} \text{(free acid)} \xrightarrow{-CO_2} \mathbf{140}$$

The thermodynamic ionization constants of bz-hydroxymonoalkyl- or polyalkylated benzofurans have been determined by ultraviolet spectroscopy at 20°C. These compounds are more acidic than the phenols, and it appears that this acidity depends markedly on the relative positions of the hydroxyl group and the hetero atom,[152] but not in a simple way. The ultraviolet spectra of methoxybenzofurans show two bands at 2600–3000 Å and 2100–2550 Å.[153]

3-Hydroxybenzofuran reacts with thioglycollic acid, in presence of hydrochloric acid, to afford *S*-(3-hydroxybenzofuranyl)thioglycollic acid which is readily oxidized to the corresponding sulfoxide.[154]

The primary adducts, obtained from Michael addition with methoxy-*p*-benzoquinone, undergo ring closure upon treatment with mineral acids to yield the corresponding quinols, for example **141**, which gives the lactone **142**. The chief crystalline product, isolated from the reaction of 2-methoxy-*p*-benzoquinone with ethyl acetoacetate, is the benzofuran **143**.[155]

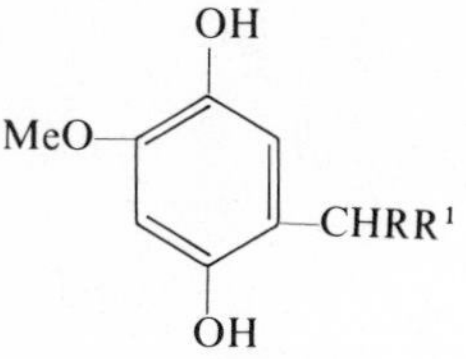

141 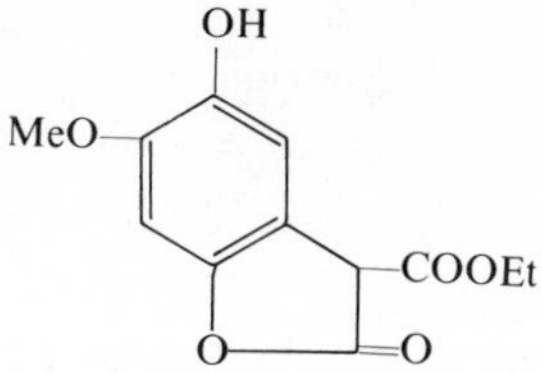 142 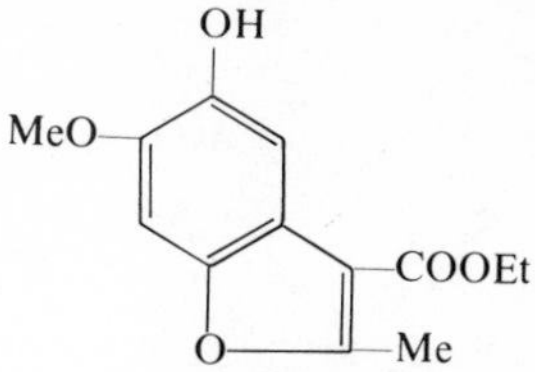 143

Treatment of 3-bromo-6,7-dimethoxycoumarin with sodium carbonate, followed by decarboxylation of the obtained acid, afforded 5,6-dimethoxybenzofuran.[156] Similarly, 3-methyl-5,6-dihydroxybenzofuran-2-carboxylic acid (**144**) has been obtained (Eq. 43), which undergoes decarboxylation to yield **145**.[157]

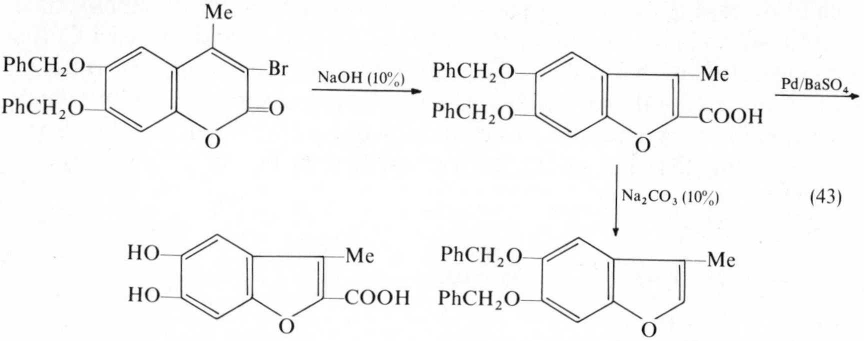

(43)

144 145

Treatment of 6,7-dihydroxy-3(2*H*)-benzofuranone (**146**) with acetic anhydride and sodium acetate in acetic acid medium gave 3,6,7-triacetoxybenzofuran (**147**). The latter yielded, upon catalytic hydrogenation, 6,7-diacetoxybenzofuran (**148**) which, by action of lithium aluminum hydride, provided 6,7-dihydroxybenzofuran (**149**) (Eq. 44).[158]

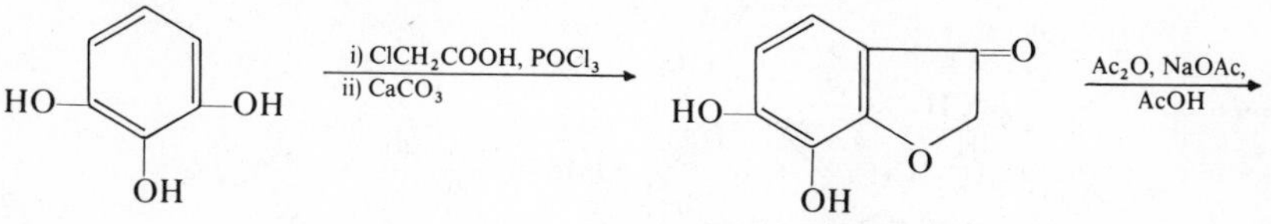

146

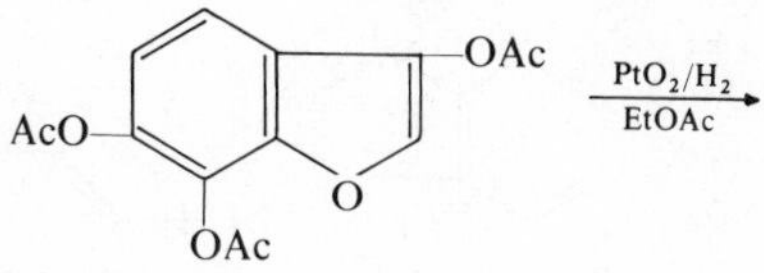

147

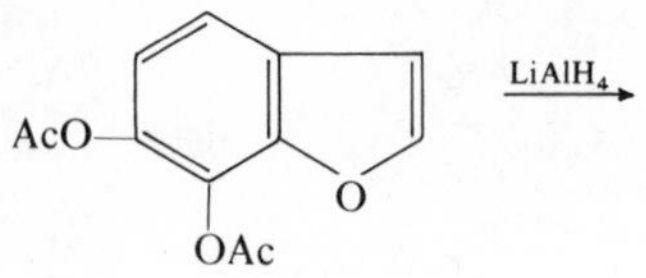

(44)

148 → ($LiAlH_4$) **149**

Cyclodehydration of 3-(mono- and/or dimethoxyaryloxy)-2-butanones to the corresponding methoxy-2,3-dimethylbenzofurans is brought out by action of phosphorus oxychloride; the compounds thus obtained are demethylated with pyridine hydrochloride.[159]

Whalley and Lloyd[160] have shown that, contrary to Freudenberg and Harder's observation,[161] attempts to prepare 2,4,6-trimethoxyphenylacetyl chloride through action of phosphorus pentachloride on the appropriate acid resulted in the formation of an anamolous acid chloride. Friedel-Crafts condensation of this acid chloride with resorcinol resulted in the formation of 3′-chloro-2,4-dihydroxy-2′,4′,6′-trimethoxydeoxybenzoin (**150**) which was readily converted into 3-(3-chloro-2,4,6-trimethoxybenzyl)-6-methoxybenzofuran (**151**) and several derivatives there of Eq. 45.[162]

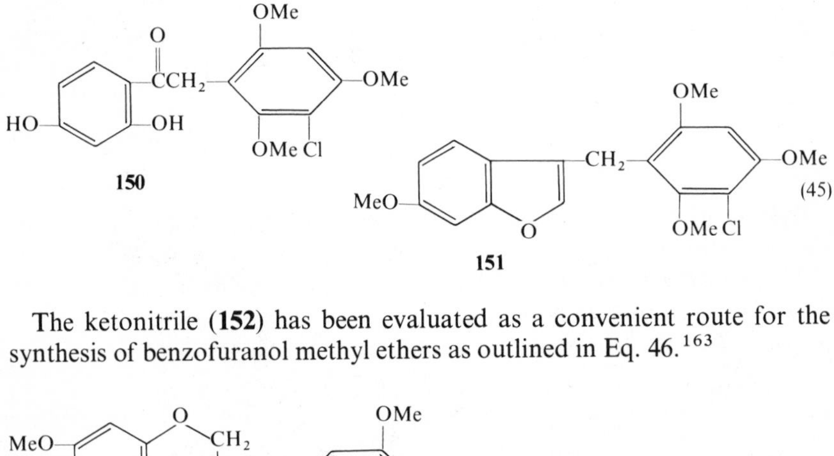

150 **151** (45)

The ketonitrile (**152**) has been evaluated as a convenient route for the synthesis of benzofuranol methyl ethers as outlined in Eq. 46.[163]

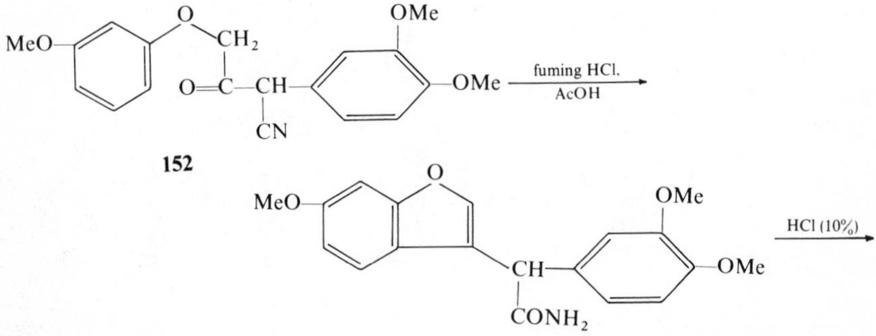

152 → (fuming HCl, AcOH) → (HCl (10%)) →

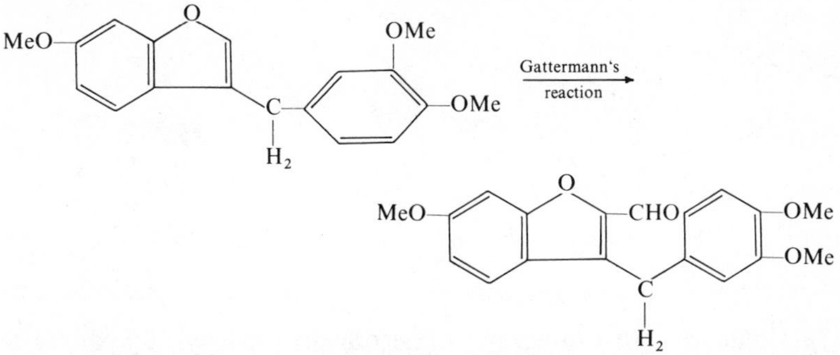

MeO– OMe –OMe Gattermann's reaction

MeO– CHO –OMe –OMe (46)

Periodic oxidation of melacacidin tetramethyl ether has been shown to give 2-(3,4-dimethoxyphenyl)-6,7-dimethoxybenzofuran (**153**). This has been synthesized as outlined in Eq. 47.[164]

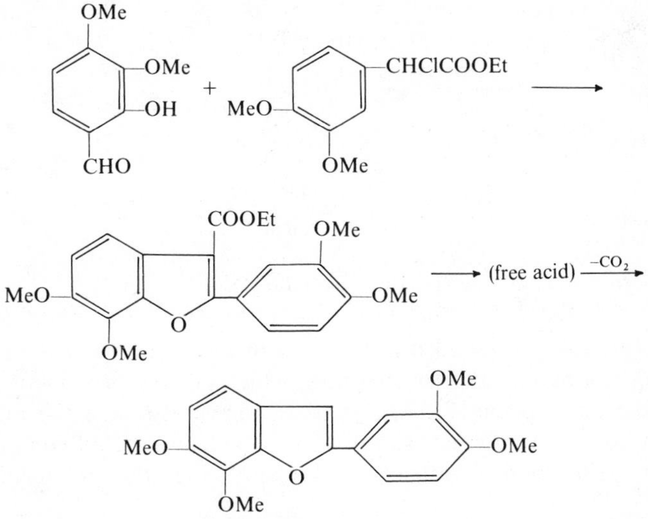

OMe –OMe –OH CHO + MeO– –CHClCOOEt OMe ⟶

COOEt OMe MeO– –OMe OMe ⟶ (free acid) $\xrightarrow{-CO_2}$

OMe MeO– –OMe OMe (47)

153

p-Methoxybenzaldehyde cyanohydrin condenses with phloroglucinol as illustrated in Eq. 48.[165]

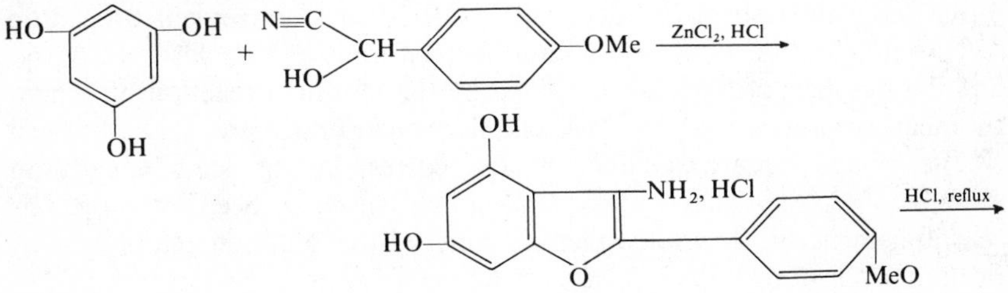

HO– –OH OH + N≡C–CH(HO)– –OMe $\xrightarrow{ZnCl_2,\ HCl}$

OH HO– –NH_2, HCl MeO $\xrightarrow{HCl,\ reflux}$

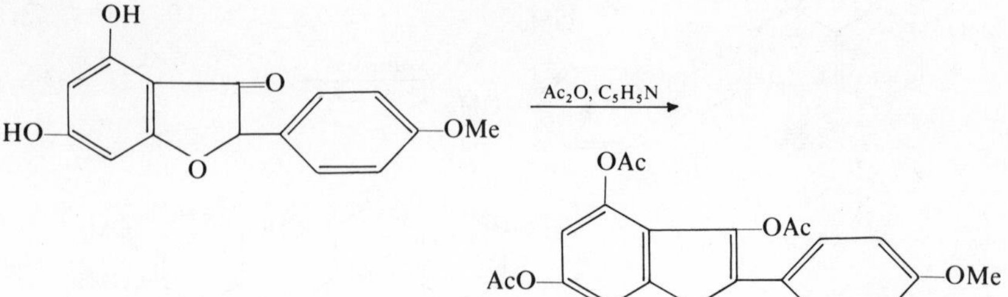

OH

HO — O — OMe

Ac_2O, C_5H_5N

OAc

OAc

AcO — O — OMe

(48)

6-(*O*-(*p*-Coumaroyl)-β-D-glucosyl-2-(3,4-dihydroxyphenyl)-6-hydroxy-4-(β-D-glucosyloxy)benzofuran-3 carboxylate (**155**) is obtained upon oxidation of cyanidin (**154**) on an acidic resin, Duolite C-25.[166]

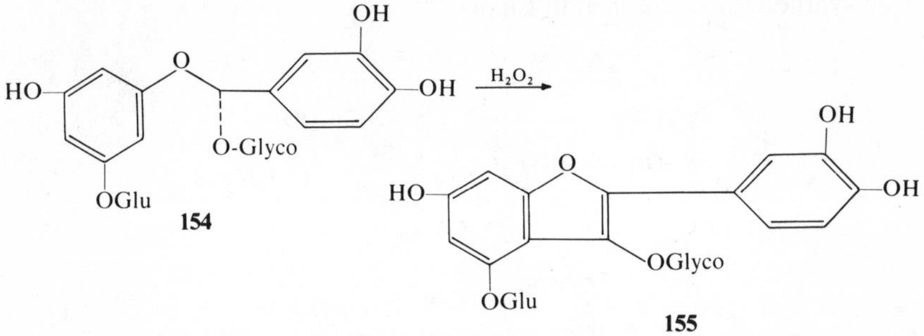

OH

HO — O — OH H_2O_2

O-Glyco

OGlu

154

OH

O

HO — OH

OGlyco

OGlu

155

Glu = β-D-glucosyl; Glyco = 6-*O*-(*p*-coumaroyl)-β-D-glucosyl

Benzofuranols are useful intermediates in the synthesis of the naturally occurring furopyrans and furopyrones, which occupy prominent position among the plant phenols. Their syntheses have attracted the interest of many authors, and reference should thus be made to the different methods cited in "Furopyrans and Furopyrones" and the references cited therein.[167,168]

7. Aminobenzofurans

Reduction of nitrobenzofurans, obtained by cyclodehydration of 3-(*m*- and *p*-nitrophenoxy)butanones, established a convenient procedure for the preparation of 4- and 5-aminobenzofurans.[169,170] Also, 4-, 5-, 6-, and 7-aminobenzofurans are obtained by the Hoffman reaction of benzofurancarboxamides.[171,176-178] Acetamidobenzofurans are also obtained by Beckmann's rearrangement of the corresponding acetylbenzofuran oximes.[171,176-178] The 5-nitro analog has recently been obtained by rearrangement of *O*-(*p*-nitrophenyl)oxime of the butanone, followed by deaminocyclization[172-175] (Charts 3–6).

Condensation of benzoin with *m*- and/or *p*-aminophenols in presence of anhydrous zinc chloride in acetic acid medium gave a mixture of 6-amino- and 6-acetamido-2,3-diphenylbenzofuran, and 5-*N*-acetyldesylamino-2,3-diphenylbenzofuran, respectively,[179] together with colored side-reaction products.

2-Aryl-3-aminobenzofurans are readily obtained by condensation of aldehyde cyanohydrin with the appropriate polyhydric phenol (Eq. 48).[180]

The reaction products of malononitrile with 1,4-benzoquinone which were considered as Michael adducts have now been identified as 2-aminobenzofurans.[181]

Active methylene compounds react with *p*-quinonemonobenzenesulfonimide (**156**) to yield the substituted phenols (**157**). When suspension of **157a** and/or **157c** were heated in constant boiling hydrochloric acid, the respective 5-sulfonamidobenzofurans (**158a, b**) are obtained. Cyclization of **157c** with sulfuric acid (70%) resulted in the formation of **158b.** The cyclization proceeds probably via the enol form[182] by the sequence of the reactions shown in Eq. 49.

NSO_2Ph — $RCOCH_2COR^1$ → $NHSO_2Ph$, C—COR¹, C—R, OH, OH

156 **157**

$PhSO_2NH$—, H, COR^1, R, O → $PhSO_2NH$—, COR^1, R, O

158 (49)

a R = R¹ = Me
b R = Me, R¹ = OEt
c R = Me, R¹ = OH

Photolysis of *N*-benzenesulfonyl-2-*t*-butyl-1,4-benzoquinone-4-imine (**159,** R = *t*-Bu) in acetic acid in sunlight yielded 2,3-dihydro-2,2-dimethyl-5-*N*-benzenesulfonaminobenzofuran (**161**). The latter compound can also be obtained upon heating a mixture of 2-(2-methylallyl)benzenesulfonaminophenol (**160b**) and aqueous hydrobromic acid. Similar treatment of 2-(2-acetoxy-2-methylpropyl)-4-benzenesulfonaminophenol (**160a**) resulted in the formation of **161**. Both **160a** and **160b** are photolytic products in the above reaction (Eq. 50).[183]

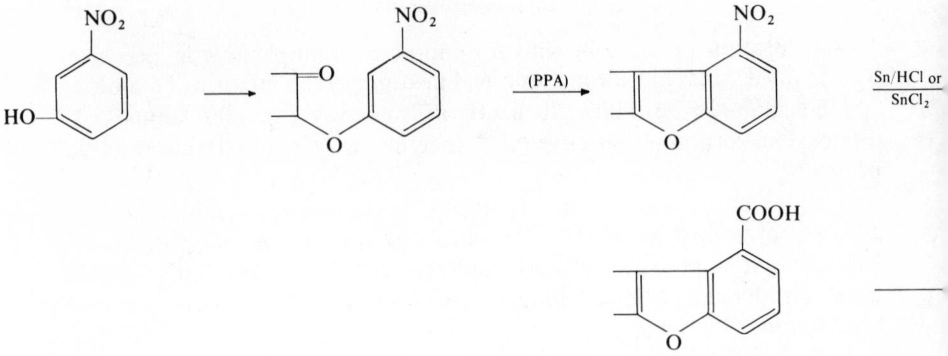

NO_2

HO

O

NO_2

O

(PPA)

NO_2

O

Sn/HCl or $SnCl_2$

COOH

O

Chart 3.

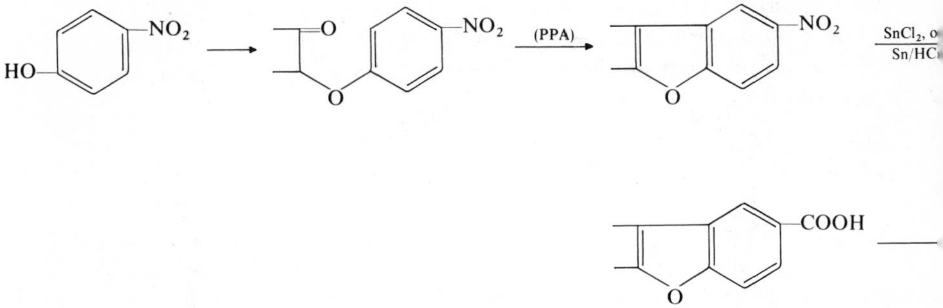

NO_2

HO

O

NO_2

O

(PPA)

NO_2

O

$SnCl_2$, o Sn/HC

COOH

O

Chart 4.

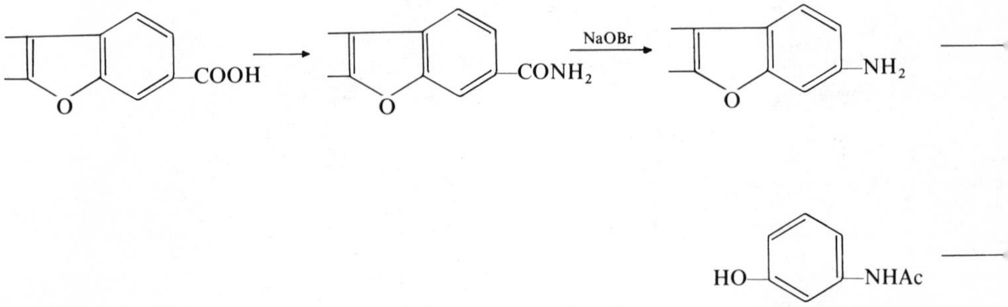

COOH

O

$CONH_2$

O

NaOBr

NH_2

O

HO

NHAc

Chart 5.

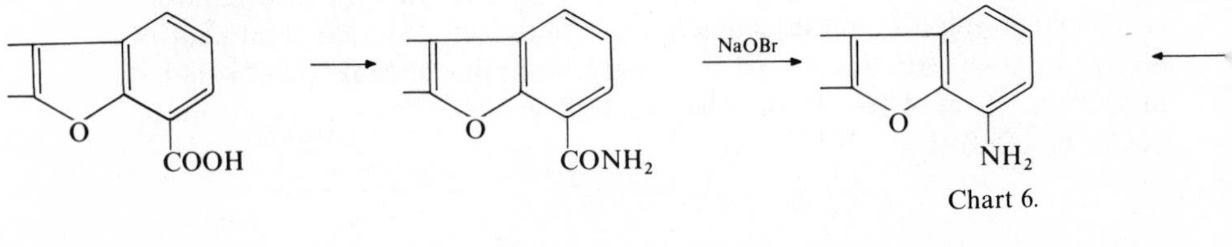

O

COOH

O

$CONH_2$

NaOBr

O

NH_2

Chart 6.

NH_2 NHAc

O O

NaOBr

$CONH_2$

O

NH_2 NHAc

O O

NaOBr

$CONH_2$ O NHAc NHAc

O O HO

NHAc

O

(PPA) or H_2SO_4

O

O NHAc

(PPA) or H_2SO_4 O

O O HO

NHAc NHAc NHAc

$$\mathbf{159} \xrightarrow[\text{AcOH}]{h\nu} \mathbf{160} + \mathbf{161} \qquad (50)$$

159: quinone imine, =O, R, =NSO$_2$Ph

160a R = $CH_2\overset{H}{C}(Me)OAc$
160b R = $CH_2C(Me){=}CH_2$

(160: OH, R, NHSO$_2$Ph)

161: PhSO$_2$NH-substituted 2,2-dimethyl(Me, Me)-2,3-dihydrobenzofuran

Phenyldiethylaminoacetylene reacts with the carbonyl group of *p*-benzoquinone anil (1,2-addition) to give the benzylidene compound **162** which, upon reduction, yields **163**. Phenylmethylamino(*n*-butyl)acetylene, which is less nucleophilic than phenyldiethylaminoacetylene, gave **164** (1,4-addition) (Eq. 51).[184]

$$\text{(O=, =NPh quinone anil)} + \xrightarrow[C_6H_6]{PhC{\equiv}CNEt_2,} \mathbf{162}\ (PhCCONEt_2, =NPh) \xrightarrow[Zr/AcOH]{KBH_4 \text{ or}} \mathbf{163}\ (HC(Ph)CONEt_2, OH)$$

$$\text{(p-benzoquinone)} + BuC{\equiv}CN(Me)Ph \longrightarrow \mathbf{164}\ (HO, Bu{-}n, N(Me)Ph \text{ benzofuran}) \qquad (51)$$

Cyclization of the Mannich bases (**165**), derived from α-phenoxyacetophenone and α-phenoxypropiophenone with polyphosphoric acid after the procedure reported by Davies and Middleton,[185] gave, in high yield, the corresponding 2-dialkylaminomethyl-3-phenylbenzofurans (**166**).[186] Reaction of **165** with Grignard reagents resulted in the formation of the corresponding tertiary carbinols (**167**), which upon treatment with polyphosphoric acid afforded 3-phenyl-2-dialkylaminodihydrobenzofurans (**168**) (Eq. 52).

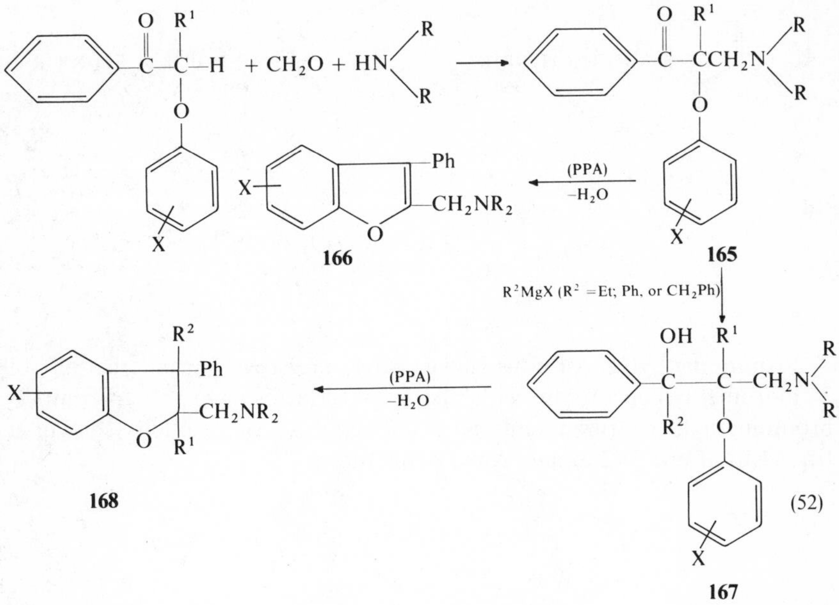

165

166

R^2MgX (R^2 = Et; Ph, or CH_2Ph)

(PPA) $-H_2O$

168

167

(52)

2-(1-Aminoethyl or dialkylaminoethyl)benzofurans of type **169** are obtained by Leuckart reaction.[187] The generalized Leuckart reaction with formylated derivatives of primary and secondary amines produces less resinous products than when formamide is used and results in formation of *N*-substituted derivatives.

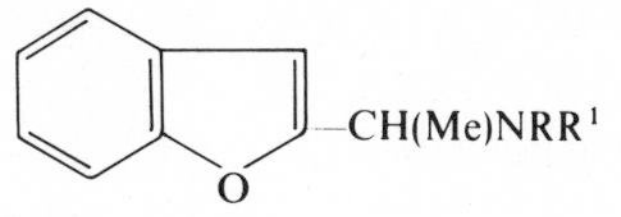

169

$R = R^1 = H$
$R = H, R^1 = Me$
$R = H, R^1 = Et$
$R = R^1 = Me$

Mannich reaction between 2-acetylbenzofurans and secondary amines gave β-aminopropionylbenzofurans (**170**) which are transformed into the corresponding secondary or tertiary alcohols (**171**), either by reduction with sodium borohydride or by Grignard reagents. The alcohols thus obtained are readily dehydrated to yield **172**.[188]

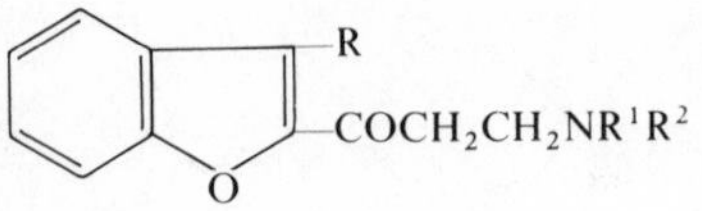

170

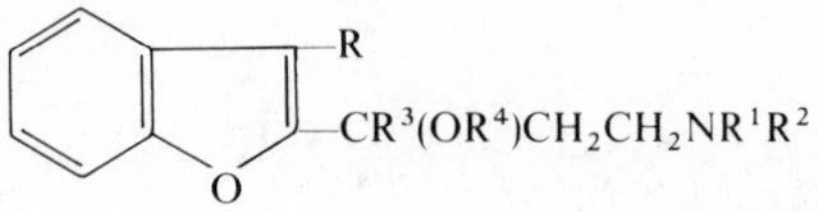

171

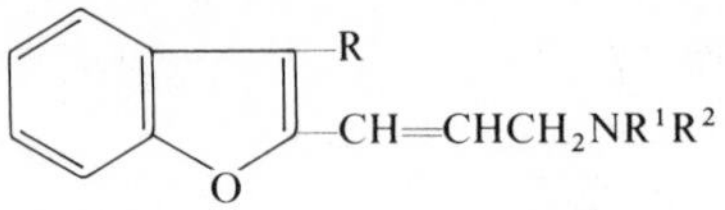

172

2-Aminomethylbenzofurans have been prepared by interaction of 2-chloromethylbenzofuran with the appropriate amine.[189] 2-Bromo-3-bromomethylbenzofuran undergoes the sequence of reactions shown in Eq. 53 to afford 3-(2-aminopropyl)benzofuran.[190]

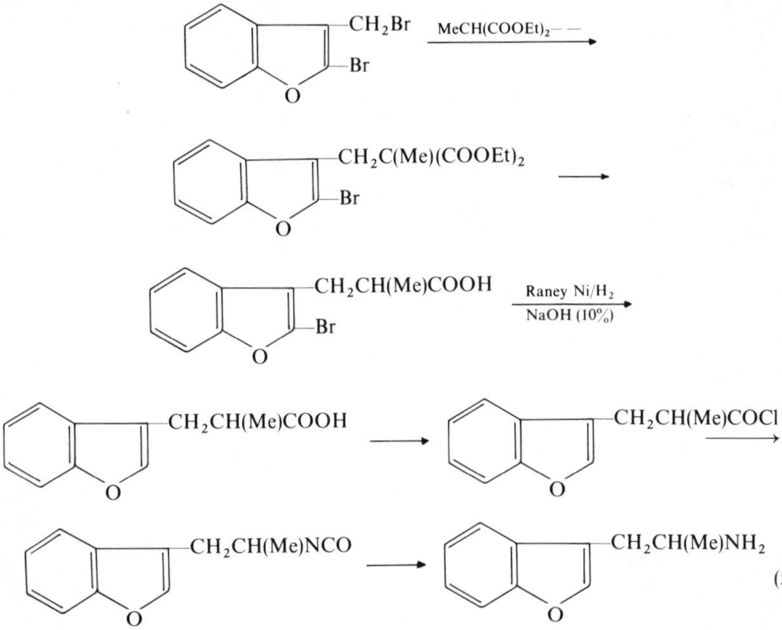

(53)

Similarly, 2-(2-methylaminopropyl)benzofuran was obtained from 2-chloromethylbenzofuran.[191] Both 3-β-(aminoethyl)- and 3-β-(acetamidoethyl)benzofurans have been synthesized[192,194] as outlined in Eq. 54. Synthesis of 3-aminoethyl (or methyl)-2-methyl (or phenyl)benzofurans has been achieved.[193]

R^1 $\rightarrow$ R^1 CN $\rightarrow$ R^1 R^2HN (54)

R^1 = H, or Me; R^2 = H or Ac

Reduction of Schiff bases, obtained from 2-ethyl-3-formyl- and 2-ethyl-3-formyl-7-methoxybenzofurans and the appropriate amine with lithium aliminum hydride establishes a convenient route for the preparation of benzofurans containing an ethylamino group[193] or an *N*-substituted aminoethyl group[195] attached to C-3, which possess hypotensive activity (Eq. 55).[193]

$CH{=}NR^1$, Et, R $\rightarrow$ CH_2NHR^1, Et, R (55)

Equation 56 outlines the synthesis of 5-hydroxy-3-(2-aminoethyl)benzofuran.[196]

$PhCH_2O$, Ac, OCH_2COOEt $\xrightarrow[C_5H_5N,\ 100°C]{SeO_2}$

$PhCH_2O$, COCOOH, OCH_2COOEt $\xrightarrow[\text{xylene}]{K_2CO_3,}$

$PhCH_2O$, COOH, COOH $\xrightarrow[\text{quinoline}]{\text{Cu,}}$

$PhCH_2O$, HOOC $\xrightarrow[HCONMe_2]{SOCl_2,}$ $\xrightarrow{Me_2NH}$

$PhCH_2O$, $CONMe_2$ $\xrightarrow{LiAlH_4}$ (crude product) $\xrightarrow{MeI}$

$$\text{PhCH}_2\text{O-[benzofuran]-CH}_2\text{NMe}_2 \xrightarrow[\text{140–150°C}]{\text{NaCN, HCONMe}_2-} \text{PhCH}_2\text{O-[benzofuran]-CH}_2\text{CN} \xrightarrow{\text{LiAlH}_4}$$

$$\text{PhCH}_2\text{O-[benzofuran]-CH}_2\text{CH}_2\text{NH}_2$$

$$\xrightarrow{\text{Pd-C/H}_2\text{, MeOH}} \text{HCl, H}_2\text{NCH}_2\text{CH}_2\text{-[2,3-dihydrobenzofuran]-OH} \qquad \xrightarrow{\text{ClCOOEt}} \text{PhCH}_2\text{O-[benzofuran]-CH}_2\text{CH}_2\text{NHCOOEt} \tag{56}$$

The ready reduction of substituted benzofurancarboxylic acid amides with lithium aluminum hydride to the corresponding substituted 2- (or 3-) aminobenzofurans and their physiological activities has been the center of interest in patent literature for the preparation of a large number of these compounds (Eq. 57).[197-201]

$$\text{MeO-[benzofuran](-COOH)(-Me)} \xrightarrow[\text{C}_6\text{H}_6]{\text{SOCl}_2,} \text{MeO-[benzofuran](-COCl)(-Me)} \xrightarrow{\text{RNH}_2}$$

$$\text{MeO-[benzofuran](-CONHR)(-Me)} \xrightarrow{\text{LiAlH}_4}$$

$$\text{MeO-[benzofuran](-CH}_2\text{NHR)(-Me)} \xrightarrow{\text{ClCH}_2\text{CH}_2\text{NEt}_2\text{, HCl}}$$

$$\text{MeO-[benzofuran](-CH}_2\text{NRCH}_2\text{CH}_2\text{NEt}_2\text{)(-Me)} \tag{57}$$

Synthesis of 3-(β-aminoethyl)-5-hydroxybenzofuran, an isosteric compound of sertinin,[202] has been achieved (Eq. 58).[203]

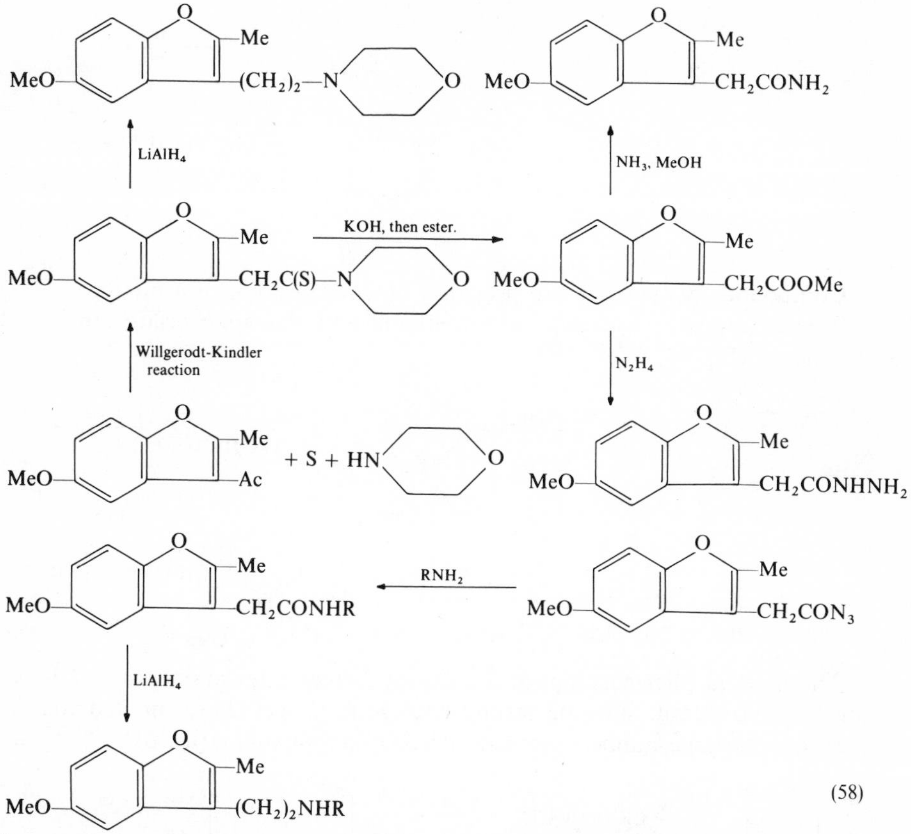

(58)

The β-Alanineamides **173** and **174**, containing the benzofuran nucleus, are obtained by heating a mixture of *N*-substituted 2-chloropropionamide and an amine, derived from substituted benzofuran or 2,3-dihydrobenzofuran.[204]

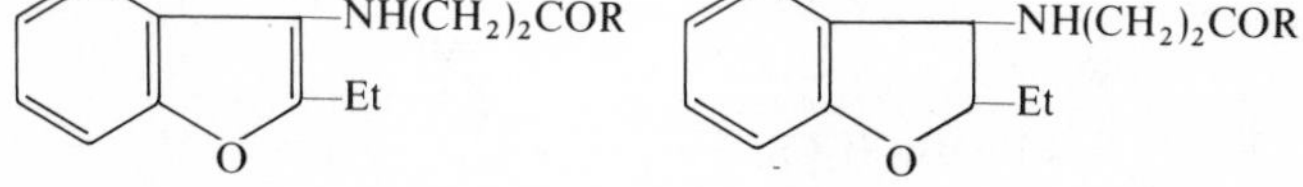

173 **174**

R = NHMe, NHEt, piperidino or morpholino

The preparation of benzofuran derivatives containing 2-(α-hydroxy-β-substituted aminoethyl) has been reported.[205-209] Reduction of 2-ethyl-3-formylbenzofuran cyanohydrin with lithium aluminum hydride yields 2-(2-ethyl-3-benzofuranyl)-2-hydroxyethylamine, (Eq. 59) (**175**).[206]

$\xrightarrow{NaHSO_3,\ KCN}$ CH(OH)CN, Et $\xrightarrow{LiAlH_4}$ $CH(OH)CH_2NH_2$, HCl; Et (59)

175

Compounds of type **176** are prepared by condensation of a mixture of 1-(2-benzofuranyl)-1-hydroxy-2-chloroethane and the appropriate amine in ethanol (Eq. 60).[205,207-209]

$COCH_2Cl$ $\xrightarrow{NaBH_4}$ $CH(OH)CH_2Cl$ $\xrightarrow{RNH_2}$ $CH(OH)CH_2NHR$ (60)

176

The marked pharmacological activity of 2-(monoalkylaminoethyl)-2,3-dihydrobenzofuran, showing strong analegesic properties, prompted the synthesis of a large number of structurally related compounds (Eq. 61).[210,211]

Br, Me, Me, CHO, OH + $ClCH_2COCH_3$ $\xrightarrow{KOH}$ Br, Me, Me, Ac $\xrightarrow{S_2Cl_2}$ Br, Me, Me, $COCH_2Cl$ $\xrightarrow{NaBH_4}$ Br, Me, Me, $CH(OH)CH_2Cl$ $\xrightarrow{iso\text{-}PrNH_2}$ Br, Me, Me, $CH(OH)CH_2NHCHMe_2$ (61)

The synthesis shown in Eq. 62 illustrates the preparation of amino-alkanols, containing the dihydrobenzofuranyl residue.[212]

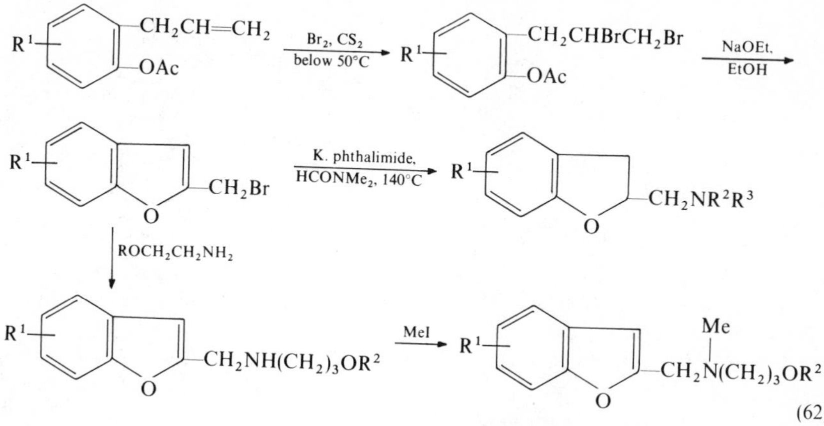

(62)

8. Benzofuranquinones

Unsubstituted 4,7-benzofuranquinone (**179**) is obtained through oxidation of 4-amino-7-methoxybenzofuran hydrochloride (**178**) which in turn is obtained by catalytic hydrogenation of 4-nitro-7-methoxybenzofuran (**177**), with sodium dichromate and sulfuric acid.[213]

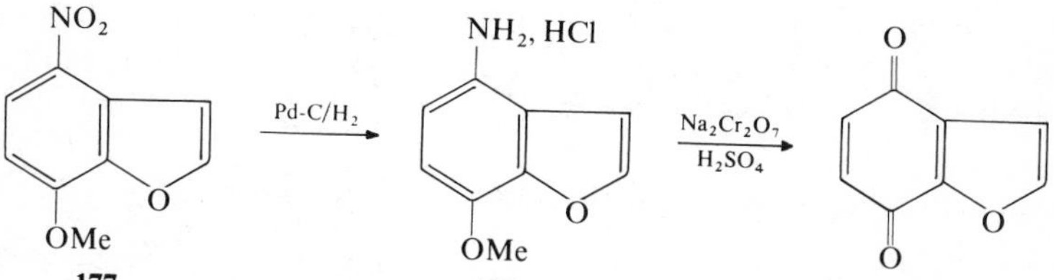

177 **178** **179**

Another approach is the treatment of 4-hydroxybenzofuran with diazobenzenesulfonic acid in presence of potassium hydroxide to yield 7-amino-4-hydroxybenzofuran; this is readily oxidized as above to give **179**. Reaction of the latter in acetic anhydride and sulfuric acid results in the formation of 4,5,7-triacetoxybenzofuran (**180**), which with sodium hydroxide shaken with air, and then acidified, yields 5-hydroxy-4,7-benzofuranquinone (**181**).[214]

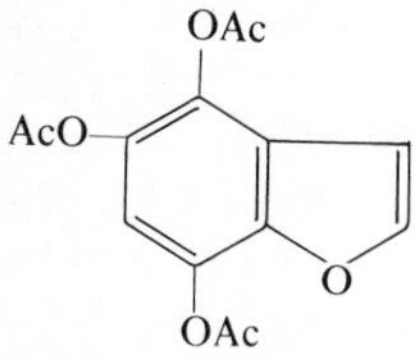

180 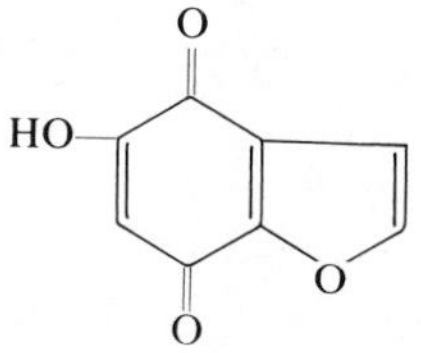 **181** 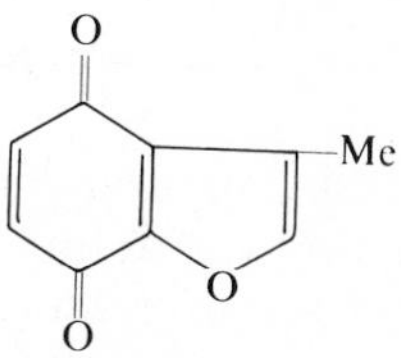 **182**

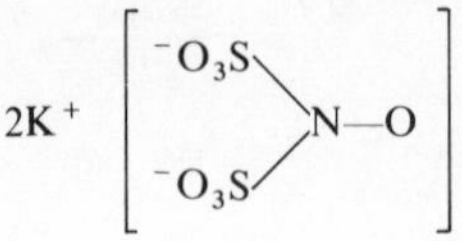

183

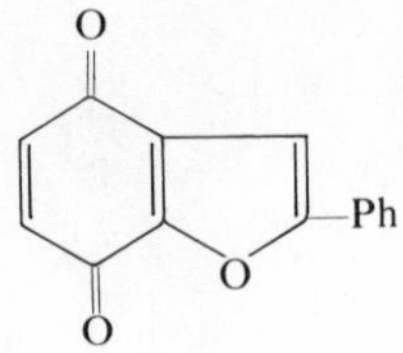

186

$182 + MeCH{=}CHCH{=}CH_2 \longrightarrow$

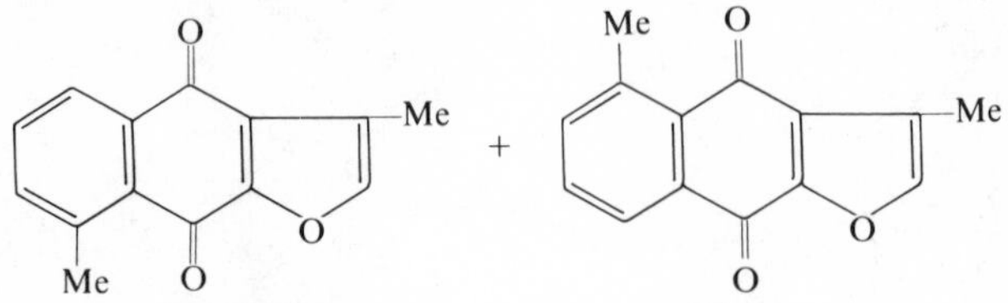

184 + 185

Oxidation of 3-methyl-4-hydroxybenzofuran with Fremy's salt[215] (**183**) afforded 3-methyl-4,7-benzofuranquinone (**182**).[216] The latter condenses with 1-methylbutadiene in ethanol and, when the initial adduct is air-oxidized in alcoholic alkaline medium, a mixture of maturinone (**184**) (a naturally occuring product isolated from Cacalia decomposita) and (**185**) is produced.[218] Synthesis of 5-acetyl-6-hydroxybenzofuran-4,7-quinone has been achieved.[217]

Benzofuran, which has a hydroxyl group at C-7 and other substituents at C-2 and/or position 3, undergoes oxidation when treated with potassium nitrosodisulfonate (**183**).[219] Thus, 2-phenyl-7-hydroxybenzofuran gives 2-phenyl-4,7-benzofuranquinone (**186**).[220,221] The same is true of 3-methyl-6-methoxy-4,7-benzofuranquinone (**188**) (Eq. 63).[222]

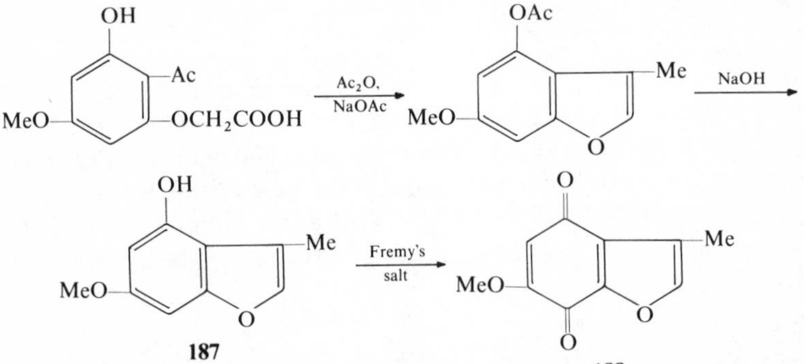

$\xrightarrow{Ac_2O,\ NaOAc}$ $\xrightarrow{NaOH}$

187 $\xrightarrow{\text{Fremy's salt}}$ **188** (63)

Toluquinone reacts with ethyl acetoacetate to give ethyl 2,6-dimethyl-5-hydroxybenzofuran-3-carboxylate. The latter undergoes the following sequence of reactions to yield 2,6-dimethyl-5-methoxy-4,7-benzofuran-quinone (**189**) (Eq. 64).[223]

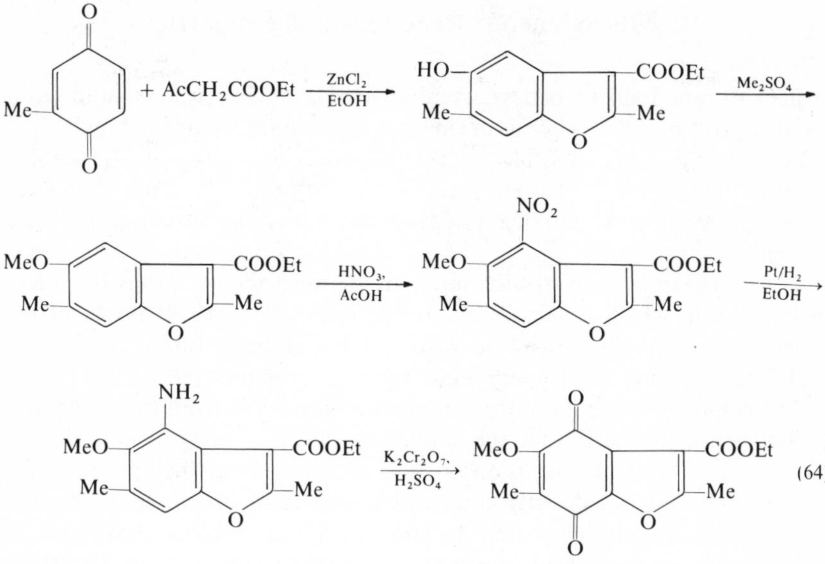

(64)

189

Oxidation of the reaction product, obtained by condensation of benzoin with pyrogallol in presence of zinc chloride with nitric acid afforded 2,3-diphenyl-6,7-benzofuranquinone (**190**).[224] However, Fremy's salt has effected the oxidation of 7-hydroxy-2,3-diphenylbenzofuran to 2,3-diphenyl-4,7-benzofuranquinone (**191**).[225] Recently **192** has been reported to undergo oxidation to 3-ethoxycarbonyl-2-methylbenzofuran-4,7-quinone (**193**).[226]

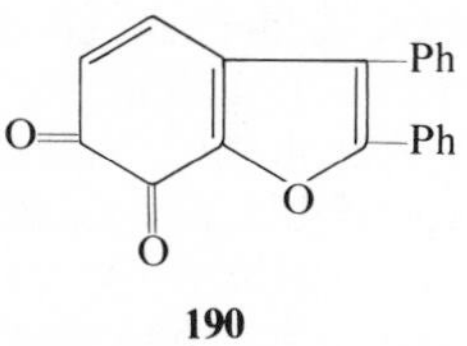

190 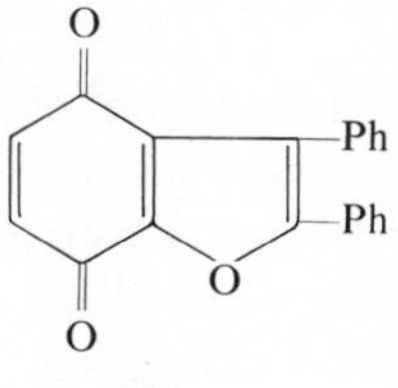 **191**

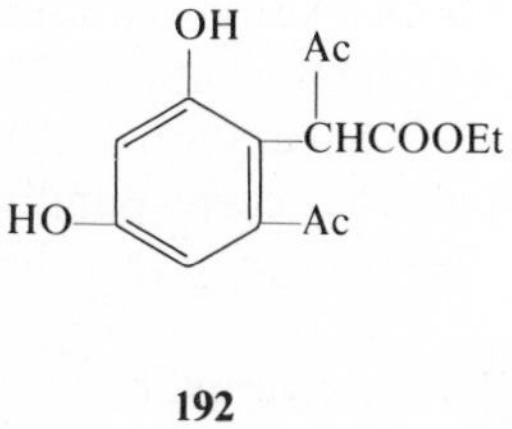

192 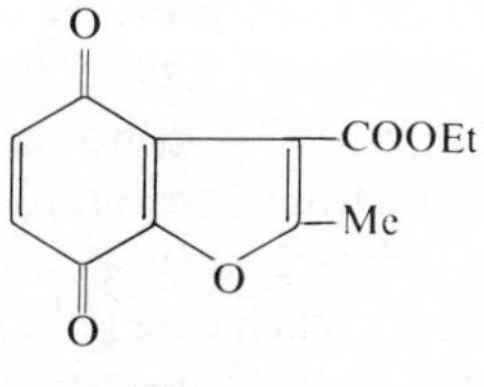 **193**

9. Miscellaneous Reactions and Properties

Benzofurans tend to behave as condensed aromatic systems and commonly afford picrates. The furan ring can, sometimes, be opened hydrolytically, but it is never possible to cause alcohols to add to the 2,3 double bond in the fashion so characteristic of dihydrofuran. The aromatic nature of benzofuran is evident in that dihydrobenzofurans may be dehydrogenated to benzofurans.

In naphthalene chemistry the interplay between the two rings is evident in the way in which a substituent in one ring affects the behavior of the other. This is equally true of benzofurans. But in benzofurans the heterocyclic ring does not have a very high degree of aromaticity.

The behavior of benzofurans can often be predicted from that of monocyclic forms, provided that the stabilizing effect of merging one double bond with a benzene ring is taken into account. Thus the spectroscopic properties[227] are not directly comparable with those of simple furans, but the susceptibility to oxidation by air, the intense colors developed in concentrated sulfuric acid, antimony trichloride, and similar properties with similar reagents, and the resistance to hydrogenation[46,228,229] are all properties found in the parent heterocycle.

Data on the reactivity of the benzofuran are mainly of the qualitative type. Benzofuran, the general reactivity of its nuclear positions being given by the sequence[231,232] $2 \rightarrow 3 \rightarrow 6 \rightarrow$, is substituted by electrophilic reagents in the two position,[116,119,233-237] except for sulfonation, which attacks position 3.[60c,237]

A. Catalytic Hydrogenation

Catalytic hydrogenation of benzofuran, under all conditions, is accompanied by partial cleavage of the furan ring and the formation of 2-ethylcyclohexanol and β-cyclohexylethyl alcohol.[230] Passing benzofuran with hydrogen over palladium on asbestos at 175°C formed up to 80% of octahydrobenzofuran together with cleavage products. With nickel catalyst, the yield of 2-ethylcyclohexanol is increased to 50% and that of octahydrobenzofuran to 21–22%. Liquid-phase hydrogenation of benzofuran in ethanol at 20–50°C in presence of platinum black and platinized charcoal proceeded analogously, but with an intermediate formation of 2,3-dihydrobenzofuran (Chart 7). Hurd and Oliver[238] have shown that in liquid ammonia, 2-methylbenzofuran (**194**) and the corresponding dihydro compound (**195**) are reduced with ring cleavage to give a high yield of *o*-propylphenol in each case (Chart 7).

Brust and Tarbell[239] have shown that **195** is converted by lithium–*t*-butyl

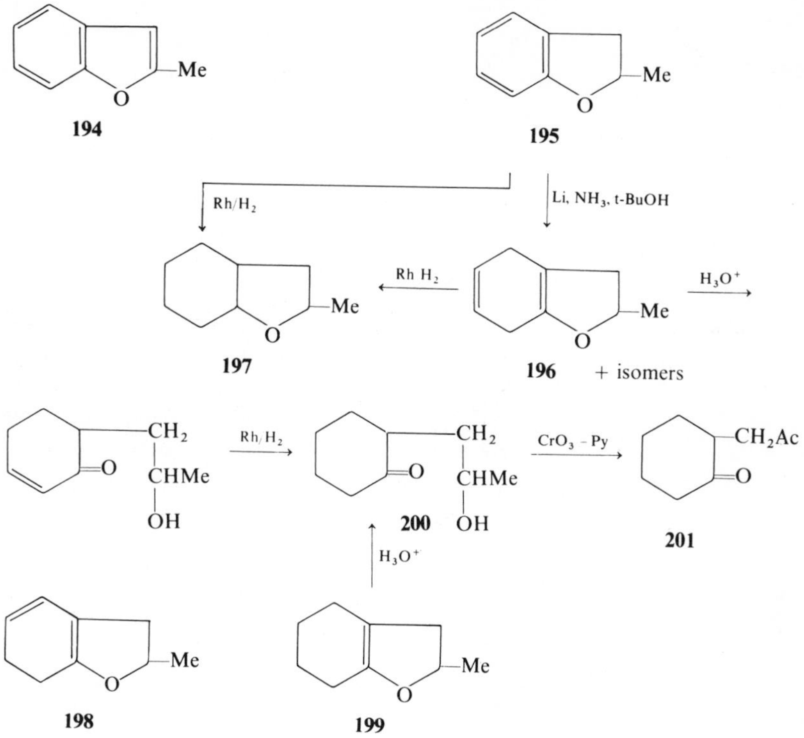

194 195

Rh/H₂ Li, NH₃, t-BuOH

197 196 + isomers

200 201

198 199

Chart 7.

alcohol–ether–liquid ammonia procedure to the tetrahydro compound (**196**) and some hexahydro compound such as **199.** The ring structure of 2,3,4,7-tetrahydro-2-methylbenzofuran (**196**) is established by catalytic reduction to the perhydro compound (**197**), and by a sequence of dilute acid hydrolysis and catalytic reduction to 2-(2-hydroxypropyl)cyclohexanone (**200**), which is apparently a mixture of diastereoisomers, as well as by an unambiguous synthesis from 2-allyloxycyclohexanone. One of the intermediates in this synthesis is converted to the known diketone **201** (Chart 8).[240]

Cleavage of the heterocyclic ring by action of sodium in liquid ammonia can be interpreted by those mechanisms offered in the case of acyclic systems.[238] Thus 2-methyl-2,3-dihydrobenzofuran (**195**) can be regarded as a β-substituted phenyl phenethyl ether in which the benzene ring is common to both moities of the ether. Hence the reaction is one of beta elimination with carbanion (**202**).

195 $\xrightarrow{NaNH_2^-}$ [202: 2-methyl-2,3-dihydrobenzofuran carbanion, $\overset{\ominus}{C}H$, CH—Me, O] $\xrightarrow{Na/NH_3}$ [o-(CH=CHMe)phenoxide, $O^{\ominus}$] $\xrightarrow{Na/NH_3}$ [o-(CH_2CH_2Me)phenol sodium salt, ONa] + NH_3

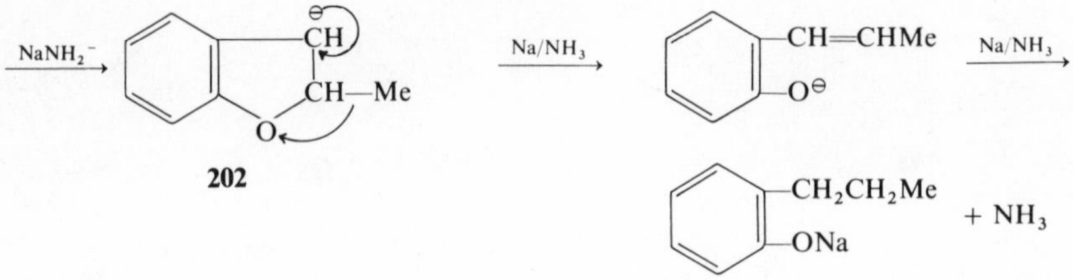

202

That benzofurans are reduced to dihydrobenzofurans, as a first step, is indicated by the rapid initial decolorization of the sodium solution as sodium was added to solution of the compound in ammonia. Thus benzofuran and its 2-methyl homolog are analogous to 2-methyl-2,3-dihydrobenzofuran. The latter has been found to suffer from cleavage with a slight excess of sodium amide to *o*-propenylphenol. A similar reaction is known to be brought out by the action of phenyllithium.[241] 2-Phenylbenzofuran behaves as a substituted benzylphenyl ether.

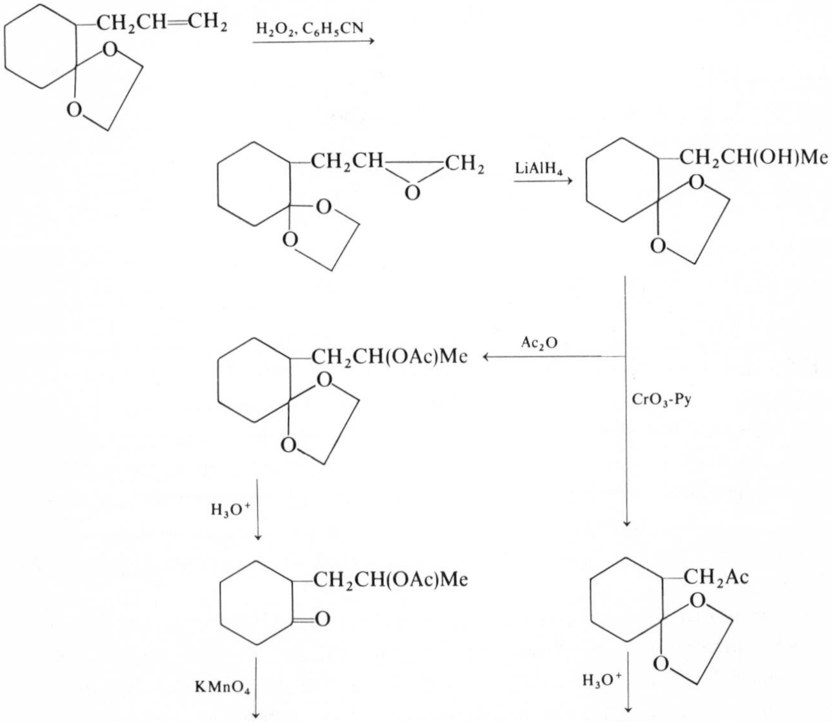

[Ketal with $CH_2CH{=}CH_2$] $\xrightarrow{H_2O_2,\ C_6H_5CN}$ [Ketal with $CH_2CH\text{—}CH_2$ epoxide (O)] $\xrightarrow{LiAlH_4}$ [Ketal with $CH_2CH(OH)Me$]

[Ketal with $CH_2CH(OH)Me$] $\xrightarrow{Ac_2O}$ [Ketal with $CH_2CH(OAc)Me$] $\xrightarrow{H_3O^+}$ [Cyclohexanone with $CH_2CH(OAc)Me$, =O] $\xrightarrow{KMnO_4}$ 200

[Ketal with $CH_2CH(OH)Me$] $\xrightarrow{CrO_3\text{-}Py}$ [Ketal with CH_2Ac] $\xrightarrow{H_3O^+}$ 201

Chart 8.

Reduction with lithium–ammonia and alcohol[239] is undoubtedly vigorous enough to cause some isomerization of **196** (Chart 8); base cleavage[238] or electron cleavage[242] of the dihydrobenzofuran **195** is averted.

Hydrogenation of 5-methoxybenzofuran (**204**) over palladium on charcoal gave 5-methoxy-2,3-dihydrobenzofuran (**206**). However reduction of the 2-methyl-5-methoxybenzofuran **203** and **204** with lithium in ammonia would be expected to yield the corresponding phenols **207** and **208.** When **203**, **204**, **205**, and **206** are reduced with lithium in solution of ammonia containing 15% absolute ethanol, the major product is the corresponding 5-methoxy-2,3,4,7-tetrahydrobenzofuran.[243] Thus compounds **203** and **205** gave **209; 204** and **206** produced **210.** Elimination of methoxide ion has been observed during more vigorous metal–ammonia reductions.[244]

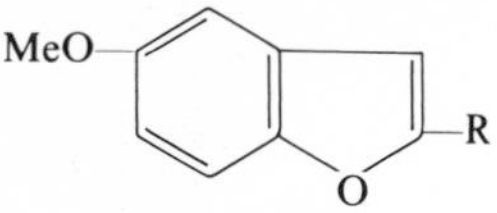

MeO R O

203 R = Me
204 R = H

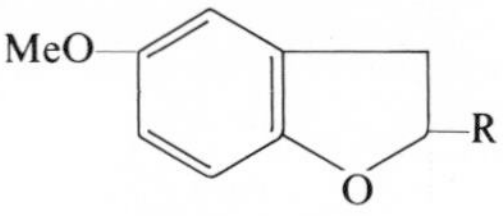

MeO R O

205 R = Me
206 R = H

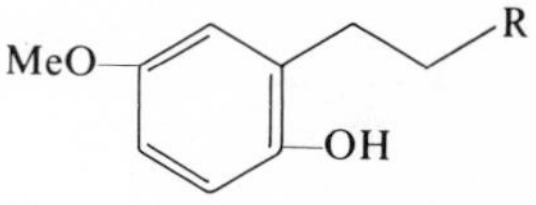

MeO R OH

207 R = Me
208 R = H

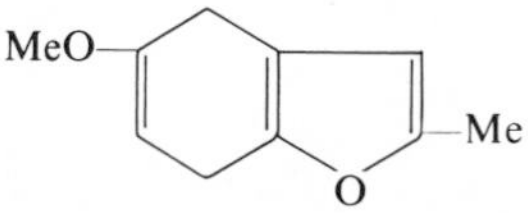

MeO Me O

209

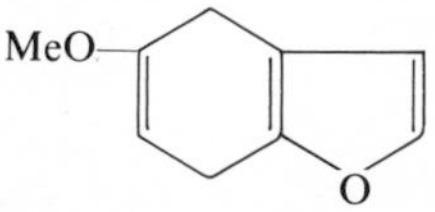

MeO O

210

B. Oxidation

Potassium permanganate or chromic acid oxidation of benzofurans was first reported by Stoermer and Richter,[245] who demonstrated either nonreaction or complete destruction of the benzofuran nucleus. However Wacek and Zeisler[246] have shown that oxidation of benzofurans with chromic acid caused scission at 2,3-position. Oxidative degradation has been referred to for structural elucidation of aryl-substituted benzofurans (Eqs. 65 and 66).[82,246,247]

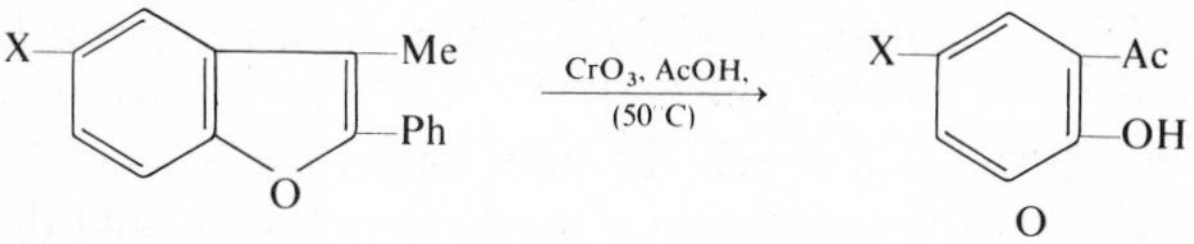

X Me Ph O $\xrightarrow[(50^\circ C)]{CrO_3,\ AcOH,}$ X Ac OH (65)

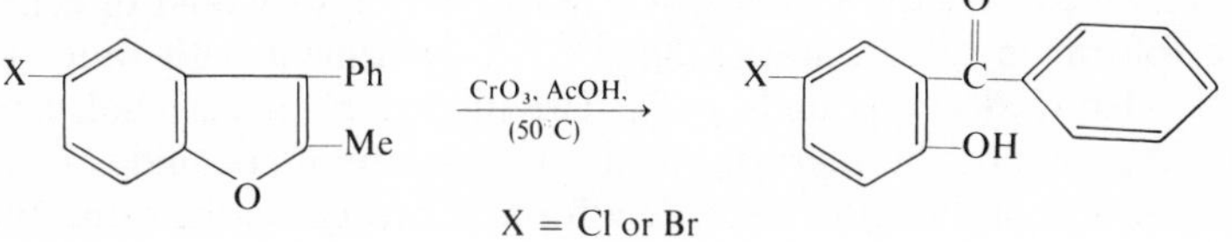

X Ph Me O $\xrightarrow[(50^\circ C)]{CrO_3,\ AcOH,}$ X C(=O) OH (66)

X = Cl or Br

Chromic acid oxidation of 6-bromo-5- (or 7) methyl- and 6-nitro-2,3-diphenylbenzofuran yielded the corresponding 4-bromo-2-benzyloxy-5-(or 3) methylbenzophenone.[248,249] Similar oxidation of 6-methyl-2-phenylbenzofuran (**211**) afforded 2,2′-bis(benzyloxy)-4,4′-dimethylbenzil (**212**), whereas 2-hydroxy-4-methylbenzil (**213**) was formed upon treatment of **211** with Jones reagent[250] (Eq. 67).

CrO_3/AcOH

Me PhCH$_2$O OCH$_2$Ph Me

212

(67)

Me Ph

211

Jones reagent

Me OH

213

C. Ozonolysis

Benzofuran and its derivatives seem to be well suited to undergo a facile reaction with ozone,[252] whether one regards ozone as a specific "double-bond reagent" or as an electrophilic reagent. Wacek and collaborators[246,251] introduced ozonolysis as a degradative method in the benzofuran series, mainly as a tool for elucidation of structures. In the case of benzofuran, salicylic acid (25%), salicylaldehyde (40%), and catechol (10%) were isolated. The homologs (2-Me, 3-Me, 3,7-di-Me) of benzofuran behaved entirely like benzofuran itself (Eq. 68). The formation of catechol can easily be explained through the anamolous ozonolysis. The type of reaction which perhaps accompanies most ozonolysis would directly lead to the diformate of catechol (**213a**).[248-250] Treatment with water would lead to identified compounds only. The diformate of catechol can also result from a normal rearrangement of benzofuran ozonide. Catalytic hydrogenation of the latter ozonide afforded catechol in the same amount

(68)

as in hydrolytic decomposition. This points clearly toward preformed catechol derivative.

The benzofuran ozonide (**214**) underwent acid-catalyzed rearrangement to $C_{14}H_{14}O_8$ which gave, on alkaline hydrolysis, one mole of oxalic acid and **218.** The rearrangement is isomeric and most probably intramolecular. The unstable structure of the conjugate acid (**214**) will begin to break down because of the positive inductive effect of the methyl group and opposite effect of the carbethoxyl group. The cleavage most probably occurs as indicated, leading to the carbonium ion **215**. The oxygen–oxygen bond is then ruptured, leaving an entity, **216**, with cationic oxygen that is apt to undergo rearrangement. The bond to the nucleus is served and the phenyl cation migrates to the oxygen giving **217** which, by expulsion of a proton, comes to rest as **218** (Eq. 69).[252]

(69)

D. Nitration

Nitration of benzofuran with N_2O_4 in benzene gives 3-nitrobenzofuran which, upon boiling with water, undergoes ring cleavage with the formation of formic acid and *o*-hydroxybenzylnitrite (Eq. 70).[253] 2,3-Dimethylbenzofuran is readily nitrated to yield 6-nitro derivative.[254] However nitration of 7-methoxybenzofuran affords 4-nitro-7-methoxybenzofuran.[130]

N_2O_4 (6–10°C); $LiAlH_4$; NO_2; NH_2; NO_2, OH; CH_2NO_2, CHO; CH_2NO_2, OH + HCOOH (70)

2-Phenylbenzofuran is readily converted, upon treatment with nitric acid and acetic acid mixture, to 3-nitro- and 6-nitro-2-phenylbenzofuran; however acetic anhydride nitration results in the formation of 3,6-dinitro-2-phenylbenzofuran.[255] Nitration of 2-phenyl- and 2,3-diphenylbenzofuran is outlined in (Chart 9).[256-258]

Nitration of 5-chloro-2,3-diphenylbenzofuran under conditions similar to 4,7-dimethyl-2,3-diphenylbenzofuran (Chart 9), but with more concentrated nitric acid, yielded 28% of 6-nitro-5-chloro-2,3-diphenyl-benzofuran. It seems that the chlorine atom deactivates the neighboring position 6 (eventually 4) of the benzofuran nucleus, as compared with the methyl group which, activates this position.

E. Halogenation

Bromination of 2-phenylbenzofuran, 5-methyl-, and 7-methyl-2,3-diphenylbenzofuran attacks first C-3 (if free) and then C-6, leading to the formation of 3-bromo-, 3,6-dibromo-2-phenylbenzofuran, 5-methyl-6-bromo-, and 7-methyl-6-bromo-2,3-diphenylbenzofuran, respectively.[259] Chloromethylation of benzofuran gives 2-chloromethylbenzofuran;[260] however 2-phenylbenzofuran yields 2-phenyl-3-chloromethylbenzofuran,[255] together with bis(2-phenyl-3-benzofuranyl)methane.

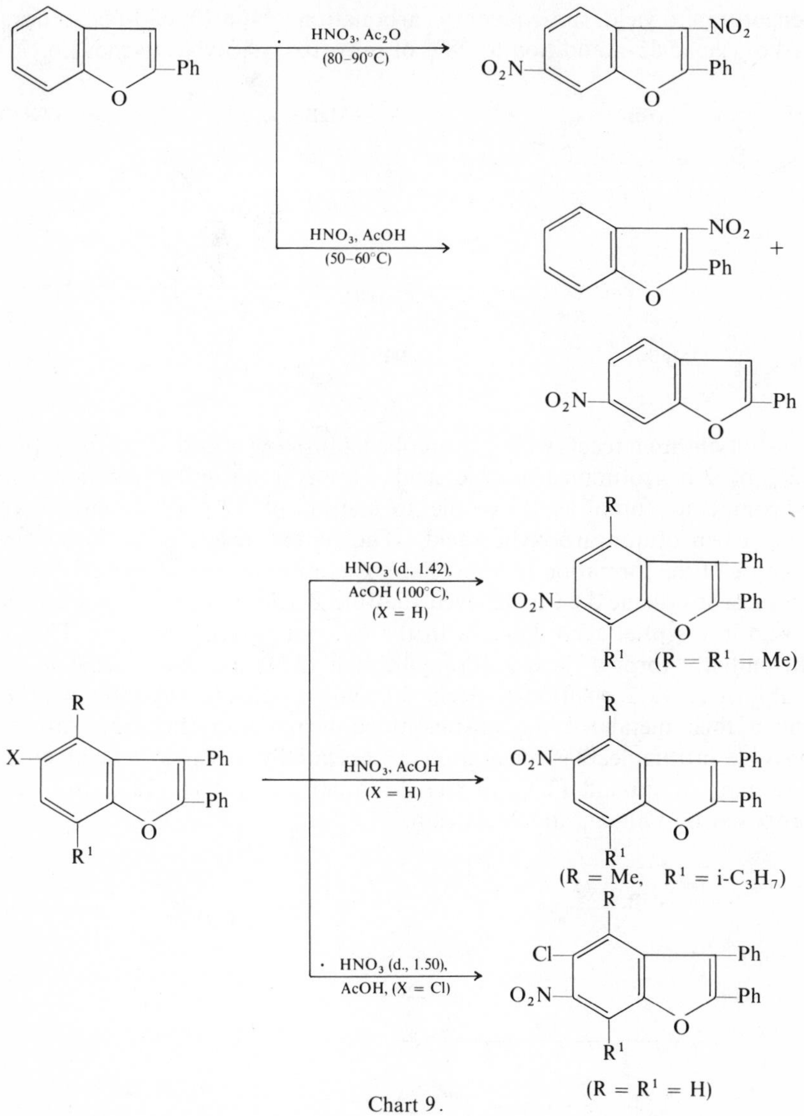

HNO_3, Ac_2O (80–90°C)

HNO_3, AcOH (50–60°C)

HNO_3 (d., 1.42), AcOH (100°C), (X = H)

(R = R[1] = Me)

HNO_3, AcOH (X = H)

(R = Me, R[1] = i-C_3H_7)

HNO_3 (d., 1.50), AcOH, (X = Cl)

(R = R[1] = H)

Chart 9.

F. Benzofuranylmetallic Compounds

2-Bromobenzofuran does not react with magnesium under conventional conditions to give Grignard reagent. However Reichstein and Baud[261] showed that the activated magnesium–copper alloy reacts with 3-bromobenzofuran to yield, subsequent to carbonation, about 1% of 3-benzofurancarboxylic acid, in addition to 28% of *o*-hydroxyphenylacetylene (Eq. 71).

Br Mg MgBr CO_2 COOH O O O Mg C≡CH OMgBr C≡CH OH

(71)

n-Butyllithium reacts with 2-bromobenzofuran to afford, on carbonation, 62% of 2-benzofurancarboxylic acid. However a similar reaction with 3-bromobenzofuran leads to the formation of 12% of 3-benzofuran- and 2-benzofurancarboxylic acid. The latter reaction is interesting because of the formation of appreciable quantities of the isomeric 2-benzofurancarboxylic acid. It is believed that the 2-acid was produced from the 3-acid in essential accordance with the following transformation. That is, the initially formed 3-benzofuranyllithium (**220**) metalated (**219**) in the highly reactive 2 position to yield 3-bromo-2-benzofuranyllithium (**221**), which then metalated the unsubstituted benzofuran (Eq. 72). Another possible intramolecular metalation, or essentially a rearrangement of the 3-lithium compound (**220**) to the 2-lithium compound (**222**), has been proposed by Gilman and Melstrom.[262]

Br n-BuLi Li + n-C_4H_9Br O O **219** **220** +219 Br + benzofuran Li CO_2, H_2O O Li O O **221** **222** HOOC

(72)

Grignard reagent (**223**), obtained from 3-chloromethyl-2-methylbenzofuran, was allowed to react with carbon dioxide and ethyl chlorocarbonate. A trace of the normal product (**224**) was isolated only with carbon dioxide.

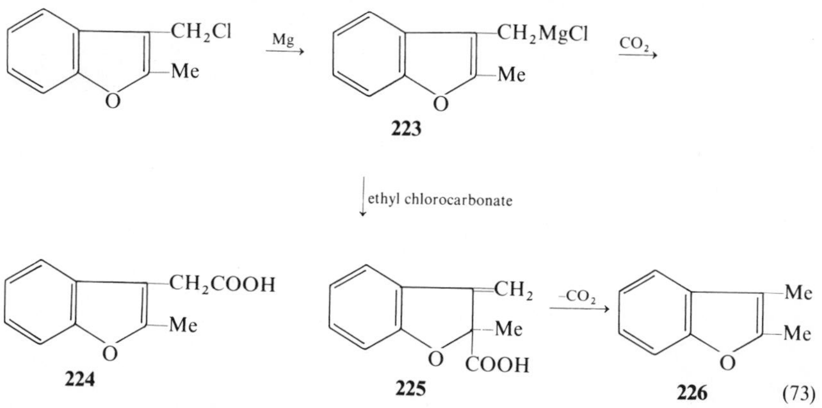

CH_2Cl, Me $\xrightarrow{Mg}$ CH_2MgCl, Me (**223**) $\xrightarrow{CO_2}$ CH_2COOH, Me (**224**)

223 $\xrightarrow{\text{ethyl chlorocarbonate}}$ CH_2, Me, COOH (**225**) $\xrightarrow{-CO_2}$ Me, Me (**226**) (73)

In both reactions the major product, after hydrolysis, was the "isoaromatic" 2-methyl-3-methylene-2,3-dihydro-2-benzofurancarboxylic acid (**225**), the result of "abnormal reaction." This acid, **225**, decarboxylated at its melting point to yield 2,3-dimethylbenzofuran (**226**), a reaction analogous to the behavior of β-keto-acids (Eq. 73).[263] 2-Chloromethylbenzofuran, however, reacts with magnesium to give *o*-allylphenol, which undergoes base-catalyzed ring closure under mild conditions to afford 2-methylbenzofuran. Gaertner[264] correlated the results by an assumption involving equilibrium between benzofuranylmethyl and *o*-allenylphenoxide anions (Chart 10).

G. Friedel-Crafts Techniques

Acylation and alkylation of benzofurans are possible by Friedel-Crafts techniques, but the catalysts have to be very carefully chosen,[176,236,265] and such reactions have not been employed extensively. Benzofuran, itself, is almost wholly resinified by Lewis acids as mild as stannic chloride and phosphoric acid, but 2-alkylbenzofurans are readily acylated at C-3 by acid chlorides in presence of stannic chloride. Again 2,3-dialkyl derivatives are unreactive; the 5-chlorine atom hampers acylation and the 7-methoxy group favors it.

Acetylation of 2-phenylbenzofuran yields a mixture of 3-acetyl-2-phenyl- (**227**), and 6-acetyl-2-phenylbenzofuran (**228**), as well as 3,6-diacetyl-2-phenylbenzofuran (**229**).[259] Under Friedel-Crafts conditions, 5-hydroxy-2,3-diphenylbenzofuran reacts with acetyl chloride in nitrobenzene to give the 4-acetyl derivative. with cinnamoyl chloride in boiling carbon tetrachloride the 5-cinnamoyloxy derivative (**230**) is obtained; but in nitrobenzene 6-cinnamoyl-5-hydroxy-2,3-diphenylbenzofuran (**231**) is produced.[266]

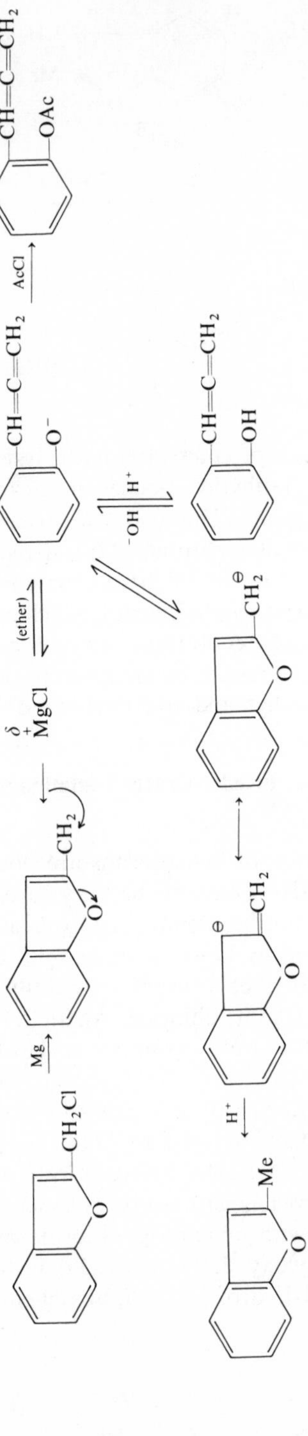

Chart 10.

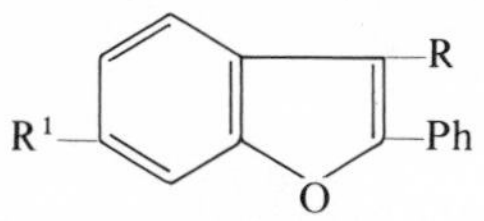

227 R = Ac, R[1] = H
228 R = H, R[1] = Ac
229 R = R[1] = Ac

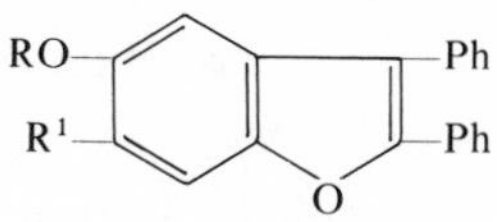

230 R = COCH=CHPh, R[1] = H
231 R = H, R[1] = COCH=CHPh

H. Hoesch and Gattermann Techniques

Hoesch and Gatterman techniques are successful,[267,268] though the latter will probably tend to lose favor, since formylation by means of dimethylformamide and phosphorus oxychloride gives an equally high yield and is much more convenient.[236] Benzofuran has been found to undergo formylation more easily than thionaphthen and, unlike the latter at position 2, the same reaction occurs at position 3 with 2-substituted benzofurans. The benzene ring is inert in this reaction as 2,3-dimethyl- and 2-ethyl-3-methylbenzofuran are not formylated. A 5-chlorine atom hinders formylation and a 7-methoxy group facilitates it. The various formylbenzofurans displayed the normal aldehyde properties, but 2-alkyl substituents caused steric hindrance as shown by the increasing difficulty in Wolff-Kishner reduction of the following sequence: 2-formyl-, 3-formyl-2-methyl, and 2-ethyl-3-formylbenzofuran. Furthermore the oxime of 2-ethyl-3-formylbenzofuran was readily dehydrated to 3-cyano-2-ethylbenzofuran, but the latter compound could not be hydrolyzed any further than to the amide.[236] Moreover 2-formyl-, but not 3-formyl-2-ethylbenzofuran, gave chalcone with acetophenone.

I. With Diazoalkanes

The behavior of benzofuran toward the action of ethyl diazoacetate and diazoacetone simulates that of vinyl ether and yields after hydrolysis the cyclopropanecarboxylic acid (**232**) and the ketone (**233**),[269-271] respectively.

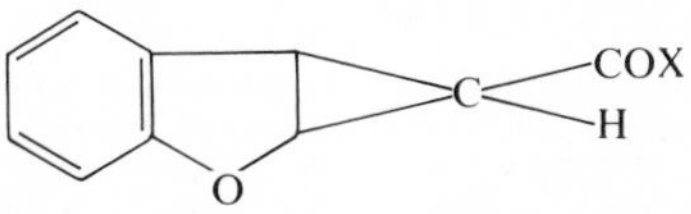

232 X = OH **233** X = Me

J. With Dihalocarbene

Benzofuran reacts with dichlorocarbene in hexane to form the initial adduct **234,** which could not be isolated, but on hydrolysis, gave the ring-expanded product **236,** probably via **235.**[272]

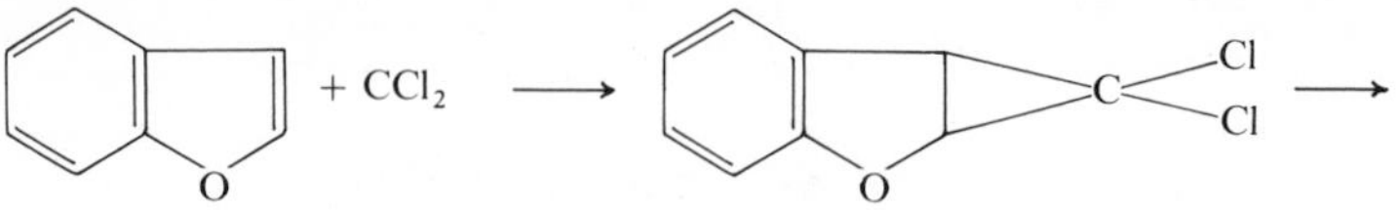

234

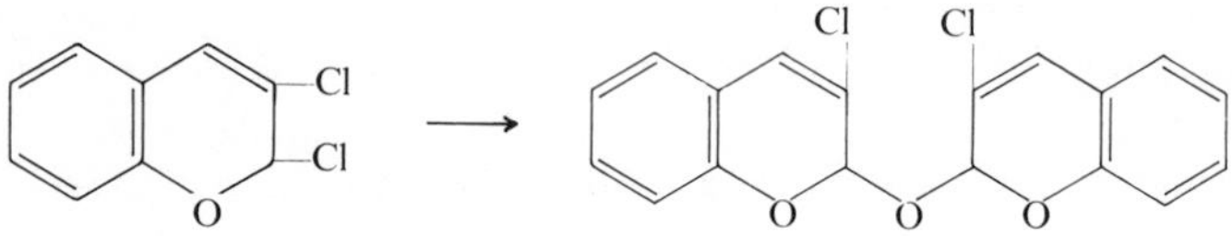

235 **236**

K. Cyclophotochemical Addition

Irradiation of benzene solution of benzofuran with acetophenone, propiophenone, acetone, and/or xanthone results in the formation of the cyclo adducts **237** and **238.** However benzofuran reacts photochemically with benzophenone, benzaldehyde, and/or thioxanthone to give the adducts **239** and **240,** respectively.[273]

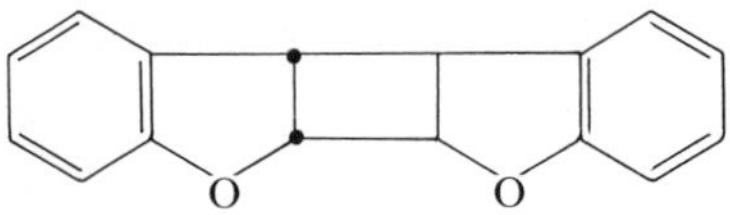

237 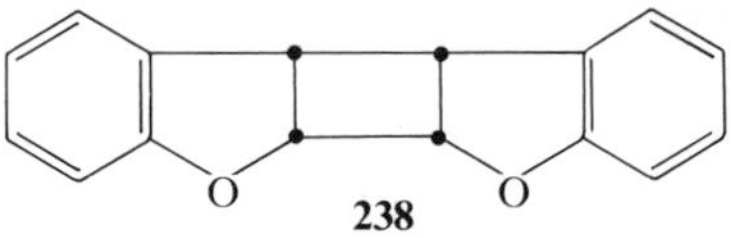 **238**

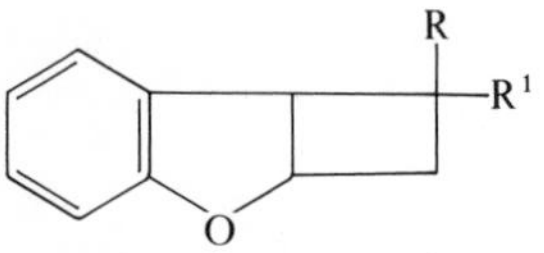

239 R = R^1 = Ph
240 R = H, R^1 = Ph

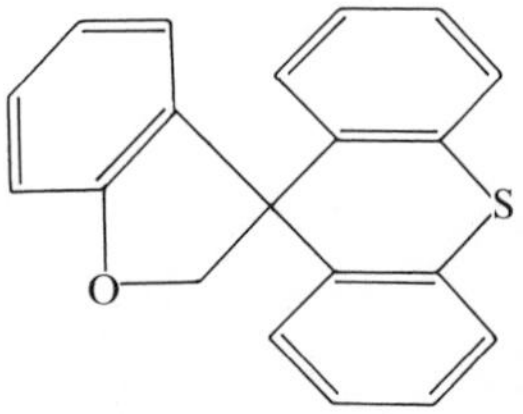

241

L. Polymerization

Benzofuran does not undergo direct sulfonation, rather resinification occurs. This is a process commercially used for the manufacture of "cumar resins" by treatment of benzofuran-containing fractions of coal tar with sulfuric acid. Allyl-and methoxy-substituted benzofurans are dimerized in the presence of sulfuric acid to the dimers **242** and **243** which are readily dissociated into the monomers.[274,275] Benzofuran is a well-known cyclic olefin which can be easily polymerized with conventional cationic catalyst.[276,277] Optically active polybenzofurans have been obtained at −80 to −100°C, using toluene as a solvent and asymmetric catalyst, for example, alkylaluminum halides with optically active acids, alcohols, hydroxy acids, and/or amino acids.[278,279]

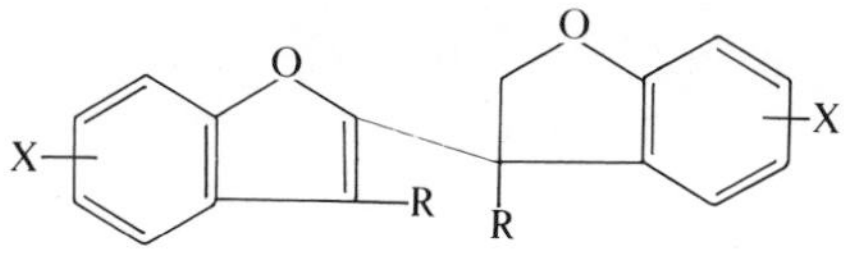

242 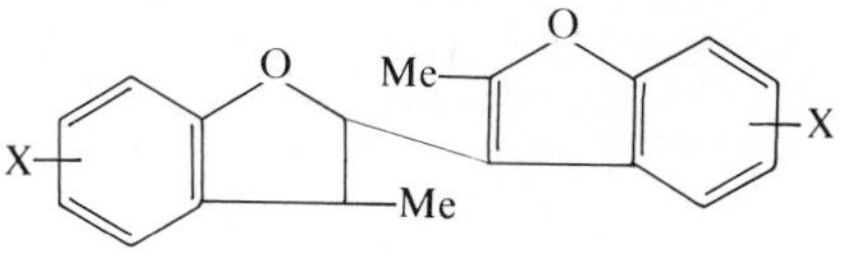 **243**

M. Miscellaneous Reactions

Benzofuran and hydrogen sulfide in presence of $CuO–Cr_2O_3$ and MoS at 500–700°C produced thionaphthen.[280] Similarly, 4-hydroxythionaphthen is obtained from 4-hydroxybenzofuran.[281]

Methylsulfonyl carbanion attacks the σ-orbital of 2 positions of the benzofuran nucleophilically and yields the intermediate **244**, which stands for the benzofuran ring-opening reaction with the quantitative formation of *o*-hydroxyphenylacetylene (**245**).[282]

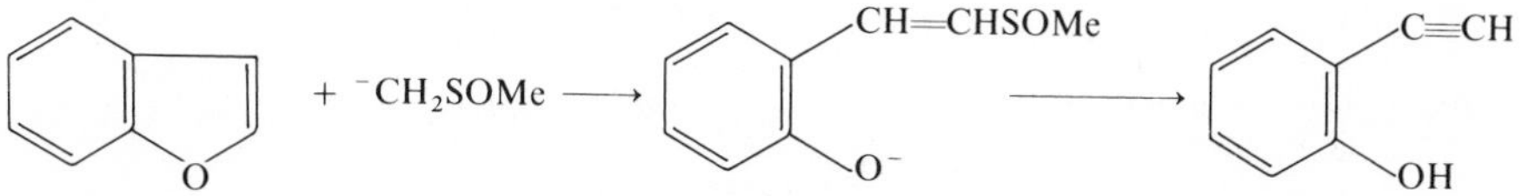

$+\ ^{-}CH_2SOMe \longrightarrow$ CH=CHSOMe, O^- $\longrightarrow$ C≡CH, OH

244 **245**

Catalytic alkylation of benzofuran with *t*-butyl chloride at 60–100°C in hexane, nitrobenzene, toluene, and/or acetic acid in presence of zinc chloride gave a mixture of 2- and 3-*t*-butylbenzofuran in 1:2 ratio.[283]

References

1. A. M. Patterson, L. T. Capell, and D. F. Walker, "The Ring Index," American Chemical Society, McGreger and Werner, Inc., Washington, 1959.
2. G. Kraemer and A. Spilker, *Chem. Ber.*, **23,** 78 (1890).
3. G. Kraemer and A. Spilker, *Chem. Ber.*, **33,** 2257 (1900).
4. K. Ouchi and J. D. Brooks, *Fuel (London)*, **46**(4–5), 367 (1967); *Chem. Abstr.*, **68,** 42132 (1968).
5. T. Lesiak, *Roczniki Chem.*, 38, 1709 (1964); *Chem. Abstr.*, **62,** 16175 (1965).
6. D. E. Boswell, P. S. Landis, E. N. Givens, and P. B. Venuto, *Ind. Eng. Chem. Process Des. Develop.*, 7(3), 215 (1968); *Chem. Abstr.*, **69,** 96365 (1968).
7. B. Sila and T. Lesiak, *Roczniki Chem.*, **35,** 1519 (1961); *Chem. Abstr.*, **57,** 5867 (1962).
8. N. I. Shuikin, E. A. Viktoreva, Shih Li, and E. A. Karakhanov, *Izv. Akad. Nauk SSSR, Otd. Khim. Nauk,* **1961,** 2054; *Chem. Abstr.,* **57,** 8478 (1962).
9. C. Hansch, C. Scott, and H. Keller, *Ind. Eng. Chem.*, **42,** 2114 (1950); *Chem. Abstr.*, **45,** 1991 (1951); C. Hansch, W. Saltonstall, and J. Settle, *J Am. Chem. Soc.*, **71,** 943 (1949).
10. B. B. Corson, H. E. Tiefenthal, J. E. Nickels, and W. J. Heintzelman, *J. Am. Chem. Soc.*, **77,** 5428 (1955).
11. B. B. Corson, W. J. Heintzelman, H. E. Tiefenthal, and J. E. Nickels, *J. Org. Chem.*, **17,** 971 (1952).
12. E. N. Givens and P. B. Venuto, *J. Catal.*, **15**(4), 319 (1969); *Chem. Abstr.*, **72,** 3298 (1970).
13. G. E. Illingworth and J. J. Lauvar, U.S. Pat. 3,285,932 (1964); *Chem Abstr.*, **66,** 28649 (1967).
14. L. v. Claisen and E. Tietze, *Justus Liebigs Ann. Chem.*, **449,** 81 (1926); L. v. Claisen and E. Tietze, *Chem. Ber.*, **59,** 2344 (1926).
15. V. I. Panservich-Kolyada and Z. B. Idel'chick, *Zh. Obshch. Khim.*, **25,** 2215 (1955); *Chem. Abstr.*, **50,** 9370 (1956); V. I. Panservich-Kolyada and Z. B. Idel'chick, *Zh. Obshch. Khim.*, **24,** 807 (1954); *Chem. Abstr.*, **49,** 8183 (1955).
16. G. C. Robinson, U.S. Pat. 3,230,237 (1966); *Chem. Abstr.*, **64,** 11176 (1966).
17. F. M. Dean, P. Halewood, S. Mongkolsuk, A. Robertson, and W. B. Whalley, *J. Chem. Soc.,* **1953,** 1250.
18. R. v. Stoermer, *Justus Liebigs Ann. Chem.*, **312,** 237 (1900).
19. R. J. S. Beer, H. F. Davenport, and A. Robertson, *J. Chem. Soc.*, **1953,** 1262.
20. R. Royer, M. Hubert-Habart, L. Rene, A. Cheutin, and M-L. Desvoye, *Bull. Soc. Chim. Fr.*, **1964,** 1259.
21. K. K. Thomas and M. M. Bokadia, *J. Indian Chem. Soc.*, **43,** 713 (1966).
22. S. Trippett, *J. Chem. Soc.*, **1957,** 419.
23. J. R. Collier, M. K. M. Dirania, and J. Hill, *J. Chem. Soc., C,* **1970,** 155.
24. Y. Kawase, *Chem. Ind. (London)*, **1970,** 687.
25. K. K. Thomas and M. M. Bekadia, *J. Indian Chem. Soc.*, **45,** 265 (1968).
26. C. K. Bradsher, *Chem. Rev.*, **38,** 447 (1946).
27. J. A. Elvidge and R. G. Foster, *J. Chem. Soc.,* **1964,** 981.
28. J. Schmitt, M. Suquet, G. Callet, J. Le Meur, and P. Comoy, *Bull. Soc. Chim. Fr.*, **1967,** 74.
29. D. D. Chapman, S. E. Cremer, R. M. Carman, M. Kunstmann, J. G. McNally, Jr., A. Rosowsky, and D. S. Tarbell, *J. Am. Chem. Soc.*, **82,** 1009 (1960).
30. D. S. Tarbell, R. W. Carman, D. D. Chapman, S. E. Cremer, A. D. Cross, K. R. Huffman, M. Kunstmann, N. J. McCorkindale, J. G. McNally, Jr., A. Rosowsky, F. H. Varino, and R. L. West, *J. Am. Chem. Soc.*, **83,** 3096 (1961).
31. A. Mooradian and P. E. Dupont, *Tetrahedron Lett.*, **1967,** 2867.
32. R. Haworth, T. Richardson, and G. Shedrick, *J. Chem. Soc.*, **1935,** 1576.

33. (a) R. Royer, P. Demerseman, J. F. Lechartier, A-M Laval Jeantet, A. Cheutin, and M-L., Desvoye, *Bull. Soc. Chim. Fr.*, **1964,** 315; (b) R. Royer, P. Demerseman, and J. P. Lechartier, *Compt. Rend.*, **254,** 2605 (1962); *Chem. Abstr.*, **57,** 3395 (1962).
34. E. Späth and M. Pailer, *Chem. Ber.*, **69,** 767 (1936).
35. T. A. Geissman and T. G. Halsall, *J. Am. Chem. Soc.*, **73,** 1280 (1951).
36. (a) G. R. Ramage and C. V. Stead, *J. Chem. Soc.*, **1953,** 3602; (b) R. v. Stoermer and E. Barthelmes, *Chem. Ber.*, **48,** 62 (1915).
37. B. Klarmann, *J. Am. Chem. Soc.*, **73,** 4476 (1951).
38. H. v. Bickel and H. Schmid, *Helv. Chim. Acta*, **36,** 664 (1953).
39. R. v. Stoermer and W. König, *Chem. Ber.*, **39,** 492 (1906).
40. P. Friedländer and F. Risse, *Chem. Ber.*, **47,** 1919 (1914).
41. E. Kamthong and A. Robertson, *J. Chem. Soc.*, **1933,** 933.
42. K. v. Horváth, *Monat. Chem.*, **82,** 901 (1951).
43. Th. Kaufmann, H. Henkler, and H. Zengel, *Angew. Chem.*, **74,** 248 (1962); *Chem. Abstr.*, **57,** 9632 (1962).
44. Y. Kawase and S. Nakamoto, *Bull. Chem. Soc. Jap.*, **35,** 1624 (1962); *Chem. Abstr.*, **58,** 1421 (1963).
45. E. D. Elliott, *J. Am. Chem. Soc.*, **73,** 754 (1951).
46. R. L. Shriner and J. Anderson, *J. Am. Chem. Soc.*, **61,** 2705 (1939).
47. I. V. Andreeva and M. M. Koton, *Zh. Obshch. Khim.*, **27,** 671 (1957); *Chem. Abstr.*, **51,** 16409 (1957).
48. G. B. Bachman and L. V. Heisey, *J. Am. Chem. Soc.*, **71,** 1985 (1949).
48a. R. Magnussen, *Acta Chem. Scand.*, **18,** 421 (1964).
49. (a) C. E. Castro, R. Halvin, V. K. Honwad, and S. Moje, *J. Am. Chem. Soc.*, **91,** 6464 (1969); (b) C. E. Castro, E. Gaughan, and D. C. Owsley, *J. Org. Chem.*, **31,** 4071 (1966).
50. H. v. Pechmann and E. Hanke, *Chem. Ber.*, **34,** 354 (1901).
51. F. Peters and H. Simonis, *Chem. Ber.*, **41,** 830 (1908).
52. F. M. Dean, E. Evans, and A. Robertson, *J. Chem. Soc.*, **1954,** 4565.
53. M. K. M. Dirania and J. Hill, *J. Chem. Soc., C*, **1968,** 1311.
54. A. Smith and H. P. Utley, *J. Chem. Soc., C*, **1970,** 1.
55. G. R. Lappin and J. S. Zannucci, *Chem. Commun.*, **1969,** 1113; *Chem. Abstr.*, **72,** 31525 (1970).
56. J. M. Holovka, P. D. Gardner, C. B. Strow, M. L. Hill, and T. V. Van Auken, *J. Am. Chem. Soc.*, **90,** 5041 (1968).
57. M. Higashi, F. Toda, and K. Akagi, *Chem. Ind. (London)*, **1969,** 491.
58. S. Kobayashi, M. Shinya, and H. Taniguchi, *Tetrahedron Lett.*, **1971,** 71.
59. O. Simonsen and C. Lohse, *Acta Chem. Scand.*, **24,** 268 (1970).
60. (a) M. L. Oftedahl, J. W. Baker, and M. W. Dietrich, *J. Org. Chem.*, **30,** 296 (1965); (b) M. L. Oftedahl, U.S. Pat. 3,256,301 (1966); *Chem. Abstr.*, **65,** 7142 (1966); (c) K. Masuda, N. Numata, and K. Ikawa, Jap. Pat. 7,006,809 (1967); *Chem. Abstr.*, **73,** 3784 (1970).
61. (a) A. P. Terent'ev and L. A. Kazitsyna, *Compt. Rend. Acad. Sci. USSR*, **55,** 625 (1947); *Chem. Abstr.*, **42,** 556 (1948); (b) L. A. Kazitsyna, *Vestn. Mosk. Gos. Univ.*, **1947,** 109; *Chem. Abstr.*, **42,** 3751 (1948); (c) E. Guenther, F. Wolf, and G. Wolter, *Z. Chem.*, **8,** 111 (1968); *Chem. Abstr.*, **68,** 104665 (1968).
62. M. D. Rausch and A. Siegel, *J. Org. Chem.*, **34,** 1974 (1969).
63. St. v. Kostanecki and J. Tambor, *Chem. Ber.*, **29,** 237 (1896).
64. S. Kawai, T. Nakamura, and N. Sugiyama, *Proc. Imp. Acad. (Tokyo)*, **15,** 45 (1939); *Chem. Abstr.*, **33,** 5394 (1939).
65. W. Davies and S. Middleton, *Chem. Ind. (London)*, **1957,** 599.
66. S. Kawai, T. Nakamura, and S. Sugiyama, *Chem. Ber.*, **72,** 1146 (1939).
67. R. Klink and K. H. Baron, U.S. Pat. 3,499,914 (1970); *Chem. Abstr.*, **72,** 121353 (1970).

68. (a) J. N. Chatterjea and S. K. Roy, *J. Indian Chem. Soc.*, **40,** 144 (1963); (b) J. N. Chatterjea, *Sci. Cult. (Calcutta)*, **24,** 40 (1958).
69. W. Davies and S. Middleton, *J. Chem. Soc.*, **1959,** 3544.
70. (a) A. N. Kost, A. Budylin, E. D. Matveeve, and D. O. Sterligov, *J. Org. Chem. USSR*, **6**(7), 1516 (1970); translated from *Zh. Org. Khim.*, **6**(7), 1503 (1970); (b) V. A. Budylin, A. N. Kost, and E. D. Matveeve, *Vestn. Mosk. Gos. Univ.*, **1969,** 121.
71. O. Dann and M. Kokorudz, *Chem. Ber.*, **91,** 172 (1958).
72. (a) J. N. Chatterjea, N. M. Sahai, and N. C. Jain, *J. Indian Chem. Soc.*, **47,** 261 (1970); (b) J. N. Chatterjea and S. K. Roy, *J. Indian Chem. Soc.*, **34,** 98 (1957).
73. H. Fiesselmann and J. Ribka, *Chem. Ber.*, **89,** 40 (1956).
74. R. Royer and E. Bisagni, *Bull. Soc. Chim. Fr.*, **1959,** 1468.
75. R. Royer and E. Bisagni, *Helv. Chim. Acta*, **42,** 2364 (1959).
76. R. Royer, E. Bisagni, and C. Hurdy, *Bull. Soc. Chim. Fr.*, **1960,** 1178.
77. R. Royer, E. Bisagni, and C. Hurdy, *J. Org. Chem.*, **26,** 4308 (1961).
78. R. Royer and E. Bisagni, *Bull. Soc. Chim. Fr.*, **1959,** 521.
79. R. Royer and C. Hurdy, *Bull. Soc. Chim. Fr.*, **1961,** 939.
80. R. Royer, J. L. Derocque, P. Demerseman, and A. Cheutin, *Compt. Rend., Ser. C*, **262,** 1286 (1966).
81. A. Hantzsch, *Chem. Ber.*, **19,** 1290 (1886); A. Hantzsch and G. Pfeiffer, *Chem. Ber.*, **19,** 1301 (1886).
82. E. Bisagni and Ch. Rivalle, *Bull. Soc. Chim. Fr.*, **1969,** 3111.
83. H. Stetter and E. Siehnhold, *Chem. Ber.*, **88,** 271 (1955).
84. J. W. Schulenberg and S. Archer, *J. Am. Chem. Soc.*, **82,** 2035 (1960).
85. W. Dilthey and F. Quint, *J. Prakt. Chem.*, **131,** 1, (1931).
86. B. Arventi'ev and H. Offenberg, *Anal. Stiint. Univ.* "A.I. Ciza," *Iasi Sect.*, **18,** 217 (1962); *Chem. Abstr.*, **59,** 7464 (1963).
87. O. Dischendorfer and A. Verdino, *Monat. Chem.*, **68,** 10 (1936); *Chem. Abstr.*, **30,** 5953 (1936).
88. O. Dischendorfer, *Monat. Chem.*, **62,** 263 (1933); *Chem. Abstr.*, **28,** 758 (1934).
89. B. Arventi'ev, H. Wexler, and M. Strul, *Acad. Rep. Populare Romine, Filiala Iasi, Studii Cercetari Stiint., Chim.*, **12,** 87 (1961); *Chem. Abstr.*, **57,** 9772 (1962).
90. R. Huisgen, H. Koenig, G. Binsch, and J. Sturm, *Angew. Chem.*, **73,** 368 (1961); *Chem. Abstr.*, **56,** 1441 (1962).
91. R. Huisgen, G. Binsch, and H. Koenig, *Chem. Ber.*, **97,** 2884 (1964).
92. G. Binsch, R. Huisgen, and H. Koenig, *Chem. Ber.*, **97,** 2893 (1964).
93. P. Yates, *J. Am. Chem. Soc.*, **74,** 5376 (1952).
94. W. Ried and A. Urschel, *Chem. Ber.*, **91,** 2459 (1958).
95. F. Wessely and E. Zbiral, *Justus Liebigs Ann. Chem.*, **605,** 98 (1957).
96. L. Hurd, *Chem. Ind. (London)*, **1963,** 1165.
97. J. B. Brown, H. B. Henbest, and E. R. H. Jones, *J. Chem. Soc.*, **1950,** 3634.
98. D. M. X. Donnelly, J. F. K. Eades, E. M. Philbin, and T. S. Wheeler, *Chem. Ind. (London)*, **1961,** 1453.
99. C. K. Bradsher and R. Wert, *J. Am. Chem. Soc.*, **62,** 280 (1940).
100. R. v. Stoermer and R. Wehln, *Chem. Ber.*, **35,** 3549 (1902).
101. C. M. Robinson and R. Robinson, *J. Chem. Soc.*, **125,** 827 (1924).
102. T. Sheradsky, *J. Heterocycl. Chem.*, **4**(3), 413 (1967); *Chem. Abstr.*, **68,** 12792 (1968).
103. J. C. Sheehan and R. M. Wilson, *J. Am. Chem. Soc.*, **86,** 5277 (1964).
104. A. L. Mndzhoyan and G. L. Papayan, *Sin. Geterotsikl. Soedin., Akad. Nauk. Arm. SSSR, Inst. Tonkoi Or. Khim.*, No. 7, 14 (1966); *Chem. Abstr.*, **68,** 78050 (1968).
105. (a) H. Singh and J. C. Verma, *J. Indian Chem. Soc.*, **40,** 817 (1963); (b) D. S. Deorha and S. K. Mukerji, *J. Indian Chem. Soc.*, **40,** 817 (1963).
106. K. B. L. Mathur and H. S. Mehra, *J. Chem. Soc.*, **1960,** 1954.

107. A. S. Angeloni, F. Delmoro, and M. Tramontini, *Ann. Chim. (Rome)*, **53,** 1751 (1963); *Chem. Abstr.*, **60,** 9123 (1964).
108. R. W. Wyn and S. A. Glickman, U.S. Pat. 2,636,885 (1953); *Chem. Abstr.*, **48,** 2779 (1954).
109. B. I. Arventi, *Bull. Soc. Chim. Fr.*, **3**(5), 598 (1936).
110. R. D. Stephens and C. E. Castro, *J. Org. Chem.*, **28,** 3313 (1963).
111. (a) C. F. H. Allen and J. A. Van Allan, *J. Org. Chem.*, **16,** 716 (1951); (b) C. F. H. Allen, *Can. J. Res.*, **4,** 264 (1931); *Chem. Abstr.*, **25,** 2992 (1931).
112. E. Cerutti and E. Laude, *Compt. Rend.*, **256,** 1122 (1963); *Chem. Abstr.*, **59,** 3858 (1963).
113. G. H. Coleman and R. H. Rigterink, U.S. Pat. 2,559,532 (1951); *Chem. Abstr.*, **46,** 3084 (1952).
114. R. Andrisano and F. Duro, *Gazz. Chim. Ital.*, **85,** 381 (1955); *Chem. Abstr.*, **50,** 5616 (1956).
115. A. P. Kuriakose and S. Sethna, *J. Indian Chem. Soc.*, **43,** 437 (1966).
116. R. v. Stoermer and B. Kahlert, *Chem. Ber.*, **35,** 1633 (1902).
117. R. Fittig and G. Ebert, *Justus Liebigs Ann. Chem.*, **216,** 162 (1882).
118. R. v. Stoermer and G. Calov, *Chem. Ber.*, **34,** 770 (1901).
119. E. W. Smith, *Iowa State Coll. J. Sci.*, **12,** 155 (1937); *Chem. Abstr.*, **32,** 2938 (1938).
120. V. S. Salvi and S. Sethna, *J. Indian Chem. Soc.*, **44,** 135 (1967).
121. W. Grubenmann and H. Erlenmeyer, *Helv. Chim. Acta*, **31,** 78 (1948).
122. V. I. Staninets and E. O. Shilov, *Dopovidi Akad. Nauk Ukr. RSR*, **1962,** 1474; *Chem. Abstr.*, **59,** 2754 (1963).
123. (a) G. M. Brooke, B. S. Furniss, and W. K. R. Musgrave, *J. Chem. Soc., C*, **1968,** 580; (b) G. M. Brooke and B. S. Furniss, *J. Chem. Soc., C*, **1967,** 869.
124. G. M. Brooke, *Tetrahedron Lett.*, **1968,** 2029.
125. D. H. R. Barton, A. K. Ganguly, R. H. Hesse, S. N. Loo, and M. M. Pechet, *Chem. Commun.*, **1968,** 806; *Chem. Abstr.*, **69,** 7781 (1968).
126. W. J. Hale, *Chem. Ber.*, **45,** 1596 (1912).
127. R. Royer, P. Demerseman, and L. Rene, *Bull. Soc. Chim. Fr.*, **1970,** 3740; *Chem. Abstr.*, **74,** 53363 (1971).
128. Y. Kawase, S. Takata, and E. Hikishima, *Bull. Soc. Chem. Japan*, **44,** 749 (1971).
129. T. Severin, R. Schmitz, and H-L. Temme, *Chem. Ber.*, **97,** 467 (1964).
130. F. Bordin, R. Bevilacqua, and F. Dalbeni-Sala, *Gazz. Chim. Ital.*, **99,** 1177 (1969); *Chem. Abstr.*, **72,** 90167 (1970).
131. B. R. Brown, G. A. Somerfield, and P. D. J. Weitzman, *J. Chem. Soc.*, **1958,** 4305.
132. B. R. Brown, W. Cummings, and G. A. Somerfield, *J. Chem. Soc.*, **1957,** 3757.
133. J. Finkelstein and S. M. Linder, *J. Am. Chem. Soc.*, **71,** 1010 (1949).
134. F. R. Japp and A. N. Meldrum, *J. Chem. Soc.*, **75,** 1035 (1899).
135. S. Tanaka, *J. Chem. Soc. Japan, Pure Chem. Sect.*, **72,** 307 (1951); *Chem. Abstr.*, **46,** 2535 (1952).
136. S. Tanaka, *J. Am. Chem. Soc.*, **73,** 872 (1951).
137. M. G. Marathey and J. M. Athavale, *J. Univ. Poona, Sci. Technol.*, **1953,** 90; Chem. *Abstr.*, **49,** 11638 (1955).
138. A. N. Grinev, I. A. Zaitsev, N. K. Venevtseva, and A. P. Terent'ev, *Zh. Obshch. Khim.*, **28,** 1853 (1958); *Chem. Abstr.*, **53,** 1299 (1959).
139. Kh. Ya. Kipper and H. Raudsepp, *Tr. Tallinsk. Politekhn. Inst., Ser. A*, No. 230, 67 (1965); *Chem. Abstr.*, **66,** 10800 (1967).
140. Kh. Ya. Kipper, I. R. Klesment, S. Ya. Saluste, O. G. Eizen, and Kh. T. Raudsepp, *Tr. Tallinsk. Politekhn. Inst., Ser. A*, No. 230, 77 (1965); *Chem. Abstr.*, **66,** 10805 (1967).
141. L. I. Smith and C. W. MacMullen, *J. Am. Chem. Soc.*, **58,** 629 (1936).
142. L. I. Smith and E. W. Kaiser, *J. Am. Chem. Soc.*, **62,** 133 (1940).
143. F. Bergel, A. Jacon, A. R. Todd, and T. S. Work, *J. Chem. Soc.*, **1938,** 1375.
144. P. Karrer, R. Escher, and H. Rentschler, *Helv. Chim. Acta*, **22,** 1287 (1939).

145. L. I. Smith and J. A. King, *J. Am. Chem. Soc.*, **65,** 441 (1943).
146. T. Reichstein and R. Hirt, *Helv. Chim. Acta*, **16,** 121 (1933).
147. M. G. Marathey and K. G. Gore, *J. Univ. Poona, Sci. Technol.*, **16,** 37 (1959); *Chem. Abstr.*, **54,** 9904 (1960).
148. R. L. Shriner and J. Anderson, *J. Am. Chem. Soc.*, **60,** 1418 (1938).
149. K. Okamoto, I. Nitta, and H. Shingu, *Bull. Soc. Chem. Japan*, **42,** 1464 (1969); *Chem. Abstr.*, **71,** 60936 (1969).
150. J. D. Brewer and J. A. Elix, *Tetrahedron Lett.*, **1969,** 4139;–*Chem. Abstr.*, **72,** 12616 (1970).
151. D. Molho and C. Mentzer, *Compt. Rend.*, **223,** 333 (1946); *Chem. Abstr.*, **41,** 442 (1947).
152. P. Demerseman, R. Reynaud, A. Cheutin, J. P. Lechartier, C. Pène, A-M. Laval-Jeantet, R. Royer, and P. Rumpf, *Bull. Soc. Chim. Fr.*, **1965,** 1464; *Chem. Abstr.*, **63,** 11321, (1965).
153. R. Andrisano, F. Duro, and G. Pappalardo, *Bull. Sci. Fac. Chim. Ind. Bologna*, **14,** 96 (1956); *Chem. Abstr.*, **52,** 19446 (1958).
154. B. Holmberg, *Chem. Ber.*, **89,** 278 (1956).
155. J. A. D. Jeffreys, *J. Chem. Soc.*, **1959,** 2153.
156. F. E. King, J. R. Housely, and T. J. King, *J. Chem. Soc.*, **1954,** 1392.
157. V. A. Zagorevski, Z. D. Kirsanova, *Khim. Geterosikl. Soedin,* **4**(4), 598 (1968); *Chem. Abstr.*, **70,** 37578 (1969).
158. O. Dann and H. G. Zeller, *Chem. Ber.*, **93,** 2829 (1960).
159. R. Royer, E. Bisagni, C. Hurdy, A. Cheutin, and M-L. Desvoye, *Bull. Soc. Chim. Fr.*, **1963,** 1003; *Chem. Abstr.*, **59,** 5106 (1963).
160. W. B. Whalley and G. Lloyd, *J. Chem. Soc.*, **1956,** 3213.
161. F. Freudenberg and M. Harder, *Justus Liebigs Ann. Chem.*, **451,** 213 (1926).
162. G. Lloyd and W. B. Whalley, *J. Chem. Soc.*, **1956,** 3209.
163. J. N. Chatterjea, *Experentia*, **7,** 374 (1951); *Chem. Abstr.*, **46,** 8083 (1952).
164. W. Bottomley, *Chem. Ind. (London)*, **1956,** 170; *Chem. Abstr.*, **50,** 13894 (1956).
165. J. Chopin, and G. Piccardi, *C. R. Acad. Sci., Paris, Ser. C*, **267,** 1336 (1968); *Chem. Abstr.*, **70,** 28769 (1969).
166. P. Niebes and J. Jadot, *Bull. Soc. Roy. Sci. Liege*, **39**(9–10), 525 (1970); *Chem. Abstr.*, **74,** 112400 (1971).
167. A. Mustafa, "Furopyrans and Furopyrones," A. Weissberger, Ed., Interscience, New York, 1967.
168. S. Tanaka, *Bull. Soc. Chem. Japan*, **42,** 1971 (1969); *Chem. Abstr.*, **71,** 91185 (1969).
169. S. Tanaka, *J. Chem. Soc. Japan, Pure Chem. Sect.*, **73,** 282 (1952); *Chem. Abstr.*, **47,** 9957 (1953).
170. B. Arventi'ev and H. Offenberg, *Anal. Stiint.* "Al. I. Cuza," *Iasi Sect. 1, Chim.* **10c,** 59 (1964); *Chem. Abstr.*, **63,** 13185 (1965).
171. A. Mooradian, U.S. Pat. 3,481,944 (1969); *Chem. Abstr.*, **72,** 53237 (1970).
172. A. Mooradian, *Tetrahedron Lett.*, **1967,** 407.
173. T. Sheradsky, *Tetrahedron Lett.*, **1966,** 5225.
174. D. Kaminsky, J. Shavel, Jr., and R. I. Meltzer, *Tetrahedron Lett.*, **1967,** 859.
175. R. Royer, Y. Kawase, M. Hubert-Habart, L. Rene, and A. Cheutin, *Bull. Soc. Chim. Fr.*, **1966,** 211; *Chem. Abstr.*, **64,** 19529 (1966).
176. E. Bisagni and R. Royer, *Bull. Soc. Chim. Fr.,* **1962,** 925
177. C. Pène, P. Demerseman, A. Cheutin, and R. Royer, *Bull. Soc. Chim. Fr.*, **1966,** 586.
178. R. Royer, P. Demerseman, C. Pène, and G. Colin, *Bull. Soc. Chim. Fr.*, **1967,** 915.
179. I. Matei, E. Cocea, and T. Lixandru, *Bull. Inst. Politeh. Iasi* (N. S.) **1**(1–2), 89 (1955); *Chem. Abstr.*, **51,** 13845 (1957).
180. J. Chopin and C. Katamna, Fr. Pat. 1,502,727 (1967); *Chem. Abstr.*, **69,** 96447 (1968).
181. H. Junek, H. Sterk, and B. Hornischer, *Monatsh. Chem.*, **99,** 2359 (1968); *Chem. Abstr.*, **70,** 57552 (1969).

182. R. Adams and L. Whitaker, *J. Am. Chem. Soc.*, **78,** 658 (1956).
183. I. Baxter and I. A. Mensah, *J. Chem. Soc., C*, **1970,** 2604.
184. J. Ficini and A. Krief, *Tetrahedron Lett.*, **1967,** 2497.
185. W. Davies and S. Middleton, *J. Chem. Soc.*, **1958,** 822.
186. J. B. Wright, *J. Org. Chem.*, **25,** 1867 (1960).
187. M. Polonovski, M. Pesson, and H. Kornowski, *Compt. Rend.*, **240,** 319 (1955); *Chem. Abstr.*, **50,** 1748 (1956).
188. A. Areschka, F. Binon, J. Mahaux, and F. Verbruggen, *Chim. Ther.*, **5**(6), 331 (1966); *Chem. Abstr.*, **67,** 73462 (1967).
189. M. Descamps and H. Inion, Belg. Pat. 704,705 (1968); *Chem. Abstr.*, **72,** 78858 (1970).
190. Smith, Kline & French Laboraties, Brit. Pat. 855,115 (1960); *Chem. Abstr.*, **55,** 12423 (1961).
191. E. Kyburz, A. Pletscher, H. Staebler, H. Besendorf, and A. Brossi, *Arzneimittel-Forsch.*, **13**(9), 819 (1963); *Chem. Abstr.*, **61,** 1124 (1964).
192. F. Hoffman-La Roche & Co., A-G., Fr. Pat. 1,339,382 (1963); *Chem. Abstr.*, **60,** 2932 (1964).
193. A. N. Grinev and N. K. Venevtseeva, *Zh. Obshch. Khim.*, **33**(5), 1436 (1963); *Chem. Abstr.*, **59,** 12739 (1963).
194. R. Landi-Vittory, F. Gatta, F. Toffler, S. Chiavarelli, and G. L. Gatta, *Farmaco (Pavia), Ed. Sci.*, **18,** 465 (1963); *Chem. Abstr.*, **59,** 9941 (1963).
195. B. Scarlata, L. Gramiceioni, L. Gambetti, and W. Pinto-Scoognamiglio, *Ann. Chim. (Rome)*, **56**(1–2), 71 (1966); *Chem. Abstr.*, **65,** 673 (1966).
196. G. Hallmann and K. Haegele, *Justus Liebigs Ann. Chem.*, **662,** 147 (1963); *Chem Abstr.*, **59,** 535 (1963).
197. L. Byk-Gulden, Fr. Pat. 1,343,073 (1963); *Chem. Abstr.*, **60,** 11986 (1964).
198. L. Byk-Gulden, Chemische Fabrik G. m. b. H Ger. Pat. 1,212,984 (1966): *Chem. Abstr.*, **64,** 19561 (1966).
199. L. Byk-Gulden, Fr. Pat. 1,343,074 (1963); *Chem. Abstr.*, **61,** 11972 (1964).
200. A. A. Aroyan and N. Kh. Khachatran, *Izv. Akad. Nauk Arm. SSR, Khim. Nauki*, **17,** 212 (1964); *Chem. Abstr.*, **61,** 8252 (1964).
201. A. L. Mndzhoyan and M. A. Kaldrikyan, *Izv. Akad. Nauk Arm. SSR, Khim. Nauki*, **15,** 85 (1962); *Chem. Abstr.*, **58,** 7894 (1963).
202. E. Shaw and D. W. Wooley, *Proc. Soc. Exptl. Biol. Med.*, **96,** 439 (1957); *Chem. Abstr.*, **52,** 4830 (1958).
203. C-T. Chou, J-Y. Chi, Hua Hsueh Hsueh Pao **29**(4), 260 (1963); *Chem. Abstr.*, **60,** 486 (1964).
204. I. R. Landi-Vittory, F. Gatta, S. Chiavarelli, and E. Ciriaci, *Rend. Ist. Super. Sanita*, **27**(1–2), 5 (1964); *Chem. Abstr.*, **62,** 1619 (1965).
205. F. Binon, C. Goldenberg, G. Deltour, and E. Gillyns, *Chim. Therap.*, **1966,** 141; *Chem. Abstr.*, **65,** 10550 (1966).
206. C. Goldenberg, F. Binon, and E. Gillyns, *Chim. Therap.*, **1966,** 221; *Chem. Abstr.*, **65,** 16925 (1966).
207. S. A. Marly, Neth. Pat. 6,506,415 (1965); *Chem. Abstr.*, **64,** 19561 (1966).
208. S. A. Marly, Belg. Pat. 663,926 (1965); *Chem. Abstr.*, **64,** 17543 (1966).
209. Laboratorio Chemico Farmaceutica, A. Menarinin S. A. S., Belg. Pat. 644,176 (1964); *Chem. Abstr.*, **63,** 8319 (1965).
210. J. M. Osbond, G. A. Fothergill, and J. C. Wickens, Brit. Pat. 1,106,059 (1968); *Chem. Abstr.*, **69,** 35924 (1968).
211. J. M. Osbond, G. A. Fothergill, and J. C. Wickens, Brit. Pat. 1,106,058 (1968); *Chem. Abstr.*, **69,** 35922 (1968).
212. S. Toyoshima, N. Hirose, T. Ogo, and A. Sugii, *Yakugaku Zasshi*, **88**(5), 503 (1968); *Chem. Abstr.*, **69,** 106373 (1968).

213. G. Rodighiero and U. Fornasiero, *Gazz. Chim. Ital.*, **91,** 90 (1961); *Chem. Abstr.*, **56,** 12843 (1962).
214. C. J. P. Spruit, *Rec. Trav. Chim. Pays-Bas*, **81,** 810 (1962); *Chem. Abstr.*, **58,** 12490 (1963).
215. H. J. Teuber and N. Götz, *Chem. Ber.*, **87,** 1236 (1954); *Chem. Abstr.*, **49,** 13956 (1955).
216. Y. Inouye and H. Kakisawa, *Bull. Chem. Soc. Japan*, **42,** 3318 (1969); *Chem. Abstr.*, **72,** 31519 (1970).
217. E. Späth and W. Gruber, *Chem. Ber.*, **71,** 106 (1938).
218. H. Kakisawa, Y. Inouye, and J. A. Romo, *Tetrahedron Lett.*, **1969,** 1929; *Chem. Abstr.*, **71,** 49678 (1969).
219. H. Zimmer, D. C. Iankin, and S. W. Horgan, *Chem. Rev.*, **71,** 229 (1971).
220. H. Ishii, T. H. Sugano, and M. Ikeda, *Yakugaku Zasshi*, **90,** 1290 (1970).
221. H. Ishii, M. Konno, M. Wakabayashi, F. Kuriyagawa, and M. Ikeda, *Yakugaku Zasshi*, **90,** 1298 (1970).
222. B. D. Cavell and J. MacMillan, *J. Chem. Soc., C*, **1967,** 310.
223. A. N. Grinev and A. P. Terent'ev, *Zh. Obshch. Khim.*, **28,** 75 (1958); *Chem. Abstr.*, **52,** 12830 (1958).
224. N. Sugiyama, *Bull. Inst. Phys. Chem. Res. (Tokyo)*, **21,** 744 (1942); *Chem. Abstr.*, **41,** 5506 (1947).
225. H. Ishii, R. Ohtake, H. Ohida, H. Mitsui, and N. Ikeda, *Yakugaku Zasshi*, **90,** 1283 (1970); *Chem. Abstr.*, **74,** 3444 (1971).
226. S. E. Fumagalli and C. H. Eugster, *Helv. Chim. Acta*, **54,** 959 (1971); *Chem. Abstr.*, **75,** 35562 (1971).
227. D. Cagniant and M. P. Cagniant, *Bull. Soc. Chim. Fr.*, **1957,** 838.
228. J. Entel, C. H. Ruof, and H. C. Howard, *J. Am. Chem. Soc.*, **73,** 4152 (1951).
229. F. Prillinger and H. Schmid, *Monatsh. Chem.*, **72,** 427 (1939); *Chem. Abstr.*, **34,** 1017 (1940).
230. N. I. Shuikin, I. I. Dmitriev, and T. P. Dobrynina, *J. Gen. Chem. (USSR)*, **10,** 967 (1940); *Chem. Abstr.*, **35,** 2508 (1941).
231. R. Adams and R. E. Rindfusz, *J. Am. Chem. Soc.*, **41,** 648 (1919).
232. H. Gilman, E. W. Smith, and L. C. Cheney, *J. Am. Chem. Soc.*, **57,** 2059 (1935).
233. M. W. Farrar and R. Levine, *J. Am. Chem. Soc.*, **72,** 4433 (1950).
234. R. Magnusson, *Acta Chem. Acad.*, **17,** 2358 (1963).
235. N. P. Buu-Hoi, N. D. Xuong, and N. V. Bac, *J. Chem. Soc., C*, **1964,** 173.
236. E. Bisagni, N. P. Buu-Hoi, and R. Royer, *J. Am. Chem. Soc.*, **1955**, 3688.
237. L. A. Kazitsyna, *Uch. Zap. Gos. Univ. im. M. V.*, **131,** 5 (1950); *Chem. Abstr.*, **47,** 10518 (1953).
238. C. D. Hurd and G. L. Oliver, *J. Am. Chem. Soc.*, **81,** 2795 (1959).
239. D. P. Brust and D. S. Tarbell, *J. Org. Chem.*, **31,** 1251 (1966).
240. H. E. Baumgarten, P. L. Creger, and C. E. Villars, *J. Am. Chem. Soc.*, **80,** 6609 (1958).
241. J. G. Traynham, *J. Sci. Lab. Denison Univ.*, **42,** 6079 (1951); *Chem. Abstr.*, **46,** 969 (1952).
242. A. Birch, *Quart. Rev. Chem. Soc.*, **4,** 69 (1950).
243. S. D. Darling and K. D. Wills, *J. Org. Chem.*, **32,** 2794 (1967).
244. A. L. Wilds and N. A. Nelson, *J. Am. Chem. Soc.*, **75,** 5360 (1953).
245. R. v. Stoermer and B. Richter, *Chem. Ber.*, **30,** 2094 (1897).
246. A. v. Wacek and F. Zeisler, *Monatsh. Chem.*, **83,** 5 (1952).
247. R. Royer, E. Bisagni, and G. Menichi, *Bull. Soc. Chim. Fr.*, **1964,** 2112.
248. H. Offenberg and B. Arventi'ev, *Anal. Stiint. Univ. "Al. I. Cuza" Iasi, Sect. Ic, Chim.*, **11**(2), 155 (1965); *Chem. Abstr.*, **65,** 5429 (1966).
249. B. Arventi'ev and H. Offenberg, *Anal. Stiint. Univ. "Al. I. Cuza", Iasi, Sect. Ic, Chim.*, **11**(1), 79 (1965); *Chem. Abstr.*, **63,** 14795 (1965).
250. H. Ishii, Y. Ishijawa, K. Mizukami, H. Mitsui, and N. Ikeda, *Chem. Pharm. Bull. Japan*, **19,** 870 (1971); *Chem. Abstr.*, **75,** 48792 (1971).

251. A. v. Wacek, H. O. Eppinger, and A. v. Bezard, *Chem. Ber.*, **73,** 521 (1940).
252. E. Bernateck, "Ozonolysis in the Naphthoquinone and Benzofuran Series," Oslo University Press, Oslo, 1960, p. 77.
253. F. Kaluza and G. Perold, *Chem. Ber.*, **88,** 597 (1955).
254. E. Bisagni, J. P. Marquet, A. Cheutin, R. Royer, and M-L. Desvoye, *Bull. Soc. Chim. Fr.*, **1965,** 1466.
255. A. S. Angeloni and M. Tramontini, *Ann. Chim. (Rome)*, **55,** 1028 (1965); *Chem. Abstr.*, **64,** 5028 (1966).
256. B. Arventi'ev and H. Offenberg, *Acad. Rep. Populare Romine, Filiala Iasi, Studii Cercetari Stiint. Chim.*, **11,** 305 (1960); *Chem. Abstr.*, **56,** 11554 (1962).
257. B. Arventi'ev and H. Offenberg, *Anal. Stiint. Univ. "Al I. Cuza," Iasi Sect. 1,* **9**(1), 225 (1963); *Chem. Abstr.*, **60,** 493 (1964).
258. O. Dischendorfer and A. Verdino, *Monatsh. Chem.*, **68,** 81 (1936).
259. A. S. Angeloni and M. Tramontini, *Bull. Sci. Fac. Chim. Ind. Bologna*, **21**(4), 243 (1963); *Chem. Abstr.*, **60,** 15808 (1964).
260. A. L. Mndzhoyan and A. A. Aroyan, *Izv. Akad. Nauk Arm. SSR Khim. Nauki*, **14,** 591 (1961); *Chem. Abstr.*, **58,** 5606 (1963).
261. T. Reichstein and J. Baud, *Helv. Chim. Acta*, **20,** 892 (1937).
262. H. Gilman and D. S. Melstrom, *J. Am. Chem. Soc.*, **70,** 1655 (1948).
263. R. Gaertner, *J. Am. Chem. Soc.*, **74,** 5319 (1952).
264. R. Gaertner *J. Am. Chem. Soc.*, **73,** 4400 (1951).
265. J. N. Chatterjea, *J. Indian Chem. Soc.*, **34,** 306 (1957).
266. A. Mustafa, W. Asker, O. H. Hishmat, M. I. Ali, A. K. E. Mansour, N. M. Abed, K. M. A. Khalil, and S. M. Samy, *Tetrahedron*, **21,** 849 (1965).
267. R. T. Foster and A. Robertson, *J. Chem. Soc.*, **1939,** 921.
268. H. F. Birch and A. Robertson, *J. Chem. Soc.*, **1938,** 306.
269. G. M. Badger, B. J. Christie, H. J. Rodda, and J. M. Pryke, *J. Chem. Soc.*, **1958,** 1179.
270. G. M. Badger, H. J. Rodda, and J. M. Sasse, *J. Chem. Soc.*, **1958,** 4777.
271. J. Nóvak, J. Ratusky, O. Sneberg, and F. Sórm, *Chem. Listy*, **51,** 479 (1957); *Chem. Abstr.*, **51,** 10508 (1957).
272. W. E. Parham, C. G. Fritz, R. W. Soeder, and R. M. Dodson, *J. Org. Chem.*, **28,** 577 (1963.)
273. C. H. Krausch, W. Metzner, and G. O. Schenck, *Chem. Ber.*, **99,** 1723 (1966).
274. T. Abe and T. Shimizu, *Nippon Kagaku Zasshi*, **91,** 753 (1970); *Chem. Abstr.*, **73,** 120436 (1970).
275. B. D. Cavell and J. MacMillan, *J. Chem. Soc.*, **1967,** 310.
276. G. Kraemer and A. Spilker, *Chem. Ber.*, **23,** 3276 (1890).
277. G. Bressan, *Chim. Ind. (Milan)*, **51,** 705 (1969); *Chem. Abstr.*, **71,** 81781 (1969).
278. G. Natta, M. Farina, M. Peraldo, and G. Bressan, *Chim. Ind. (Milan)*, **43,** 161 (1961); *Chem. Abstr.*, **60,** 15986 (1964).
279. A. Mizote, T. Tanaka, T. Higashumura, and S. Okamura, *J. Polym. Sci.*, **Part A-I** (4), 869 (1966); *Chem. Abstr.*, **64,** 19795 (1966).
280. T. Lesiak, *Roczniki Chem.*, **39,** 589 (1965); *Chem. Abstr.*, **63,** 16293 (1965).
281. H. M. Foster, U.S. Pat. 3,381,018 (1968); *Chem. Abstr.*, **69,** 51981 (1968).
282. O. Oyamamoto, H. Kato, and T. Yonezawa, *Bull. Chem. Soc. Japan*, **40,** 1580 (1967).
283. E. A. Karakhanov, G. V. Drovyannikova, E. A. Viktorova, *Khim. Geterosikl. Soedin*, 7(2), 156 (1971); *Chem. Abstr.*, **75,** 33677 (1971).

CHAPTER II

Acylbenzofurans

1. Formylbenzofurans

The fact that 3-methyl- and 6-hydroxy-3-methylbenzofuran invariably form α-formyl derivatives in the Gattermann aldehyde synthesis is well established.[1] Moreover, it seems clear that the carbethoxy group in the α-position can be used to protect the furan residue in the application of the Gattermann reaction for synthesis of formylbenzofurans having the formyl group in the benzene nucleus. Furthermore, it seems reasonable to expect that, in general, replacement of the carboxyl group with an alkyl group yields similar results.[2] Thus, by Gattermann's procedure with the aid of aluminum chloride, an excellent yield of ethyl 4,6-dimethoxy-7-formylbenzofuran-2-carboxylate is obtained.[2] Orientation of formylation of bz-methoxy-2,3-dimethylbenzofurans, using phosphorous oxychloride and dimethylformamide, has been thoroughly studied.[3,4] The ready reduction of aldehydes, by Huang-Minlon modification of the Wolff-Kishner procedure, to the methyl derivatives established the orientation. 5-Hydroxy-3-methylbenzofuran failed to undergo the Gattermann aldehyde reaction.[2]

Alkylated benzofurans are readily formylated, either by boiling with phosphorus oxychloride in dimethylformamide, or by saturation of benzofuran solution with hydrogen cyanide and hydrogen chloride.[5-8] A comparative study of the ease of formylation of some alkylated benzofurans has been recently reported.[9] Vilsmier-Haack formylation of 4,6-dimethoxybenzofuran afforded the 7-formyl compound (**1**).[10,11]

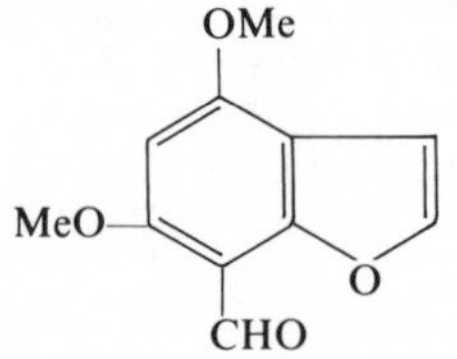

OMe

MeO

O

CHO

1

Another approach for the preparation of substituted 2-formylbenzofurans is exemplified by treatment of 3-methyl-7-methoxybenzofurancarboxylic acid chloride with a mixture of palladium catalyst and sulfur–quinoline inhibitor in xylene saturated at 140°C with hydrogen leading to the formation of **2**.[13]

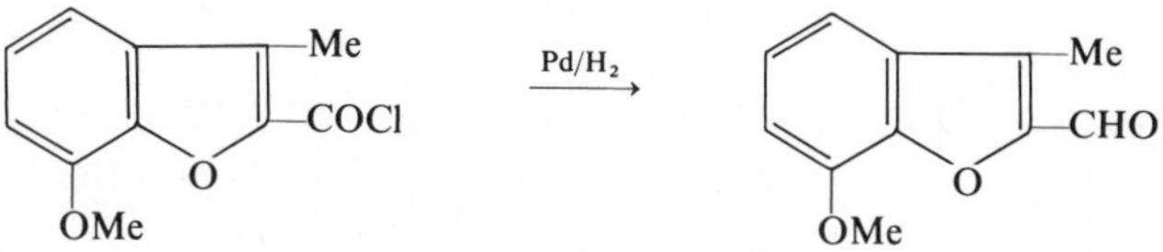

Me COCl OMe O $\xrightarrow{Pd/H_2}$ Me CHO OMe O

2

3-Formylbenzofuran has been obtained via oxidation of 3-benzofuranyl alcohol, which is readily produced upon treatment of methyl-3-benzofurancarboxylate with lithium aluminum hydride (Eq. 1).[14]

$$RCOOH \xrightarrow[(90\%)]{CH_2N_2} RCOOMe \xrightarrow{LiAlH_4} RCH_2OH \xrightarrow{MnO_2} RCHO \quad (1)$$

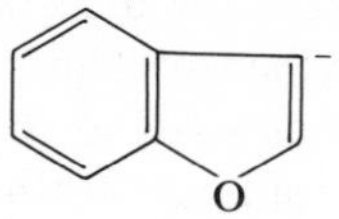

R =

2. Acylbenzofurans

It has been reported that alkylbenzofurans are readily acetylated in the furan ring,[15-23] and in the benzene ring when the furan ring is fully substituted (the 6 position is the most reactive site in the benzene ring),[24-28] and that further acetylation furnishes diacetylbenzofurans readily.[29,30] These acylation patterns of alkylbenzofurans seem to be due to the great contributions of extreme resonance structures **3**–**8**, and especially **3**–**6**.

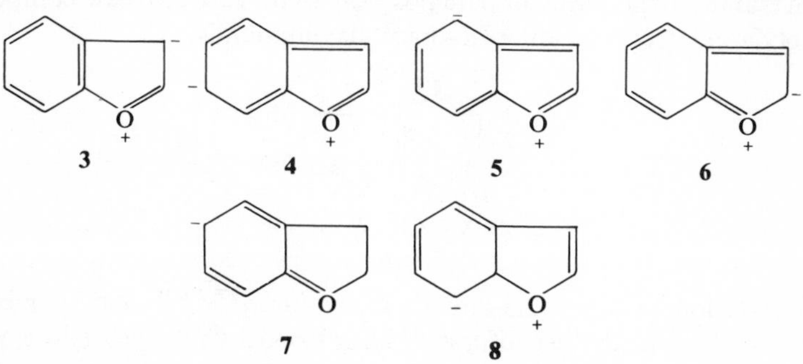

3 **4** **5** **6**

7 **8**

Treatment of 2-benzofuranecarboxylic acid cyanide with aniline resulted in the formation of 2-formylbenzofuran (Eq. 2).[12]

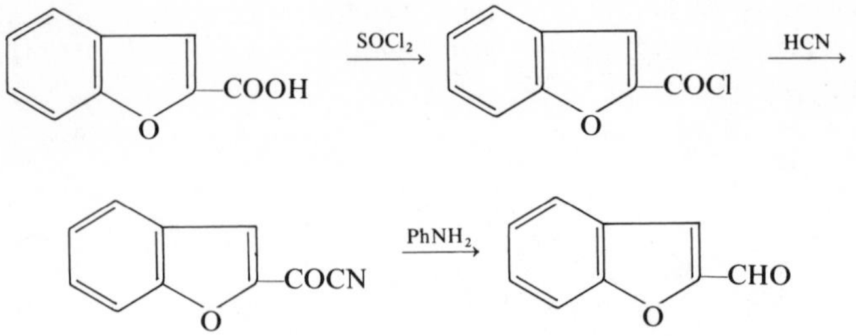

SOCl$_2$ HCN

—COOH —COCl

O O

PhNH$_2$

—COCN —CHO (2)

O O

Acetyl derivatives of 2,3-dimethylbenzofuran and their methyl homologs are derived into their corresponding carboxylic acids by the haloform reaction, and thence acetylation of their esters by Friedel-Crafts reaction is possible.[31] The keto esters thus obtained are hydrolyzed to yield the keto acids which, upon oxidation by the haloform reaction into the dicarboxylic acids and followed by decarboxylation, afford acetylbenzofurans. Of interest is the observation that the dicarboxylic acid can also be obtained via oxidation of the diacetylbenzofurans. However the haloform reaction of hindered acetyl compounds leads to the formation of only halogenated ketones. Acetylation of 5-alkoxycarbonyl-2,3-dimethylbenzofuran furnished the 6-acetylated compounds; that of 6-alkoxycarbonyl compounds and their methyl homologs yielded, under the same conditions, the 4-acetylated compounds. Thus acetylbenzofurans can be obtained by: (1) acetylation of benzofurans; (2) dehydrocyclization of phenylbutanones with polyphosphoric acid (PPA) or sulfuric acid; and (3) decarboxylation of keto acids (Chart 1).

Acetylation of bz-chloro- (**25**), and bz-bromo-2,3-dimethylbenzofuran (**26**) with acetyl chloride in presence of aluminum chloride in carbon disulfide has been thoroughly investigated,[32,33] Acetylation occurs at C-6 and/or C-4 and is activated by the furan ring, regardless of whether the halogen atom is at the ortho, meta, or para position. In the case of 7-halo compounds, diacetylation occurs yielding 4,6-diacetyl compounds.

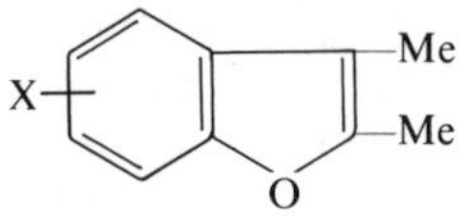

X— —Me

—Me

O

25 X = Cl **26** X = Br

Acetylation of acetyl-2,3-dimethylbenzofuran[27,29-31] and of methoxycarbonyl-2,3-dimethylbenzofuran,[31] after Friedel-Crafts procedure, yielded the acetylated compounds, generally meta to the acetyl group or methoxycarbonyl group except that methyl 2,3-dimethyl-5-benzofurancarboxylate afforded the 6-acetylated compound. Isopropylation of 2,3-dimethyl-6-

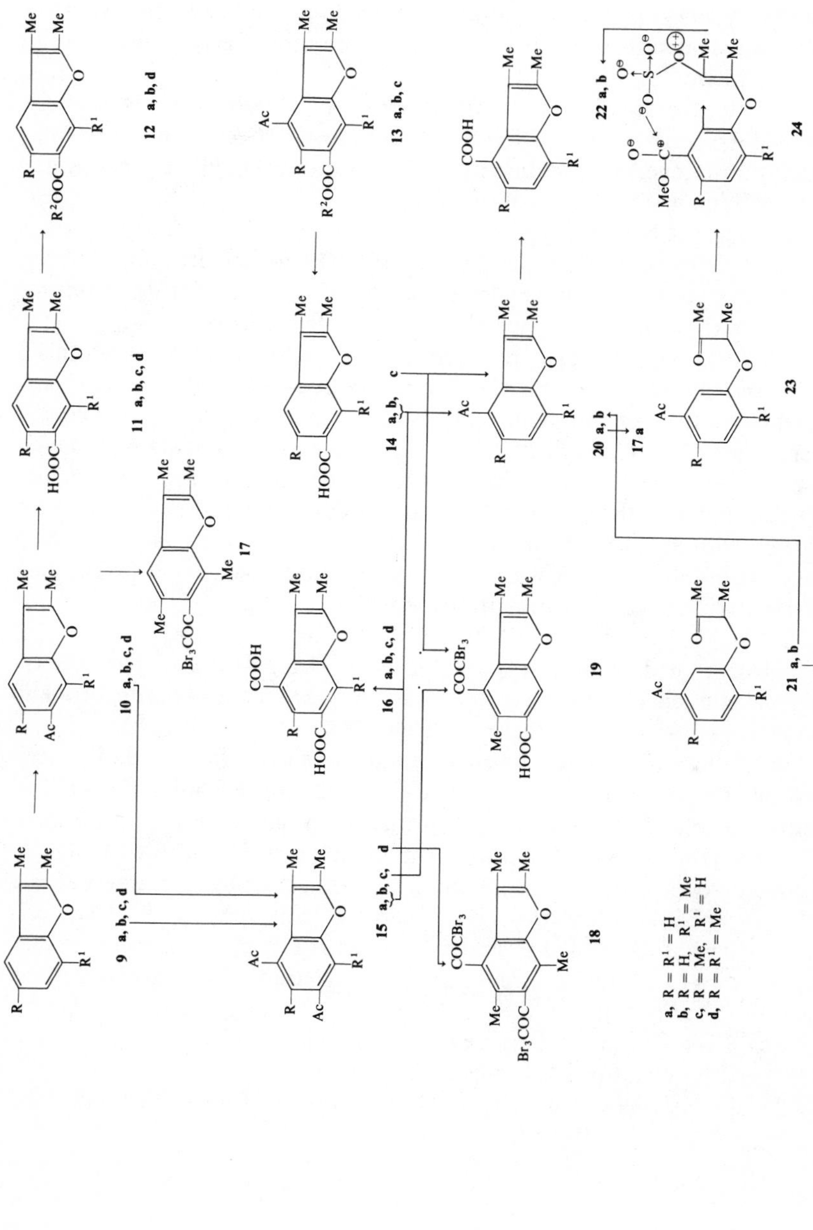

a, R = R^1 = H
b, R = H, R^1 = Me
c, R = Me, R^1 = H
d, R = R^1 = Me

Chart 1.

benzofuranyl methyl ketone furnished the 4-isopropyl compound.[27,29] Kawase et al.[34] have concluded that in electrophilic substitution of bz-acetyl- or bz-methoxycarbonyl-2,3-dimethylbenzofurans all positions, but especially the 6- and 4-positions, are activated by the effect of the furan ring, and that the meta position relative to the acetyl or methoxycarbonyl group is less deactivated than the ortho or para position by effect of the group. In case of methyl 2,3-dimethyl-4- and methyl 2,3-dimethyl-6-benzofurancarboxylate, the reaction occurred at the 6 and 4 positions, respectively, and was favored by both effects. The furan-ring effect predominated in acetylation of methyl 2,3-dimethyl-7-benzofurancarboxylate, yielding the corresponding 6-acetylated compound.

Polyalkylated benzofurans are acylated at C-6, provided this position is unoccupied. Thus, for example, 2,3,6,7- and 2,3,4,6-tetramethylbenzofuran are acylated in presence of acyl chloride and aluminum chloride or stannic chloride to the yield 4- and 7-acetyl compounds,[26] respectively.

2-Alkylated benzofurans are considerably more stable than benzofuran itself, and undergo normal aluminum chloride-catalyzed Friedel-Crafts acylation with both aliphatic and aromatic acid chlorides to give the appropriate 3-acyl-2-alkylbenzofurans.[35] Phosphoric acid-catalyzed acylations (in the 3 position) are also possible. Benzofuran can readily be converted into 2-acetylbenzofuran by means of acetic anhydride in acetic acid medium with 85% phosphoric acid as a catalyst. When it was extended to higher acid anhydrides, it was necessary to replace acetic acid by the acid corresponding to the acid anhydride used.[35]

Acylation of 2,3-dimethyl-4-hydroxybenzofuran (**27**) with acetic acid, phenylacetic acid, 2-methoxyphenylacetic acid, and/or 2,4,5-trimethoxyphenylacetic acid in presence of polyphosphoric acid gave a mixture of 5- and 7-acyl compounds, **28** and **29**, respectively (Chart 2).[36]

Application of Hoesch ketone synthesis with trifluoro- and/or trichloromethyl cyanides to 3-methyl-, 6-hydroxy-, 5,6-dimethoxy-, 5-hydroxy-, and 5-methoxy-3-methylbenzofuran gave the corresponding 2-haloacetyl derivatives. 7-Methoxy-3-methylbenzofuran afforded a mixture of 7-methoxy-3-methyl-2-trifluoroacetylbenzofuran and an isomeric ketone, which is probably either a 4- or 5-trifluoroacetyl compound.[20] 5-Hydroxy-3-methylbenzofuran failed to underg Hoesch reaction, but its benzyl ether produced the respective ketones; debenzylation occurred during the reaction. Modification of the usual reaction conditions has effected the successful condensation of trifluoro- and trichloromethyl cyanides with a number of 2,3-unsubstituted benzofurans.[1,2,37-40]

Friedel-Crafts acylation of 4,6-dimethoxybenzofuran (**32**) gave the 7-acetyl compound (**33**) (Eq. 3).[10,11]

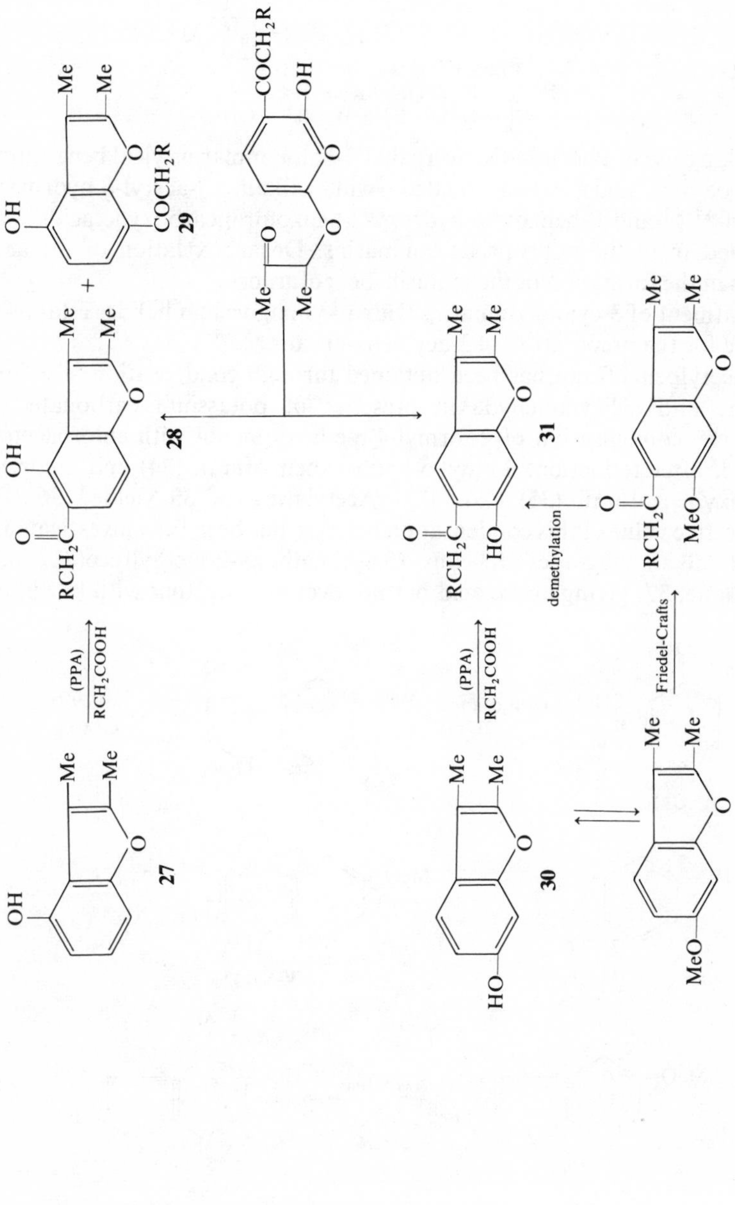

Chart 2.

(3)

32 → **33**

R = R^2 = OMe, R^1 = H
R = R^2 = OH, R^1 = H
R = OMe, R^1 = H, R^2 = OH

Making use of Perkin's discovery that 3-halocoumarins yield benzofuran-2-carboxylic acids when treated with alkali, 7-acetyl-6-hydroxy-3-methyl-[41-43] and 7-benzoyl-6-hydroxy-2-benzofurancarboxylic acids[44] are obtained from the appropriate coumarins. Decarboxylation of the acids results in the formation of the requisite benzofurans.

Treatment of 3-cyanocoumarins with alkylmagnesium halides establishes a route for the preparation of 3-acylbenzofurans.[45,46]

2-Acetylbenzofuran has been obtained through condensation of chloroacetone and salicylaldehyde in presence of potassium carbonate.[46,48] Similarly, condensation of 2-formyl-4-methoxyphenol with chloroacetone yielded, after reduction, 2-ethyl-5-hydroxybenzofuran (**34**) and 2-ethyl-5-methoxybenzofuran (**35**) (Eq. 4).[47] Acetylation of **35** yielded **36**. The preparative value of this condensation reaction has been fully investigated.[48]

Hot alkali degrades Khellin (5,8-dimethoxy-2-methylfuro(3′,2′-6,7)-chromone, **37**) giving acetic acid but not acetone, together with khellinone

$ClCH_2COCH_3$, K_2CO_3, Me_2CO; N_2H_4, H_2O, $(CH_2OH)_2$

34 + **35**

AcCl, $SnCl_4$; $AlCl_3$

36 demethylation

(4)

(5-acetyl-4,7-dimethoxy-6-benzofuranol, **39**).[49] Similarly, visnaginone (5-acetyl-4-methoxy-6-benzofuranol, **40**) is obtained upon alkaline degradation of visnagin (**38**).[50] Both **39** and **40** react with ethyl acetate in presence of sodium to afford the β-diketones **41**[51] and **42**,[52] respectively (Eq. 5).

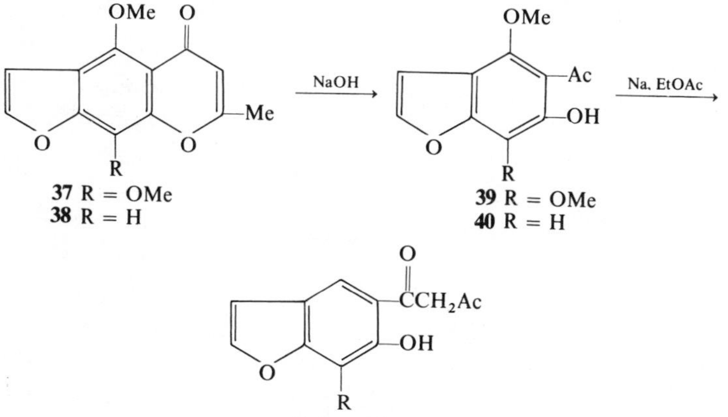

37 R = OMe
38 R = H

NaOH →

39 R = OMe
40 R = H

Na, EtOAc →

41 R = OMe
42 R = H

(5)

2-Aroylbenzofurans were prepared by Rap-Stoermer condensation,[53,54] involving condensation of ω-bromoacetophenone with the sodio derivative of salicylaldehyde; this procedure was extended to derivatives of each component.[55] Thus ω-bromo derivatives of 3-chloro-5 and 3-fluoro-4-methoxyacetophenone afforded 2-(3-chloro-4-methoxybenzoyl)- and 2-(3-fluoro-4-methoxybenzoyl)benzofuran, both of which were demethylated in the usual way to the corresponding hydroxy ketones; 2-(2-hydroxy-5-methylbenzoyl)- and 2-(*p*-bromobenzoyl)benzofuran were similarly prepared.

An alternative route to 2-aroylbenzofurans consisted of Friedel-Crafts acylations.[56-60] 3-Aroylbenzofurans are prepared by Friedel-Crafts acylations of 2-substituted benzofurans and reactive aroyl derivatives.[61] 3-Benzoyl-2-benzylbenzofuran[62] is similarly prepared, and the same is true of 3-methyl-2-benzofuranylpyridyl ketone synthesis.[63]

Treatment of 3-haloflavones with sodium hydroxide in presence of ethanol resulted in the formation of 2-aroyl-3-hydroxybenzofurans.[64-67] Similar results were obtained in the case of 4′-methoxy- and 4′-methoxy-6-methyl-3,3′-dibromoflavanone.

Oxidation of 4′-hydroxyflavylium chloride (**43**) in boiling methanol, containing a buffer solution (pH 5.8) with hydrogen peroxide yielded 2-(4-hydroxyphenyl)-3-acetylbenzofuran (**44**) (Eq. 6).[68]

Jurd[69,70] has shown that in aqueous methanolic solution peroxide oxidation of 3-methyl-, and 3-methoxyflavylium salts primarily yielded

(6)

43 **44**

3-acetyl- and 3-carbmethoxy-2-(4-hydroxyphenyl)benzofuran, for example **45** yields **46** (Eq. 7). Investigation of this reaction, both in aqueous methanol

R = Me or OMe (7)

45 **46**

and/or aqueous acetic acid, has confirmed that 3-acetyl-2-arylbenzofurans are formed only in alcoholic media.[71] In absence of methanol (or ethanol) a new type of oxidation product, an enol benzoate of structure **47**, is formed.

47

Such oxidation in aqueous acetic acid medium resembles the Baeyer-Villiger oxidation of ketones to esters, thus a similar reaction mechanism has been proposed:

Since the susceptibilty of methanol greatly exceeds that of water, nucleophilic attack of methanol on the carbonium ion may compete with the above reaction in aqueous methanol solutions to yield 3-acetyl-2-phenylbenzofurans:

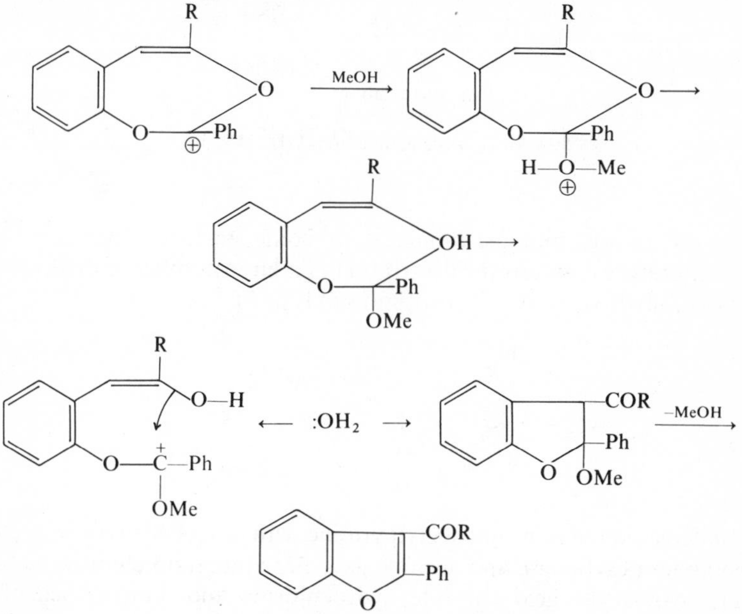

Condensation of *p*-benzoquinone with *β*-(diethylamino)vinyl ketones ($Et_2NCR{=}CHCOR^1$) in glacial acetic acid medium at room temperature provided a large number of substituted 5-hydroxy-3-acylbenzofurans (**48**).[72,73]

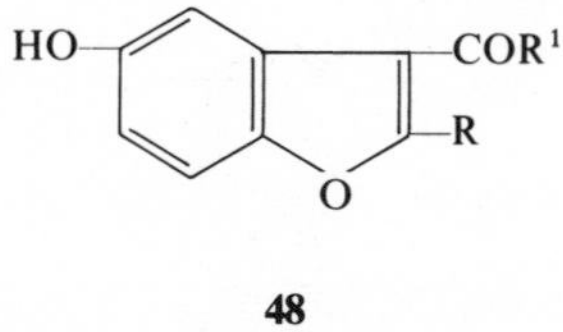

48 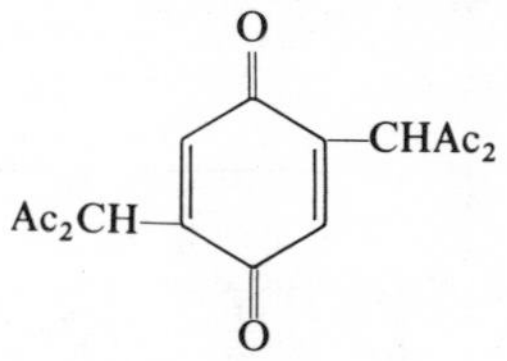 **49**

Bernatek et al.[74-77] have shown that the base-catalysed reaction between *p*-benzoquinone and acetylacetone leads to the formation of $\omega,\omega,\omega',\omega'$-tetraacetyl-*p*-benzoquinone (**49**) and 2-methyl-3-acetyl-5-hydroxybenzofuran (**48**, R = R′ = Me). The latter has also been obtained in 60% yield, upon treatment of *p*-benzoquinone and acetylacetone with absolute EtOH-$ZnCl_2$[77] 4-Hydroxy-2,6-di-*t*-butyl-3-acetylbenzofuran (**50**), an auto-oxidation product of 5-*t*-butylresorcinol, has been synthesized (Eq. 8).[78,79]

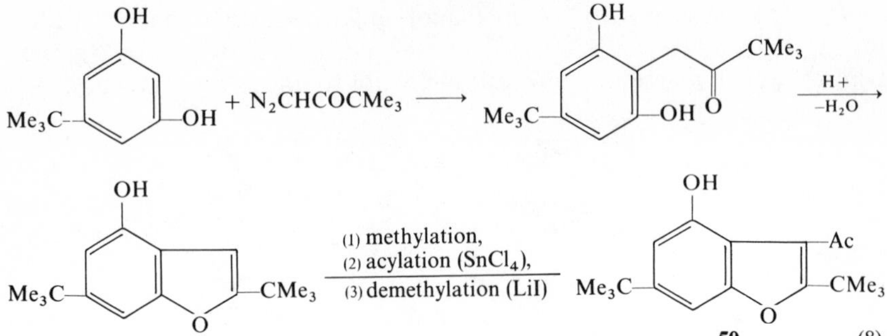

(8)

50

Because of the biological interest of some acylbenzofurans,[56,57] the patent literature contains the description of a large number of dialkylamino ethers of 3-hydroxy-2-aroylbenzofurans of type **51**.[80]

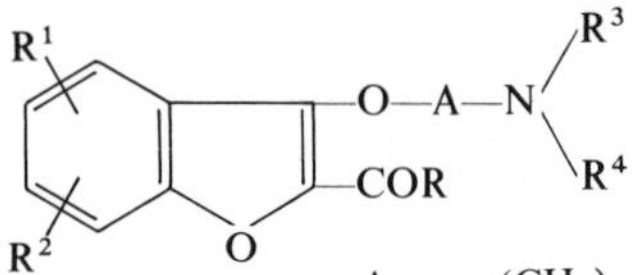

51 $A = -(CH_2)_n-$

Condensation of ω-bromoacetylpyridine with salicylaldehyde in presence of potassium carbonate and acetone gave **52**; aluminum chloride added to pyridinecarboxylic acid chloride hydrochloride and 3-methylbenzofuran in carbon disulfide yielded 3-methyl-2-benzofuranylpyridyl ketone (**53**).[63] 2-Hydroxy-ω-isonicotinylacetophenone, treated with bromine solution in chloroform, then with hydrochloric–acetic acid mixture produced 2-isonicotinoyl-3-benzofuranone hydrochloride.[63] Synthesis of **54** has been achieved.[81]

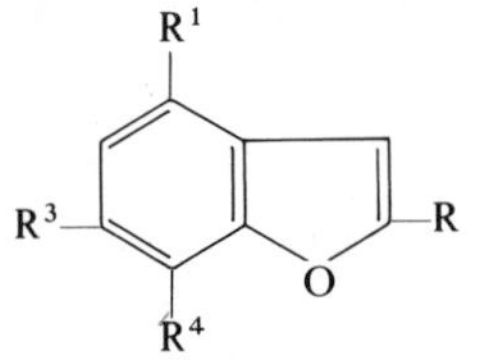

52 R = R = 2-, 3-, or 4-pyridyl

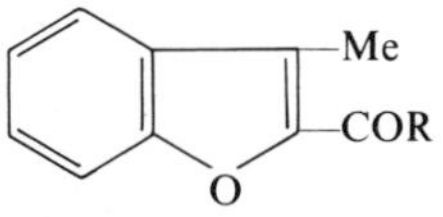

53

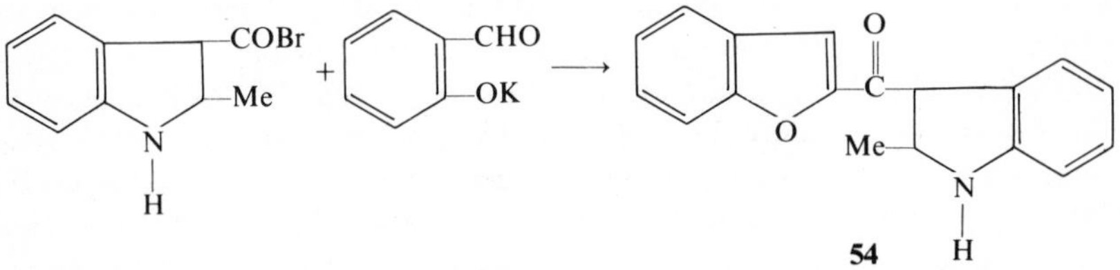

54

Synthesis of 4-hydroxy-5-benzofuranylmethyl ketone (**55**) has been achieved (Eq. 9).[82]

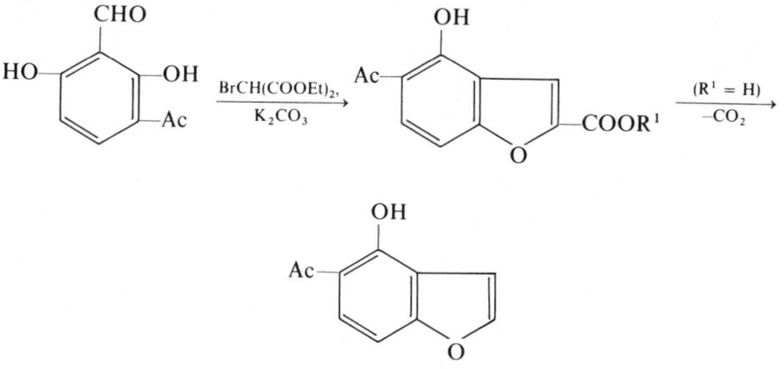

$$\xrightarrow[K_2CO_3]{BrCH(COOEt)_2,} \qquad \xrightarrow[-CO_2]{(R^1 = H)} \qquad (9)$$

55

Treatment of dihydrobenzofuran-2,3-dione with ω-bromo-2-acetonaphthalene in presence of sodium and methanol resulted in the formation of 2-(2-naphthoyl)benzofuran-3-carboxylic acid.[83]

3. Miscellaneous Reactions

A. Reduction

2-Formylbenzofuran is reduced by the Huang-Minlon modification of the Wolff-Kishner reduction to 2-methylbenzofuran.[3,4] Catalytic hydrogenation of 2-acetylbenzofuran[84,85] is thoroughly studied (Chart 3).

B. Oxidation

Oxidation of 3-acetyl-5-methoxybenzofuran with morpholine at 110–120°C yields 5-methoxybenzofuran-3-acetic acid.[72] However, treatment of 2-acetylbenzofuran with selenium dioxide gives 2-benzofuranylglyoxal, which is converted readily into the dioxime. The latter, upon treatment of its nickel salt with dilute alkali solution yields the oxime of 2-benzofuranyl formamide. Further, 2-benzofuranylglyoxal reacts with semicarbazide and/or thiosemicarbazide in presence of potassium carbonate to afford the 3-hydroxy and 3-mercapto derivative of 5-(2-benzofuranyl)-1,2,4-triazine.[86]

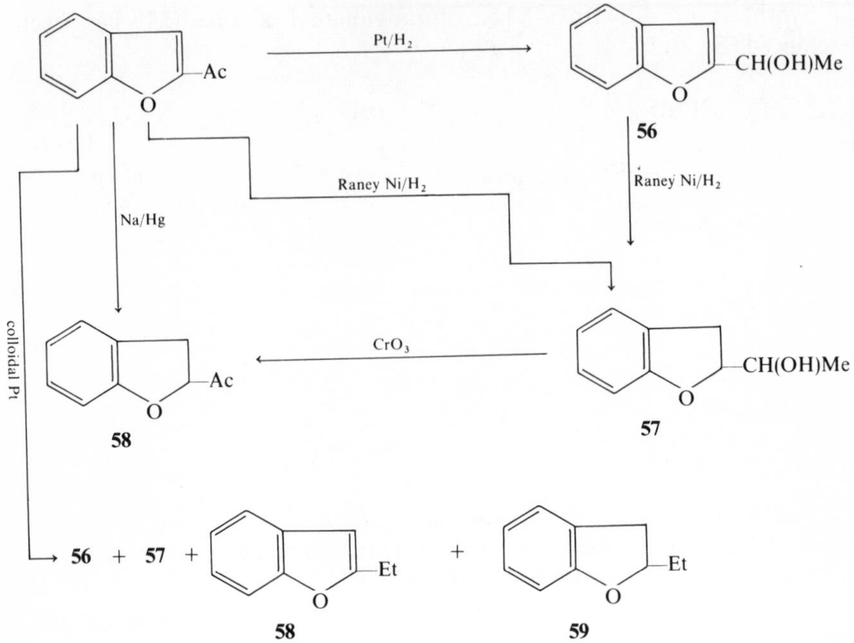

Pt/H_2
-Ac
O
-CH(OH)Me
O
56
Raney Ni/H_2
Raney Ni/H_2
Na/Hg
colloidal Pt
CrO_3
-Ac
O
58
-CH(OH)Me
O
57
56 + 57 +
-Et
O
58
+
-Et
O
59

Chart 3.

C. Alkaline Degradation

2-Ethyl-3-formyl-, 2-ethyl-3-propionyl-, and 2-ethyl-3-(4-methoxyphenyl) -benzofuran are readily converted in alkaline media to the same 2-hydroxy-benzylethyl ketone which, when heated with isatin and potassium hydroxide, underwent the Pfitzinger condensation to afford a mixture of 2-ethyl-3-(2-hydroxyphenyl)cinchoninic acid (**60**) and 2-(2-hydroxybenzyl)-3-methyl-cinchoninic acid (**61**).[87] Under alkaline conditions, the hetero ring of 2-alkyl-3-acylbenzofurans opens and degradation then leads to the formation of acids and hydroxybenzyl alkyl or aryl ketones, depending on the substituents of the heterocyclic ring. On acidification, the ketones cyclize to give 2-alkylbenzofurans (Chart 4).[88] It is concluded that benzofuran compounds with 3-carbonyl substituents are as unstable toward alkalies as the corresponding chromone molecules; thus 3-acylbenzofurans can be used to prepare 2-alkyl- or 2-arylbenzofurans.

Degradation of benzofurans substituted at C-3 by electron-attracting groups not containing a carbonyl residue is more difficult than degradation of formyl or acyl analogs. Consequently, the 1,2 bond of 3-formyl-, or acylbenzofuran (**62**) is broken with ammonia to give (2-hydroxyphenyl)-

β-enamino ketones (**63**), which regenerate the initial compound (**62**) by loss of ammonia either spontaneously, upon heating, and/or upon treatment with dilute mineral acid.[89]

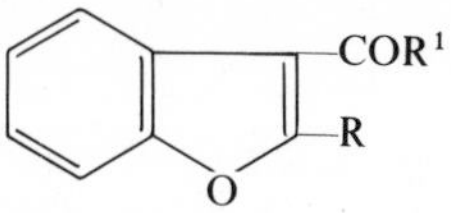

62

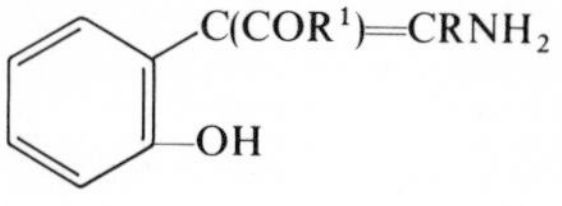

63

2-Ethyl-3-acetylbenzofuran (**62**, R = Et, R^1 = Me) and the isomer (62, R = Me, R^1 = Et) gave two different β-enamino ketones; however, 2-ethyl-3-formylbenzofuran (**62**, R = Et, R^1 = H) and its isomer (**62**, R = H, R^1 = Et) afforded the same compound (**63**) (Chart 5).

However 3-cyano- (**64**), 2-ethyl-3-cyano- (**65**), and 2-ethyl-3-carbethoxy-benzofuran (**66**) do not undergo this degradation reaction. Benzofurans bearing nitrile, formyl, and/or acyl groups at C-3 are degraded in alkaline medium in the same way as the 3-carboxybenzofuran. The reaction is carried out in aqueous alcohol with 3 moles sodium hydride. Alternatively, 2-ethyl-3-carboxy- (**67**) and 2-ethyl-3-amidobenzofuran (**68**) are not degraded by alkali, but rather saponified.

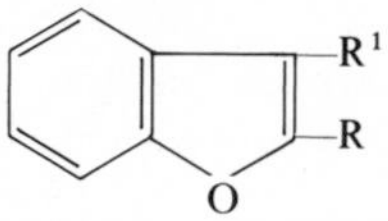

64 R = H, R^1 = CN
65 R = Et, R^1 = CN
66 R = Et, R^1 = COOEt
67 R = Et, R^1 = COOH
68 R = Et, R^1 = $CONH_2$

Thus **65** gave 2′-hydroxyphenylacetic acid, while 2-ethyl-3-formyl- and 3-propionylbenzofuran gave the same 2-hydroxybenzyl ethyl ketone (**A,** Chart 5). With hydroxylamine in neutral or alkaline medium 2-ethyl-3-acetylbenzofuran and 2-methyl-3-propionylbenzofuran gave two different isoxazoles, **B** and **C**, respectively. **69** (R = CHO, Ac, COEt, COPh)reacts with *N*-(guanidinomethyl)morpholine to afford **70** (R = CHO, Ac, COEt, or COPh). **69** (R = CHO, Ac, COEt, or COPh) yields **71** (R = CHO, Ac, COEt, or COPh); **69** (R = CN) provides **70** (R = NH_2).[90]

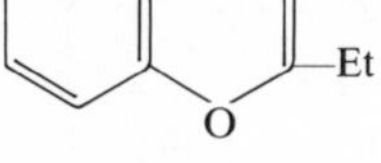

69

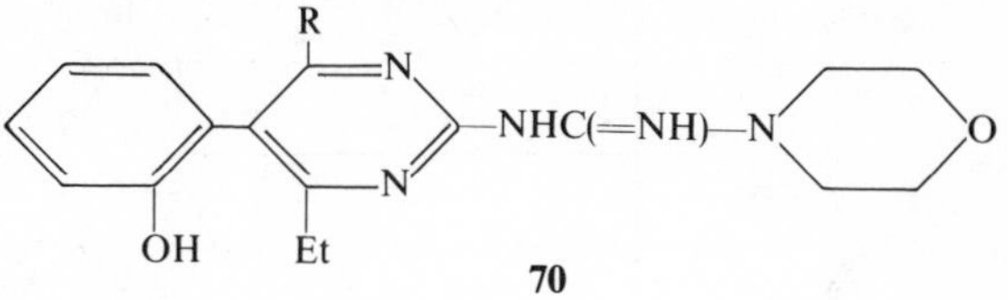

70

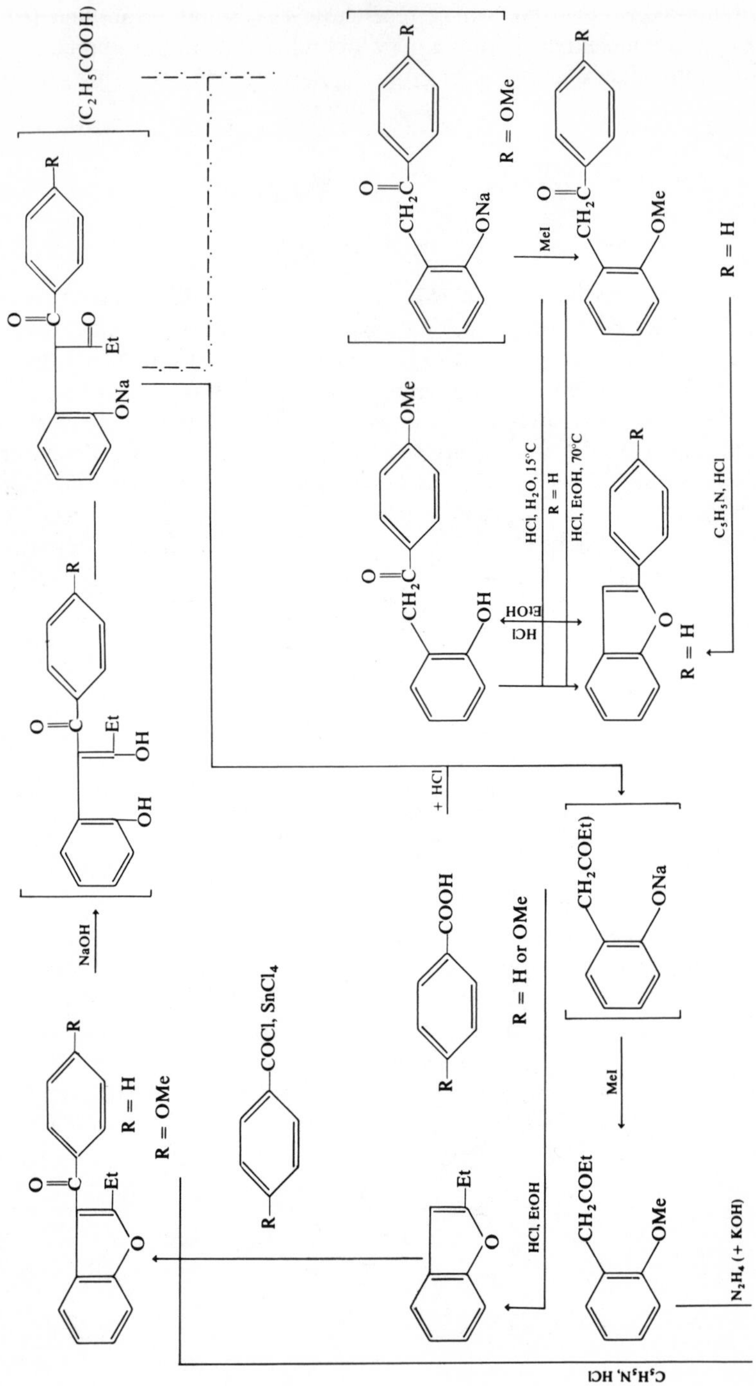

NaOH
(C_2H_5COOH)
R = H
R = OMe
COCl, $SnCl_4$
COOH
R = H or OMe
+ HCl
MeI
HCl, EtOH
N_2H_4 (+ KOH)
C_5H_5N, HCl
HCl, H_2O, 15°C
R = H
HCl, EtOH, 70°C
HCl
EtOH

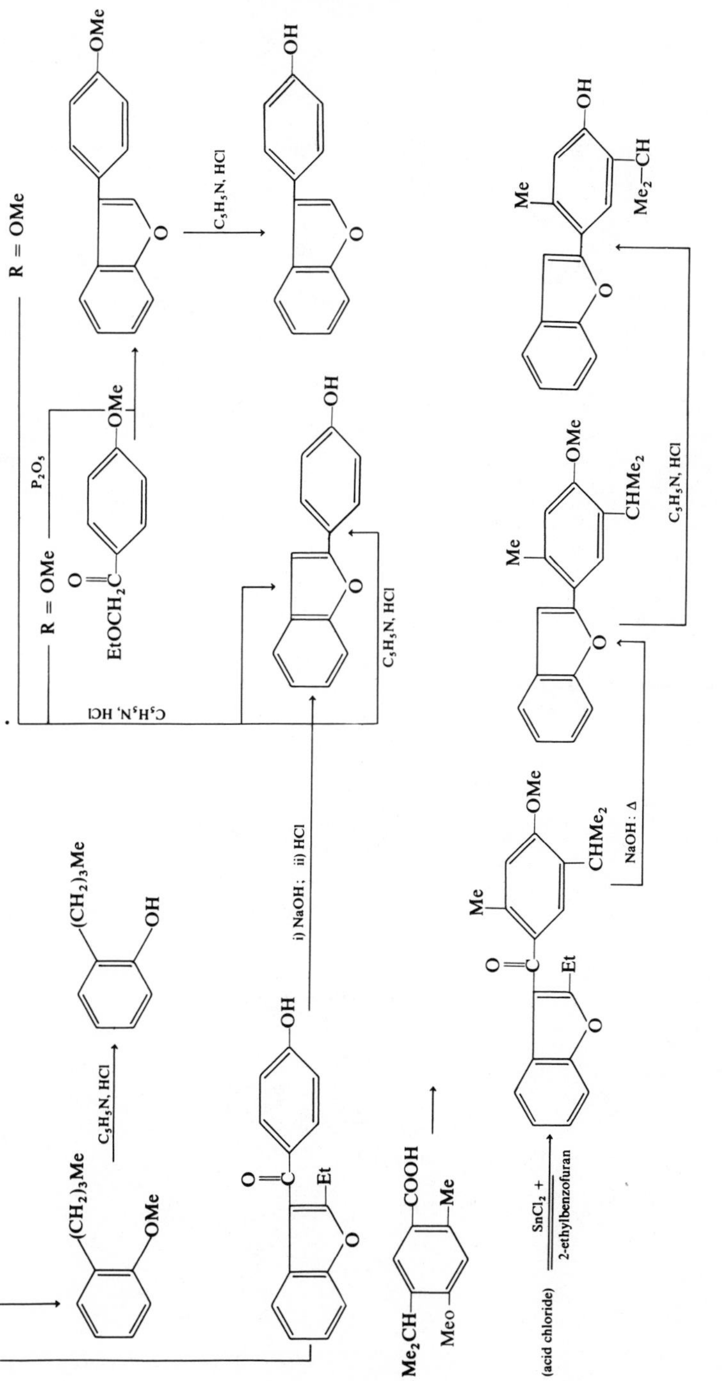
R = OMe
R = OMe
P_2O_5
EtOCH$_2$C
C_5H_5N, HCl
i) NaOH; ii) HCl
(acid chloride)
SnCl$_2$ +
2-ethylbenzofuran
NaOH : Δ
Me
CHMe$_2$
OMe
OH
Et
Me$_2$CH
Meo
COOH
(CH$_2$)$_3$Me
Me$_2$—CH

Chart 4.

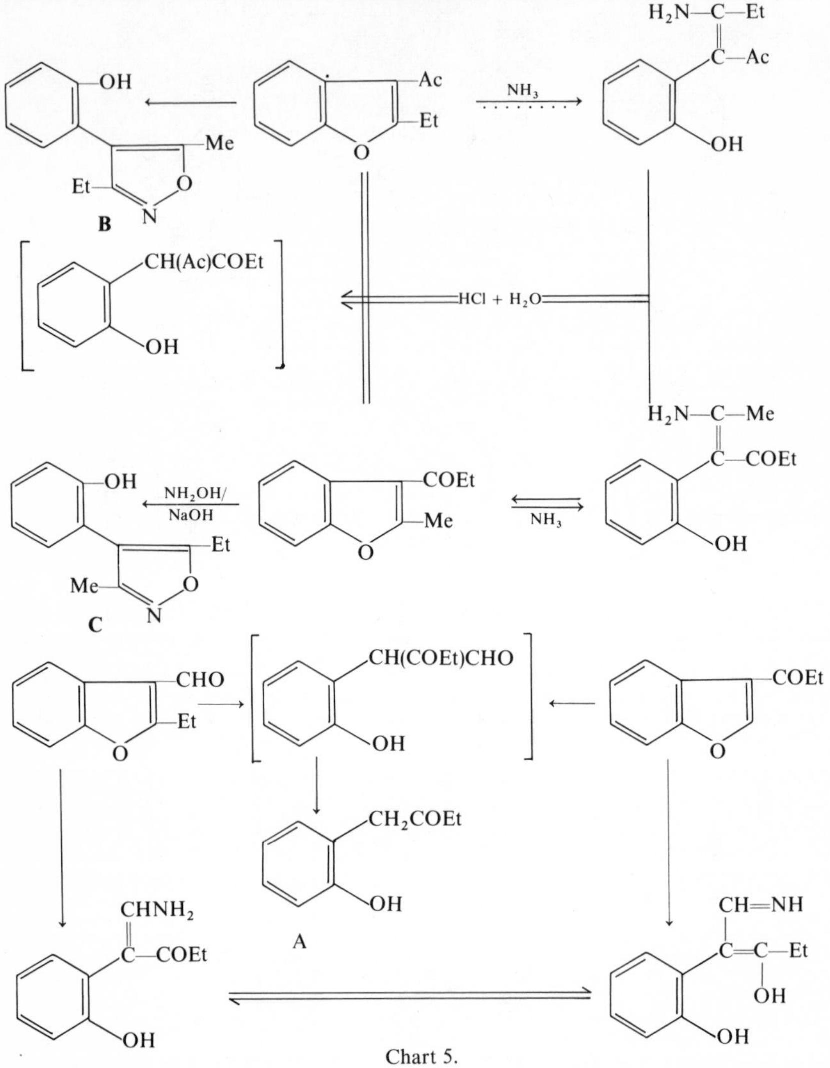

OH
Me
Et
O
N
B
Ac
Et
O
NH3
H2N—C—Et
C—Ac
OH
CH(Ac)COEt
OH
HCl + H2O
H2N—C—Me
C—COEt
OH
NH2OH/
NaOH
COEt
Me
O
NH3
OH
Et
Me
O
N
C
CHO
Et
O
CH(COEt)CHO
OH
COEt
O
CH2COEt
OH
A
CHNH2
C—COEt
OH
CH=NH
C=C—Et
OH
OH

Chart 5.

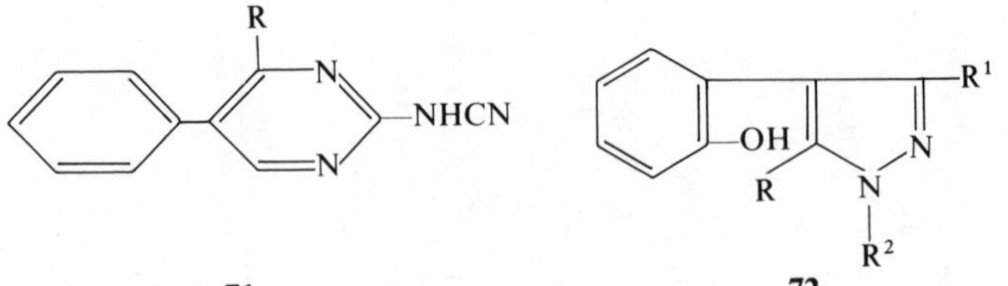

R
N
NHCN
N
R1
OH
N
R
N
R2

71 **72**

The 3-acetylbenzofurans **65** and **66** are degraded by the action of NH_2NHR^2 (R^2 = H, Me, $CONH_2$ or $CSNH_2$) and yield the corresponding pyrazoles (**72**). In the case of hydrazine hydrate, each pair of 2-ethyl-3-acetyl-, 2-methyl-3-propionylbenzofuran, 2-ethyl-3-formyl-, and 3-propionylbenzofuran gave the same pyrazole derivative; however, they gave different pyrazoles with methylhydrazine. Compounds **64**, **67**, and **68** are not degraded by hydrazine hydrate. Guanidine carbonate or hydrochloride, urea and/or thiourea react with 3-acetylbenzofuran, **64**, and/or **65** to afford pyrimidines. Thus guanidine, even in the absence of another alkaline reagent, yields aminopyrimidine (**73**); with urea and/or thiourea, the reaction should be carried out in a dry medium in the absence of sodium ethoxide and ethanol to afford 1,2-dihydropyrimidines (**74**). Acetamidine reacts with benzofuran derivatives having a carbonyl group at position 3, for example 2-methyl-3-formylbenzofuran to furnish 2-methyl-5-(2-hydroxyphenyl)pyrimidines (75).[91]

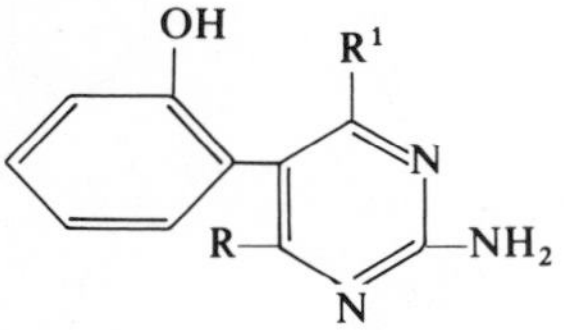

OH R¹ N R NH₂ N

73

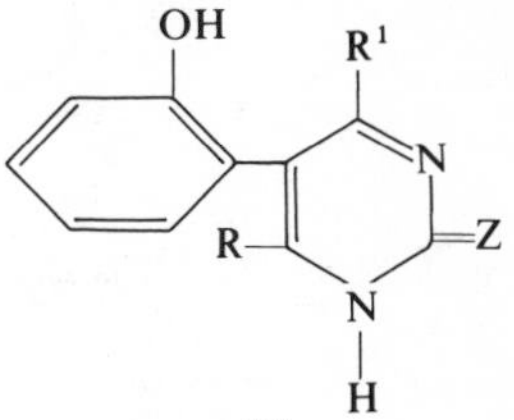

OH R¹ N R Z N H

74

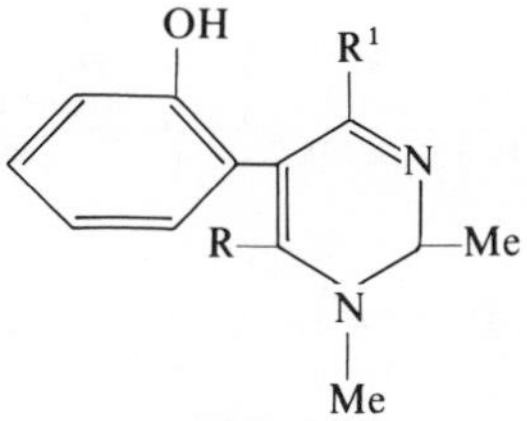

OH R¹ N R Me N Me

75

The reaction of a series of 3-acylbenzofurans toward action of hydroxylamine to yield either the expected oxime or the corresponding isoxazoles (**76**) has been thoroughly investigated.[92] The ratio of the two products depends on structure of the benzofuran and reaction conditions, and particularly on the pH. The resulting **76** is substituted in the 3 and 5 positions by an alkyl group at position 2 and an alkyl (or aryl) group of the 3-acetyl group of the starting material (Chart 6).[92]

1-(4,6-Dimethoxybenzofuran-5-yl)butane-1,2,3-trione 2-arylhydrazones (**78**) are readily transformed to the corresponding 3-(benzofuran-5-yl)-4-arylazo-5-methoxypyrazoles (**79**).[93] Thus 5-acetoacetylbenzofurans are

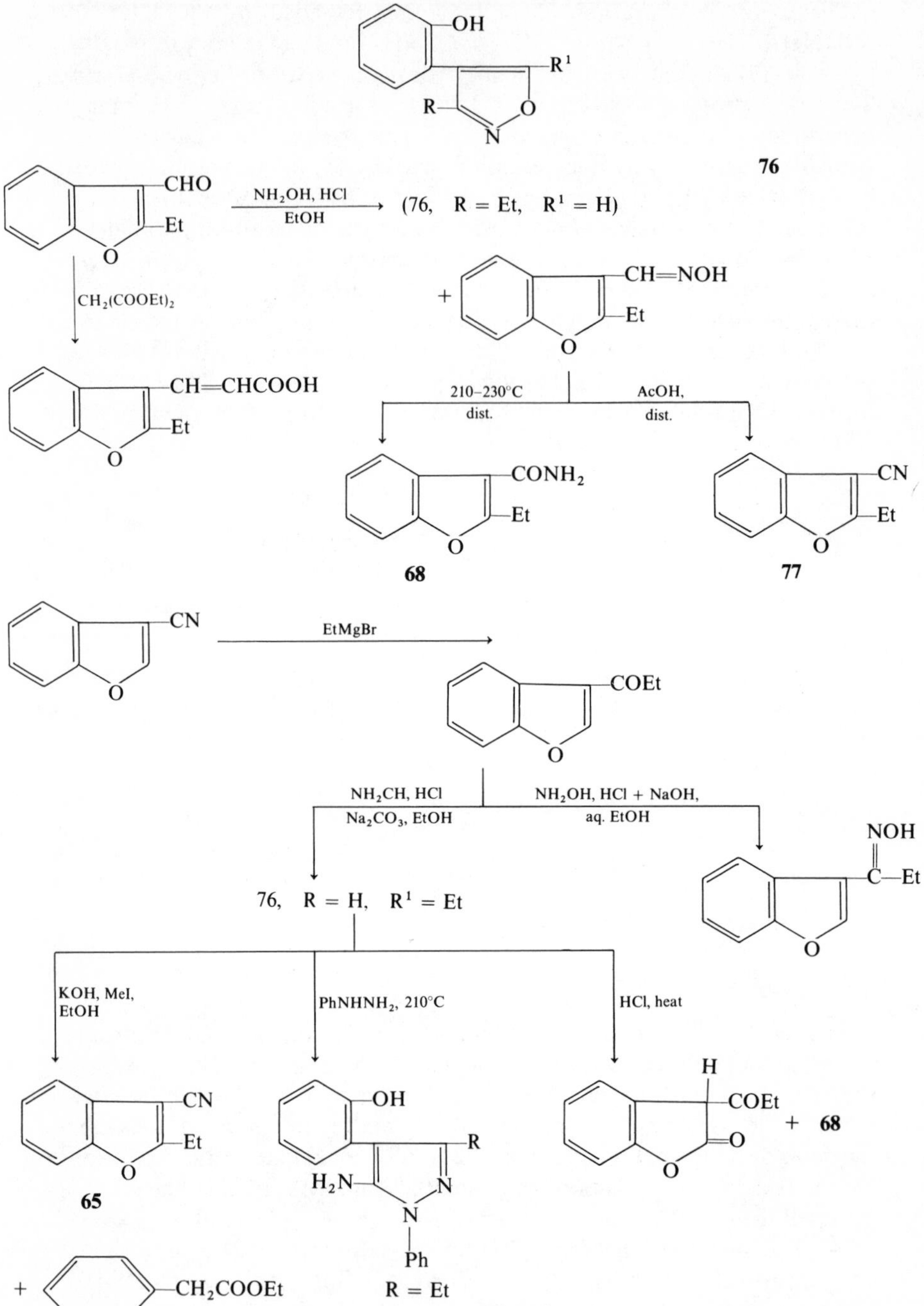

OH
R^1
R
O
N
76

CHO
Et
O
NH_2OH, HCl
EtOH
(76, R = Et, R^1 = H)

CH=NOH
+
Et
O

$CH_2(COOEt)_2$

CH=CHCOOH
Et
O

210–230°C
dist.
AcOH,
dist.

$CONH_2$
Et
O
68

CN
Et
O
77

CN
O
EtMgBr

COEt
O

NH_2CH, HCl
Na_2CO_3, EtOH
NH_2OH, HCl + NaOH,
aq. EtOH

NOH
C—Et
O

76, R = H, R^1 = Et

KOH, MeI,
EtOH
$PhNHNH_2$, 210°C
HCl, heat

CN
Et
O
65

OH
R
H_2N
N
N
Ph
R = Et

H
COEt
=O
O
+ **68**

+ CH_2COOEt

Chart 6.

treated with aryldiazonium chloride to give **78** (Ar = Ph or substituted phenyl) which, when treated with formamide, yields 3-(benzofuran-5-yl)-4-arylazo-5-methylisoxazole (**80**). 1-Carbamoyl-3-(benzofuran-5-yl)-4-arylazo-5-methylpyrazoles (**81**) are obtained upon treatment of **78** with semicarbazide.[93]

OMe, O, CC(=NNHAr)Ac, OMe, O, R — **78**

H, N, N, Me, N=NAr, OMe, OMe, O, R — **79**

O, N, Me, N=NAr, OMe, OMe, O, R — **80**

$CONH_2$, N, N, Me, N=NAr, OMe, OMe, O, R — **81**

2-Bromoacetylbenzofuran, obtained by bromination of 2-acetylbenzofuran,[72] reacts with thioacetamide to give 2-methyl-4-(2-benzofuranyl)-thiazole;[94] with thiourea the thiazole derivatives (**82**) are obtained.[95]

O, N, NHR^1, S — **82**

D. Rearrangement of Acylbenzofuran Oximes

2,3-Dimethyl-5-methoxy-6-acetylbenzofuran oxime (**83**) undergoes, in presence of pyridine hydrochloride, rearrangement to a mixture of compounds of the general type **84** and 2,6,7-trimethylfuro(2,3-f) benzoxazole (**89**) (Chart 7).[96]

Vargha et al.[97] reported the rearrangement of the tosylate of 2-acetylbenzofuran oxime (**90**) to 2-methylchromonol-3 (**91**) and the acetal (**92**) (Eq. 10). Later, Geissman and Armen[98] studied the mechanism of the reaction (Eq. 11).

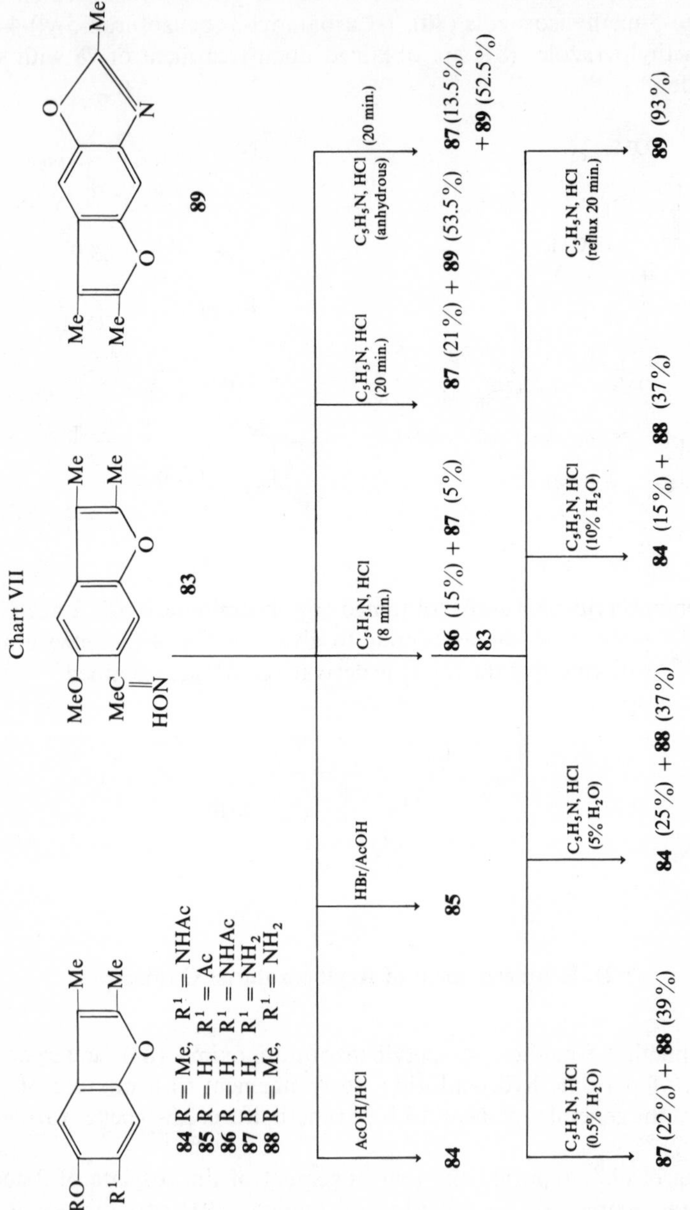

Chart VII
RO
R¹
Me
Me
O
84 R = Me, R¹ = NHAc
85 R = H, R¹ = Ac
86 R = H, R¹ = NHAc
87 R = H, R¹ = NH₂
88 R = Me, R¹ = NH₂
MeO
MeC
HON
83
89
N
AcOH/HCl
84
HBr/AcOH
85
C₅H₅N, HCl (8 min.)
86 (15%) + 87 (5%)
83
C₅H₅N, HCl (20 min.)
87 (21%) + 89 (53.5%)
C₅H₅N, HCl (anhydrous) (20 min.)
87 (13.5%) + 89 (52.5%)
C₅H₅N, HCl (0.5% H₂O)
87 (22%) + 88 (39%)
C₅H₅N, HCl (5% H₂O)
84 (25%) + 88 (37%)
C₅H₅N, HCl (10% H₂O)
84 (15%) + 88 (37%)
C₅H₅N, HCl (reflux 20 min.)
89 (93%)

Chart 7.

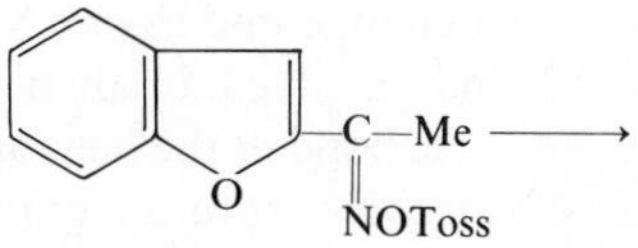

C—Me
NOTOss
O
90
(10)

O
C
OH
Me
O
91
+
O
C
OMe
OMe
Me
O H
92
90

Tos⊖
C—Me
O N
MeOH
OMe
C—Me
O N
MeOH,
HA
+ HOTos

OMe
C—OMe
O N
Me H
Me
HA
OMe
C—Me
O NH_2
⊕ OMe

OMe
⊕
NH_2
Me
O OMe
H_2O
OMe
O
Me
O OMe

OMe
⊕
OH
Me
O OMe
O
OH
Me
O OMe
HA

O
C H
OH
Me
O OMe
–MeOH
91
⊕ OH
OH
Me
O OMe
COOH
Me
O

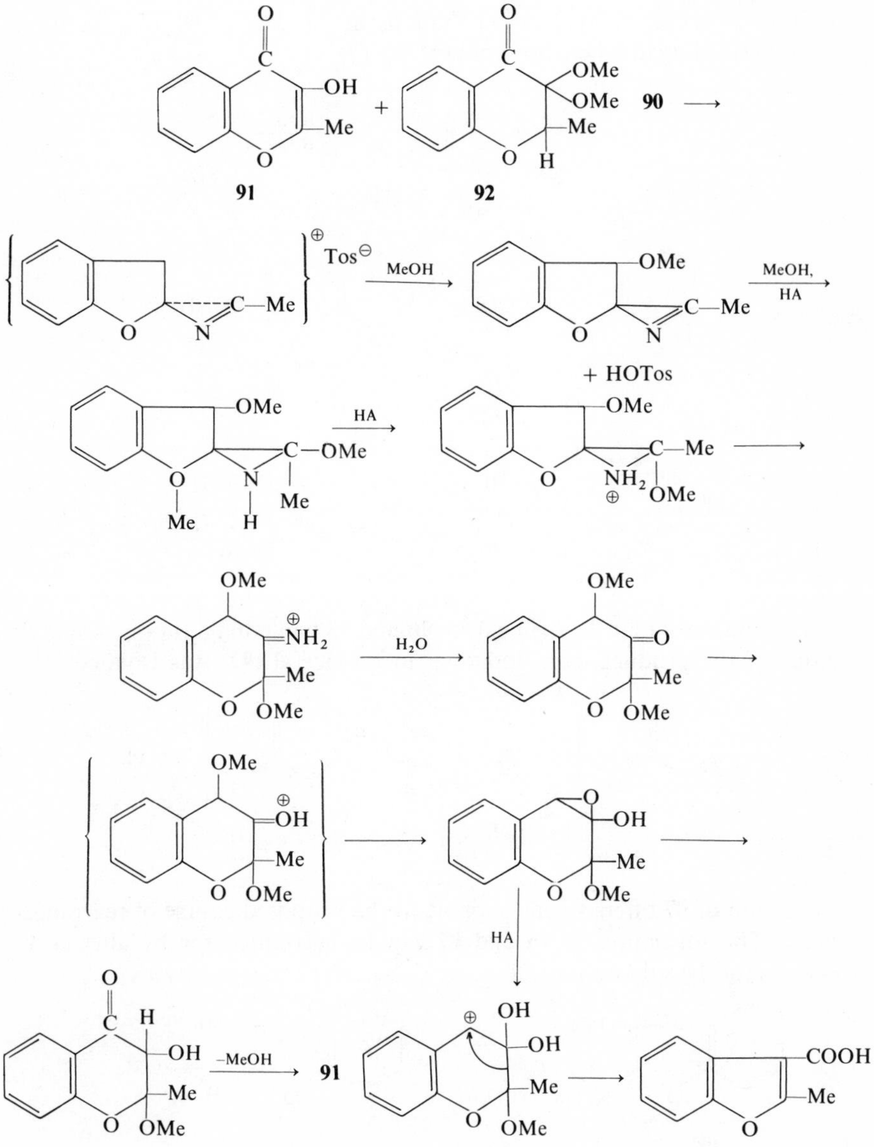

It appeared that the initial change undergone by **90** is probably allied to the Neber reaction[99-103] and involved in displacement of the solvated tosyloxy anion by the 2,3 double bond of the benzofuran ring.

Tosyl derivatives of 2-benzoylbenzofuran oxime (**93**), when heated in ethanol, are rapidly transformed into the rearranged products (chiefly resulting from a Beckmann rearrangement); they are derivatives of benzofuran-2-carboxylic acid rather than, as in the case of **90**, of aminobenzofuran (leading to 2-benzofuranone) (Eq. 12).

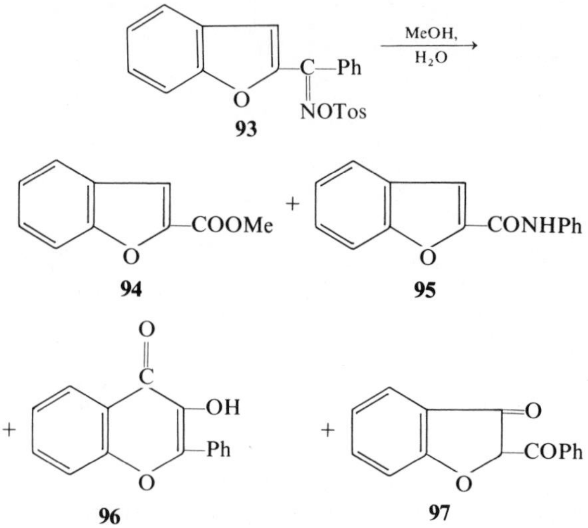

C—Ph, NOTos, O, **93**; MeOH, H_2O →

—COOMe, O, **94** + —CONHPh, O, **95**

+ O=C, —OH, —Ph, O, **96** + =O, —COPh, O, **97** (12)

This indicates that the oxime from **90** and **93** had configurations **A** and **B**. None of the products corresponding to the acetal (**92**) was isolated. The

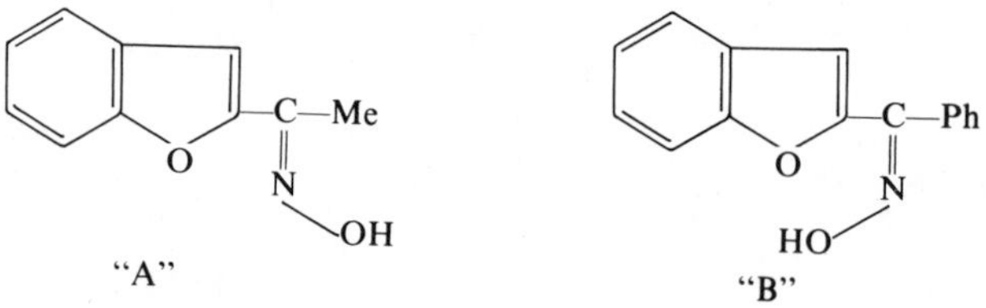

C—Me, O, N, OH "A"   C—Ph, O, N, HO "B"

formation of **97** offers strong support to the proposed course of rearrangement. The formation of **96** and **97** can be accounted for by alternative courses for the solvolysis of **98**.

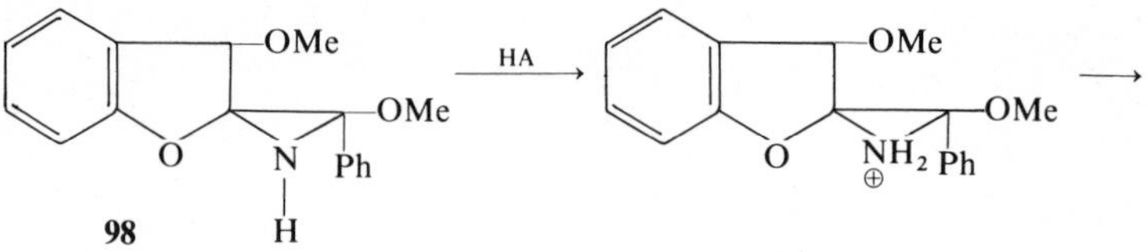

—OMe, —OMe, O, N—H, Ph, **98** —HA→ —OMe, —OMe, O, $NH_2^{\oplus}$, Ph →

OMe | OMe | C⊕—Ph | O NH$_2$ | OMe | O | Ph | O OMe | 96 | 97

E. Rearrangement (Migration) in Acylbenzofurans

Kawase et al.[27,29] have reported that 2,3-dimethyl-5-ethyl- and 5-isopropyl-6-benzofuranyl methyl ketones **99a** and **99c** afforded the corresponding 4-alkyl ketones (**100a–c**) by analogous migration of the 5-alkyl group to the 4 position. Migration of the C-5 alkyl group of 2,3-dimethyl-5-alkyl-6-benzofuranyl methyl ketones by aluminum chloride was further studied and an analogous migration has been observed in the case of methyl 2,3-dimethyl-5-alkyl-6-benzofurancarboxylates.[104] 5-Ethyl-, 5-*n*-propyl-, and 5-isopropylbenzofuranyl ketones are converted to the corresponding carboxylic acids by the bromoform reaction, and further to the esters. 5-Alkyl groups are less likely to migrate in the esters than the corresponding alkyl groups in the ketones; however *n*-propyl groups of the ketone and the ester are less likely to migrate than other alkyl groups, and migration seems to be intramolecular. Moreover, it seems that migration of 5-alkyl groups of 5-alkyl-6-benzofuranyl ketones and esters by aluminum chloride occurs through the cationic cleavage of the alkyl group caused by the electron-attracting effect of the carbonyl group, followed by the synchronous attack of the cationic alkyl groups at the adjacent 4 position (Eq. 13).

R | Ac | Me | Me | O | 99 | HOOC | MeOOC | 100

a, R = Et, b, R = n-Pr, c, R = iso-Pr (13)

Aqueous pyridine hydrochloride transforms 2-ethyl-3-(4-methoxybenzoyl)benzofuran (**101**) to 2-ethyl-3-(4-hydroxybenzoyl)benzofuran (**102**) and 2-(4-hydroxyphenyl)-3-propionylbenzofuran (**103**). However demethylation of **101** with anhydrous pyridine hydrochloride leads to **102**, which is partly converted to **103** by the action of aqueous pyridine hydrochloride (Chart 8).[22]

F. Willgerodt-Kindler Reaction

2-Acylbenzofurans, for example, 2-acetylbenzofuran, undergoes the Willgerodt-Kindler reaction in presence of sulfur and morpholine to give 2-benzofuranylacetic acid (Eq. 14).[105]

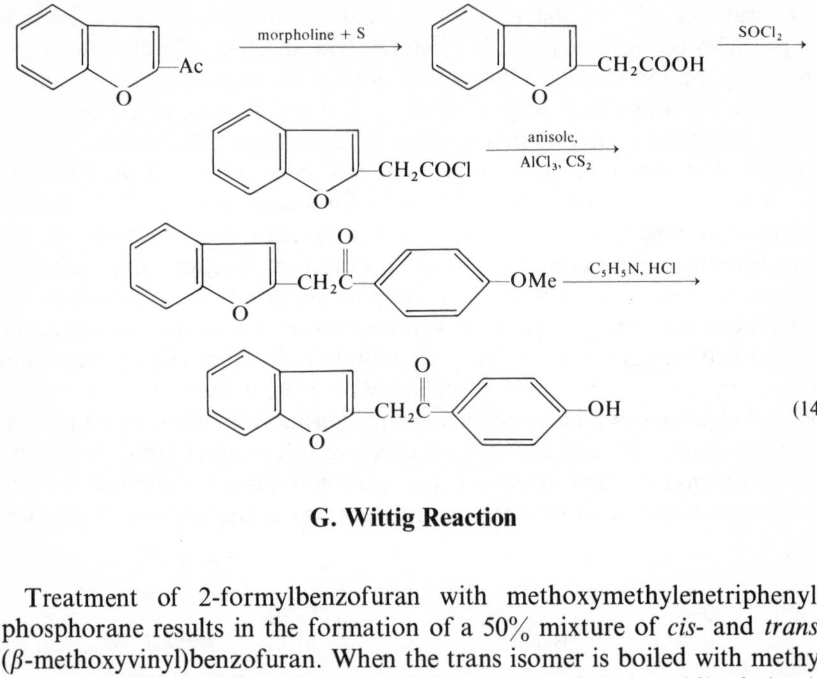

Ac → (morpholine + S) → CH_2COOH → ($SOCl_2$) → CH_2COCl → (anisole, $AlCl_3$, CS_2) → $CH_2C(=O)C_6H_4OMe$ → (C_5H_5N, HCl) → $CH_2C(=O)C_6H_4OH$ (14)

G. Wittig Reaction

Treatment of 2-formylbenzofuran with methoxymethylenetriphenylphosphorane results in the formation of a 50% mixture of *cis*- and *trans*-(β-methoxyvinyl)benzofuran. When the trans isomer is boiled with methyl acetylenedicarboxylate, only 34% of **104** is obtained. It is oxidized slowly with air to afford **105** which is also obtained by boiling **104** with *N*-bromosuccinimide in carbon tetrachloride.[106]

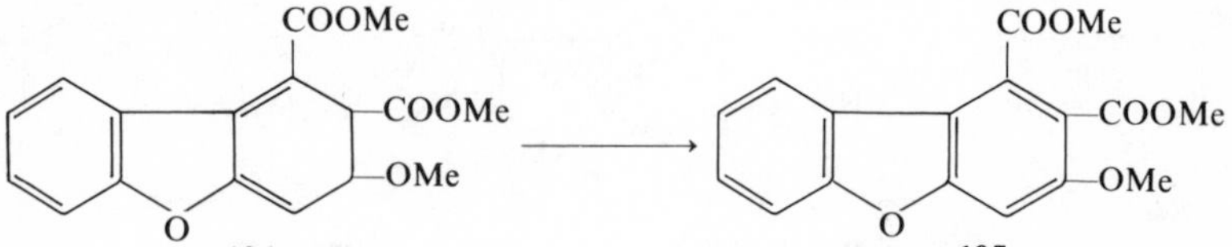

COOMe, COOMe, OMe → COOMe, COOMe, OMe

104 **105**

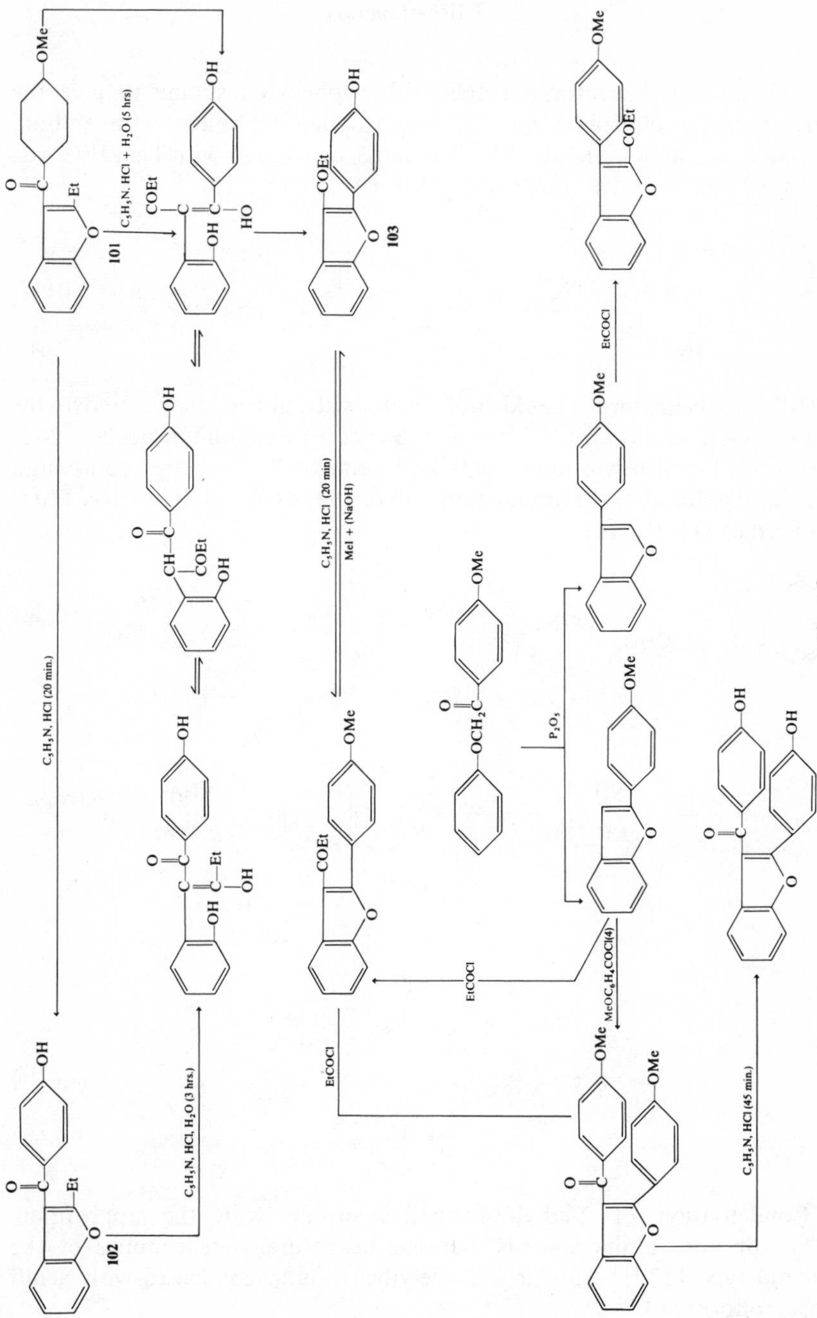
101
102
103
C5H5N, HCl + H2O (5 hrs)
C5H5N, HCl (20 min.)
C5H5N, HCl, H2O (3 hrs.)
C5H5N, HCl (20 min)
MeI + (NaOH)
P2O5
EtCOCl
MeOC6H4COCl(4)
C5H5N, HCl (45 min.)

Chart 8.

H. Miscellaneous

2-Bromoacetylbenzofuran reacts with triphenylphosphine to give the corresponding phosphonium salt which, upon treatment with sodium ethoxide in ethanol, yielded **106**. The latter condensed with benzaldehyde to afford benzofuranyl-2-phenylvinyl ketone (**107**) (Eq. 15).[107,108]

$$\text{2-benzofuranyl}-\overset{O}{\overset{\|}{C}}-\underset{H}{\underset{|}{C}}=PPh_3 \;(\mathbf{106}) \xrightarrow{PhCHO} \text{2-benzofuranyl}-\overset{O}{\overset{\|}{C}}-CH=CHPh \;(\mathbf{107}) \qquad (15)$$

2-Formylbenzofuran could not react with nitromethane under the conventional conditions,[109] but in presence of ammonium acetate 2-(2-nitroprop-1-enyl)benzofuran (**108**) is obtained.[110] The latter compound is readily reduced to the amino compound, **109**; its acetyl derivative, **110** is cyclized to **111** (Eq. 16).

$$\text{2-benzofuranyl}-CHO + EtNO_2 \xrightarrow{AcONH_4} \text{2-benzofuranyl}-CH=C(NO_2)-Me \;(\mathbf{108}) \xrightarrow{LiAlH_4}$$

$$\text{2-benzofuranyl}-CH_2-CH(NH_2)-Me \;(\mathbf{109}) \xrightarrow{Ac_2O} \text{2-benzofuranyl}-CH_2-CH(NHAc)Me \;(\mathbf{110}) \xrightarrow{PCl_5,\ C_6H_6}$$

$$\text{1,3-dimethyl-3,4-dihydroisoquinoline} \;(\mathbf{111}) \qquad (16)$$

Condensation of 2-ethyl-3-formylbenzofuran with the appropriate alkyl (or aryl) amine affords 2-ethyl-3-benzofuranylideneamines of the general type **112**.[111] Similarly, 2-acetylbenzofuran condenses with Schiff bases to form **113**.[112]

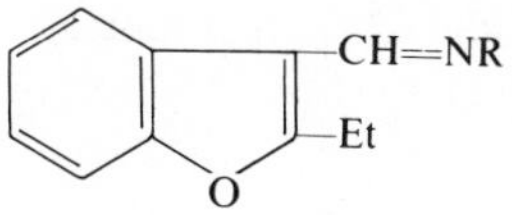

112

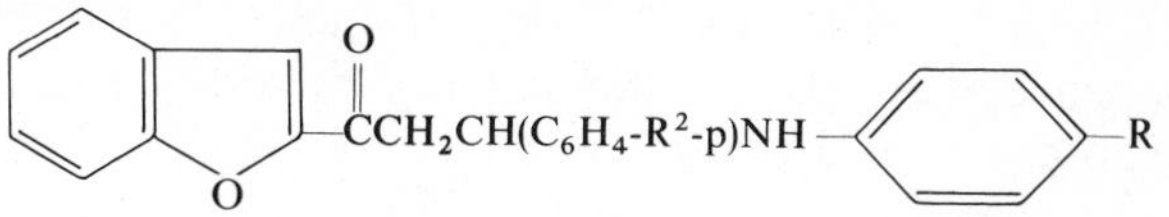

113

In presence of piperidine or hydrochloric acid, 2-acetylbenzofuran condenses with aromatic aldehydes, for example, benzaldehyde, and 2-cinnamoylbenzofuran (**107**) is obtained.[113] Treatment of 2-(cyanoacetyl)- benzofuran with chlorosulfonic acid gave 2-(cyanoacetyl)benzofuran-5-sulfonyl chloride which, with aniline, provided 2-(cyanoacetyl)-5-benzenesulfonanilide (**114**).[114]

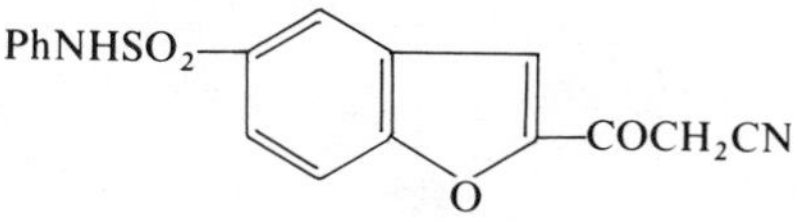

114

2-Formylbenzofuran reacts with diazomethane to afford 2-acetylbenzofuran and 1,3-bis(2-benzofuranyl)propan-1-ol-3-one (**115**).[14] However addition of the formyl compound, without cooling, to a mixture of the formyl compound and ethereal diazomethane solution yields **115** and 1,3-bis(2-benzofuranyl)-2-propen-1-one (**116**). Further treatment of the formyl compound with diazomethane results in the formation of 2-acetylbenzofuran and **117** (Eq. 17).

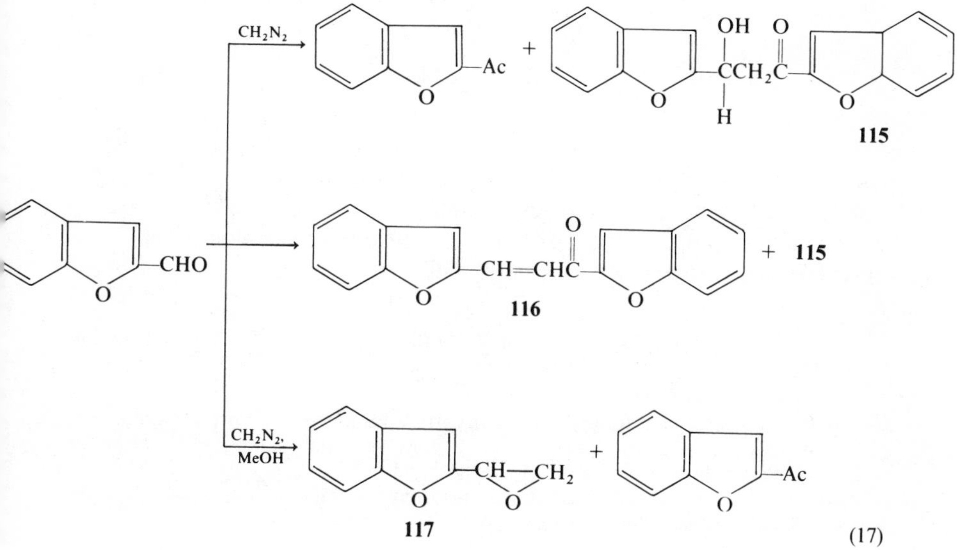

115

116

117

(17)

2,3,8-Trimethylfuro(2,3-g)-6(5*H*)quinolone[115] (**118** and **119**) are obtained upon heating (**120**) with sodium hydroxide. A mixture of **120** and polyphosphoric acid yields **119**.

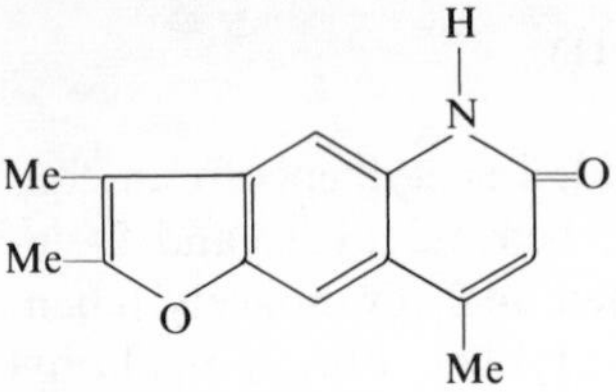

118

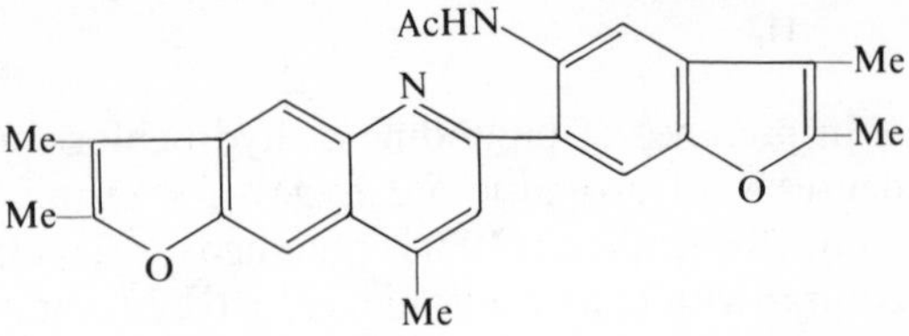

119

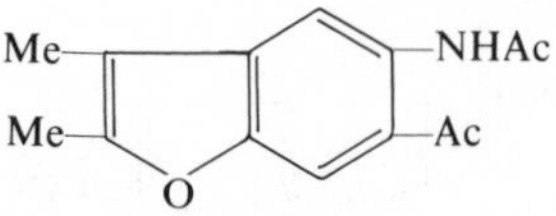

120

Photochemical addition of olefins and *o*-quinones in sunlight has been reported.[116-119] 5-Acetyl-4,7-dimethoxy-6-benzofuranol (**39**) and 5-acetyl-4-methoxy-6-benzofuranol (**40**) react with phenanthraquinone in sunlight to give the photo-adducts **121** and **122**, respectively.[120]

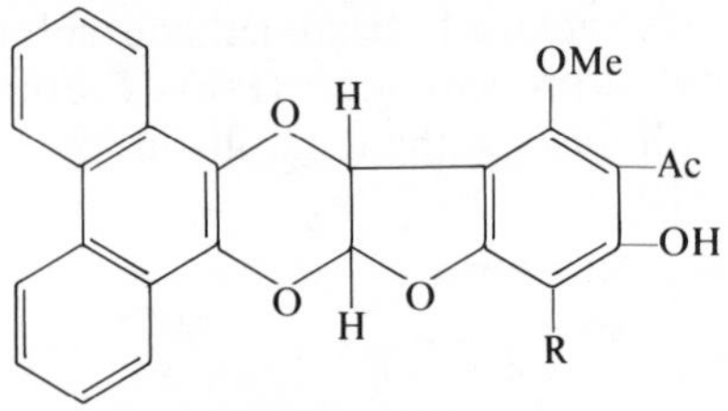

121 R = OMe
122 R = H

Mustafa et al.[121] have reported a color reaction for **39** and **40**, as well as ω-acetokhellinone (**41**), ω-acetovisnaginone (**42**) and 7-bromovisnaginone (**40**, R = Br) with aqueous uranyl acetate. It is accompanied by separation of the metal complex upon dilution with water. The color developed is destroyed with mineral acids or acetic acid.

References

1. J. B. D. Mackenzie, A. Robertson, and (in part) A. Bushra and R. Towers, *J. Chem. Soc.*, **1949,** 2057.
2. R. T. Foster and A. Robertson, *J. Chem. Soc.*, **1939,** 921.
3. R. Royer, E. Bisagni, A. M. Laval-Jeantet, and J-P. Marquet, *Bull. Soc. Chim. Fr.*, **1965,** 2607; *Chem. Abstr.*, **63,** 16319 (1965).

4. R. Royer, J. L. Derocque, P. Demerseman, and A. Cheutin, *Compt. Rend., Ser. C,* **262,** 1286 (1966); *Chem. Abstr.,* **65,** 3819 (1966).
5. F. M. Dean, D. S. Deorha, J. C. Knight, and T. Francis, *J. Chem. Soc.,* **1961,** 327.
6. F. Pan and G. A. Wiese, *J. Am. Pharm. Assoc., Sci. Ed.,* **49,** 259 (1960); *Chem. Abstr.,* **54,** 24635 (1960).
7. V. T. Suu, N. P. Buu-Hoi, and N. D. Xuong, *Bull. Soc. Chim. Fr.,* **1962,** 1875.
8. J. N. Chatterjea, *J. Indian Chem. Soc.,* **34,** 347 (1957).
9. R. Royer and L. Rene, *Bull. Soc. Chim. Fr.,* **1970,** 1037.
10. R. Royer, P. Demerseman, A. M. Laval-Jeantet, J. F. Rossignol, A. Cheutin, and M. L. Desvoye, *Bull. Soc. Chim. Fr.,* **1968,** 1026; *Chem. Abstr.,* **69,** 96364 (1968).
11. P. Demerseman, J. P. Lechartier, C. Pene, A. Cheutin, and R. Royer, *Bull. Soc. Chim. Fr.,* **1965,** 1473; *Chem. Abstr.,* **63,** 11321 (1965).
12. T. v. Tadeus and I. Reichstein, *Helv. Chim. Acta,* **13,** 1275 (1930).
13. B. Sila, *Rocz. Chem.,* **42**(3), 553 (1968); *Chem. Abstr.,* **69,** 43707 (1968).
14. L. Capuano, *Chem. Ber.,* **98,** 3659 (1965).
15. A. Hantzsch, *Chem. Ber.,* **19,** 1294 (1886).
16. R. v. Stoermer, *Chem. Ber.,* **28,** 1253 (1895).
17. R. v. Stoermer, *Justus Liebigs Ann. Chem.,* **312,** 274 (1900).
18. E. W. Smith, *Iowa State Coll. J. Sci.,* **12,** 155 (1937); *Chem. Abstr.,* **32,** 2938 (1938).
19. H. F. Birch and A. Robertson, *J. Chem. Soc.,* **1938,** 306.
20. W. B. Whalley, *J. Chem. Soc.,* **1953,** 3479.
21. E. Bisagni, N. P. Buu-Hoi, and R. Royer, *J. Chem. Soc.,* **1956,** 625.
22. R. Royer, P. Demerseman, and E. Bisagni, *Bull. Soc. Chim. Fr.,* **1960,** 685.
23. E. Bisagni and R. Royer, *Bull. Soc. Chim. Fr.,* **1960,** 1968.
24. H. Gilman, E. W. Smith, and L. C. Cheney, *J. Am. Chem. Soc.,* **57,** 2095 (1935).
25. E. Bisagni and R. Royer, *Bull. Soc. Chim. Fr.,* **1962,** 925.
26. R. Royer, M. Hubert-Habart, L. Rene, A. Cheutin, and M-L. Desvoye, *Bull. Soc. Chim. Fr.,* **1964,** 1259.
27. Y. Kawase, R. Royer, M. Hubert-Habart, A. Cheutin, L. Rene, J-P. Desvoye, *Bull. Soc. Chim. Fr.,* **1964,** 3131.
28. R. Royer, E. Bisagni, M. Hubert-Habart, L. Rene, and J-P. Marquet, *Bull. Soc. Chim. Fr.,* **1965,** 1794.
29. Y. Kawase, M. Hubert-Habart, J-P. Buisson, and R. Royer. *Compt. Rend.,* **258,** 5007 (1964).
30. R. Royer, Y. Kawase, M. Hubert-Habart, L. Rene, and A. Cheutin, *Bull. Soc. Chim. Fr.,* **1966,** 211.
31. Y. Kawase and M. Takashima, *Bull. Chem. Soc. Japan,* **40,** 1224 (1967); *Chem. Abstr.,* **67,** 82013 (1967).
32. Y. Kawase and S. Hori, *Bull. Chem. Soc. Japan,* **43,** 3496 (1970); *Chem. Abstr.,* **74,** 53364 (1971).
33. R. Royer and L. Rene, *Bull. Soc. Chim. Fr.,* **1970,** 3601; *Chem. Abstr.,* **74,** 53358 (1971).
34. Y. Kawase, T. Okada, and T. Miwa, *Bull. Chem. Soc. Japan,* **43,** 2884 (1970).
35. N. P. Buu-Hoi, N. D. Xuong, and N. V. Bac, *J. Chem. Soc.,* **1964,** 173.
36. K. Yoshiyuki, N. Mutsumu, and M. Fumihisha, *Bull. Chem. Soc. Japan,* **41,** 2676 (1968); *Chem. Abstr.,* **70,** 47229 (1969).
37. W. B. Whalley, *J. Chem. Soc.* **1951,** 3229.
38. P. Karrer and F. Widmer, *Helv. Chim. Acta,* **2,** 454 (1919).
39. F. Prillinger and H. Schmid, *Monatsh. Chem.,* **72,** 427 (1939).
40. W. Gruber and F. Traub, *Monatsh. Chem.,* **77,** 414 (1947).
41. D. B. Limaye and T. B. Panse, *Rasayanam,* **2,** 27 (1950); *Chem. Abstr.,* **45,** 6620 (1951).
42. R. D. Dessai and M. Ekhlas, *Proc. Indian Acad. Sci.,* **8A,** 194 (1938); *Chem. Abstr.,* **33,** 2119 (1939).

43. D. B. Limaye and N. R. Sathe, *Rasayanam*, **1**, 48 (1936); *Chem. Abstr.*, **31**, 2212 (1937).
44. M. G. Marathey and J. M. Athavale, *J. Univ. Poona, Sci. Technol.*, **1953**, 90; *Chem. Abstr.*, **49**, 11638 (1955).
45. M. Martynoff, *Compt. Rend.*, **233**, 878 (1951); *Chem. Abstr.*, **47**, 8725 (1953).
46. M. Martynoff, *Compt. Rend.*, **232**, 2454 (1950); *Chem. Abstr.*, **46**, 3998 (1952).
47. M. Hubert-Habart, G. Menichi, K. Takagi, A. Cheutin, L-M. Desvoye, and R. Royer, *Chim. Ther.*, **3**(4), 280 (1968); *Chem. Abstr.*, **70**, 37765 (1969).
48. R. Royer, G. Menichi, J. P. Buisson, M. Hubert-Habart, A. Cheutin, and M-L. Desvoye, *Bull. Soc. Chim. Fr.*, **1967**, 2405; *Chem. Abstr.*, **68**, 2756 (1968).
49. E. Späth and W. Gruber, *Chem. Ber.*, **71**, 106 (1938).
50. E. Späth and W. Gruber, *Chem. Ber.*, **74**, 1492 (1941).
51. T. A. Geissman, *J. Am. Chem. Soc.* **71**, 1498 (1949).
52. J. R. Clarke, G. Glaser, and A. Robertson, *J. Chem. Soc.*, **1948**, 2260.
53. E. Rap, *Gazz. Chim. Ital.*, **25**(2), 285 (1895).
54. R. v. Stoermer, *Justus Liebigs Ann. Chem.*, **312**, 237 (1900); R. v. Stoermer and M. Schaeffer, *Chem. Ber.*, **36**, 2864 (1903).
55. E. Bisagni, N. P' Buu-Hoi, and R. Royer, *J. Chem. Soc.*, **1955**, 3693.
56. N. P. Buu-Hoi, E. Bisagni, R. Royer, and C. Routier, *J. Chem. Soc.*, **1957**, 625.
57. N. P. Buu-Hoi, G. Saint-Ruf, T. B. Loc, and N. D. Xuong, *J. Chem. Soc.*, **1957**, 2593.
58. F. Zwayer and St. v. Kostanecki, *Chem. Ber.*, **41**, 1338 (1908).
59. A. B. Sen and S. Saxena, *J. Indian Chem. Soc.*, **36**, 283 (1959); A. B. Sen and S. Saxena, *J. Indian Chem. Soc.*, **35**, 136 (1958).
60. C. Routier, N. P. Buu-Hoi, and R. Royer, *J. Chem. Soc.*, **1956**, 4276.
61. W. B. Whalley and G. Lloyd, *Sci. Proc. Dublin Soc.*, **27**, 105 (1956); *Chem. Abstr.*, **51**, 8082 (1957).
62. K. R. Huffman and E. F. Ullman, U.S. Pat. 3,331,854 (1967); *Chem. Abstr.*, **69**, 43785 (1968).
63. F. Binon and H. Inion, *Chim. Ther.*, **2**(2), 113 (1967); *Chem. Abstr.*, **68**, 59388 (1968).
64. H. K. Pendse, *Rasayaṇam*, **2**, 121 (1956); *Chem. Abstr.*, **51**, 5062 (1957).
65. H. K. Pendse and N. D. Patwardhan, *Rasayanam*, **2**, 117 (1956); *Chem. Abstr.*, **51**, 5062 (1957).
66. H. K. Pendse and K. S. Moghe, *Rasayanam*, **2**, 114 (1956); *Chem. Abstr.*, **51**, 5062 (1957).
67. S. D. Limaye, H. K. Pendse, K. R. Chandorkar, and G. V. Bhide, *Rasayanam*, **2**, 97 (1956); *Chem. Abstr.*, **51**, 5064 (1957).
68. L. Jurd, *Chem. Ind. (London)*, **1963**, 1165; *Chem. Abstr.*, **59**, 7458 (1963).
69. L. Jurd, U.S. Pat. 3,165, 537 (1965); *Chem. Abstr.*, **62**, 7728 (1965).
70. L. Jurd, *J. Org. Chem.*, **29**, 2602 (1964).
71. L. Jurd, *Tetrahedron*, **22**, 2913 (1966).
72. F. A. Trofimov, T. L. Mukhanova, A. N. Grinev, and V. I. Shvedov, *Zh. Org. Khim.*, **3**(12), 2185 (1967); *Chem. Abstr.*, **68**, 68809 (1968).
73. F. A. Trofimov, T. I. Mukhanova, N. G. Tsyshkova, A. N. Grinev, and K. S. Shadurskii, *Farm Zh.*, **1**(9), 14 (1967); *Chem. Abstr.*, **69**, 2769 (1968).
74. E. Bernatek, *Acta Chem. Scand.*, **6**, 160 (1952).
75. E. Bernatek and S. Ramstad, *Acta Chem. Scand.*, **7**, 1351 (1953).
76. E. Bernatek, *Acta Chem. Scand.*, **7**, 677 (1953).
77. A. N. Grinev, L. A. Bukhtenko, and A. N. Terent'ev, *Zh. Obshch. Khim.*, **29**, 945 (1959); *Chem. Abstr.*, **54**, 1481 (1960).
78. H. Cuntze and H. Musso, *Chem. Ber.*, **103**, 62 (1970).
79. H. Alles, D. Bormann, and H. Musso, *Chem. Ber.*, **103**, 2526 (1970).
80. A. Hassle, Neth. Pat. 6,413,996 (1965); *Chem. Abstr.*, **63**, 18043 (1965).
81. G. Sanna, *Gazz. Chim. Ital.*, **59**, 694 (1929); *Chem. Abstr.*, **24**, 850 (1930).
82. Ch. B. Rao, G. Subramanyam, and V. Venkateswarlu, *J. Org. Chem.*, **24**, 683 (1959).

83. J. N. Chatterjea and V. N. Mehrotra, *J. Indian Chem. Sod.*, **39**, 599 (1962); **40**, 203, (1963).
84. R. L. Shriner and J. Anderson, *J. Am. Chem. Soc.*, **61**, 2705 (1939).
85. R. v. Stoermer, C. W. Chydenius, and E. Schinn, *Chem. Ber.*, **57**, 72 (1924).
86. S. Fatutta, *Univ. Studi Trieste Fac. Sci., Ist. Chim.*, **31**, 33 (1961); *Chem. Abstra.*, **58**, 526 (1963).
87. K. Takagi, M. Hubert-Habart, A. Cheutin, and R. Royer, *Bull. Soc. Chim. Fr.*, **1966**, 3136.
88. R. Royer and E. Bisagni, *Bull. Soc. Chim. Fr.*, **1960**, 395.
89. M. Hubert-Habart, K. Takagi, A. Cheutin, and R. Royer, *Bull. Soc. Chim. Fr.*, **1966**, 1587.
90. K. Takagi and M. Hubert-Habart, *Chim. Ther.*, **5**(4), 264 (1970); *Chem. Abstr.*, **74**, 3578 (1971).
91. M. Hubert-Habart, K. Takagi, and R. Royer, *Bull. Soc. Chim. Fr.*, **1967**, 356; M. Hubert-Harbert, K. Takagi, and R. Royer, *Compt. Rend.*, **260**, 5302 (1965).
92. R. Royer and E. Bisagni, *Bull. Soc. Chim. Fr.*, **1963**, 1746.
93. A. Mustafa, O. H. Hishmat, and M. M. Younes, *J. Prakt. Chem.*, **312**, 1011 (1970); *Chem. Abstr.*, **75**, 48794 (1971).
94. E. B. Middleton, U.S. Pat. 2,524,674 (1950); *Chem. Abstr.*, **45**, 2348 (1951).
95. F. Russo and M. Ghelardoni, *Ann. Chim. (Rome)*, **54**(10), 987 (1964); *Chem. Abstr.*, **62**, 5265 (1965).
96. R. Royer, P. Demerseman, and G. Colin, *Bull. Soc. Chim. Fr.*, **1967**, 4210; *Chem. Abstr.*, **69**, 103316 (1968).
97. L. Vargha, J. Ramonczai, and J. Bathory, *J. Am. Chem. Soc.*, **71**, 2652 (1949).
98. T. A. Geissman and A. Armen, *J. Am. Chem. Soc.*, **77**, 1623 (1955).
99. P. W. Neber and A. v. Friedolscheim, *Justus Liebigs Ann. Chem.*, **449**, 109 (1926).
100. P. W. Neber and A. Burgard, *Justus Liebigs Ann. Chem.*, **493**, 281 (1932).
101. P. W. Neber and G. Huh, *Justus Liebigs Ann. Chem.*, **515**, 283 (1935).
102. P. W. Neber, A. Burgard, and W. Thier, *Justus Liebigs Ann. Chem.*, **526**, 277 (1936).
103. D. J. Cram and M. J. Hatch, *J. Am. Chem. Soc.*, **75**, 33, 38 (1953).
104. Y. Kawase, S. Takata, and T. Miwa, *Bull. Chem. Soc. Japan*, **43**, 1796 (1970).
105. E. Bisagni and R. Royer, *Bull. Soc. Chim. Fr.*, **1962**, 86; *Chem. Abstr.*, **56**, 14192 (1962).
106. W. J. Davidson and J. A. Elix, *Tetrahedron Lett.*, **1968**, 4589.
107. M. I. Shevchuck, A. S. Antonyuk, and A. V. Dombrovskii, *Zh. Obshch. Khim.*, **39**(4), 860 (1969); *Chem. Abstr.*, **71**, 61479 (1969).
108. M. I. Shevchuk, A. S. Antonyuk and A. V. Dombrovskii, *Zh. Org. Khim.*, **6**(12), 2579 (1970); *Chem. Abstr.*, **74**, 64276 (1971).
109. M. Descamps and F. Binon, *Bull. Soc. Chim. Belges*, **71**, 579 (1962); *Chem. Abstr.* **59**, 1605 (1963).
110. C. J. Cattanach and R. G. Rees, *J. Chem. Soc., C*, **1971**, 53.
111. R. Landi-Vittory, F. Gatta, F. Toffler, S. Chiavarelli, and G. L. Gatta, *Farmaco (Pavia) Ed. Sci.*, **18**, 465 (1963); *Chem. Abstr.*, **59**, 9941 (1963).
112. N. S. Kozlov and I. P. Sonich, *Izv. Vysslikh Uchebn. Zavedenii Khim. i. Khim. Tekhnol.*, **6**(6), 970 (1963); *Chem. Abstr.*, **60**, 15810 (1964).
113. M. Poonvski, M. Pesson, and G. Polmanse, *Bull. Soc. Chim. Fr.*, **1953**, 200; *Chem. Abstr.*, **48**, 663 (1954).
114. I. A. Solov'eva and G. I. Arbuzov, *Zh. Obshch. Khim.*, **21**, 765 (1951); *Chem. Abstr.*, **45**, 9524 (1951).
115. Y. Kawase and S. Kondo, *Bull. Chem. Soc. Japan*, **43**, 3268 (1970); *Chem. Abstr.*, **74**, 22734 (1971).
116. A. Mustafa, *Nature (London)*, **175**, 992 (1955).
117. A. Mustafa, A. K. Mansour, and H. A. A. Zaher, *J. Org. Chem.*, **25**, 949 (1960).

118. A. Mustafa, "Advances in Photochemistry", W. A. Noyes, Jr., G. S. Hammond, and J. N. Pitts, Jr., Eds., Interscience, New York 1964, p. 104.
119. A. Mustafa and A. M. Islam, *J. Chem. Soc. Suppl.*, 81 (1949).
120. A. Schönberg and M. M. Sidky, *J. Org. Chem.*, **21,** 476 (1956); C. H. Krausch, S. Farid, and G. O. Schenck, *Chem. Ber.*, **98,** 3102 (1965).
121. A. Mustafa, N. A. Starkovsky, and E. Zaki, *J. Org. Chem.*, **25,** 794 (1960).

CHAPTER III

Benzofurancarboxylic Acids

1. Benzofuran monocarboxylic Acids

A. 2-Benzofurancarboxylic Acids

The name "coumarilic acid" was coined as a consequence of an early preparation of 2-benzofurancarboxylic acid (**3b**) from coumarin; this was made possible by Perkin's discovery[1] that 3-halocoumarin (**1b**) when heated with alkali gave **3b**. The reaction of 3-chlorocoumarin with sodium methoxide yields methyl 2-benzofurancarboxylate only if methanol is present.[2] It is postulated that the methoxide ion attacks the carbonyl group of **1a** to yield a phenoxide ion (**2a**). A Michael addition of methanol to the double bond of **2a** (or its conjugate acid) gives an intermediate **2b**, which has an sp^3 carbon-bearing chlorine which could then undergo an intramolecular displacement easily to yield **2c**. This would be followed by loss of methanol to afford **3a** (Eq. 1).

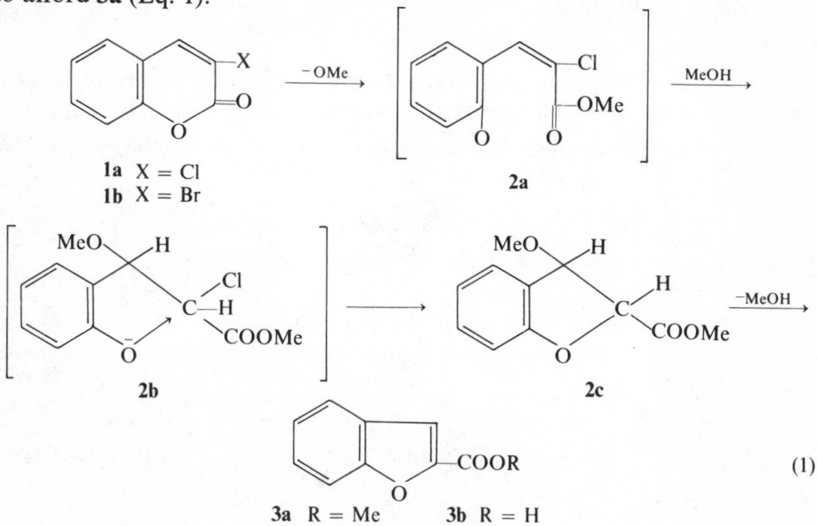

1a X = Cl
1b X = Br

2a

2b

2c

3a R = Me **3b** R = H (1)

The ultraviolet spectra of compounds possessing the benzofuran-2-carboxylate structure very rarely absorb at wavelengths above 300 mμ, while coumarins almost invariably have one or more absorption bands above 300 mμ.[2]

Treatment of **1b** with diethylamine affords the diethylamide of 2-benzofurancarboxylic acid (**4**).[3] The acid amide (**6**) has been similarly obtained upon treatment of the coumarin (**5**) with ammonia.[4-6]

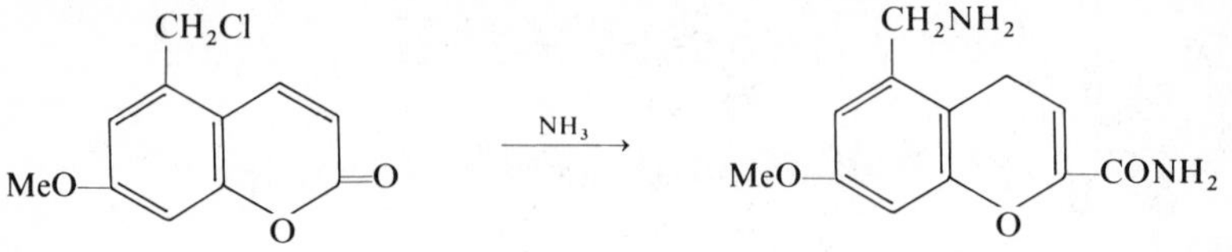

4-Phenylcoumarins exhibit a marked tendency to undergo ring contraction upon treatment with mercuric oxide in alkaline mediums to yield the corresponding 3-phenyl-2-benzofurancarboxylic acids,[7,8] for example, **7**.

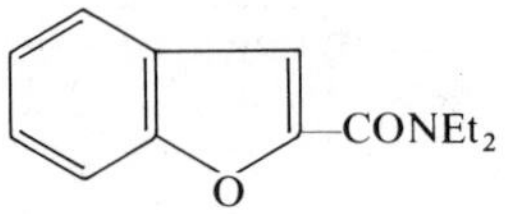

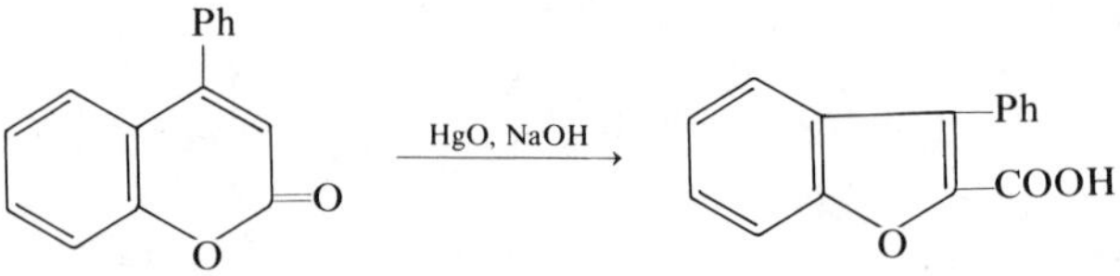

A convenient route for the preparation of 2-benzofurancarboxylic acid and its substituted derivatives is the condensation of *o*-hydroxybenzaldehydes with ethyl bromomalonate in presence of potassium carbonate (Eq. 2).[9]

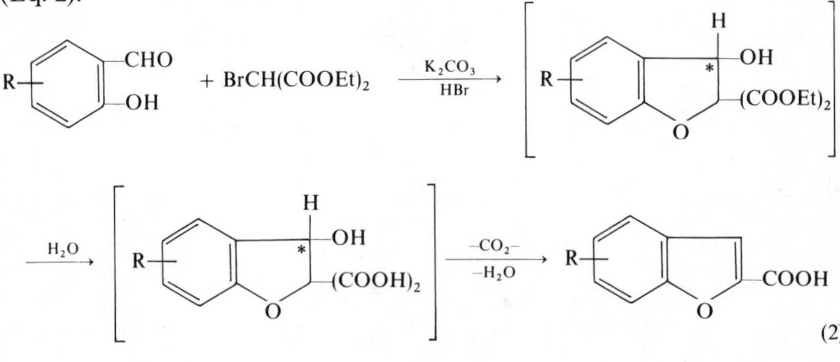

(2)

The interaction of ethyl bromoacetate with *o*-hydroxybenzaldehydes has been thoroughly studied and the scope of the reaction is stressed.[10-18]

5,2-$R_2NCH_2(OH)C_6H_3CHO$, prepared from 5,2-$ClCH_2(OH)C_6H_3CHO$ with ethyl bromomalonate, afforded substituted 2-benzofurancarboxylic acids, which are esterified and ammonolyzed to yield 5-aminomethylbenzofuran-2-carboxamides. 3-Piperidinomethyl-5-methylsalicylaldehyde is, similarly, converted to 7-piperidinomethyl-5-methylbenzofurancarboxamide. 4-Aminomethyl-7-methoxybenzofuran-2-carboxamide has been obtained upon treatment of ethyl 4-chloromethyl-7-methoxybenzofuran-2-carboxylate with amines and then ammonia.[17] Ethyl bromomalonate condenses with 4,2-$O_2N(OH)C_6H_3CHO$ in presence of potassium carbonate and ethyl methyl ketone to yield 6-nitro-2-benzofurancarboxylic acid.[18]

3-Methylbenzofuran-2-carboxylic acid was obtained by the sequence of reactions shown in Eq. 3.[19]

Me; Br_2, CS_2 (0°C) → Me, Br; CuCN, quinoline →

Me, CN; Na/MeOH → Me, COOH; EtOH/HCl →

Me, COOEt; NBS (130°C) → CH_2Br, COOEt; NaOEt →

CH_2OEt, COOH (3)

Metalation of bis(*o*-methoxyphenyl)-1-chloroethylene (**8**), followed by carboxylation, resulted in the formation of neutral material which was chromatographically separated into **9** and **10** (Eq. 4).[20]

OMe, C=CHCl, OMe **8**; BuLi (−72°C) → Me, O, Li, C=C, Cl, O, Me →

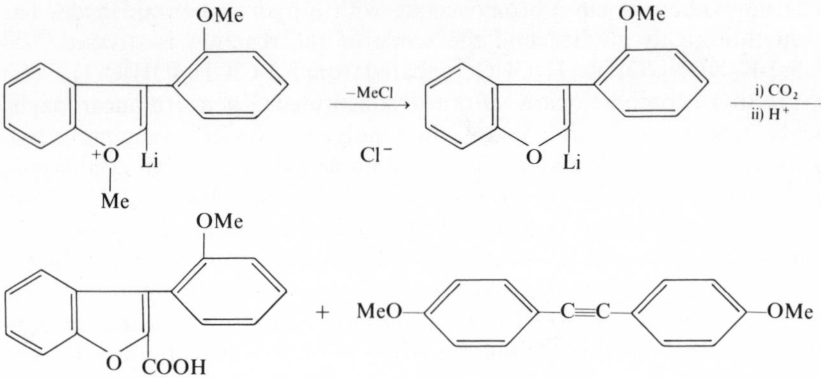

(4)

B. 3-Benzofurancarboxylic Acids

Thermolysis of 3,4,5,6-tetrachlorophenyl-2-diazo-1-oxide (**11**) to the oxocarbene (**12**) in presence of ethyl phenylpropiolate establishes a convenient approach for the synthesis of esters of 3-benzofurancarboxylic acids, for example, **13** (Eq. 5).[21,22]

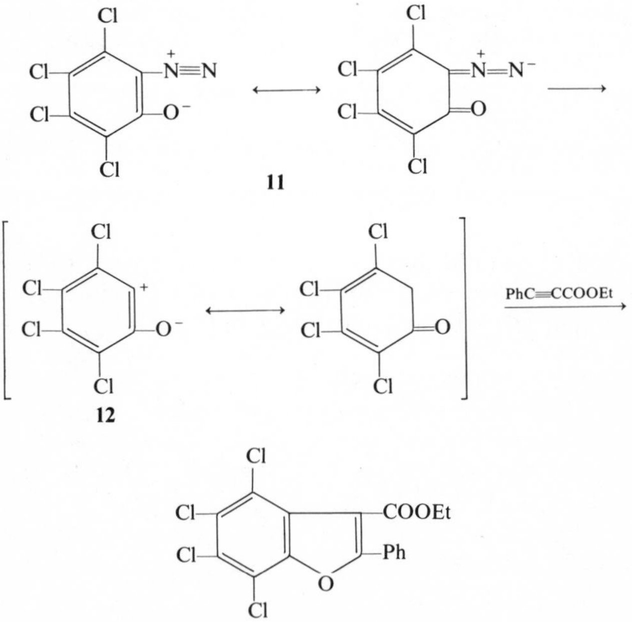

(5)

The 1,3 Cycloaddition of ethyl cinnamate to oxocarbene results in the formation of ethyl 2-phenyl-2,3-dihydrobenzofuran-3-carboxylate (**14**).[22]

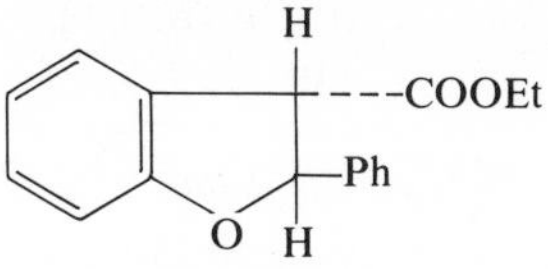

H
---COOEt
Ph
O H

14

2-Methyl-3-carbethoxy-4,5,6,7-tetrafluorobenzofuran (**15**), obtained by condensation of hexafluorobenzene with ethyl acetoacetate in presence of sodium hydride, is readily hydrolized to the acid (**16**). Catalytic hydrogenation of **15** gave 2-methyl-3-carbethoxybenzofuran (**17**) (Eq. 6).[23] Similar condensation of pentafluorobenzene with ethyl acetoacetate resulted in the formation of ethyl 4,5,7-trifluorobenzofuran-3-carboxylate.[24]

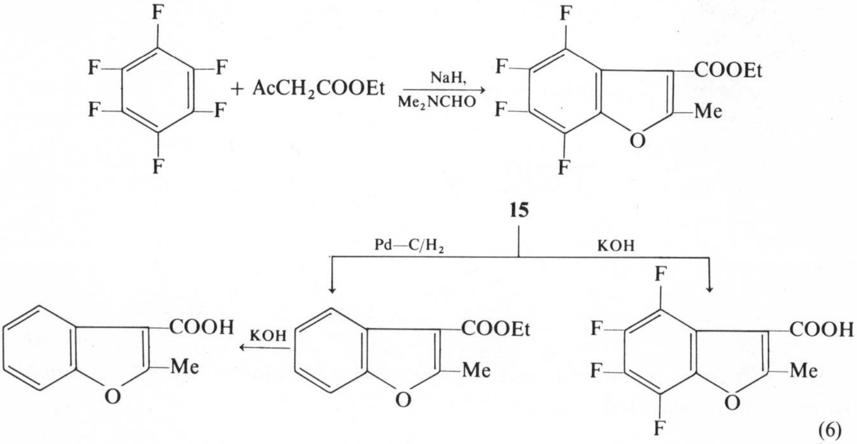

$$C_6F_6 + AcCH_2COOEt \xrightarrow[Me_2NCHO]{NaH,} \textbf{15}$$

15 —Pd—C/H_2→ **17** —KOH→ 2-methylbenzofuran-3-COOH; **15** —KOH→ **16** (6)

17 **16**

2-Aryl-3-benzofurancarboxylic acid and its substituted derivatives have been obtained by alkaline oxidation, with hydrogen peroxide and sodium hydroxide (AFO reaction), of 2′-hydroxychalcones (**18**).[25-27] Wheeler[27] has suggested that a benzylic acid-type rearrangement of the postulated intermediate, α-diketone (**19**) is involved here.[28] This rearrangement may proceed through either a benzyl or phenyl group migration. The occurrence of labeled C-atoms in 2-phenylbenzofuran, isolated in the ^{14}C tracer synthesis of 2-phenyl-3-benzofurancarboxylic acid, **20** from **18**, indicates that the reaction proceeds predominantly through benzyl migration (Eq. 7).

$$\text{PhOH} + (\text{Me}\overset{*}{\text{C}}\text{O})_2\text{O} \longrightarrow \text{C}_6\text{H}_5\text{O}\overset{*}{\text{C}}\text{OMe} \longrightarrow o\text{-HOC}_6\text{H}_4\overset{*}{\text{C}}\text{OMe} \longrightarrow$$

$$o\text{-HOC}_6\text{H}_4\overset{*}{\text{C}}\text{OCH=CHPh}\ (\mathbf{18}) \xrightarrow{\text{AFO}} [o\text{-}^{-}\text{OC}_6\text{H}_4\overset{*}{\text{C}}\text{OCOCH}_2\text{Ph}]\ (\mathbf{19}) \longrightarrow [o\text{-}^{-}\text{OC}_6\text{H}_4\text{C(OH)(COO}^-)\text{CH}_2\text{Ph}]$$

(19) ↓ $o\text{-}^{-}\text{OC}_6\text{H}_4{}^{+}\text{C(OH)(COO}^-)\text{CH}_2\text{Ph}$ ⟶ 3-carboxy(*)-2-phenylbenzofuran **20** (85 ± 2 c.p.m) ⟶ 2-phenylbenzofuran (* at C-3) (77 ± c.p.m.)

(dashed) ↓ 3-(*COOH)-2-phenylbenzofuran ⟶ 2-phenylbenzofuran (7)

Synthesis of **20**[29] is outlined in Eq. 8:

$$o\text{-HOC}_6\text{H}_4\text{CH}_2\text{CN} + \text{PhCOOEt} \longrightarrow o\text{-HOC}_6\text{H}_4\text{CH(CN)COPh} \longrightarrow \text{3-cyano-2-phenylbenzofuran} \xrightarrow{\text{H}^+} \mathbf{20} \quad (8)$$

Unsubstituted 3-benzofurancarboxylic acid was obtained by the sequence of reactions illustrated in Eq. 9.[30]

$$\text{2,3-benzofurandione} + \text{BrCH}_2\text{COOR} \xrightarrow{\text{NaOEt}} o\text{-(COCOOH)C}_6\text{H}_4\text{OCH}_2\text{COOH} \xrightarrow{\text{CH}_2\text{N}_2}$$

$$o\text{-(COCOOMe)C}_6\text{H}_4\text{OCH}_2\text{COOMe} \xrightarrow{\text{Na/EtOH}} \text{2,3-benzofurandicarboxylic acid (COOH, COOH)} \xrightarrow[\text{quinoline}]{\text{Cu,}} \text{3-benzofurancarboxylic acid (HOOC)} \quad (9)$$

Benzofurancarboxylic acids, having the carboxy substituent in the benzene nucleus, are obtained either by: (1) haloform reaction, (2) dehydrocyclization of butanones, and/or (3) hydrolysis of the appropriate esters[31] (see chart 1, Chapter II, p. 81. 5-Carboxy- and 6-carboxy-2,3-

dimethylbenzofuran, obtained from 5-acetyl-- and 6-acetyl-2,3-dimethylbenzofuran by the haloform reaction, are fluorescent whitening agents for organic materials.[32]

C. Hydroxybenzofurancarboxylic Acids

In 1888 von Pechmann[33] allowed *p*-benzoquinone to react with ethyl acetoacetate in presence of anhydrous zinc chloride and obtained a product with a high melting point; Ikuta[34] proved this to have the difuran structure (**21**), and was able to isolate the benzofuran derivative (**22**).

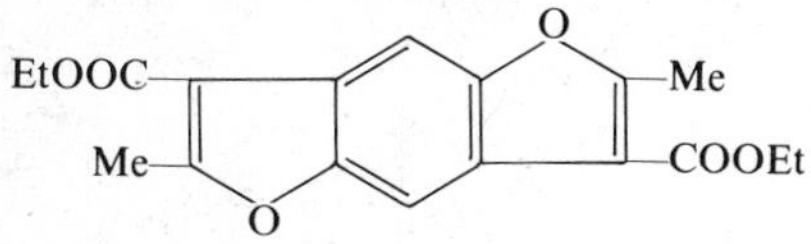

EtOOC, Me, O, Me, COOEt, O

21

HO, COOEt, Me, O

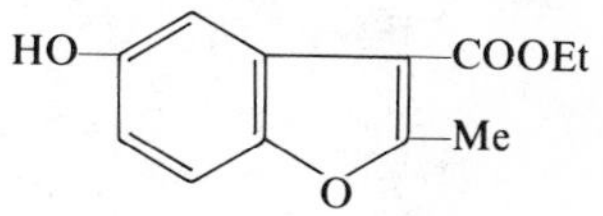

22

This reaction involved the initial formation of a quinol, which was cyclized to **22** or oxidized by action of unreacted benzoquinone to its corresponding quinone. This then added a second molecule of ethyl acetoacetate to form disubstituted quinol and cyclized to **21**.[35] In a similar manner, 2,3-dichloro-*p*-benzoquinone condensed with ethyl acetoacetate and/or ethyl benzoylacetate to yield ethyl 6,7-dichloro-2-methyl- (**23**) and ethyl 6,7-dichloro-2-phenyl-5-hydroxybenzofuran-3-carboxylate (**24**), respectively.[36,37]

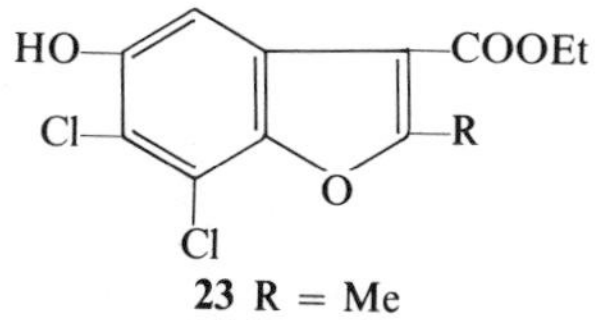

HO, COOEt, Cl, R, O, Cl

23 R = Me
24 R = Ph

Ethyl 3-amino-2-(2,5-dihydroxyphenyl)crotonate (**25**), one of the products of Nenitzescu condensation with *p*-benzoquinone and ethyl 3-aminocrotonate, upon acid-catalyzed hydrolysis yielded 2-(2,5-dihydroxyphenyl)acetoacetate (**26**). Cyclization of **26** gave 2-methyl-3-carbethoxy-5-hydroxybenzofuran (**27**) (Eq. 10).[38]

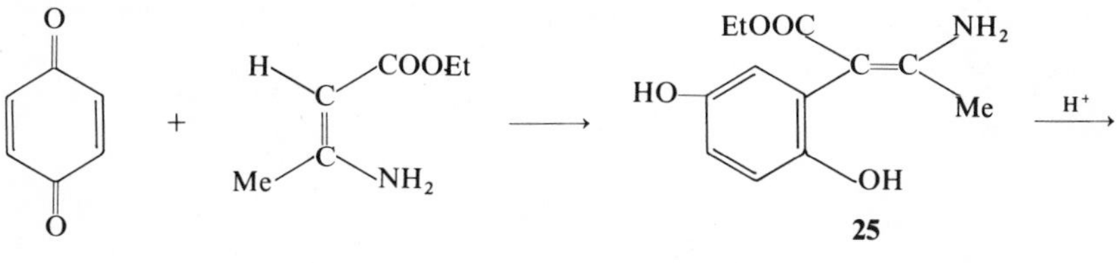

O, O + H, COOEt, C, C, Me, NH_2 ⟶ EtOOC, NH_2, C=C, HO, Me, OH $\xrightarrow{H^+}$

25

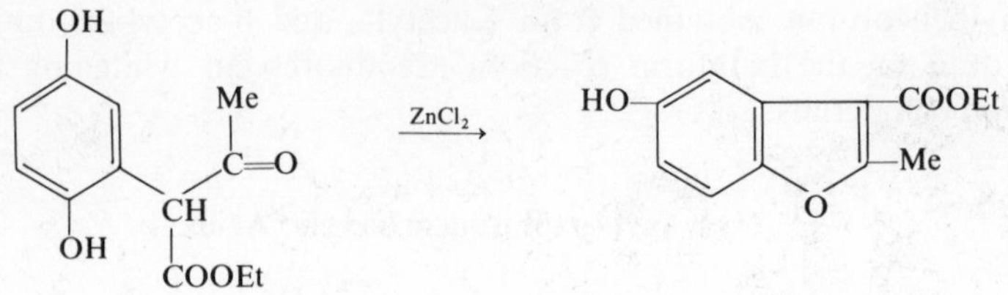

(10)

The ethyl esters of dimethylamino- (**28**) and β-piperidino-crotonic acid (**29**) gave, under normal Nenitzescu conditions[39] with an inert solvent in presence of acetic acid, 2-methyl-3-ethoxycarbonyl-5-hydroxybenzofuran (**27**) as the main product[40] and **31** (Eq. 11).

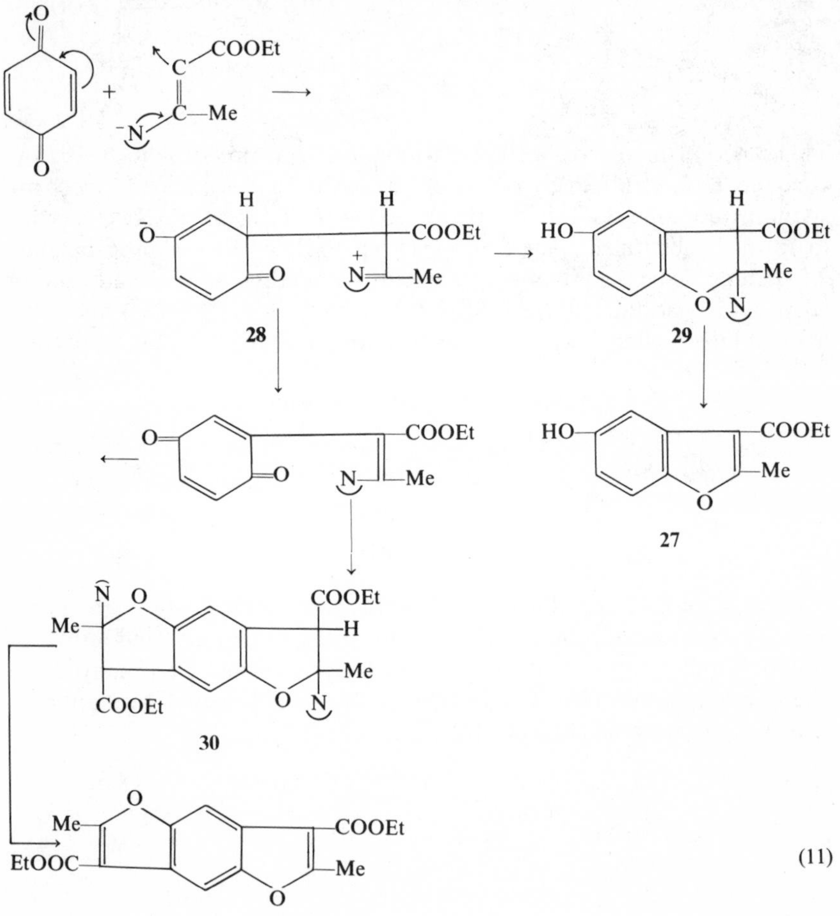

(11)

o-Hydroxybenzofurancarboxylic acids are obtained by treating the alkali salts of 5-hydroxy- and/or 6-hydroxybenzofuran with carbon dioxide under pressure.[41]

Another approach for the synthesis of hydroxybenzofurancarboxylic acid is the treatment of hydroxycoumarins, for example, 3-bromo5-hydroxy-4,7-dimethylcoumarin, with sodium carbonate to yield 3,6-dimethyl-4-hydroxy benzofuran-2-carboxylic acid.[42]

Condensation of 2,6-dihydroxy-3-methoxycarbonylbenzaldehyde with ethyl bromomalonate in presence of potassium carbonate yields a mixture of **32** and **33**.[43] However 4-benzyloxy-2-hydroxybenzaldehyde gave, under the same conditions, 2,2-dicarbethoxy-3-hydroxy-6-benzyloxy-2,3-dihydrobenzofuran (**34**) which, upon hydrolysis, afforded 6-benzyloxy-benzofuran-2-carboxylic acid (**35**) and the 6-hydroxy isomer **36** (Eq. 12).

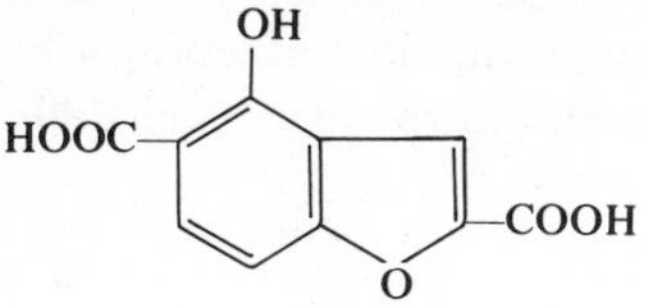

OH HOOC COOH O **32** OH COOH O **33**

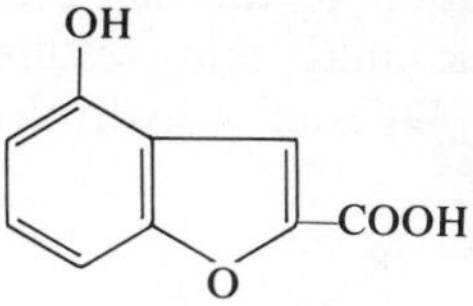

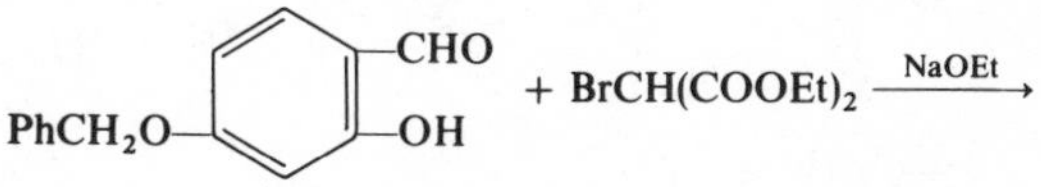

CHO PhCH$_2$O OH + BrCH(COOEt)$_2$ $\xrightarrow{\text{NaOEt}}$

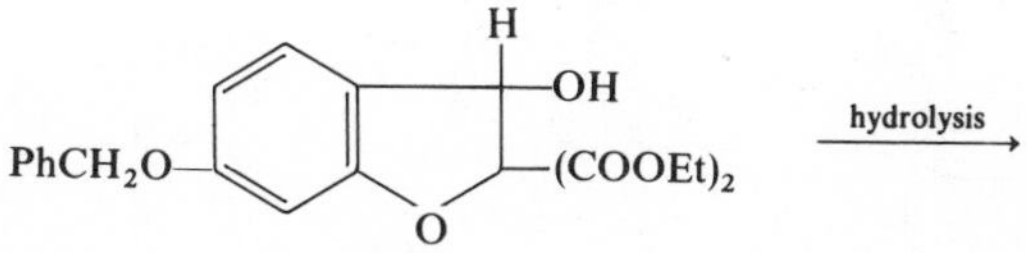

H OH PhCH$_2$O (COOEt)$_2$ O **34** $\xrightarrow{\text{hydrolysis}}$

PhCH$_2$O COOH O **35** + HO COOEt O **36** (12)

The scope of this synthesis[44] has been further extended to a variety of substituted *o*-hydroxybenzaldehydes (Eq. 13).

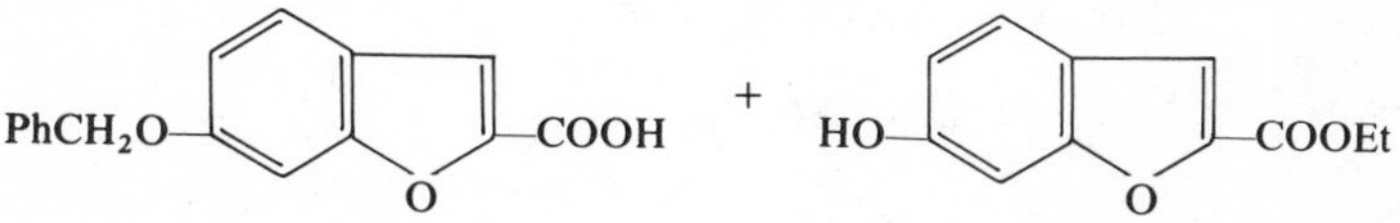

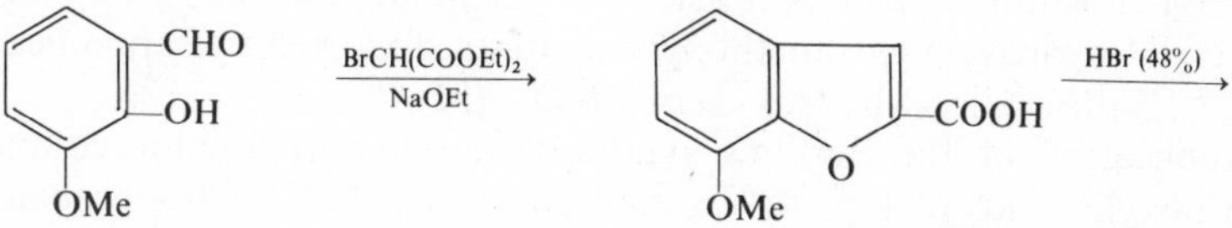

CHO OH OMe $\xrightarrow[\text{NaOEt}]{\text{BrCH(COOEt)}_2}$ COOH O OMe $\xrightarrow{\text{HBr (48\%)}}$

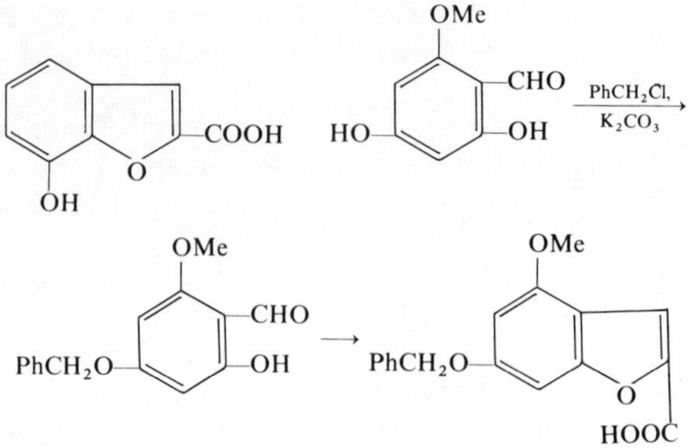

(13)

Cyclization of the half-esters **37** and **38** by action of sodium acetate and acetic anhydride, resulted in the formation of ethyl 4-hydroxy-7-phenyl- (**39**) and 4-hydroxybenzofuran-6-carboxylic acid (**40**), respectively.[45,46]

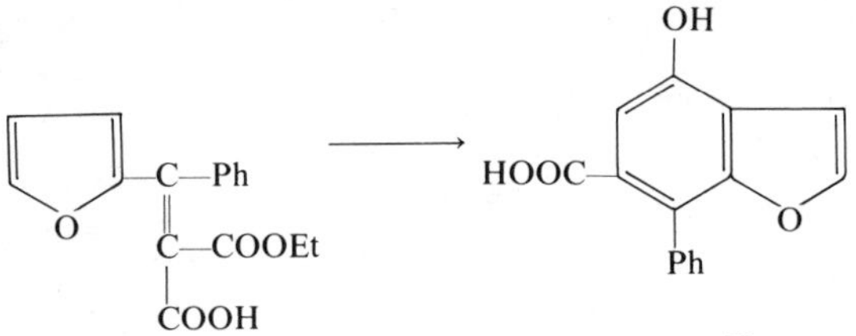

37 **39**

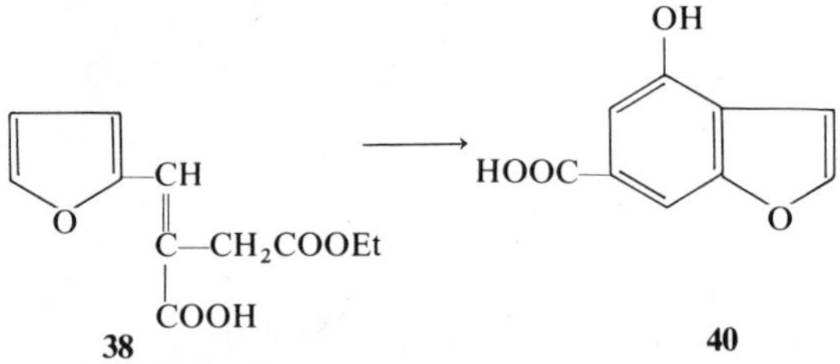

38 **40**

2. Benzofuran Dicarboxylic Acids

Benzofuran dicarboxylic acids are obtained by haloform oxidation of diacetyl-substituted benzofurans.[31,46] Selenium dioxide oxidation of 2-acetyl-4-hydroxy-5-carbmethoxybenzofuran, followed by saponification, gave 2,5-dicarboxy-4-hydroxybenzofuran (**32**).[46]

Application of the Berkin's synthesis to esters of 3-chlorocoumarin-4-carboxylic acids of type **41**[47] establishes a pathway for the preparation

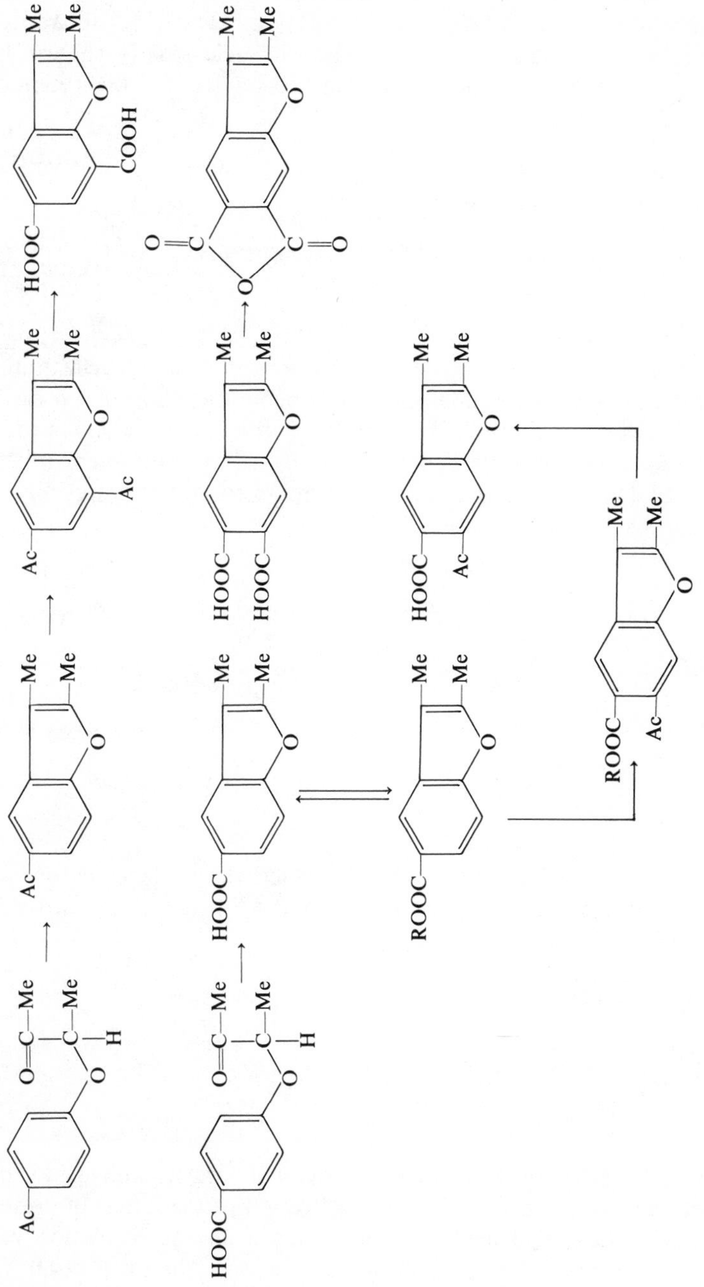

Chart 1.

of benzofuran dicarboxylic acids. Furthermore, treatment of the esters of 3-methoxycoumarin carboxylic acids of type **43** with alkali afforded the corresponding benzofuran-2,3-dicarboxylic acids. This process furnishes a close analogy to the mechanism postulated for the rotenone–rotenonic acid cleavage.[48]

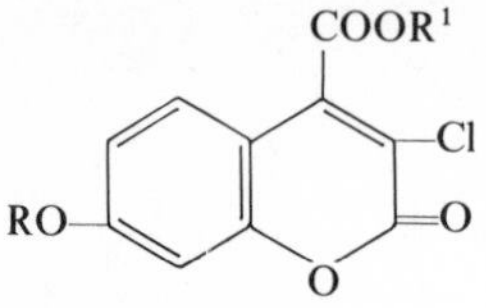

COOR^1 ... Cl ... O ... RO ... O — **41**

COOR^1 ... COOR^1 ... RO ... O — **42**

COOEt ... OMe ... O ... RO ... O — **43**

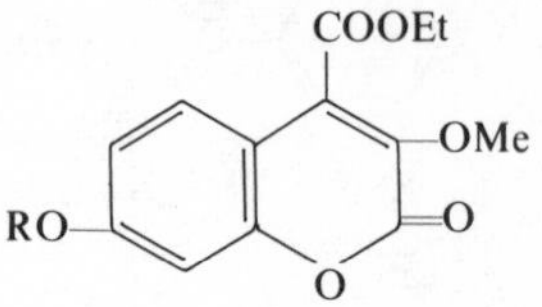

Koelsch and Whitney[49] reported sulfuric–acetic acid cyclization of the product, obtained by condensation of phenoxyacetic acid and diethyl oxalate. Huntress and Olsen[50] designated the condensation product as diethyl oxalophenoxyacetate (**44**), which reacted as a tautomeric mixture of structures **44a** and **44b**. The ester thus obtained underwent cyclization to 2,3-dicarboxybenzofuran and **45** (Eq. 14).

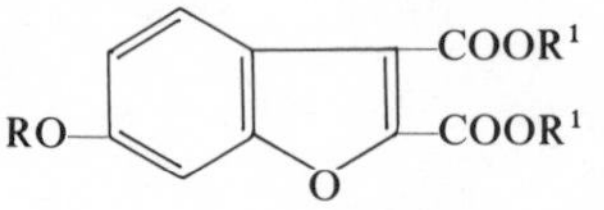

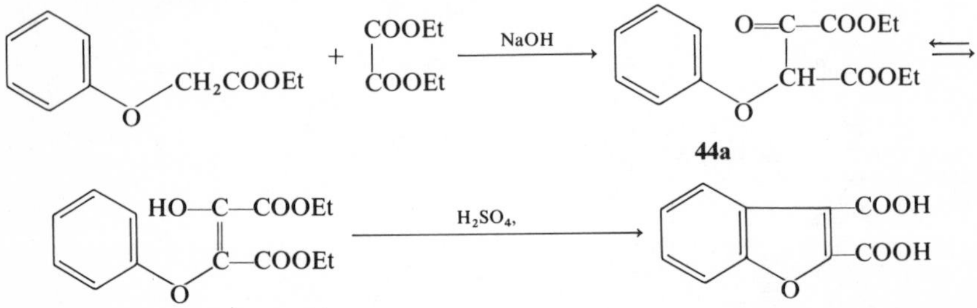

$$\text{PhOCH}_2\text{COOEt} + \text{COOEt–COOEt} \xrightarrow{\text{NaOH}} \text{PhO–CH(COOEt)–C(=O)–COOEt}\ (\mathbf{44a}) \rightleftharpoons$$

$$\text{PhO–C(COOEt)=C(OH)–COOEt} \xrightarrow{\text{H}_2\text{SO}_4,} \mathbf{45}\ (\text{benzofuran-2,3-dicarboxylic acid; COOH, COOH})$$

(14)

Addition of 2-vinylbenzofuran (**46**) to dimethyl acetylenedicarboxylate at room temperature gave a mixture of **47** and 4,5-benzofurandicarboxylic acid dimethyl ester (**48**).[51]

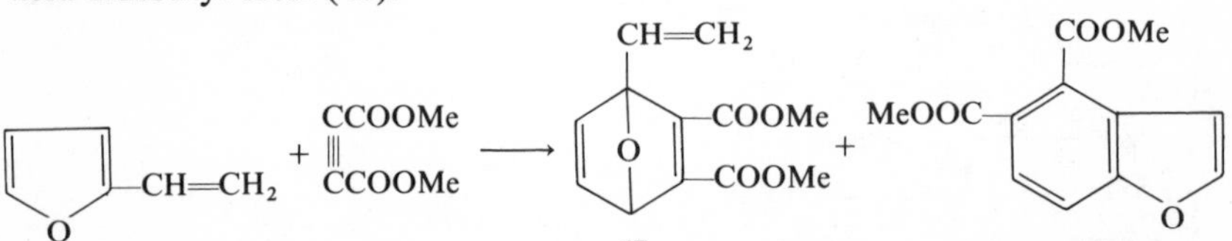

$$\mathbf{46}\ (\text{CH=CH}_2) + \text{CCOOMe} \equiv \text{CCOOMe} \longrightarrow \mathbf{47}\ (\text{CH=CH}_2,\ \text{COOMe},\ \text{COOMe}) + \mathbf{48}\ (\text{MeOOC},\ \text{COOMe})$$

5-Bromo-2-furylpropiolic acid was converted into its anhydride upon reflux with acetic anhydride; thermal cyclization of the latter, in presence of aqueous sodium hydroxide solution, results in the formation of 2-(bromo)-4-(5)bromo-2-furylbenzofuran-5,6-dicarboxylic acid (**49**) (Eq. 15).[52] Compound **50** was similarly prepared.[53]

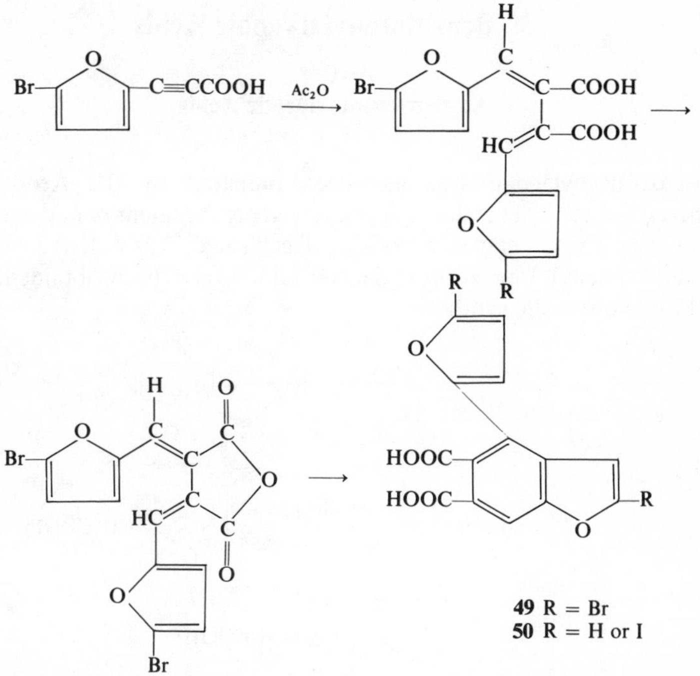

(15)

49 R = Br
50 R = H or I

Interaction of the methyl ester of carboxysalicylaldehyde in methyl ethyl ketone with ethyl bromomalonate in presence of potassium carbonate afforded dicarboxylic acids of type **51**, for example, 5-, 6-, and 7-carboxybenzofuran-2-carboxylic acid (Eq. 16).[54]

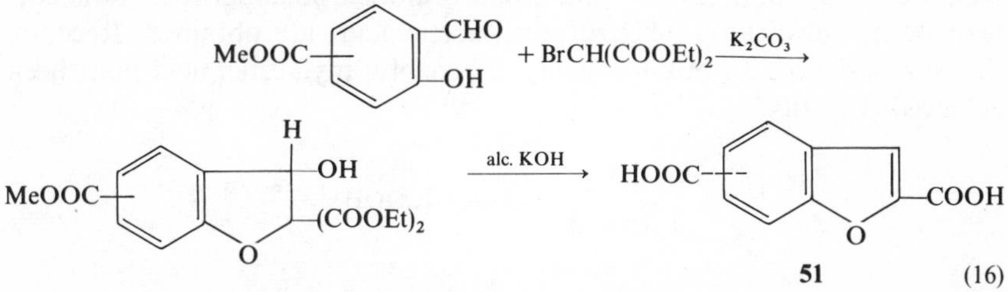

51 (16)

The 1,3 Cycloaddition of dimethyl fumarate and/or diethyl maleate to the oxocarbene results in the formation of **52**.[22]

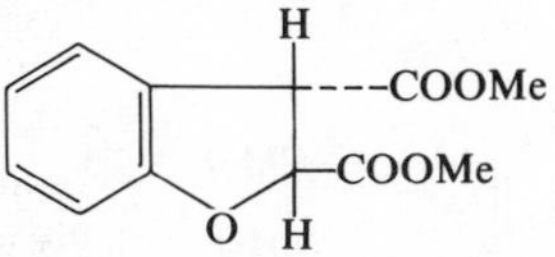

52

3. Benzofuranylalkanoic Acids

A. Benzofuranylacetic Acids

2-Benzofuranylacetic acid has been prepared by the Arndt-Eistert synthesis[55] in 29% yields through catalytic rearrangement of ω-diazoacetylbenzofuran. Thus 2- and 3-benzofuranylacetic acid,[56] as well as 5-methoxy-2-[57] and 2-phenyl-3-benzofuranylacetic acid,[58] have been obtained. Equation 17 illustrates the synthesis.

R–(benzofuran)–COOH $\xrightarrow{SOCl_2}$ R–(benzofuran)–COCl $\xrightarrow{CH_2N_2}$

R–(benzofuran)–$COCHN_2$ $\xrightarrow{Ag_2O}$ R–(benzofuran)–CH_2COOEt $\xrightarrow{H^+}$

R–(benzofuran)–CH_2COOH (17)

Benzofurancarboxaldehydes are readily transformed into the corresponding azlactones which provide a convenient route for the preparation of benzofuranylacetic acid and substituted derivatives. Thus starting with 3-methyl-,[59] 3-methyl-5-chloro-,[60] 3,6-dimethyl-5-chloro-,[61] 3-ethyl-5-chloro-,[62] 3,5-dimethyl-,[63] and 2-benzofurancarboxaldehyde,[65] the corresponding substituted 2-benzofuranylacetic acids are obtained. Recently the synthesis of 3-*t*-butyl-6-methyl-2-benzofuranylacetic acid has been achieved (Eq. 18).[66]

Me–(benzene)(OH)($COCMe_3$) + $ClCH_2COOH$ $\xrightarrow{K_2CO_3}$

Me–(benzene)(OCH_2COOH)($COCMe_3$) $\longrightarrow$ Me–(benzofuran)–CMe_3 + $\xrightarrow[POCl_3]{HCONMe_2,}$

Me–(benzofuran)(CMe_3)(CHO) $\xrightarrow[NaOAc,\ Ac_2O]{PhCONHCH_2COOH}$

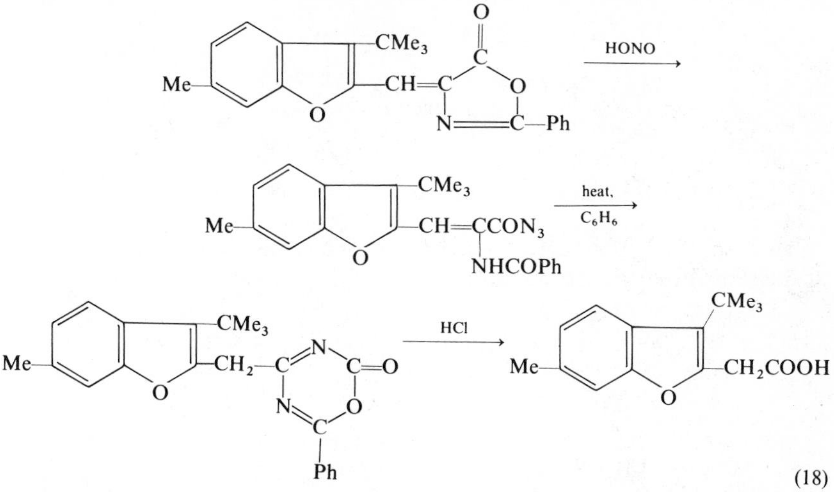

(18)

The azlactones obtained in the synthesis, upon treatment with potassium hydroxide (20%), establish a pathway for the preparation of substituted 2-benzofuranylpyruvic acids.[64] This holds true when the azlactones are treated with hydrazine hydrate, yielding α-benzamido-β-(2-benzofuranyl) acrylohydrazides. 3-Benzofuranylpyruvic acid has also been obtained as outlined in (Eq. 19).[56]

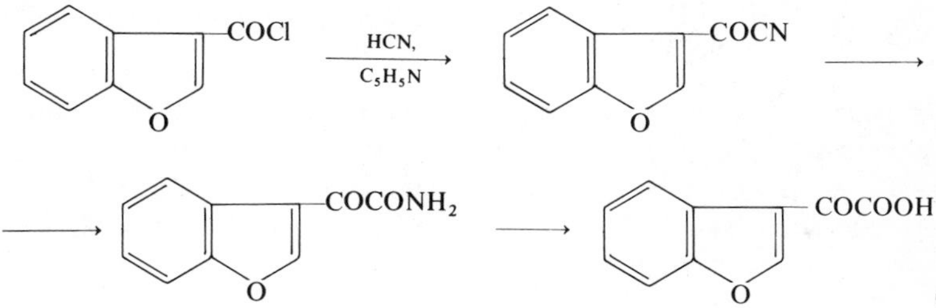

(19)

Acid-catalyzed cyclization of 3-phenylbenzofuran-2-acetic acid may have the cyclobutanone structure **53** instead of the formerly claimed structure **54**, since treatment of the reaction product with sodium carbonate gave a mixture of 2-phenyl-3-benzofuranylacetic acid (**55**) and naphthobenzofuran (**56**).[67]

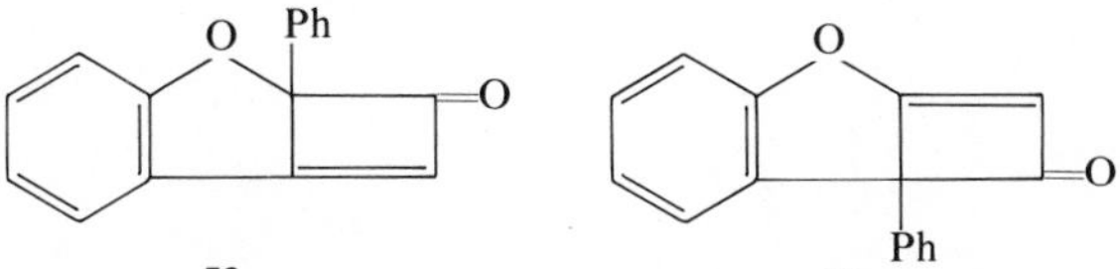

53 **54**

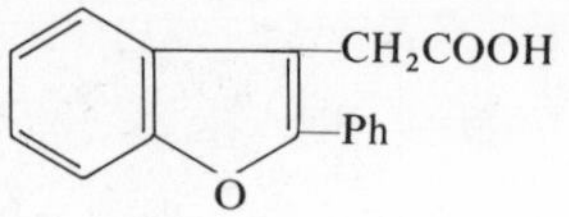

55 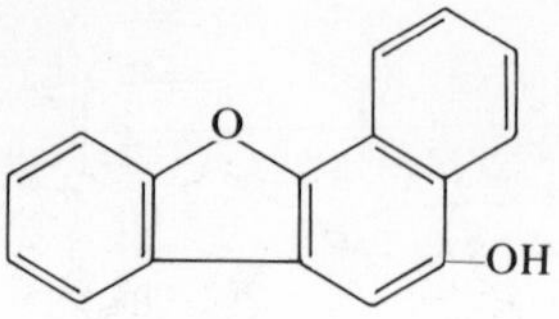 **56**

Unsubstituted 3-benzofuranylacetic acid has been obtained according to the sequence of reactions in Eq. 20.[6]

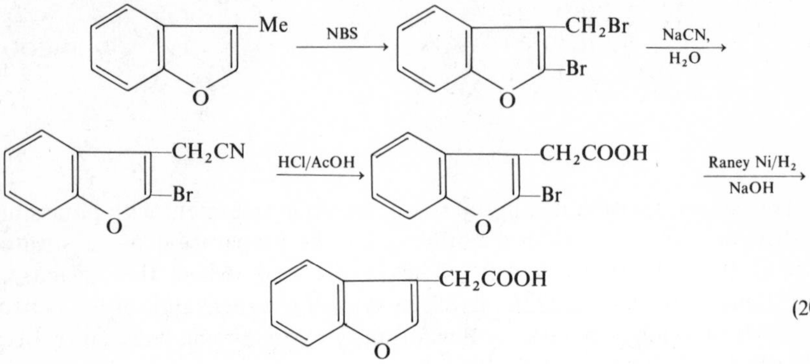

(20)

B. Benzofuranylpropionic Acids

Catalytic hydrogenation of β-(2-benzofuranyl)acrylic acid (**57**), obtained by condensation of 2-benzofurancarboxaldehyde with malonic acid in presence of pyridine[68] or its substituted derivatives, affords 2-benzofuranylpropionic acids (**58**) (Eq. 21).[69] Compound **58** (R = H) is also

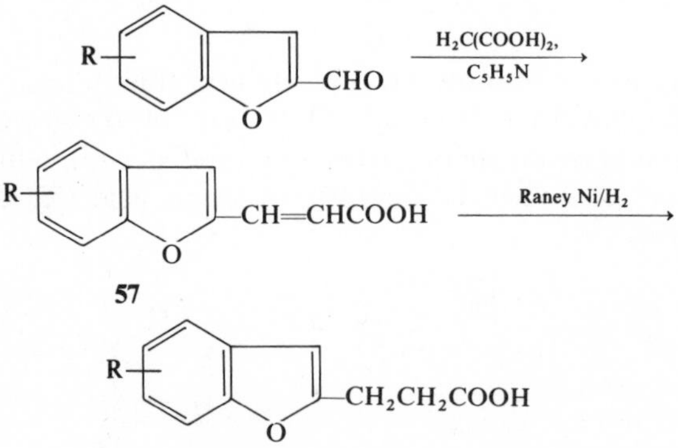

57

58 (21)

obtained upon treatment of benzofuran with acrylic acid in presence of acetic anhydride and acetic acid.[69] The reaction has been shown to be of wide scope (Eq. 22).

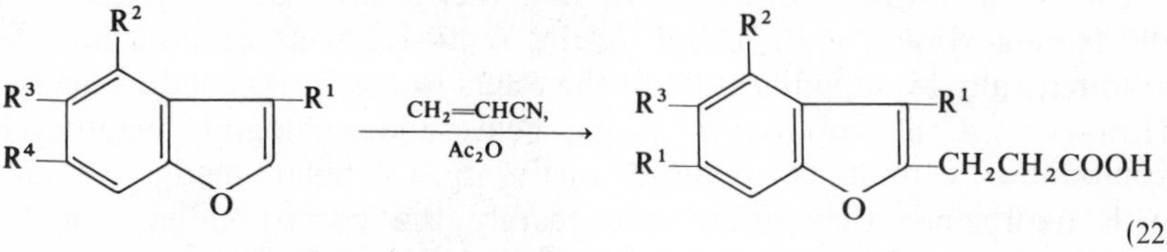

(22)

β-(3-Benzofuranyl)propionic acid and its substituted derivatives are obtained by condensation of 2-bromo-3-bromomethylbenzofuran with sodioethyl malonate (Eq. 23).[70,71]

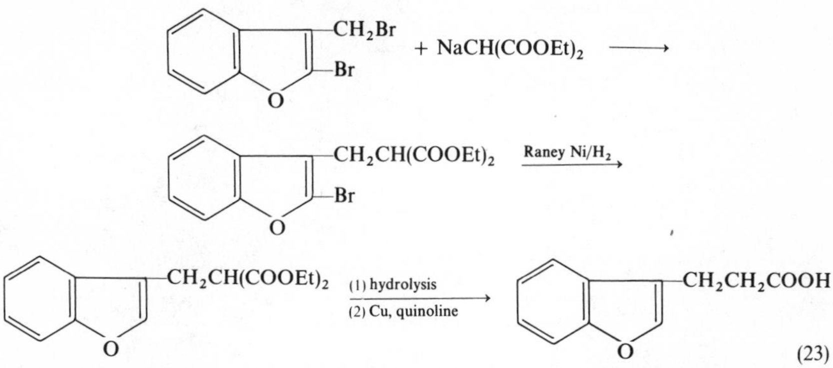

(23)

Oxidative coupling of substituted β-(2-benzofuranyl)propionic acids (**59**)[72] with manganese dioxide results in the formation of spiro-2,3-dihydrobenzofurans (**60**) which, when treated with oxygen, yielded **61**.

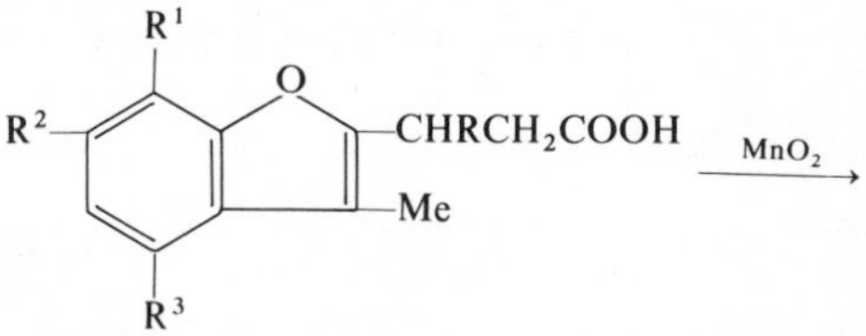

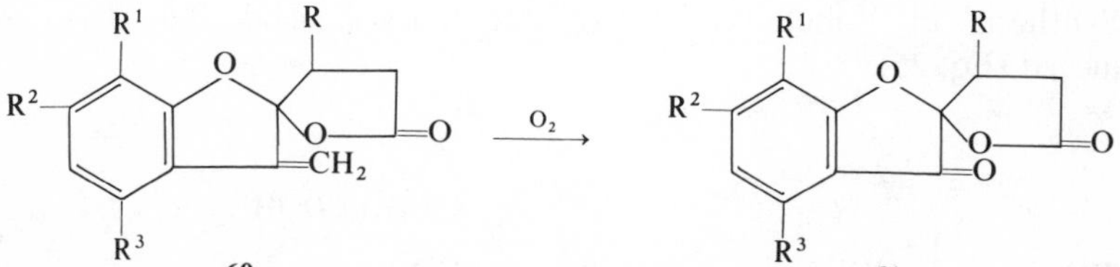

60

61

C. Benzofuranylbutyric Acids

Esters of β-(3-benzofuroyl)propionic acids are readily reduced by the Huang-Minlon modification of the Wolff-Kishner reaction, followed automatically by saponification of the esters to the corresponding acids.[73] However β-(3-benzofuroyl)propionic acids and analogous heterocyclic compounds cannot be reduced easily in a similar manner because with hydrazines they afford too readily the corresponding dihydropyridazinones. Thus reduction of β-(2-ethyl-3-benzofuroyl)propionic acid (**62**) yields 6-(2-ethyl-3-benzofuranyl)-3-oxo-2,3,4,5-tetrahydropyridazine (**63**), whereas the ester (**64**) on reduction gives γ-(2-ethyl-3-benzofuranyl)-butyric acid (**65**) (Eq. 24).[74]

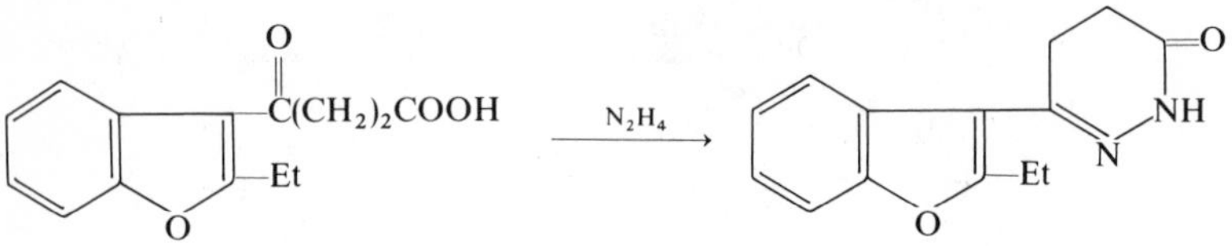

O ‖ $C(CH_2)_2COOH$, Et, O — $\xrightarrow{N_2H_4}$ — O, NH, N, Et, O

62 **63**

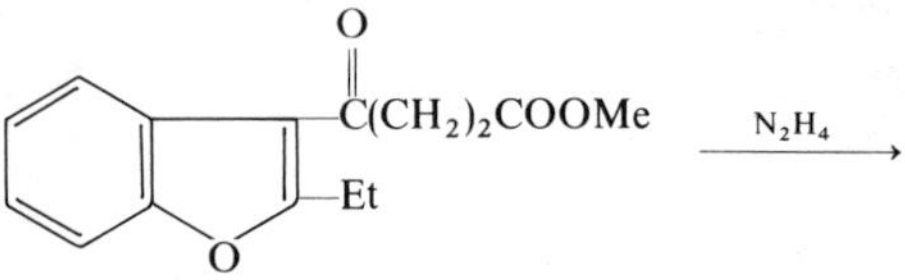

O ‖ $C(CH_2)_2COOMe$, Et, O — $\xrightarrow{N_2H_4}$

64

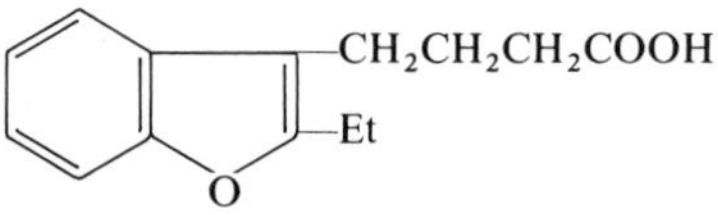

$CH_2CH_2CH_2COOH$, Et, O (24)

65

Synthesis of 6-methoxy-3-benzofuranylbutyric acid (**66**) has been achieved (Eq. 25).[75]

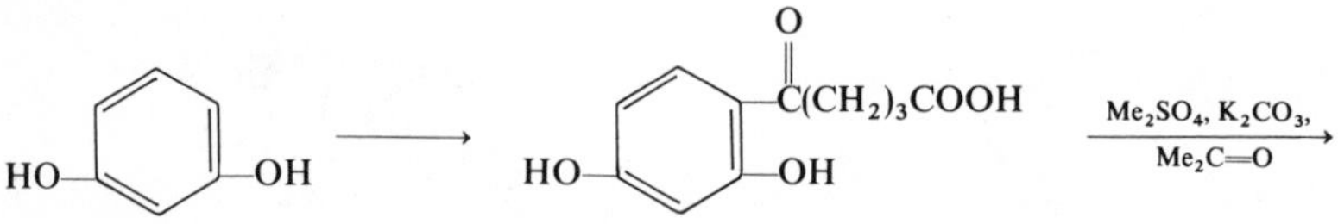

HO, OH — $\longrightarrow$ — O ‖ $C(CH_2)_3COOH$, HO, OH — $\xrightarrow[Me_2C{=}O]{Me_2SO_4,\ K_2CO_3,}$

$$\text{MeO-C}_6\text{H}_3\text{(OH)-C(=O)(CH}_2)_3\text{COOMe} \xrightarrow[K_2CO_3,\ MeCOEt]{BrCH_2COOEt} \left[\text{MeO-C}_6\text{H}_3\text{(OCH}_2\text{COOEt)-C(=O)(CH}_2)_3\text{COOMe}\right]$$

$$\xrightarrow[(10\%)]{aq.\ NaOH} \text{MeO-C}_6\text{H}_3\text{(OCH}_2\text{COOH)-C(=O)(CH}_2)_3\text{COOH} \xrightarrow{Ac_2O,\ NaOAc} \mathbf{66}\ \text{(MeO-benzofuranyl-(CH}_2)_3\text{COOH)}$$

$$\left[\ldots\right] \xrightarrow{Ac_2O,\ NaOAc} \text{MeO-benzofuranyl-(CH}_2)_3\text{COOEt}; \quad \mathbf{66} \xrightarrow[EtOH]{Ba(OH)_2} \text{MeO-benzofuranyl-(CH}_2)_3\text{COOEt} \qquad (25)$$

Substituted benzofuranylbutyric acids, which have a high choleretic activity, are prepared by converting the chloromethylbenzofurans into the corresponding nitriles by interaction with alkali cyanides. The nitriles are alkylated with the appropriate alkyl halide to form compounds, which are transformed into the requisite acids by saponification.[76,77]

Intramolecular dehydration of γ-(2-ethyl-3-benzofuranyl)butyric acids (with or without a second ethyl group in the benzene ring) has been achieved by action of polyphosphoric acid. Similar cyclization of the lower and higher homologous ω-(2-ethyl-3-benzofuranyl)alkanoic acids cannot be accomplished. The resulting cyclanones are reduced by the Huang-Minlon procedure to the corresponding mono- and diethyl-3,4-butanobenzofurans. 2-Ethyl-3,4-butanobenzofuran (**67**) can be acylated in the 6 position in the same manner as 2,3-dialkylbenzofurans. Reduction of **67** in liquid ammonia yields a mixture of the corresponding coumarin and hydroxypropyl suberane. 2-Ethyl-3,4-(4-oxobutano)benzofuran (**68**) is transformed into the carbinol (**69**), which can be dehydrated to 2-ethyl-3,4-(3-butano)benzofuran (**70**) (Eq. 26).[78]

Et

O

67

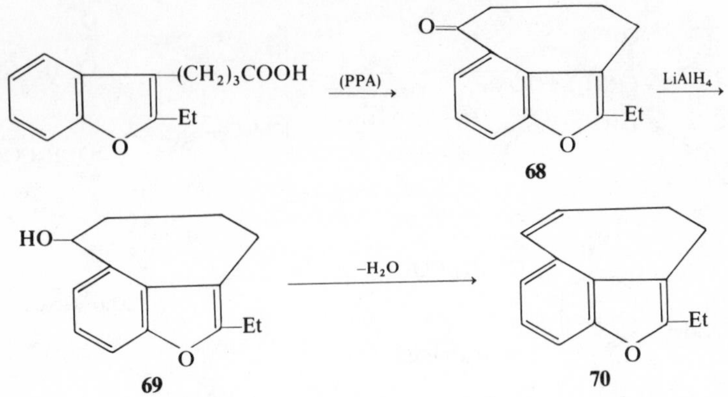

$$\text{(26)}$$

68

69

70

D. Miscellaneous Benzofuranylalkanoic Acids

Application of Willgerodt-Kindler reaction to 2- or 3-acylbenzofurans establishes an efficient procedure for the preparation of a large number of benzofuranylalkanoic acids.[74]

Synthesis of ethyl 2-(3-benzofuranyl)cyclopropanecarboxylate (**71**), useful as an antidepressive and hypotensive agent, has been achieved (Eq. 27).[79]

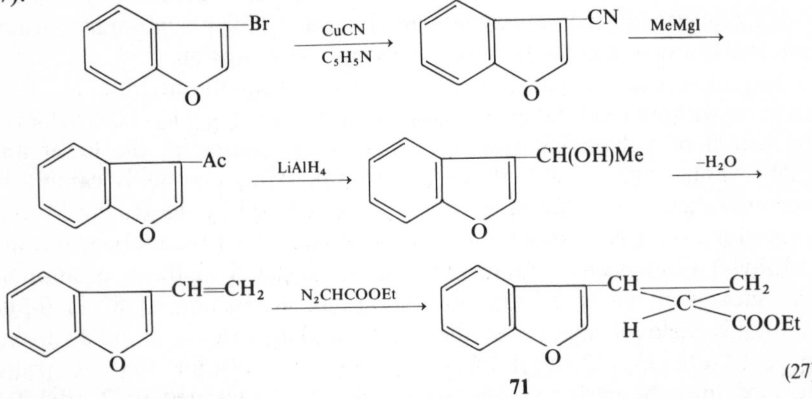

(27)

71

Sila[80-83] has reported the preparation of benzofurans containing amino acid residues, for example, α-amino-β-(3-methyl-2-benzofuranyl)propionic acid (**72**) (Eq. 28).[80,81]

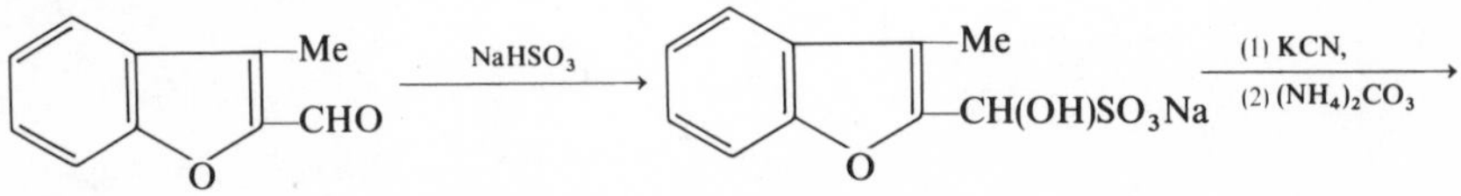

HCl → 72 $CH_2CH(NH_2)COOH$ (28)

Erlenmeyer and Grubenmann[84] have reported the preparation of α-amino-β-(3-benzofuranyl)propionic acid (**73**) according to the sequence of reactions shown in Eq. 29).

NBS, Bz_2O_2 → CH_2Br, Br → $AcNHCNa(COOEt)_2$, heat (5 hrs.) → NHAc, $CH_2C(COOEt)_2$, Br → aq. KOH, EtOH → NHAc, $CH_2C(COOH)_2$, Br → PhBr, 130°C (10 min.) → $CH_2CH(NHAc)COOH$, Br → Raney Ni/H_2, EtOH → $CH_2CH(NHAc)COOH$ → HCl, reflux (10 min.) → $NH_2.HCl$, $CH_2CHCOOH$ → C_5H_5N, EtOH → $CH_2CH(NH_2)COOH$ **73** (29)

Synthesis of a number of alkyl-substituted benzofuranylglycine derivatives has been reported.[82,83]

4. Miscellaneous Reactions of Benzofurancarboxylic Acids

A. Halogenation

Bromination of the potassium salt of 2-benzofurancarboxylic acid results in the formation of 2-bromobenzofuran with elimination of the carboxyl group.[85] However Smith[86] found that 2-benzofurancarboxylic acid did not undergo bromination reaction, but its ethyl ester gave the 5-bromo derivative. Bromination of methyl 3-methyl-6-hydroxy-2-benzofurancarboxylate (**74**) in acetic acid yields methyl 3-methyl-6-hydroxy-7-bromo-2-benzofurancarboxylate (**75**).[87]

Me
HO COOMe
O
74
$\xrightarrow{Br_2}$
Me
HO COOMe
O
Br **75**

The Bromination reaction with methyl 3-methyl-5-methoxy-2-benzofurancarboxylate is illustrated in Chart 2.[88] Bromination and iodination of ethyl 2-methyl-5-hydroxy-3-benzofurancarboxylate (**77**) yield the corresponding 6-halo derivatives, but chlorination affords primarily the 4-chloro compound.[89] Mercuration also occurs at C-4, thus affording a route for the preparation of the corresponding 4-iodo compound. Deiodination is readily effected by the action of zinc and acetic acid without ring cleavage; chlorine is not removed by this procedure. Iodine at the 4 position is especially labile toward the base and, by use of potassium cyanide in dimethyl sulfoxide, a remarkably selective removal of the iodine atom takes place. However the 6-iodo compound undergoes no change. 4,6-Dichloro derivative forms methanol and water-insoluble salts with alkyl and alicyclic amines as do many simple dihalophenols (Chart 3).

B. Chloromethylation

Electrophilic attack on methyl- or ethyl 2-benzofurancarboxylate takes place at C-5 during chloromethylation.[90-92]

C. Nitration

Direct nitration of 2-benzofurancarboxylic acid and its derivatives, under different nitration conditions, is illustrated in Chart 4.[93]

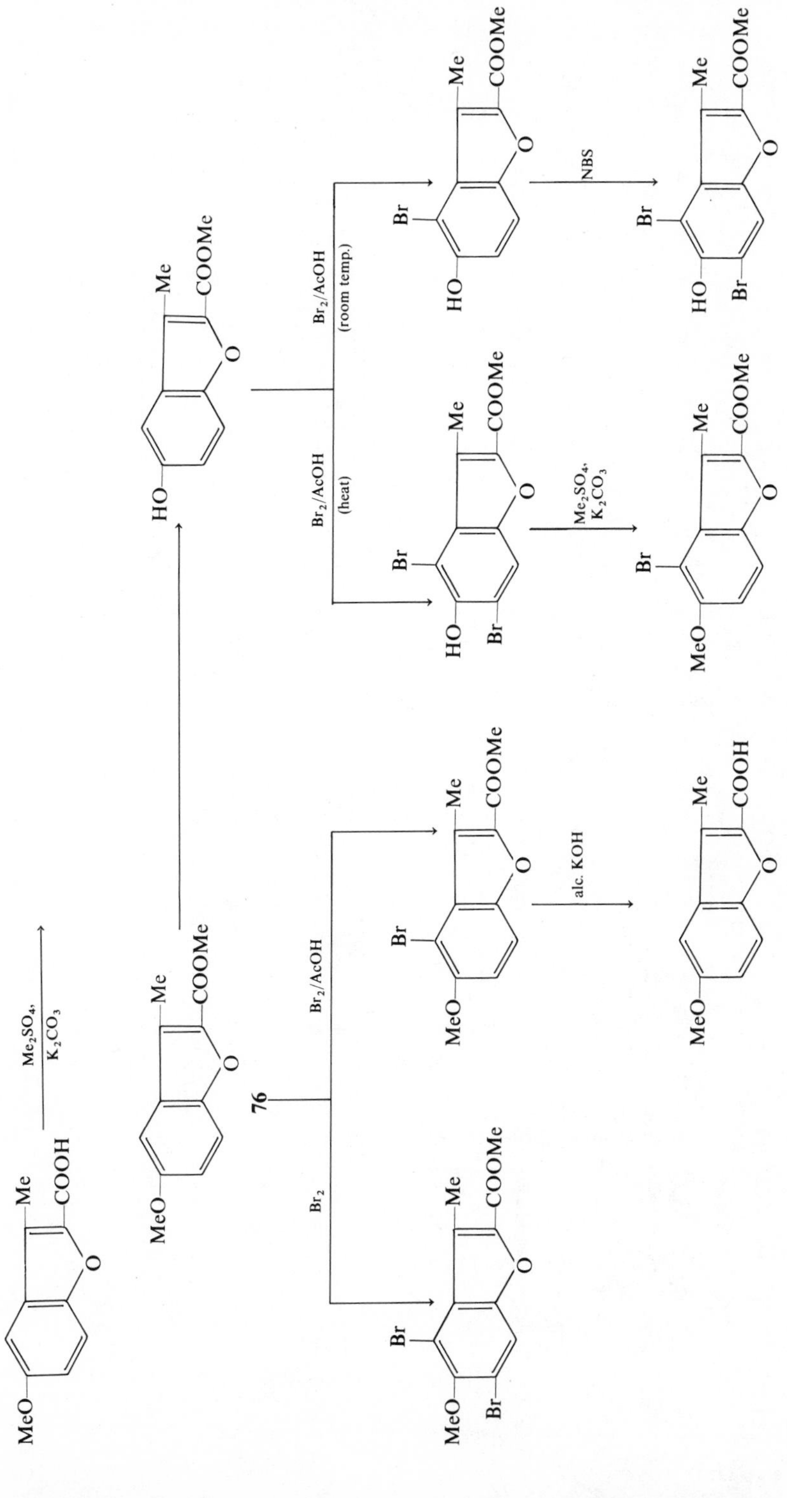

Me
COOH
MeO
O
Me_2SO_4,
K_2CO_3
76
Br_2
Br_2/AcOH
alc. KOH
Br_2/AcOH
(heat)
Br_2/AcOH
(room temp.)
Me_2SO_4,
K_2CO_3
NBS
HO
Br
COOMe

Chart 2.

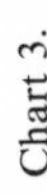

Chart 3.

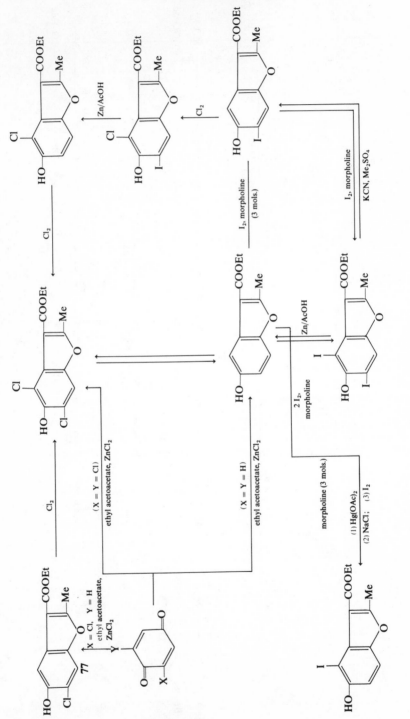

D. Saponification

The rate of alkaline saponification of methyl carboxylic esters in the heterocyclic series in dioxan (70%) has been shown to follow the sequence: 3-benzofuranyl, 2-benzofuranyl, 2-benzothienyl, indol-2-yl, 3-benzothienyl.[94] In alkaline hydrolysis (a nucleophilic reaction) the rate will be governed by electron-withdrawal from the reaction site. The hetero atoms exert this reaction in the sequence: O > S > NH, that is, in the order of electronegativities (− I effect). In the benzofuran series, the 3-ester reacts more quickly than the 2-ester (at 0°C), and then other factors must operate. A neighboring benzo group is rate-enhancing in the case of 3-furyl position and rate-diminishing with the 3-thienyl or phenyl groups.

E. Catalytic Hydrogenation

Catalytic hydrogenation of 2-benzofurancarboxylic acid in presence of Ni–Al alloy and aqueous alkali produces *β*-(*o*-hydroxyphenyl)propionic acid (**78**), isolated in the form of lactone (**79**).[95] Sodium amalgam reduction of 2-benzofurancarboxylic acid, followed by submitting to the action of Raney alloy, produces the dihydro compound (**80**) (Eq. 30).

CH_2CH_2COOH

COOH

OH

O

78

O

O

COOH

O

79

80

(30)

F. Peroxide Formation and Ozonolysis

Benzofuranyl peroxide (**81**) decomposes[96] in benzene, carbon tetrachloride, chlorobenze, or nitrobenzene; but it appears not to yield radicals capable of attacking the solvents. In each case, 2-benzofurancarboxylic acid is isolated together with small amounts of **82**.

COOH

$HNO_3, Ac_2O,$ H_2SO_4

NO_2 (4%) + NO_2 COOH (1.4%) + O_2N COOH (6%)

COOEt

$HNO_3, Ac_2O,$ H_2SO_4

COOEt NO_2 + O_2N COOEt

alc. KOH

O_2N COOH $\xrightarrow{-CO_2}$ O_2N

OMe $\xrightarrow{HNO_3, H_2SO_4}$ NO_2 OMe $\xleftarrow{Cu, quinoline}$ NO_2 OMe COOH $\xrightarrow{Sn, HCl}$ NH_2 OMe COOH

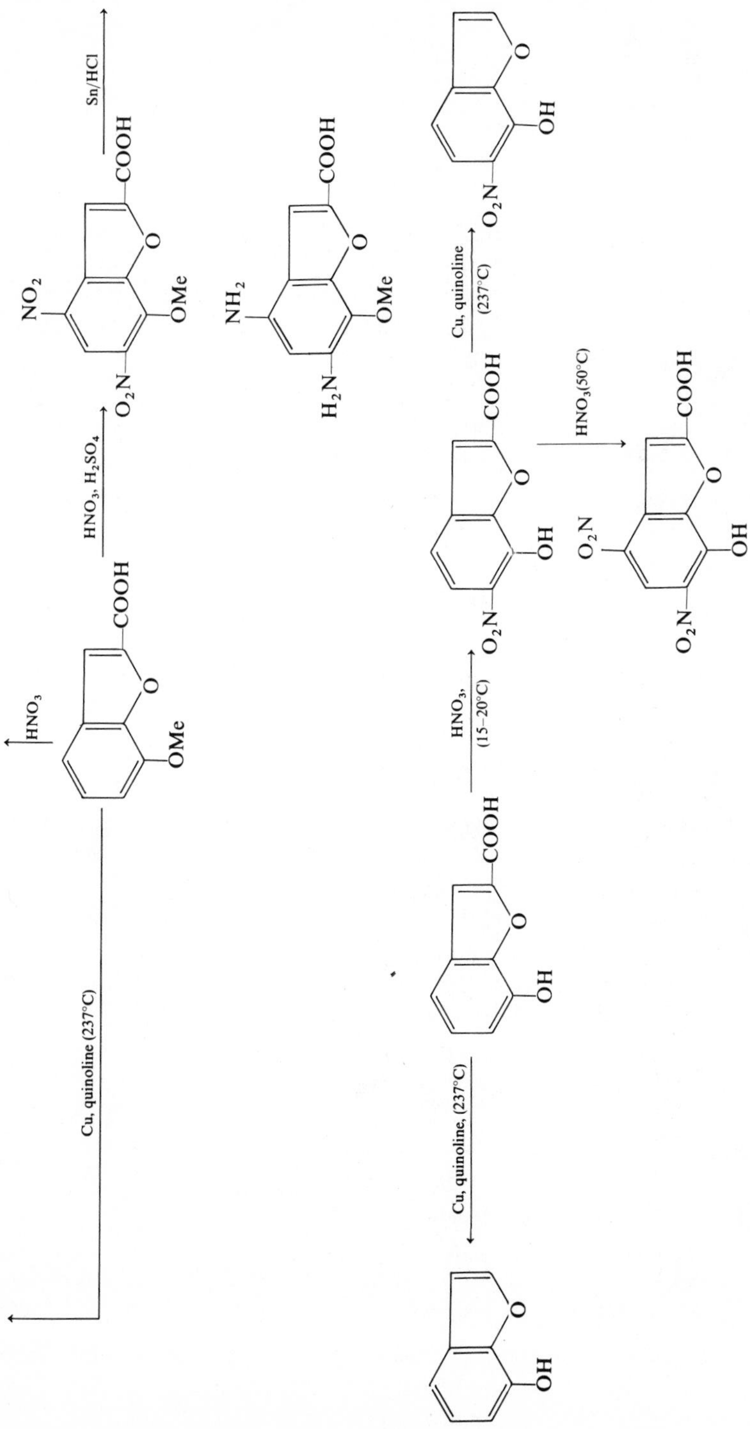
HNO3
HNO3, H2SO4
Sn/HCl
Cu, quinoline (237°C)
HNO3, (15–20°C)
Cu, quinoline (237°C)
HNO3(50°C)
Cu, quinoline, (237°C)

Chart 4.

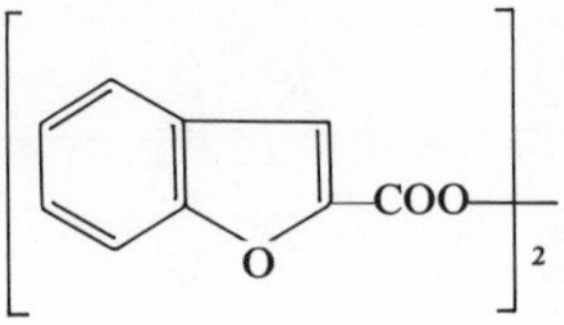

81

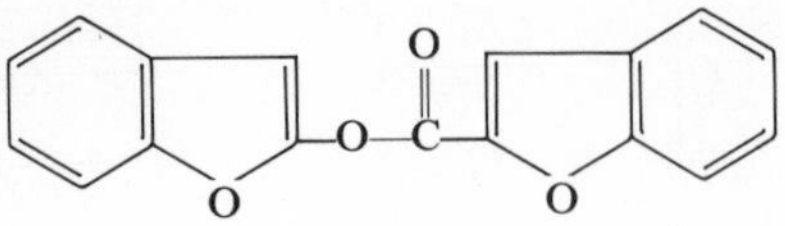

82

The stable crystalline ozonides[97] **83** and **84** have been isolated in methyl acetate, upon ozonization of 2-methyl- and 2,7-dimethyl-5-acetoxy-3-carbethoxybenzofuran, respectively. The ozonide **84** is readily reduced with sodium iodide in glacial acetic acid to yield the monoacetyl derivative of ethyl 2,5-dihydroxy-3-methylphenylglyoxalate (**85**).

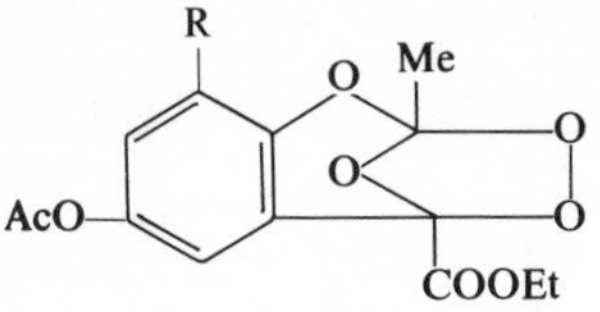

83 R = H
84 R = Me

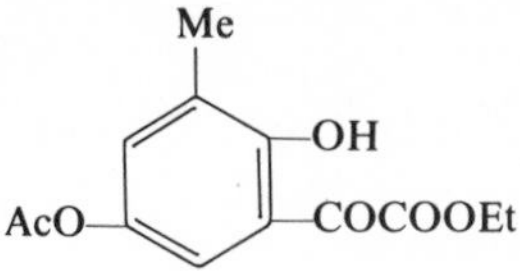

85

G. Acylation

Investigation of the point of attack in acetylation of methyl 4-, 5-, and 7-methoxy-2-benzofurancarboxylates has shown that only in the case of **86** and **87** did acetylation occur ortho to the methoxyl group.[98] Acetylation of methyl 5-methoxy-2-benzofurancarboxylate (**87**) in presence of boron trifluoro–acetic anhydride–acetic acid mixture[99] occurs at C-4. Similarly, methyl 4-acetyl-7-methoxy-2-benzofurancarboxylate is obtained upon acetylation of **89**. The choice of position 4 as the point of attack arose from the elimination of position 6 because of the retention of the methoxy group in the acetylation reaction.

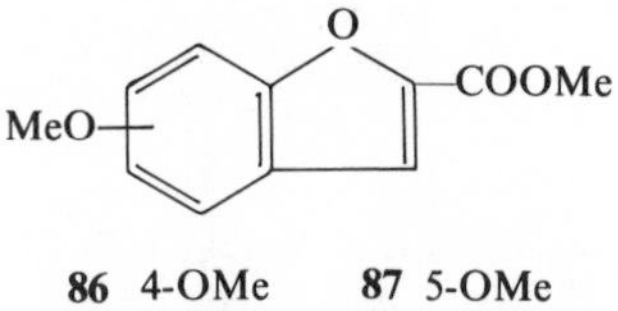

86 4-OMe **87** 5-OMe
88 6-OMe **89** 7-OMe

In contrast to the reported observation[100] that 2-phenyl-2,3-dihydro-2-benzofurancarboxylic acid (**90**) was formed upon treatment of benzofurancarboxylic acid with benzene under Friedel-Crafts conditions, Krishnaswamy and Seshadri[101] have shown that the acid reacts with benzene in presence of aluminum chloride under reflux to afford 3-phenyl-2-benzofurancarboxylic acid (**91**).

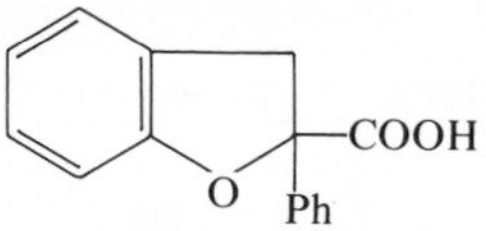

90 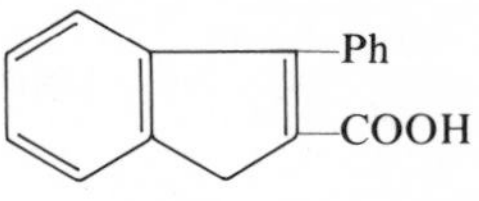 **91**

Treatment of methyl 6-hydroxy-3-methyl-2-benzofurancarboxylate with hexamine in glacial acetic acid results in the formation of 7-formyl-6-hydroxy-3-methyl-2-benzofurancarboxylate; similar treatment of methyl 5-hydroxy-3-methyl-2-benzofurancarboxylate affords the 4-formyl compound.[102]

H. Alkylation

Ethyl 3-hydroxybenzofuran-2-carboxylate is methylated with methyl iodide in presence of lithium, sodium, and/or potassium; with ethanol, *n*-propanol and/or butanol, lithium is used as a catalyst. The rate of *C*-alkylation decreases more rapidly than *O*-alkylation in the solvent series: MeOH, EtOH, PrOH, BuOH. In methanol the rate of *C*-alkylation increases in the series: Li, Na, K, and *O*-alkylation in the series Li, Na or K.[103]

I. Miscellaneous Reactions

Interaction of ethyl 6-benzyloxybenzofuran-2-carboxylate and methylmagnesium iodide results in the formation of 6-benzyloxy-2-(1-hydroxyisopropyl)benzofuran;[104] however treatment of 6-benzyloxy-2-acetylbenzofuran with an excess of the same reagent affords 6-benzyloxy-2-isopropenylbenzofuran, which on simultaneous hydrogenation and debenzylation yields 6-hydroxy-2-isopropyl-2,3-dihydrobenzofuran.

Mannich bases have been obtained upon treatment of ethyl 6-hydroxy-3-methyl-2-benzofurancarboxylate with paraformaldehyde and secondary amines in ethanol (**92**).[105]

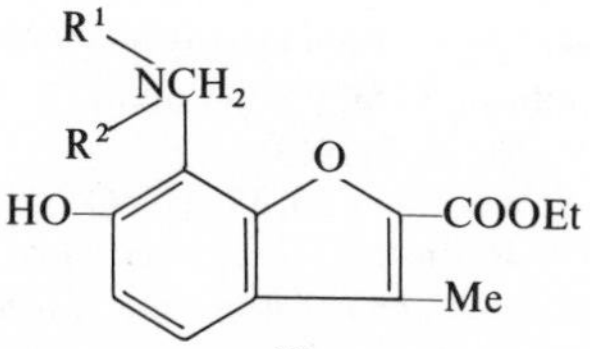

92

Degradation of 2-benzofurancarboxamide with sodium hypochlorite in presence of methanol resulted in the formation of methyl benzofuran-carbamate which, on heating with concentrated hydrochloric acid, yielded *o*-hydroxyphenylacetic acid lactone.[106]

References

1. W. H. Perkin, *J. Chem. Soc.*, **8,** 368 (1870); **9,** 37 (1871).
2. M. S. Newman and C. K. Dalton, *J. Org. Chem.*, **30,** 4126 (1965).
3. V. A. Zagorevskii, V. L. Savel'ev, N. V. Dudykina, and S. L. Portnova, *Zh. Org. Khim.*, **3**(3), 568 (1967); *Chem. Abstr.*, **67,** 11396 (1967).
4. M. Descamps, J. van der Elst, and F. Binon, *Chim. Therap.*, **87** (1966); *Chem. Abstr.*, **64,** 17518 (1966).
5. A. L. Mndzhoyan and M. A. Kaldrikyan, *Izv. Akad. Arm. SSSR, Khim. Nauki*, **13**(1), 55 (1960); *Chem. Abstr.*, **57,** 8524 (1962).
6. A. L. Mndzhoyan and M. A. Kaldrikyan, *Izv. Arm. SSSR, Khim. Nauki*, **13,** 425 (1960); *Chem. Abstr.*, **55,** 27254 (1961).
7. V. K. Ahluwalia, A. C. Mehta, and T. R. Seshadri, *Tetrahedron*, **4,** 271 (1958).
8. T. R. Seshadri, *Tetrahedron*, **6,** 169 (1959).
9. S. Tanaka, *J. Am. Chem. Soc.*, **73,** 872 (1951).
10. St. v. Kostanecki and J. Tambor, *Chem. Ber.*, **42,** 901 (1909).
11. A. v. Graffenried and St. v. Kostanecki, *Chem. Ber.*, **43,** 2155 (1910).
12. A. F. Shepard, N. R. Winslow, and J. R. Johnson, *J. Am. Chem. Soc.*, **52,** 2083 (1930).
13. R. v. Stoermer and M. Schäffer, *Chem. Ber.*, **36,** 2863 (1903).
14. S. Kawai, T. Nakamura, and N. Sugiyama, *Chem. Ber.*, **72,** 1146 (1939).
15. S. Kawai, T. Nakamura, and M. Yoshida, *Chem. Ber.*, **73,** 581 (1940).
16. R. T. Foster, A. Robertson, and A. Bushra, *J. Chem. Soc.*, (**1948**), 2254.
17. M. Descamps, J. van der Elst, and F. Binon, *Ind. Chim. Belge*, (**1967**), 32; *Chem. Abstr.*, **70,** 68028 (1969).
18. R. Rumpf and Ch. Gansser, *Helv. Chim. Acta*, **37,** 435 (1954).
19. W. Grubenmann and H. Erlenmeyer, *Helv. Chim. Acta*, **31,** 78 (1948).
20. G. Köbrich and H. Trapp, *Chem. Ber.*, **99,** 680 (1966).
21. R. Huisgen, G. Binsch, and H. König, *Chem. Ber.*, **97,** 2884 (1964).
22. G. Binsch, R. Huisgen, and H. König, *Chem. Ber.*, **97,** 2893 (1964).
23. G. G. Yakobson, T. D. Petrova, L. I. Kann, T. I. Savchenko, A. K. Petrov, and N. N. Vorozhisov, Jr., *Dokl. Akad. Nauk SSSR*, **158**(4), 926 (1964); *Chem. Abstr.*, **62,** 2755 (1965).
24. T. D. Petrova, L. I. Kann, V. A. Barkhash, and G. G. Yakobson, *Khim. Geterotsikl. Soedin*, (**1969**), 778; *Chem. Abstr.*, **72,** 111180 (1970).
25. B. Cummins, D. M. X. Donnelly, E. M. Philbin, J. Swirski, T. S. Wheeler, and R. K. Wilson, *Chem. Ind. (London)*, (**1960**), 348.
26. D. M. X. Donnelly, J. F. K. Eades, E. M. Philbin, and T. S. Wheeler, *Chem. Ind. (London)*, (**1961**), 1453.
27. T. S. Wheeler, *Rec. Chem. Prog.*, **18,** 133 (1957); *Chem. Abstr.*, **52,** 376 (1958).
28. J. C. O'Connor, W. I. O'Sullivan, E. M. Philbin, and T. S. Wheeler, *Chem. Ind. (London)*, (**1962**), 1864; *Chem. Abstr.*, **58,** 2332 (1963).
29. J. N. Chatterjea, *J. Indian Chem. Soc.*, **33,** 175 (1956).
30. V. Titoff, H. Müller, and T. Reichstein, *Helv. Chim. Acta*, **20,** 883 (1937).
31. Y. Kawase and M. Takashima, *Bull. Chem. Soc. Japan*, **40**(5), 1224 (1967); *Chem. Abstr.*, **67,** 82013 (1967).

32. Y. Kawase and S. Takashima, Japan. Pat. 6,924,897 (1969); *Chem. Abstr.*, **72,** 43425 (1970).
33. H. v. Pechmann, *Chem. Ber.*, **21,** 3005 (1888).
34. M. Ikuta, *J. Prakt. Chem.*, **45**(2), 65 (1892).
35. E. Bernatek and T. Ledaal, *Acta Chim. Scand.*, **12,** 2053 (1958).
36. A. B. Terent'ev and A. P. Terent'ev, *Zh. Obshch. Khim.*, **26,** 560 (1956); *Chem. Abstr.*, **50,** 13860 (1956).
37. A. N. Grinev, N. K. Venevtseva, and A. P. Terent'ev, *Zh. Obshch. Khim.*, **28,** 1850 (1958); *Chem. Abstr.*, **53,** 1296 (1959).
38. S. A. Monti, *J. Org. Chem.*, **31,** 2669 (1966).
39. C. D. Nenitzescu, *Bull. Soc. Chim. Romania*, **II,** 37 (1929); *Chem. Abstr.*, **24,** 110 (1930).
40. G. Domschke, *J. Prakt. Chem.*, **32**(4), 144 (1966); *Chem. Abstr.*, **65** 12156 (1966).
41. I. G. Farbenindustrie, A-G., Brit. Pat. 461,189 (1937); *Chem. Abstr.*, **31,** 4830 (1937).
42. R. D. Desai and M. M. Gaitonde, *Proc. Indian Acad. Sci.*, **25A,** 364 (1947); *Chem. Abstr.*, **42,** 1915 (1948).
43. T. Matsumoto and K. Fukui, *Bull. Chem. Soc. Japan*, **30,** 3 (1957); *Chem. Abstr.*, **51,** 12891 (1957).
44. T. Reichstein, R. Oppenauer, A. Grüssner, R. Hirt, L. Rhyner, and C. Glatthaar, *Helv. Chim. Acta*, **18,** 816 (1935).
45. E. B. Knott, *J. Chem. Soc.*, (**1945**), 189.
46. M. Miyano, S. Muraki, T. Nishikubo, and M. Matsui, *Nippon Nogeikagaku Kaishi*, **34**(8), 678 (1960); *Chem. Abstr.*, **59,** 13928 (1963).
47. G. W. Holton, G. Parker, and A. Robertson, *J. Chem. Soc.*, (**1949**), 2049.
48. F. B. LaForge and L. E. Smith, *J. Am. Chem. Soc.*, **52,** 1091 (1930).
49. C. F. Koelsch and A. G. Whitney, *J. Am. Chem. Soc.*, **63,** 1762 (1941).
50. E. H. Huntress and R. T. Olsen, *J. Am. Chem. Soc.*, **70,** 2856 (1948).
51. W. J. Davidson and J. A. Elix, *Tetrahedron Lett.*, (**1968**), 4589; *Chem. Abstr.*, **69,** 96362 (1968).
52. L. I. Vereshchagin, S. P. Korshunov, S. L. Aleksandrova, and R. L. Bol'shedvorskaya, *Zh. Org. Khim.*, **1**(5), 960 (1965); *Chem. Abstr.*, **63,** 6944 (1965).
53. L. I. Vereshchagin, S. P. Korshunov, R. L. Bol'shedvorskaya, and T. V. Lipovich, *Zh. Org. Khim.*, **2**(3), 522 (1966); *Chem. Abstr.*, **65,** 7123 (1966).
54. F. Duro and P. Condorelli, *Ann. Chim. (Rome)*, **53**(11), 1582 (1963); *Chem. Abstr.*, **60,** 9123 (1964).
55. B. Eistert, "Newer Methods of Preparative Organic Chemistry," Interscience, New York, 1948, p. 513.
56. R. B. Wagner and J. M. Tome, *J. Am. Chem. Soc.*, **72,** 3477 (1950).
57. A. Mcgookin, A. Robertson, and W. B. Whalley, *J. Chem. Soc.*, (**1940**), 787.
58. J. N. Chatterjea, *Experientia*, **12,** 18 (1956); *Chem. Abstr.*, **50,** 15504 (1956).
59. D. S. Deorha and P. Gupta, *Chem. Ber.*, **97,** 3577 (1964).
60. D. S. Deorha and P. Gupta, *Indian J. Chem.*, **2**(11), 459 (1964).
61. D. S. Deorha and P. Gupta, *Rec. Trav. Chim. Pays-Bas*, **83**(9/10), 1056 (1964).
62. D. S. Deorha and P. Gupta, *J. Indian Chem. Soc.*, **41,** 371 (1964).
63. B. Sila and K. Seyda, *Rocz. Chem.*, **44**(6), 1319 (1970); *Chem. Abstr.*, **73,** 87709 (1970).
64. J. N. Chatterjea, *J. Indian Chem. Soc.*, **30,** 103 (1953).
65. F. M. Dean, D. S. Deorha, J. C. Knight, and T. Francis, *J. Chem. Soc.*, (**1961**), 327.
66. D. S. Deorha and P. Gupta, *Bull. Chem. Soc. Japan*, **39**(12), 2768 (1966); *Chem. Abstr.*, **66,** 65342 (1967).
67. J. N. Chatterjea and S. Srivastava, *Experientia*, **26**(12), 1291 (1970); *Chem. Abstr.*, **74,** 42234 (1971).
68. F. Pan and T-C. Wang, *J. Chinese Chem. Soc. (Taiwan), Ser. II*, No. 8, 374 (1961); *Chem. Abstr.*, **58,** 1388 (1963).

69. D. S. Deorha and P. Gupta, *Chem. Ber.*, **99,** 2063 (1966).
70. F. Boyer, *C. R. Acad. Sci. Paris, Ser. C,* **264**(18), 1546 (1967); *Chem. Abstr.*, **67,** 64142 (1967).
71. F. Boyer and C. Fournier, *C. R. Acad. Sci. Paris, Ser. C*, **270**(11), 1027 (1970); *Chem. Abstr.*, **72,** 132416 (1970).
72. D. S. Deorha and P. Gupta, *Chem. Ind. (London)*, (**1964**), 1862; *Chem. Abstr.*, **62,** 508 (1965).
73. N. P. Buu-Hoi and G. Saint-Ruf, *Bull. Soc. Chim. Fr.*, (**1966**), 624; *Chem. Abstr.*, **64,** 17524 (1966).
74. E. Bisagni and R. Royer, *Bull. Soc. Chim. Fr.*, (**1962**), 86; *Chem. Abstr.*, **56,** 14192 (1962).
75. G. V. Bhide, N. L. Tikotkar, and B. D. Tilak, *Tetrahedron*, **10,** 223 (1960).
76. F. Lauria, P. N. Giraldi, P. N. Francavilla, G. Ninnini, and W. Logeman, U.S. Pat. 3,452,085 (1969); *Chem. Abstr.*, **71,** 70480 (1969).
77. S. P. A. Carlo Erba, Brit. Pat. 1,109,527 (1968); *Chem. Abstr.*, **69,** 35925 (1968).
78. R. Royer, E. Bisagni, A. Cheutin, J. P. Marquet, and M. L. Desvoye, *Bull. Soc. Chim. Fr.*, (**1964**), 309.
79. C. Kaiser and Ch. L. Zirkle, U.S. Pat. 3,010,971 (1960); *Chem. Abstr.*, **56,** 15484 (1962).
80. B. Sila, *Rocz. Chem.*, **38**(9), 1387 (1964); *Chem. Abstr.*, **63,** 3030 (1965).
81. B. Sila, *Rocz. Chem.*, **39**(10), 1535 (1965); *Chem. Abstr.*, **64,** 19756 (1966).
82. B. Sila, *Rocz. Chem.*, **41**(1), 157 (1967); *Chem. Abstr.*, **68,** 13337 (1968).
83. B. Sila, *Rocz. Chem.*, **41**(2), 399 (1967); *Chem. Abstr.*, **68,** 40046 (1968).
84. H. Erlenmeyer and W. Grubenmann, *Helv. Chim. Acta*, **30,** 297 (1947); *Chem. Abstr.*, **41,** 3087 (1947).
85. R. v. Stoermer and G. Calov, *Chem. Ber.*, **34,** 770 (1901).
86. E. W. Smith, *Iowa State Coll. J. Sci.*, **12,** 155 (1937); *Chem. Abstr.*, **32,** 2938 (1938).
87. V. S. Salvi and S. Sethna, *J. Indian Chem. Soc.*, **44**(2), 135 (1967); *Chem. Abstr.*, **67,** 82016 (1967).
88. V. S. Salvi and S. Sethna, *J. Indian Chem. Soc.*, **45**(2), 433 (1968); *Chem. Abstr.*, **69,** 106375 (1968).
89. C. A. Giza and R. L. Hinman, *J. Org. Chem.*, **29,** 1453 (1964).
90. A. L. Mndzhoyan and A. A. Aroyan, *Izv. Akad. Nauk. Arm. SSSR, Khim. Nauki*, **11**(1), 45 (1958); *Chem. Abstr.*, **53,** 3185 (1959).
91. E. L. Martin, U.S. Pat. 2,754,286 (1957); *Chem. Abstr.*, **51,** 919 (1957).
92. E. I. du Pont de Nemours Co., Brit. Pat. 705, 950 (1954); *Chem. Abstr.*, **49,** 2233 (1955).
93. F. Bordin, R. Bevilaequa, and F. Dalbeni-Sala, *Gazz. Chim. Ital.*, **99**(11), 1177 (1969); *Chem. Abstr.*, **72,** 90167 (1970).
94. A. Feinstein, P. H. Gore, and G. L. Reed, *J. Chem. Soc., B*, (**1969**), 205.
95. D. Papa, E. Schwenk, and H. F. Ginsberg, *J. Org. Chem.*, **16,** 253 (1951).
96. M. C. Ford and W. A. Waters, *J. Chem. Soc.*, (**1951**), 824.
97. E. Bernatek and I. Bϕ, *Acta Chem. Scand.*, **13,** 337 (1959); *Chem. Abstr.*, **54,** 24621 (1960).
98. P. K. Ramachandran, A. T. Tefteller, G. O. Paulson, T. Cheng, C. T. Lin, and W. J. Horton, *J. Org. Chem.*, **28,** 398 (1963).
99. W. J. Horton and M. G. Stout, *J. Org. Chem.*, **26,** 1221 (1961).
100. E. J. King, *J. Am. Chem. Soc.*, **49,** 562 (1927).
101. B. Krishnaswamy and T. R. Seshadri, *Proc. Indian. Chem. Sci.*, **16A** 151 (1942).
102. V. S. Salvi and S. Sethna, *J. Indian Chem. Soc.*, **45,** 439 (1968); *Chem. Abstr.*, **69,** 106429 (1968).
103. A. Brändström and I. Forsblad, *Acta Chem. Scand.*, **11,** 914 (1957).
104. J. B. D. MacKenzie, A. Robertson, and (in part) A. Bushra and R. Towers, *J. Chem. Soc.*, (**1949**), 2057.
105. S. Kumar and S. S. Joshi, *J. Indian Chem. Soc.*, **41,** 737 (1964); *Chem. Abstr.*, **62,** 11756 (1965).
106. I. J. Rinkes, *Rec. Trav. Chim. Pays-Bas*, **51,** 349 (1932).

CHAPTER IV

Hydrogenated Benzofurans

1. Dihydrobenzofurans

A. Alkyl- (or Aryl-) Substituted 2,3-Dihydrobenzofurans

The system followed in naming the substituted dihydrobenzofurans is that of Stoermer and Göhl;[1] numbering the positions or their respective nuclei. The inconsistency of starting to number from the α-carbon in 2,3-dihydrobenzofuran is obvious; and since there can be no substitution on the oxygen, it seems preferable to do this. However, both methods are in use (see chroman).

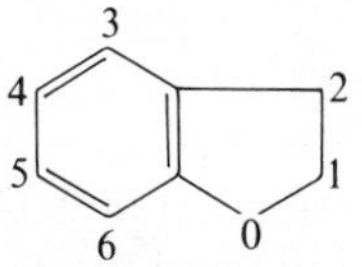

"Dihydrobenzofuran"

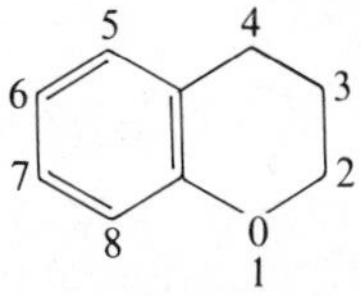

"Chroman"

a. Preparation

(i) CONDENSATION OF PHENOLS WITH UNSATURATED ALCOHOLS. Various methods have been applied for the preparation of 2,3-dihydrobenzofurans.[2-10] Many of the methods reported for preparation of alkyl- and aryl-substituted 2,3-dihydrobenzofurans do not appear to be completely satisfactory from the standpoint of yield, directions, and generality.

Niederl and Storch[11] have reported the condensation of an unsaturated alcohol and phenol as a simple process that is widely applicable for the preparation of 2,3-dihydrobenzofurans. The synthesis involves heating of allyl alcohol with phenols or cresols in presence of a sulfuric–acetic acid mixture for 5 hr; the polymer thus obtained is decomposed pyrolytically and the alkyl-substituted dihydrobenzofurans are formed along with small quantities of isopropenylcresol. Equation 1 has been presented to illustrate a possible mechanism for the formation of (4). Assuming alpha addition of the acid radical has occurred (the structure of the final products rules out the possibility of beta addition), the resulting ether would be *β*-(*p*-cresoxy)propylene glycol (**1**). In the presence of sulfuric acid this transition is accompanied by loss of water from the side chain. This loss may either precede or follow the rearrangement. In any case the resulting product, an alkali-insoluble polymer (**2**), was obtained. On distillation, depolymerization takes place and 3,5-dimethyl-2,3-dihydrobenzofuran (**4**), together with a small quantity of **3** are obtained.

$$CH_2{=}CHCH_2OH + H_2SO_4 \longrightarrow HOCH_2CH(Me)\boxed{OSO_2OH + H}OC_6H_4Me\text{-p} \longrightarrow$$

$$\underset{\mathbf{1}}{HOCH_2CH(Me)OC_6H_4Me\text{-p}} \xrightarrow[\text{rearrangement}]{-H_2O} \underset{\mathbf{2}}{(C_{10}H_{20}O)_x} \xrightarrow{\text{pyrolysis}}$$

$$\mathbf{3} + \mathbf{4} \qquad (1)$$

(ii) Cyclization of *o*-Allylphenols. Formation of 2,3-dihydrobenzofurans through cyclization of *o*-allylphenols[12,13] and their methyl ethers[14] by acids has long been known. Cyclization catalysts used are pyridinium chloride,[15] hydrobromic–acetic mixture,[16] sulfuric acid–hydrochloric acid,[17] and potassium bisulfate.[18] In general, electrophilic reagents would be expected to effect such cyclizations; for example, mercuric acetate does react with *o*-allylphenol to afford 2-acetoxymercurimethyl-2,3-dihydrobenzofuran.[19] There is also some evidence that bromine[20] and hypochlorous acid[21] can bring about the formation of 2-halomethyl-2,3-dihydrobenzofurans in similar fashion. Tarbell et al.[22-24] have studied cyclization of 2-(α-methylallyl)-3-methyl-4,6-dichlorophenol (**5**) and its methyl ether (**6**) by acids and iodine monochloride. Both are cyclized to give a mixture of *trans*-**7** and *cis*-**8** forms by sulfuric acid (96.7%) or $0.02M$ solution of *p*-toluenesulfonic acid in benzene at reflux. Addition of iodine monochloride to **6** in carbon tetrachloride produced *trans*-iodomethyl compound (**9**). Treatment of the latter with alcoholic potassium hydroxide solution resulted in the formation of 5,7-dichloro-2,3,4-trimethylbenzofuran (**10**) (Eq. 2).

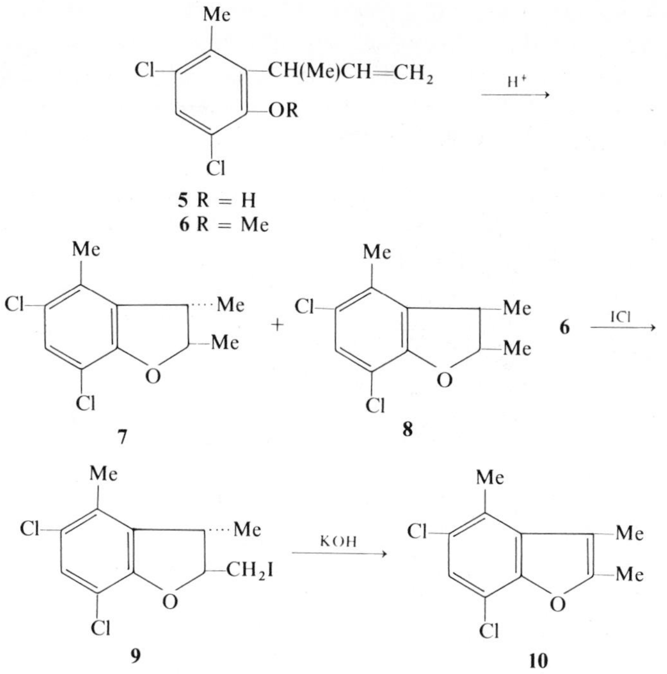

5 R = H
6 R = Me

$\xrightarrow{H^+}$ **7** + **8**

6 $\xrightarrow{ICl}$ **9** $\xrightarrow{KOH}$ **10** (2)

Cyclization of *o*-allyphenols and related compounds to 2,3-dihydrobenzofurans and chromans may be regarded as special case of addition to an olefinic double bond.[25] Since Kharasch et al.[26] have found that a trace of peroxide often governs the direction of addition, the attempt is made to direct ring closure by means of the "peroxide effect." Formation of dihydrobenzofuran ring would be anticipated under peroxide-free conditions and since *o*-allylphenols may act similarly to quinol, it forms with hydrobromic acid 2-methyl-2,3-dihydrobenzofuran, whether or not peroxide is present. On the other hand, air, ascaridole, and/or benzoyl peroxide failed to favor the formation of chroman. The acetate derivative of *o*-allylphenol and hydrobromic acid yielded 2-methyl-2,3-dihydrobenzofuran; however in the presence of ascaridole or other peroxides chroman is produced. In the same way, the acetate of *o*-allylphenol gave 2,5-dimethyl-2,3-dihydrobenzofuran in the absence, and 6-methylchroman in the presence of peroxide.

Alkyl-substituted 2,3-dihydrobenzofurans can be readily produced upon reflux of allyl bromides with phenols in presence of potassium carbonate followed by rearrangement of the allylphenyl ethers thus obtained and ring closure with hydrobromic acid. In large scale reactions, substitution of allyl chlorides for the corresponding bromides is economically desirable.

Claisen rearrangement of allylphenyl ethers (**11**)[27] establishes a convenient route for the preparation of *o*-allylphenols (**12**). Solvent effects on Claisen rearrangement of β-methylallyl ether (**13**) have been studied.[28] Neutral solvents afford the expected allylic compound **14** as the principal product, contaminated with both the isobutenyl (**15**) and the cyclic isomer (**16**). An acidic solvent, 2,6,-xylenol, led to the formation of the dihydrobenzofuran (**16**). *o*-(β-Methylallyl)phenol is an intermediate in the formation of **16**; this is supported by complete conversion of **14** to **16** by using 2,6-xylenol as a solvent (Eq. 3).

$$C_6H_5OH \xrightarrow[K_2CO_3,\ Me_2CO]{RCH{=}CHCH_2X} C_6H_5OCH_2CH{=}CHR\ (\mathbf{11}) \xrightarrow{heat} o\text{-}HOC_6H_4CH(R)CH{=}CH_2\ (\mathbf{12})$$

$$C_6H_5OCH_2C(Me){=}CH_2\ (\mathbf{13}) \longrightarrow o\text{-}HOC_6H_4CH_2C(Me){=}CH_2\ (\mathbf{14}) + o\text{-}HOC_6H_4CH{=}CMe_2\ (\mathbf{15}) + \text{2,2-dimethyl-2,3-dihydrobenzofuran}\ (\mathbf{16}) \tag{3}$$

Basic solvents, for example, 2,6-xylidine, led to the expected phenol (**14**) after the first half-hour of reflux at the end of the reaction; however **15** and **16** were the principal products.

Cyclization of *o*-allyl-*p*-*t*-butyl- (or *t*-amyl)phenol with pyridine hydrochloride or hydrobromic acid in acetic acid solution affords 5-*t*-butyl- (**17**) and 5-*t*-amyl-2-methyl-2,3-dihydrobenzofuran (**18**), respectively.[29,30]

$$4\text{-}R\text{-}2\text{-}(CH_2CH{=}CH_2)C_6H_3OH \longrightarrow \text{5-R-2-}(CH_2HgAc)\text{-2,3-dihydrobenzofuran} \longrightarrow \text{5-R-2-Me-2,3-dihydrobenzofuran}$$

17 R = t-C_4H_9
18 R = t-C_5H_{11}

In the reaction of rearranged methylallylphenyl ethers, alkali-insoluble products are always present. If heating is continued too long, appreciable quantities of dimethyl-2,3-dihydrobenzofurans are produced by isomerization of methylallylphenol; these are identical with those formed by action of pyridine hydrochloride on methylallylphenol.[31]

Aqueous sodium hydroxide solution has been successfully used to bring about the conversion of the rearranged products, obtained upon heating with ascanatite, 2-methyl-6-allylphenol, *o*-allylguiacol, and *o*-allyl-6-ethoxyphenol, to 2,7-dimethyl-, 2-methyl-7-methoxy-, and 2-methyl-7-ethoxy-2,3-dihydrobenzofuran, respectively.[32] Similar cyclization of the acid phthalate of **19** with aqueous sodium hydroxide solution has been reported to afford 3-phenyl-2,3-dihydrobenzofuran (**20**).[33]

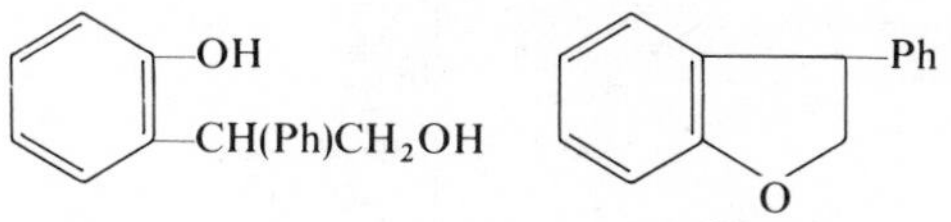

OH
CH(Ph)CH_2OH
19

Ph
O
20

Allylic alcohols, with formic acid alone or with zinc chloride, afford allylation without rearrangement; this is followed by cyclization.[34] However methylvinylcarbinol and methylvinylcarbinyl chloride are exceptions; they undergo rearrangement during the course of reaction (Eq. 4).[35]

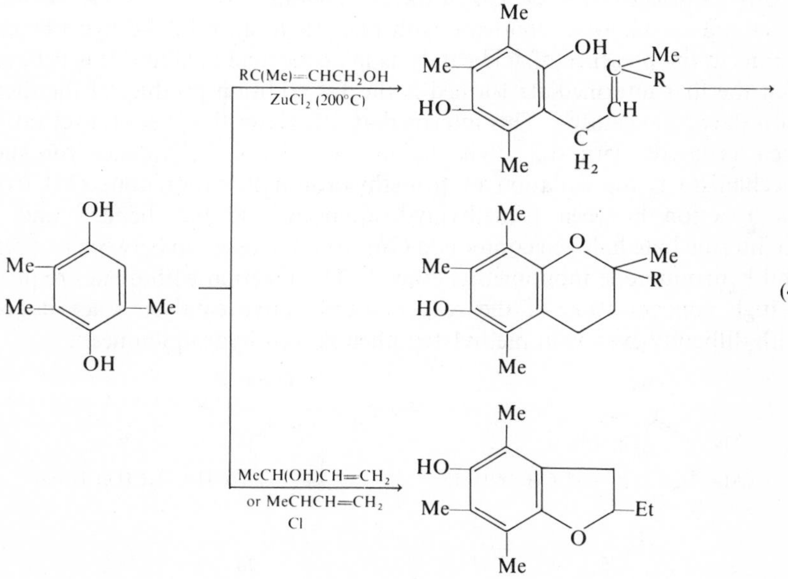

RC(Me)=CHCH_2OH
$ZuCl_2$ (200°C)

MeCH(OH)CH=CH_2
or MeCHCH=CH_2
Cl

(4)

Cresols condense with vinylcarbinols in presence of phosphoric acid to give monosubstituted cresols with isomerized γ-substituted allyl groups. The neutral fractions afford 2-isopropyl-6-methyl-, 2-isobutyl-, and 2-cyclohexyl-2,3-dihydrobenzofuran.[36]

Compound **21**, in boiling diethylaniline, undergoes the so-called "anomalous Claisen rearrangement," to furnish **22** and **23**. The former product (**22**) is readily cyclized by action of hydrobromic–acetic acid mixture to yield **24** (Eq. 5).[37]

$OCH_2C(Me)=CH_2$; Me; Me — **21** —[$PhNEt_2$ (185°C, 16 hr)]→ Me; $CH_2C(Me)=CH_2$; OH; Me — **22** + $CH=CMe_2$; Me; OH; Me — **23**

22 —[HBr (48%)/AcOH, (20°C, 24 hrs.)]→ Me; Me; Me; Me; O — **24** (5)

(iii) Condensation of Conjugated Dienes with Phenols. Conjugated dienes, likewise, condense with phenols to afford 2,3-dihydrobenzofurans in the presence of acid catalysts in acetic acid medium. It is believed that the first intermediate formed is the 1,4 addition product of the diene with the acidic catalyst. This intermediate alkylates the phenol; cyclization then proceeds with the allylic halide or alcohol.[38] Evidence for such mechanism is the isolation of trimethylcrotonylhydroquinone (**25**) from the reaction between trimethylhydroquinone and butadiene,[35] and of an intermediate halogen compound (**26**) from the reaction between isoprene and hydroquinone monomethyl ether.[39] The reaction with dienes requires a high concentration of the reagents and active catalyst, since it goes with difficulty even with methylated phenols and hydroquinones.

OH; Me; Me; Me; $CH_2CH=CHMe$; OH — **25**

OMe; $CH_2CH_2CHXCHMe_2$; OH — **26**

5-Methyl-2-ethyl-2,3-dihydrobenzofuran constitutes the principal reaction product of the catalytic alkenylation of *p*-cresol with butadiene in presence of zinc chloride at 120 to 150°C.[40] Similarly, with *m*-cresol, 6-methyl-2-ethyl-2,3-dihydrobenzofuran is obtained. Piperylene in presence of ethyl hydrogen sulfate alkenylates *p*- and *m*-cresol to yield 3,5-dimethyl- and 3,6-dimethyl-2-ethyl-2,3-dihydrobenzofuran, respectively.[41]

(iv) Interaction of Substituted Styrene Oxides with Phenoxide Ion. The acid-catalyzed reaction of *p*-methoxystyrene oxide with phenoxide ion[42] simulates that with styrene oxide,[43] nitrostyrene oxide,[44,45] and/or 2,4,6-trimethylstyrene oxide, and proceeds as illustrated in Eq. 6 for the latter case.

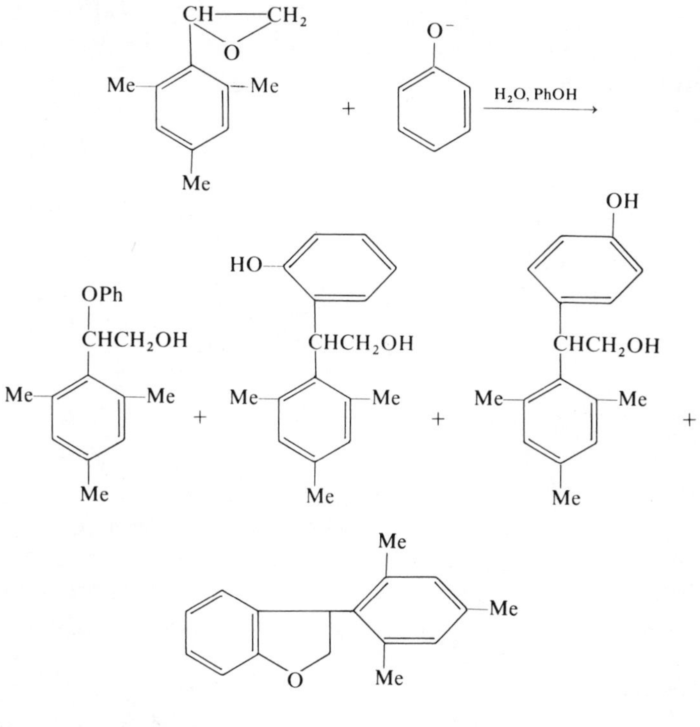

$\xrightarrow{H_2O,\ PhOH}$ (6)

(v) Alkenylation of Phenols with Substituted Acetylenes. Substituted acetylenes react with phenol to bring about alkenylation of the latter; cyclization of the alkenylated products results in the formation of 2,3-dihydrobenzofurans, for example, 3-*t*-butyl-[46] and 3-isopropyl-2,3-dihydrobenzofuran.[47] 3-Propargyl-4-hydroxy-6-methylpyron-(2)-5-carboxylic acid ethyl ester (**27**) condenses with salicylaldehyde to yield **28**.[48]

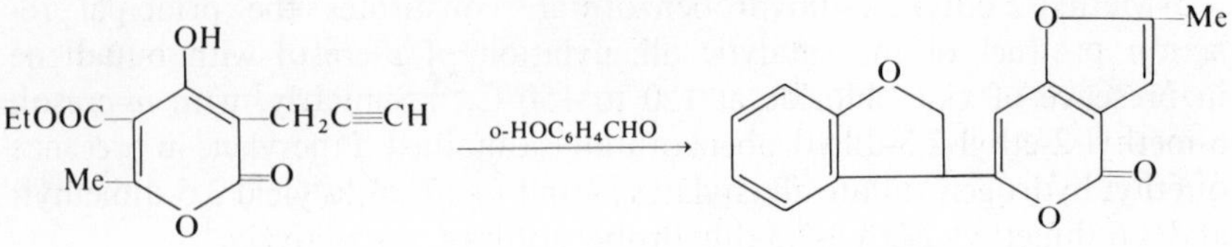

27 → (o-HOC_6H_4CHO) 28

(vi) INTERACTION OF ALLYL DIPHENYLPHOSPHATE WITH PHENOL. Allyl diphenylphosphate (**29**), when heated with excess of phenol, gives diphenylhydrogenphosphate (**30**) together with 2-methyl-2,3-dihydrobenzofuran (**31**) (Eq. 7).[49] It has been shown that the intermediate *o*-allylphenol (**32**), in presence of **30**, cyclizes readily to give **31**. Moreover, rearrangement and cyclization of the ether and of *p*-allylphenol to yield the same **31** appear to be examples of acid-catalyzed aromatic rearrangements. When allylphenyl ether is heated with **30**, it gives only **31**.

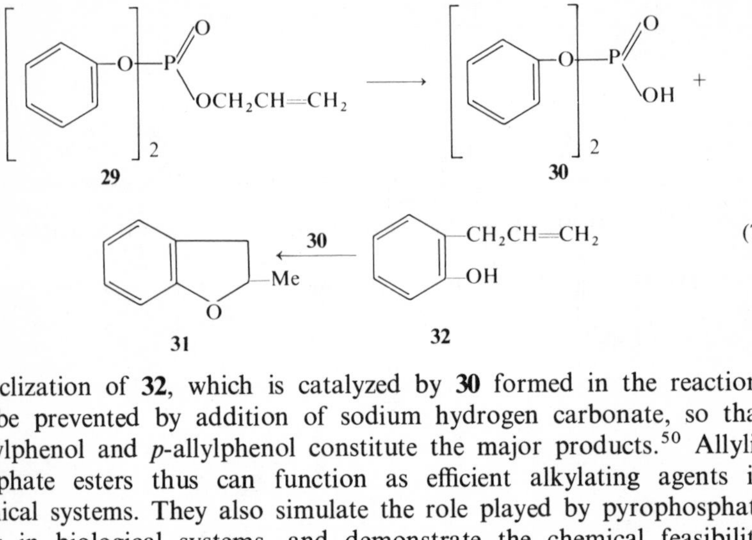

$$(PhO)_2P(=O)OCH_2CH{=}CH_2 \;(\mathbf{29}) \longrightarrow (PhO)_2P(=O)OH \;(\mathbf{30}) \; + \quad (7)$$

$$\mathbf{31} \xleftarrow{\mathbf{30}} o\text{-}CH_2CH{=}CH_2\text{-}C_6H_4OH \;(\mathbf{32})$$

Cyclization of **32**, which is catalyzed by **30** formed in the reaction, can be prevented by addition of sodium hydrogen carbonate, so that *o*-allylphenol and *p*-allylphenol constitute the major products.[50] Allylic phosphate esters thus can function as efficient alkylating agents in chemical systems. They also simulate the role played by pyrophosphate esters in biological systems, and demonstrate the chemical feasibility of the biogenic hypotheses which have been suggested.[51]

(vii) OXIDATION OF *o*-ALLYLPHENOLS. Oxidation of *o*-allylphenol with thallium acetate in acetic acid medium for short periods gives almost quantitative yields of 2,3-dihydrobenzofuranylthallium diacetate; however when the reaction mixture was heated (80°C), 2-acetoxymethyl-2,3-dihydrobenzofuran is obtained.[52]

(viii) CYCLIZATION OF *β*-ARYLETHANOLS. The parent compound, 2,3-dihydrobenzofuran, is obtained either by cyclization of *o*-methoxyphenyl-ethanol with hydrobromic acid or cyclodehydration of β-phenoxyethanol by action of phosphorus pentoxide.[53] Reflux of β-(*m*-chlorophenyl)ethanol (**33**) and diethylamine in ether with phenyllithium, followed by hydrolysis of the reaction product, results in the formation of 2,3-dihydrobenzofuran which, when refluxed with Et_2NLi and followed by hydrolysis, yields *o*-vinylphenol (Eq. 8).[54]

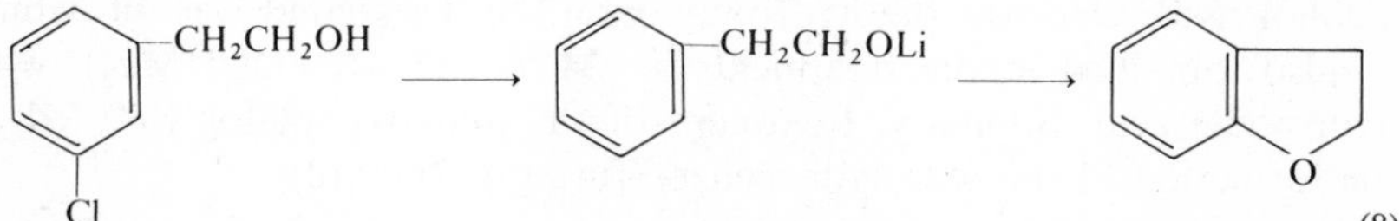

CH_2CH_2OH (Cl) **33** → CH_2CH_2OLi → O (8)

Decomposition of 1-phenoxy-2-propanol over activated alumina yields 2-methyl-2,3-dihydrobenzofuran.[55] Dehydration of 2-(*o*-hydroxyphenyl)-1-butanol with *p*-toluenesulfonic acid in benzene gives 3-ethyl-2,3-dihydrobenzofuran.[56] Dropwise addition of ω-chloro-4-methyl-2-hydroxyacetophenone in tetrahydrofuran to methylmagnesium iodide solution resulted in an abnormal Grignard reaction which led to the formation of 2,2,6-trimethyl-2,3-dihydrobenzofuran (**34**) (Eq. 9).[57]

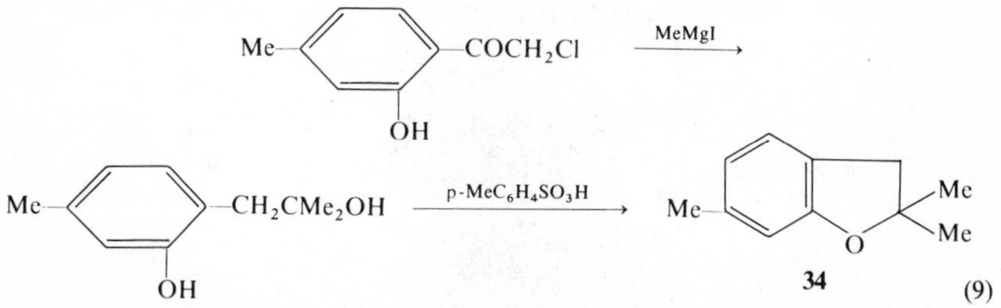

Me— —$COCH_2Cl$ (OH) $\xrightarrow{MeMgI}$

Me— —CH_2CMe_2OH (OH) $\xrightarrow{p\text{-}MeC_6H_4SO_3H}$ Me— O Me Me **34** (9)

(ix) Photochemical Formation. Photochemical intramolecular addition of 2,4-($H_2C{=}CHCH_2$)-*t*-BuC_6H_4OH or its formate gave 2-methyl-5-*t*-butyl-2,3-dihydrobenzofuran (**35**); however 2-($Me_2C{=}CHCH_2$)C_6H_4OH yielded 2,2-dimethylchroman (**36**).[58]

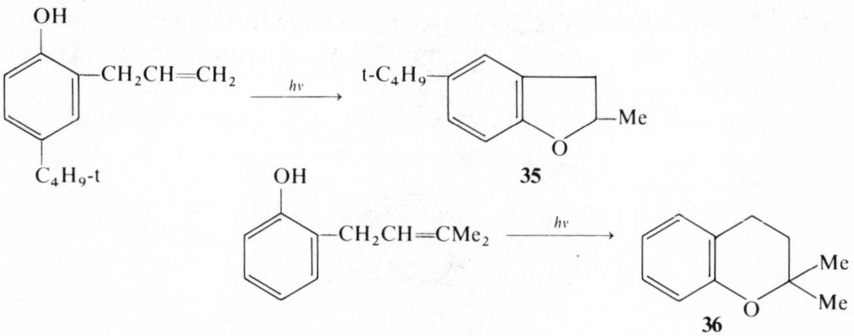

OH —$CH_2CH{=}CH_2$ (C_4H_9-t) $\xrightarrow{h\nu}$ t-C_4H_9— O —Me **35**

OH —$CH_2CH{=}CMe_2$ $\xrightarrow{h\nu}$ O Me Me **36**

A comparison of cyclization of *o*-allyl phenols with hydrobromic acid through irradiation with ultraviolet in benzene at 20°C for 10 hr has been reported.[59]

Irradition of 2,5-di-*t*-butylbenzoquinone[60,61] in methanol afforded **38** and **39**. Preparative gas–liquid chromatography of the reaction mother liquor gave two additional components, independent of the alcohol used. One was the hydroquinone (**37**), the second was **40**, which is also obtained upon treatment of **38** or **37** (R=$OCHMe_2$) with hydroiodic acid. Similarly, treatment of the dialkoxy analog (**39**) led to the formation of the tetrahydrobenzodifuran **41** (Eq. 10).

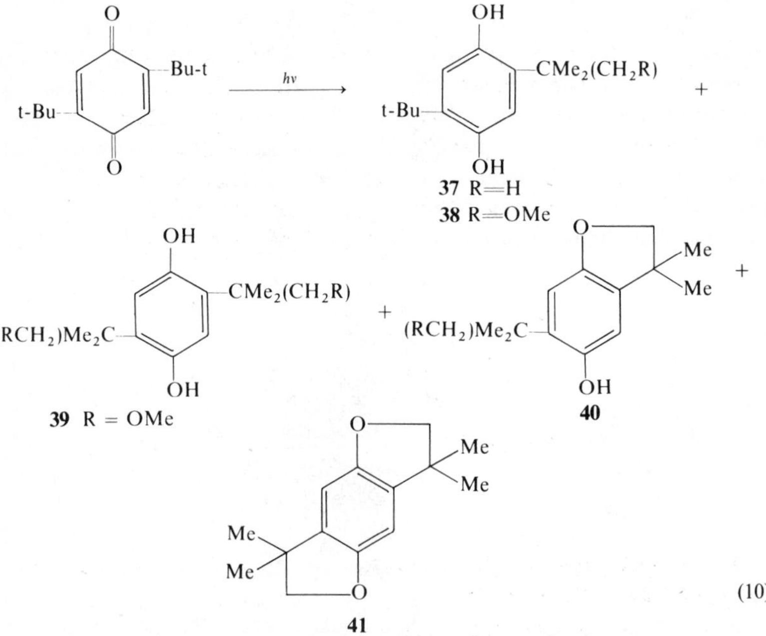

37 R=H
38 R=OMe
39 R = OMe
40
41

(10)

Miller[62] has reported that irradiation of 4-(2,3-dibromopropyl)-4-methyl-2,6-di-*t*-butylcyclohexa-(2,5)-dien-1-one (**42**) results in the formation of 2-bromomethyl-2,3-dihydro-4,7-di-*t*-butylbenzofuran (**43**) as the main product.

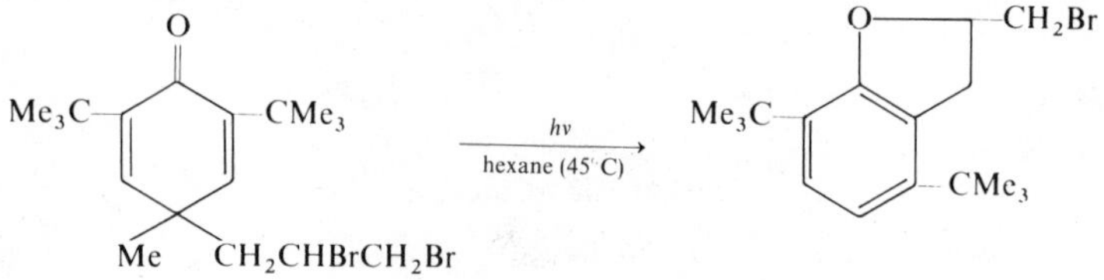

hv, hexane (45°C)

42 **43**

Synthesis of **43** has been achieved (Eq. 11).

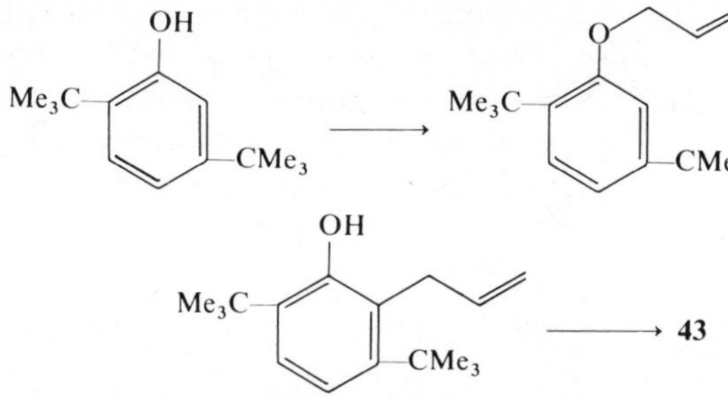

$\longrightarrow$ **43** (11)

Although **43** was the first product which could be isolated from photolysis of **42**, the linear dienone was the intermediate in its formation. This was confirmed by addition of bromine to the dienone (**45**) resulting from photorearrangement of 4-vinyl-4-methyl-2,6-di-*t*-butylcyclohexa-2,5-dienone (**44**).[64] The intermediate product of bromine addition retained the characteristic ultraviolet and infrared bands of a linearly conjugated cyclohexadienone, but these were lost on gentle heating. Methyl bromide and **43** were obtained. Formation of **43** from **42** appears to involve photochemical rearrangement of **42** to **46**, followed by thermal conversion of **46** to **43**, undoubtedly via the intermediate ketonium salt (**47**). The unusually easy cyclization to **47** is presumably facilitated by steric interference between the dibromopropyl group of **46** and the adjacent *t*-butyl group. Finally, attack of a bromide ion at the angular methyl group of **47** gives methyl bromide and **43** (Eq. 12).

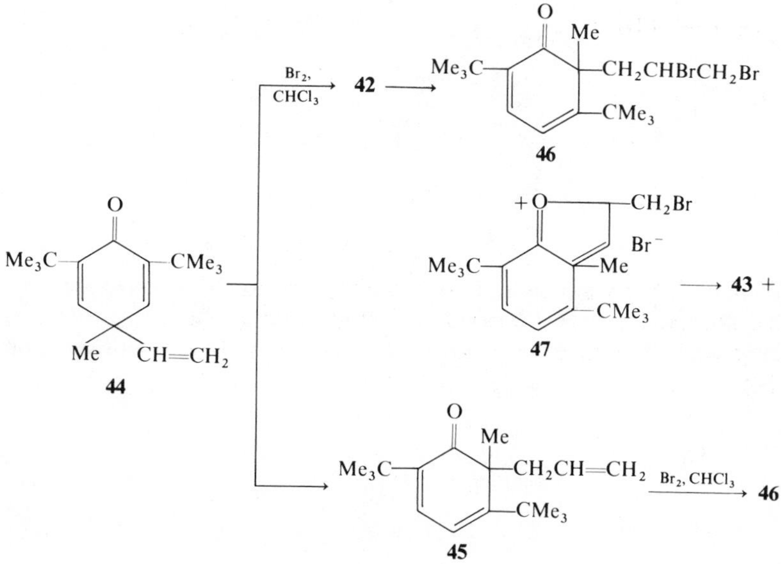

$\xrightarrow{Br_2, CHCl_3}$ **42** $\longrightarrow$ **46**

44 $\longrightarrow$ **47** $\longrightarrow$ **43** + MeBr (12)

45 $\xrightarrow{Br_2, CHCl_3}$ **46**

Photolysis of 2-(benzoyloxy)-4-(dodecyloxy)benzophenone and/or 2-isopropoxy-4-methoxybenzophenone[63] proceeded mainly by ring closure between the carbonyl carbon and the α-carbon of the 2 substituent to give 6-(dodecyloxy)-2,3-dihydro-3-benzofuranol and 2,2-dimethyl-6-methoxy-3-phenyl-2,3-dihydro-3-benzofuranol (**48**), respectively. Further photolysis of **48** gave only 2-hydroxy-4-methoxybenzophenone.

OH Ph MeO Me O Me

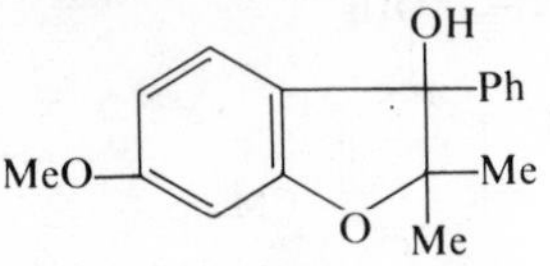

48

(x) MISCELLANEOUS. Mixtures of **50** and **51** are obtained by reduction of **52**. Treatment of **49** with titanium chloride, MeO^-, and/or iso-PrO^- affords **50** and **51** (Eq. 13).[65]

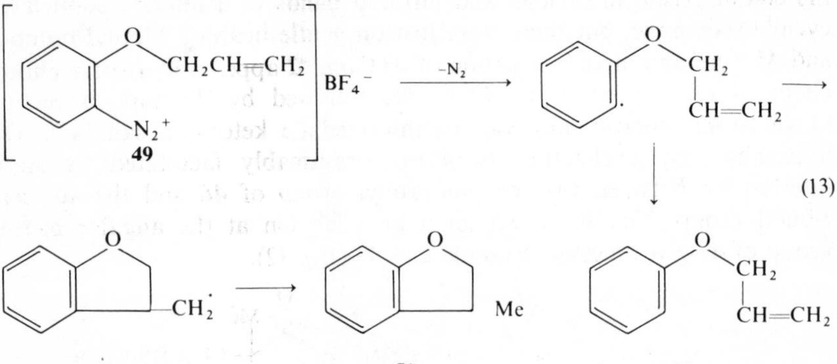

$$[\text{49: } o\text{-}(CH_2CH{=}CH_2O)C_6H_4N_2^+ \; BF_4^-] \xrightarrow{-N_2} \text{aryl radical } (OCH_2CH{=}CH_2) \longrightarrow \text{2,3-dihydrobenzofuran-3-}\dot{C}H_2 \longrightarrow \mathbf{51}\ (\text{3-Me}) ; \quad \text{aryl radical} \downarrow \mathbf{50}\ (C_6H_5OCH_2CH{=}CH_2) \tag{13}$$

OCH₂CH=CH₂ X

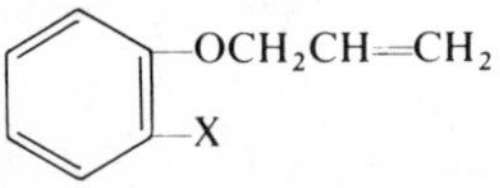

52

o-Dimethylallylphenol is readily transformed to 2-hydroxyisopropyl-2,3-dihydrobenzofuran,[66-68] upon treatment with acetic anhydride and peracetic acid. The epoxide seems to be an intermediate in this reaction (Eq. 14).

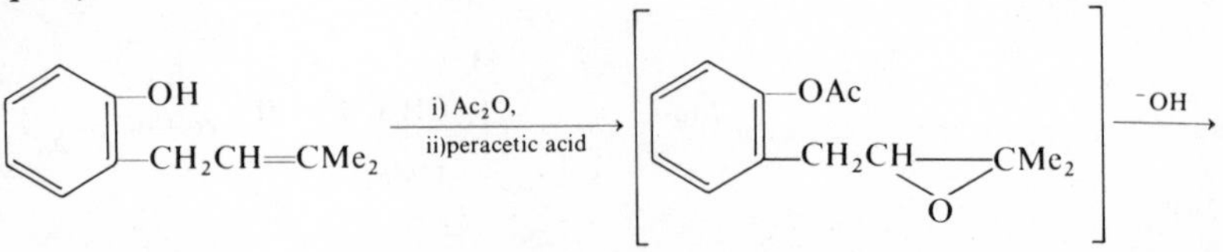

$$o\text{-}HOC_6H_4CH_2CH{=}CMe_2 \xrightarrow[\text{ii) peracetic acid}]{\text{i) } Ac_2O,} \left[o\text{-}AcOC_6H_4CH_2CH\overset{O}{—}CMe_2 \right] \xrightarrow{^-OH}$$

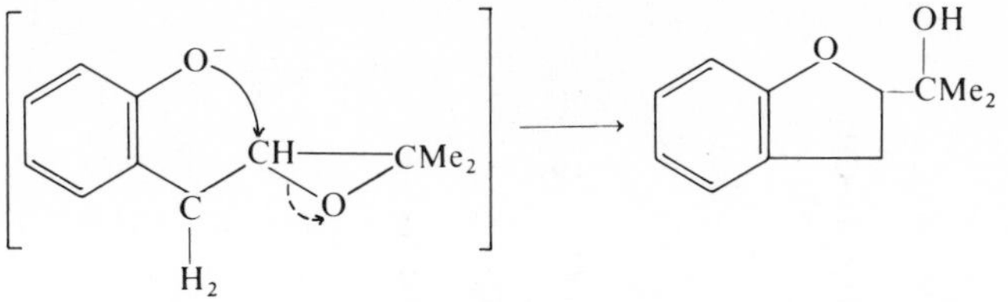

(14)

B. Halogen-Substituted 2,3-Dihydrobenzofurans

Zinc chloride-catalyzed rearrangement of allyl-2,6-dichlorophenyl ether (**53**) in nitrobenzene resulted in the formation of 2-allyl-4,6-dichlorophenol (**54**) along with some 2,6-dichlorophenol and 2-methyl-5,7-dichloro-2,3-dihydrobenzofuran (**55**). The yield of **55** is increased at the expense of **54**. This is in contrast to the thermal rearrangement of this ether, which gave mainly 4-allyl-2,6-dichlorophenol along with smaller amounts of 2-allyl-6-chlorophenol and 2-allyl-4,6-dichlorophenol (Chart 1).[69,70] Zinc chloride–allyl-2,6-dichlorophenylether complex (**56**) cleaved in the usual manner via route 1 under the influence of Lewis acids[71] to form 2,6-dichlorophenol. Simultaneously, there occurred a preferential intramolecular rearrangement of **56** via route 2 to form the intermediate dienone **57**.[72] The dienone then underwent migration of halogen through an allylic shift to a small extent via route 4, but chiefly via the agency of zinc chloride bridge (**58**) (route 3). Aromatization of **59** gave 2-allyl-4,6-dichlorophenol, which cyclized to the 2,3-dihydrobenzofuran in presence of zinc chloride. Halogen migration and elimination have also been observed in the Claisen rearrangement of 2,6-dichlorophenyl ether.[70]

Sulfuryl chloride reacts with flavone and 3-chloroflavone to give 2,3,3-trichloroflavanone (**60**) which, upon hydrolysis, yields 2,3-dichloro-2-phenyl-2,3-dihydrobenzofuran-3-carboxylic acid (**61**). Formation of **61** from **60** is a benzylic type of rearrangement.[73,74] Similarly, 6-chloroflavone reacts with sulfuryl chloride to yield 2,3,3,6-tetrachloroflavanone which, upon mild alkali treatment, yields 2,3,6-trichloro-2-phenyl-2,3-dihydrobenzofuran-3-carboxylic acid (**62**). However, under the same conditions, 2,3-dibromoflavanone afforded 2-benzoyl-3(2*H*)benzofuranone (Eq. 15).

2-(Halomethyl)-2,3-dihydrobenzofurans are obtained by addition of halogen to the double bond of the ester (**63**), followed by alcoholysis of the ester and cyclization with potassium hydroxide.[75,76] Treatment of 2-chloromethyl-2,3-dihydrobenzofuran (**64**) with magnesium in boiling ether and/or with metallic sodium in boiling toluene yields 60% *o*-allylphenol. However 2-methylbenzofuran (15%) and 2-(hydroxymethyl)-2,3-dihydrobenzofuran (50%) are obtained upon treatment of 2-bromo-

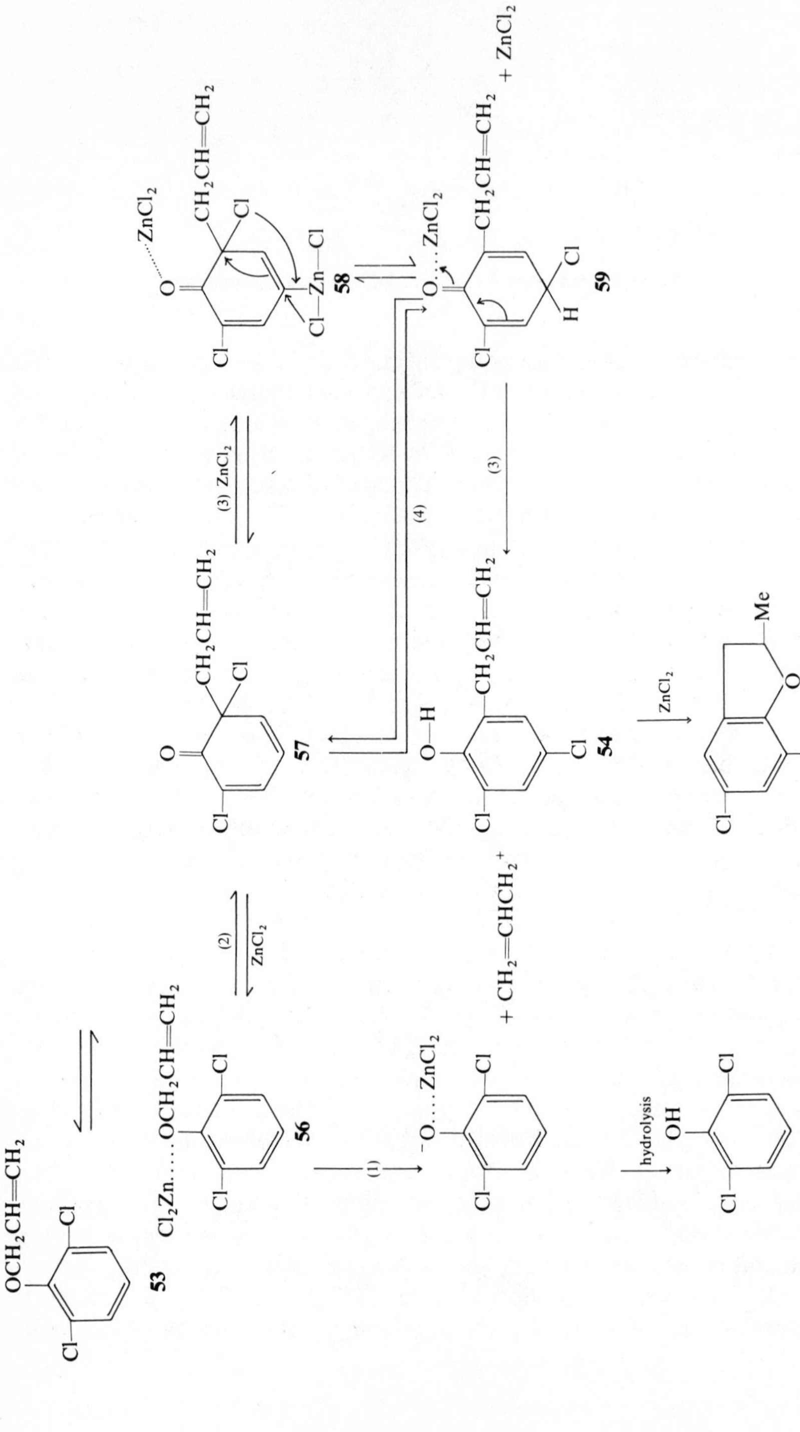

Chart 1

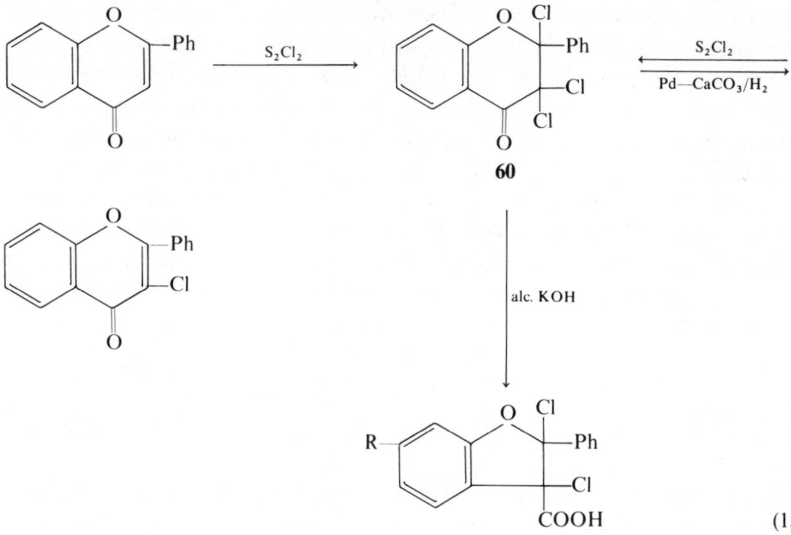

(15)

61 R = H
62 R = Cl

methyl-2,3-dihydrobenzofuran with potassium acetate and subsequent reaction with potassium hydroxide.

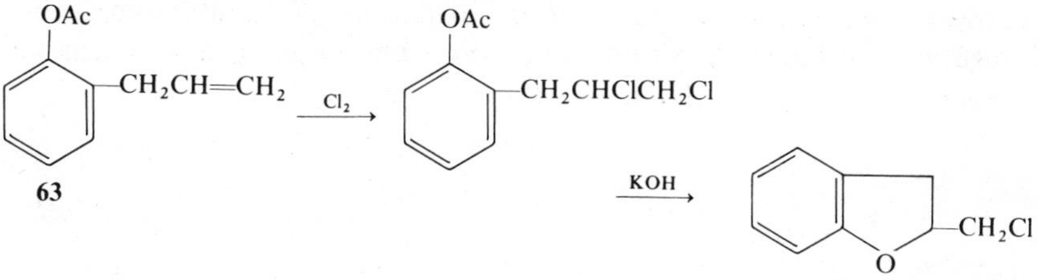

63

64

Iodination of 4,6-dichloro-2-propenylphenol yielded 5,7-dichloro-3-iodo-2-methyl-2,3-dihydrobenzofuran. Similarly, 2,4-dimethyl-6-propenylphenol gave 2,5,7-trimethyl-3-iodo-2,3-dihydrobenzofuran.[77]

C. Nitro-Substituted 2,3-Dihydrobenzofurans

The literature has very little to say about nitration of 2,3-dihydrobenzofurans or rules of orientation in this reaction. Nitro-substituted 2,3-dihydrobenzofurans, as a class of compounds, are virtually unknown. The only examples of nitration of 2,3-dihydrobenzofurans are those reported by Arnold and McCool,[78] and by Chatelus.[79] The first authors[78] nitrated 5-acetamido-2-methylchroman expecting to obtain 5-acetamido-2-methyl-6-nitro-2,3-dihydrobenzofuran; instead they obtained the 4-nitro derivative. Chatelus,[79] on the other hand, obtained a dinitrodihydrobenzofuran on nitration of 2,3-dihydrobenzofuran; the same was true with chroman, which resulted in isolating also dinitrochroman. Chatelus[79] proposed 5,7-dinitro-2,3-dihydrobenzofuran without support for the given structure. Two other arguments support the 5,7 orientation assignment for the dinitro 2,3-dihydrobenzofuran: one is the fact that the resonance hybrid of 2,3-dihydrobenzofuran shows high electron density at positions 5 and 7; the second is the observation that 2,3-dihydrobenzofuran is acetylated at position 5.[80,81]

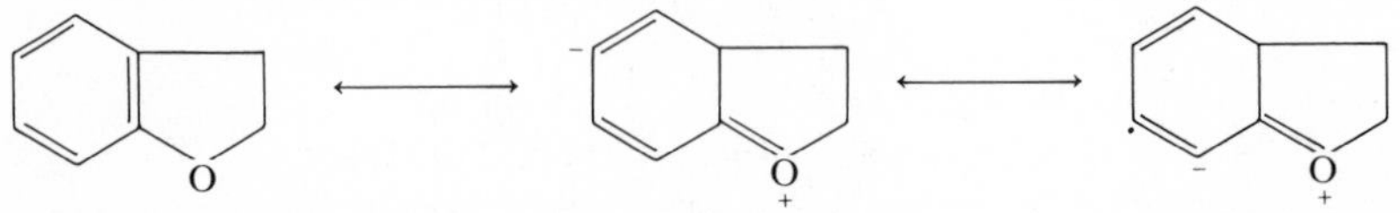

A number of alkylated 2,3-dihydrobenzofurans has been nitrated to give the 5,7-dinitro derivatives; however 3-phenyl-2,3-dihydrobenzofuran on nitration yielded 5,7-dinitro-3-(*p*-nitrophenyl)-2,3-dihydrobenzofuran.[82] Degradation of the nitration product of 2-methyl-2,3-dihydrobenzofuran afforded further evidence for the 5,7 orientation. This involved dehydrogenation, followed by oxidation of the dinitrobenzofuran thus obtained to the known dinitrosalicylic acid, for example, the formation of 3,5-dinitro-2-hydroxybenzoic acid upon oxidation of dinitro-substituted 2-methyl-2,3-dihydrobenzofuran with potassium permanganate in acetone in the presence of a small amount of water (Eq. 16).

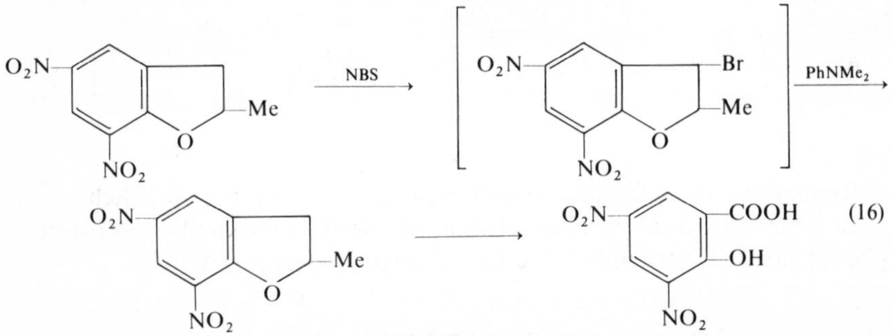

(16)

N-Bromosuccinimide reacts readily with simple 2,3-dihydrobenzofurans substituted in the dihydrofuran ring to give satisfactory yields of bromo-substituted 2,3-dihydrobenzofurans or their dehydrobromination products.[83,84] However it was found that 5,7-dinitro-2-methyl-2,3-dihydrobenzofuran reacted very sluggishly with this reagent except when the reaction was carried out with benzene as the solvent and for longer reaction periods.

2,4-Dinitrophenol reacts with acetone in presence of sodium ethoxide with the formation of the salt (**65**), which with methylamine, formaldehyde, and acetic acid, yields the bicyclic amine (**66**). Compound **65** is reduced by the action of sodium borohydride to the alcohol (**67**), which is converted by action of acid to 2,3-dihydrobenzofuran derivative (**68**) (Eq. 17).[85]

NO_2 HO NO_2 + O MeCMe $\xrightarrow{EtO^- Na^+}$ [NO_2^- H O CH$_2$COMe $=NO_2^-$ **65**] $\xrightarrow{NaBH_4}$

NO_2^- H O CH$_2$CH(OH)Me $=NO_2^-$ **67** 2 Na$^+$ $\xrightarrow{H^+}$ (17)

NO_2 RO Me O **68** R = H, R = Me

65 $\xrightarrow{MeNH_2,\ HCHO,\ H^+}$ O_2N NMe H O CH$_2$COMe NO_2 **66**

Treatment of the adduct **69** (M = Na or K), obtained from *sym*-trinitrobenzene, acetone, and alkali, with sodium borohydride effected the reduction of the carbonyl group and addition of the hydride ions to the aromatic nucleus. The resulting salt, **70**, upon treatment with bromine, gave, with reoxidation of the nucleus and removal of one nitro group, **71** (R = Me).[86] Treatment of **71** in pyridine with iodine resulted in the formation of 4,6-dinitro-2-methylbenzofuran (**72**) (Eq. 17). An analogous reaction, starting with cyclohexanone and/or cyclopentanone, afforded **73a** and **73b**, respectively (Eq. 18).[86]

NO_2 O_2N NO_2 $\longrightarrow$ [O_2N H CH$_2$COMe $^{\ominus}O_2N=$ NO_2 **69**] $\xrightarrow{M^+,\ NaBH_4}$

[$NO_2^{\ominus}$ H CH$_2$CH(OH)Me $^{\ominus}O_2N=$ $=NO_2^{\ominus}$ **70**] 3 Na$^{\oplus}$ $\xrightarrow{Br_2}$

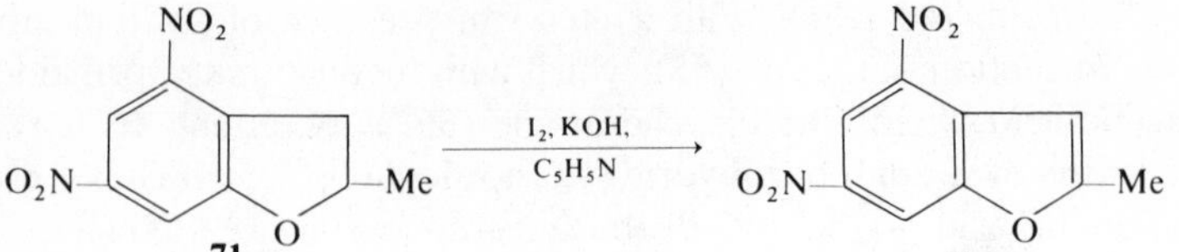

(18)

71 → (I_2, KOH, C_5H_5N) → **72**

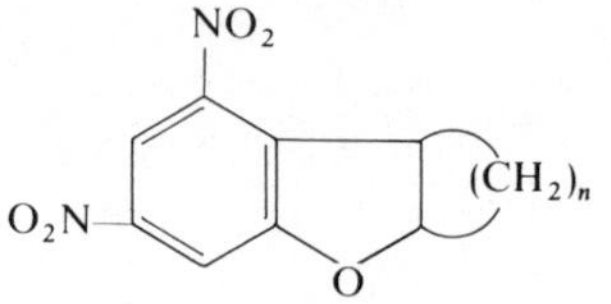

73a n = 4
73b n = 3

Mononitration of 2,3-dihydrobenzofuran by the action of acetyl chloride and silver nitrate in acetonitrile leads to the formation of 5- and 7-nitro derivatives. Attempted nitration with nitronium tetrafluoroborate results in intensive ring opening.[87]

5-Acetamido-2-methyl-2,3-dihydrobenzofuran undergoes the sequence of reactions illustrated in Eq. 19.[78]

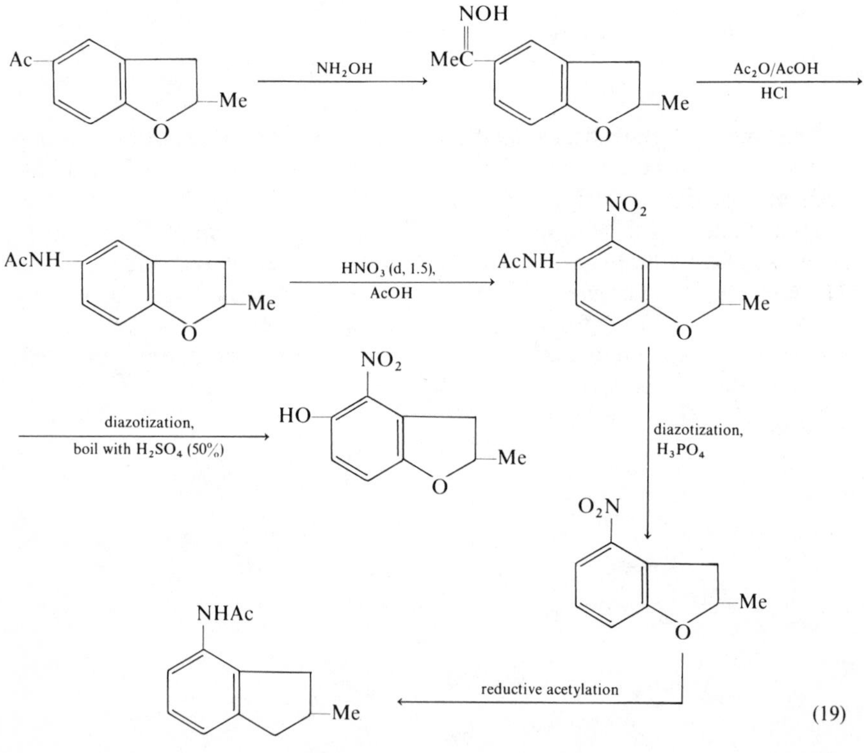

NH_2OH; Ac_2O/AcOH, HCl; HNO_3 (d, 1.5), AcOH; diazotization, boil with H_2SO_4 (50%); diazotization, H_3PO_4; reductive acetylation

(19)

D. Amino-Substituted 2,3-Dihydrobenzofurans

Because of the marked physiological activities of nuclear or side-chain substituted 2,3-dihydrobenzofurans, the patent literature covers the report of the preparation of a large number of compounds in this class. Karrer and Fritzsche[88] have reported the preparation of 2-methyl-5-amino-2,3-dihydrobenzofuran by catalytic reduction of the azo dye, obtained by treatment of 2-methyl-2,3-dihydrobenzofuran with diazonium salt of 2,4-dinitroaniline. Similarly, 2,4,7-trimethyl-5-amino-2,3-dihydrobenzofuran was obtained.

3-Amino-substituted 2,3-dihydrobenzofurans[89,90] have been successfully prepared from the corresponding substituted 3(2*H*)-benzofuranone oximes, for example, treatment of 5-chloro-, and 7-methyl-3(2*H*)-benzofuranone oximes with sodium amalgam gave 5-chloro- and 7-methyl-3-amino-2,3-dihydrobenzofuran, respectively. These compounds show potent adrenergic nerve-blocking activities, especially in the form of quaternary salts.[90]

o-Hydroxyanils are stable or very resistant toward ethereal diazomethane solution, but react in the ketonic form to yield dihydrobenzofuran derivatives when treated with ethereal diazomethane solution in presence of methanol (Chart 2).[91,92]

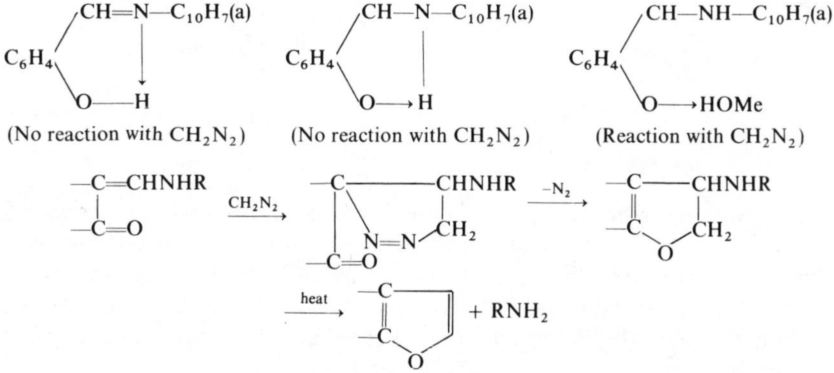

C_6H_4 (CH=N—$C_{10}H_7$(a) / O—H) (No reaction with CH_2N_2)

C_6H_4 (CH—N—$C_{10}H_7$(a) / O→H) (No reaction with CH_2N_2)

C_6H_4 (CH—NH—$C_{10}H_7$(a) / O→HOMe) (Reaction with CH_2N_2)

—C=CHNHR / —C=O $\xrightarrow{CH_2N_2}$ —C——CHNHR / N=N—CH$_2$ / —C=O $\xrightarrow{-N_2}$ —C——CHNHR / —C—O—CH$_2$ $\xrightarrow{\text{heat}}$ furan + RNH_2

Chart 2

Photolysis of *N*-benzenesulfonyl-2-*t*-butyl-1,4-benzoquinone-4-imine (**74**) in acetic acid in sunlight yields 5-benzenesulfonylamino-2,3-dihydro-2,2-dimethylbenzofuran (**77**) together with **75** and **76**. Compound **77** can also be obtained upon heating a mixture of the olefin, namely 2-(2-methylallyl)benzenesulfonylaminophenol (**76**) and/or 2-(2-acetoxy-2-methylpropyl)-4-benzenesulfonylaminophenol (**75**) with hydroiodic acid (Eq. 20).[93]

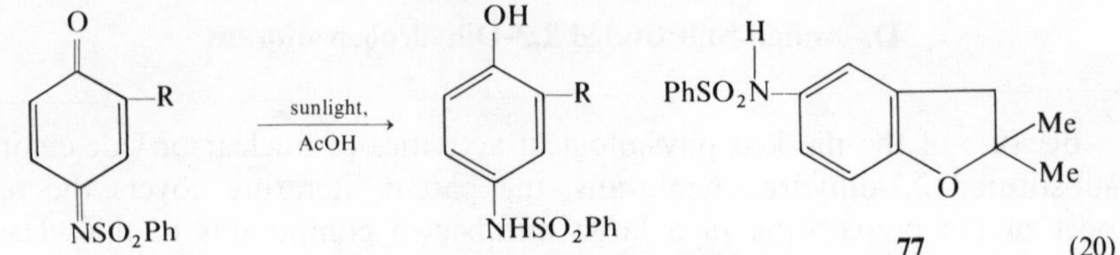

74 R = t-Bu

75 R = CH_2CMe_2OAc

76 R = $CH_2CMe{=}CH_2$

77 (20)

2-Alkylaminoalkyl-substituted 2,3-dihydrobenzofurans are obtained:[94-99] (1) by reaction of 2-bromomethyl-2,3-dihydrobenzofuran with alkylamine in an inert solvent to give, for example, 2-aminomethyl-2,3-dihydrobenzofuran; (2) by lithium aluminum hydride reduction of 2-acetyl-2,3-dihydrobenzofuran oxime to yield, for example, 2-(1-aminoethyl)-2,3-dihydrobenzofuran; and (3) by reaction of 2-acetyl-2,3-dihydrobenzofuran with formic acid and formamide, followed by lithium aluminum hydride reduction to afford 2-(1-methylaminoethyl)-2,3-dihydrobenzofuran.

Amino-2-alkyl-2,3-dihydrobenzofurans are readily obtained by catalytic hydrogenation of amino-2-alkylbenzofurans under pressure in presence of Raney nickel.[100] Similarly, **78** (R = NO_2, R^1 = H or R = H, R^1 = NO_2) are catalytically reduced to the corresponding amino compounds [**79**, R = NH_2 (or H), R^1 = H (or NH_2)].[101]

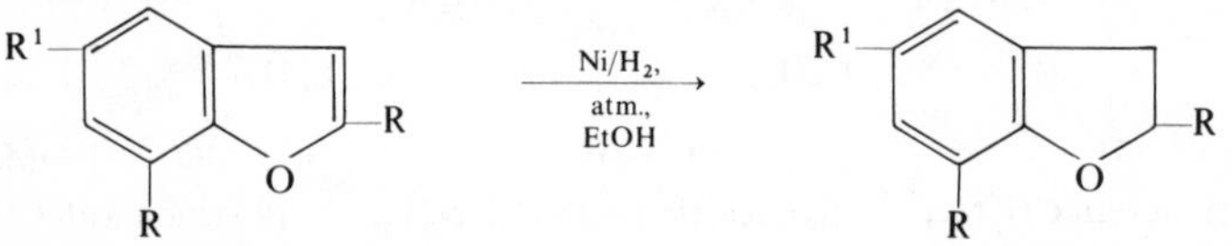

78 → (Ni/H_2, atm., EtOH) **79**

Substituted 2-aminomethyl-2,3-dihydrobenzofurans (**80**) are prepared by interaction of substituted 2-bromomethyl-2,3-dihydrobenzofuran with the appropriate amine.[102] Condensation of 1-(2-chlorophenyl)piperazine with 2-bromo-5-methoxy-2,3-dihydrobenzofuran in presence of sodium carbonate gave the hypotensive 2-(1-piperazinyl)methyl-5-methoxy-2,3-dihydrobenzofuran (**81**).[103,104]

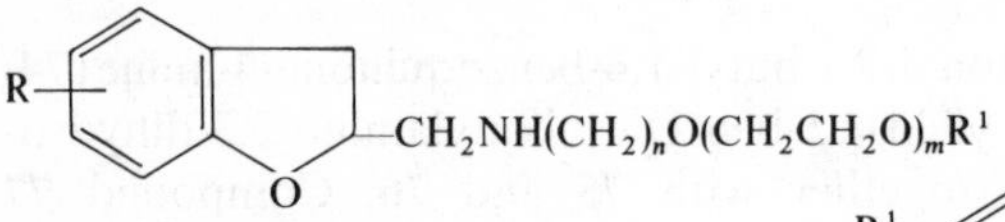

80 R = H, or alkoxy or alkyl (1–6, C-atoms), R^1 = alkyl (1–6, C-atoms), n = 0, 1, 2, or 3, m = 0, 1, or 2

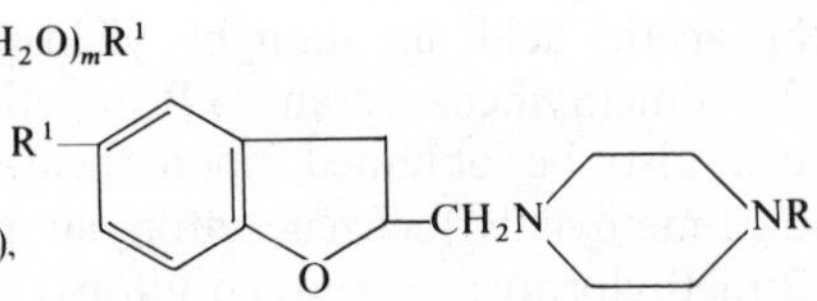

81 R = H, R^1 = OMe

Synthesis of 2-guanidinomethyl-2,3-dihydrobenzofuran (**82**) has been achieved (Eq. 21).[105]

$$\text{2-(bromomethyl)-2,3-dihydrobenzofuran (}CH_2Br\text{)} \xrightarrow[(110°C,\ 6\ hr)]{PhCH_2NH_2,} CH_2NHCH_2Ph \xrightarrow[AcOH]{Pd\text{—}C/H_2,} CH_2NH_2 \longrightarrow CH_2NHC(=NH)\text{—}NH_2\ \textbf{82} \qquad (21)$$

2-Ethyl-3-aminomethyl-2,3-dihydrobenzofuran (**83**) condenses with *N*-substituted 2-chloropropionamide to yield **84**.[106]

$$\textbf{83}\ (3\text{-}CH_2NH_2,\ 2\text{-Et}) + ClCH_2CH_2C(=O)NHMe \longrightarrow \textbf{84}\ (3\text{-}CH_2NH(CH_2)_2CONHMe,\ 2\text{-Et})$$

Study of the significance of the cyclic ether linkage on the analgesic activity necessitates the preparation of a large number of dihydrobenzofurans containing aminoalcohol groups. Reduction of the amino ketones of type **86**, obtained by interaction of 2-bromoacetylbenzofuran (**85**) with the appropriate amine and aluminum isopropoxide, afforded (**87**). Catalytic hydrogenation of the latter in the presence of palladized charcoal yielded 2,3-dihydrobenzofuranylaminoalcohols (**88**) (Eq. 22).[107]

$$\textbf{85}\ (COCH_2Br) + NHRR^1 \longrightarrow \textbf{86}\ (COCH_2NRR^1) \xrightarrow{Al(iso\text{-}C_3H_7)_3} \textbf{87}\ (CH(OH)CH_2NRR^1) \xrightarrow{Pd\text{—}C/H_2} \textbf{88}\ (CH(OH)CH_2NRR^1) \qquad (22)$$

Nuclear halogen is more readily attacked by hydrogen on palladized charcoal than the furanoid double bond.[107] Thus 1-(5-bromo-2-benzofuranyl)-2-piperidinoethanol (**89**) loses its bromine atom and affords **90**.

Br— —$CH(OH)CH_2NC_5H_{10}$ $\xrightarrow{Pd-C/H_2}$ —$CH(OH)CH_2NC_5H_{10}$

89 90

2-Aminomethyl- and 2-aminoethyl-2,3-dihydrobenzofuran (**91**) react with acrylonitrile to yield β-(*N*-alkyl-2,3-dihydrofuranyl)propionitriles (**92**) which, upon hydrogenation, furnish γ-(*N*-alkyl-2,3-dihydrobenzofuranyl)trimethylenediamines (**93**, R = H) (Eq. 23).[108]

R —$CHNH_2$ + $CH_2{=}CHCN$ $\longrightarrow$

91

R —$CHNHCH_2CH_2CN$ $\xrightarrow[Ni-Cr_2O_3/H_2]{MeOH,\ NH_3}$

92

R —$CHNHCH_2CH_2CH_2NH_2$ (23)

93

The following exemplifies the procedure for the preparation of 2-dimethylaminomethyl-2,3-dihydrobenzofurans (**94**) (Eq. 24).[109,110]

R— $CH_2CH{=}CH_2$, OAc $\xrightarrow[below\ 5°C]{Br_2,\ CS_2}$ R— $CH_2CHBrCH_2Br$, OAc $\xrightarrow[EtOH]{NaOEt,}$

R— —CH_2Br $\xrightarrow[K.\ phthalimide]{HCONMe_2}$ R— —$CH_2NR^1R^2$ (24)

94 $R^1 = R^2 = H$

E. 2,3-Dihydrobenzofuranols

Michael addition and intermediate cyclization of enamines with several aldehydes or ketones result in the formation of substituted 2-piperidino-(or morpholino)-5-hydroxy-2,3-dihydrobenzofuran, for example Michael addition of enamine, in the appropriate solvent, to *p*-benzoquinone (Eq. 25).[111] In general, the reaction of enamine with unsubstituted quinone affords a cyclic addition product (**95**);[112] no reaction of acyloxyquinone with an enamine has yet been reported.[113]

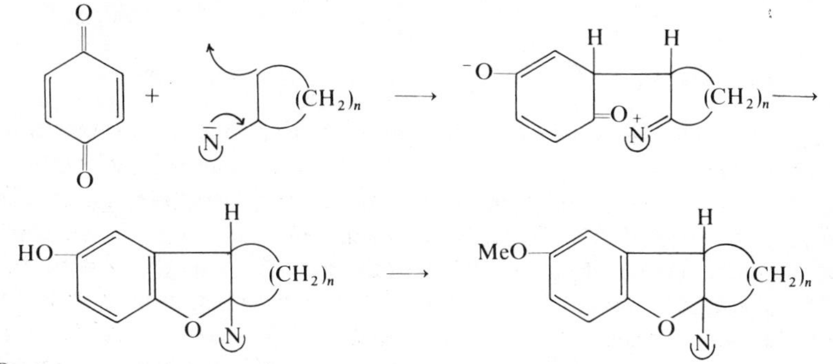

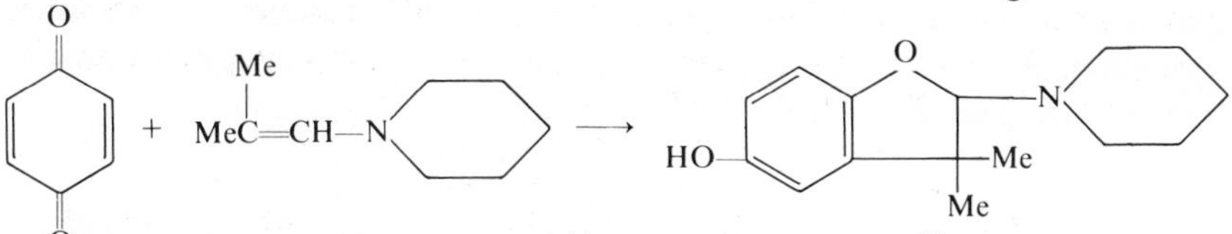

95

Substituted *p*-benzoquinones, for example, 2-phenyl- and 2-methoxy-*p*-benzoquinone, condense with enamines,[114] for example *N*-(1-alken-1-yl)morpholines and 1-(1-alken-1-yl)piperidines, to yield the appropriate dihydrobenzofuranols of type **96**. 3,3-Dialkyl-2-alkoxy-2,3-dihydrobenzofuran-5-ols (**96**) are generally obtained by condensation of 1,4-quinones with enol ethers of type $R^1{}_2$ C═CHOR^2, for example, Me_2C═CHOEt, in presence of boron trifluoride. When an excess of the enol ether is used, acetals are formed, which can be converted into the appropriate dihydrobenzofuranols by acid hydrolysis (Eq. 25).[115]

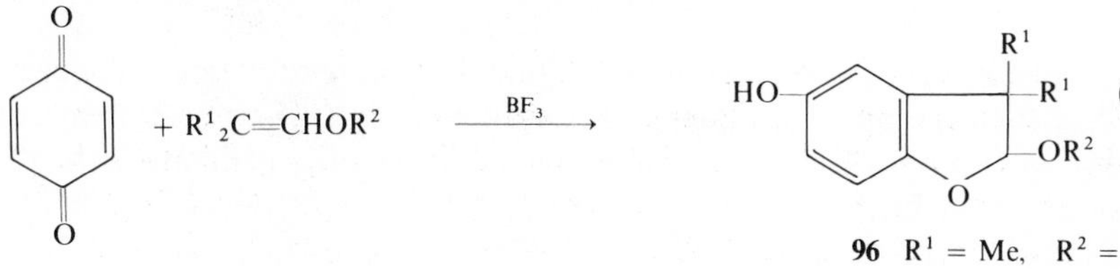

(25)

96 R^1 = Me, R^2 = Et

Thermal treatment of benzenesulfonyl-*p*-benzoquinone (**97** and **98**) in dichloroethylene (or acetic acid) medium at 40 to 50°C effects the formation of **99** (Eq. 26).[116] However 2-acetyl-1,4-benzoquinone condenses with enol ethers to give 2-alkoxy-4-acetyl-5-hydroxy-2,3-dihydrobenzofurans, and with acetylenic ethers to yield the corresponding benzofurans.[117]

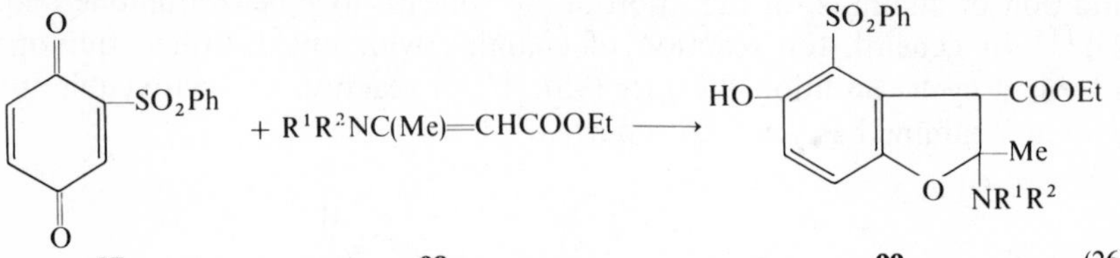

$$\mathbf{97} + R^1R^2NC(Me){=}CHCOOEt\ (\mathbf{98}) \longrightarrow \mathbf{99} \quad (26)$$

Synthesis of 6-hydroxy-2,3-dihydrobenzofuran via an organolithium intermediate (**100**) has been achieved.[118] Treatment of **100** with allyl bromide yields **102**. Demethylation of the latter proceeds easily to afford 6-hydroxy-2-methyl-2,3-dihydrobenzofuran (**103**, R = Me). Treatment of **100** with ethylene oxide gives **101**. Compound **102** is oxidized with osmic acid–periodate to the aldehyde, which is reduced with sodium borohydride to yield **101**. Demethylation of the latter is achieved only with methylmagnesium iodide with high yield. The phenolic product, so obtained by demethylation of **101**, is then cyclized to give 6-hydroxy-2,3-dihydrobenzofuran (**103**, R = H) (Eq. 27).

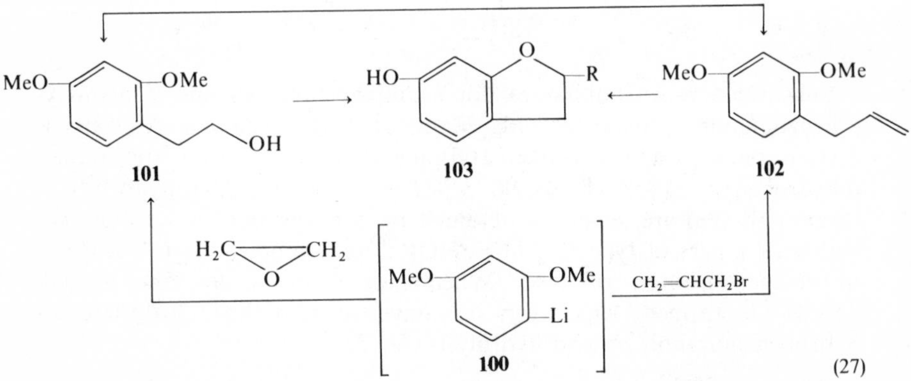

$$\mathbf{100} \xrightarrow{H_2C\text{—}CH_2\ (\text{O})} \mathbf{101} \longrightarrow \mathbf{103} \longleftarrow \mathbf{102} \xleftarrow{CH_2{=}CHCH_2Br} \mathbf{100} \quad (27)$$

Dehydration of 1,4-dimethoxy-2,5-dimethyl-3β-hydroxypropylbenzene (**105**), obtained upon treatment of 1,4-dimethoxy-2,3,5-trimethylbenzene (**104**) with *n*-butyllithium, followed by addition of acetaldehyde to the reaction product, resulted in the formation of 5-methoxy-2,4,7-trimethyl-2,3-dihydrobenzofuran (**106**) (Eq. 28).[119]

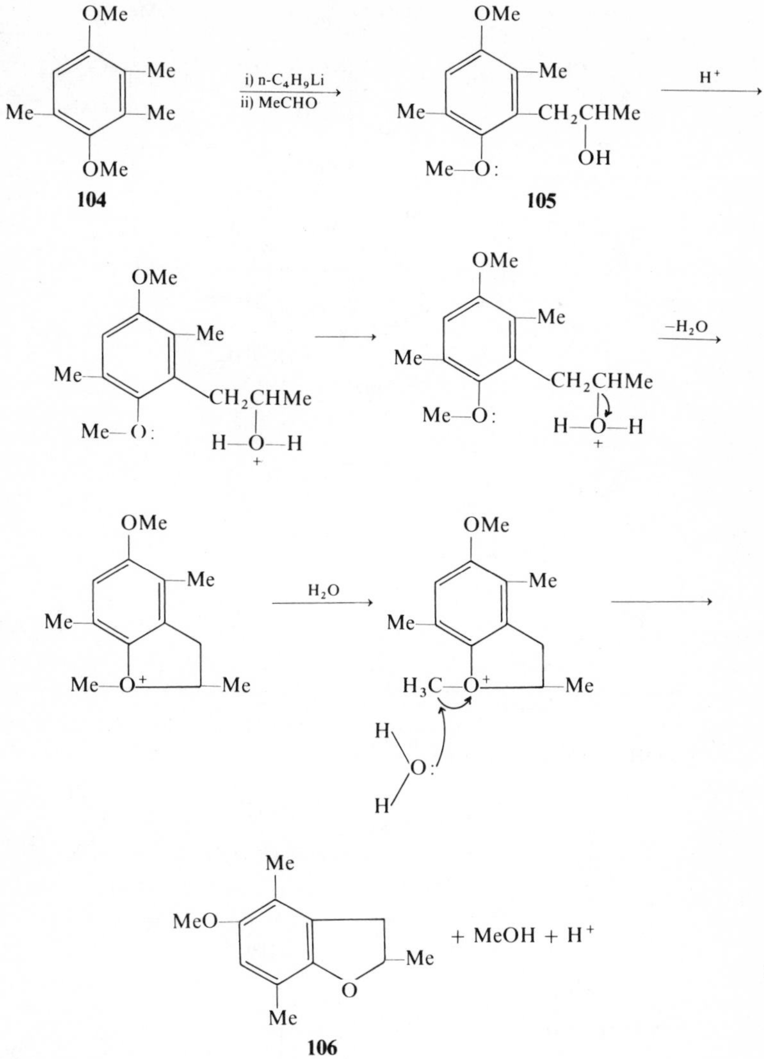

(28)

Claisen rearrangement of 2-allyloxy-6-methoxy-4′-chlorobenzophenone (**107**) yields 2-methyl-6-methoxy-7-(*p*-chlorobenzoyl)-2,3-dihydrobenzofuran (**108**) (Eq. 29).[120]

Double Claisen rearrangement of *α,α-bis*(aryloxy)isobutenes, for example, 1,3-di-(*o*-tolyl)-2-methylenepropane (**109**), is effected on heating with *N,N*-dimethylaniline to give a mixture of **110** and 2-(3′-methyl-2′-hydroxybenzyl)-2,7-dimethyl-2,3-dihydrobenzofuran (**111**) (Eq. 30).[121]

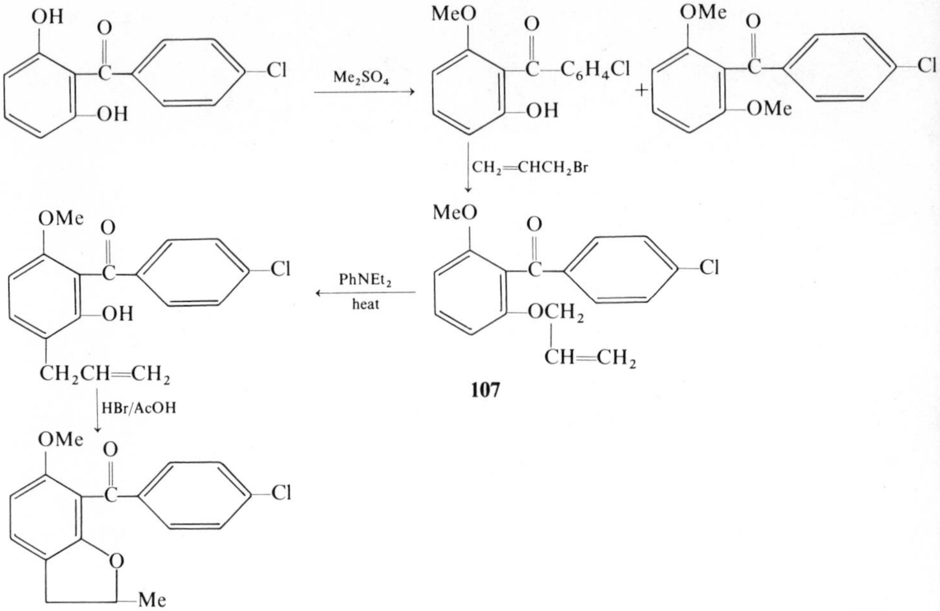

(29)

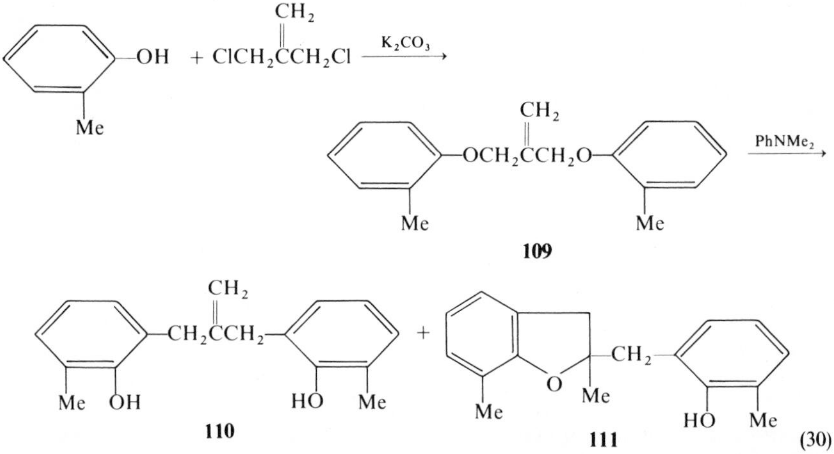

(30)

Dehydration of *meso*-3,4-*bis*(*o*-hydroxyphenyl)-3,4-dihydroxyhexane (**112**), obtained by pinacolization of *o*-propionylphenol by action of aluminum amalgam in ether with acetic anhydride–acetyl chloride mixture, yields 2,3-diethyl-2-acetoxy-3-(*o*-acetoxyphenyl)-2,3-dihydrobenzofuran (**113**). The latter gives, upon hydrolysis, **114** (Eq. 31).[122]

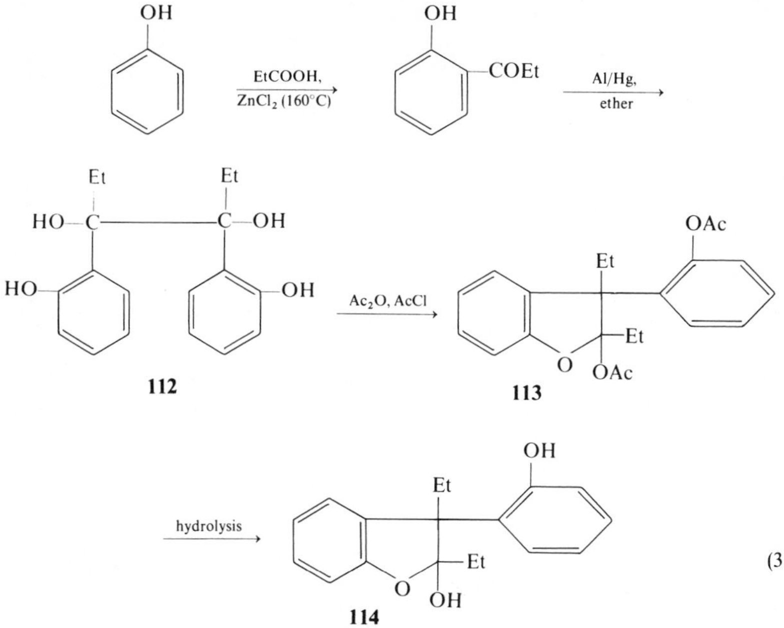

(31)

Dimethylsulfoxonium methylide[123,124] reacts with salicylaldehyde to yield 3-hydroxy-2,3-dihydrobenzofuran (**115**) together with small amounts of benzofuran. Compound **115** decomposes readily to yield benzofuran; on oxidation it affords 3(2*H*)-benzofuranone (**116**), which in turn can be reduced to **115** (Eq. 32).

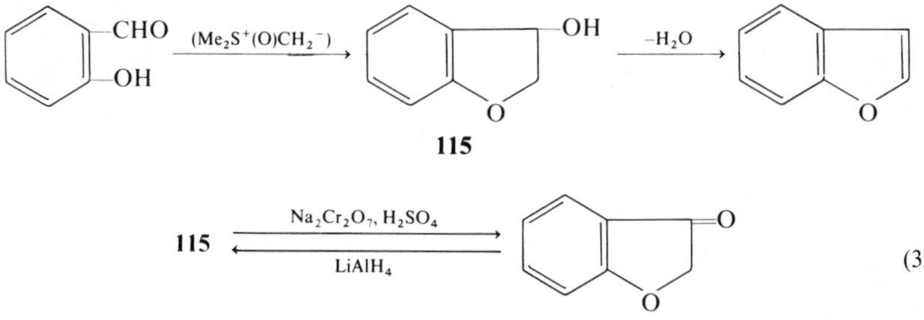

(32)

Chart 3 outlines the synthesis of 2-methyl- and 2,4,6,7-tetramethyl-5-hydroxy-2,3-dihydrobenzofuran from the corresponding *o*-allylphenols; it involves diazo coupling, cleavage of the azo compound to the aminophenol, oxidation to the allylic quinone, reduction to the hydroquinone, and cyclization.[125]

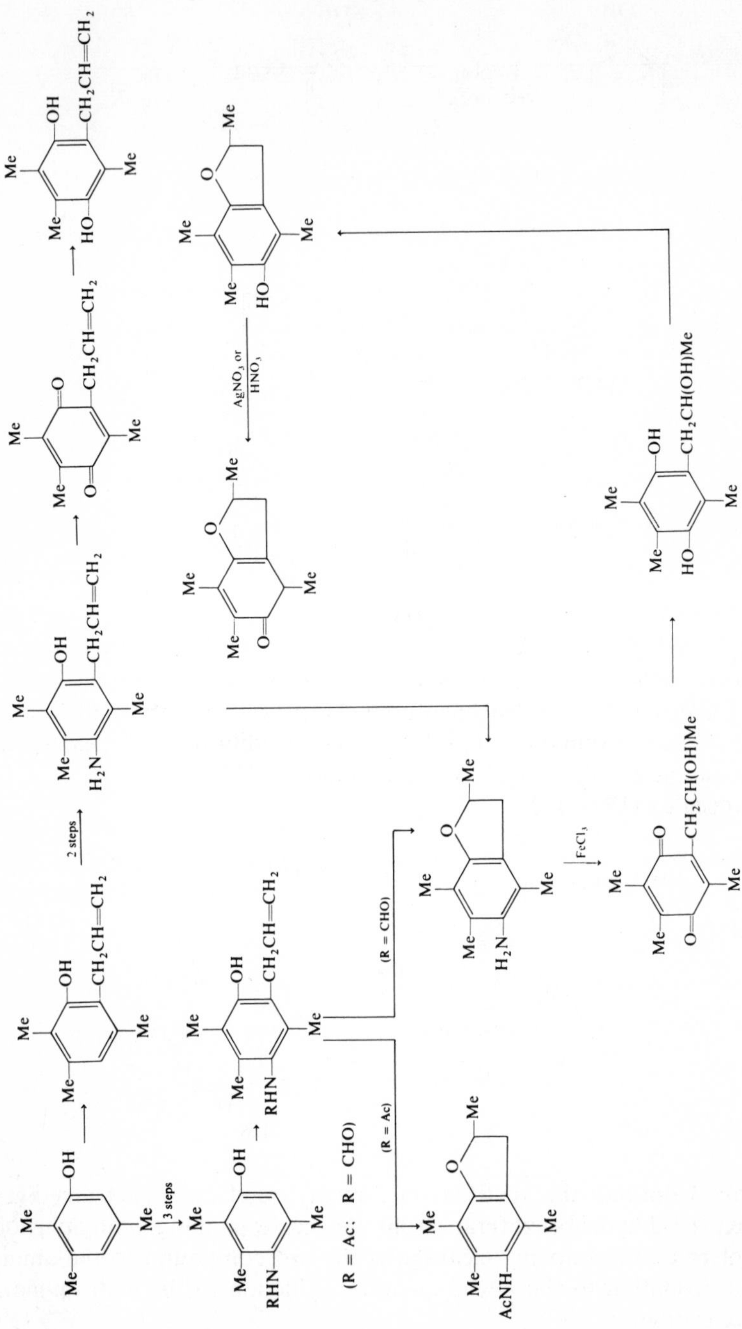

Chart 3

The 2,3-dihydro-5-benzofuranols, related to β- and γ-tocopherol and to toxol, have been prepared (Chart 4) and their oxidation with alkaline ferricyanide has been studied.[126] Oxidation of 2,2-dimethyl-2,3-dihydrobenzofuran-5-ol yields a spiroketal trimer (**121**) by C–C and C–O coupling by the annelated 5-membered heterocyclic ring; this is in accord with the predictions based on the Mills-Nixon effect. Similar oxidation of 2,2,4,7-tetramethyl- (**118**), and of 2,2,6,7-tetramethyl-2,3-dihydrobenzofuran-5-ol (**119**, analogous to β- and γ-tocopherol), gives the hydroquinones **122** and **123**; the reactions involve opening of the heterocyclic ring.

2-(α-Methylallyl)-4,6,7-trimethyl-2,3-dihydrobenzofuran-5-ols (**124**) form a class of compounds which has attracted great interest because the members of this class are isomeric with, and closely related to, 2,2-dialkyl-5,7,8-chroman-6-ol structural-type tocopherols, for example, **125**. Methyl *n*-amyl ketone reacts with 4,6,7-trimethyl-3(2*H*)-benzofuranone (**126**) to yield 2-(2′-heptylidene)-4,6,7-trimethyl-3(2*H*)-benzofuranone (**127b**), which is catalytically reduced in presence of Raney nickel to **128a**. The latter is converted to the bromo compound (**128b**), which forms the Grignard reagent (**128c**). The latter is oxidized to yield the bromomagnesium phenolate (**128d**), and is readily converted on hydrolysis to the dihydrobenzofuran (**124b**).[127-131]

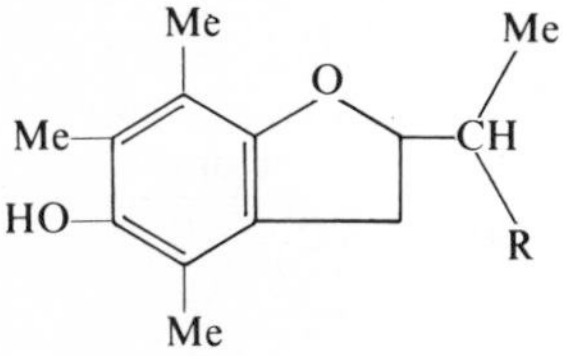

124a R = Me
124b R = H—C_5H_{11}
124c R = $C_{15}H_{31}$
(4, 8, 12-trimethyltridecyl)

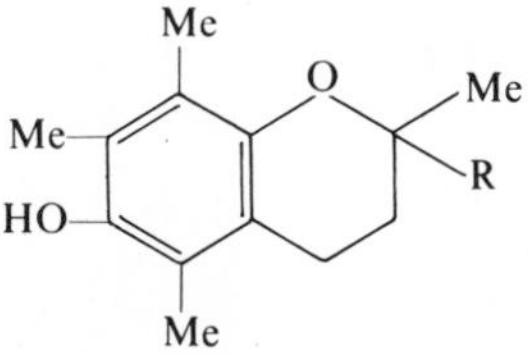

125

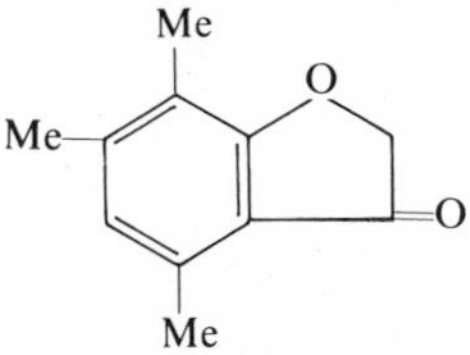

126

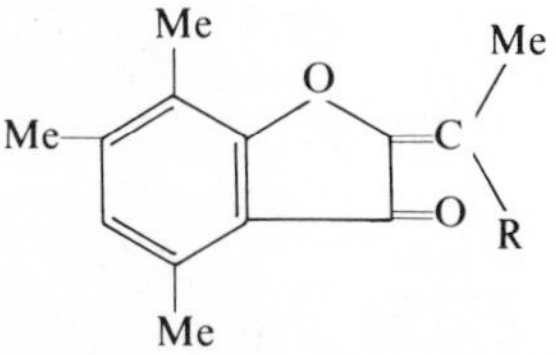

127 (a, b, c, R as in **124**)

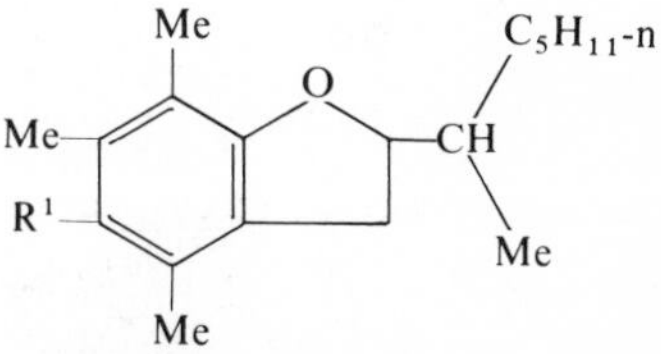

128a R^1 = H
128b R^1 = Br
128c R^1 = MgBr
128d R^1 = OMgBr

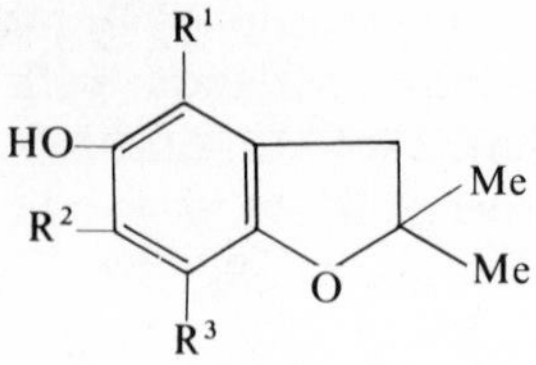

117 $R^1 = R^2 = R^3 = H$
118 $R^1 R^3 = Me, \quad R^2 = H$
119 $R^1 = H, \quad R^2 = R^3 = Me$
120 $R^1 = R^3 = H, \quad R^2 = Me$

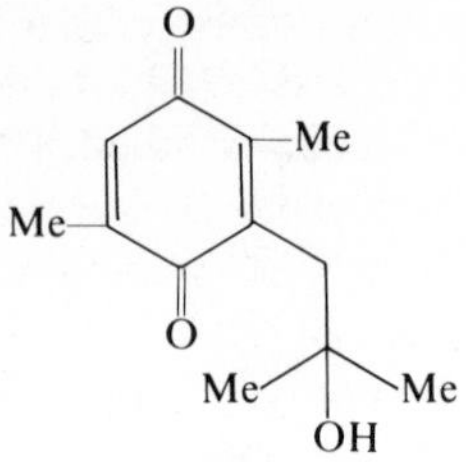

122

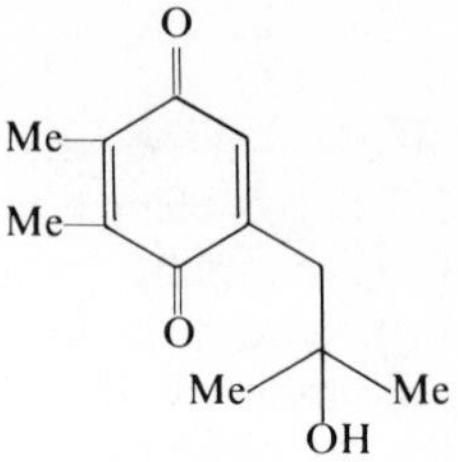

123

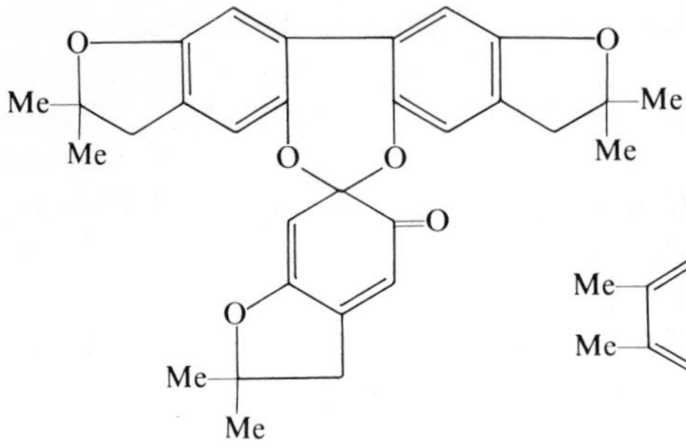

121

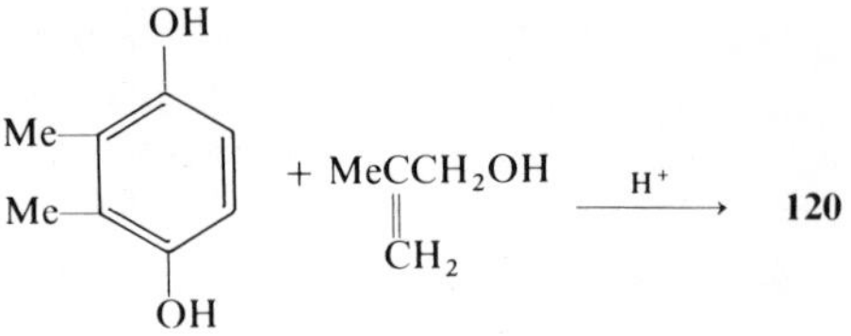

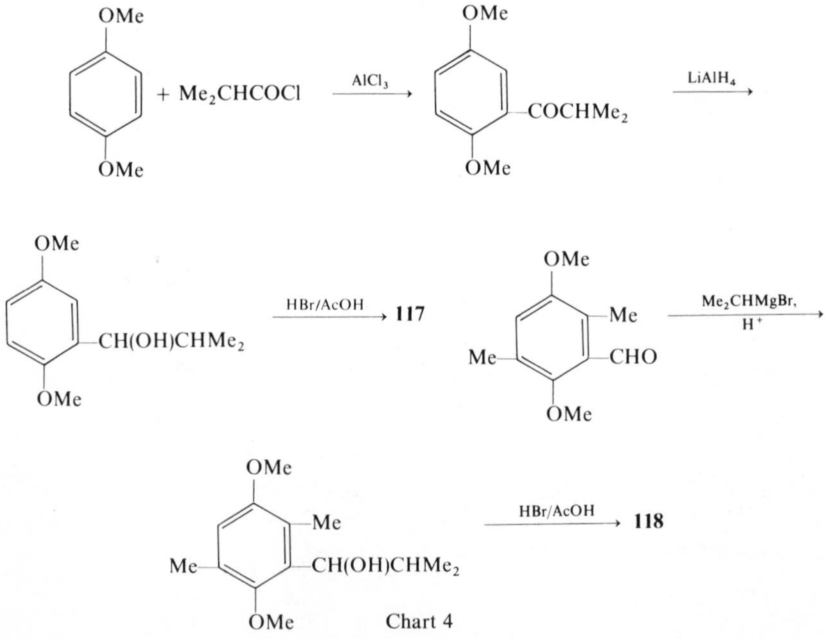

$+ \ MeC(=CH_2)CH_2OH \xrightarrow{H^+}$ **120**

$+ \ Me_2CHCOCl \xrightarrow{AlCl_3}$ COCHMe$_2$ $\xrightarrow{LiAlH_4}$

CH(OH)CHMe$_2$ $\xrightarrow{HBr/AcOH}$ **117**

$\xrightarrow{Me_2CHMgBr, \ H^+}$

CH(OH)CHMe$_2$ $\xrightarrow{HBr/AcOH}$ **118**

Chart 4

The dihydrobenzofuran, namely, 2-(6′,10′,14′-trimethyl-2′-pentadecyl)-4,6,7-trimethyl-5-hydroxy-2,3-dihydrobenzofuran possesses only about 5% as much as vitamin E activity and simulates that of **124c**.[129]

5-Hydroxy-4,6,7-trimethyl-2,3-dihydrobenzofuran[130] is obtained by cyclization of 2-(2-hydroxyethyl)pseudocumohydroquinone with hydrobromic–acetic acid mixture.

In general, 2-alkyl-2,3-dihydrobenzofuranols are prepared conveniently by: (1) condensing a quinone, particularly 2,3,5-trimethyl-*p*-benzoquinone, with the alkali metal enolate of a symmetrical *β*-diketone to form a hydroquinone with the diacylmethyl residue of the *β*-diketone attached to the ring in the 6 position, cyclization of the latter with loss of one acyl group, and hydrogenation of the resulting 2-alkylbenzofurans to the corresponding 2,3-dihydro compounds;[131] (2) interaction of conjugated dienes (or the diene reaction material is produced simultaneously in the reaction in which it is used or in a preliminary reaction) and hydroquinone compounds, leading to the formation of dihydrobenzofuranols,

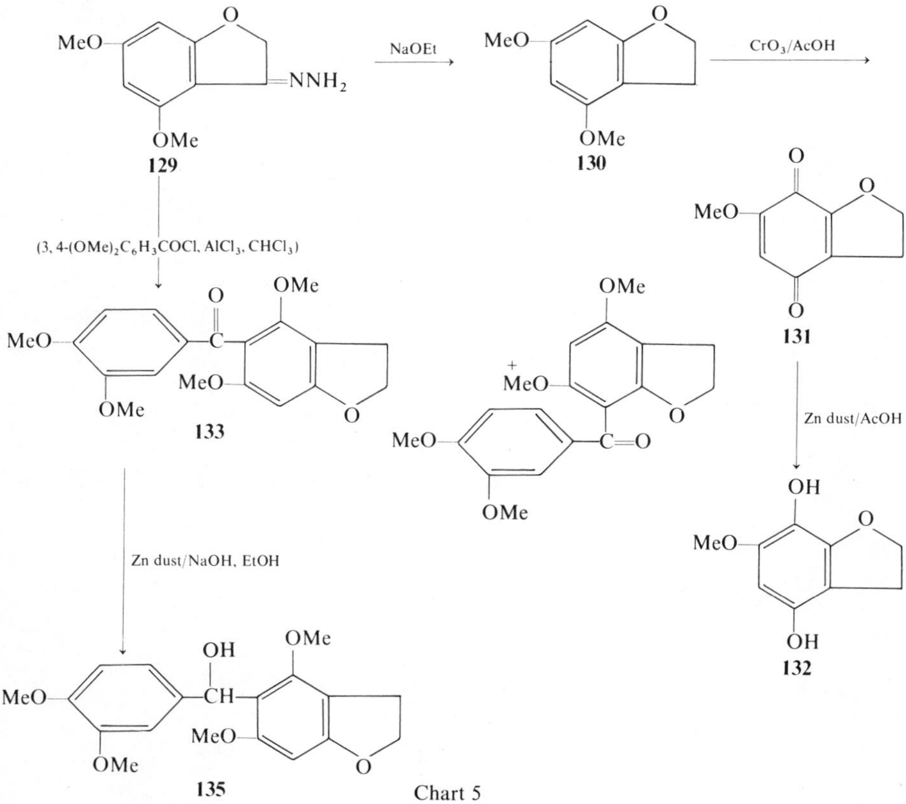
MeO
NNH2
OMe
129
NaOEt
MeO
OMe
130
CrO3/AcOH
MeO
O
131
Zn dust/AcOH
OH
OH
132
(3, 4-(OMe)2C6H3COCl, AlCl3, CHCl3)
MeO
OMe
133
+
MeO
C=O
OMe
Zn dust/NaOH, EtOH
OH
CH
135

Chart 5

for example, 2,4,6,7-tetramethyl-3-ethyl-5-hydroxy-2,3-dihydrobenzofuran;[132-141] (3) the action of Grignard reagents on alkyl-substituted 5-hydroxy-3(2*H*)-benzofuranones.[142] 4,6-Dimethoxy-2,3-dihydrobenzofuran (**130**), obtained by reduction of 4,6-dimethoxy-3(2*H*)-benzofuranone hydrazone (**129**), is readily oxidized to yield 6-methoxy-2,3-dihydrobenzofuran-4,7-quinone (**131**), which is reduced to 4,7-dihydroxy-6-methoxy-2,3-dihydrobenzofuran (**132**).[143] Condensation of **129** with 3,4-dimethoxybenzoyl chloride, under Friedel-Crafts conditions, gave a mixture of **133** and **134**. Reduction of **133** furnished **135** (Chart 5).[143]

Sodium amalgam reduction of the oxime **136** yields **137**;[144] reduction of substituted 3(2*H*)-benzofuranones,[145] for example, **138**, gives the corresponding 2,3-dihydrobenzofuran (**140**) via **139** (Eq. 33).[146]

$\xrightarrow[\text{NaOAc}]{NH_2OH,\ HCl}$ **136** $\xrightarrow{Na/Hg}$ **137**

$\xrightarrow[\text{(2) NaOAc}]{\text{(1) HBr.}}$ **138** $\xrightarrow{LiAlH_4}$ **139** $\xrightarrow{Pd—C/H_2}$ **140** (33)

2,3-Dihydro-2,2-dimethylbenzofuran-7-ol (**141**, R = H),[147] an intermediate in the preparation of the insecticide 2,3-dihydro-2,2-dimethylbenzofuran-7-yl methylcarbamate,[148] is obtained by rearrangement and cyclization of 2-methylallyloxyacetophenone, followed by oxidation and hydrolysis. Equation 34 illustrates the synthesis of this potent paraciticide (**141**, R = Me).[149]

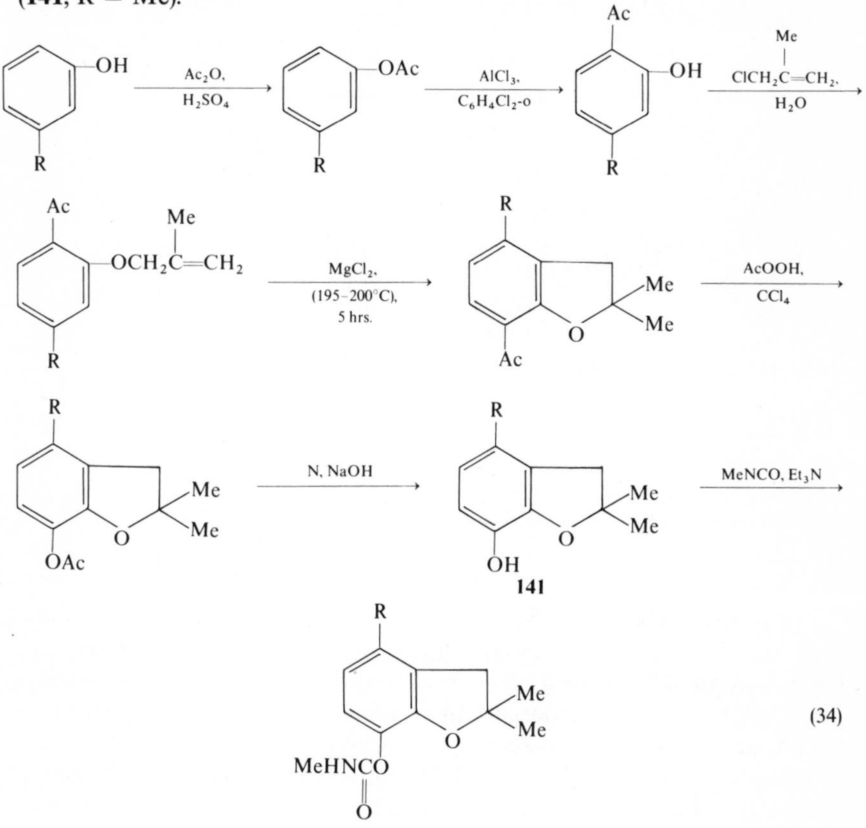
OH
Ac_2O, H_2SO_4
R
OAc
$AlCl_3$, $C_6H_4Cl_2$-o
Ac
Me
$ClCH_2C{=}CH_2$, H_2O
$OCH_2C{=}CH_2$
$MgCl_2$, (195–200°C), 5 hrs.
Me
O
AcOOH, CCl_4
N, NaOH
MeNCO, Et_3N
MeHNCO

141

(34)

"Furadan" (NIA-10242)
(R = H)

Different synthetic routes have been reported in the patent literature for the preparation of **141** (R = H or Me).[148,150,151]

Photochemical formation of substituted 2,3-dihydrobenzofuranols has been extensively studied during the last decade. Photolysis of 2-*t*-butyl-, 2,5-di-*t*-butyl-, and 2,6-di-*t*-butyl-*p*-benzoquinone in a variety of solvents, besides reduction to the corresponding hydroquinone and rearrangement of the side-chain, takes place and leads to an ether and a dihydrobenzofuranol.[152,153] Thus, photolysis of 2-*t*-butyl-*p*-benzoquinone

(**142**), for example, affords 2,2-dimethyl-5-hydroxy-2,3-dihydrobenzofuran (**144**), which is obtained in high yield upon heating for a few minutes 2-(2-ethoxy-2-methyl-1-propyl)hydroquinone (**143**) with hydroiodic acid, or by action of the latter acid on 2-(β-methylallyl)hydroquinone (**145**) (Eq. 35).

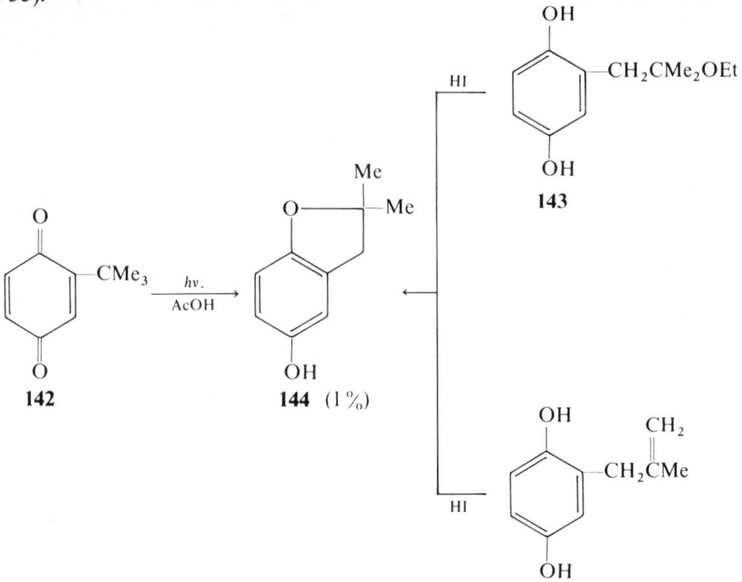

142 $\xrightarrow{h\nu,\ AcOH}$ **144** (1%) $\xleftarrow{HI}$ **143** (CH_2CMe_2OEt); **144** $\xleftarrow{HI}$ **145** ($CH_2C(=CH_2)Me$) (35)

o-Benzyloxybenzaldehyde undergoes light-induced cyclization to furnish *cis*-2-phenyl-3-hydroxy-2,3-dihydrobenzofuran and a closely related compound, possibly the trans isomer.[154] Methyl *o*-benzyloxyphenylglyoxalate (**146**) undergoes light-induced intramolecular cyclization to afford an isomeric mixture of 2-phenyl-3-carbmethoxy-3-hydroxy-2,3-dihydrobenzofuran (**147a** and **147b**) in high yield. The stereochemistry of photocyclization is significantly dependent on both the solvent and the temperature: one of the isomers is formed almost exclusively in nonpolar solvents at low temperature.[155] The results of photoisomerization and quenching experiments indicate that the reaction occurs predominantly via the triplet state; this suggests that the stereochemical fate of the triplet depends on both solvent and temperature.

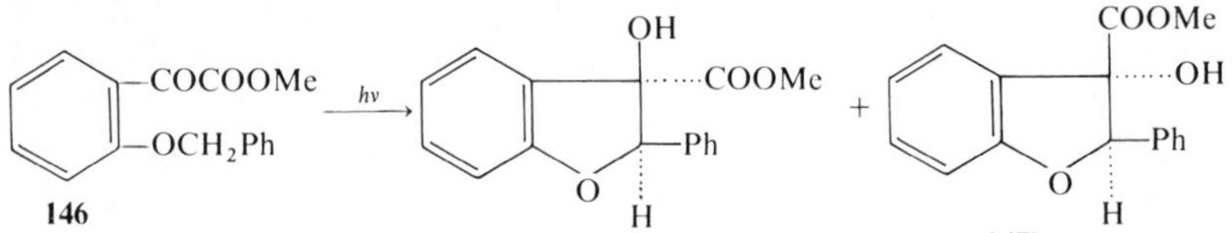

146 $\xrightarrow{h\nu}$ **147a** + **147b**

The quinol (**149**), obtained by irradiation 2,3-dichloro-*p*-benzoquinone and **148**, is transformed by the action of zinc dust in boiling aqueous methanol to a major product, **150**, and a minor one: 4,6,7-trichloro-5-hydroxy-2-methyl-2-isopropenyl-2,3-dihydrobenzofuran (**151**) (Eq. 36).[156,157]

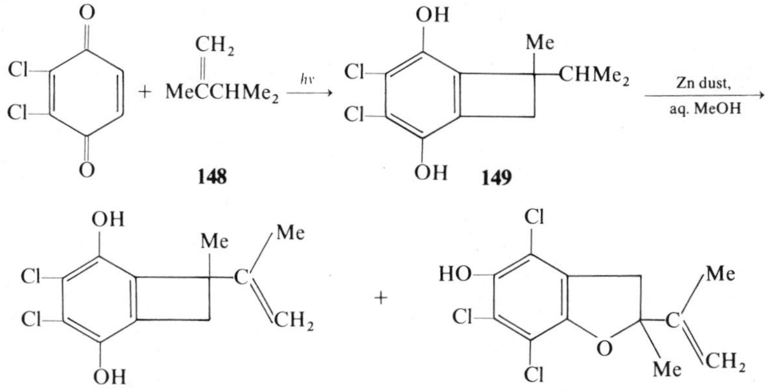

(36)

148 **149** **150** **151**

Irradiation of 1,4-benzoquinone in outgassed acetaldehyde with light from a tungsten filament lamp affords 70% acetylquinol, but its formation is largely suppressed when 1,1-diphenylethylene is present. The proton magnetic spectrum of the solution indicates that only one major product, the enedione (**152**), accumulates. In ethanolic solution this enolizes to the corresponding quinol which, on heating, decomposes quantitatively to a mixture of quinol, olefin (**153**), and dihydrobenzofuranol (**154**); the relative proportions of **153** and **154** depend on the decomposition temperature.[158] Treatment of **153** with *p*-toluenesulfonic acid in boiling toluene affords **154**.

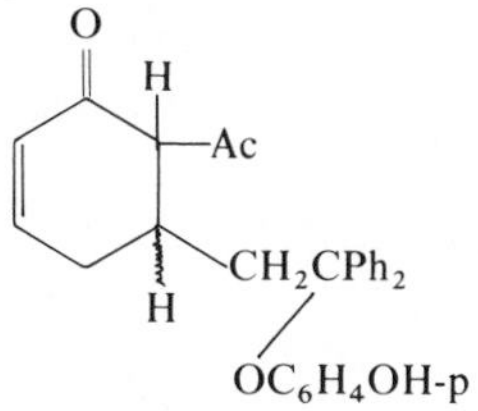

152 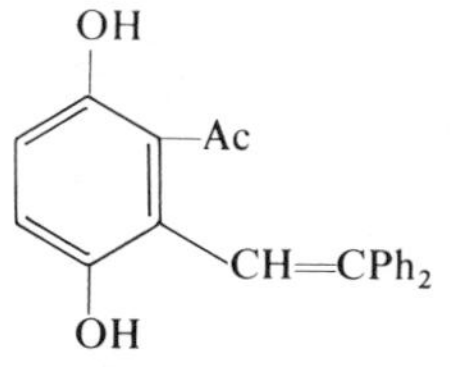 **153** 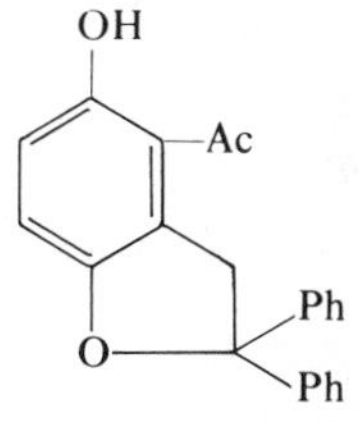 **154**

When colupulone (**155**) is treated with moist air at 20°C in light petroleum, the dihydrobenzofuran (**156**) is produced. Hydrogenation of the latter compound over palladized charcoal yields the benzenoid derivative (**157**) which, upon dehydration with *p*-toluenesulfonic acid, affords 2-iso Pr-7-isobutynoyl-4,6-dihydroxy-5-isopentylbenzofuran (**158**) (Eq. 37).[159,160]

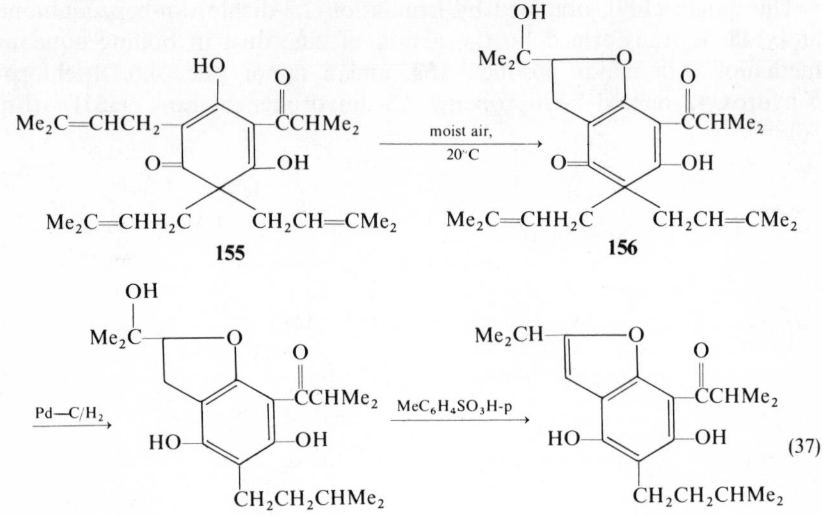

(37)

Synthesis of **158** has been achieved[160] after the procedure adopted by Nickel (Eq. 38).[161]

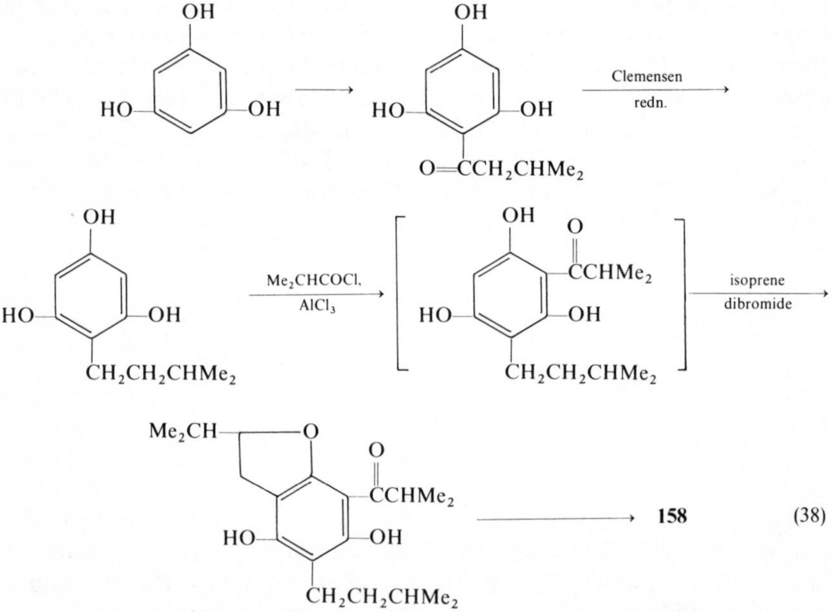

(38)

Chart 6 illustrates different syntheses of polyhydroxy-substituted 2,3-dihydrobenzofurans.[162-164]

(1)

$\xrightarrow[\text{EtOH}]{\text{NaOAc,}}$

$\xrightarrow{\text{PhCOCl}}$

$\xrightarrow[\text{AcOH/HCl}]{\text{3, 4-(PhCOO)}_2\text{C}_6\text{H}_3\text{CHO,}}$

$\xrightarrow[\text{37 lbs pressure}]{\text{PtO}_2\text{/H}_2\text{; 25°C}}$

$\xrightarrow{\text{KOH}}$

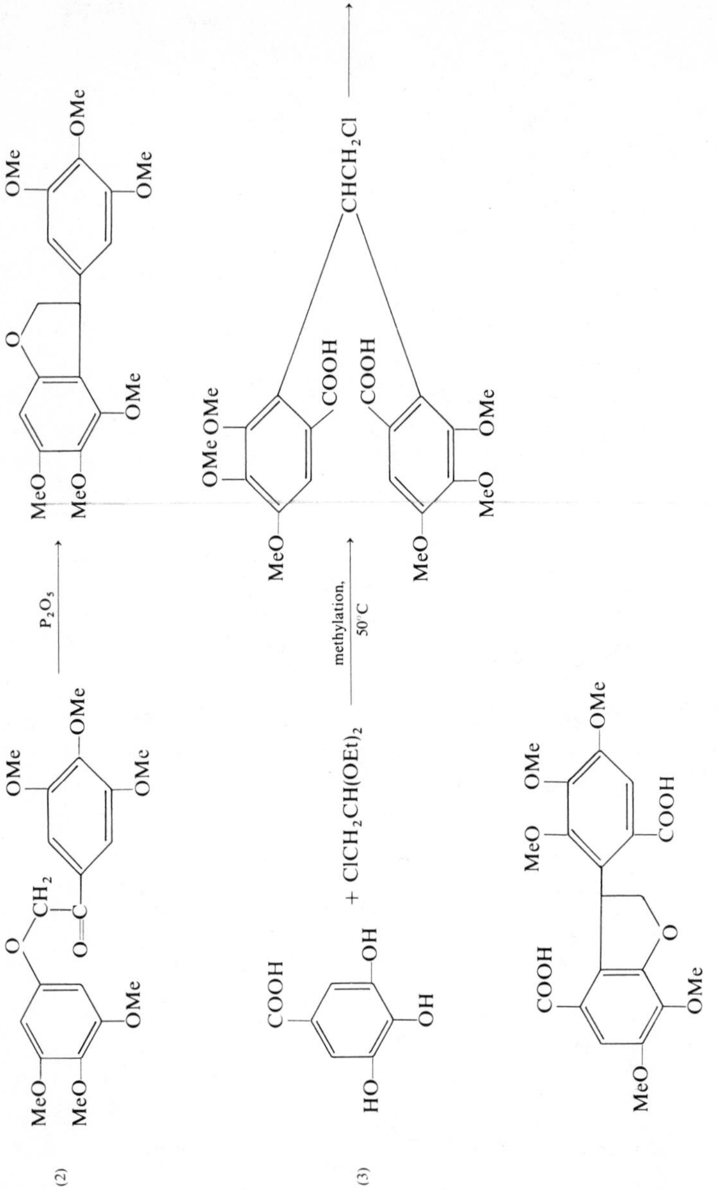

(2) $\xrightarrow{P_2O_5}$

(3) + $ClCH_2CH(OEt)_2$ $\xrightarrow[50^\circ C]{\text{methylation,}}$ $CHCH_2Cl$ $\longrightarrow$

Chart 6

F. Geometrical Isomers of 2,3-Dihydrobenzofurans

Cis- and *trans*-dihydrobenzofurans (**159**) were prepared[165] by nitration of **160**, followed by subsequent separation of the isomers of **161**, reduction to **162**, deaceylation, and treatment with methyl iodide. 5-Dimethylamino-2-methylbenzofuran-3(2*H*)-one (**163**) and the 7 isomer (**164**) were obtained from 5- and 7-nitro-3(2*H*)-benzofuranones, respectively.

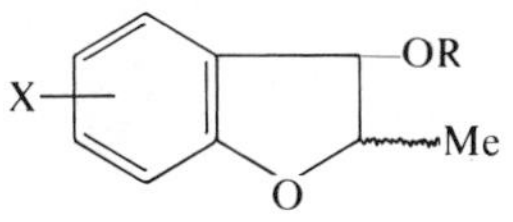

159 R = H, X = 5-, and 7-N^+MeI
160 R = Ac, X = H
161 R = Ac, X = 5-, and 7-NO_2
162 R = Ac, X = 5-, and 7-NMe_2

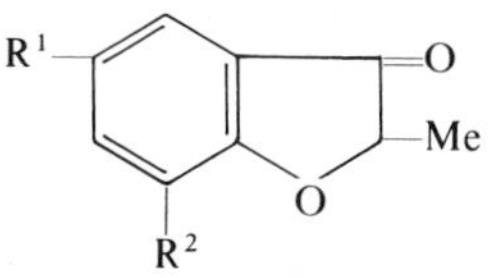

163 R^1 = $NMe_2 \cdot MeI$, R^2 = H
164 R^1 = H, R^2 = $NMe_2 \cdot MeI$

Stereospecific formation of *trans*-phenyl-2,3-dihydrobenzofurans by dehydrogenative dimerization of *trans*- and *cis*-isoeugenol has been recently established.[166a] *Trans*- and *cis*-isoeugenol were treated with hydrogen peroxide in aqueous acetone containing peroxidase to yield dimeric products, which were separated on silica gel to the *trans*-phenyl-dihydrobenzofuran (**165**) (65% yield) from *trans*-eugenol, and **166** (22% yield) from *cis*-isoeugenol. Hydrogenation of the two products with palladium over charcoal (5%) in ethanol brought out the formation of identical hydrogenated products (**167**). Lack of formation of *cis*-phenyl-dihydrobenzofurans from *trans*- and *cis*-isoeugenols is due to the rotation about the C-α C-β bond of an initially formed intermediate prior to cyclization, rather than to the rapidity of the cyclization process.[166]

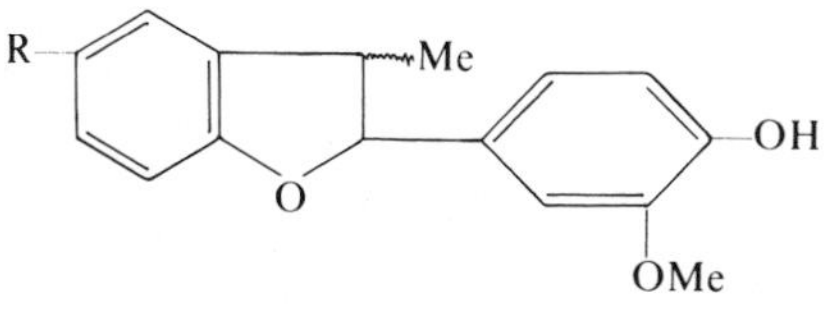

165 R = *trans*-CH=CHMe
166 R = *cis*-CH=CHMe
167 R = CH_2CH_2Me

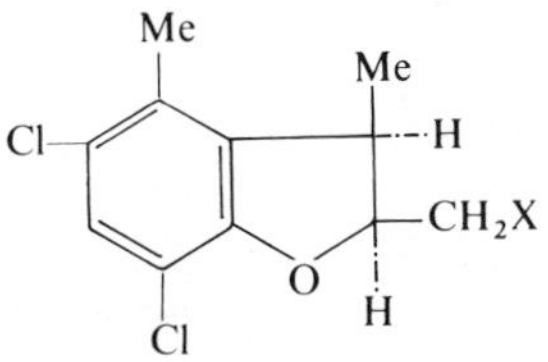

168a X = H
168b X = I

Synthesis and configuration of *cis*- and *trans*-2,3,4-trimethyl-5,7-dichloro-2,3-dihydrobenzofuran (**168a** and **169a**), as well as the corresponding 2-iodomethyl compounds (**168b** and **169b**), were established.[167]

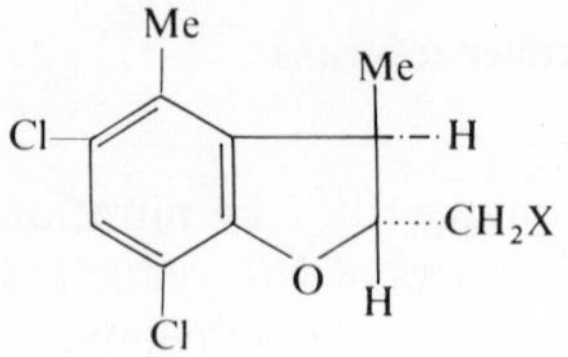

169a X = H
169b X = I

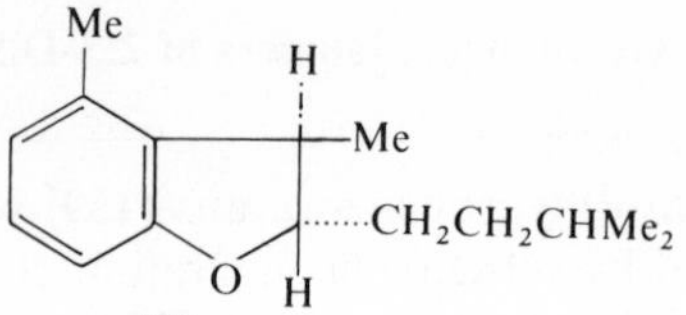

170

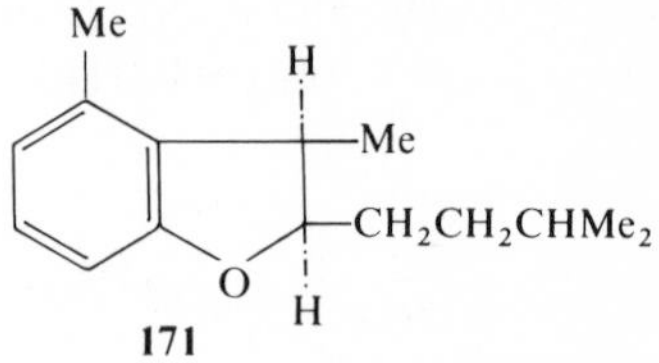

171

Cyclization of 2,4-dichloro-5-methyl-6-(α-methylallyl)phenol (**172**) and its methyl ether gave *cis*- and *trans*-2,3-dihydro-2,3,4-trimethyl-5,7-dichlorobenzofurans (**173** and **174**). Treatment of **172** and its methyl ether with ICl gave the *cis*- and *trans*-iodomethyl compounds (**168b** and **169b**). The latter is converted to **175** (Eq. 39).[166]

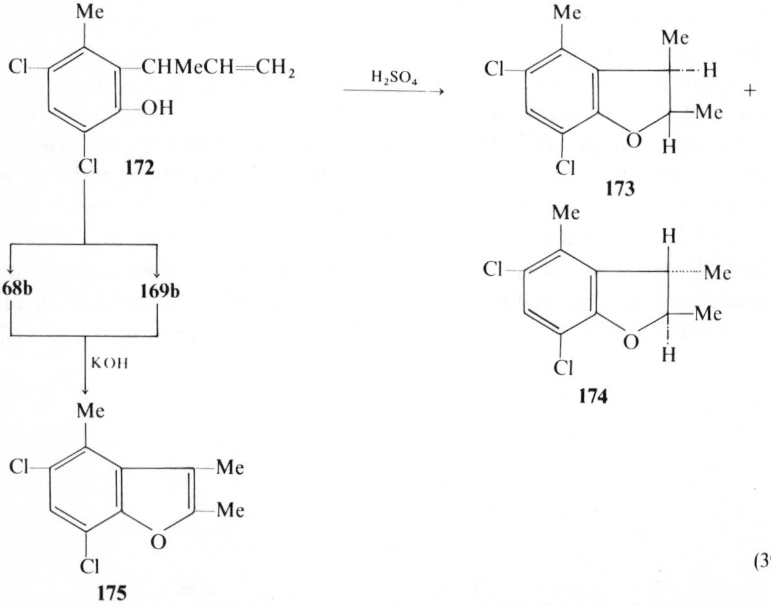

172 $\xrightarrow{H_2SO_4}$ **173** + **174**

172 → **168b**, **169b** $\xrightarrow{KOH}$ **175** (39)

The configuration of the *cis*- and *trans*-2-aroyl-3-hydroxy-2,3-dihydrobenzofurans (**176** and **177**) is established by nuclear magnetic resonance (NMR) spectroscopy and by the rate of dehydration with 2 N-hydrochloric acid in dioxan to **178**; **176** has the lower dehydration rate and **177** has the higher.[168]

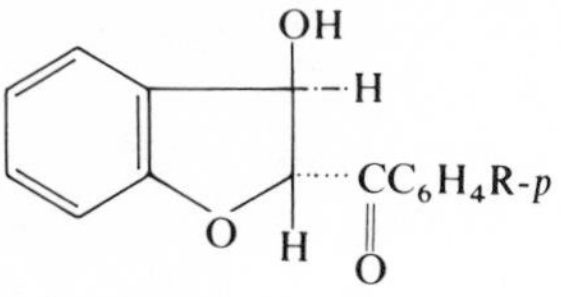

176 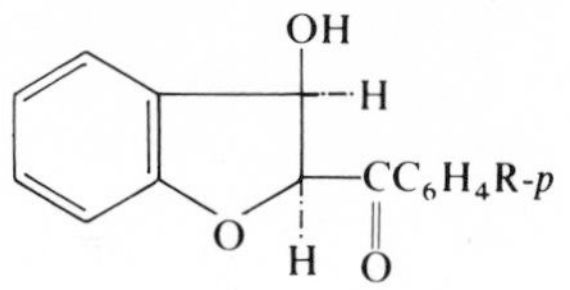 **177** 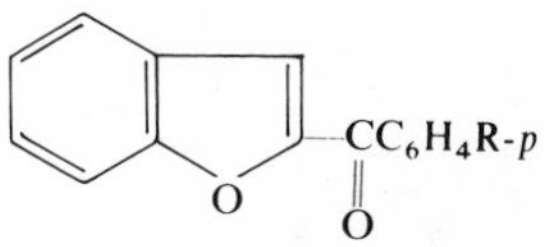 **178**

Trans-2-isopropyl-3-hydroxy-5-acetyl-2,3-dihydrobenzofuran (**180a**) is obtained by treatment of 2-hydroxy-2,5′-dibromo-3-methylbutyrophenone (**179**) with sodium borohydride in aqueous ethanolic potassium hydroxide, butyllithium, carbon dioxide, and finally methyllithium.[169,170] When **179** was reduced with sodium borohydride and the product was then treated with potassium hydroxide, *cis*-2-isopropyl-3-hydroxy-5-bromo-2,3-dihydrobenzofuran (**181b**) was obtained. The latter was converted by an identical sequence of reactions into *cis*-2-isopropyl-3-hydroxy-5-acetyl-2,3-dihydrobenzofuran (racemic dihydrotubanol) (*cis*-**181a**). Reduction of **179** with sodium borohydride in an alkaline medium is believed to lead first, by cyclization, to 2-isopropyl-5-bromo-3(2*H*)-benzofuranone, followed by reduction to give the thermodynamically more stable *trans*-**180b**. In absence of alkali, **179** is first reduced to give the erythro isomer (steric approach) which then leads to *cis*-**180b** by backside displacement of the α-bromine atom by the phenolic OH group. The unexpected observation, that J_{trans}-2,3 J_{cis}-2,3 in the two families of 2,3-dihydrobenzofurans, is believed to arise from a structural stereochemical dependence of the electronegativity effect of the OH group in the *cis* series.[169,170]

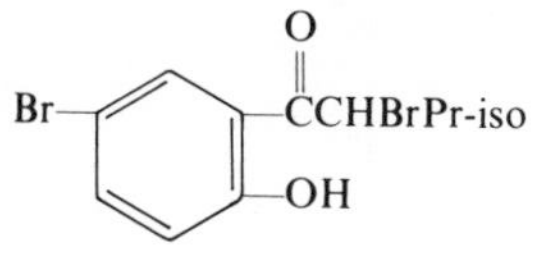

179

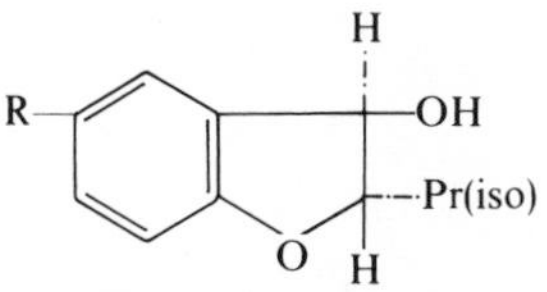

180a R = Ac (trans)
180b R = Br

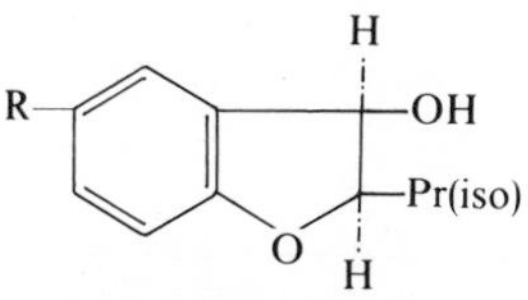

181a R = Ac (cis)
181b R = Br

Synthesis of *trans*- and *cis*-3-cyanomethyl-2-carbethoxy-3,5-dimethyl-2,3-dihydrobenzofuran has been achieved (Chart 7).[171]

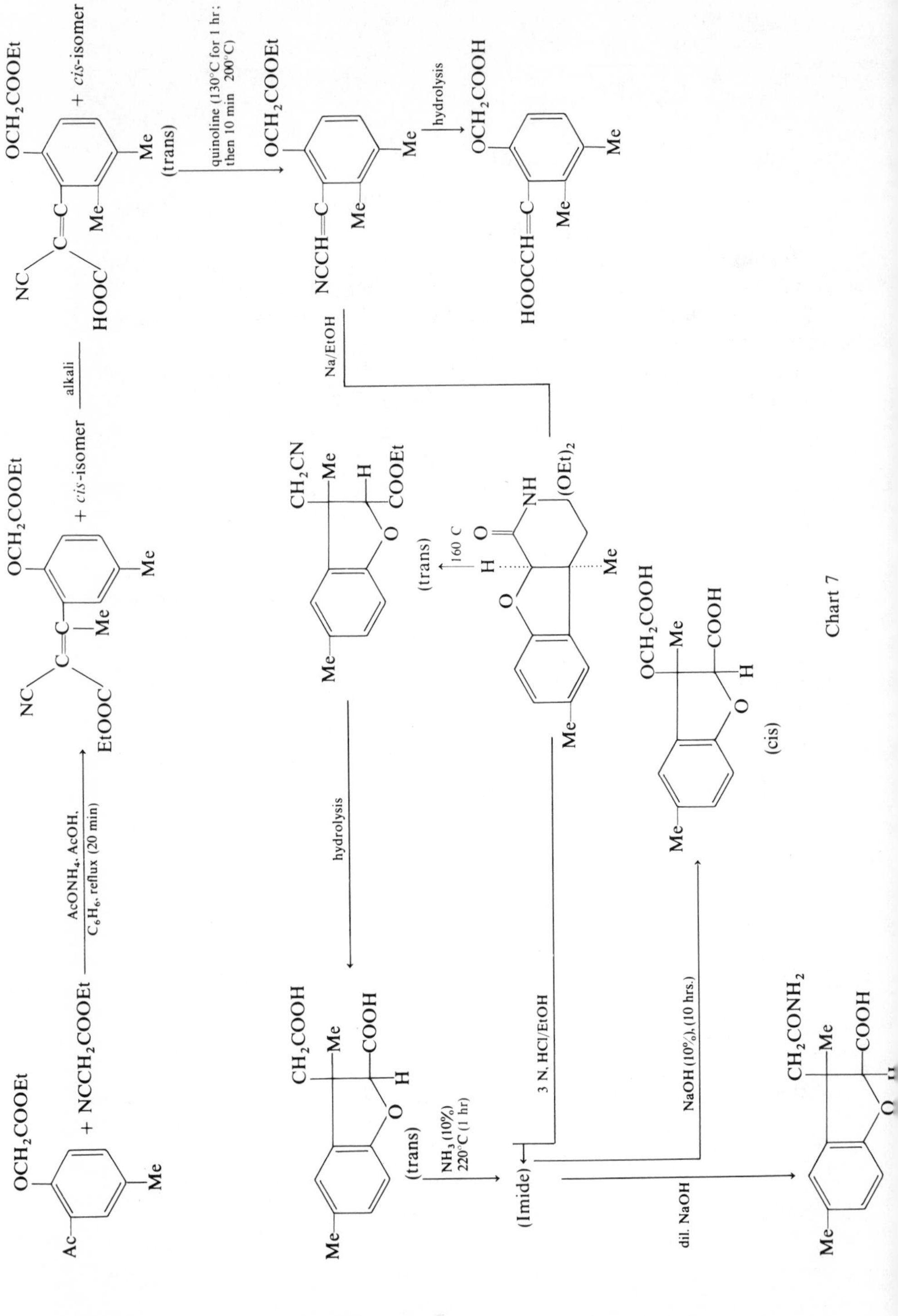

Ac, OCH$_2$COOEt, Me + NCCH$_2$COOEt

AcONH$_4$, AcOH, C_6H_6, reflux (20 min)

NC, EtOOC, C=C, Me, OCH$_2$COOEt, Me + *cis*-isomer

alkali

NC, HOOC, C=C, Me, OCH$_2$COOEt, Me (trans) + *cis*-isomer

quinoline (130°C for 1 hr; then 10 min 200°C)

NCCH=C, Me, OCH$_2$COOEt, Me

hydrolysis

HOOCCH=C, Me, OCH$_2$COOH, Me

Na/EtOH

Me, CH$_2$CN, Me, H, COOEt, O (trans)

160 C

Me, O, H, O, NH, (OEt)$_2$, Me

hydrolysis

Me, CH$_2$COOH, Me, COOH, H, O (trans)

NH$_3$ (10%) 220°C (1 hr)

3 N. HCl/EtOH

(Imide)

NaOH (10%), (10 hrs.)

Me, OCH$_2$COOH, Me, COOH, H, O (cis)

dil. NaOH

Me, CH$_2$CONH$_2$, Me, COOH, O, H

Chart 7

Trans- and *cis*-5-nitro- and 7-nitro-3-acetoxy-2-methyl-2,3-dihydrobenzofurans (**182–185**) have been recently prepared through nitration of a trans and cis mixture of 3-acetoxy-2-methyl-2,3-dihydrobenzofuran. The 7-nitro isomer (**184**) is reduced to the dimethylamino derivative (**186**); the acetoxy group is saponified to the aminoalcohol (**187**), which is converted to *cis*-7-dimethylamino-3-hydroxy-2-methyl-2,3-dihydrobenzofuran methiodide (**188**). X-ray diffraction analysis of **188** has shown the 2-methyl- and 3-hydroxy groups to be cis.[172] The cis isomers **182**, **184**, **186**, **187**, and **188** show $J_{2,3} = 6$ Hz, and the trans isomers **183** and **185** give the coupling constant $J_{2,3} = 2$ Hz. This agrees with assignments made on the basis of the Karplus equation.

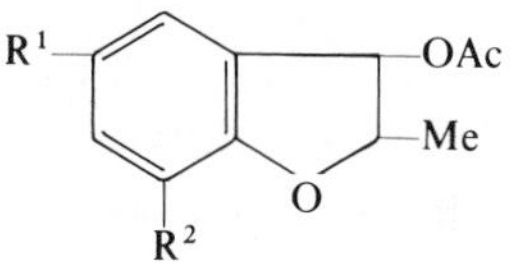

182 $R^1 = NO_2$, $R^2 = H$
184 $R^1 = H$, $R^2 = NO_2$

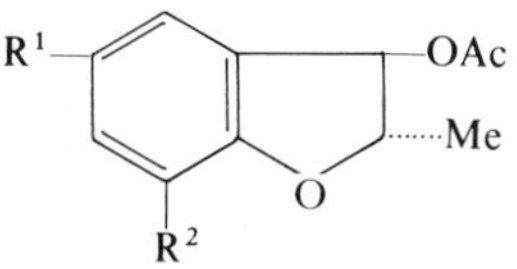

183 $R^1 = NO_2$, $R^2 = H$
185 $R^1 = H$, $R^2 = NO_2$

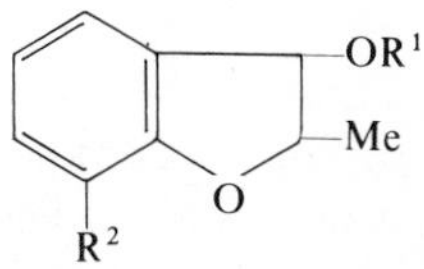

186 $R^1 = Ac$, $R^2 = NMe_2$
187 $R^1 = H$, $R^2 = NMe_2$
188 $R^1 = H$, $R^2 = NMe_3I$

Isolation and identification of the cis-trans stereoisomers of substituted 3-hydroxy- (or acetoxy)-2-methyl-2,3-dihydrobenzofurans have been achieved.[173] The cis-trans stereoisomers of 5-nitro- (**189a**) and 7-nitro-3-acetoxy-2-methyl-2,3-dihydrobenzofuran (**189b**) were separated by a combination of column chromatography and fractional crystallization. *cis*-7-Nitro-3-acetoxy-2-methyl-2,3-dihydrobenzofuran was converted to the dimethylamino isomer by catalytic reduction of the nitro group in methanol–formaldehyde. Hydrolysis of the ester and quaternization of the aminoalcohol gave *cis*-7-dimethylamino-3-hydroxy-2-methyl-2,3-dihydrobenzofuran methiodide (**190**). The stereoisomers of **189**, where $J_{2,3} = 2$ Hz, were assigned trans stereochemistry, and those where $J_{2,3} = 6$ Hz were assigned cis stereochemistry. These assignments were confirmed by determining the x-ray diffraction patterns of **190**, and this confirms the availability of the Karplus equation in predicting the stereochemistry of these substituted 2,3-dihydrobenzofurans.

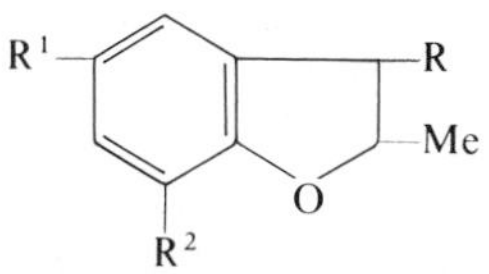

189a R = OAc, $R^2 = H$; $R^1 = NO_2$
189b R = OAc, $R^2 = NO_2$, $R^1 = H$

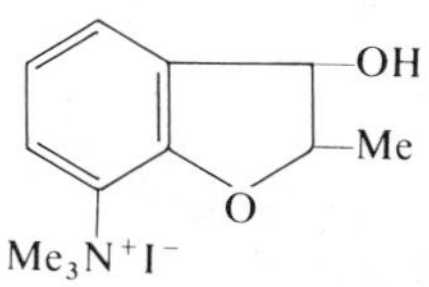

$Me_3N^+I^-$

190

G. Miscellaneous Reactions of 2,3-Dihydrobenzofurans

a. Heterocyclic Ring Opening

2,3-Dihydrobenzofuran behaves as a typical alkyl aryl ether; its treatment with hydroiodic acid effects the cleavage of the dihydrofuran ring[174] with the formation of *o*-ethylphenol (Eq. 40). Treatment of

(1) HI, (2) Zn/HCl; Et, OH (40)

2- and/or 3-aryl-2,3-dihydrobenzofuran with strong bases causes ring opening, accompanied in some cases by disproportionation. 3-Phenyl-2,3-dihydrobenzofuran gives, upon treatment with alcoholic potassium hydroxide at 200°C, a mixture of **191** and **192** (Eq. 41).[175]

Ph; alc. KOH, 200°C; CHCOOH + C=CH_2; OH; OH

191 **192** (41)

Reflux of 2-chloromethyl-2,3-dihydrobenzofuran solution in toluene with sodium brought out the formation of *o*-allylphenol (**193**) and 2-methylbenzofuran. However, with magnesium, only *o*-allylphenol was formed (Eq. 42).[176]

$CH_2CH{=}CH_2$; OH; Mg; CH_2Cl; Na/toluene

193

Me + **193** (42)

Hurd and Oliver[177] have reported the cleavage of the dihydrofuran ring in 2-methyl-2,3-dihydrobenzofuran (**194**). Similar cleavage is outlined[178] in Eq. 43.

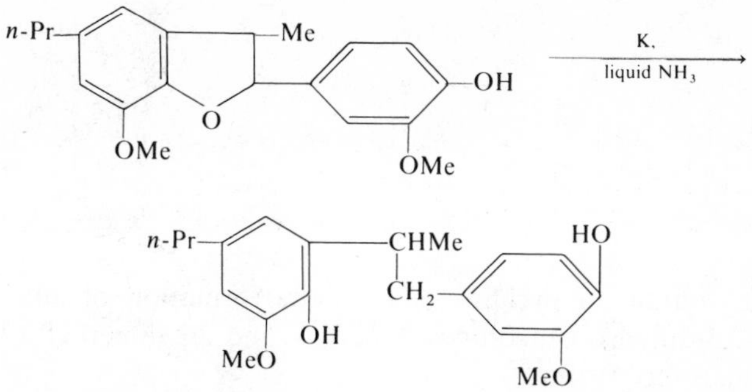

(43)

Treatment of **194** with lithium-ammonia and alcohol results in the formation of 2-methyl-2,3,4,7-tetrahydro- (**195**), and 2,3,4,5,6,7-hexahydrobenzofuran (**196**)[179] (Eq. 44). While the latter conditions are undoubtedly enough to cause some isomerization of **195**, base cleavage[177] or electron cleavage[180] of the dihydrobenzofuran is averted. Lithium, in a solution of ammonia that contains 15% absolute ethanol, effects the conversion of **197** and **198** to the corresponding 5-methoxy-2,3,4,7-tetrahydrobenzofurans. Thus **197** gave **199**; while **198** produced 98% 5-methoxy-2,3,4,7-tetrahydrobenzofuran (**200**). Reduction of **199** yielded **196**.[181] The elimination of methoxide ion has been observed during more vigorous metal-ammonia reduction.[182]

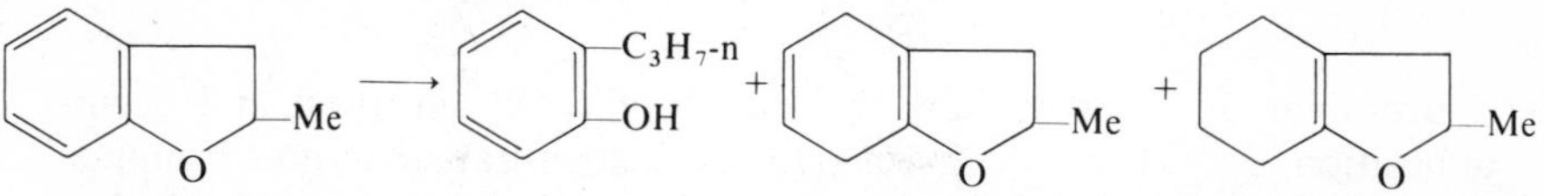

194 **195** **196** (44)

Birch reduction of *cis*-2-isoamyl-3,4-dimethyl-2,3-dihydrobenzofuran (**204**), using 5 g-atoms of lithium gave a mixture, after hydrolysis, of about 75% α,β-unsaturated ketones; acetylation, followed by chromatography on alumina, yielded **204** (10%) and, also, a mixture containing three parts of α,β-unsaturated keto acetate (**205**) and one part of saturated keto acetal (**206**).[183]

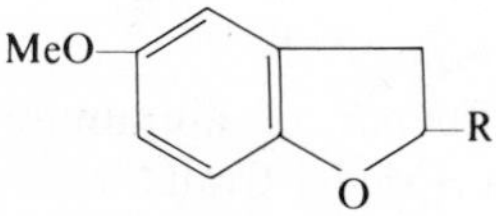

197 R = Me
198 R = H

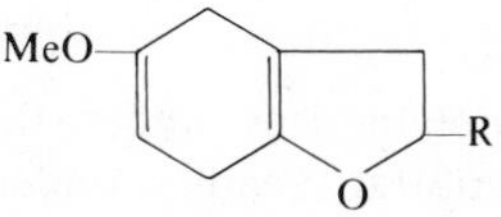

199 R = Me
200 R = H

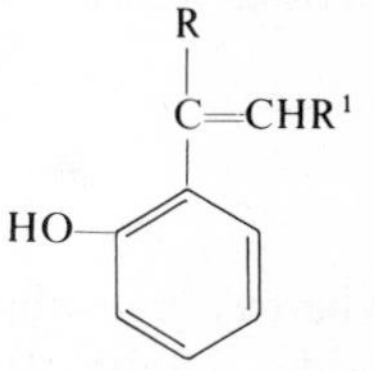

201 R = R^1 = H
202 R = Me, R^1 = H
203 R = R^1 = Me

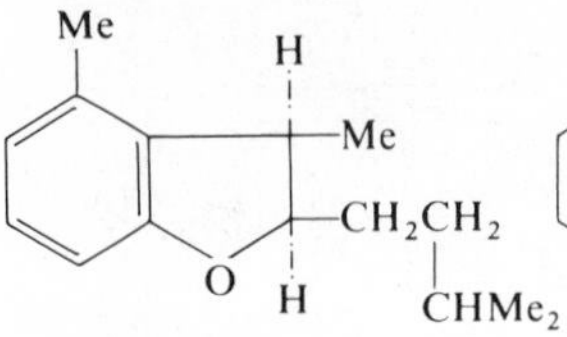

204 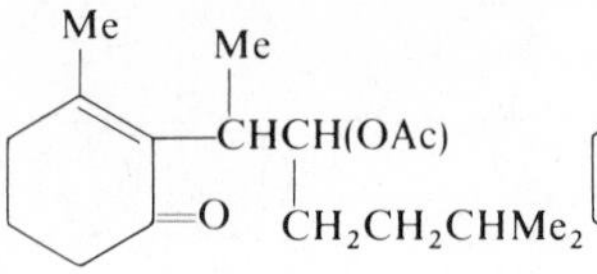 205 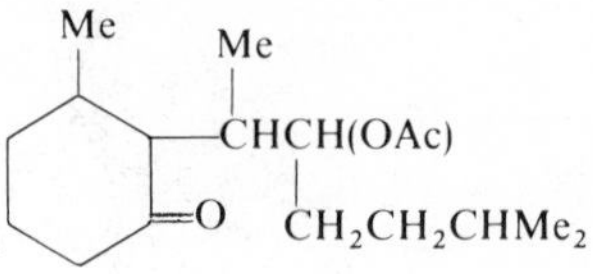 206

Metallic sodium in pyridine effected the formation of **201**, **202**, and **203** from 2,3-dihydro-benzofuran 2-methyl- and 2,3-dimethyl-2,3-dihydrobenzofuran, respectively.[184]

Another type of cleavage has been brought about by the action of tritylsodium–triphenylboron on 2,3-dihydrobenzofuran, yielding triphenylmethane derivatives **207** and **208**. Tritylsodium alone gives similar cleavage products, along with triphenylcarbinol.[185]

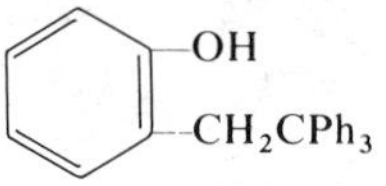

207 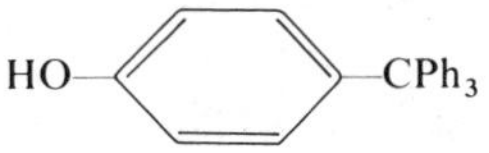 208

b. Halogenation and Chloromethylation

Bromination is directed by the heterocyclic oxygen atom and occurs at position 7.[186] Thus 5-methyl-2,3-dihydrobenzofuran undergoes bromination in carbon bisulfide solution to yield 5-methyl-7-bromo-2,3-dihydrobenzofuran (**209**).

Chloromethylation of 2-methyl-2,3-dihydrobenzofuran (**194**) affords 2-methyl-5-chloromethyl-2,3-dihydrobenzofuran together with *bis*(2-methyl-2,3-dihydrobenzofuran-5-yl)methane.[81] However chloromethylation of 5-methyl-2,3-dihydrobenzofuran gives 5-methyl-7-chloromethyl-2,3-dihydrobenzofuran which, when subjected to Wolff-Kishner-Minlon reduction, yields 5,7-dimethyl-2,3-dihydrobenzofuran.[186]

c. Acylation

Whereas benzofuran polymerizes under the influence of aluminum chloride, 2,3-dihydrobenzofuran readily undergoes Friedel-Crafts reaction.[78,81] 2,3-Dihydrobenzofuran reacts with acid chloride and/or half-esters of dicarboxylic acid chlorides in presence of aluminum chloride to yield the corresponding 5-acyl derivatives.[53,79,187] However 5-alkyl-substituted 2,3-dihydrobenzofurans on acylation give the 7-acyl derivatives. When the 5 and 7 positions are occupied, the acyl group is introduced in the 6 position.[186]

Acylation of 4-hydroxy-2,3-dihydrobenzofuran leads, after Seetharamiah[188] under the Hoesch reaction conditions, to the formation of the 7-acyl compound; however Miyano et al.[189-191] have reported the formation of a mixture of 5- and 7-acyl compounds on acylation of 2-alkyl-2,3-dihydrobenzofurans. Kawase et al.[192] have shown that acylation of 4-hydroxy-2,3-dichlorobenzofuran with phenylacetic acid in presence of polyphosphoric acid afforded a mixture of the 5- and 7-acylated compounds. A Hoesch reaction between 4,6-dihydroxy-2,3-dihydrobenzofuran (**210**) and the nitrile (**211**) resulted in the formation of 5-(*o*-carboxymethoxyphenylacetyl)-4,6-dihydroxy-2,3-dihydrobenzofuran (**212**) and small amount of the 7 isomer (**213**) (Eq. 43).[193]

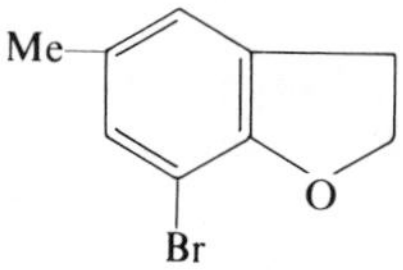

Me, Br

209

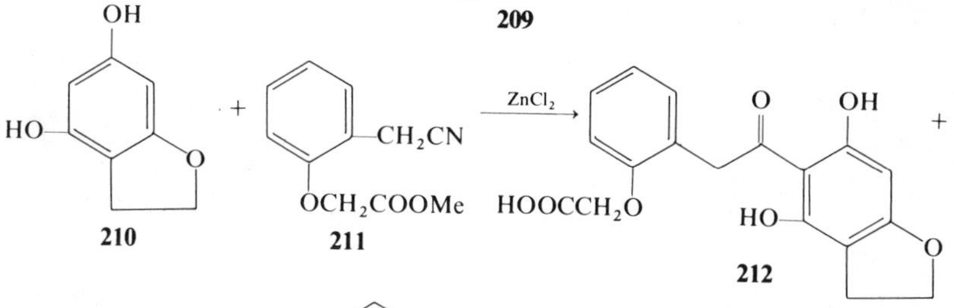

OH, HO, O

210

+ CH_2CN, OCH_2COOMe

211

$\xrightarrow{ZnCl_2}$

O, OH, $HOOCCH_2O$, HO, O

212

+

O, O, $HOOCCH_2O$, HO, OH

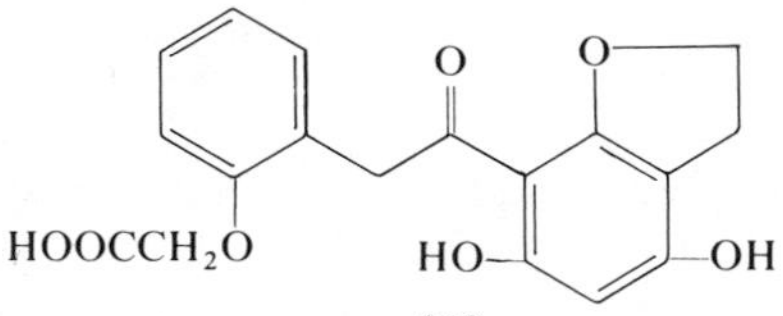

213

Introduction of a formyl group in position 2 is effected as outlined in Eq. 44.[194] Reduction of 5-formyl-2,3-dihydrobenzofuran with lithium aluminum hydride yielded the 5-hydroxymethyl compound.[195]

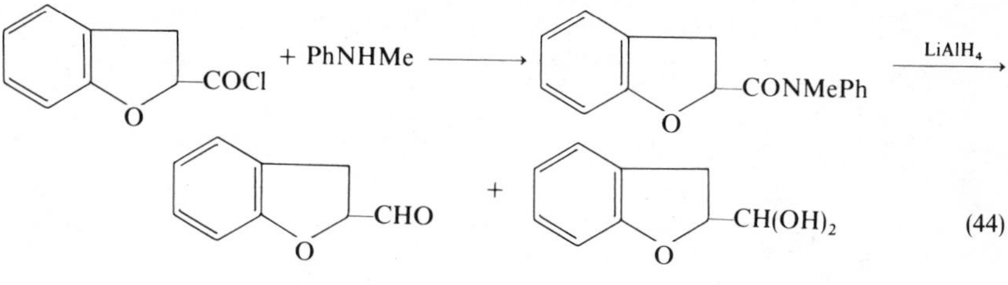

COCl, O + PhNHMe ⟶ CONMePh, O $\xrightarrow{LiAlH_4}$

CHO, O + $CH(OH)_2$, O (44)

d. Dehydrogenation

2,3-Dihydrobenzofurans are dehydrogenated via halogenation, followed by dehydrohalogenation in presence of suitable basic reagent, preferably tertiary amine. This yields the corresponding benzofurans. This procedure is applicable to dihydrobenzofurans in which the benzene nucleus is substituted with lower alkoxy, acyl, or acyloxy radical.[19,83] Halogenation reagents may be *N*-bromosuccinimide, *N*-chloro isomer, *N*-bromophthalimide and/or *N*-bromoacetamide. Tertiary bases, for example, *N*,*N*-dimethylaniline, collidine, quinoline, and/or pyridine, may as well be used.

e. Mercuration

Mercury salts react with *o*-allylphenol to give mercurated 2-methyl-2,3-dihydrobenzofuran (**214**),[19] which exhibits stability toward acids and shows reactions typical of any of the known hydroxymercuric compounds (Eq. 45).[196]

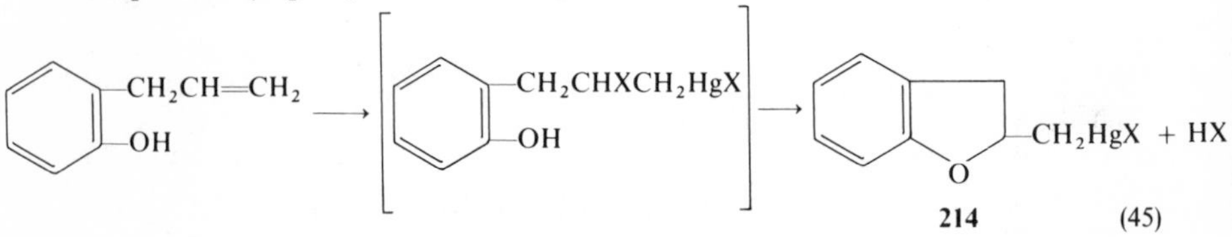

$CH_2CH{=}CH_2$ / OH → [CH_2CHXCH_2HgX / OH] → CH_2HgX + HX

214 (45)

f. Miscellaneous

Treatment of 2,4,6,7-tetramethyl-2,3-dihydrobenzofuran with chlorosulfonic acid, followed by reduction of the reaction product with lithium aluminum hydride, effected the formation of the 5-mercapto compound (**215**).[197] However treatment of 2,3-dichloro-2,3-dihydrobenzofuran with *O*,*O*-diethyldithiophosphate yields 3-chloro-2,3-dihydrobenzofuran-2-thio-*S*-(*O*,*O*-diethylphosphorodithioate).[198]

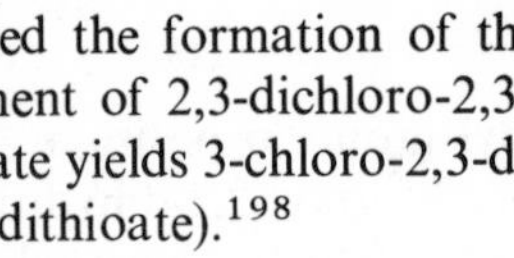

Me, HS, Me, Me, O, Me

215

2-Methylthiomethyl-2,3-dihydrobenzofuran (**216**) undergoes oxidation with hydrogen peroxide to afford 2-methylsulfonylmethyl- (**217**),[200] and 2-methylsulfinylmethyl-2,3-dihydrobenzofuran (**218**).[199] Sulfur monochloride reacts with 2,3-dihydrobenzofuran to yield di-(2,3-dihydrobenzofuranyl)trisulfide.[201]

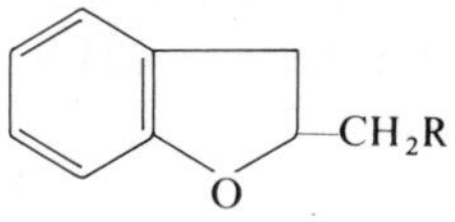

216 R = SMe **217** R = SO_2Me **218** R = SOMe

6-Hydroxy-2,3-dihydrobenzofuran undergoes facile carbonylation with potassium bicarbonate and carbon dioxide[202,203] in glycerin medium to yield the 5-carboxy compound (**219**). The acid chloride of the latter acid is readily converted to the diazoketone which undergoes smooth transformation to 5-(ω-methoxy)acetyl-6-hydroxy-2,3-dihydrobenzofuran (**220**). Compound **219** was also obtained by oxidation of 5-formyl-6-hydroxy-2,3-dihydrobenzofuran.[203]

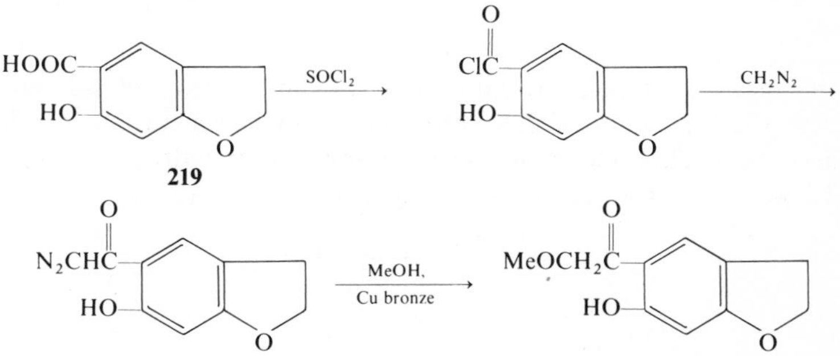

219 $\xrightarrow{SOCl_2}$ $\xrightarrow{CH_2N_2}$ $\xrightarrow{\text{MeOH, Cu bronze}}$ **220**

2,4,7-Trimethyl-5-hydroxybenzofuran couples with diazotized *o*-anisidine to give a purple azo dye oriented at the 6 position; however 6-methyl-5-hydroxy-2,3-dihydrobenzofuran does not undergo the coupling reaction.[204]

The ultraviolet and infrared spectra of a variety of alkyl-substituted 2,3-dihydrobenzofurans have been studied.[205] In addition to the usual bands due to aromatic and aliphatic C–H bonds, and the aromatic C=C bonds, the dihydrofuran ring system is characterized by three strong bands: one between 1200 and 1260 cm^{-1} (attributed to the carbonyl group), another between 980 and 1000 cm^{-1} and the third at 944 cm^{-1}, the latter two appear to be influenced by substitution.

2. Bz-Dihydrobenzofurans

The literature contains few examples of bz-dihydrobenzofurans. Reaction of olefins with active methylene compounds in the presence of thallic acetate has been recently investigated.[206] Equation 46 exemplifies the synthesis of 3-acetyl-2,4-dimethyl- (**221**), and 3-carbethoxy-2-methyl-4-phenyl-4,5-dihydrobenzofuran (**222**).

$$Ac_2CH_2 + MeCH{=}CH_2 \xrightarrow[HClO_4\ (1\text{-}2^\circ C)]{(AcO)_3Tl/AcOH} \mathbf{221} + Me_2CO + AcOCH_2CH(OAc)Me$$

$$AcCH_2COOEt + PhCH{=}CH_2 \xrightarrow[BF_3\text{-}Et_2O]{(AcO)_3Tl,} \mathbf{222} \qquad (46)$$

Condensation of benzoin with pyrogallol in presence of sulfuric acid leads to the formation of 2,3-diphenyl-6,7-dihydrobenzofuran[207] (**223a**); similarly pyrogallol and *p*-anisoin in presence of anhydrous zinc chloride yield 2,3-di-(*p*-methoxyphenyl)-6,7-dihydrobenzofuran (**223b**).[207]

223a $R = R^1 = Ph$
223b $R = R^1 = C_6H_4OMe\text{-}p$

3. Tetrahydrobenzofurans

Alkylation of enolate ions, followed by hydrolysis–dehydration of the resulting *β*-chloroallyl ketone (**224**) with sulfuric acid, gave substituted tetrahydrobenzofurans (**225**) (Eq. 47).[208]
The "one-pot" hydrolysis–dehydration of **224a** proceeded in about 80% yield to the known 2-methyl-4,5,6,7-tetrahydrobenzofuran (**225a**). In like manner, **224b** proceeded in high yield to 2,7-dimethyl-4,5,6,7-tetrahydrobenzofuran (**225b**).

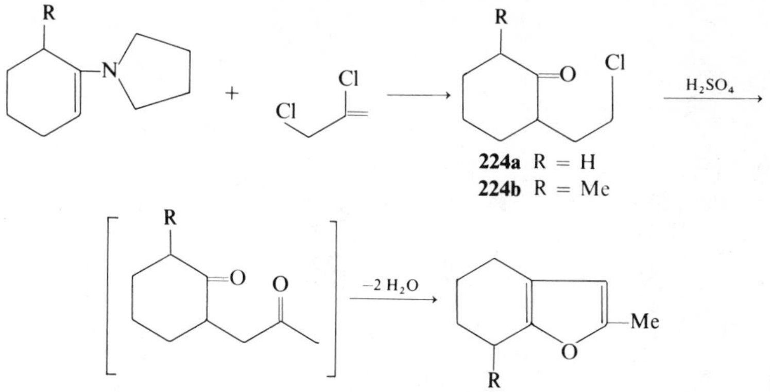

224a R = H
224b R = Me

225a R = H
225b R = Me

(47)

Cyclization of 1,3-dioxo-2-phenacylcyclohexanone, obtained by treatment of phenacyl bromide and dihydroresorcinol in the presence of potassium hydroxide with sulfuric acid, yields 4,5,6,7-tetrahydro-4-oxo-2-phenylbenzofuran (**226**).[209,210] On Wolff-Kishner reduction, this affords 4,5,6,7-tetrahydro-2-phenylbenzofuran (**227**). The latter dehydrogenates readily with sulfur to yield 2-phenylbenzofuran (**228**). Treatment of **227** with mercuric chloride, sodium acetate, and ethanol gives the mercuri compound (**229**), (Eq. 48).[211]

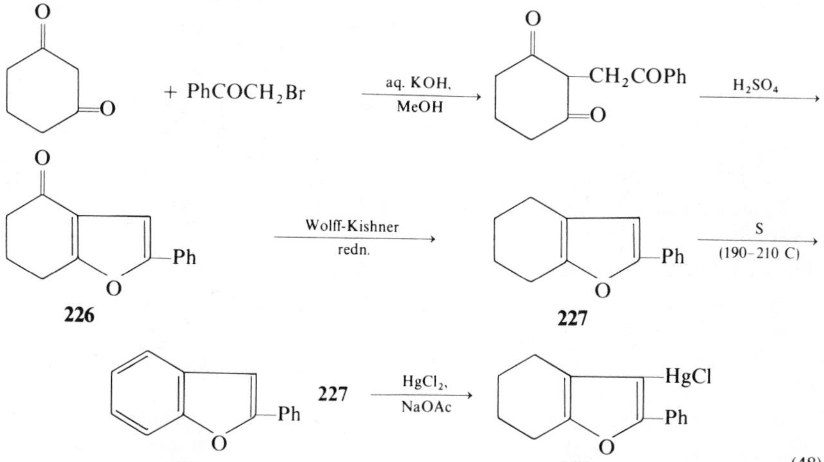

226 227 228 229

(48)

2-Phenyl-3-fluoro-4,5,6,7-tetrahydrobenzofuran (**231**) is obtained upon treatment of 2-benzylidenecyclohexanone (**230**) with difluorocarbene; the latter adds to the double bond by 1,2 addition, and the adduct thus obtained rearranges with elimination of hydrogen fluoride.[212]

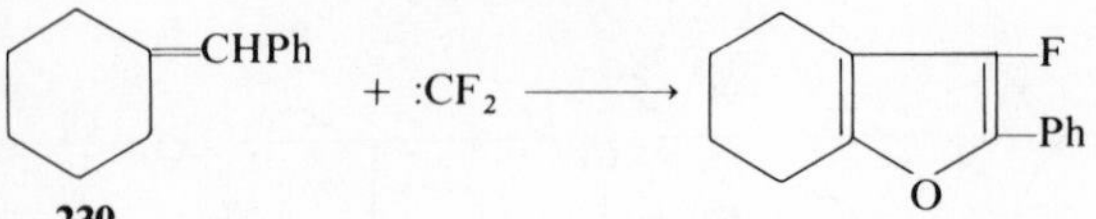

230 + $:CF_2$ → 231

Pyrolysis of *d*-sultone (**233**), obtained from pulegone and/or isopulegone (**232**), results in the formation of menthofuran (**234**) (Eq. 49).[213-215]

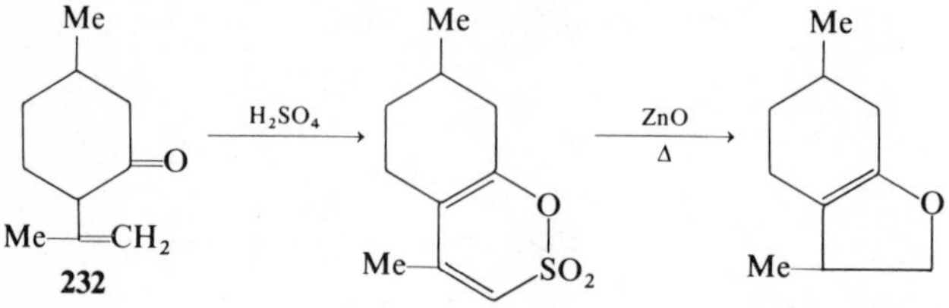

$$\mathbf{232} \xrightarrow{H_2SO_4} \mathbf{233} \xrightarrow[\Delta]{ZnO} \mathbf{234} \qquad (49)$$

Condensation of 3-bromo-2-acetoxy-1-cyclohexene with ethyl malonate in presence of sodium chloride affords 2-ethoxy-3-carbethoxy-4,5,6,7-tetrahydrobenzofuran (**235a**);[216] with ethyl acetoacetate, 2-methyl-3-carbethoxy-4,5,6,7-tetrahydrobenzofuran, **235b**, is produced.[217-220]

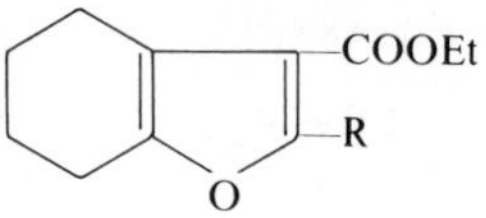

235a R = oEt
235b R = Me

Hexahydro-3-methyl-2-benzofuranone (**236**) is reduced by action of dialkylaluminum hydride to give 4,5,6,7-tetrahydro-3-methylbenzofuran (**237**);[221] this readily undergoes oxidation by atmospheric oxygen to afford carbonyl compounds, thus rendering the yield of **237** unsatisfactory. Rearrangement of **236** with isobutylaluminum hydride affords **237**.[223]

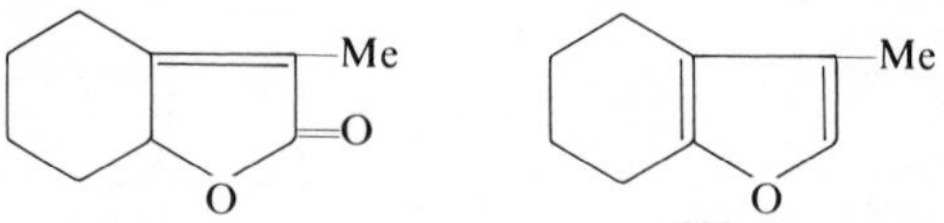

236 **237**

3-Acetyl-2,4,5,6,7,7α-hexahydrobenzofuran-2-one (**238**) undergoes rearrangement in presence of hydrochloric–acetic acid mixture or upon slow distillation, and affords 4,5,6,7-tetrahydro-2-methyl-3-benzofurancarboxylic acid (**239**) (Eq. 50).[222]

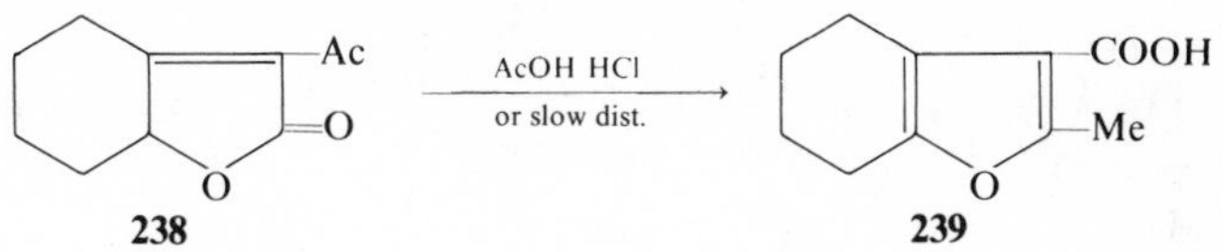

238 $\xrightarrow[\text{or slow dist.}]{\text{AcOH HCl}}$ **239** (50)

α-Haloketones (**240**) are transformed in one step to the corresponding benzofurans upon reflux with cuprous acetylides.[224] Cyclization through the copper-coordinated enol is consistent with the ready preparation of benzofurans (Eq. 51).[225]

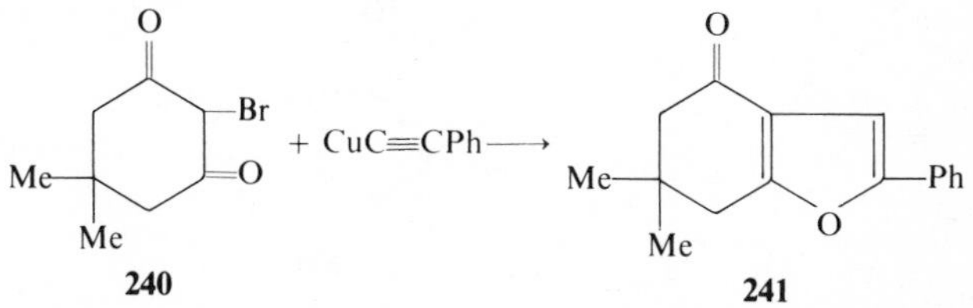

240 + CuC≡CPh ⟶ **241** (51)

α-Bromomethylacetylene[226] reacts with 1,3-cyclohexanedione to give 2-propargylcyclohexane-1,3-dione (**242**); the latter, when treated with zinc carbonate, yields 2-methyl-4-oxo-4,5,6,7-tetrahydrobenzofuran (**243**, R = Me) (Eq. 52).

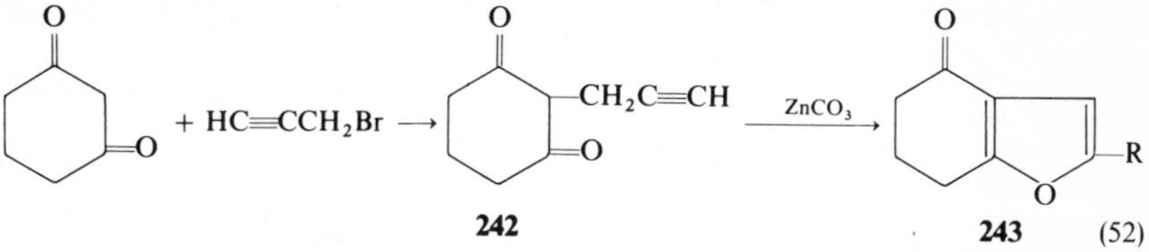

+ HC≡CCH$_2$Br ⟶ **242** $\xrightarrow{\text{ZnCO}_3}$ **243** (52)

The benzofuranone (**244**) reacts with vinylmagnesium bromide to yield the carbinol (**245**) which, when allowed to react with thiourea, yields *S*-(2-(2-alkyl (or aryl)-4,5,6,7-tetrahydrobenzofuran-4-ylidene)ethyliso-thiouran acetate (**246**) (Eq. 53).[227]

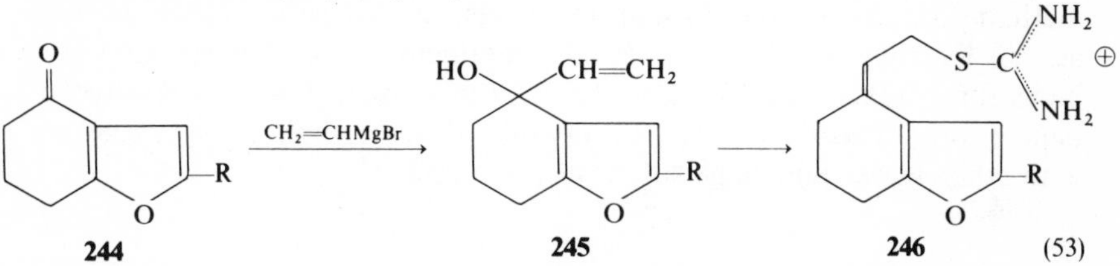

244 $\xrightarrow{\text{CH}_2\text{=CHMgBr}}$ **245** ⟶ **246** (53)

Dimedone, in presence of triethylamine, undergoes condensation with α-(*p*-nitrophenyl)-β-nitro-β-bromoethylene to yield 2-nitro-3-(*p*-nitro-phenyl)-4-oxo-6,6-dimethyl-2,3,4,5,6,7-hexahydrobenzofuran (**247**); when refluxed with triethylamine, it gives 3-(*p*-nitrophenyl)-4-oxo-6,6-dimethyl-4,5,6,7-tetrahydrobenzofuran (**248**) (Eq. 54).[228,229]

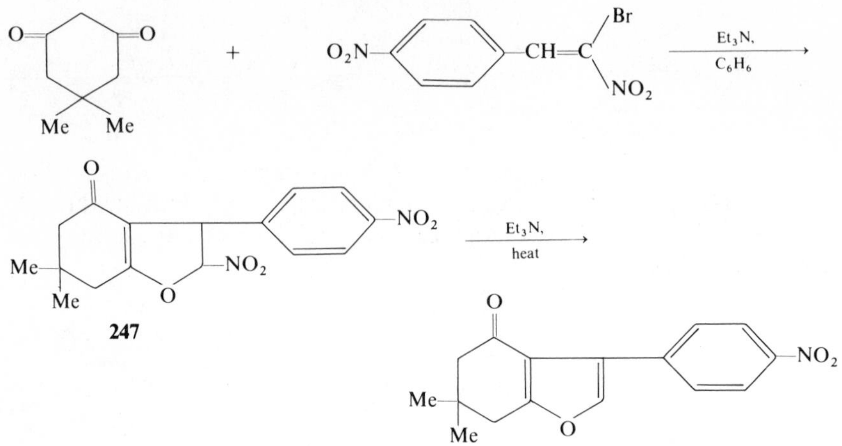

(54)

Dimedone condenses with chloroacetone[230] to afford **249** as outlined in Eq. 55.

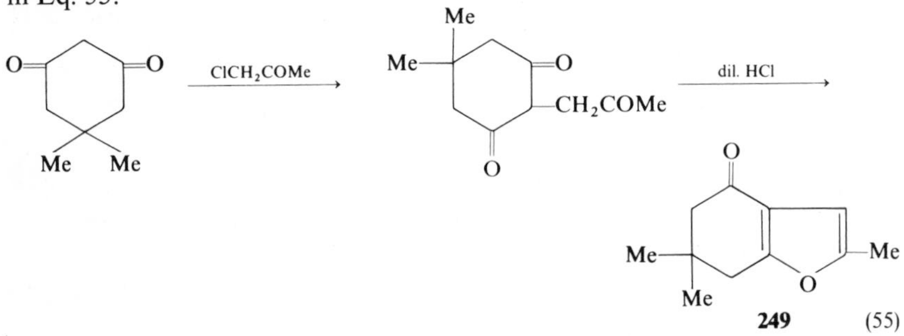

(55)

Methylation of tetra-*c*-methylphloroacetophenone (**250**) with diazomethane results in the formation of its methyl ether (**251**) as well as 3-hydroxy-4,6-dioxo-3,5,5,7,7-pentamethyl-2,3,4,5,6,7-hexahydrobenzofuran (**252**) and 4,6-dioxo-3,5,5,7,7-pentamethyl-4,5,6,7-tetrahydrobenzofuran (**253**); similarly, 2-acetyldihydroresorcinol, upon treatment with ethereal diazomethane solution, yields **254**.[231]

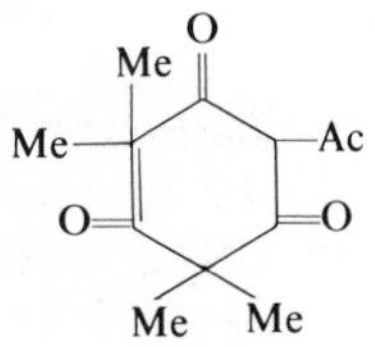

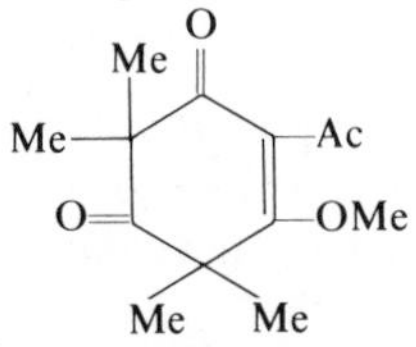

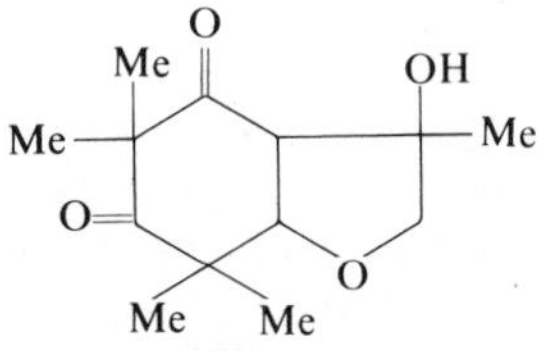

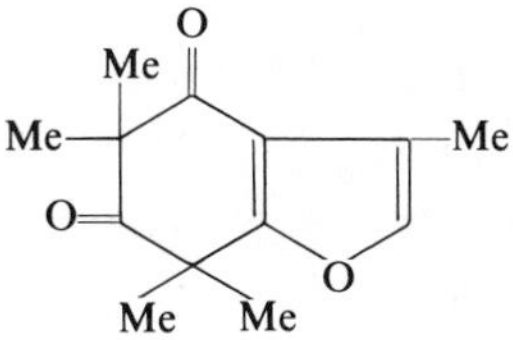

253

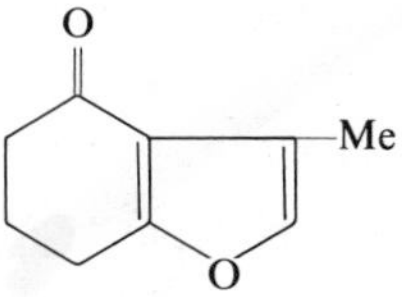

254

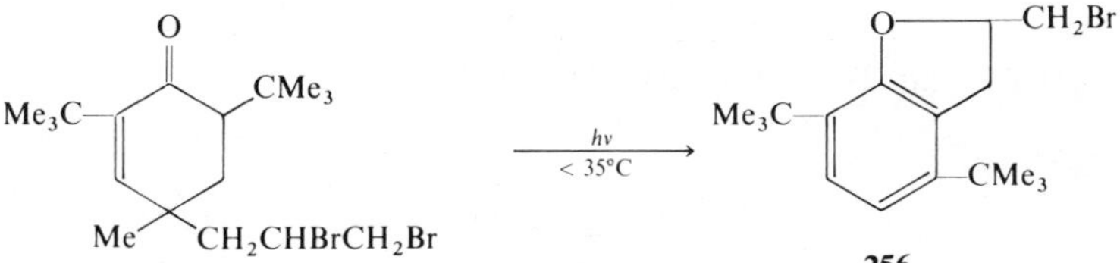

255

256

Photochemical rearrangement of 2,6-di-*t*-butyl-4-methyl-4-(2,3-dibromopropyl)cyclohexen-2,5-dienone (**255**) results in the formation of the dihydrobenzofuran (**256**).[232] If care is taken to keep the temperature of the solution below 35°C during irradiation of **255** and if the solvent is evaporated at low temperature, the ultraviolet and infrared spectra of the photolysis product show peaks that would be expected of a structure like **257**, which loses the elements of methyl bromide and is converted to **256**.[232]

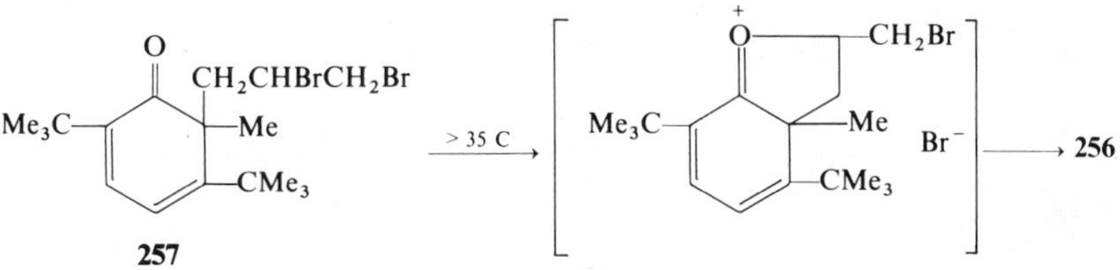

257

Cyclization of β-(5-methyl-2-furyl)butyric acid with phosphorus pentachloride and stannic chloride in benzene has been shown[233] to afford 4,5,6,7-tetrahydro-2-methyl-4-oxobenzofuran (**243**, R = Me).

4-Oxo-4,5,6,7-tetrahydrobenzofurans (**258**) react readily with amines to give the corresponding tetrahydroindoles (**259**).[210]

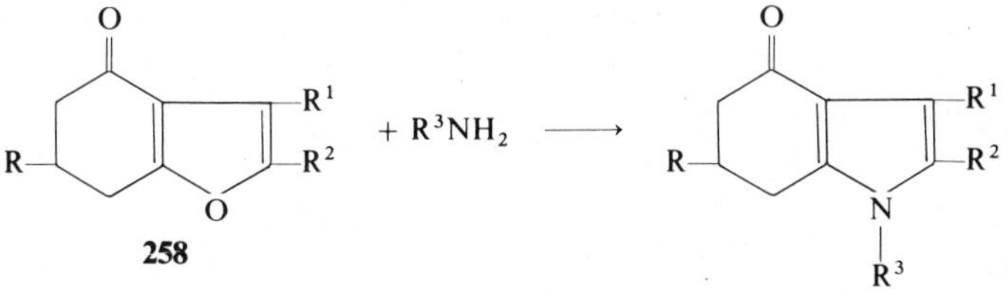

258

259

4. Hexahydrobenzofurans

Cyclic enamines react with styrene to give cyclic *O*,*N*-ketals, having a furan nucleus.[234] Thus treatment of 2-phenyl-8-(*N*-pyrrolidyl)perhydrobenzofuran (**260**), obtained from 1-pyrrolidino-1-cyclohexene with oxalic acid, gave 2-phenyl-2,3,4,5,6,7-hexahydrobenzofuran (**261**) which, on acidic hydrolysis, produced **262**, and, by aromatization with sulfur, was transformed into 2-phenylbenzofuran (**263**) (Eq. 56).

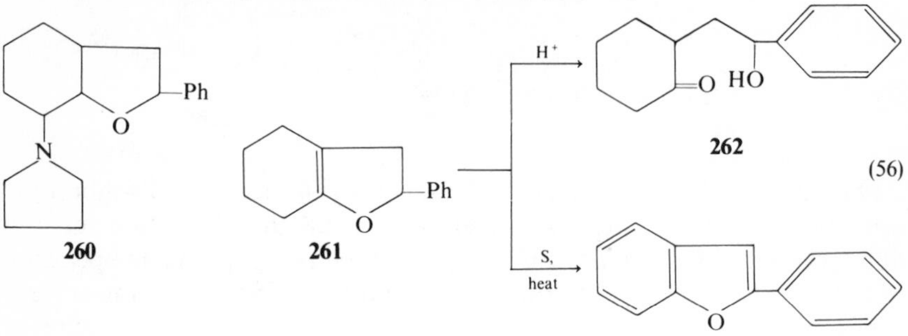

Ph O N **260** Ph O **261** H^+ O HO **262** S, heat O **263** (56)

Dihydroresorcinol reacts readily with 2-(β-bromo-β-nitro)-α-phenylethylene in presence of triethylamine to yield 2-nitro-4-oxo-2,3,4,5,6,7-hexahydrobenzofuran (**264**).[228]

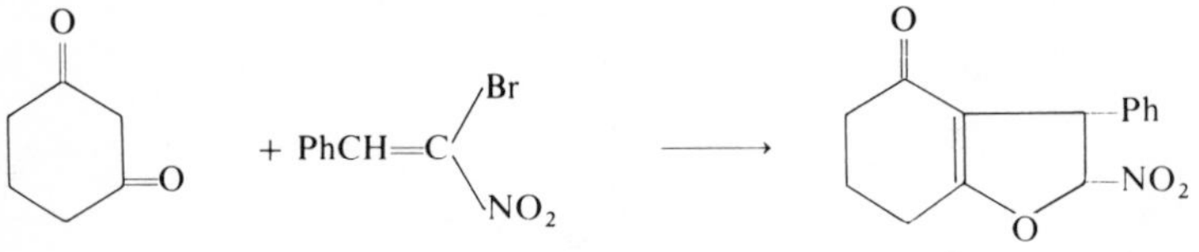

O O + PhCH=C(Br)(NO_2) → O Ph NO_2 O **264**

The base-catalyzed condensation product of dimedone and β-nitrostyrene results in the formation of **265,** which has also been obtained in 40% yield when the reaction was carried out in ether in the presence of triethylamine.[235]

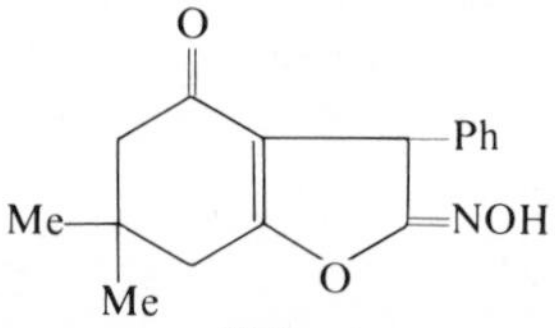

O Ph Me =NOH Me O **265**

Denitration may take place when the condensations are effected with a fourfold excess of triethylamine and the reaction time is extended (Eq. 57).[229]

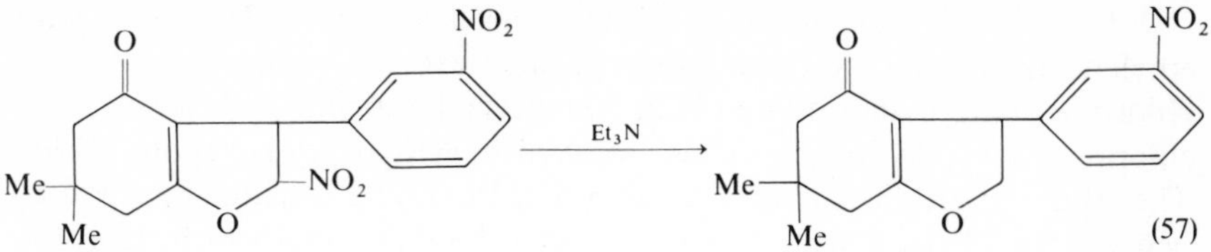

Et$_3$N (57)

Dimedone reacts with the oxymercurial of propylene to form a mixture of 2,6,6-trimethyl- (**266**), and 3,6,6-trimethyl-4-oxo-2,3,4,5,6,7-hexahydrobenzofuran (**267**) under simultaneous demercuration (Eq. 58).[236]

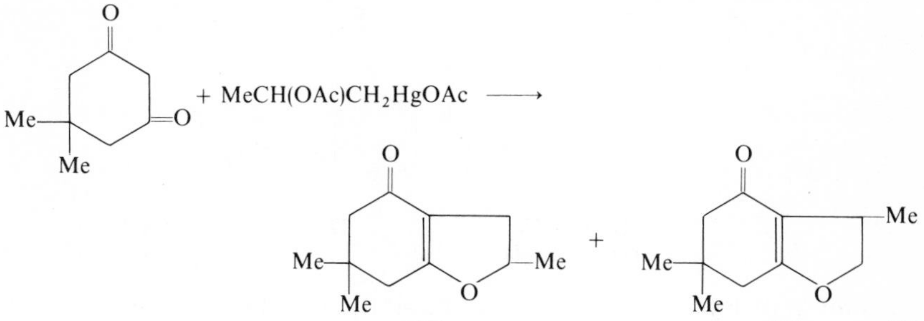

+ MeCH(OAc)CH_2HgOAc ⟶

266 + **267** (58)

5. Octahydrobenzofurans

Oxonin (**268**) undergoes thermal bond-relocation to what is believed to be 8,9-dihydrobenzofuran (**269**).[237] The stereochemistry of **269** has been shown to be the cis form. Catalytic hydrogenation of the thermal tautomer **270** produced *cis*-octahydrobenzofuran (**269**), identical with that obtained by catalytic hydrogenation of benzofuran or via a known, stereochemically consistent, multistep procedure employing *o*-methoxybenzoic acid as starting material.[238]

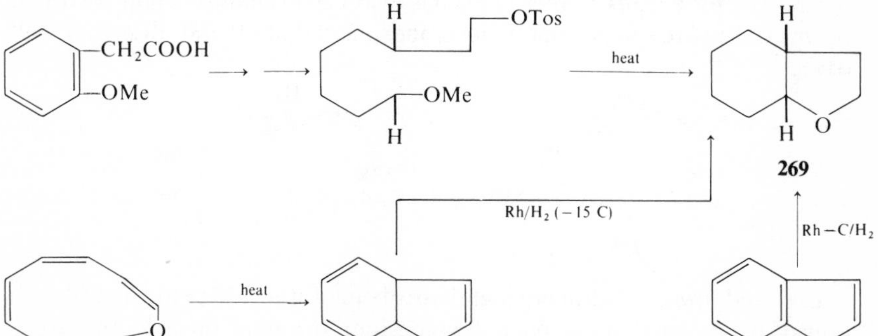

heat; Rh/H_2 (−15 C); Rh−C/H_2

269 **268** **270**

N-Cyclohexylcyclohexylimine bromomagnesium salt (**271**) reacts with ethylene oxide at 0°C to yield, after treatment with hydrochloric acid, 2-(2′-chloromethyl)cyclohexanone (**272a**); however treatment with ammonium chloride prior to hydrolysis yields 2-(2′-hydroxyethyl)cyclohexanone (**272b**). The latter, on distillation in vacuo is cyclized to give 2,3,4,5,6,7-hexahydrobenzofuran (**273**), which is readily reduced to *cis*-octahydrobenzofuran (**274**).[238,239] Analogously, *cis*-2-methyloctahydrobenzofuran (**275**) has been obtained.[240]

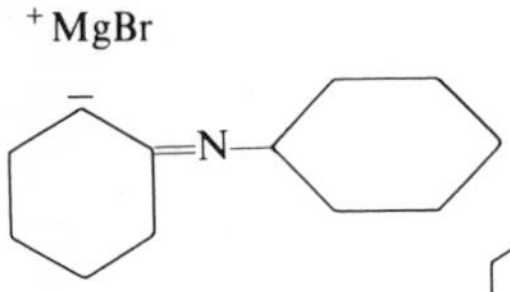

$^+$MgBr

271

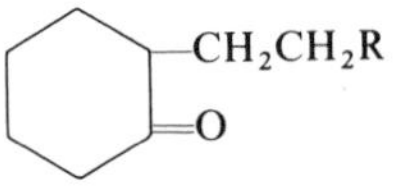

272a R = Cl
272b R = OH

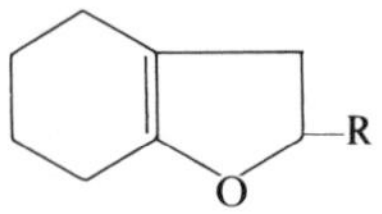

273

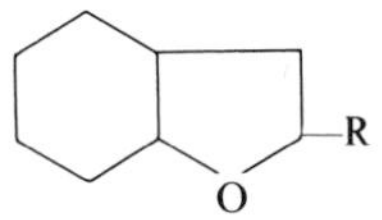

274 R=Me
275 R=Me

Trans-2-allylcyclohexanol, on isomerization with phosphoric acid, yields 2-methyloctahydrobenzofuran.[241] *Cis*-2-(2-methoxycyclohexyl)-ethanol (**276**), upon treatment with *p*-toluenesulfonyl chloride and pyridine, yields *cis*-perhydrobenzofuran (**274**). However *trans*-**278** is obtained from *trans*-2-(2-hydroxycyclohexyl)ethanol (*trans*-**277**).[242]

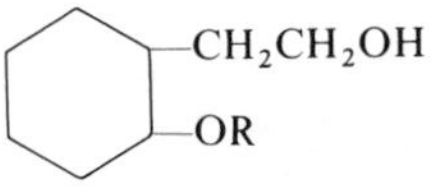

276 R = Me
277 R = H

Hydrogenation 2,2-dimethyl-2,3-dihydrobenzofuran in acetic acid over Rh–Al^2O^3 catalyst under pressure yields *cis*-2,2-dimethyloctahydrobenzofuran (**278**); the trans isomer (**279**) has also been obtained upon treatment of *trans*-2-hydroxy-α,α-dimethylcyclohexane ethanol and dimethyl sulfoxide.[243,244]

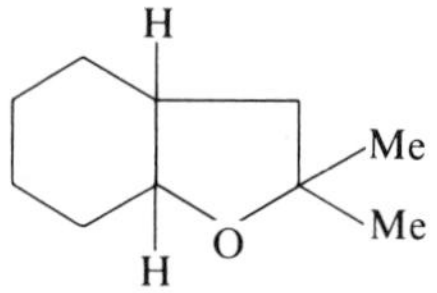

278

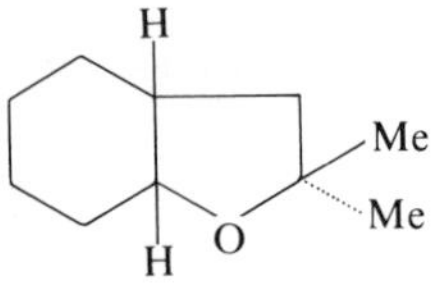

279

Cis- and *trans*-2,2-diphenyloctahydrobenzofuran (**281**) are readily obtained by the action of phenylmagnesium bromide on *cis*- and *trans*-2-hydroxycyclohexylacetic acid lactone (**280**).[245]

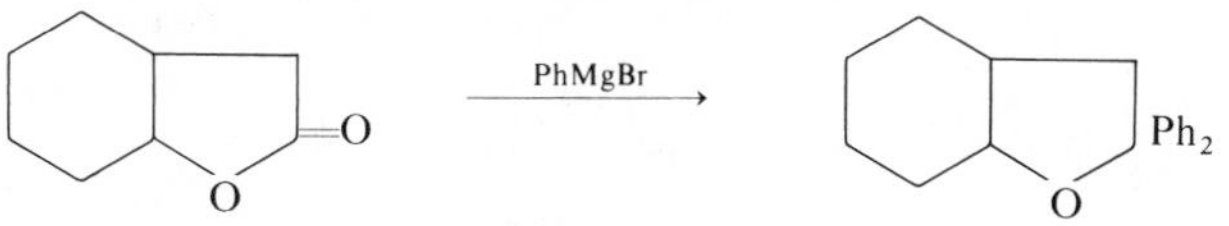

PhMgBr

280 281

The synthesis of 2-keto-3-substituted octahydrobenzofurans has attracted the interest of many authors. Photochemical addition of cyclohexene to ethyl pyruvate resulted in the formation of the stereoisomeric hydroxy ester (**282**, R = Et) mixture which, upon treatment with *p*-toluene-sulfonlc acid gave a mixture of neutral and acidic products. From the neutral material the lactone (**283**) has been isolated.[246]

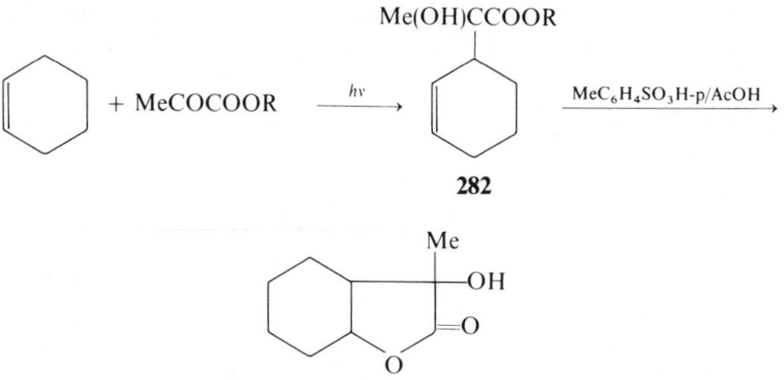

Me(OH)CCOOR

+ MeCOCOOR $\xrightarrow{h\nu}$ $\xrightarrow{MeC_6H_4SO_3H\text{-p/AcOH}}$

282

Me OH O O

283

Benzylpyruvic acid reacts with cyclohexanone in alkaline medium to furnish two compounds; γ-phenyl-α-hydroxy-α-(2-oxocyclohexyl)butyric acid (**284**) which, upon treatment with potassium borohydride, yields **285.** Acid dehydration of **284** is accompanied by rearrangement (equivalent to lactonization of the hydrated ketone) to form **286.**[247]

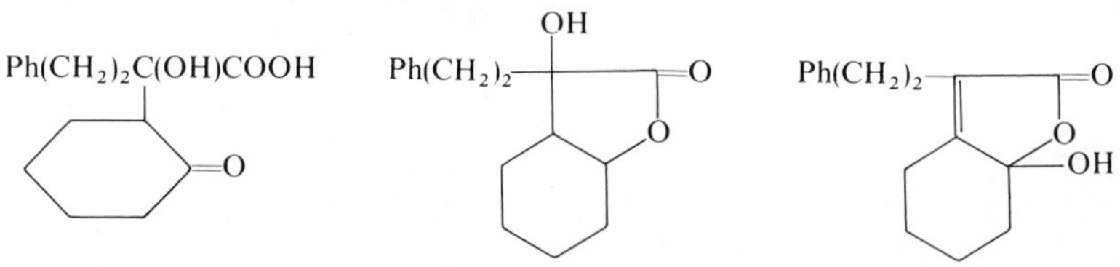

$Ph(CH_2)_2C(OH)COOH$ OH $Ph(CH_2)_2$ $Ph(CH_2)_2$

284 285 286

Synthesis of **287** has been achieved (Eq. 59).[248,249] The last step in the sequence is thought to proceed first by displacement of the carboxylate anion by iodide in a manner similar to the use of lithium iodide in the cleavage of methyl esters, then by decarboxylation with concurrent β-elimination of trimethylamine.

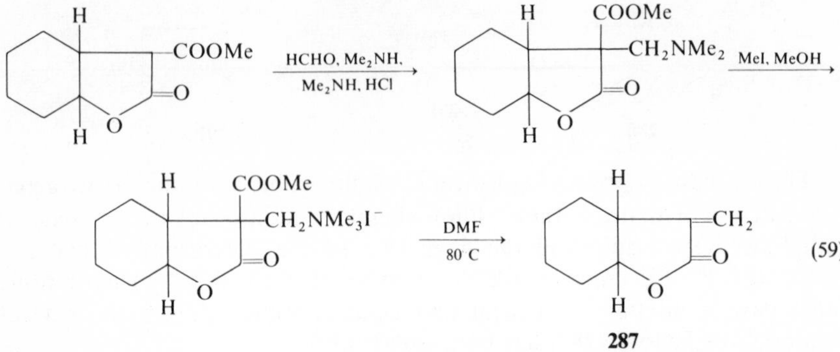

(59)

287

Lactonization of the sodium derivative of the product obtained by the interaction of ethyl malonate with cyclohexene oxide, upon acidification, results in the formation of ethyl 2-keto-octahydrobenzofuran-3-carboxylate (**288a**). Condensation of the above sodium derivative with β-bromopropionic acid furnished ethyl 2-keto-octahydrobenzofuran-3-propionate (**288b**).[250]

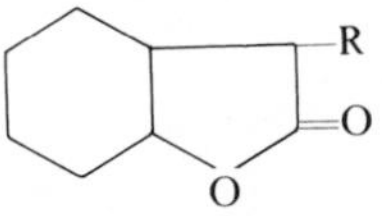

288a R = COOEt
288b R = CH_2CH_2COOEt

Catalytic thermal decomposition of octahydrobenzofuran has recently been reported to yield (besides 2,3-dihydrobenzofuran and benzofuran), the furan ring cleavage products: *o*-ethylphenol, phenol, cyclohexylacetaldehyde, toluene, ethylbenzene, and benzene.[251]

References

1. R. v. Stoermer and F. Göhl, *Chem. Ber.*, **36,** 2873 (1903).
2. H. Alexander, *Chem. Ber.*, **25,** 2409 (1892).
3. R. v. Stoermer and B. Kahlert, *Chem. Ber.*, **34,** 1810 (1901).
4. R. v. Stoermer and M. Schäffer, *Chem. Ber.*, **36,** 2863 (1903).
5. R. E. Rindfusz, *J. Am. Chem. Soc.*, **41,** 665 (1919).
6. R. E. Rindfusz, P. M. Ginnings, and V. I. Harnack, *J. Am. Chem. Soc.*, **42,** 157 (1920).
7. K. Fries and P. Moskopp, *Justus Liebigs Ann. Chem.*, **372,** 187 (1910).
8. R. v. Stoermer, C. W. Chydenius, and E. Schinn, *Chem. Ber.*, **57,** 72 (1924).
9. L. Claisen and E. Tietze, *Chem. Ber.*, **58,** 275 (1925).
10. L. Claisen and E. Tietze, *Chem. Ber.*, **59,** 2344 (1926).
11. J. B. Niederl and E. A. Storch, *J. Am. Chem. Soc.*, **55,** 4549 (1933).
12. R. C. Elderfield and V. B. Meyer, "Heterocyclic Compounds," Vol. 2, R. C. Elderfield, Ed., Wiley, New York, 1950, p. 10.

13. S. Wawzoneck, "Heterocyclic Compounds," Vol. 2, R. &. Elderfield, Ed., Wiley, New York, 1950, p. 393.
14. R. T. Arnold and J. Moran, *J. Am. Chem. Soc.*, **64,** 2986 (1942).
15. L. Claisen and F. Kremers, *Justus Liebigs Ann. Chem.*, **418,** 69 (1919)
16. F. Kremers, F. Roth, and L. Claisen, *Justus Liebigs Ann. Chem.*, **442,** 210 (1925).
17. L. F. Fieser, *J. Am. Chem. Soc.*, **48,** 3201 (1926); **49,** 857 (1927).
18. L. Claisen, Ger. Pat. 394,797 (1922); *Chem. Abstr.*, **18,** 2175 (1924).
19. R. Adams, F. L. Roman, and W. N. Sperry, *J. Am. Chem. Soc.*, **44,** 1781 (1922).
20. R. Adams and R. E. Rindfusz, *J. Am. Chem. Soc.*, **41,** 648 (1919).
21. H. Normant, *Compt. Rend.*, **219,** 163 (1944).
22. D. P. Brust, D. S. Tarbell, and S. M. Hecht, *Proc. Natl. Acad. Sci., U.S.*, **53,** 233 (1965); *Chem. Abstr.*, **62,** 14607 (1965).
23. S. T. Young, J. R. Turner, and D. S. Tarbell, *J. Org. Chem.*, **28,** 928 (1963).
24. D. S. Tarbell, R. M. Carman, D. D. Chapman, S. E. Cremer, A. D. Cross, K. R. Huffman, M. Kunstmann, N. J. McCorkindale, J. G. McNally, Jr., A. Rosowsky, F. H. L. Varino, and R. L. West, *J. Am. Chem. Soc.*, **83,** 3096 (1961).
25. C. D. Hurd and W. A. Hoffman, *J. Org. Chem.*, **5,** 212 (1940).
26. M. S. Kharasch and F. R. Mayo, *J. Am. Chem. Soc.*, **55,** 2468 (1933).
27. D. S. Tarbell, "Organic Reactions," Vol. 2, R. Adams, Ed., Wiley, New York, 1944, p. 1.
28. A. T. Shulgin and A. W. Baker, *J. Org. Chem.*, **28,** 2468 (1963).
29. A. B. Sen and R. P. Rostogi, *J. Indian Chem. Soc.*, **30,** 355 (1953).
30. A. B. Sen and R. P. Rostogi, *J. Indian Chem. Soc.*, **30,** 556 (1953).
31. Q. R. Bartz, R. F. Miller, and R. Adams, *J. Am. Chem. Soc.*, **57,** 371 (1935).
32. E. D. Laskina, T. A. Rudol'fi, and V. N. Belov, *Zh. Obshch. Khim.*, **33**(8), 2513 (1963); *Chem. Abstr.*, **60,** 491 (1964).
33. C. O. Guss, *J. Am. Chem. Soc.*, **73,** 608 (1951).
34. F. Bergel, A. M. Copping, A. Jacob, A. R. Todd, and T. S. Work, *J. Chem. Soc.*, **1938,** 1382.
35. L. I. Smith and J. A. King, *J. Am. Chem. Soc.*, **63,** 1887 (1941).
36. A. I. Kakhazava, G. Sh. Glonti, G. D. Bagratisshavi, D. G. Kitiashvili, and I. I. Abkhazava, *Soobsch. Akad. Nauk. Gruz. SSSR*, **36**(3), 565 (1964); *Chem. Abstr.*, **62,** 11756 (1965).
37. A. Habich, R. Barmer, W. v. Phillipsborn, and H. Schmid, *Helv. Chim. Acta*, **48,** 1297 (1965).
38. L. I. Smith, H. E. Ungnade, J. R. Stevens, and C. C. Christman, *J. Am. Chem. Soc.*, **61,** 2615 (1939).
39. L. I. Smith, H. E. Ungnade, H. H. Hoehn, and S. Wawzonek, *J. Org. Chem.*, **4,** 311 (1939).
40. E. A. Viktorova, N. I. Shuikin, and E. A. Karakhanov, *Izv. Akad. Nauk. SSSR Ser. Khim.*, **1966,** 915; *Chem. Abstr.*, **65,** 13587 (1966).
41. E. A. Viktorova, N. I. Shuikin, and S. A. Karakhanov, *Izv. Akad. Nauk. SSSR Ser. Khim.*, **1963,** 1281; *Chem. Abstr.* **59,** 13927 (1963).
42. C. O. Guss, *J. Am. Chem. Soc.*, **74,** 2561 (1952).
43. C. O. Guss, *J. Am. Chem. Soc.*, **71,** 3460 (1949).
44. C. O. Guss and H. G. Mautner, *J. Org. Chem.*, **16,** 887 (1951).
45. C. O. Guss, *J. Am. Chem. Soc.*, **75,** 3177 (1953).
46. A. K. Sopkina and V. D. Ryabov, *Zh. Org. Khim.*, **1**(12), 2164 (1965); *Chem. Abstr.*, **64,** 11156 (1966).
47. V. D. Ryabov, A. K. Sopkina, and V. S. Vesova, *Neftekhimyia*, **5**(3), 335 (1965); *Chem. Abstr.*, **63,** 11407 (1965).

48. K. E. Schulte, J. Reisch, and K. H. Kauder, *Arch. Pharm.*, **295,** 801 (1962); *Chem. Abstr.*, **58,** 11337 (1963).
49. J. A. Miller and H. C. S. Wood, *J. Chem. Soc., C*, **1968,** 1837.
50. J. A. Miller and H. C. S. Wood, *Chem. Commun.*, **1965,** 39.
51. W. D. Ollis and I. O. Sutherland, "Recent Developments in the Chemistry of Natural Phenolic Products," W. D. Ollis, Ed., Pergamon, London, 1961, p. 74.
52. H. J. Kabbe, *Justus Liebigs Ann. Chem.*, **656,** 204 (1962).
53. G. Chatelus and P. Cagniant, *Compt. Rend.*, **224,** 1777 (1947); *Chem. Abstr.*, **42,** 1260 (1948).
54. R. Huisgen, H. König, and A. R. Lepley, *Chem. Ber.*, **93,** 1496 (1960).
55. B. B. Corson, W. J. Heintzelman, H. E. Tiefenthal, and J. E. Nickels, *J. Org. Chem.*, **17,** 971 (1952).
56. J. Colonge and G. Descartes, *Compt. Rend.*, **246,** 777 (1958); *Chem. Abstr.*, **52,** 18386 (1958).
57. F. Bohlmann and C. Zodero, *Tetrahedron Lett.*, **1968,** 3683; *Chem. Abstr.*, **69,** 76793 (1968).
58. W. M. Horspool and P. L. Lauson, *Chem. Commun.*, **1967,** 196; *Chem. Abstr.*, **66,** 115546 (1967).
59. Gy. Fràter and H. Schmid, *Helv. Chim. Acta*, **50,** 255 (1967).
60. A. T. Shulgin and H. O. Kerlinger, *Tetrahedron Lett.*, **1965,** 3355; *Chem. Abstr.*, **63,** 16251 (1965).
61. J. Petranek and O. Ryba, *Chem. Ind. (London)*, **1965,** 225; *Chem. Abstr.*, **62,** 8560 (1965).
62. B. Miller, *Chem. Commun.*, **1966,** 327; *Chem. Abstr.*, **65,** 7124 (1966).
63. G. R. Lappin and J. S. Zannucci, *J. Org. Chem.*, **36,** 1808 (1971).
64. B. Miller, *J. Org. Chem.*, **30,** 1964 (1965).
65. A. L. J. Beckwith and W. B. Gara, *J. Am. Chem. Soc.*, **91,** 5691 (1969).
66. S. W. Tinsley, *J. Org. Chem.*, **24,** 1197 (1959).
67. V. I. Pansevich-Kolyda and Z. B. Idel'chick, *Zh. Obshch. Khim.*, **24,** 807 (1954); *Chem. Abstr.*, **49,** 8183 (1955).
68. R. Aneja, S. K. Mukerjee, and T. R. Seshadri, *Tetrahedron*, **2,** 203 (1958).
69. E. Piers and R. K. Brown, *Can. J. Chem.*, **41,** 329 (1963).
70. E. Piers and R. K. Brown, *Can. J. Chem.*, **41,** 2917 (1963).
71. R. L. Burwell, Jr., *Chem. Rev.*, **54,** 615 (1954).
72. H. Conroy and R. A. Firestone, *J. Am. Chem. Soc.*, **78,** 2290 (1956).
73. J. R. Merchant and D. V. Rege, *Chem. Commun.*, **1970,** 380.
74. J. R. Merchant and D. V. Rege, *Tetrahedron Lett.*, **1969,** 3589.
75. H. Normant, *Ann. Chim.*, **17,** 335 (1942); *Chem. Abstr.*, **38,** 3282 (1944).
76. H. Normant, *Compt. Rend.*, **218,** 683 (1944).
77. V. I. Staninets, E. A. Shilov, and E. N. Koryak, *Zh. Org. Khim.*, **4**(2), 268 (1968); *Chem. Abstr.*, **68,** 104886 (1968).
78. R. T. Arnold and J. C. McCool, *J. Am. Chem. Soc.*, **64,** 1315 (1942).
79. G. Chatelus, *Ann. Chim.*, **4,** 505 (1949); *Chem. Abstr.*, **44,** 1975 (1950).
80. C. Marschalk, *Chem. Ber.*, **43,** 1695 (1910).
81. St. v. Kostanecki, V. Lampe, and C. Marschalk, *Chem. Ber.*, **40,** 3665 (1907).
82. C. D. Hurd and R. Dowbenko, *J. Am. Chem. Soc.*, **80,** 4711 (1958).
83. T. A. Geissman, T. G. Halsall, and E. Hinreiner, *J. Am. Chem. Soc.*, **72,** 4326 (1950).
84. M. P. Grundon and N. J. McCorkindale, *J. Chem. Soc.*, **1957,** 2177.
85. Th. Severin and H. L. Temme, *Chem. Ber.*, **98,** 1159 (1965).
86. Th. Severin, R. Schmitz, and H. L. Temme, *Chem. Ber.*, **97,** 467 (1964).
87. N. R. Raulins, W. G. Kruggel, D. D. Titus, and D. C. Van Landuyt, *J. Heterocycl. Chem.*, **5**(1), 1 (1968); *Chem. Abstr.*, **68,** 78074 (1968).

88. P. Karrer and H. Fritzsche, *Helv. Chim. Acta*, **22,** 657 (1939).
89. M. W. Baines, R. Fielden, and W. Tertiuk, U.S. Pat. 3,153,057 (1964); *Chem. Abstr.*, **62,** 14633 (1965).
90. M. W. Baines, R. Fielden, A. M. Roe, G. L. Willey, and W. Teriuk, Brit. Pat. 962,856 (1964); *Chem. Abstr.*, **61,** 9467 (1964).
91. A. Schönberg and A. Mustafa and (in part) M. K. Hilmy, *J. Chem. Soc.*, **1947,** 1045.
92. A. Schönberg and A. Mustafa, *J. Chem. Soc.*, **1946,** 746.
93. I. Baxter and I. A. Mensah, *J. Chem. Soc., C*, **1970,** 2604.
94. G. L. Willey and A. M. Roe, Brit. Pat. 1,120,763 (1968); *Chem. Abstr.*, **69,** 96449 (1968).
95. L. H. Werner, Fr. Pat. 1,344,997 (1963); *Chem. Abstr.*, **60,** 14475 (1964).
96. H. Zaugg, R. W. DeNet, and R. J. Michaels, Jr., Belg. Pat. 618,672 (1962); *Chem. Abstr.*, **59,** 3897 (1963).
97. Smith, Kline, and French Laboratories, Brit. Pat. 855,115 (1960); *Chem. Abstr.*, **55,** 12423 (1961).
98. A. M. Roe and G. L. Willey, U.S. Pat. 3,131,199 (1964); *Chem. Abstr.*, **61,** 4315 (1964).
99. H. E. Zaugg, R. W. DeNet, and R. J. Michaels, Jr., U.S. Pat. 3,156,688 (1964); *Chem. Abstr.*, **62,** 2764 (1965).
100. A. Eitel, Ger. Pat. 1,903,201 (1970); *Chem. Abstr.*, **73,** 77037 (1970).
101. A.-G. Farben industrie Bayer, Fr. Pat. 2,003,825 (1969); *Chem. Abstr.*, **70,** 111286 (1969).
102. Ward Blinkinsop and Co. Ltd., Fr. Pat. 1,505,676 (1967); *Chem. Abstr.*, **70,** 19921 (1969).
103. C. F. Huebner and L. H. Werner, U.S. Pat. 3,470,185 (1969); *Chem Abstr.*, **71,** 112978 (1969).
104. C. F. Huebner and L. H. Werner, *S. African Pat.* 6,803,709 (1968); *Chem. Abstr.*, **71,** 39004 (1969).
105. J. Augstein, A. M. Monro, G. W. H. Potter, and P. Scholfield, *J. Med. Chem.*, **11**(4), 844 (1968); *Chem. Abstr.*, **69,** 34421 (1968).
106. I. R. Landi-Vittory, F. Gatta, S. Chiavarelli, and E. Ciriaci, *Rend. Ist. Super. Sanita*, **27**(1–2), 5 (1964); *Chem. Abstr.*, **62,** 1619 (1965).
107. A. Berger and A. J. Deinet, *J. Am. Chem. Soc.*, **67,** 566 (1945).
108. A. L. Mndzhoyan and M. A. Kaldrikyan, *Izv. Akad. Nauk. Arm. SSSR Khim. Nauki*, **14,** 495 (1961); *Chem. Abstr.*, **58,** 3377 (1963).
109. E. Fourneau, P. Mederni, and Y. de Lestrange, *J. Pharm. Chim.*, **18,** 185 (1933); *Chem. Abstr.*, **27,** 5738 (1933).
110. S. Toyoshima, N. Hirose, T. Ogo, and A. Sugii, *Yakugaku Zasshi*, **88**(5), 503 (1968); *Chem. Abstr.*, **69,** 106373 (1968).
111. G. Domschke, *J. Prakt. Chem.*, **32**(3–4), 144 (1966); *Chem. Abstr.*, **65,** 12156 (1966).
112. K. C. Brannock, R. D. Burpitt, H. E. Davis, and H. S. Pridgen, *J. Org. Chem.*, **29,** 2579 (1964).
113. H. Karkisawa and M. Tateishi, *Bull. Chem. Soc. Japan*, **43,** 824 (1970).
114. G. R. Allen, Jr., *J. Org. Chem.*, **33,** 3346 (1968).
115. K. C. Brannock and R. D. Burpitt, U.S. Pat. 3,285,937 (1966); *Chem. Abstr.*, **66,** 37762 (1967).
116. F. A. Trofinov, N. G. Tsyshkova, V. I. Nozdrich, and A. N. Grinev, *Khim-Fram. Zh.*, **5**(1), 30 (1971); *Chem. Abstr.*, **75,** 3567 (1971).
117. P. Kuser, E. F. Frauenfelder, and C. H. Eugster, *Helv. Chim. Acta*, **54**(4), 969 (1971); *Chem. Abstr.*, **75,** 35557 (1971).
118. N. S. Narasimhan and M. V. Paradkar, *J. Chem. Soc.*, **1969,** 1004.
119. E. A. Kun and H. G. Cassidy, *J. Org. Chem.*, **27,** 841 (1962).
120. A. A. Shanshurin and L. P. Singavskya, *Zh. Org. Khim.*, **6**(8), 1682 (1970); *Chem. Abstr.*, **73,** 98547 (1970).
121. F. Weiss and A. Isard, *Bull. Soc. Chim. Fr.*, **1967,** 2033; *Chem. Abstr.*, **68,** 21630 (1968).

122. R. Devis and P. Depovere, *Bull. Soc. Chim. Fr.*, **1967,** 3185; *Chem. Abstr.*, **68,** 77358 (1968).
123. B. Holt and P. A. Lowe, *Tetrahedron Lett.*, **1966,** 683; *Chem. Abstr.*, **64,** 14156 (1966).
124. M. Pierre, *C. R. Akad. Sci. Paris, Ser. C*, **266**(25), 1712 (1968); *Chem. Abstr.*, **69,** 106372 (1968).
125. L. I. Smith, H. H. Hoehn, and A. G. Whitney, *J. Am. Chem. Soc.*, **62,** 1863 (1940).
126. J. Lars, G. Nilson, H. Selander, H. Sievertsson, and I. Skänberg, *Tetrahedron*, **26,** 879 (1970).
127. L. I. Smith and G. A. Boyack, *J. Am. Chem. Soc.*, **70,** 2687 (1948).
128. L. I. Smith, J. A. King, W. I. Guss, and J. Nichols, *J. Am. Chem. Soc.*, **65,** 1594 (1943).
129. L. I. Smith and G. A. Boyack, *J. Am. Chem. Soc.*, **70,** 2690 (1948).
130. L. I. Smith, S. Wawzoneck, and H. C. Miller, *J. Org. Chem.*, **6,** 229 (1941).
131. L. I. Smith and J. A. King, U.S. Pat. 2,388,579 (1945); *Chem. Abstr.*, **40,** 904 (1946).
132. L. I. Smith and H. E. Ungnade, U.S. Pat. 2,411,942 (1946); *Chem. Abstr.*, **41,** 2447 (1947).
133. L. I. Smith, H. E. Ungnade, H. H. Hoehn, and S. Wawzonek, *J. Org. Chem.*, **4,** 305 (1939).
134. Wm. K. T. Gleim, U.S. Pat. 2,546,499 (1951); *Chem. Abstr.* **45,** 7600 (1951).
135. P. Karrer and H. Rentschler, *Helv. Chim. Acta*, **22,** 1287 (1939).
136. L. I. Smith and C. W. MacMullen, *J. Am. Chem. Soc.*, **58,** 629 (1936).
137. L. I. Smith and J. A. King, *J. Am. Chem. Soc.*, **63,** 1887 (1941).
138. W. M. Lauer and E. E. Renfrew, *J. Am. Chem. Soc.*, **67,** 808 (1945).
139. F. Hoffman-La Roche and Co., Swiss Pat. 205,529 (1939); *Chem. Abstr.*, **35,** 2681 (1941).
140. F. Hoffman-La Roche and Co., Swiss Pat. 219,859 (1942); *Chem. Abstr.*, **42,** 6856 (1948).
141. F. Bergel, A. Jacob, A. R. Todd, and T. S. Work, *J. Chem. Soc.*, **1938,** 1375.
142. P. Karrer, M. Favarger, A. Merz, and G. Milhaud, *Helv. Chim. Acta*, **31,** 1505 (1948).
143. H. F. Dean and M. Nierstein, *J. Am. Chem. Soc.*, **46,** 2798 (1924).
144. O. Dann, U.S. Pat. 2,855,406 (1958); *Chem. Abstr.*, **54,** 579 (1960).
145. W. J. Horton and E. G. Paul, *J. Org. Chem.*, **24,** 2000 (1959).
146. Y. Kawase and S. Nakamoto, *Bull. Chem. Soc. Japan*, **35,** 1624 (1962); *Chem. Abstr.*, **58,** 1421 (1963).
147. D. L. Towns, U.S. Pat. 3,419,579 (1968); *Chem. Abstr.*, **70,** 68119 (1969).
148. R. L. Pelley and E. F. Orwell, Fr. Pat. 1,545,645 (1968); *Chem. Abstr.*, **72,** 43209 (1970).
149. H. W. Weber, Jr., Fr. Pat. 1,511,331 (1968); *Chem. Abstr.*, **70,** 77767 (1969).
150. E. F. Orwell, U.S. Pat. 3,356,690 (1967); *Chem. Abstr.*, **68,** 104964 (1968).
151. F M C Corp., Neth. Pat. 6,602,601 (1966); *Chem. Abstr.*, **66,** 46319 (1967).
152. C. M. Orlando, Jr., H. Mark, A. K. Bose, and M. S. Manhas, *J. Am. Chem. Soc.*, **89,** 6527 (1967).
153. C. M. Orlando, Jr., H. Mark, A. K. Bose, and M. S. Manhas, *Tetrahedron Lett.*, **1966,** 3003.
154. S. P. Pappas and E. Blackwell, Jr., *Tetrahedron Lett.*, **1966,** 1171.
155. S. P. Pappas, B. C. Pappas, and J. E. Blackwell, Jr., *J. Org. Chem.*, **32,** 3066 (1967).
156. J. A. Barltrop and B. Hesp, *J. Chem. Soc., C*, **1967,** 1625.
157. J. A. Barltrop and B. Hesp, *J. Chem. Soc.*, **1965,** 5182.
158. J. M. Bruce and E. Cutts, *Chem. Commun.*, **1965,** 2.
159. P. V. R. Shannon, *Chem. Ind. (London)*, **1968,** 720.
160. D. M. Cahill and P. V. R. Shannon, *J. Chem. Soc., C*, **1969,** 938.
161. J. Nickel, *Chem. Ber.*, **91,** 553 (1958).
162. R. L. Shriner and F. Grosser, *J. Am. Chem. Soc.*, **64,** 382 (1942).
163. W. Baker, W. M. Morgans, and R. Robinson, *J. Chem. Soc.*, **1933,** 374.
164. R. Quelet, C. Broquet, and M. Francoise, *Compt. Rend.*, **254,** 1811 (1962); *Chem. Abstr.*, **57,** 712 (1962).

165. M. P. Mertes, L. J. Powers, and M. M. Hava, *J. Med. Chem.*, **14**(4), 361 (1971); *Chem. Abstr.*, **74**, 125309 (1971).
166. D. P. Brust, D. S. Tarbell, S. M. Hecht, E. C. Hayward, and L. D. Colebrook, *J. Org. Chem.*, **31**, 2192 (1966).
166a K. V. Sarkanen and A. F. A. Wallis, *Chem. Commun.*, **1969**, 298; *Chem. Abstr.*, **70**, 114905 (1969).
167. E. C. Hayward, D. S. Tarbell, and L. D. Colebrook, *J. Org. Chem.*, **33**, 399 (1968).
168. M. Ghelardoni, V. Pestellini, and C. Musant, *Gazz. Chim. Ital.*, **99**, 1273 (1969); *Chem. Abstr.*, **73**, 24801 (1970).
169. L. H. Zalkow and M. Ghosal, *J. Org. Chem.*, **34**, 1646 (1969).
170. L. H. Zalkow and M. Ghosal, *Chem. Commun.*, **1967**, 922; *Chem. Abstr.*, **68**, 38865 (1968).
171. T. Shimizu, *Nippon Kagaku Zasshi*, **83**, 203 (1962); *Chem. Abstr.*, **59**, 5121 (1963).
172. M. P. Mertes and L. J. Powers, *Chem. Commun.*, **1970**, 620.
173. M. P. Mertes, L. J. Powers, and E. Shafter, *J. Org. Chem.*, **36**, 1805 (1971).
174. A. Baeyer and O. Seuffert, *Chem. Ber.*, **34**, 40 (1901).
175. R. v. Stoermer and O. Kippe, *Chem. Ber.*, **36**, 3992 (1903).
176. H. Normant, *Bull. Soc. Chim. Fr.*, **12**(5), 609 (1945); *Chem. Abstr.*, **40**, 3754 (1946).
177. C. D. Hurd and G. L. Oliver, *J. Am. Chem. Soc.*, **81**, 2795 (1959).
178. K. Freudenberg, W. Lautsch, and G. Piazolo, *Chem. Ber.*, **74**, 1879 (1941).
179. D. P. Brust and D. S. Tarbell, *J. Org. Chem.*, **31**, 1251 (1966).
180. A. J. Birch, *Quart. Rev. (London)*, **4**, 69 (1950).
181. S. D. Darling and K. D. Wills, *J. Org. Chem.*, **32**, 2794 (1967).
182. A. L. Wilds and N. A. Nelson, *J. Am. Chem. Soc.*, **75**, 5360 (1953).
183. L. H. Brannigan and D. S. Tarbell, *J. Org. Chem.*, **35**, 2339 (1970).
184. J. Gripenberg and T. Hase, *Acta Chem. Scand.*, **20**(6), 1561 (1966); *Chem. Abstr.*, **66**, 28463 (1967).
185. G. Wittig and G. Kolb, *Chem. Ber.*, **93**, 1469 (1960).
186. M. P. Cagniant and P. Cagniant, *Bull. Soc. Chim. Fr.*, 827 (1957); *Chem. Abstr.*, **51**, 16410 (1957).
187. J. S. H. Davies, P. A. McCrea, W. L. Norris, and G. R. Ramage, *J. Chem. Soc.*, **1950**, 3206.
188. A. Seetharamiah, *J. Chem. Soc.*, **1948**, 894.
189. M. Miyano and M. Matsui, *Proc. Acad.*, **35**, 175 (1959).
190. M. Miyano and M. Matsui, *Chem. Ber.*, **92**, 2487 (1959); **93**, 54 (1960).
191. M. Miyano, A. Kobayashi, and M. Matsui, *Bull. Agr. Chem. Soc., Japan*, **24**, 540 (1960); *Chem. Abstr.*, **61**, 2630 (1964).
192. Y. Kawase, M. Nanbu, and H. Yangihara, *Bull. Chem. Soc. Japan*, **41**(5), 1201 (1968).
193. D. J. Ringshaw and H. J. Smith, *J. Chem. Soc., C*, **1968**, 102.
194. D. Mütsiti, F. De Marchi, and V. Rosnati, *Gazz. Chim. Ital.*, **93**, 52 (1963); *Chem. Abstr.*, **59**, 606 (1963).
195. G. Baddeley, N. H. P. Smith, and M. A. Vickars, *J. Chem. Soc.*, **1956**, 2455.
196. L. E. Mills and R. Adams, *J. Am. Chem. Soc.*, **45**, 1842 (1923).
197. O. Hromatka and I. Kirnig, *Monatsh. Chem.*, **85**, 235 (1954); *Chem. Abstr.*, **49**, 9637 (1955).
198. G. A. Buntin, U.S. Pat. 2,749,272 (1956); *Chem. Abstr.*, **50**, 14173 (1956).
199. K. Masuda, M. Numata, and K. Ikawa, Japan. Pat. 7,006,809 (1967); *Chem. Abstr.*, **73**, 3784 (1970).
200. H. Hirano, K. Masuda, M. Numata, and Y. Oka, Japan. Pat. 6,806,943 (1965); *Chem. Abstr.*, **70**, 19919 (1969).
201. Z. S. Ariyan and L. A. Wiles, *J. Chem. Soc.*, **1962**, 4709.
202. S. K. Pavanaram, L. Ramachandra Row, and G. S. R. Subba, *J. Sci. Ind. Res. (India)*, **19B**, 57 (1960); *Chem. Abstr.*, **55**, 24732 (1961).

203. J. Eisenbeiss and H. Schmid, *Helv. Chim. Acta*, **42,** 61 (1969).
204. J. Green, D. McHale, S. Marcinkiewicz, P. Mamalis, and P. R. Watt, *J. Chem. Soc.*, **1959,** 3362.
205. P. Cagniant and M. P. Cagniant, *Bull. Soc. Chim. Fr.*, **1957,** 838.
206. S. Uemura, T. Nakano, and K. Ichikawa, *Nippon Nagaku Zasshi,* **89**(2), 203 (1968); *Chem. Abstr.*, **69,** 51913 (1968).
207. N. Sugiyama, *Bull. Inst. Phys. Chem. Res., (Tokyo)*, **21,** 744 (1942); *Chem. Abstr.*, **41,** 5506 (1947).
208. E. J. Nienhouse, R. M. Irwin, and G. R. Finni, *J. Am. Chem. Soc.*, **89,** 4557 (1967).
209. H. Stetter and E. Siehnhold, *Chem. Ber.*, **88,** 271 (1955).
210. H. Stetter and R. Lauterbach, *Justus Liebigs Ann. Chem.*, **655,** 20 (1962); *Chem. Abstr.*, **57,** 15056 (1962).
211. P. E. Verkade, Th. Morel, and H. G. Gerritsen, *Rec. Trav. Chim. Pays-Bas*, **74,** 763 (1955).
212. P. Hodge and M. Derenberg, *J. Chem. Soc., D*, **1971,** 233.
213. Th. Morel and P. E. Verkade, *Rec. Trav. Chim. Pays-Bas*, **70,** 35 (1951).
214. Th. Morel and P. E. Verkade, *Rec. Trav. Chim. Pays-Bas*, **67,** 539 (1948).
215. W. Triebs, *Chem. Ber.*, **70,** 85 (1937).
216. I. V. Machinskaya and V. A. Barkhash, *Zh. Obshch. Khim.*, **28,** 2873 (1958); *Chem. Abstr.*, **53,** 9002 (1959).
217. I. V. Machinskaya and V. A. Barkhash, *Zh. Obshch. Khim.*, **27,** 1978 (1957); *Chem. Abstr.*, **52,** 4615 (1957).
218. M. Kuhn, *J. Prakt. Chem.*, **156,** 103 (1940); *Chem. Abstr.*, **35,** 1762 (1941).
219. F. Ebel, F. Huber, and A. Brunner, *Helv. Chim. Acta*, **12,** 16 (1929).
220. W. Cooker and S. Hornsby, *J. Chem. Soc.*, **1947,** 1157.
221. H. Minato and T. Nagasaki, *J. Chem. Soc. Org.*, **1966,** 377.
222. R. N. Lacey, *J. Chem. Soc.*, **1954,** 822.
223. Shioogi and Co., Ltd., Brit. Pat. 1,081,647 (1967); *Chem. Abstr.*, **68,** 69209 (1968).
224. K. Gump, S. W. Moje, and C. E. Castro, *J. Am. Chem. Soc.*, **89,** 6770 (1967).
225. C. E. Castro, E. J. Gaughan, and D. C. Owsley, *J. Org. Chem.*, **31,** 4071 (1966).
226. K. E. Schulte, J. Reisch, and A. Mock, *Arch. Pharm.*, **295,** 645 (1962); *Chem. Abstr.*, **59,** 1575 (1963).
227. G. Lehmann and B. Lücke, *Justus Liebigs Ann. Chem.*, **727,** 88 (1969).
228. A. S. Sopova, O. V. Perekalin, and V. M. Lebendnova, *Zh. Obshch. Khim.*, **34**(8), 2638 (1964); *Chem. Abstr.*, **61,** 14618 (1964).
229. V. B. Berestovitskaya, A. S. Sopova, and O. V. Perekalin, *Khim. Geterotsikl. Soedin,* **1967,** 396; *Chem. Abstr.*, **68,** 69209 (1968).
230. H. J. Schaeffer and R. Vince, *J. Org. Chem.*, **27,** 4502 (1962).
231. G. Nowy, W. Riedel, and H. Simon, *Chem. Ber.*, **99,** 2075 (1966).
232. B. Miller, *J. Am. Chem. Soc.*, **89,** 1685 (1967).
233. D. A. H. Taylor, *J. Chem. Soc.*, **1959,** 2767.
234. P. Jakobson and S. O. Lawesson, *Tetrahedron,* **24,** 3671 (1968).
235. S. J. Dominianni, M. O. Chaney, and N. D. Jones, *Tetrahedron Lett.*, **1970,** 4735; *Chem. Abstr.* **74,** 68923 (1971).
236. K. Ichikawa, O. Itoh, and T. Kawamura, *Bull. Chem. Soc. Japan,* **41,** 1240 (1968); *Chem. Abstr.*, **69,** 77404 (1968).
237. A. G. Anastassiou and R. P. Cellura, *Chem. Commun.*, **1969,** 1521.
238. S. E. Cantor and D. S. Tarbell, *J. Am. Chem. Soc.*, **86,** 2902 (1964).
239. N. I. Shuikin, I. I. Demitriev, and T. P. Dobrynina, *J. Gen. Chem. USSR,* **10,** 967 (1940); *Chem. Abstr.*, **35,** 2508 (1941).
240. W. E. Harvey and D. S. Tarbell, *J. Org. Chem.*, **32,** 1679 (1967).
241. J. Colonge and F. Collomb, *Bull. Soc. Chim. Fr.*, **1951,** 241; *Chem. Abstr.*, **46,** 443 (1952).

242. R. J. Gargiulo, Y. Yamamoto, and D. S. Tarbell, *J. Org. Chem.*, **36,** 846 (1971).
243. A. I. Meyers and K. Baburao, *J. Heterocycl. Chem.*, **1**(4), 203 (1964); *Chem. Abstr.*, **62,** 1618 (1965).
244. N. Belorizky and D. Gagnaire, *C. R. Acad. Sci. Paris, Ser. C,* **268**(8), 688 (1969); *Chem. Abstr.*, **70,** 96035 (1969).
245. J. Klein and E. D. Bermann, *J. Chem. Soc.*, **1964,** 3484.
246 P. W. Jolly and P. De Mayo, *Can. J. Chem.*, **42**(1), 170 (1964).
247. P. Cordier and A. Haberzettl, *Pharm. Acta Helv.*, **38** (7–8), 522 (1963); *Chem. Abstr.*, **60,** 4052 (1964).
248. E. S. Behare and R. B. Miller, *Chem. Commun.*, **1970,** 402.
249. E. S. Behare and R. B. Miller, *J. Chem. Soc., D,* **1970,** 402
250. E. C. Kendall, A. E. Osterberg and B. F. MacKenzie, *J. Am. Chem. Soc.*, **48,** 1384 (1926).
251. E. A. Karakhanov, L. G. Saginova and E. A. Viktorova, *Vestn. Mosk. Univ. Khim.*, **11**(4), 456 (1970).

CHAPTER V

Benzofuranones

1. 3 (2*H*)-Benzofuranones

A. Preparation

3-Hydroxybenzofurans, tautomers of 3(2H)-benzofuranones, exist by preference in the ketonic state. In contrast to 2(3*H*)-benzofuranones, they enolize readily and are therefore often soluble in alkali and form *O*- and *C*-acyl- and alkyl derivatives with equal facility.[1-4] Thè trivial name for 3(2*H*)-benzofuranones is "β-coumaranones" or "coumaran-3-ones."

Many different syntheses for 3(2*H*)-benzofuranones have been reported.[5-15] Schroeder et al.[5] have found that unsatisfactory results are often obtained and believe that the most effective method is a Dieckmann reaction performed on ethyl *o*-carboxymethylsalicylate (Eq. 1).[16]

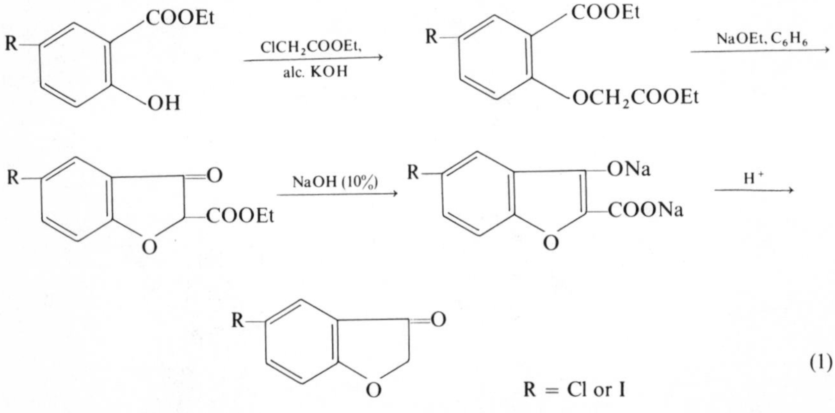

R—, COOEt, OH $\xrightarrow[\text{alc. KOH}]{ClCH_2COOEt,}$ R—, COOEt, OCH_2COOEt $\xrightarrow{NaOEt,\ C_6H_6}$

R—, =O, COOEt, O $\xrightarrow{NaOH\ (10\%)}$ R—, ONa, COONa, O $\xrightarrow{H^+}$

R—, =O, O (1)

R = Cl or I

In a comparative study of the different syntheses, Kalinowski and Kalinowski[14] have investigated three methods for the preparation of 3(2*H*)-benzofuranones **1, 2,** and **3:** (a) cyclization of the corresponding phenoxyacetyl chloride in the presence of aluminum chloride;[13] (b) cyclization of the corresponding 2-hydroxy-ω-chloroacetophenone by dilute alkali;[17] and (c) cyclodehydration of the corresponding phenoxyacetic acid in benzene solution with phosphorus pentoxide.[6]

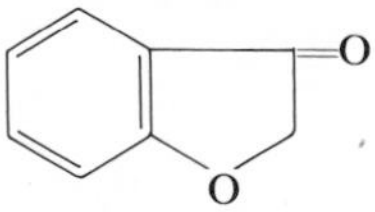

1 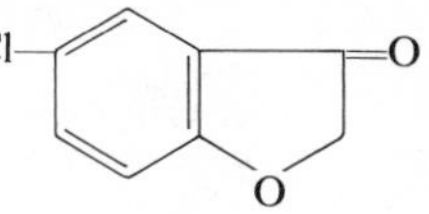 **2**

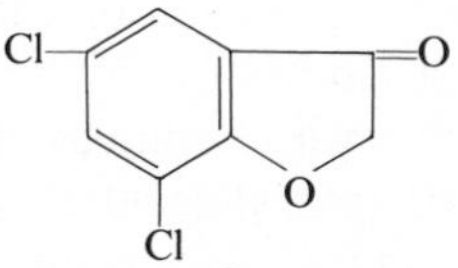

3

Whereas **2** was obtained in fair yield by method (b) and in traces by methods (a) and (c); compound **1** was prepared equally well by all three methods. Compound **3** could not be obtained by any of the three cited methods. Fries rearrangement of 2,4-dichlorophenyl chloroacetate to 3,5-dichloro-2-hydroxy-ω-chloroacetophenone was unsuccessful. 5,7-Dichloro-3(2*H*)-benzofuranone (**3**) has been synthesized by an unambiguous method starting with salicyclic acid[18] and/or 2,4-dichlorophenol[9–11] (Eqs. 2 and 3).

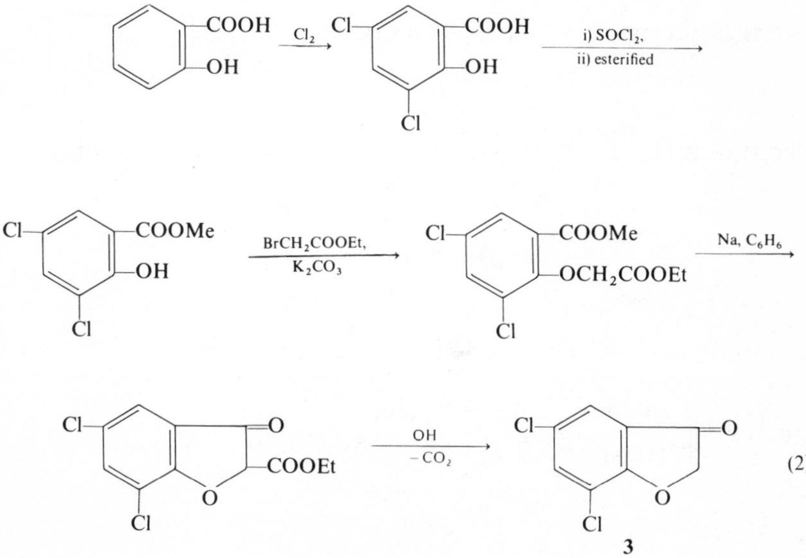

3 (2)

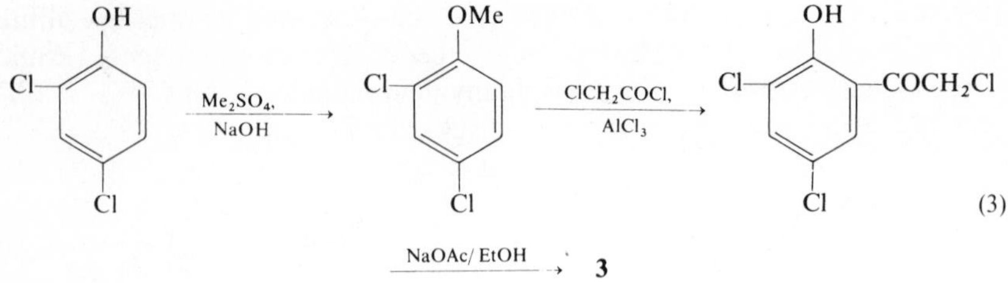

OH, Cl, Cl $\xrightarrow[NaOH]{Me_2SO_4,}$ OMe, Cl, Cl $\xrightarrow[AlCl_3]{ClCH_2COCl,}$ OH, Cl, $COCH_2Cl$, Cl (3)

$\xrightarrow{NaOAc/EtOH}$ **3**

Unsubstituted 3(2*H*)-benzofuranone **1** is readily obtained by cyclization of phenoxyacetyl chloride in presence of aluminum chloride, but the nature of the byproducts is uncertain.[13,19-22] Interaction of substituted aryloxyacetyl chlorides [19-22] with aluminum chloride in benzene solution leads to intramolecular and intermolecular acylation, diphenylmethane, the corresponding phenol, and *o*-benzylphenol. The product ratio varies with the substituent. The alkylation products arise from unimolecular decarbonylation of the aryloxyacetylium ion, followed by attack on the solvent and cleavage of the benzyl ether. Electron-releasing substituents assist decarbonylation and cyclization to the 3(2*H*)-benzofuranone while electron-attracting substituents lead to a high proportion of the 2-aryloxyacetophenones (Chart 1).[23]

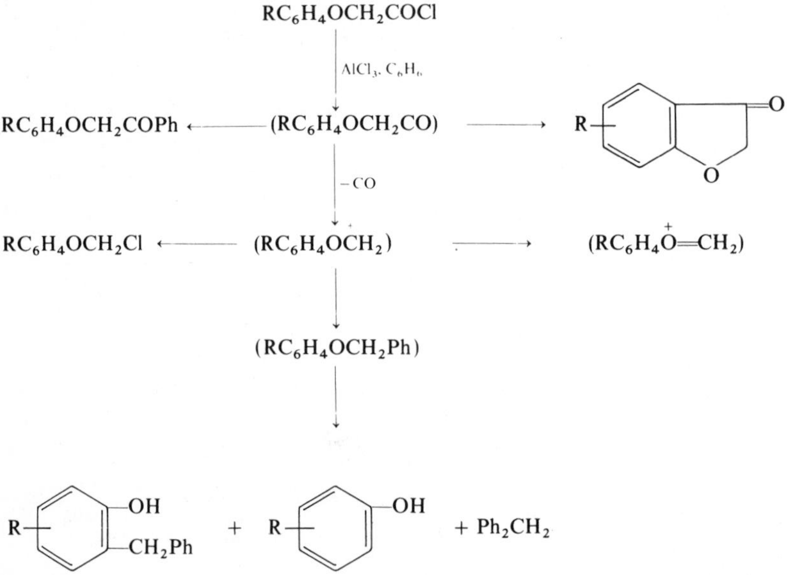

$RC_6H_4OCH_2COCl$

$AlCl_3, C_6H_6$

$RC_6H_4OCH_2COPh$ ← $(RC_6H_4OCH_2\overset{+}{C}O)$ → R, O, =O

$-CO$

$RC_6H_4OCH_2Cl$ ← $(RC_6H_4O\overset{+}{C}H_2)$ → $(RC_6H_4\overset{+}{O}{=}CH_2)$

$(RC_6H_4OCH_2Ph)$

R, OH, CH_2Ph + R, OH + Ph_2CH_2

Chart 1

The interaction of alkyl-substituted phenoxyacetyl halides with aluminum chloride has been shown to afford cyclic ketones as the dominant products; thus 3-methyl-4-chloro- and 3,4-dimethylphenoxyacetyl chloride undergoes cyclization to the corresponding 5,6-disubstituted 3(2*H*)-benzofuranones.[24] Cyclization of *o*-cresyloxyacetic acid and/or thymyloxyacetic acid with phosphorus pentachloride and aluminum chloride results in the formation of 3(2*H*)-benzofuranones **4** and **5**, respectively.[25] Hydrofluoric acid effects cyclization of *p*-cresol-*β*-chloropropionate to 2,5-dimethyl-3(2*H*)-benzofuranone (**6**).[26]

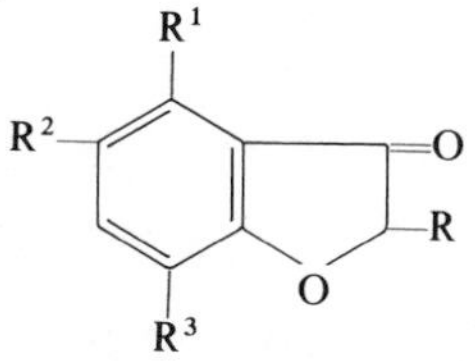

R^1, R^2, =O, R, O, R^3

4 R = R^1 = R^2 = H, R^3 = Me
5 R = R^2 = H, R^1 = Me, R^3 = *n*-Pr
6 R = R^2 = Me, R^1 = R^3 = H

Kuhn and Hensel[27] have announced the ready conversion of 4-chloro-*ω*-diazoacetophenone to 6-chloro-3(2*H*)-benzofuranone (**7**). The half-ester of hemipinic acid (**8**), has been successfully transformed, in a similar manner, to **10**.[28] The ammonium ion (**9**) is formed as an intermediate of nucleophilic attack on the lone pair of oxygen atoms (Eq. 4). In **9**, the diazoacetyl group is flanked on both sides by an ester and an ether function; it is the latter that reacts preferentially because it is a more powerful nucleophile than the carbonyl oxygen.

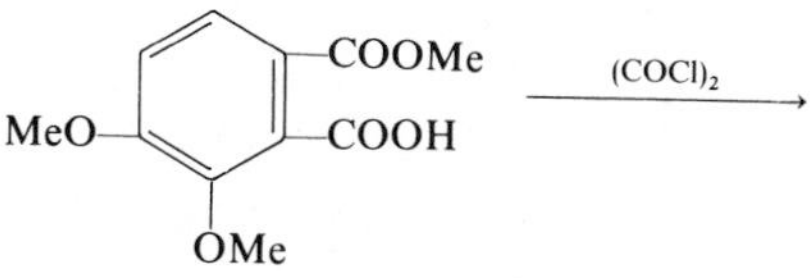

COOMe, COOH, MeO, OMe — $(COCl)_2$ →

8

COOMe, COCl, MeO, OMe — CH_2N_2 →

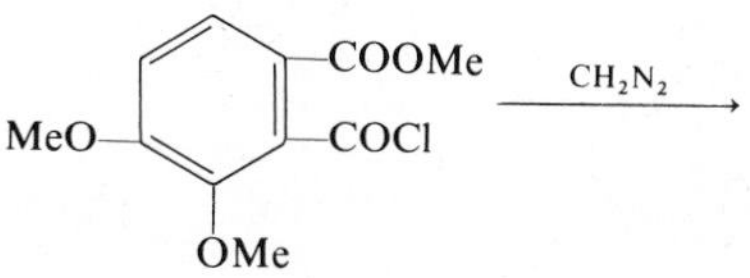

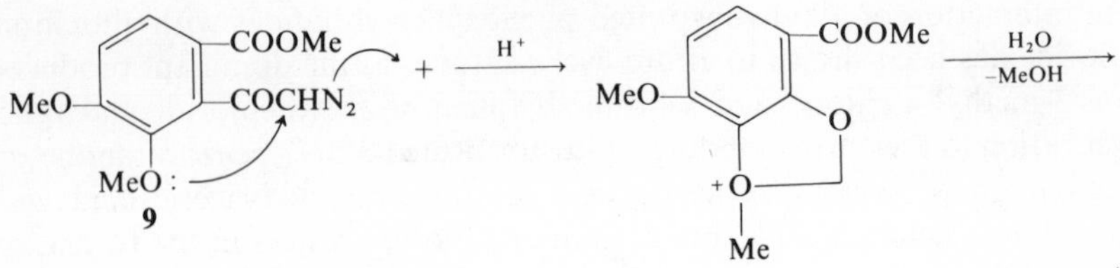

(4)

10 **7**

Dallacker and Korb[29] have determined the optimum conditions for the formation of 3(2*H*)-benzofuranones from *o*-alkoxy-ω-diazoacetophenones, the different reactivities of methylenedioxy- and alkoxy groups in the ortho position to the diazocarbonyl functions have also been investigated. Bose and Gates[30] have suggested the reaction scheme shown in Eq. 5 to be compatible with the catalytic nature of acetic[27,31] and hydrochloric acid[30,32]

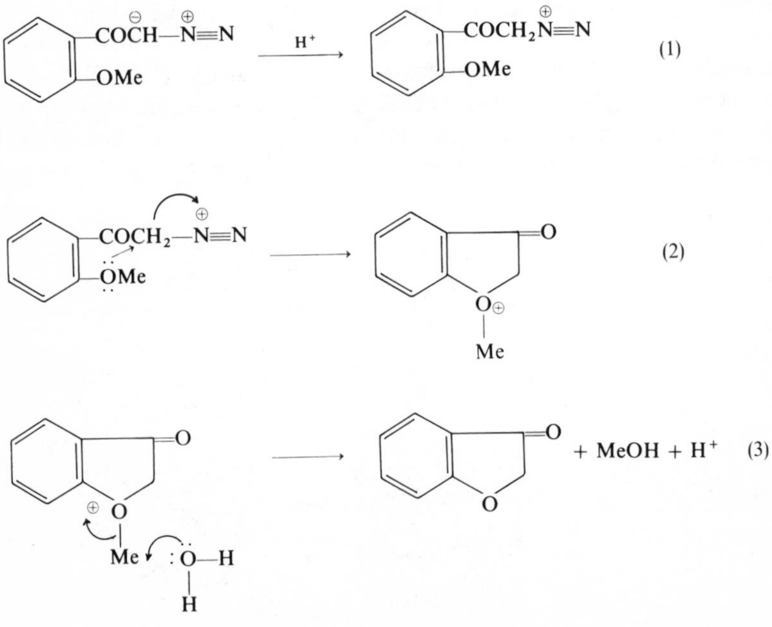

(1)

(2)

+ MeOH + H^+ (3)

(5)

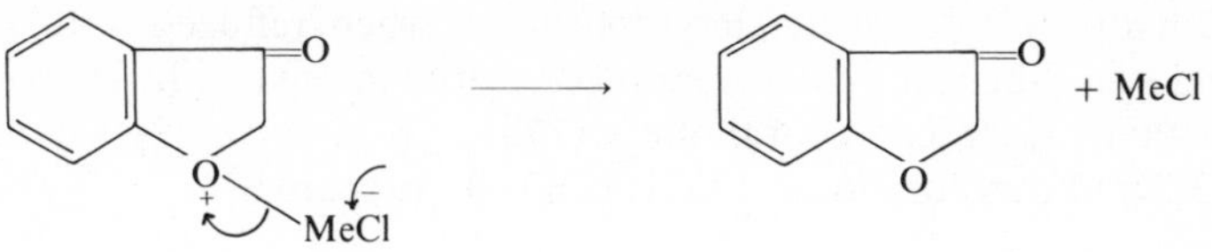

+ MeCl

In reaction (1) the nucleophilic methine carbon of the diazoketone is attacked by a proton to give an aliphatic diazonium ion which loses nitrogen by a displacement reaction on carbon involving an unshared pair on oxygen (2). The oxonium ion thus formed is then attacked by the solvent to give 3(2*H*)-benzofuranone and methanol at the same time the proton is regenerated. Reactions (2) and (3) might well be concerted. Although this scheme conforms with the catalytic nature of hydrochloric acid, the acid will be partially consumed by an alternative final step—the attack of a chloride ion on the oxonium intermediate. However, this will only occur as a side reaction since the presence of water in large excess ensures more frequent attack by solvent than by chloride ion.

ω-Diazo-*o*-methoxyacetophenone reacts with **11** and/or **12** in the presence of catalytic amounts of boron trifluoride ethereate to yield **13** and/or **14**, respectively.[33] Pyrolysis of **14** in the presence of potassium hydrogen sulfate yields **15**. 3(2*H*)-Benzofuranone reacts with trialkyl orthoformate or **15** to give **16** (R = Ac), which is hydrolyzed with alkali to afford **16** (R = H) (also obtained by pyrolysis of **13**) (Eq. 6).

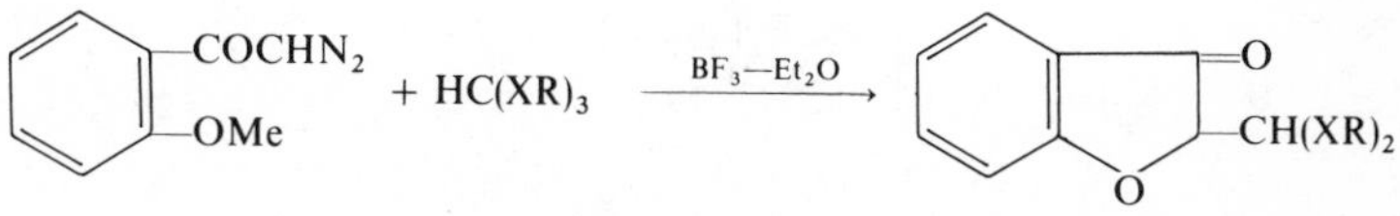

$COCHN_2$, OMe + $HC(XR)_3$ $\xrightarrow{BF_3\text{—}Et_2O}$ $CH(XR)_2$

11 X = O, R = Me
12 X = S, R = Et

13 X = O, R = Me
14 X = S, R = Et

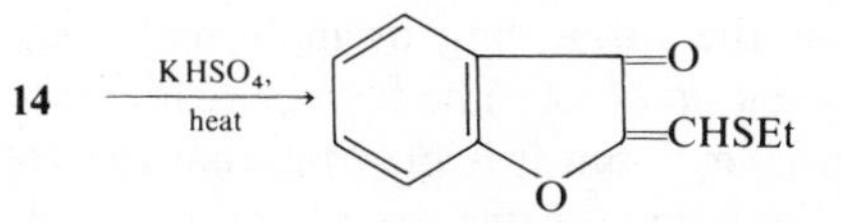

14 $\xrightarrow{KHSO_4,\ \text{heat}}$ =CHSEt

15

$\xrightarrow{HC(OR)_3}$ OR, CH=

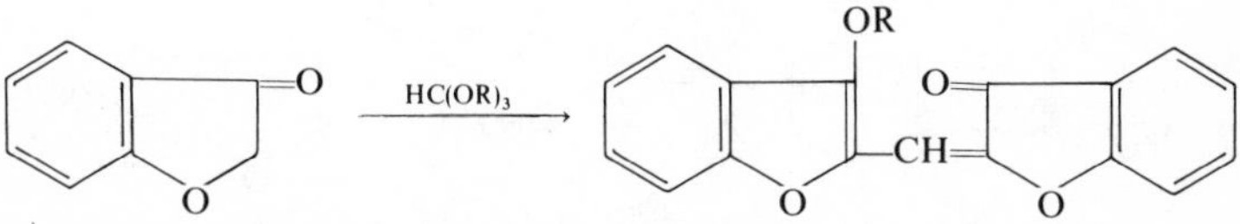

16 (6)

2′,4′-Dihydroxy-3-(2-pyridyl)acrylophenone, when refluxed with phosphoric acid in butanol, underwent cyclization to give 6-hydroxy-2-(2-pyridylmethyl)-3(2*H*-benzofuranone (**17a**). 6-Hydroxy-2-(4-pyridylmethyl)-3(2*H*)-benzofuranone, **17b**, is similarly prepared.[34]

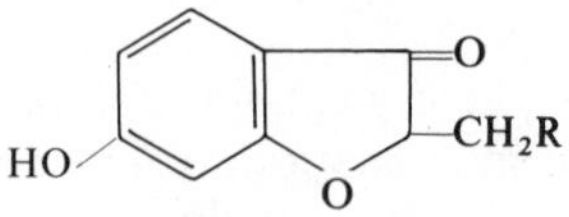

17a R = 2-C_5H_4N
17b R = 4-C_5H_4N

The cyclohexenyl diketone **18** has accessible and reactive β-hydrogens, but the biradical resulting from intramolecular hydrogen abstraction would cyclize in a photo process to give a highly strained product. Nevertheless **18** in cyclohexane underwent rapid photochemical reaction to give the benzofuranone **20**. This product probably does not result from hydrogen abstraction. Rather the diketone **18** may be pictured as an analog of cross-conjugated dienones, which frequently undergo light-induced bonding between their terminal carbon atoms. Rearrangement of the intermediate **19**, possibly by way of the 1,2 hydrogen shifts, would not be unexpected.[35]

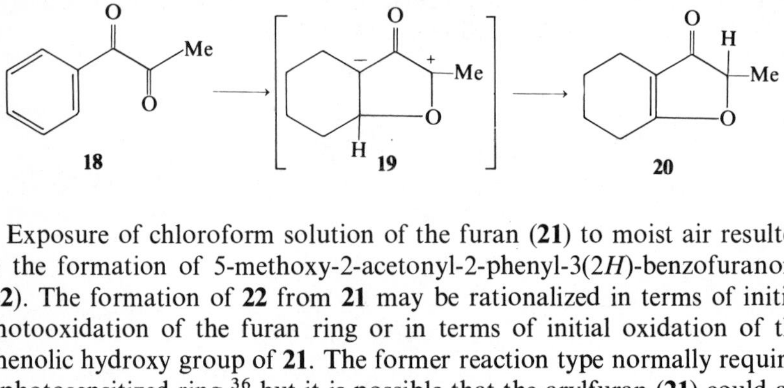

18 **19** **20**

Exposure of chloroform solution of the furan (**21**) to moist air resulted in the formation of 5-methoxy-2-acetonyl-2-phenyl-3(2*H*)-benzofuranone (**22**). The formation of **22** from **21** may be rationalized in terms of initial photooxidation of the furan ring or in terms of initial oxidation of the phenolic hydroxy group of **21**. The former reaction type normally requires a photosensitized ring,[36] but it is possible that the arylfuran (**21**) could act as the photosensitized ring for its own oxidation.[37]

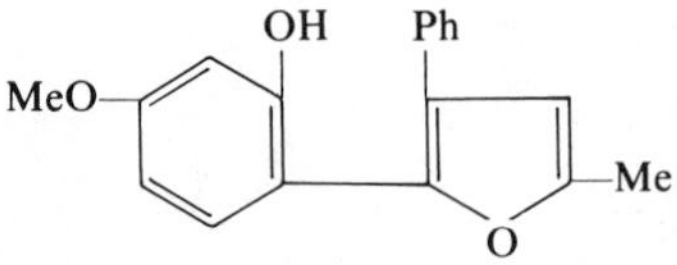

21

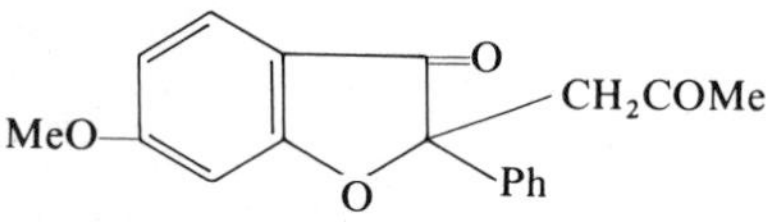

22

Oxidation of 2-hydroxyacetophenone[38-40] results in the formation of a minor product; 2-hydroxy-3(2*H*)-benzofuranone (**24**), the cyclic hemiacetal of 2-hydroxyphenylglyoxal, as well as the major product **23**, (Eq. 7).[40]

$$\text{2-Ac-C}_6\text{H}_4\text{OH} \xrightarrow{SeO_2} \text{2-(COCOOH)-C}_6\text{H}_4\text{OH } (\mathbf{23}) + \text{2-hydroxy-3(2}H\text{)-benzofuranone } (\mathbf{24}) \qquad (7)$$

The formation of 2-vinyl-3(2*H*)-benzofuranones (**25**) has been observed during the interaction of several isoflavones with dimethyl sulfoxonium methylide, and in all cases these compounds can be readily identified from their molecular formulas and spectroscopic properties.[37]

$$\text{8-OMe-isoflavone }(R^1, R^2) \xrightarrow{(Me_2\overset{\oplus}{S}(O)\overset{\ominus}{C}H_2)} \text{7-OMe-2-}R^2\text{-2-}(C(R^1){=}CH_2)\text{-3(2}H\text{)-benzofuranone } (\mathbf{25})$$

2-Azido-3(2*H*)-benzofuranone (**26**) is obtained by interaction of sodium azide with 2-bromo-3(2*H*)-benzofuranone in ethanol.[41] It is an unstable oil; however 5-bromo-6-methoxy-2-azido-3(2*H*)-benzofuranone (**27**) is stable. Compound **26** undergoes decomposition under the slight alkalinity that results from hydrolysis or alcoholysis of the unreacted sodium azide (Eq. 8).

$$\text{2-Br-3(2}H\text{)-benzofuranone} \xrightarrow{NaN_3} \text{2-}N_3\text{-3(2}H\text{)-benzofuranone } (\mathbf{26}) \xrightarrow[EtOH]{NaN_3,} \text{2-(COOEt)-C}_6\text{H}_4\text{OH} + N_2 + NaBr + HCN$$

$$\text{5-Br-6-MeO-2-Br-3(2}H\text{)-benzofuranone} \xrightarrow{NaN_3} \text{5-Br-6-MeO-2-}N_3\text{-3(2}H\text{)-benzofuranone } (\mathbf{27}) \xrightarrow{NaN_3,\ NaOAc} \text{Br, MeO-C}_6\text{H}_2\text{(COOEt)(OH)} \qquad (8)$$

Ethyl 2-(carbmethoxy)-4-bromophenoxymalonate (**28**) undergoes intramolecular cyclization in presence of sodium methoxide to give 2-(carbmethoxy)-5-bromo-3(2*H*)-benzofuranone (**30**) instead of the expected Dieckmann product (**29**). The identity of **30** was established by cyclizing ethyl 4-bromo-2-(carbmethoxy)phenoxyacetate (**31**) with powdered sodium in xylene. Interaction of the potassium salt of **30** with ethyl chlorocarbonate in polar solvent, for example, acetonitrile, at room temperature provides a high yield of *C*-alkylated keto-ester (**29**); this is also obtained by alkylation of the sodium salt of **30** in benzene without appreciable *O*-alkylation (Eq. 9).[42]

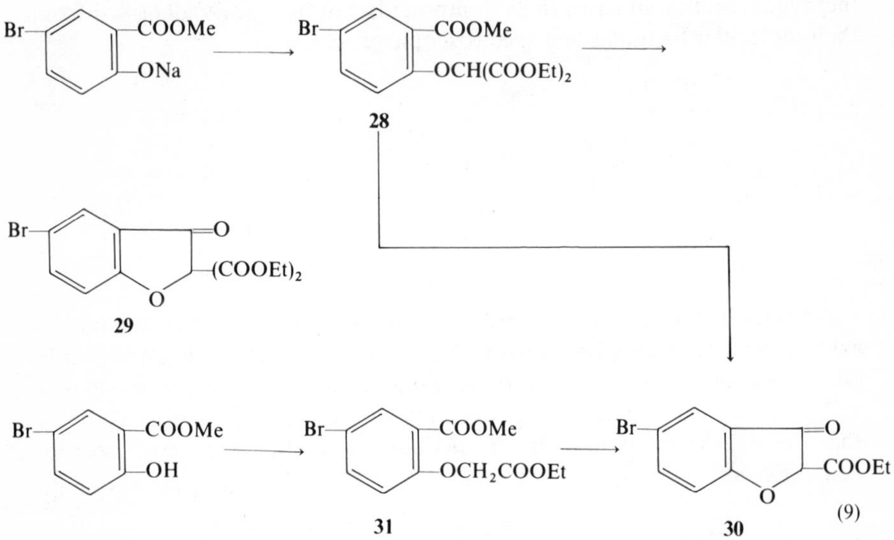

Br, COOMe, ONa → Br, COOMe, $OCH(COOEt)_2$ **28** →

Br, =O, $(COOEt)_2$, O **29**

Br, COOMe, OH → Br, COOMe, OCH_2COOEt **31** → Br, =O, COOEt, O **30** (9)

Synthesis of 5,7-dinitro-3(2*H*)-benzofuranone (**32**) has been achieved[43] by two successful methods (Chart 2).

Dean et al.[44] have shown that the previously reported condensation of maleic anhydride with 3,5-xylenol[45] under Friedel-Crafts conditions leads to the formation of colorless **33** and yellow **34**.

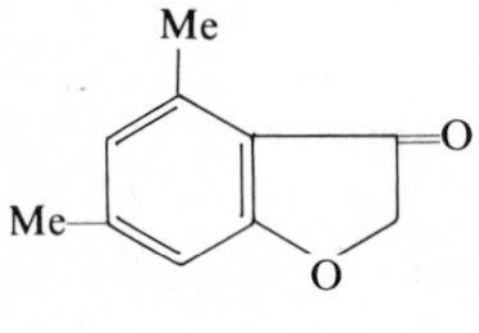

Me, Me, =O, O **33** HO, Me, COCH=CHCOOH, Me **34**

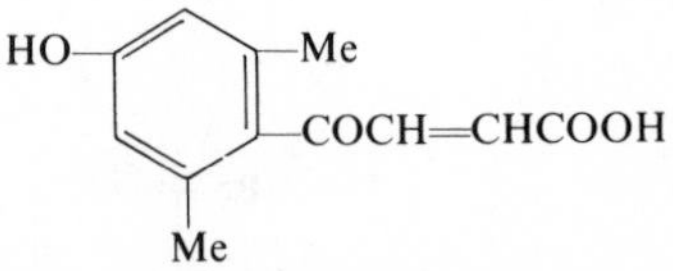

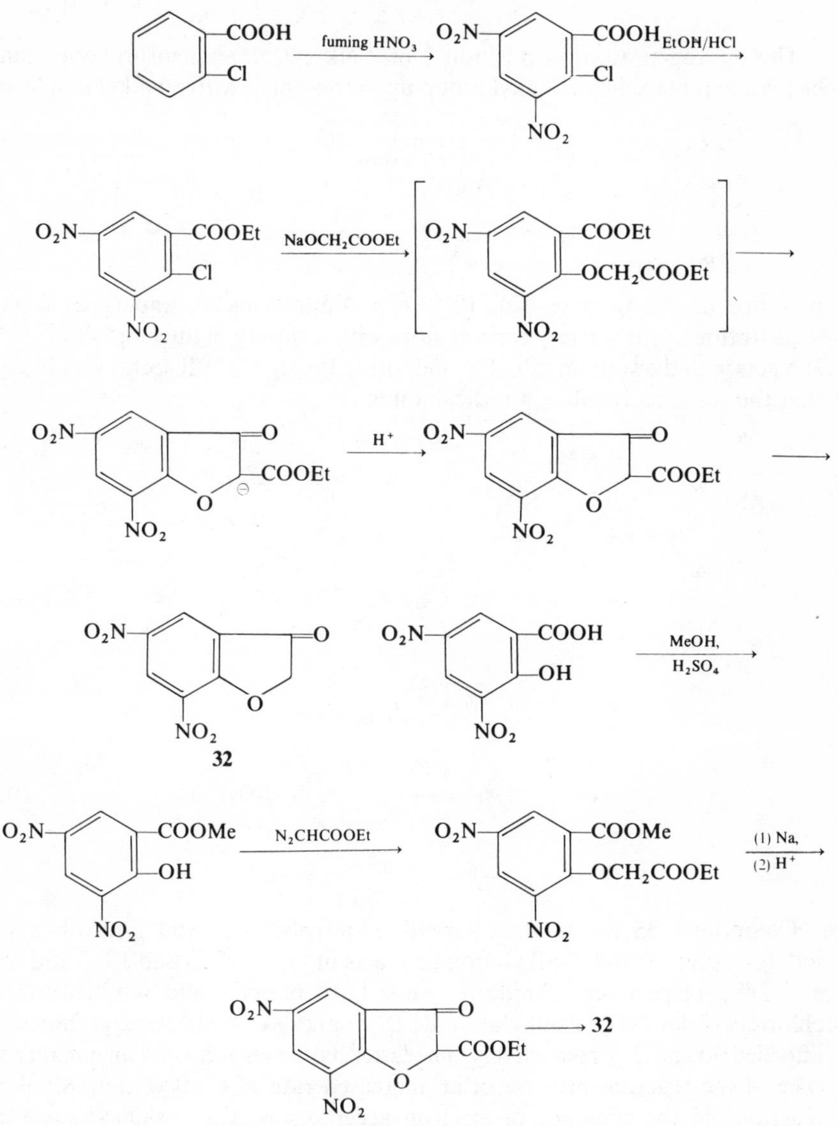

COOH Cl fuming HNO_3 → O_2N COOH Cl NO_2 EtOH/HCl →

O_2N COOEt Cl NO_2 $NaOCH_2COOEt$ → [O_2N COOEt OCH_2COOEt NO_2] →

O_2N O COOEt O ⊖ NO_2 H^+ → O_2N O COOEt O NO_2 →

O_2N O O NO_2 **32**

O_2N COOH OH NO_2 MeOH, H_2SO_4 →

O_2N COOMe OH NO_2 $N_2CHCOOEt$ → O_2N COOMe OCH_2COOEt NO_2 (1) Na, (2) H^+ →

O_2N O COOEt O NO_2 → **32**

Chart 2

B. Reactions

The hydrogen atom in position 2 of 2-alkyl-3(2*H*)-benzofuranones can be readily replaced by an alkyl group upon treatment with an alkyl halide in

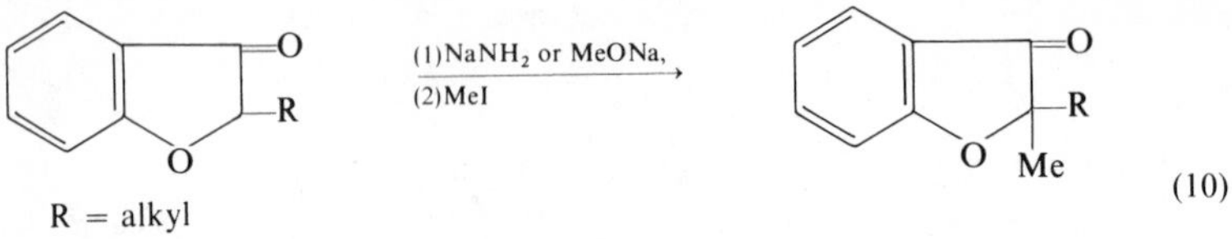

$$\text{2-R-3(2H)-benzofuranone} \xrightarrow[\text{(2) MeI}]{\text{(1) NaNH}_2\text{ or MeONa,}} \text{2-R-2-Me-3(2H)-benzofuranone} \qquad (10)$$

R = alkyl

presence of strong base (Eq. 10)[2]; α, *p*-Dinitrocumene, undergoes facile substitution at the tertiary carbon atom with a variety of nucleophiles.[46–48] It reacts with the sodium salt of **35** and forms **36** (Eq. 11).[49] It seems very likely that the reaction involves a radical anion.[50]

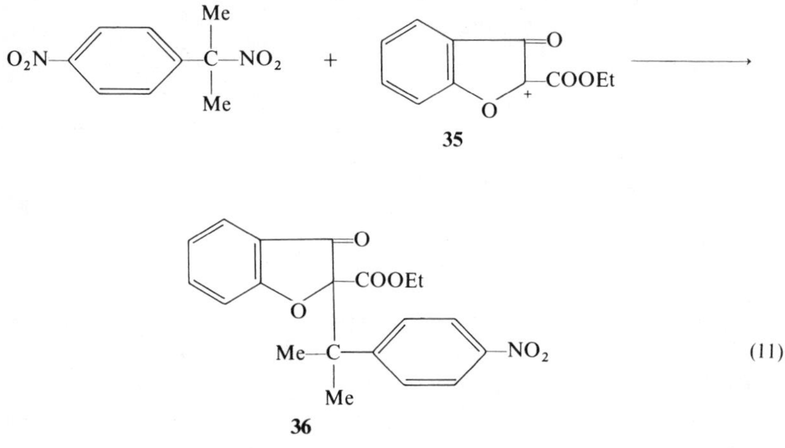

$$O_2N\text{–}C_6H_4\text{–}C(Me)_2\text{–}NO_2 + \mathbf{35} \longrightarrow \mathbf{36} \qquad (11)$$

35 (2-COOEt-3(2H)-benzofuranone sodium salt); **36** (2-COOEt-2-[C(Me)₂–C₆H₄–NO₂]-3(2H)-benzofuranone)

Compound **35** reacts with benzyl-, *m*-nitrobenzyl-, and *p*-nitrobenzyl iodide to give *C*- and *O*-alkylation products in yields of 73 and 18%, and 65 and 27%, respectively. Similarly, while both benzyl-, and *m*-nitrobenzyl chlorides yielded 40% *C*-alkylate and 51% *O*-alkylate, *p*-nitrobenzyl chloride afforded 90 and 2%, respectively. The latter halide also showed an enhanced take of the reaction because of an increased rate of *C*-alkylation. Similar reactions in the presence of electron acceptors or Cu^{2+} salts showed a supression of *C*-alkylation only in the case of *p*-nitrobenzyl radical anions compared with an S_N2 mechanism in the other cases.[51]

Acetylation of 3(2*H*)-benzofuranone with acetic anhydride and sodium acetate offers 3-acetoxybenzofuran (**37**) but, if sulfuric acid is used as catalyst, there is a danger that a second acyl group may enter position 2 by a kind of Friedel-Crafts or Fries reaction.[52] However when 5-chloro-3(2*H*)-benzofuranone (**38**), is heated with acetic anhydride and sodium acetate, **39** and **40** are obtained.[53]

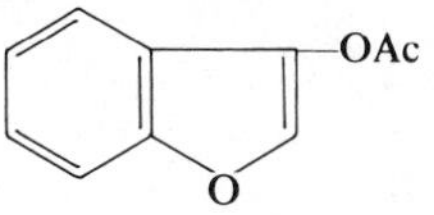

37

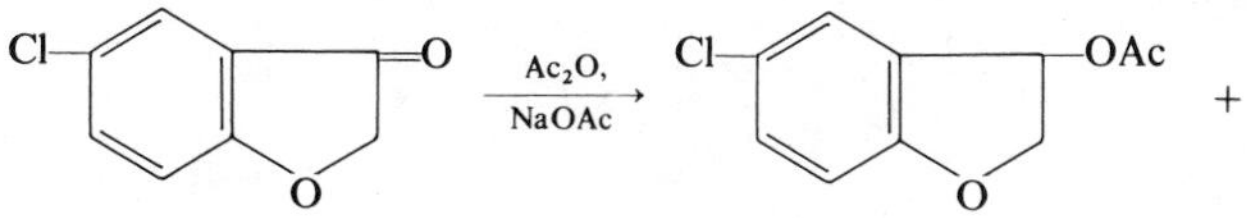

Ac_2O, NaOAc

38 → **39** +

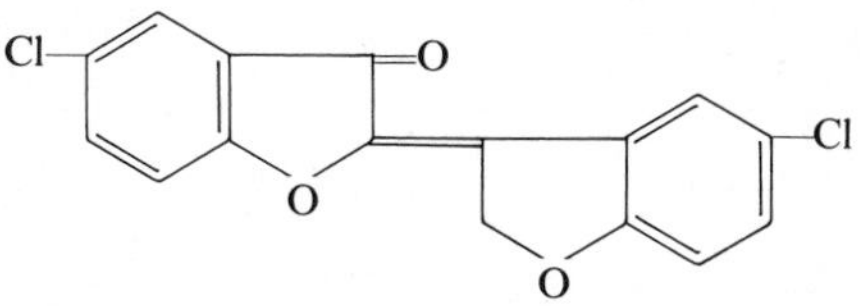

40

3(2*H*)-Benzofuranone is soluble in cold alkali solutions and can be recovered unchanged by acids, although warm alkali solutions open the furan ring to yield salicylaldehyde.[25,52] Cleavage of the furan ring between C-2 and C-3 is illustrated in the reactions of 3(2*H*)-benzofuranones. Thus, on catalytic oxidation of 2,5-dimethyl-3(2*H*)-benzofuranone (**41**), **42** is produced.[54] Hydrogen proxide oxidation of **43** in alkaline medium affords **44** (Eq. 12).[55]

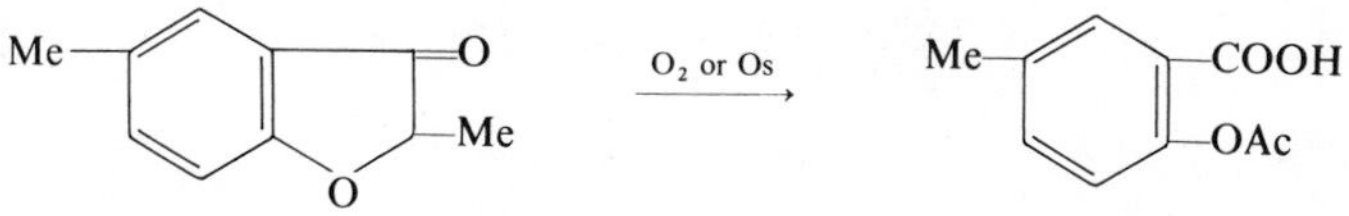

41 $\xrightarrow{O_2 \text{ or Os}}$ **42**

43 $\xrightarrow[OH^-]{H_2O_2,}$ **44** (12)

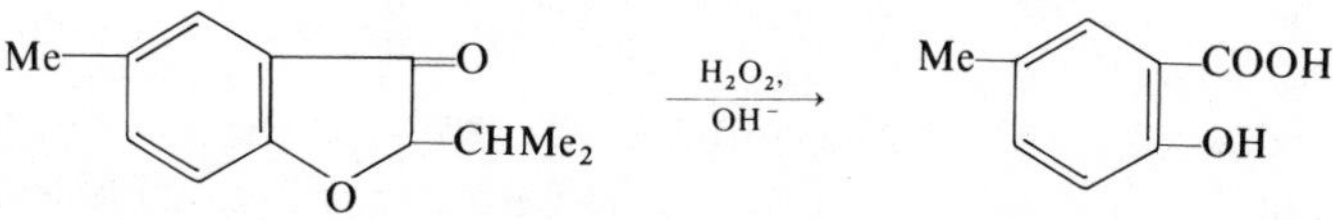

3(2*H*)-Benzofuranones behave normally if somewhat sluggishly, toward carbonyl reagents. Even moderately bulky substituents at position 2 and position 4 tend to suppress reactions altogether for steric reasons. 3(2*H*)-Benzofuranone and its 2-alkyl derivatives form the corresponding oximes readily. However, with such reagents as phenylhydrazine and semicarbazide, the normal carbonyl derivatives are obtained only with difficulty because of the susceptibility of the molecule toward basic reagents.

Hydrazine and methylhydrazine react with 3(2*H*)-benzofuranones[57] to give a series of trisubstituted pyrazoles (**45**) having a phenolic character. A series of phenolic isoxazoles (**46**) was prepared from the appropriate 3(2*H*)-benzofuranone by the method after Royer and Bisagni.[56]

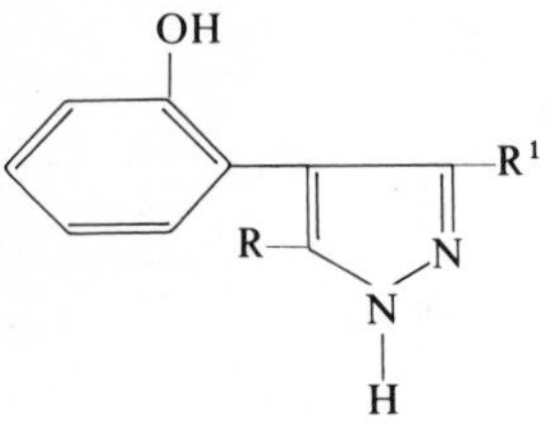

OH R R¹ N N H **45**

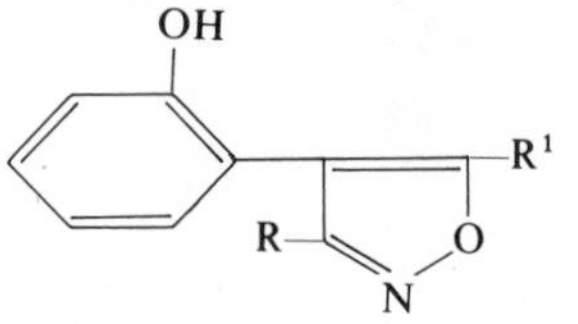

OH R R¹ O N **46**

When **47** is treated with an excess of phenylhydrazine in ethanolic solution, **48** is obtained. Considerable heat is evolved when **47** is mixed with hydrazine hydrate in ethanol, and the hydrazinium salt of the enolic form **49** is obtained (Eq. 13).[5]

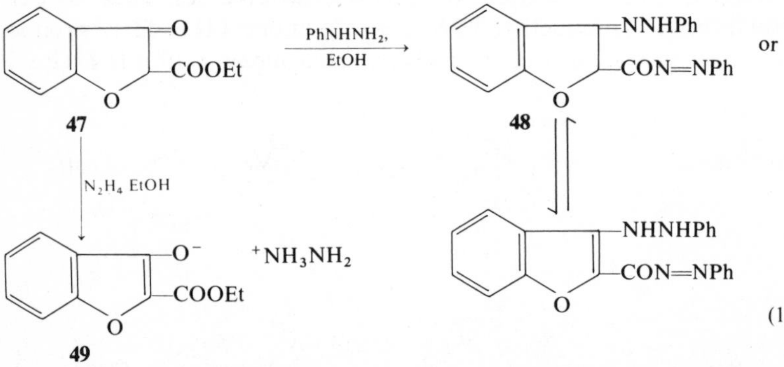

47 =O, COOEt, O —$PhNHNH_2$, EtOH→ **48** =NNHPh, CON=NPh, O or ⇌ NHNHPh, CON=NPh, O

47 —N_2H_4 EtOH→ **49** O^-, COOEt, O $^{+}NH_3NH_2$

(13)

With hydrazine sulfate, 3(2*H*)-benzofuranones (**50**) yield the corresponding azines.[58] The Fischer-indole reaction of **50** with phenylhydrazine leads to the formation of benzo(3,2-b)indoles (**0 l**) (Eq. 14)[5,58]

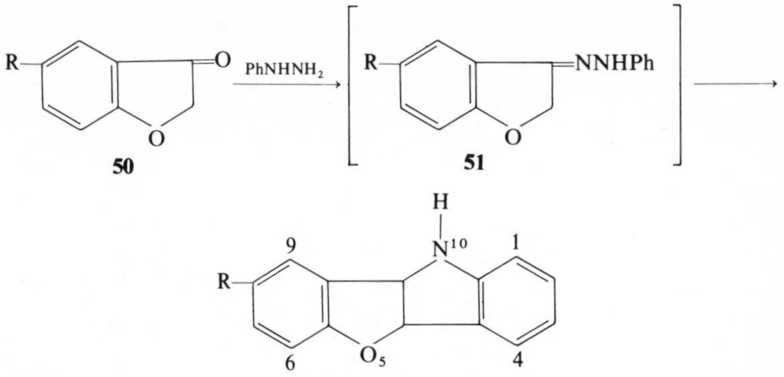

R, 50, PhNHNH$_2$, [R, 51, =NNHPh] →

R, 52, H, N^{10}, 1, 4, 6, 9, O$_5$

(14)

Although **51** (R = H) has been claimed to be an intermediate in the preparation of **52** (R = H),[59] Schroeder et al.[5] were unable to find support for its formation. Moreover, they have shown that **50** (R = H) could not form the corresponding thiosemicarbazone or the semicarbazone.[60]

The transposition of a furan ring to an oxazine ring is a general reaction. It was applied to the transformation of various phenols via phenoxyacetic acids to 3(2*H*)-benzofuranones and the conversion of the corresponding 2-isonitroso derivatives to the diketobenzisoxazines (**53**) (Eq. 15).[61-63]

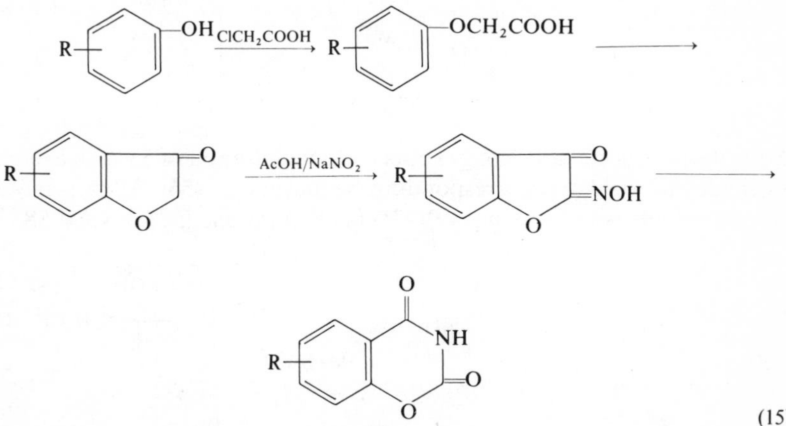

R–C$_6$H$_4$–OH $\xrightarrow{ClCH_2COOH}$ R–C$_6$H$_4$–OCH$_2$COOH →

R–benzofuranone $\xrightarrow{AcOH/NaNO_2}$ R–(2-=NOH)benzofuranone →

53

(15)

3(2*H*)-Benzofuranone behaves as a typical ketone toward Grignard reagents. The carbinol, which is the initial product, generally undergoes dehydration during decomposition of the magnesium complex to yield 3-substituted benzofuran (Eq. 16).[64] However 2,2,5-trimethyl-3(2*H*)-benzofuranone (**54**) gave stable carbinols (**55**), which dissolved in strong acids to afford colored salts (**56**) with exception of **55** (R = Me) (Eq. 17).[65]

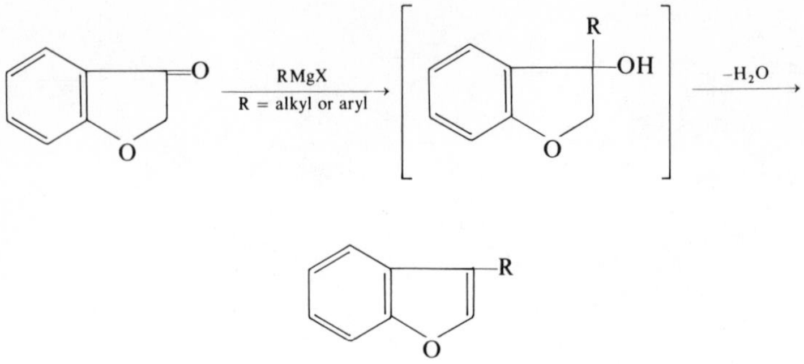

RMgX, R = alkyl or aryl; $-H_2O$

(16)

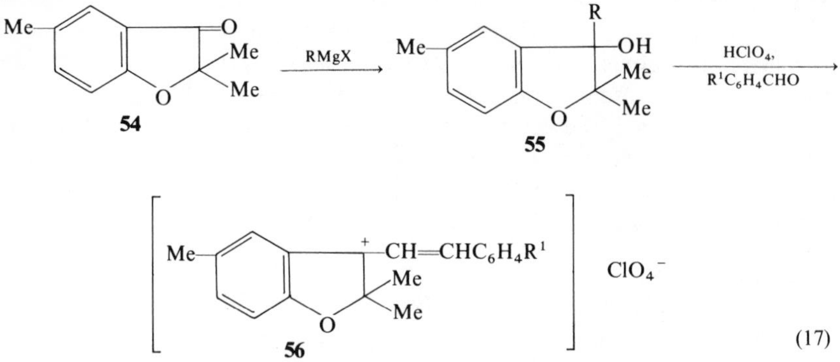

54; RMgX; 55; $HClO_4$, $R^1C_6H_4CHO$

56; $CH{=}CHC_6H_4R^1$; ClO_4^-

(17)

Aqueous perchloric acid brings about the condensation of **55** with aromatic aldehydes to yield the corresponding hemicyanines (**56**). Allylmagnesuim bromide reacts with 6-methoxy-3(2*H*)-benzofuranone (**57**) to yield **58**.[66]

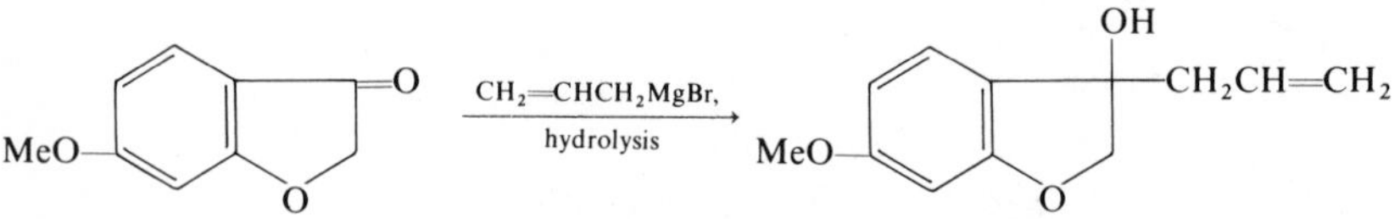

57; $CH_2{=}CHCH_2MgBr$, hydrolysis; 58

The methylene group at position 2 in 3(2*H*)-benzofuranones is very reactive;[52] condensation reactions with aldehydes, ketones, nitrous acid and/or *p*-nitrosodimethylaniline have been reported.[67,68] The behavior of 3(2*H*)-benzofuranones toward halogenation is shown in Eq. 18.[1,2,12,70–73] Because these compounds are both α-haloketones α-haloethers, it is not surprising that the halogen atoms are very reactive. They readily undergo

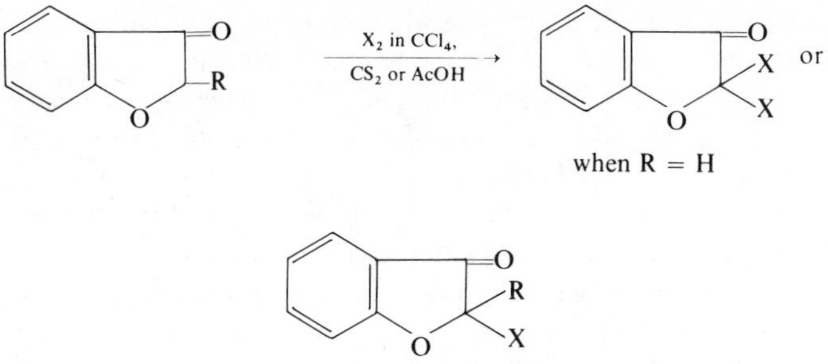

X$_2$ in CCl$_4$, CS$_2$ or AcOH; when R = H — or (18)

displacement reactions by such ions as hydroxyl, acetate, or alkoxyl.[2] Thus treatment of 2-bromo-3(2*H*)-benzofuranone with thioacetamide in ethanol yielded the 3-hydroxy-2-benzofuranyl ester of thioacetimidic acid (**59**), which is easily decomposed in sulfuric acid to afford 2-methylbenzofuro (3,2-d)thiazole (**60**).[69]

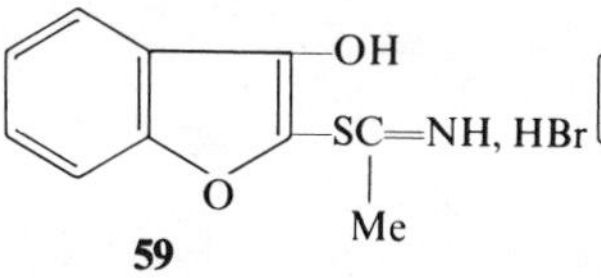

59 —SC=NH, HBr; Me

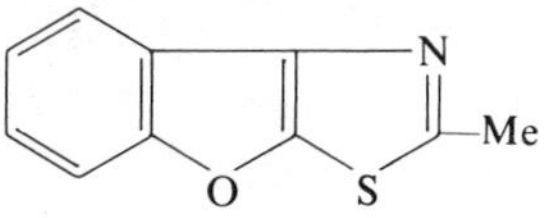

60

2,2-Dichloro-3(2*H*)-benzofuranones undergo hydrolysis to afford the corresponding benzofuran-2,3-diones.[71] When 3(2*H*)-benzofuranone is treated with phosphorus pentachloride, the ketonic oxygen is replaced by two chlorine atoms,[72] and the resulting compound, on dehydrohalogenation, gave 3-chlorobenzofuran.

Whereas the greater amount of chemical evidence favors the ketonic form for 3(2*H*)-benzofuranones, a few reactions are best explained on basis of the enolic structure. A coupling reaction with diazonium salts leads to the formation of **61** or the tautomer;[74,75] with thioglycollic acid, **62** is obtained.[76]

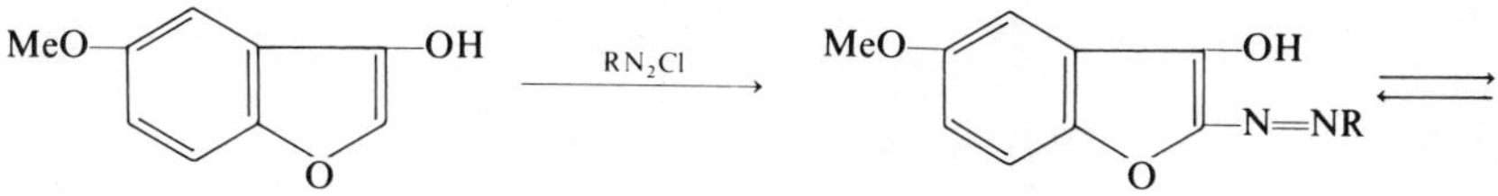

RN$_2$Cl

61

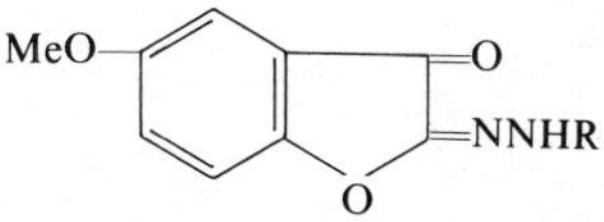

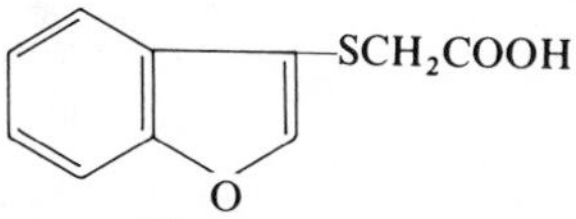

62

3(2*H*)-Benzofuranone undergoes aldol condensation with itself, followed by a subsequent oxidation to yield a series of oxygen analogs of indirubin. In absence of air, **63** and **64** are obtained;[73] however in presence of air the leuco compound, **65** is formed. Oxindirubin (**66**) itself is produced through condensation of 3(2*H*)-benzofuranone with *o*-hydroxyphenylglyoxylic acid, which probably proceeds via 2,3-dihydrobenzofurandione (Eq. 19).

3(2*H*)-Benzofuranone gives brilliant colors with alkaline nitroprusside and Zimmermann's reagent, it suffers oxidation by air[77] to peroxides and yellow resins,[78] and by periodates or Fehling's solution to the corresponding salicylic acid. It is not surprising, then, that 3(2*H*)-benzofuranones polymerize by self-condensation, are hard to store for long periods, and that

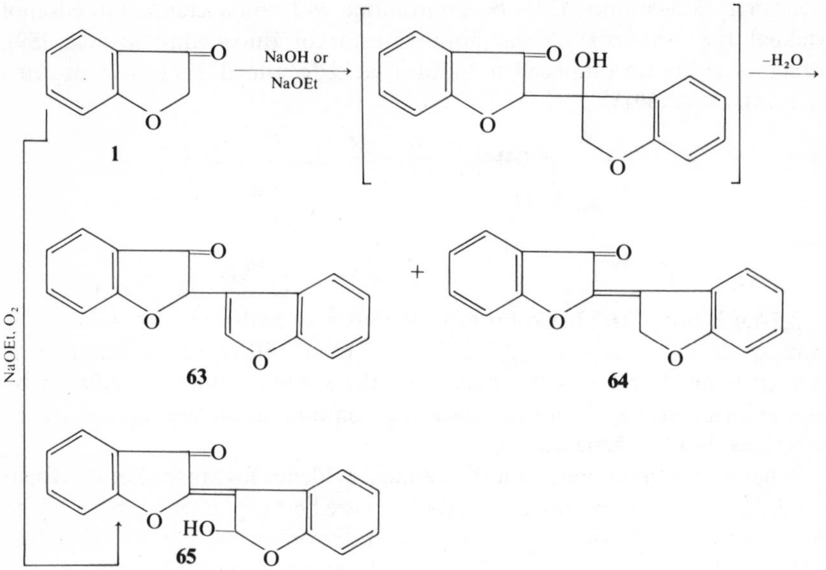

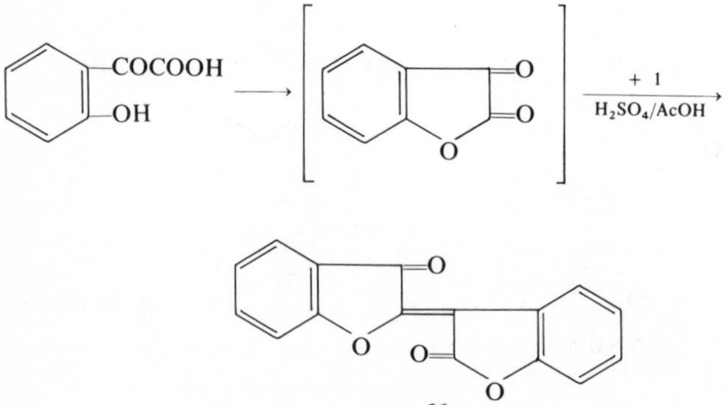

NaOH or NaOEt; $-H_2O$; NaOEt, O_2

1 **63** **64** **65**

COCOOH; OH; + **1**, H_2SO_4/AcOH

66

(19)

when both hydrogen atoms of the 2-methylene group are replaced, there is a marked increase in stability; in fact, no naturally occurring 3(2*H*)-benzofuranone has hydrogen at position 2.

2. Substituted 3 (2*H*)-Benzofuranones

A. 2-Hydroxy-2-Benzyl-3 (2*H*)-Benzofuranones

Algar and Flynn[79] and independently Oyamada[80,81] have found that 2′-hydroxychalcones (**67**) with alkaline hydrogen peroxide give flavonols (**71**) in satisfactory yield; dihydroflavonols (**69**) are intermediate in this reaction (AFO reaction).[82,83] It was later shown that if a methoxy[84] or methyl[85] substituent is present in the 6′ position in the chalcone, aurones (**72**) rather than flavonols are obtained, provided the 2,[86] or 4 position[87] does not carry a hydroxyl group.

Conflicting statements in the literature[88-90] led to the need for reinvestigation of the course of the reaction.[93] Besides aurones and flavonols, 2-benzyl-2-hydroxy-3(2*H*)-benzofuranones (**75**) and 2-arylbenzofuran-3-carboxylic acids (**81**) are sometimes formed.[91,92]

The mechanism of formation of flavonols (**71**) and aurones (**72**) (Chart 3) has been considered by a number of authors.[88] It involves the formation of ketone epoxide (**68**) from which (via route 1) dihydroflavonol (*trans*-2,3-dihydro-) (**69**),[94,95] or (via route 2) a "hydrated aurone" (**70**) is formed. The presence of a substituent (Me or OMe) in the 6′ position of the chalcone displaces the keto group from the plane of the nucleus. This produces steric inhibition from the 2-O ion, and promotes activation of the α-carbon of the chalcone.[84,96,97] A hydroxyl group in the 2 or 4 position causes resonance expulsion of the β-epoxide bond with consequent dihydroflavonol formation.[87]

A route to the production of 2-benzyl-2-hydroxy-3(2*H*)-benzofuranones (**75**) and 2-arylbenzofuran-3-carboxylic acids (**81**) in the AFO reaction is based on a mechanism suggested for the production of compounds (**75**) and **77** by the action of alkali on dihydroflavonols.[96-99]

As the Chart 3 shows, products of the AFO reaction may be divided into two classes: (a) Dihydroflavonols (**69**) and flavonols (**71**), where attack of the O^- ion of compound **68** on the β-carbon atom of the propane unit is involved; and (b) aurones (**72**), the production of these compounds is related to the O^- ion on the α-carbon atom of the epoxide (**68**).

It will be noted that, as Gripenberg[96] has pointed out, the hypothetical "hydrated aurone" (**70**) is unstable and immediately loses the elements of water to form the aurone (**72**). On the other hand, compound **75** is stable

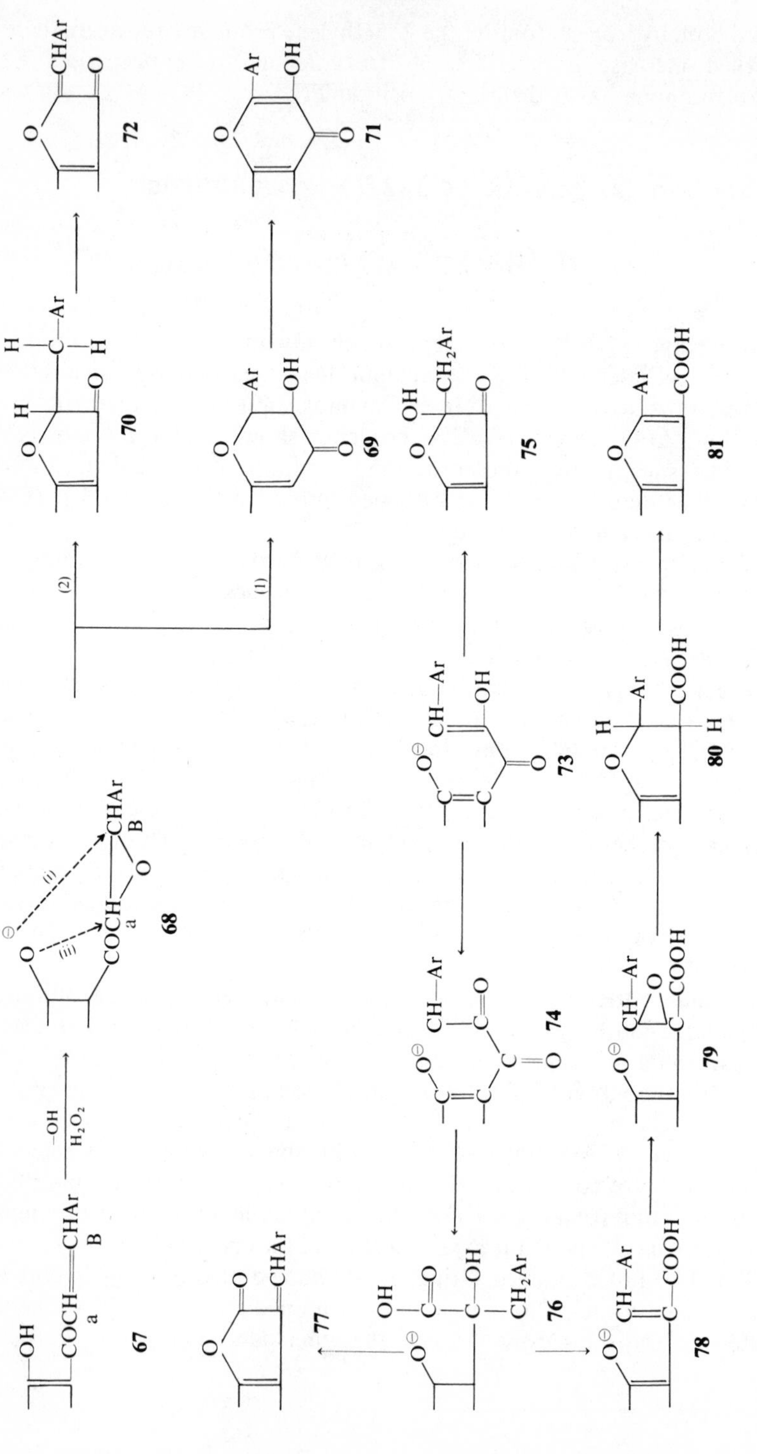

Chart 3

and requires treatment with dehydrating agent to form **72**. It is unlikely to be an intermediate in the AFO synthesis of aurones.

Donnelly et al.[100] have shown that application of the AFO reaction to phloroglucinol-type α-methoxychalcones (**82**) results in the production of a new structure, 2-α-hydroxybenzyl-2-methoxy-3(2*H*)-benzofuranone, **83**, and not, as previously reported,[101] the corresponding 3-methoxyflavones.

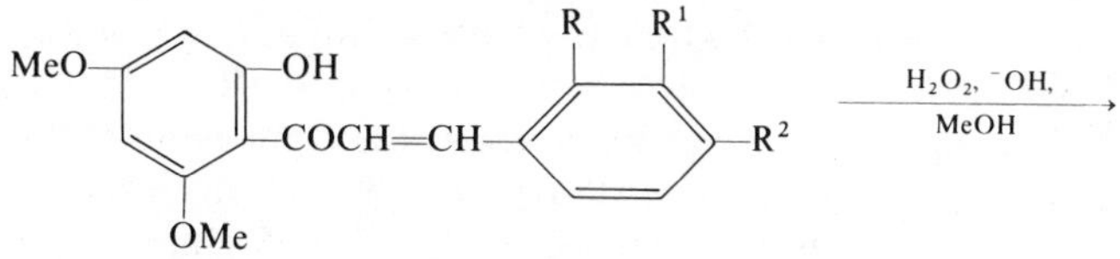

82 $\xrightarrow{H_2O_2,\ ^-OH,\ MeOH}$ **83** $\xrightarrow{MnO_2,\ CH_2Cl_2}$ **84**

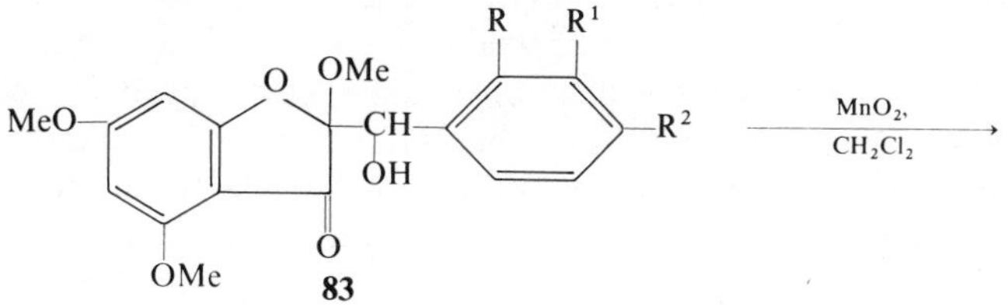

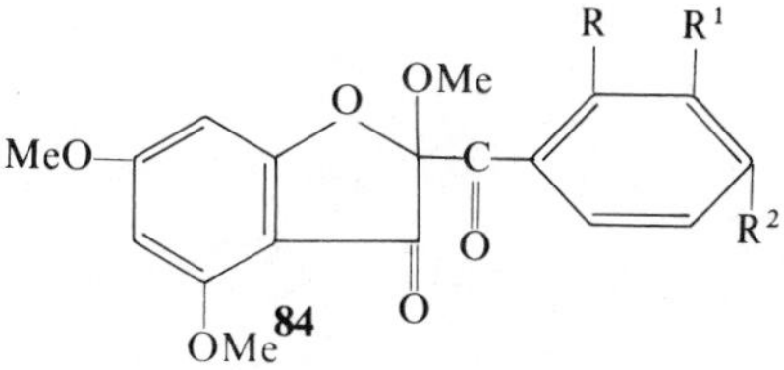

a, $R = R^1 = R^2 = H$
b, $R = OMe$, $R^1 = R^2 = H$
c, $R = R^1 = H$, $R^2 = OMe$
d, $R = H$, $R^1 = R^2 = OMe$

Further oxidation of **83a** with sodium dichromate[102] or manganese dioxide[100] gave **84a**.

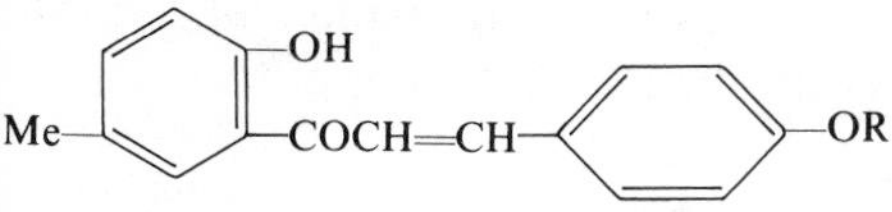

85 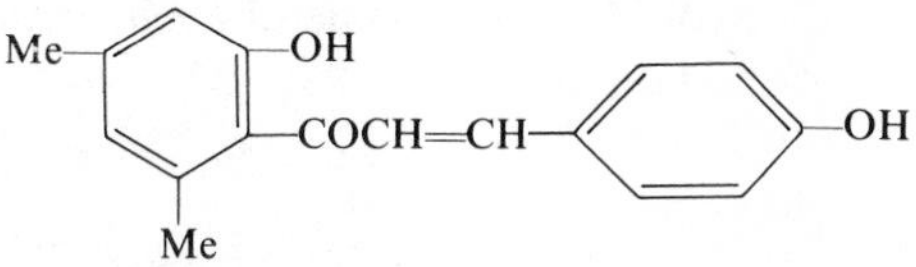 **86**

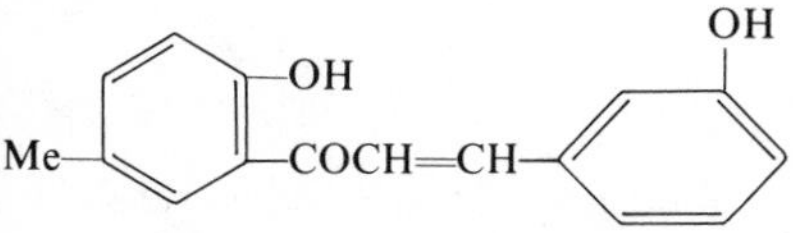

87 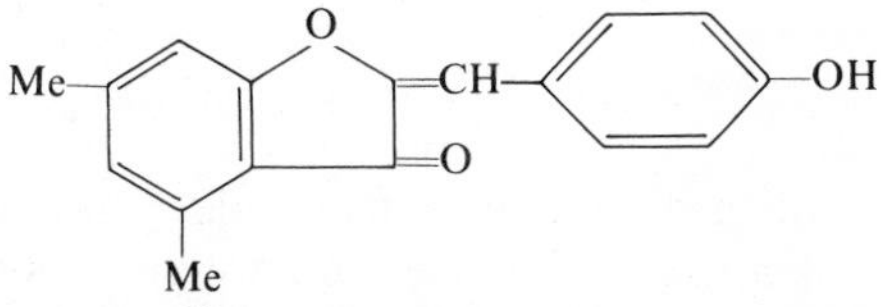 **90**

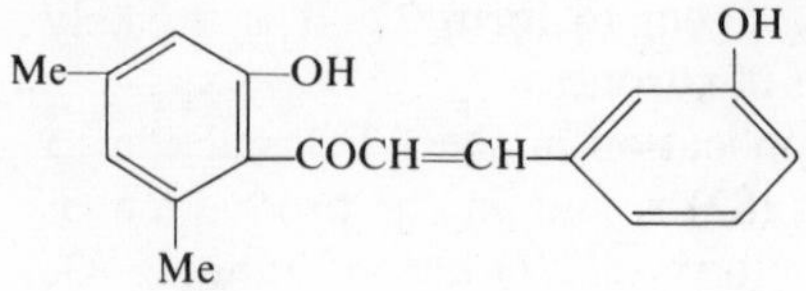

88 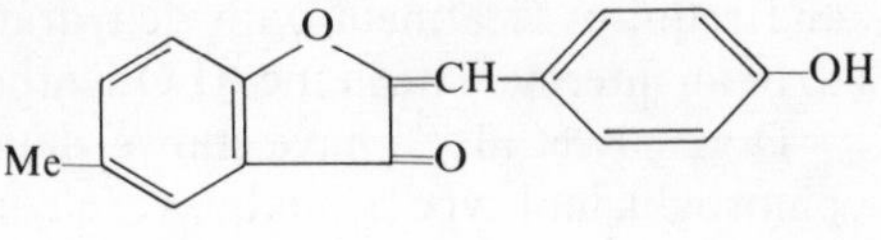 **89**

In contrast to alkaline hydrogen peroxide, alkaline ferricyanide failed to attack the chalcone derivatives **85** (R = Me), **86** (R = Me), **87** and **88**. However this reagent[103] has brought about the conversion of **85** (R = H) and **86** (R = H) to the aurone derivatives **89** and **90**, respectively. This was done presumably through route 3, which is analogous to that assumed to underlie the conversion of polyhydroxybenzophenones to derivatives of "grisan".[104,105] Thus ferricyanide oxidations do not occur unless there is a free hydroxyl group at the 4 position in the 2′-hydroxychalcone, and the products are always aurones. The characteristics are completely different from those of the AFO reaction which, in particular, never affords 4′-hydroxyaurones; the new method is, therefore, a useful synthetic supplement to the old.[103]

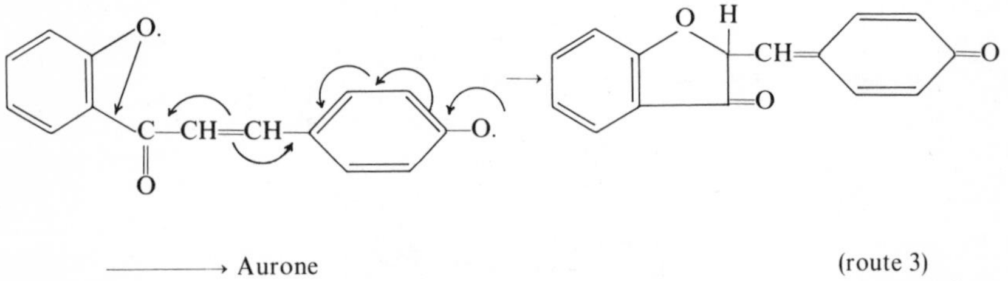

⟶ Aurone (route 3)

It is believed that favorable mode of cyclization of 2′-hydroxychalcones, having an α-methoxy group under both acidic and basic conditions, leads to the formation of a 5-membered ring rather than a 6-membered ring.[96-98,106] However Dean et al.[107] have found that cyclization of 2′-hydroxy-α,4,4′-trimethoxychalcone resulted in the formation of 7,4′-dimethoxyflavonol (**91**). Moreover, Gripenberg[96] has shown that the

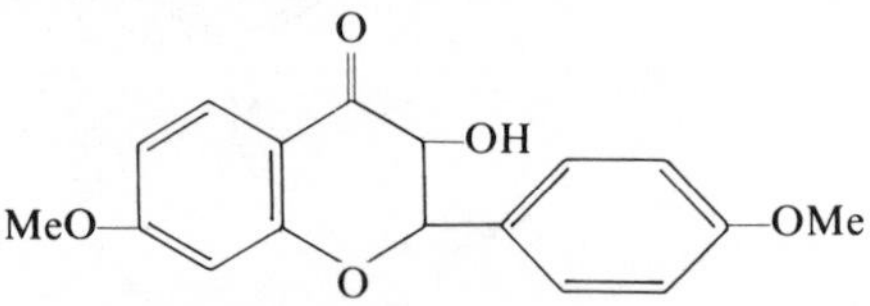

91

alleged 3-hydroxyflavonols, obtained by Kimura,[108] are 2-benzyl-2-hydroxy-3(2*H*)-benzofuranones, which can be produced by rearrangement of the true 3-hydroxyflavanones (Eq. 20).

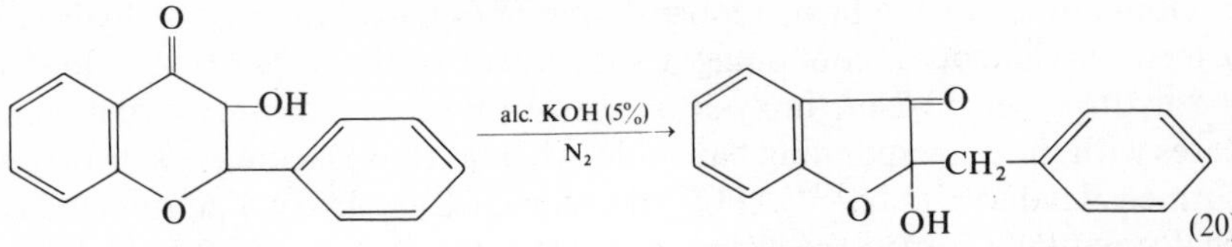

alc. KOH (5%), N_2 (20)

The reported oxidation product of 2′-hydroxy-α,4′6′-trimethoxychalcone (**92**)[101] has been found to be 2-hydroxybenzyl-2,4,6-trimethoxy-3(2*H*)-benzofuranone (**94**). The mechanism of the formation of **94** could involve the initial formation of chalcone epoxide (**93**), followed by ring closure. An alternative procedure would be earlier ring closure to **95**, followed by oxidation at the benzylic position (Eq. 21).[109]

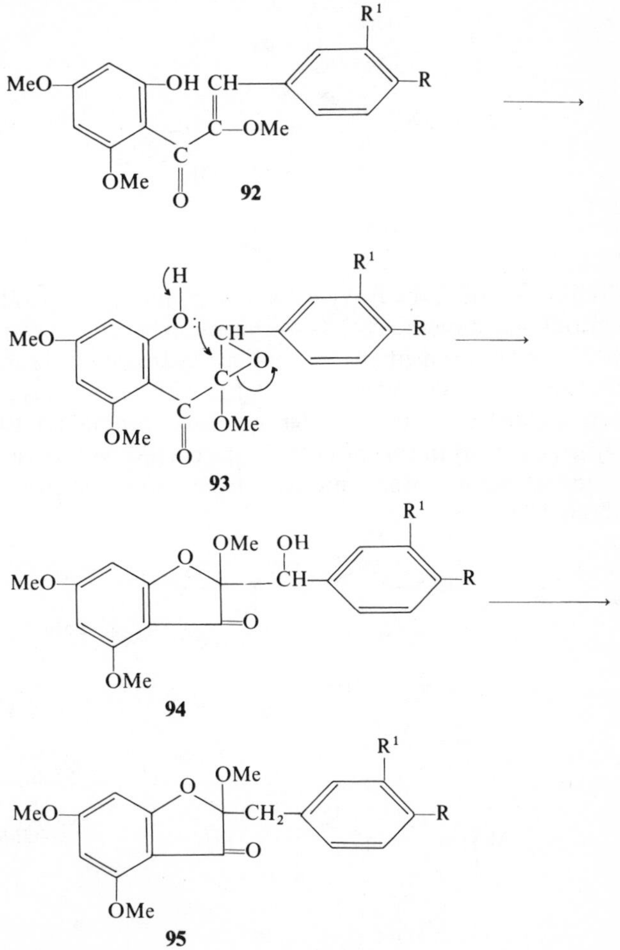

92

93

94

95

(21)

Donnelly et al.[110] have reported that AFO oxidation of 2′-hydroxy-α-methoxychalcones, containing a substituent in the 6′ position, afforded 2-α-hydroxybenzyl-2-methoxy-3(2*H*)-benzofuranones accompanied in some cases with the corresponding flavonols, whereas those lacking 6′ substituent formed flavonols only.[100] AFO oxidation of 2′-hydroxy-4′,6′-dimethoxychalcone (**96**) afforded the expected 4,6-dimethoxy-2-methyl-2-(α-hydroxybenzyl)-3(2*H*)-benzofuranone (**97**).

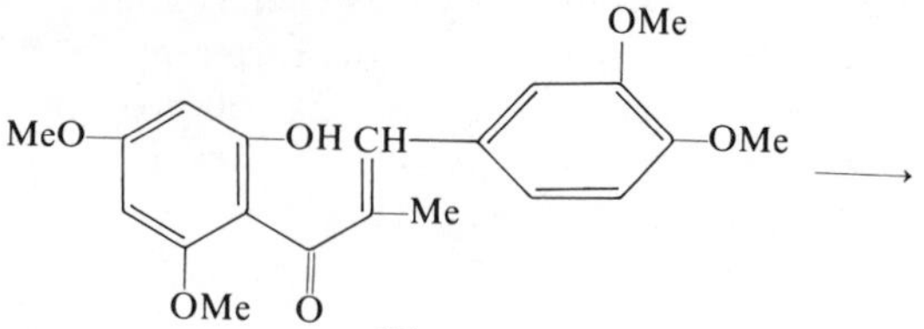

96

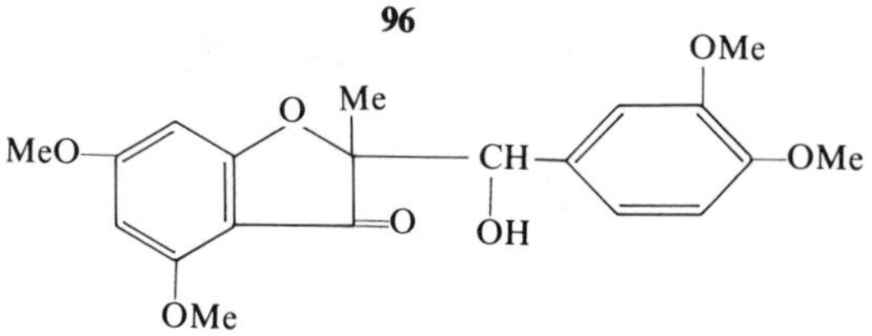

97

Cyclization of 2′-hydroxy-α,3,4,4′,6′-pentamethoxychalcone (**98**) gave 2-hydroxy-4,6-dimethoxy-2-(3,4-dimethoxybenzyl)-3(2*H*)-benzofuranone (**99**),[111] which in methanolic sodium hydroxide solution afforded the sodium salt of **98**. Treatment of the latter with acetic anhydride and sodium acetate yielded a mixture of **100** and **101**. Compound **101** is formed by benzylic acid rearrangement of **98** to the corresponding carboxylate anion, and its structure was inferred from spectrometric data and its synthesis.[111,112]

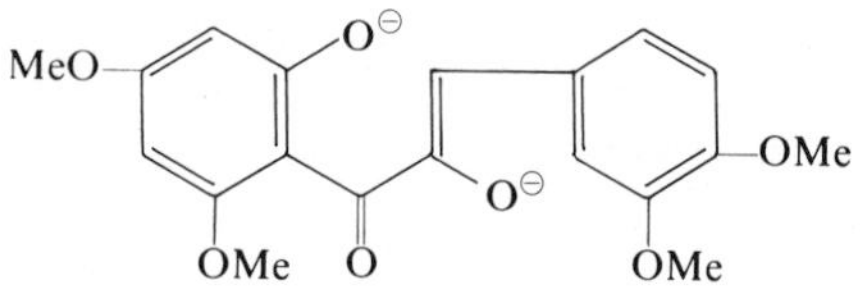

98

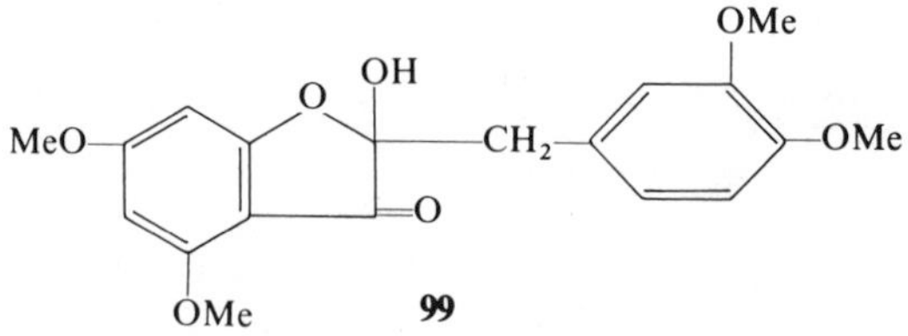

99

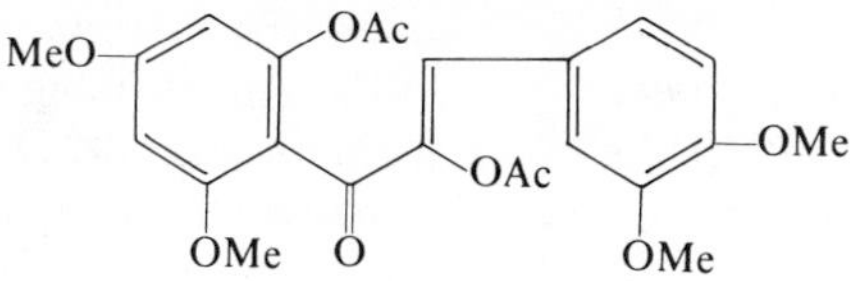

100

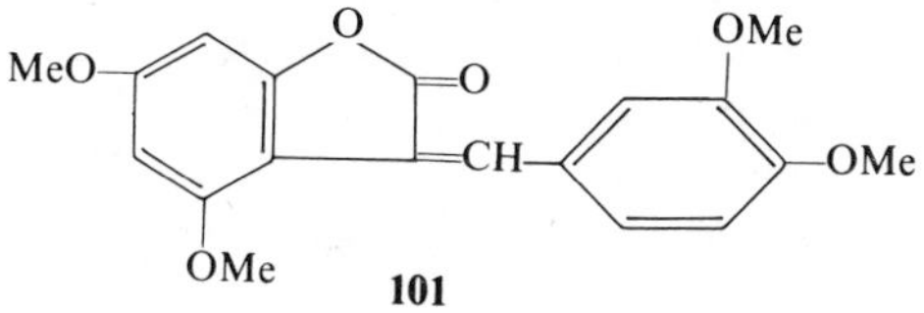

101

Treatment of **102** (R = R^1 = alkoxy) with sodium acetate–acetic acid mixture yielded **103**. However **102** ($R^1 = R^2 = H$) did not isomerize under these conditions, and it was necessary to convert them into the corresponding chlorohydrins (**103**) by action of stannic chloride in benzene, followed by treatment with sodium acetate–ethanol mixture to yield **104**.[113] Debenzylation of **104**, upon treatment with hydrochloric-acetic acid mixture, afforded **105**, which can also be obtained from **106** under the same conditions. In alcoholic medium, only **106** was demethylated to give **105**.

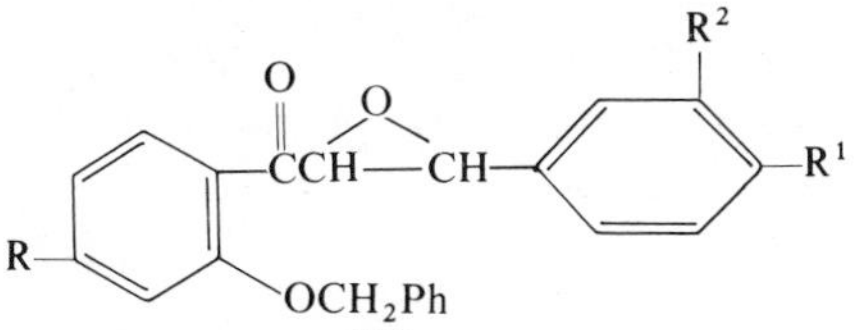

102

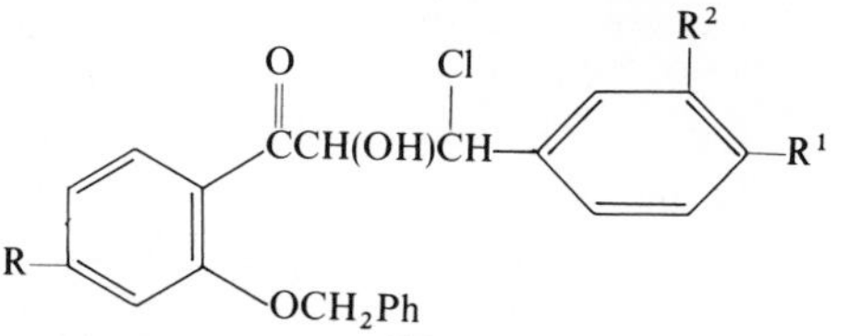

103

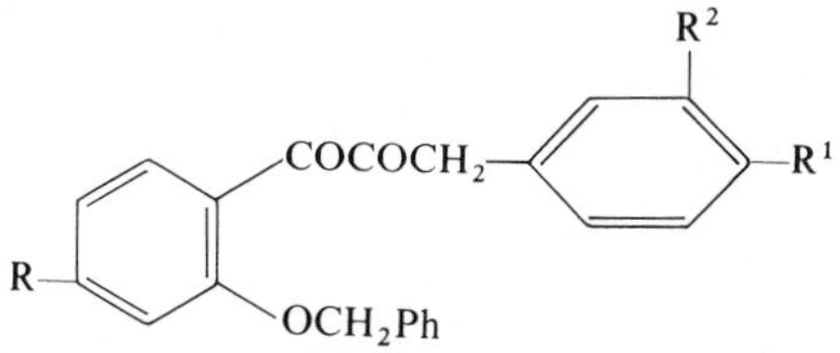

104

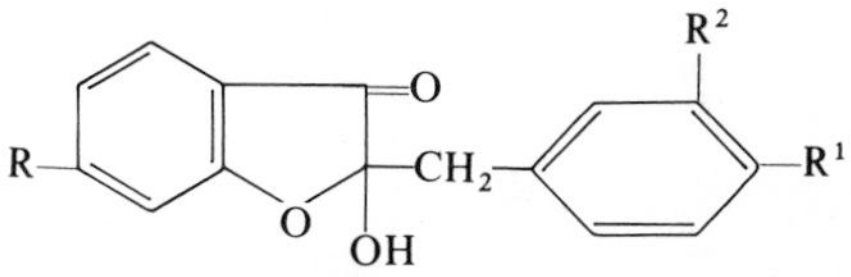

105

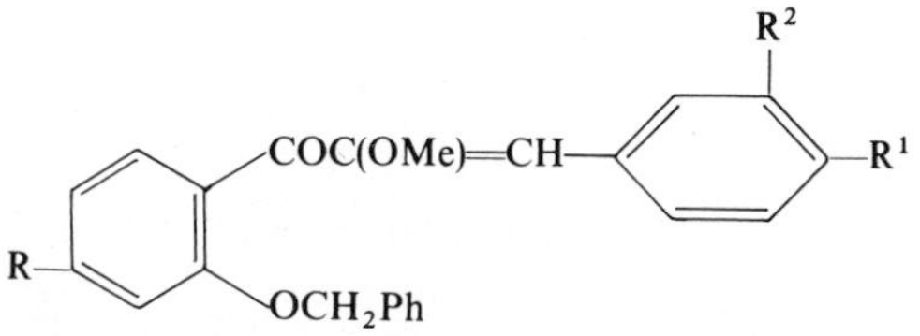

106

Pyridinium perbromide and trimethylphenylammonium perbromide react with 2′-hydroxychalcones in different fashions. Thus treatment of **107** (R = R^1 = H) results in the formation of **107** (R = H, R^1 = Br). However, under the same conditions **107** (R = OMe, R^1 = H) afforded **108** (R = Br, R^1 = Me), which was produced by thermal cyclization of **107** (R = OMe, R^1 = Br) and obtained by the action of pyridinium perbromide or of *N*-bromosuccinimide on **108** (R = H, R^1 = Me). Treatment of an acetic acid solution of **107** (R = OMe, R^1 = H) with hydrobromic acid effects the formation of **108** (R = R^1 = H).[114] In the presence of hydrobromic acid at cold temperatures, both demethylation and cyclization occur together whether the α-methoxychalcone contains bromine or not.

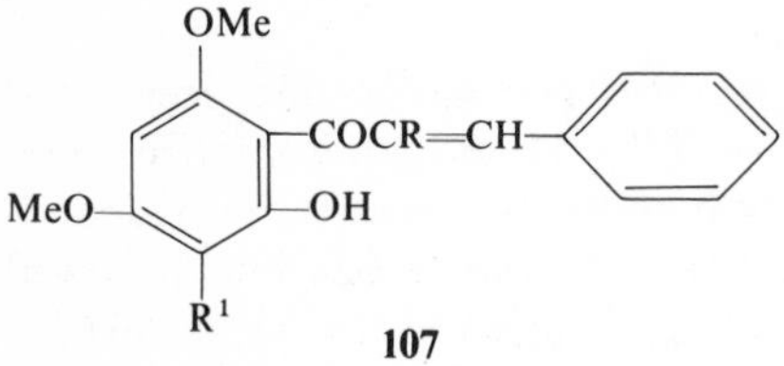

107 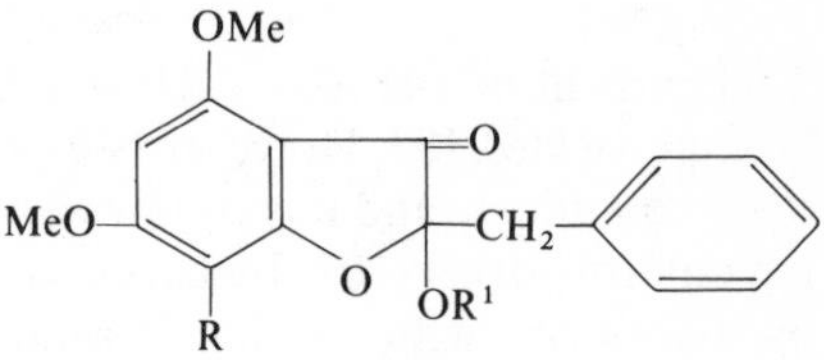 **108**

Chlorinated 6-methoxychalcones in the AFO machine undergo oxidation to yield the corresponding aurones and flavonols. Thus 2-hydroxy-3,4,6-trimethoxy-2′-chlorochalcone (**109**) gives a mixture of **110** and **111** (Eq. 22).[115] However 2′-, 3′-, and 4′-nitro-substituted 2-hydroxy-3,4,6-trimethoxychalcones do not undergo oxidation; this may be attributed to the stabilizing effect of the nitro group on the chalcone molecule.

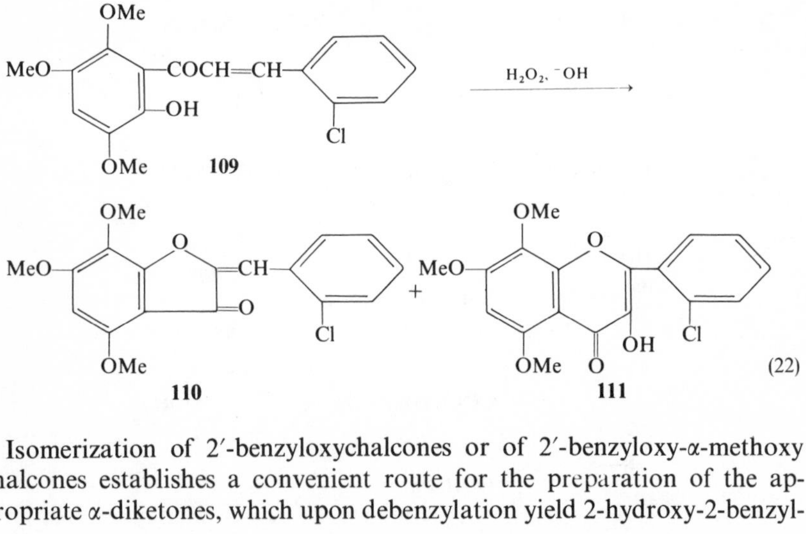

109 $\xrightarrow{H_2O_2,\ ^-OH}$ **110** + **111** (22)

Isomerization of 2′-benzyloxychalcones or of 2′-benzyloxy-α-methoxy chalcones establishes a convenient route for the preparation of the appropriate α-diketones, which upon debenzylation yield 2-hydroxy-2-benzyl-

3(2*H*)-benzofuranones.[116] The existence of an equilibrium between the cyclic and open forms is determined by their magnetic resonance spectra and is in good agreement with the results obtained by alkaline rearrangement of 3-hydroxyflavanones. Isomerization of 2′-methoxymethoxy-4-methoxychalcone epoxide, upon reflux in alcoholic potassium hydroxide solution to 1-(2-methoxymethoxy-4-methoxy)-3-phenylpropane-1,2-dione followed by treatment with sulfuric acid, afforded 2-hydroxy-2-benzyl-6-methoxy-3(2*H*)-benzofuranone.[117]

Thermal cyclization of 2′-hydroxy-α-4′,6′-trimethoxychalcone results in rearrangement to 2-benzyl-2,4,6-trimethoxy-3(2*H*)-benzofuranone; this is identical with the product obtained by methylation of 2-hydroxy-2-benzyl-4,6-dimethoxy-3(2*H*)-benzofuranone.[118]

Ollis and co-workers[119] have recently shown that thallium acetate reacts with chalcones via the intermediate enol-thallium derivatives. Thus oxidation of angolensin methyl ether (**112**)[120] with methanolic thallium(**111**) acetate gave the benzofuranone (**113**). Oxidation of 4,4′-dimethoxy-2′-hydroxychalcone (**114**) yields, in addition to the flavone (**115**), a product which is formulated like **116**. Formation of **116** by oxidation of **114** suggests that the intermediate acetal **117** is formed, but that **117** is then oxidized further to **116** in a process similar to oxidation of angolensin methyl ether **112** to **113**.

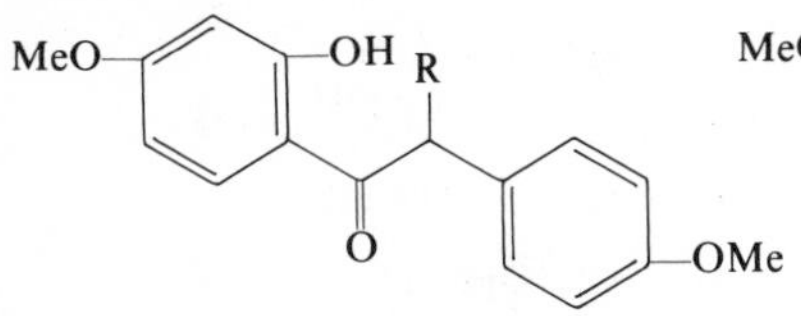

MeO OH R O OMe

112 R = Me
117 R = CH(OMe)$_2$

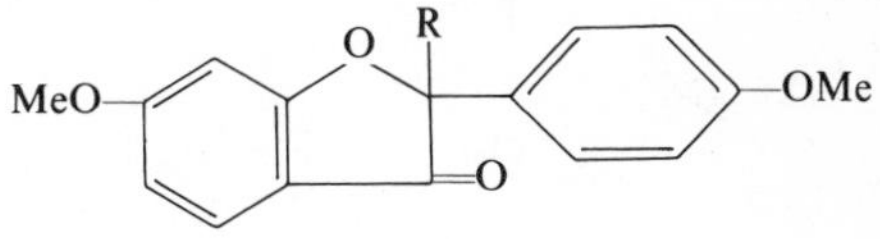

MeO O R OMe O

113 R = Me
116 R = CH(OMe)$_2$

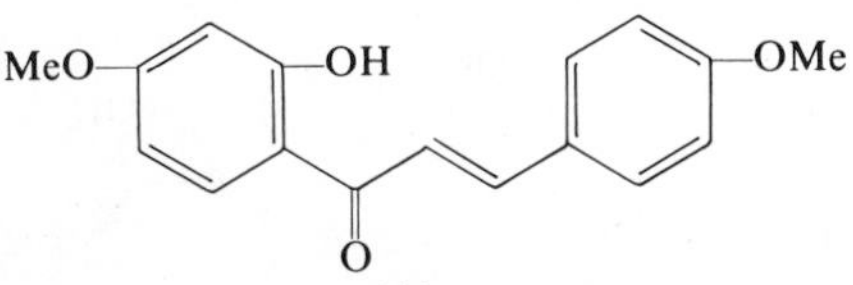

MeO OH OMe O

114

MeO O OMe O

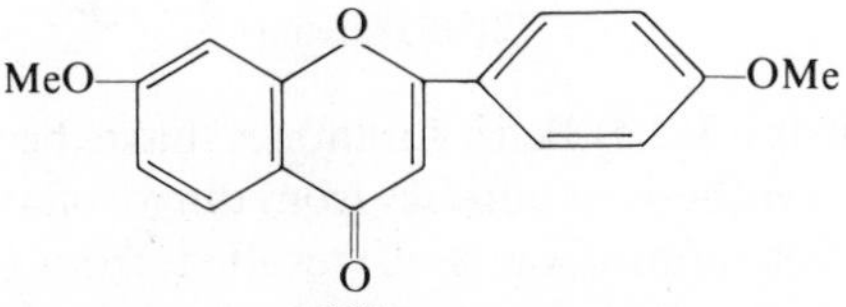

115

The existence of 2-benzyl-2-hydroxy-3(2*H*)-benzofuranone in Quebrach tannin extract (aqueous heartwood extract of *Schinopsis balansae* and *S. lorentzii*) has thrown light on the biogenetic relations of C_6-C_3-C_6 compounds.[121,122] The heartwood contains two C_{15} compounds of a type not previously reported as natural products; 2-benzyl-2,6,3′,4′-tetrahydroxy-3(2*H*)-benzofuranone (**120**) and its 4-methyl ether,[121] besides 3,7,4′-trihydroxyflavone and fisetin (**119**).[123] The structural analogy of these two compounds (**120** and its 4-methyl ether) to **119** and 4′-methoxyfisetin suggests a biogenetic metabolic relation between **119** and fustin **118**, and 2-benzyl-2-hydroxy-3(2*H*)-benzofuranone. This is also supported by observation of the hydroxyaurone sulfuretin, **121**, in the extract of the heartwood of *Rhus cotinus* (Anacardiaceae) which is also a known source of fustin, since sulfuretin is readily derived from benzyltetrahydrobenzofuranone by dehydration.

The general reactivity of the hydroxylated 2-benzyl-2-hydroxy-3(2*H*)-benzofuranones resembles that of the catechins and leucoanthocyanidins; this new group of polyphenols must, therefore, be included in consideration of the condensed tannins.

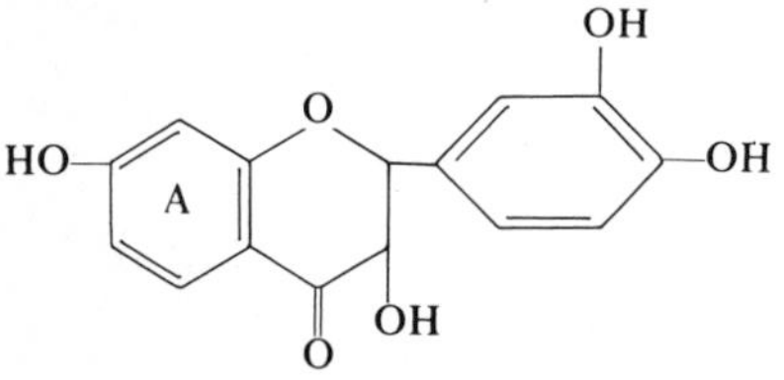

118 (Fustin)

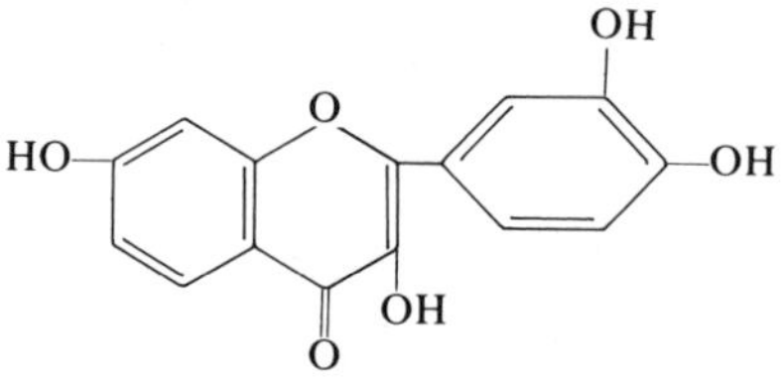

119 (Fisetin)

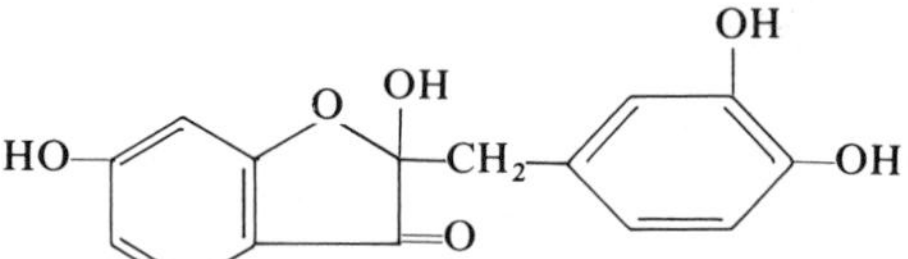

120

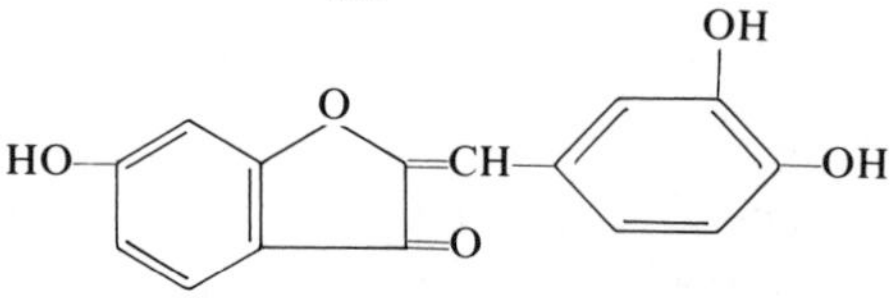

121 (Sulfuretin)

2-Benzyl-2-hydroxy-3(2*H*)-benzofuranones have been suggested as intermediates in the synthesis of aurones from dihydroflavonols.[124] 2-Benzyl-2-hydroxy-3(2*H*)-benzofuranone itself resulted from 3-hydroxyflavanone upon treatment with alcoholic potassium hydroxide. Lindstedt[125] has obtained 2-benzyl-2-hydroxy-4,6-dimethoxy-3(2*H*)-benzofuranone "apoalpinone methyl ether" from pinobanksin dimethyl ether (3-hydroxy-5,7-dimethoxyflavanone).

Epiampeloptin pentamethyl ether was considered as a stereoisomer of ampelopting (3′,4′,5,5′,7-pentahydroxyflavonol pentamethyl ether); recent investigations indicate the absence of a flavanone or chalcone moiety in its revised structure (**122**),[97] which loses water readily to form 2-(3′,4′,5′-trimethoxybenzal)-4,6-dimethoxy-3(2*H*)-benzofuranone (**123**).[126]

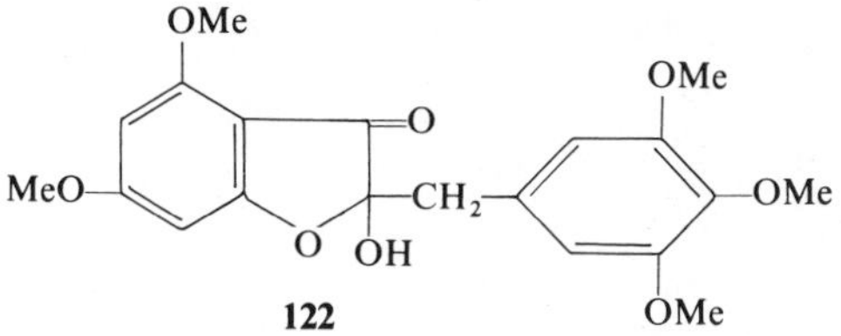

122

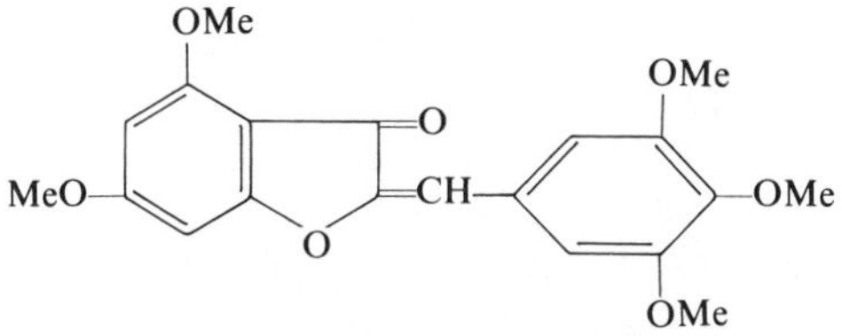

123

Treatment of fustin (**118**) with boiling alcoholic potassium hyroxide solution gave 2-benzyl-2,6,3′,4′-tetrahydroxy-3(2*H*)-benzofuranone (**120**).[127] Similar treatment to taxifolin (dihydroquercetin) yielded 2-benzyl-4,6,3′,4′-tetrahydroxy-3(2*H*)-benzofuranone.[128] Using Wheeler's procedure[129] for the preparation of chalcones, it was possible to bring about the condensation of ω-hydroxyacetophenone with isovanillin in presence of aqueous potassium hydroxide to give, among other products, 2-benzyl-2,6,3′-trihydroxy-4′-methoxy-3(2*H*)-benzofuranone directly; the expected α-hydroxychalcone is presumably unstable and rearranges immediately.[127] The synthesis, although unexpected, served to establish the first record of the existence in nature of compounds of this general structure, since 2-benzyl-2,6,3′,4′-tetrahydroxy-3(2*H*)-benzofuranone (**120**) is also present in tannin extracts.

Dihydrorobinetin (**124**), upon methylation to tetramethyl ether (**125**), followed by treatment with aqueous sodium hydroxide and acidification, resulted in the formation of 3-(3,4,5-trimethoxybenzal)-6-methoxy-2(3*H*)-benzofuranone (**126**) (Eq. 23).[99]

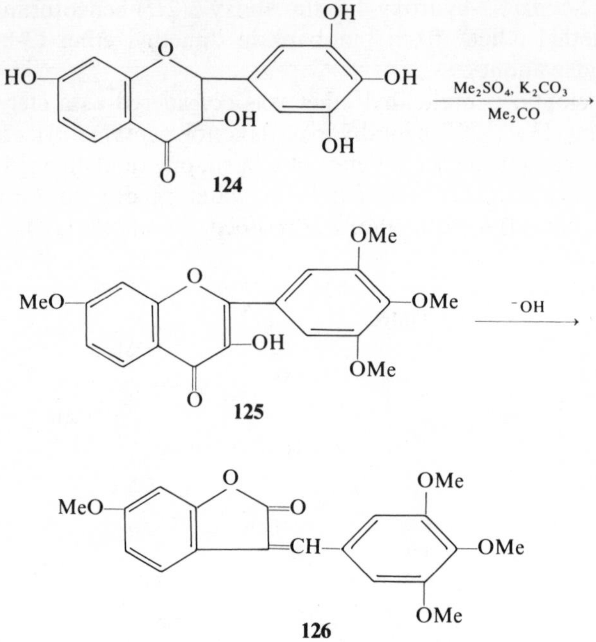

124

$\xrightarrow[\text{Me}_2\text{CO}]{\text{Me}_2\text{SO}_4,\ \text{K}_2\text{CO}_3}$

125

$\xrightarrow{^{-}\text{OH}}$

126

(23)

Quercetin (**127**), upon treatment with aqueous potassium hydroxide, gave 3′,4,4′,6-tetrahydroxy-2-phenyl-3(2*H*)-benzofuranone (**128**) (Eq. 24).[130]

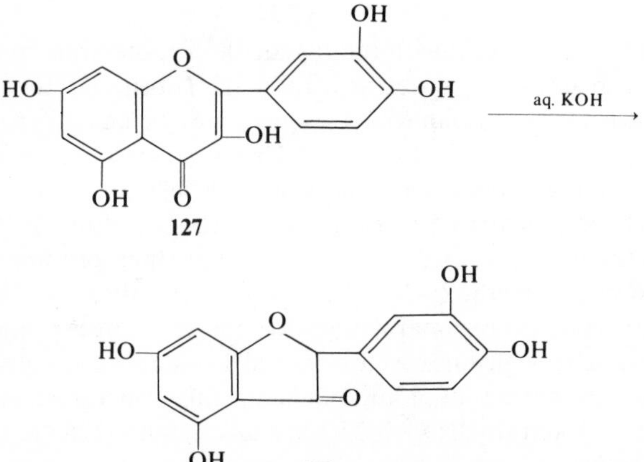

127

$\xrightarrow{\text{aq. KOH}}$

128

(24)

Pew[131] succeeded in reducing **127** to taxifolin (**129**) with sodium dithionite in the presence of aqueous sodium carbonate solution. Geissman and Lischner[132] obtained, upon treatment of **127** with sodium hydrosulfite, dihydroquercetin (**129**) and 4,6,3′,4′-tetrahydroxy-2-benzyl-3(2*H*)-benzofuranone (**130**), which in some experiments exceeded **129**. This involved ring opening and formation of the benzofuranone with further reduction. The reaction is reminiscent of the formation of flavonols and 2-benzal-3(2*H*)-benzofuranones by the alkaline hydrogen peroxide oxidation of 2′-hydroxychalcones (Eq. 25).[84]

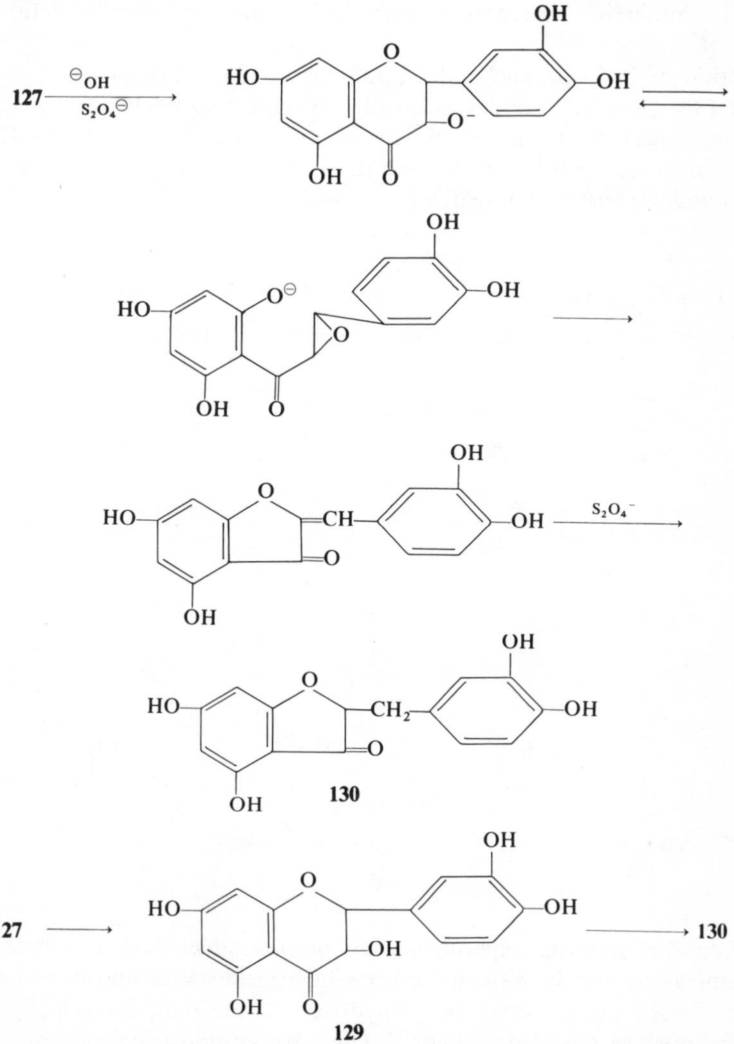

127 $\xrightarrow[S_2O_4^{\ominus}]{^{\ominus}OH}$ HO, OH, O, OH, OH, O^-, O $\rightleftharpoons$

HO, $O^{\ominus}$, OH, OH, O, OH, O $\longrightarrow$

HO, O, CH, OH, OH, O, OH $\xrightarrow{S_2O_4^-}$

HO, O, CH_2, OH, OH, O, OH

130

127 $\longrightarrow$ HO, O, OH, OH, OH, OH, O $\longrightarrow$ **130**

129

Methylation of **129** in methanol with methyl sulfate resulted primarily in a mixture of **99** and **95** (R = R^1 = OMe). The tetramethyl ether (**99**) could also be obtained upon treatment of **131** with potassium hydroxide.[133]

Later it was shown that prolonged treatment of **131** or **99** with alkali gave **132** contrary to expectations.[96] Moreover, treatment of **129** with methyl sulfate in the presence of sodium hydroxide at high temperatures effected the formation of **99, 133** and **134**. One would expect that such treatment should, through a benzylic rearrangement, finally lead to **101**, isomeric with **132**.[135-137] Compound **101** has been synthesized, making use of the fact that it should be formed from taxifolin tetramethyl ether by action of alkali (Chart 4).[134,137]

Action of hydrobromic acid on **135** yields **136**, which on methylation gives **139**. However, condensation of **135** with benzaldehyde in alkaline medium leads to the formation of 2-hydroxy-2-benzyl-4,6-dimethoxy-3(2*H*)-benzofuranone (**137**).[138,139] Dehydration of **137** affords the formation of 4,6-dimethoxyaurone **138** (Eq.26).

MeO, OH, $COCH_2OMe$, OMe **135** —HBr→ MeO, OH, $COCH_2OH$, OMe **136** —Me_2SO_4, K_2CO_3→ MeO, OH, CO, CH_2OH, MeO **139**

136 —PhCHO, ^-OH→ MeO, O, OH, CH_2, =O, OMe **137** —H_2SO_4→ MeO, O, =CH, =O, OMe **138**

(26)

A parallel series of experiments has been conducted with ω-methoxyacetophenone and its 4-methyl ether which yield fisetol and its 4-methyl ether, respectively, not the 6-hydroxy- and 6-methoxy-3(2*H*)-benzofuranones as reported earlier.[139] Thus the action of hydrobromic acid offers a convenient route and direct method for preparation of fisetol derivatives and consequently 2-hydroxy-2-benzyl-3(2*H*)-benzofuranones.[140-142]

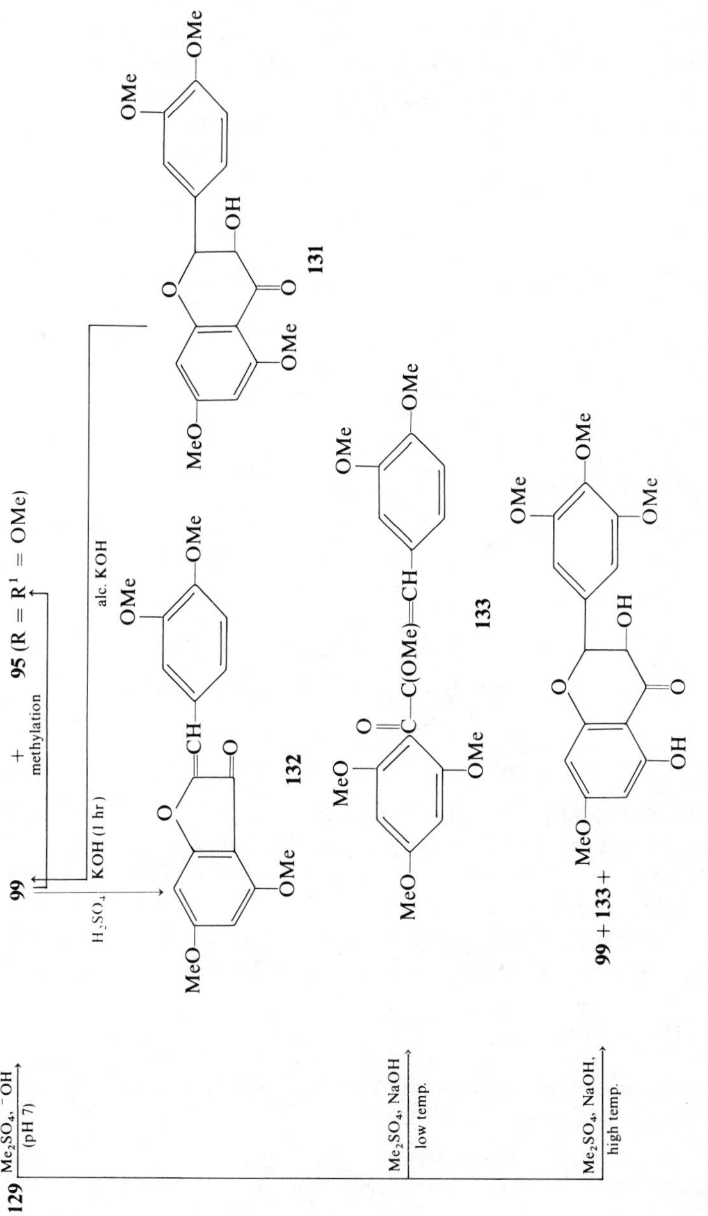
129
Me2SO4, −OH
(pH 7)
99
+
methylation
95 (R = R1 = OMe)
H2SO4
KOH (1 hr)
alc. KOH
132
131
Me2SO4, NaOH
low temp.
133
C(OMe)=CH
Me2SO4, NaOH,
high temp.
99 + 133 +
MeO
OMe
OH
O

Chart 4

B. 2-Benzyl-3 (2*H*)-Benzofuranones

Catalytic reduction of 2-benzal-5- (or 6-) methoxy-3(2*H*)-benzofuranones furnishes a convenient method for the preparation of 2-benzyl-3(2*H*)-benzofuranones;[96,143-147] it involves the formation of an aurone by condensation of benzaldehyde with the appropriate 3(2*H*)-benzofuranone (Eq. 27).[146,147]

MeO–C$_6$H$_3$(OH)–C(=O)CHBrCH$_2$Ph ← [$PhCH_2CHBrCOCl$, $AlCl_3$] MeO–C$_6$H$_4$–OMe →

MeO–C$_6$H$_3$(OH)–COCH$_2$Cl → [NaOAc/EtOH] → MeO-3(2*H*)-benzofuranone → [PhCHO, $^{\ominus}$OH] → MeO-2-benzal-3(2*H*)-benzofuranone (=CHPh) → [Pt/H_2, AcOH] → MeO-2-benzyl-3(2*H*)-benzofuranone (CH$_2$Ph)

(27)

An alternative approach is through the Friedel-Crafts reaction, using α-bromo-β-phenylpropionyl chloride and the appropriate phenol methyl ether, followed by ring closure.[148] The preparation would be simple if benzylation of 3(2*H*)-benzofuranone itself could be carried out. But this does not take place, although the methylene group is active enough to undergo condensation with aldehydes and ketones. It was, therefore,

MeO–C$_6$H$_3$(OH)–OCH$_2$Cl → MeO–C$_6$H$_3$(OAc)–COCH$_2$Cl →

MeO-2-acetyl-2-H-3(2*H*)-benzofuranone (**140**) → [MeO-2-acetyl-2-benzyl-3(2*H*)-benzofuranone (Ac, CH$_2$Ph)] →

MeO— [structure 141: 2-benzyl-6-methoxy-3(2H)-benzofuranone, O, H, —CH_2Ph, =O] ⟶ MeO— [structure 142: =CHPh, =O]

141 **142** (28)

considered necessary to activate the methylene group by introduction of an acetyl group. Thus 2-acetyl-6-methoxy-3(2*H*)benzofuranone (**140**) undergoes smooth benzylation with benzyl chloride and sodium ethoxide (Eq. 28).[149,150] Dehydrogenation of **141** via oxidation with neutral permanganate in acetone to 2-hydroxy-2-benzyl-3(2*H*)-benzofuranone and subsequent dehydration with sulfuric acid yielded **142**.[96,140]

C. 2-Acyl-3 (2*H*)-Benzofuranones

3-Chloroflavones undergo ring contraction upon treatment with alcoholic potassium hydroxide to yield 2-benzoyl-3(2*H*)-benzofuranones (**143**) (Eq. 29). The latter are readily transformed to the corresponding enol acetate (**144**).[151-153]

[flavone, —Ph, O] $\xrightarrow{SOCl_2}$ [3-chloroflavone, —Ph, —Cl, O] $\xrightarrow{KOH (1\%)/EtOH}$

[—OK, C(OH)—Ph, C—OH, O] $\xrightarrow{HCl}$ [**143**: =C(—OH)—Ph, =O] $\xrightarrow{Ac_2O}$ [**144**: =C(OAc)Ph, =O] (29)

Direct synthesis of 2-benzoyl-3(2*H*)-benzofuranones has been achieved in only a few instances. von Auwers[149] prepared 2-benzoyl-5-methyl-3(2*H*)-benzofuranone by treatment of 2-benzoyloxy-ω-chloroacetophenone (**145**) with a base. Among others, **147** (R = H) was also obtained by bromination of 2-hydroxydibenzoylmethane (**146**, R = H) in presence of base (Eq. 30).[152,154]

R— [benzene, —OCOPh, —$COCH_2Cl$] $\xrightarrow{base}$ [R— [benzene, —OH, —COCHClCOPh]] ⟶

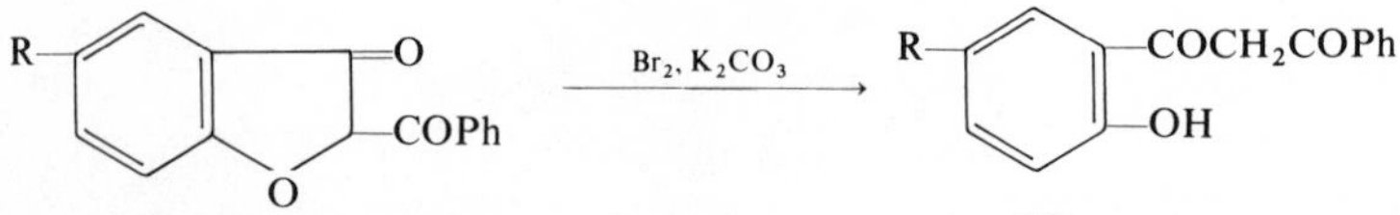

(30)

147 **146**

Condensation of methyl salicylate with monochloroacetone in the presence of potassium carbonate results in the formation of 2-acetyl-3(2*H*)-benzofuranone (**148**).[152] This is the earliest example of intramolecular acyl migration and cyclization taking place in a single operation.[149,150] The migration of the acetyl group takes place with ease because the concerned methylene group is activated with the chlorine atom. The mechanism of such migration involves internal Claisen condensation, and is analogous to Baker-Venkataraman transformations (Eq. 31).[150,154]

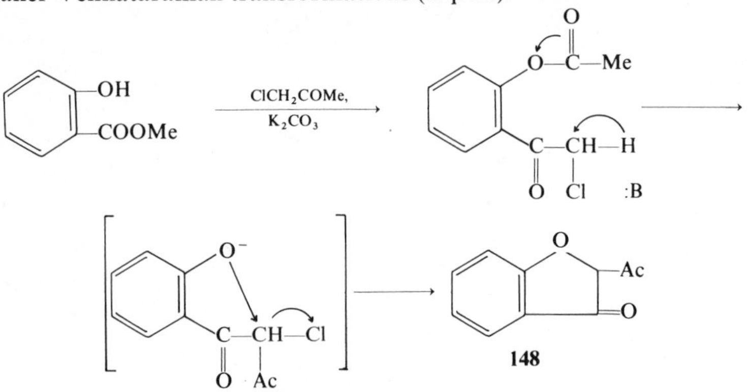

(31)

148

Other preparations of 2-benzoyl-3(2*H*)-benzofuranones include the action of alkali on 2-benzal-3(2*H*)-benzofuranone dibromide,[75] on 3,3-dibromoflavanones,[155-159] and the action of sodium peroxide on 2-benzal-3(2*H*)-benzofuranone.[160] Bryant and Haslam[161,162] have shown that condensation of ω-chloro-2,4,6-trimethoxyacetophenone with the sodium salt of methyl salicylate in xylene results in the formation of 2-(2,4,6-trimethoxybenzoyl)-3 (2*H*)-benzofuranone (**149**) (Eq. 32).

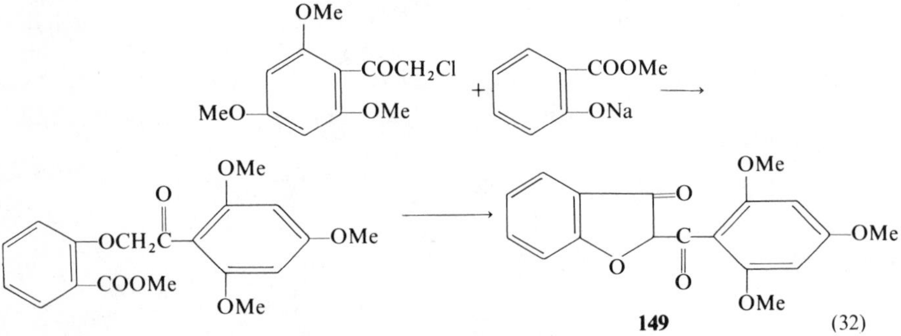

149 (32)

Synthesis of 2-aroyl-3(2*H*)-benzofuranones has been achieved more directly from ω-chloro-*o*-hydroxyacetophenone and the appropriate acid chloride.[163] The use of acid anhydrides led to a high yield in the case of 2-benzoyl-6-benzoyloxy- and of 2-anisoyl-6-anisoyloxy-3(2*H*)-benzofuranones, **150a** and **150b**, respectively.

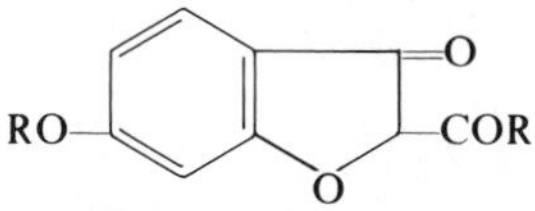

150a R = Ph
150b R = C_6H_4OMe-*p*

D. Hydroxy Bz-Substituted 3 (2*H*)-Benzofuranones

Of all benzofuran derivatives, it is the 3(2*H*)-benzofuranones which are the most easily prepared. Phenols are readily transformed into chloroketones by Friedel-Crafts and/or Hoesch reactions and cyclization is then effected by sodium acetate or pyridine.

6-Hydroxy-3(2*H*)-benzofuranone and its alkyl derivatives have been obtained as shown in Eqs. 33[164-168] and 34.[169-171]

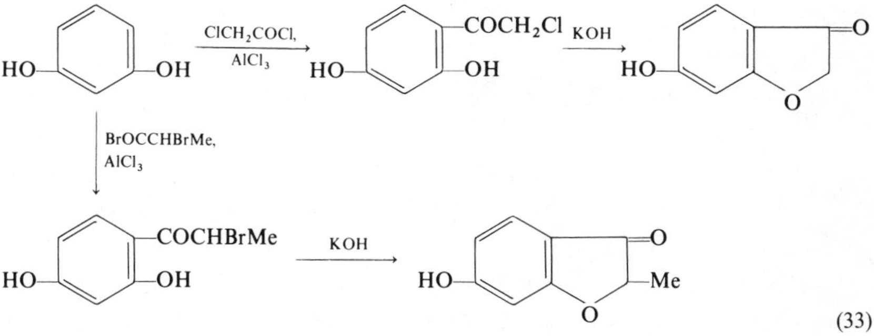

$ClCH_2COCl$, $AlCl_3$; KOH; $BrOCCHBrMe$, $AlCl_3$; KOH

(33)

$ClCH_2CN$, HCl, $ZnCl_2$; KOAc

$ZnCl_2$/HCl, $NCCH(OH)C_6H_4OMe$-p

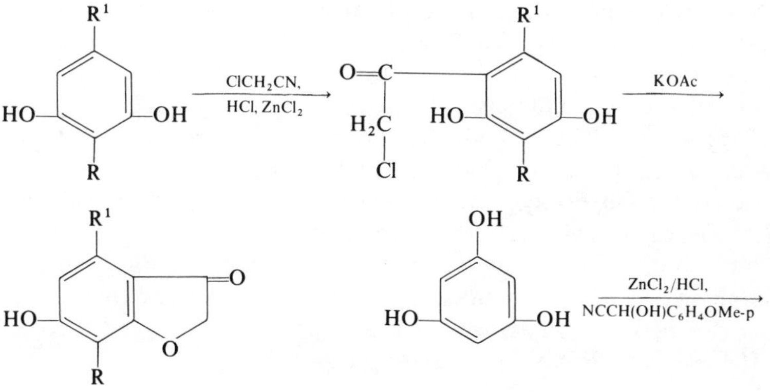

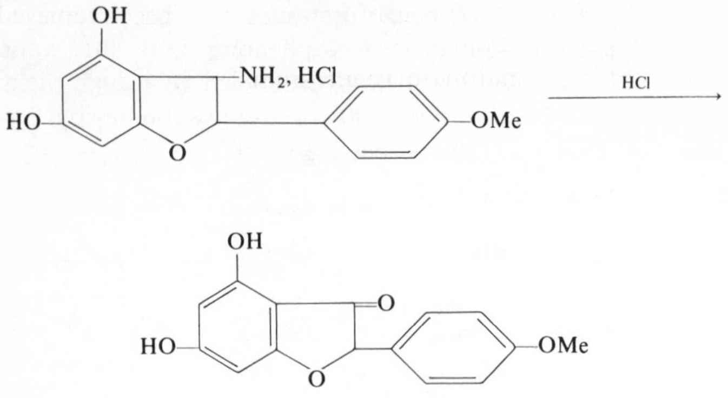

(34)

Cyclization of ω-chloro-2,4-dihydroxy-5-methoxyacetophenone affords 5-methoxy-6-hydroxy-3(2*H*)-benzofuranone (**151**).[172] Similarly prepared are 4,6,7-trimethoxy-[173] and 4,5,6-trimethoxy-3(2*H*)-benzofuranone.[174,175]

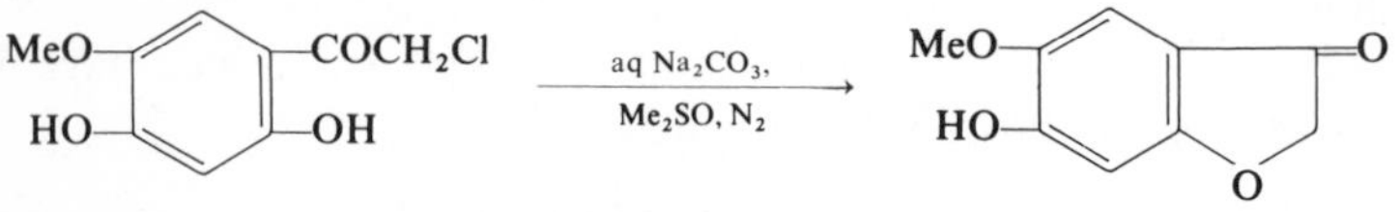

151

The ultraviolet spectral data of 3(2*H*)-benzofuranone and its hydroxy bz-substituted derivatives have been reported.[176]

E. Reactions of Hydroxy-Substituted 3 (2*H*)-Benzofuranones

Partial methylation of 4,6-dihydroxy-3(2*H*)-benzofuranone (**152**) has been reported to be brought out by the action of methyl sulfate in presence of potassium carbonate.[139] Treatment of **152** with diazomethane, however, yields 4-methoxy-6-hydroxy-3(2*H*)-benzofuranone.[177] This behavior is noteworthy; the reluctance of the hydroxyl group ortho to the carboxyl group, in such compounds as *o*-hydroxyacetophenone derivatives, 5-hydroxyflavones, and 1-hydroxyanthraquinones, to undergo diazomethane methylation under these conditions is well known.[178] The preferential methylation of the 4-hydroxy group in **152** has been attributed to its greater acidity.[177] The same argument applied to 5,7-dihydroxychromanone (**153**) would imply preferential methylation of the 5-hydroxy group; because only a monomethyl ether is formed, it must be the 7-methoxy compound. The difference between the two homologs is due to a stronger hydrogen bond in **153**; this is confirmed by infrared spectra.[179]

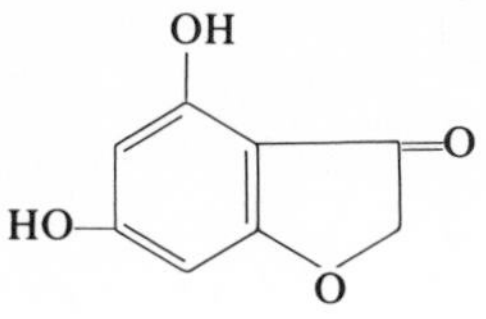

152 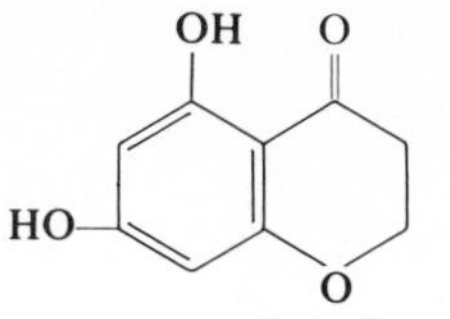 **153** 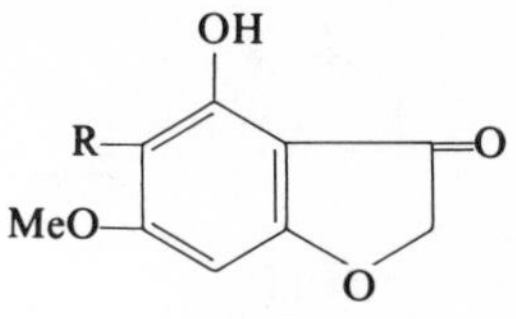 **154a** R = H **154b** R = Me

Mulholland and Ward[180] have shown that methylation of **152** after Drumm et al.[143] yielded **154a**. Compound **154a** was also readily obtained when the methylation was effected with methyl sulfate in dioxan–benzene medium.[140]

Several 2-(α-hydroxybenzyl)-2,4,6-trimethoxy-3(2*H*)-benzofuranones (**155**), substituted in the benzyl moiety, were found to undergo transformations in presence of boron trifluoride-ethereate to the respective 3-phenyl-4-hydroxycoumarins (**160**), thus constituting a novel synthesis of **160** since **155** is readily accessible by oxidation of α-methoxychalcones.[181] It is believed that **155**, being a ketal, first, undergoes demethylation in presence of a Lewis acid of type **156**. This, being a benzylic alcohol, loses an OH anion in favorable cases giving a carbonium ion (**157**). Subsequently, rearrangement takes place to give **158**, followed by deprotonation to the isoflavonol structure **159**, which is tautomeric with **160** (Eq. 35).

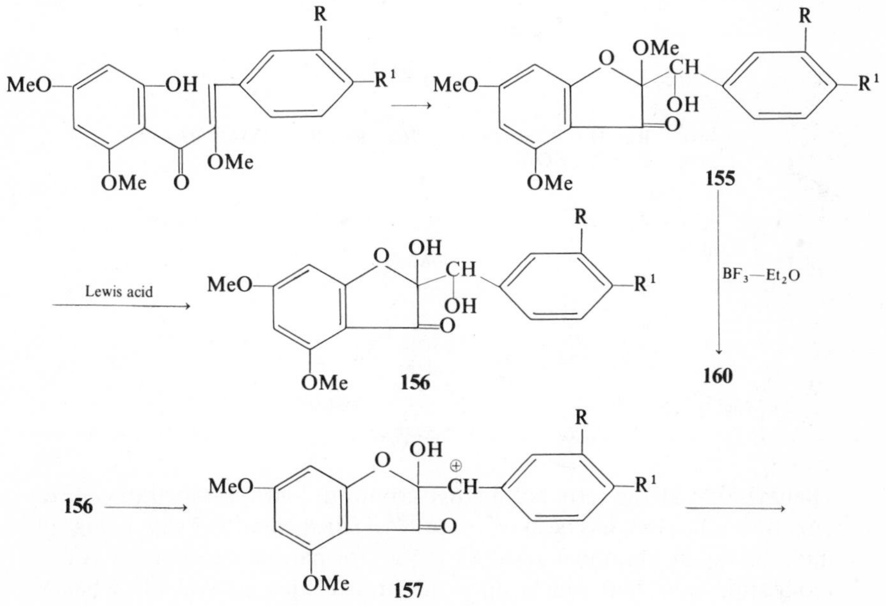

155

$\xrightarrow{\text{Lewis acid}}$ **156**

155 $\xrightarrow{BF_3—Et_2O}$ **160**

156 ⟶ **157** ⟶

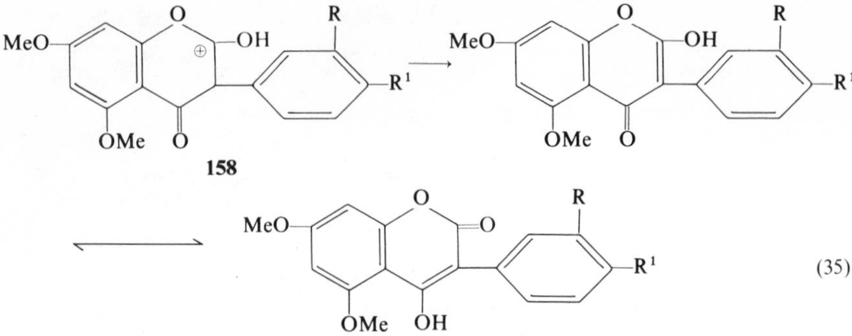

(35)

158

160

Hydroxy-3(2*H*)-benzofuranones are acetylated with ketene only on the phenolic hydroxy groups, without the formation of an enol acetate,[182] thus 6-acetoxy- (**161a**) and 4,6-diacetoxy-3(2*H*)-benzofuranone (**161b**) are obtained upon treatment with the appropriate hydroxy-3(2*H*)-benzofuranone with ketene at 25°C; but in the presence of methyl sulfuric acid ($MeSO_3H$), a mixture of 3,4,6-triacetoxy-3,4,6-triacetoxybenzofuran (**162**) and **161b** is obtained. Acetylation of 6-methoxy-3(2*H*)-benzofuranone with acetic anhydride yields 3-acetoxy-5-methoxybenzofuran which, on catalytic reduction followed by loss of acetic acid, affords 6-methoxybenzofuran (Eq. 36).[183]

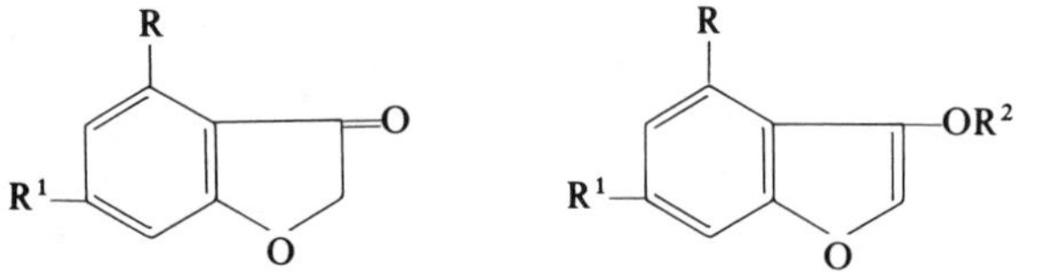

161a R = H, R^1 = OAc
161b R = R^1 = OAc

162 R = R^1 = OAc, R^2 = Ac

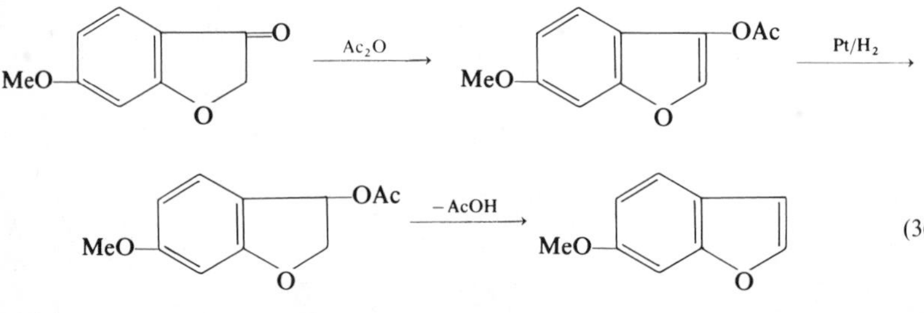

(36)

Benzylation at the terminal methyl group in 2-acetyl-6-methoxy-3(2*H*)-benzofuranone (**163**) has been reported.[184] Treatment of **163** with potassium amide in liquid ammonia resulted in the formation, presumably, of the dipotassium salt (**164**) which, upon addition of one equivalent of benzyl-chloride followed by copper acetate, gave the copper chelate of 2-dibenzylacetyl-6-methoxy-3(2*H*)-benzofuranone (**165**). The possible dibenzyl derivative that might have resulted from alkylation at both the methyl and methinyl groups of **163** could not have formed a copper chelate.

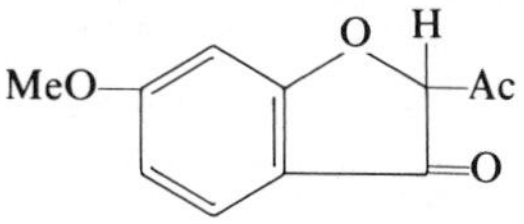

163 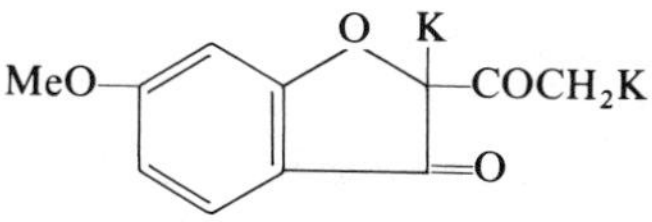 **164**

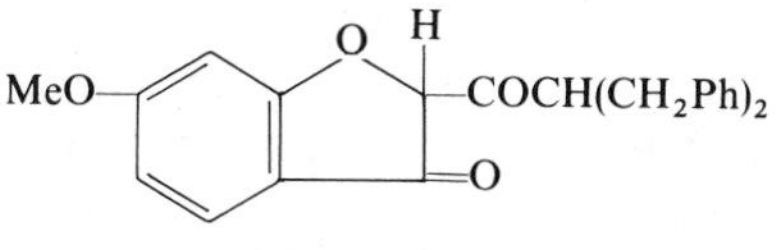

165

Under acidic conditions (a variety of Lewis acids) 6-methoxy-2-(2,4,6-trimethoxybenzoyl)-3(2*H*)-benzofuranone (**166a**) did not undergo demethylation to yield the expected 6-methoxy-2-(2-hydroxy-4,6-dimethoxybenzoyl)-3(2*H*)-benzofuranone (**166b**), but cleaved to give 1,3,5-trimethoxybenzene and 6-methoxy-3(2*H*)-benzofuranone (**167**). The cleavage presumably took place by electrophilic attack of the Lewis acid (here denoted by H^+) at the aromatic carbon atom (see **168**) giving a carbonium ion **169**.[185]

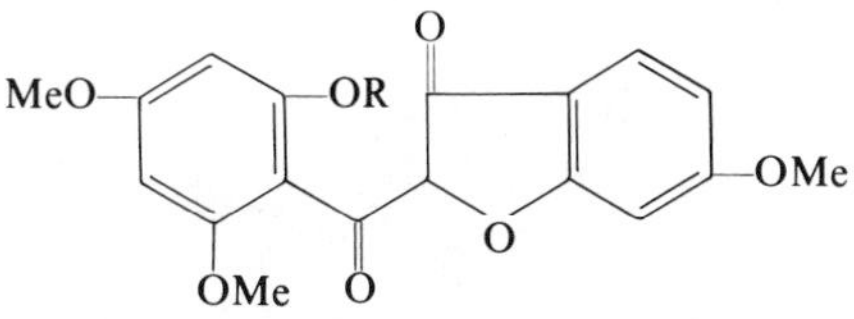

166a R = Me
166b R = H

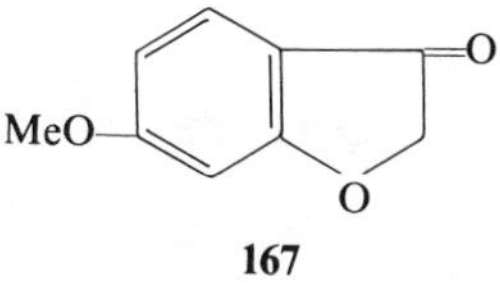

167

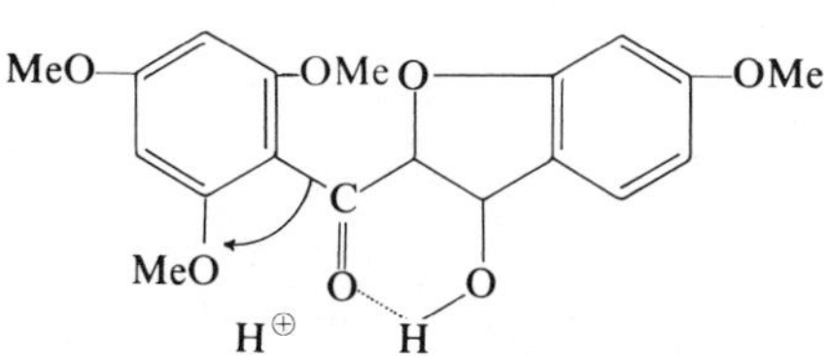

168 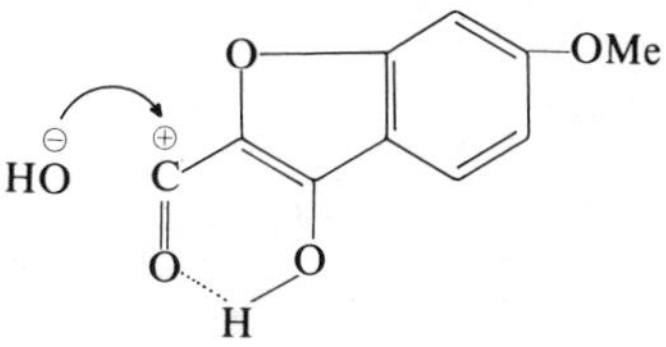 **169**

It has been found that aliphatic ketones and aldehydes do not react with the active methyene in 3(2*H*)-benzofuranones in a manner analogous to the well-known Knoevenagel condensation.[186] In general, acetic-hydrochloric acid mixture is used as a catalyst to bring about the condensation of 6-methoxy-3(2*H*)-benzofuranone (**167**) to yield 2,2′-*bis*(6-methoxy-3(2*H*) benzofuranyl)dialkylmethane, **170** and **171**. Similar condensation between **167** and formaldehyde and/or acetaldehyde, and so on in presence of zinc chloride–hydrochloric acid mixture took place.[140,183,187] However, with aromatic aldehydes in presence of alcoholic alkali hydroxides, the appropriate 2-arylidene derivatives (aurones) are obtained.[144,188,189]

2-Benzyl-2,6,3′-trihydroxy-4′-methoxy-3(2*H*)-benzofuranone (**172**) and the related 2-benzylidene-6,3′-dihydroxy-4′-methoxy-3(2*H*)-benzofuranone (**173**) produces an intense immediate cherry-red color with concentrated sulfuric acid.[190] Where the overall structure is chromogenic, successive substitution of methoxy- or benzyloxy groups for hydroxy groups leads to intensification of the developed color; this is accompanied by a change from yellow to red (partially substituted) to purple (fully methylated), but the specific structure responsible for color production is not entirely clear. The presence of a double bond conjugated with a keto group, as part of a chain, is a contributing factor.

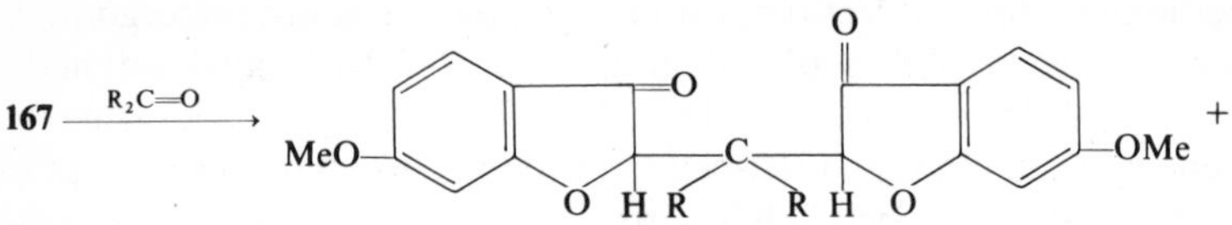

167 $\xrightarrow{R_2C{=}O}$ MeO, O, O H R, C, R H O, O, OMe +

170

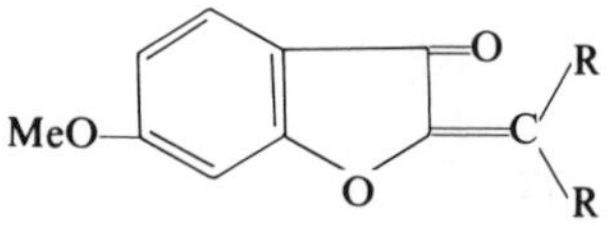

MeO, O, O, C, R, R

171

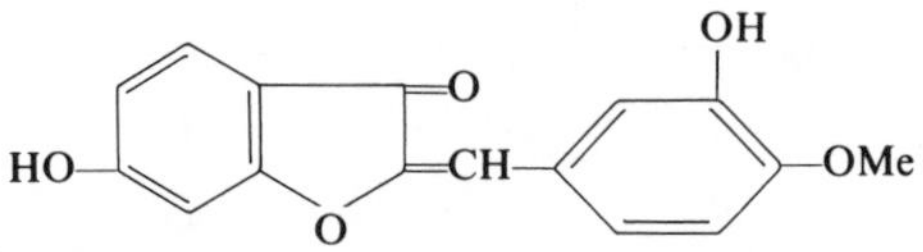

HO, O, O, CH, OH, OMe

173

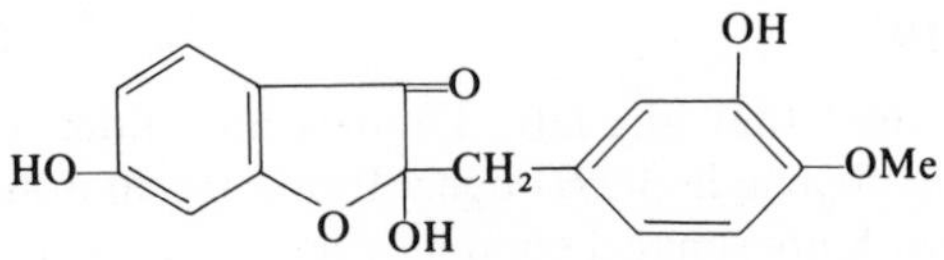

HO, O, O, CH_2, OH, OH, OMe

172

3. 2(3*H*)-Benzofuranones

A. Preparation

The trivial names for 2(3*H*)-benzofuranones are α-coumaranones, coumaran-2-ones, and/or isocoumaranones. 2-Hydroxybenzofurans are quite unknown except for certain derivatives unable to tautomerize to 2(3*H*)-benzofuranones (**173**). They invariably behave as lactones and rarely occur as natural products; **1** is the lactone of *o*-hydroxyphenylacetic acid. A number of authors[191-196] have shown that α-mono- and disubstituted *o*-hydroxyphenylacetic acids cyclize spontaneously to 2(3*H*)-benzofuranones in acid solution. Thermal cyclization of 2-hydroxy-4-methoxyphenylacetic acid to 6-methoxy-2(3*H*)-benzofuranone has been reported.[195] 4,6-Dimethoxyphthalide 3-carboxylic acid, upon treatment with hydroiodic acid, is readily transformed into 4-carboxy-6-hydroxy-2(3*H*)-benzofuranone (**174**) (Eq. 37).[197]

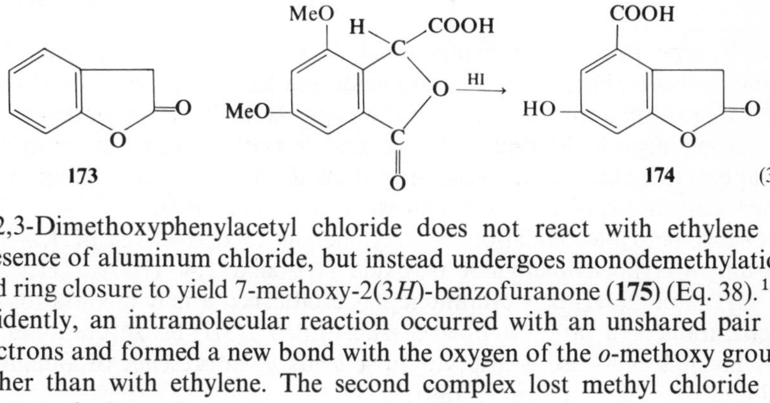

MeO H COOH C O HI MeO C O → COOH HO O O

173 **174** (37)

2,3-Dimethoxyphenylacetyl chloride does not react with ethylene in presence of aluminum chloride, but instead undergoes monodemethylation and ring closure to yield 7-methoxy-2(3*H*)-benzofuranone (**175**) (Eq. 38).[198] Evidently, an intramolecular reaction occurred with an unshared pair of electrons and formed a new bond with the oxygen of the *o*-methoxy group, rather than with ethylene. The second complex lost methyl chloride to generate the benzofuranone.

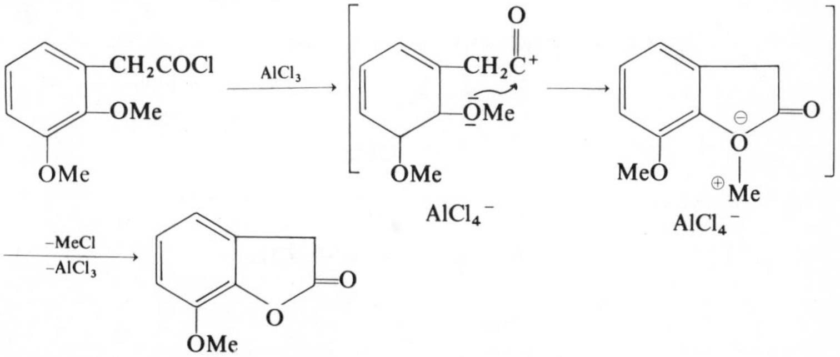

CH_2COCl OMe OMe $\xrightarrow{AlCl_3}$ [O CH_2C^+ OMe OMe $AlCl_4^-$ → O MeO O ⊖ ⊕ Me $AlCl_4^-$]

$\xrightarrow[-AlCl_3]{-MeCl}$ O O OMe

175 (38)

Mandelic acid condenses with phenol in the presence of concentrated sulfuric acid with the resulting formation of 3-phenyl-2(3*H*)-benzofuranone (**176**) together with *p*-hydroxyphenylphenylacetic acid (**177**).[199-202] This condensation has also been successfully effected by direct fusion of mandelic acid with the appropriate phenol or *m*-hydroxybenzoic acid.[203,204] Mandelonitrile[202] can replace mandelic acid in this reaction. The interaction of the sodium salt of **178** with *N,N*-aminoalkyl chlorides affords **179**.[205]

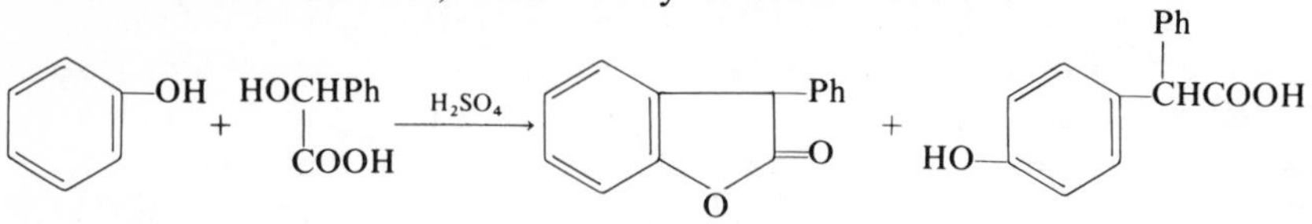

176 **177**

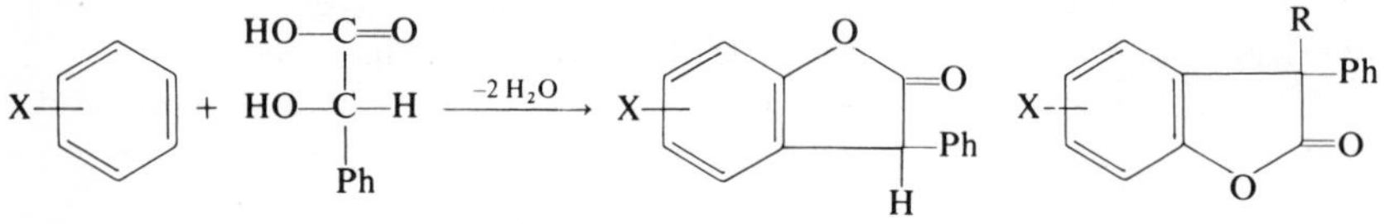

178 **179**

The lactone ring of 3,3-disubstituted 2(3*H*)-benzofuranones (e.g., **179**) shows considerable stability toward acids and bases, a property of distinct advantage in therapeutic agents of this type. Actually, lactonization of the corresponding *o*-hydroxyphenylacetic acid derivatives occurs in strong acid solution. The ring is also stable toward alkali; however no opening of the ring takes place upon refluxing with excess bases or amines.

Diacyl peroxide oxidation of alkylphenols to derivatives of 1,2- and 1,4-dihydroxyphenylmethanol peroxide (1,2- and 1,4- $(HO)_2C_6H_3CH_2$-COOH), and 2(3*H*)-benzofuranone has established a new method for the preparation of a large number of substituted 2(3*H*)-benzofuranones.[206] Polyphenols condense with arylpyruvic acids[207] in presence of aluminum chloride to yield, for example, **180** (Eq. 39).

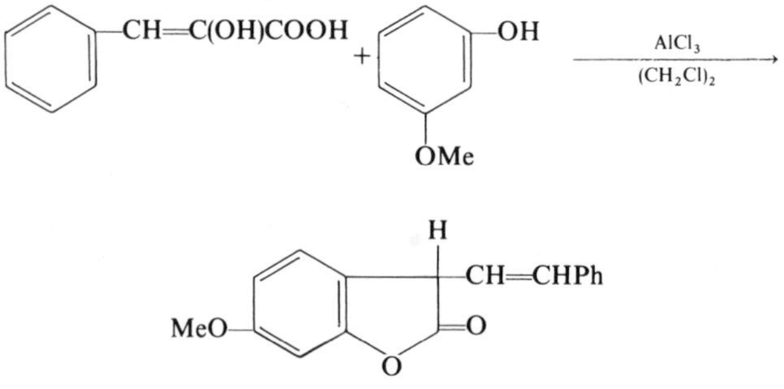

180 (39)

Borsche et al.[208] have shown that the compounds obtained by Marsh and Stephen,[209] upon treatment of arylglyoxylic nitriles with resorcinol after Hoesch's procedure, have a structure of type **181**. The iminolactones (**182**) have been isolated in the case of phloroglucinol.

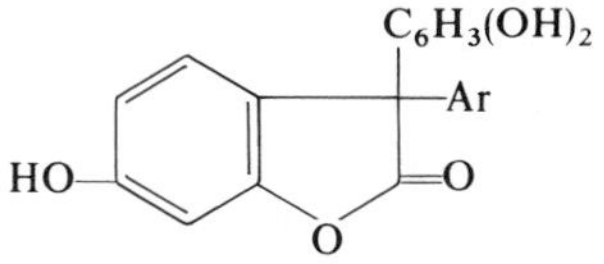

$C_6H_3(OH)_2$ / Ar / HO / O / O

181

$C_6H_3(OH)_2$ / Ar / HO / NH / O

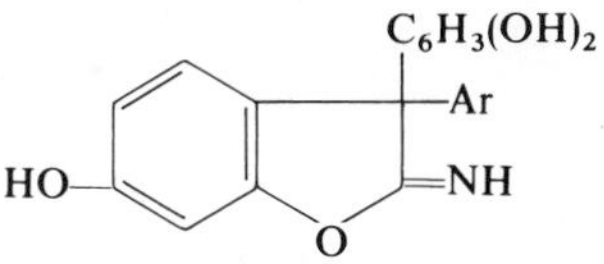

182

Reflux of a mixture of the sodium salt of 2-hydroxyphenylacetic acid with furfural in the presence of acetic anhydride led to the formation of β-(2-furyl)-α-(2-hydroxyphenyl)acrylic acid, on reduction with sodium amalgam, this gave 3-(2-furfuryl)-2(3*H*)-benzofuranone (**183**).[210]

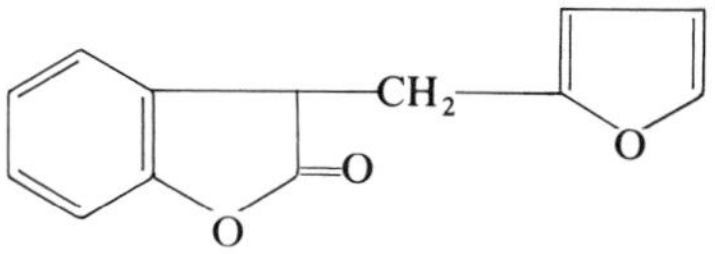

CH_2 / O / O / O

183

Synthesis of 3-methyl-7-benzyl-2(3*H*)-benzofuranone (**184**) has recently been achieved (Eq. 40).[211]

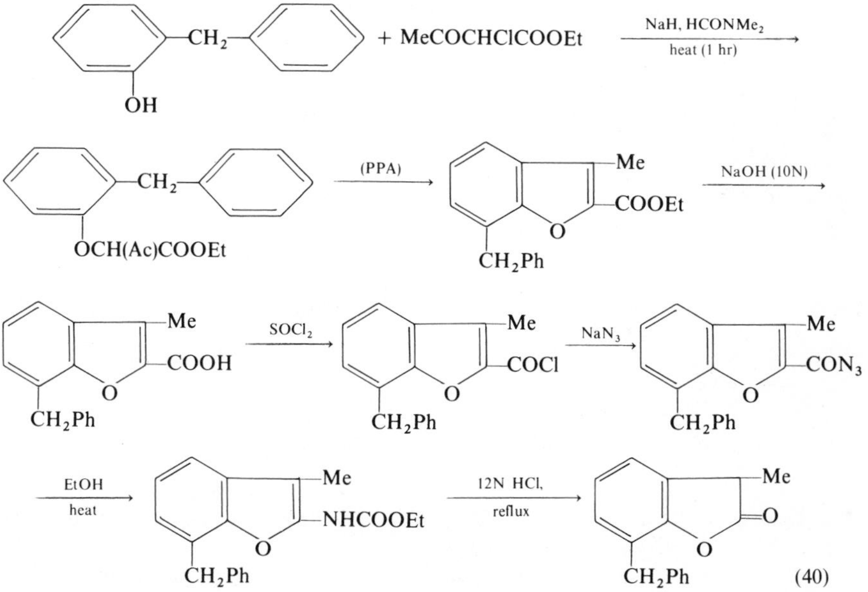

CH_2 / OH + MeCOCHClCOOEt —NaH, $HCONMe_2$, heat (1 hr)→

CH_2 / OCH(Ac)COOEt —(PPA)→ Me / COOEt / O / CH_2Ph —NaOH (10N)→

Me / COOH / O / CH_2Ph —$SOCl_2$→ Me / COCl / O / CH_2Ph —NaN_3→ Me / CON_3 / O / CH_2Ph

—EtOH, heat→ Me / NHCOOEt / O / CH_2Ph —12N HCl, reflux→ Me / O / O / CH_2Ph

184 (40)

Anisilic acid (**185**, R = OMe) is thermally decomposed to anisilide (**186**, R = OMe), which then rearranges to the benzofuranone (**187**, R = OMe).[212] The latter can also be obtained by condensation of resorcinol monomethyl ether and mandelonitrile (Eq. 41).

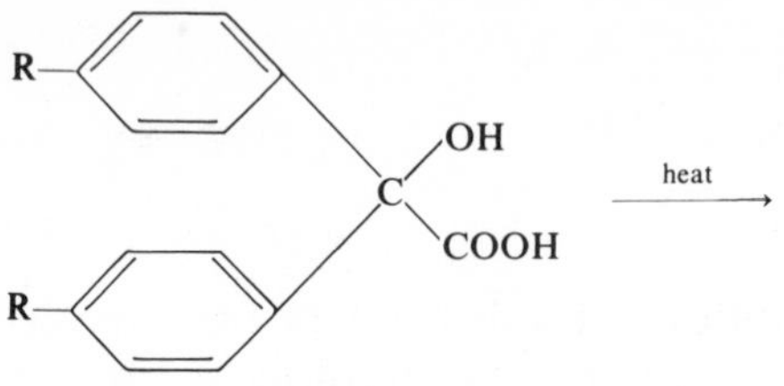

185

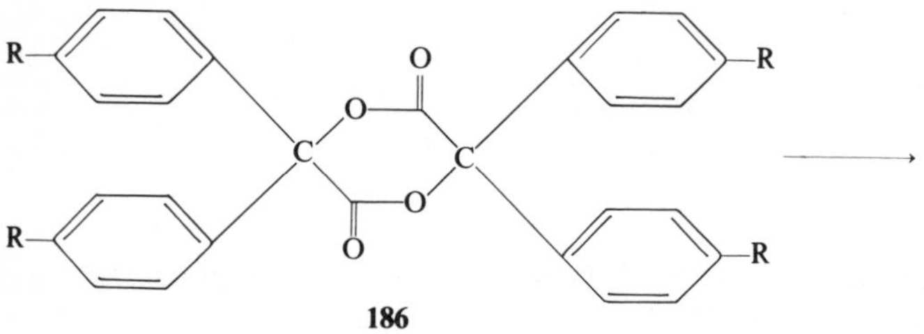

186

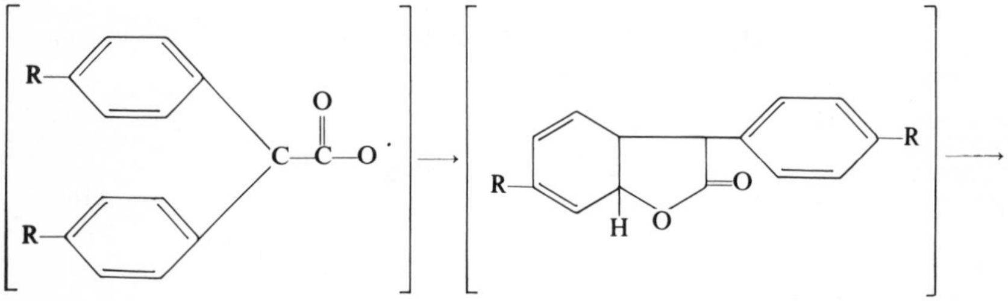

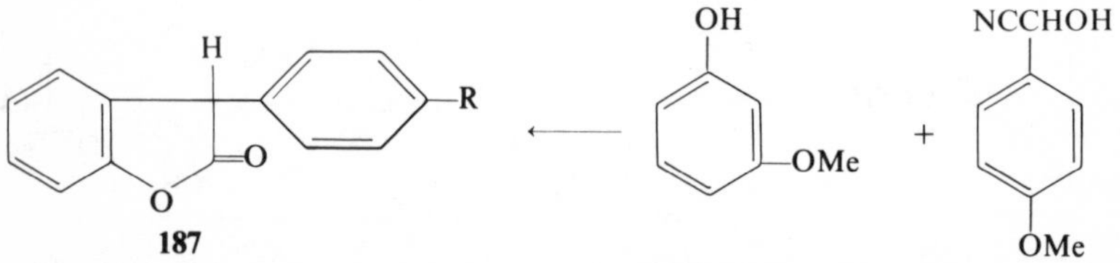

187

(41)

Demethylation of 2,4′-dimethoxybenzilic acid with hydroiodic acid gives 2,4′-dihydroxydiphenylacetic acid lactone.[213]

Cyclization of the Stobbe condensation product (**188**) with methyl β-phenyl-γ-benzoylbutyrate leads to the formation of **189**.[214]

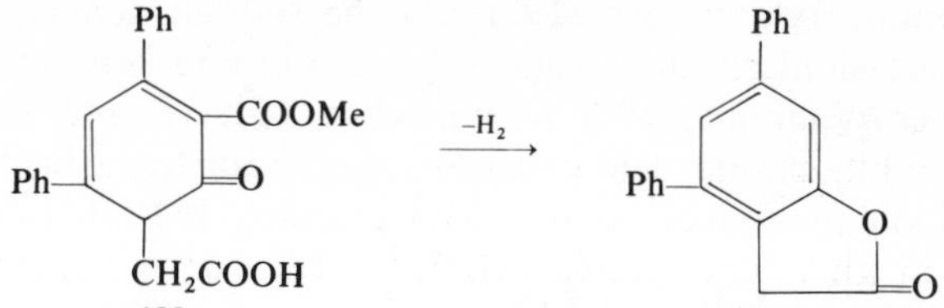

An interesting series of condensations leading to 2(3*H*)-benzofuranones, for example, **190** and **191**, proceeds from 1,4-benzoquinones such as 2,5,6-trimethyl-1,4-benzoquinone and the sodio derivative of active methylene compounds such as diethylmalonate or ethyl acetoacetate (Eq. 42).[215,216,224]

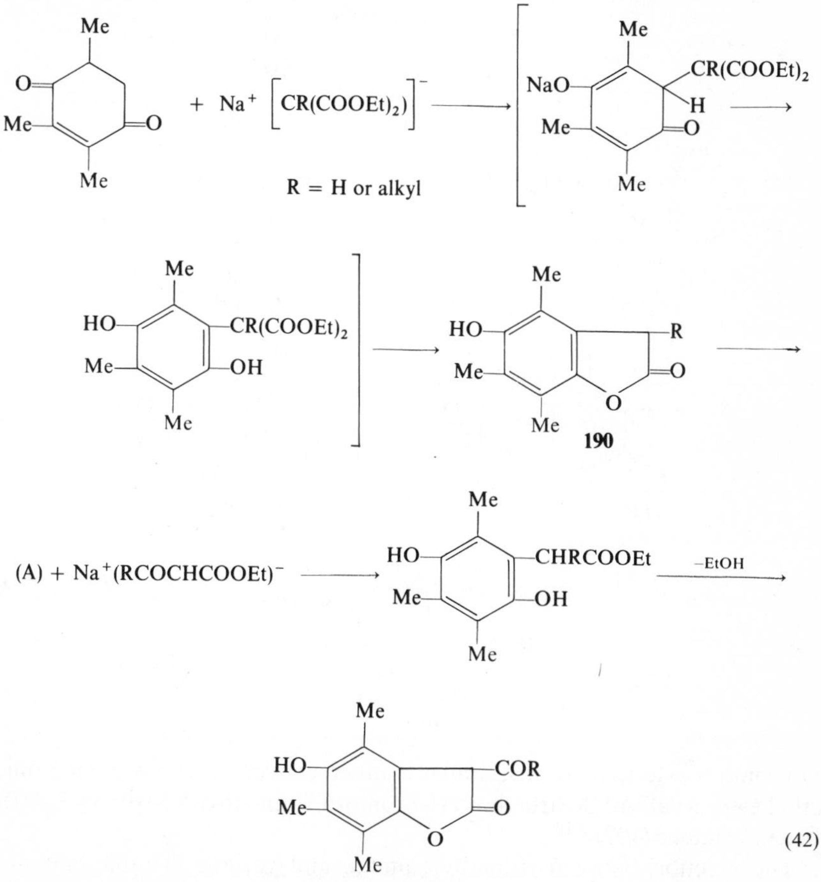

R = H or alkyl

(42)

2,5-Dibromo-*p*-xyloquinone (**192**) and the sodium derivative of ethyl malonate react similarly in dioxan to yield 5-bromo-2-dicarbethoxy-3,6-dimethyl-*p*-benzoquinone (**193**) which, when reduced and subjected to the action of dilute sulfuric acid undergoes cyclization to the benzofuranone (**194**). *p*-Xyloquinone gives, as the main product, **194** (H instead of Br) accompanied with a minor product (**195**).[217] The isomer 3-carbethoxy-4,6-dimethyl-7-bromo-5-hydroxy-2(3*H*)-benzofuranone (**196**) is obtained through reaction of dibromo-*m*-xyloquinone and sodium derivative of ethyl malonate (Eq. 43).[218]

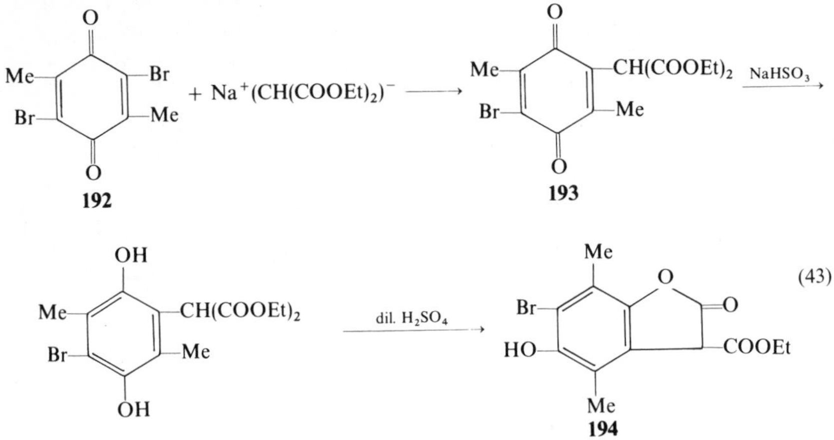

$$\mathbf{192} + Na^+(CH(COOEt)_2)^- \longrightarrow \mathbf{193} \xrightarrow{NaHSO_3} \text{hydroquinone} \xrightarrow{\text{dil. } H_2SO_4} \mathbf{194} \quad (43)$$

192 **193** **194** **195** **196** 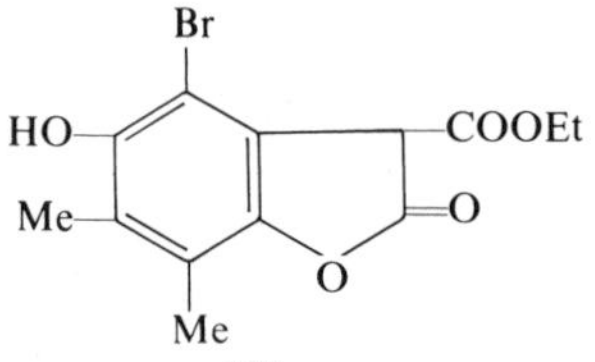 **197**

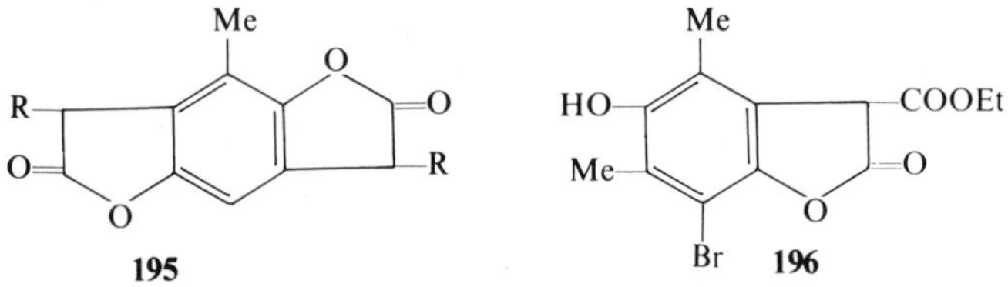

Dibromo-*o*-xyloquinone, in a similar manner, condenses with sodio malonic ethyl ester to afford 3-carbethoxy-4-bromo-6,7-dimethyl-5-hydroxy-2(3*H*)-benzofuranone (**197**).[223]

The reaction between trimethylquinone and various metallic enolates involves a primary 1,4 addition to the open conjugated system in the quinone to give the hydroquinone **198** (R = COOEt, R^1 = H). Often the hydroquinones[215,216,219–221,224] can be isolated only with difficulty; in these cases the chief product is one derived from the hydroquinone (**198**) and the benzofuranone (**199**, R = COOEt or H) derived by loss of elements of ethanol from **198**. This reaction may or may not be accompanied by loss of a carbethoxyl group.[222]

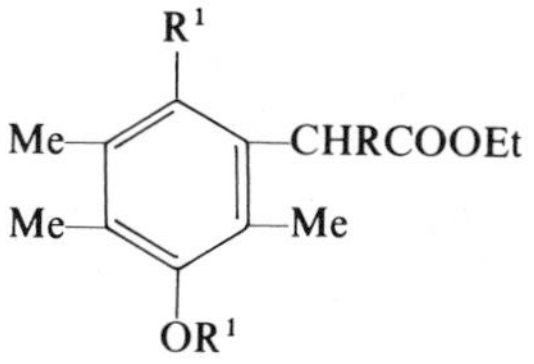

198, R = COOEt, R^1 = H

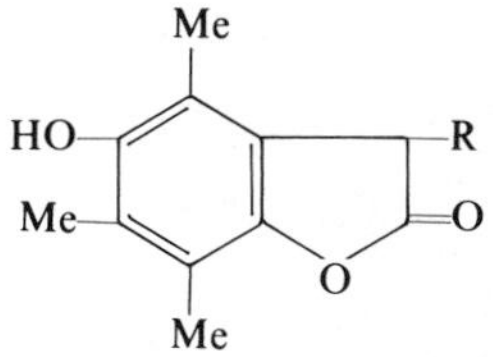

199, R = COOEt or H

The hydroquinone **198** (R = CN, R^1 = H), obtained by the interaction of sodio cyanoacetate and trimethylquinone, gave, upon treatment with acetic anhydride and sulfuric acid, the enol acetate (**200**); with benzoyl chloride and pyridine, the enol benzoate (**201**) and the corresponding hydroquinone dibenzoate (**202**) are obtained. When **198** (R = CN, R^1 = H) is treated with sulfuric acid, the main product is **199** (R = H) and is formed by cyclization and elimination of the cyano group. The benzofuranone (**203**) was the only product isolated upon treatment of the enolate of benzyl cyanide with trimethylquinone in methanol.

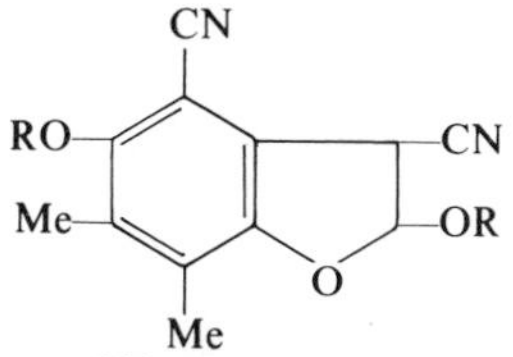

200 R = Ac
201 R = COPh

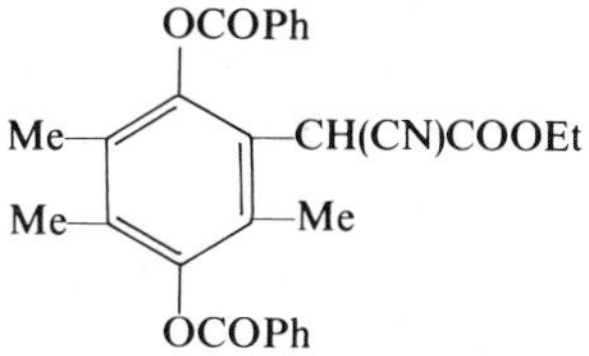

202

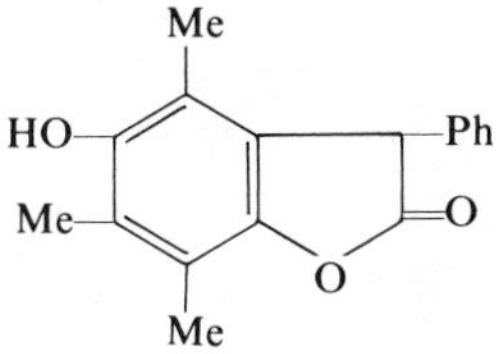

203

Condensation of 5-chlorotoluoquinone and sodio malonic ethyl ester in dioxan yields 3-carbethoxy-4-chloro-5-hydroxy-7-methyl-2(3*H*)-benzofuranone (**204**) which, upon treatment with chlorine, affords 4,6-dichloro-5-hydroxy-7-methyl-2(3*H*)-benzofuranone (**205**); this is also obtained by the interaction of 3,5-dichlorotoluoquinone and the sodio malonic ethyl ester. Decarbethoxylation of **204** yields 4-chloro-5-hydroxy-7-methyl-2(3*H*)-benzofuranone (**206**) (Eq. 44).[225]

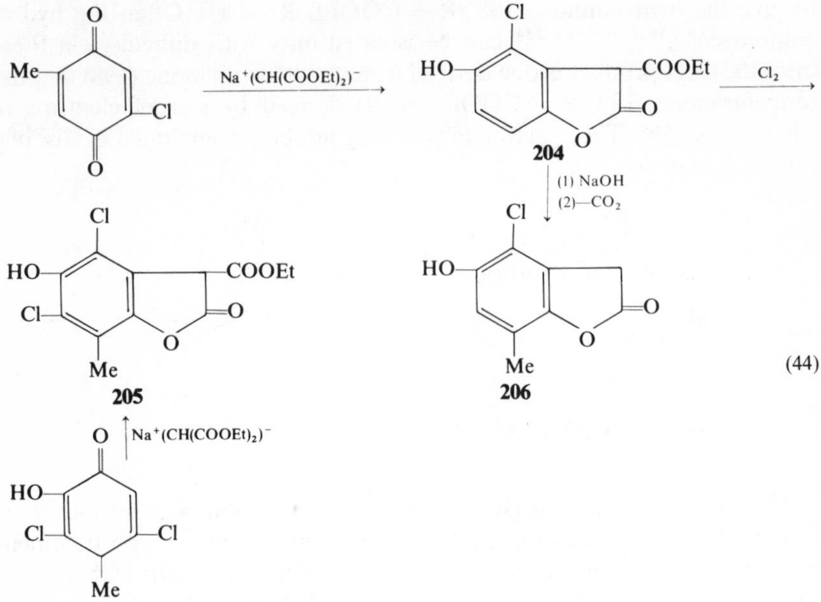

(44)

Dioxosuccinate condenses with *p*-cresol, *p*-methoxyphenol, 2,4-xylenol and/or 3,4-xylenol in the presence of anhydrous zinc chloride and acetic acid to yield **207a–d** which, when refluxed with acetic–hydrobromic acid mixture gave the benzofuranones **208a–d**.[226]

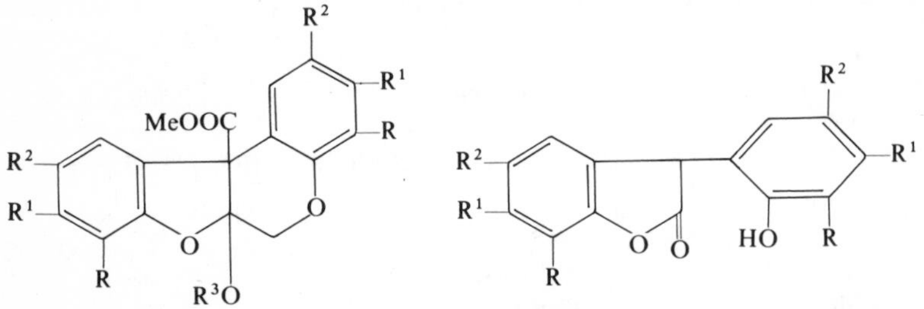

207a $R = R^1 = R^3 = H$, $R^2 = Me$
207b $R = R^1 = R^3 = H$, $R^2 = OMe$
207c $R^1 = R^3 = H$, $R = R^2 = Me$
207d $R^1 = R^2 = Me$, $R = R^3 = H$

208a $R = R^1 = H$, $R^2 = Me$
208b $R = R^1 = H$, $R^2 = OMe$
208c $R = R^2 = Me$, $R^1 = H$
208d $R^1 = R^2 = Me$, $R = H$

The reaction of ketene diethylacetal with *p*-benzoquinone results in the formation of **209** since it is readily converted to 5-ethoxy-2(3*H*)-benzofuranone (**212**).[227] Compound **209** is converted to hydroquinone–acetic acid (homogentisic acid) by hydrolysis in acid medium.[228] While neither **209** nor **210** (homogentisic lactone) could be directly ethylated by either ketene acetal or diazoethane, their sodium salts in isoamyl alcohol are alkylated with ethyl iodide. The benzofuran **209** gives **212** after hydrolysis of the intermediately formed oil (**211**); the lactone (**210**) produces **212** directly (Eq. 45).[227]

HO— ... —OEt **209** $\xrightarrow{HBr}$ HO— ... —OEt, Br $\xrightarrow{-EtBr}$ **210**

209 $\xrightarrow{EtI,\ C_5H_{11}ONa}$ EtO— ... —OEt **211** $\xrightarrow{H_2O}$ EtO— ... =O **212**

210 (HO— ... =O) $\xrightarrow{EtI,\ C_5H_{11}ONa}$ **212**

OH, OEt $\xrightarrow{CH_2{=}CHCH_2Br}$ $OCH_2CH{=}CH_2$, OEt $\longrightarrow$ OH, $CH_2CH{=}CH_2$, OEt $\longrightarrow$

(45)

OAc, $CH_2CH{=}CH_2$, OEt $\longrightarrow$ OAc, CH_2COOH, OEt $\longrightarrow$ **212**

m-Xyloquinone reacts much more slowly with ketene acetal than does benzoquinone at 150°C to yield 2-ethoxy-5-hydroxy-4,6-dimethylbenzofuran (**213**). However **209** and **213** are readily hydrolyzed with dilute aqueous acid solutions to the benzofuranones **210** and **214**, respectively.

Me, HO—, Me— ... —OEt **213**

Me, HO—, Me— ... =O **214**

Claisen rearrangement of γ-phenoxycrotonic acid esters at 220°C produces 3-isopropylidene-2(3*H*)-benzofuranone (**215**).[229] Treatment of glycollic acid with 2,4,5-trichlorophenol and chlorosulfonic acid yields bis(3,5,6-trichloro-2-hydroxyphenyl)acetic acid lactone (**216**, R = R^1 = Cl). The same is true for **216** (R = R^1 = H).[230]

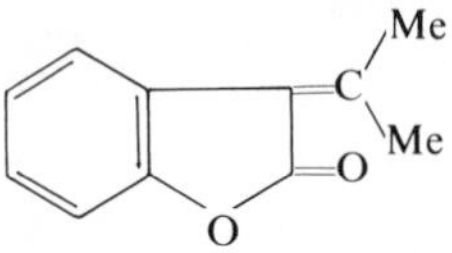

215

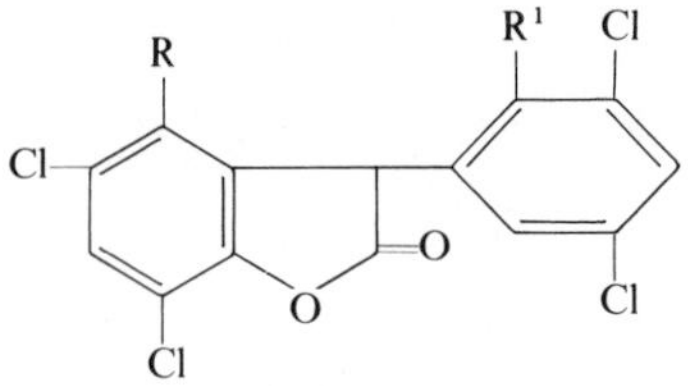

216

Self-condensation of 2(3*H*)-benzofuranone and its substituted derivatives has been achieved.[12,231,232] Chatterjea[233] has proposed structure **217** for the condensation product, which largely exists in the enolic form (**218**). When **218** is boiled with hydrochloric acid in acetic acid medium, an isomeric phenolic acid (**219**) is formed.[234] The formation of **219** from **218** includes opening of the furan ring and then reconstitution. This transformation is found to be general. Thus self-condensation of 6-methoxy-2(3*H*)-benzofuranone is isomerized by action of hydrochloric acid in acetic acid medium to **220**. Again, 3-acetyl-2(3*H*)-benzofuranone (**221**), which was originally regarded by Pfeiffer and Enders[15] as 2-methylbenzofuran-3-carboxylic acid (**222**, R = Me), is indeed converted to the latter in a similar way.

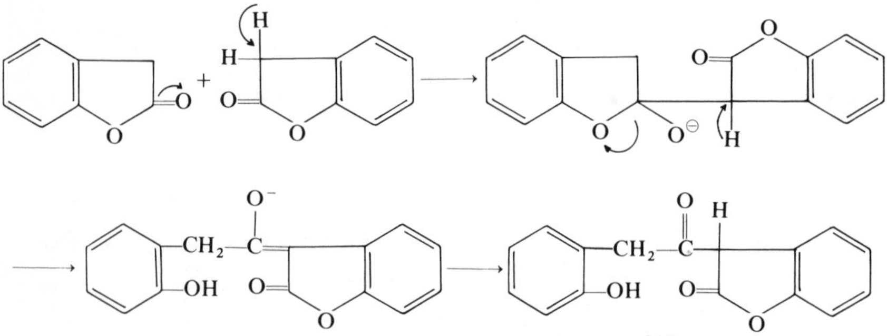

217

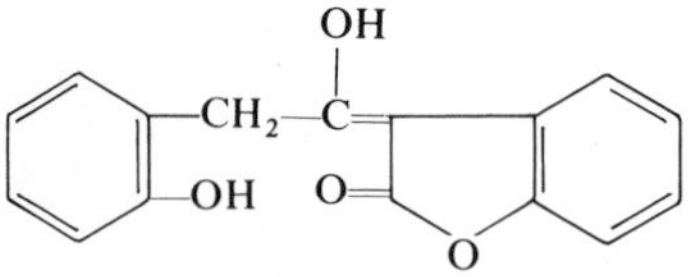

218

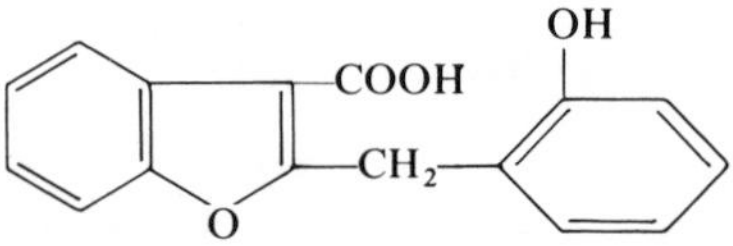

219

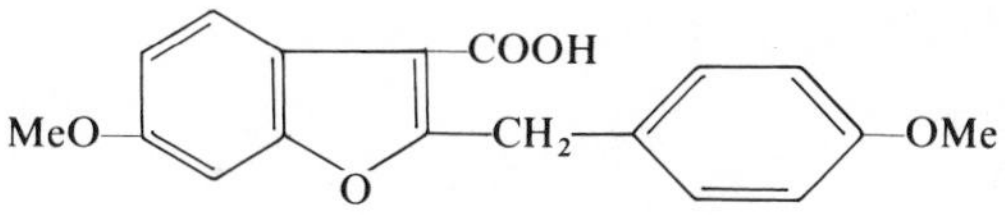

220 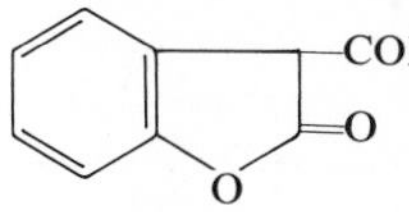 **221**

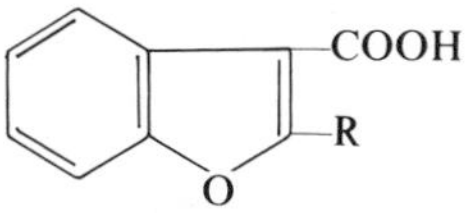

222 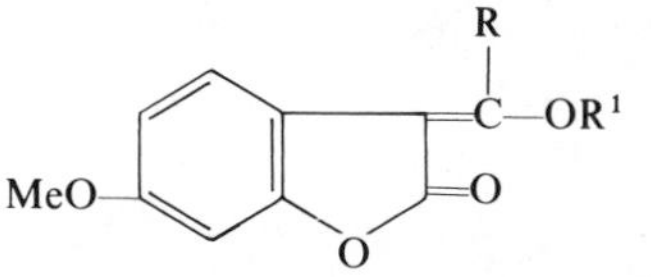 **223** R = Me, R[1] = OMe

3-Propionyl- and 6-methoxy-3-acetyl-2(3*H*)-benzofuranone isomerize to 2-ethyl- and 2-methyl-,6-methoxybenzofuran-3-carboxylic acid, respectively. 3-Benzoyl-2(3*H*)-benzofuranone, however, affords 2-phenylbenzofuran.

6-Methoxy-2(3*H*)-benzofuranone undergoes self-condensation, but with less reactivity with sodium hydride and the condensation product isomerizes with acetic anhydride and sodium acetate to afford **223**.[233]

The chalcones (**224**), obtained by condensation of ω-methoxy-*o*-hydroxyacetophenones with aromatic aldehydes, rearrange to 2-methoxy-2-benzyl-3(2*H*)-benzofuranones (**226**) on thermal treatment without a catalyst or in presence of dilute alkali. Compound **224**, when heated with alcoholic hydrochloric acid, undergoes demethylation at the methoxy group alpha to the oxo group, followed by cyclization to **225**. Acid bentonites, used for hydrochloric acid, play the role of Lewis acids and give rise to rearrangement forming **227** after dehydration. Similar treatment effected the conversion of **225** to **227**; however **226** yielded **228** without rearrangement.[236] It is therefore assumed that on heating with bentonite demethylation takes place prior to any cyclization, and that the dioxo derivatives thus obtained are rearranged (Eq. 46).

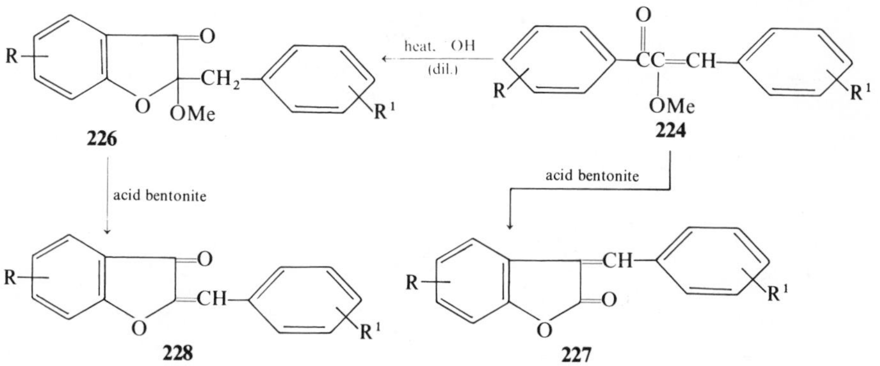

226 ← heat, OH⁻ (dil.) — **224**

226 → acid bentonite → **228**

224 → acid bentonite → **227**

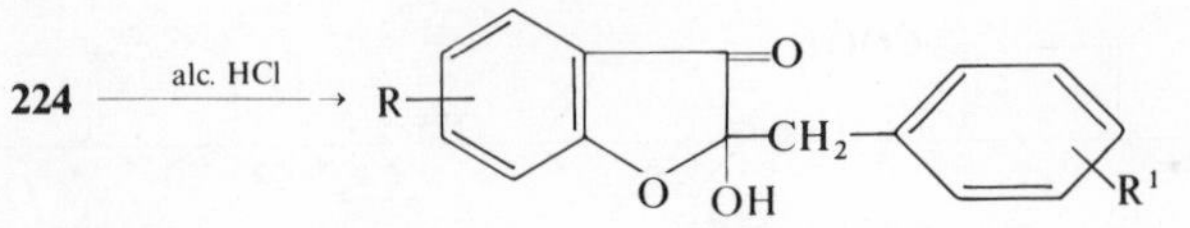

224 $\xrightarrow{\text{alc. HCl}}$ 225 (46)

Alkaline rearrangement of 3-hydroxyflavanones,[235] for example, **229** ($R = R^2 = R^3 = R^4 = H$, $R^1 = OMe$) and **229** ($R = R^2 = R^4 = H$, $R^1 = R^3 = OMe$), results in two fractions; one is soluble in aqueous sodium bicarbonate (5%) affording **230** ($R = R^2 = R^3 = R^4 = R^5 = R^6 = H$, $R^1 = OMe$) and **230** ($R = R^2 = R^4 = R^5 = R^6 = H$, $R^1 = R^3 = OMe$) which, when heated, yielded **231** ($R = R^2 = R^3 = R^4 = H$, $R^1 = OMe$) and **231** ($R = R^2 = R^4 = H$, $R^1 = R^3 = OMe$), respectively.[237] The second fraction, insoluble in sodium bicarbonate solution, affords **232** ($R = R^2 = R^3 = R^4 = H$, $R^1 = OMe$) and **232** ($R = R^2 = R^4 = H$, $R^1 = R^3 = OMe$).

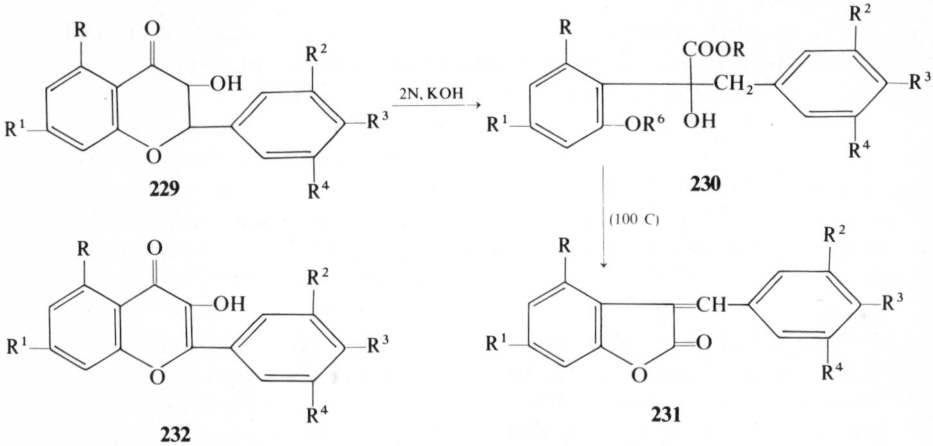

229 $\xrightarrow{\text{2N, KOH}}$ 230 $\xrightarrow{(100\ C)}$ 231

232

Irradiation of 1,4-benzoquinoneacetaldehyde[238] with visible light results in the formation of 5-hydroxy-2(3*H*)-benzofuranone (**233**). Similarly, 2-(1,4-benzoquinonyl)-2-methylpropionaldehyde yields its cyclic hemiacetal, namely, 5-hydroxy-3,3-dimethyl-2(3*H*)-benzofuranone (**234**) and 5-hydroxy-3-methyl-2(3*H*)-benzofuranone (**235**).

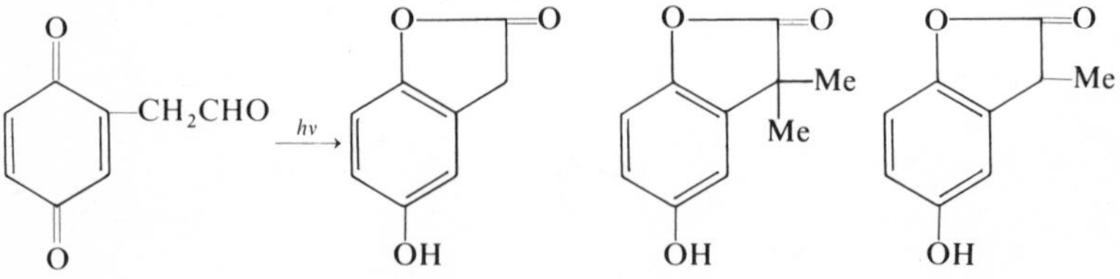

$\xrightarrow{h\nu}$ 233 234 235

A possible route for the photochemical formation of **235** (Eq. 47) is abstraction of formyl hydrogen. This would be favored and would be expected to occur intramolecularly, since the acyl radical, which would result from intermolecular abstraction, is the probable precursor of the resulting tertiary allylic radical, and an identical driving force would be available from concomitant hydrogen transfer to give the isopropenyl group. Tautomerism and oxidation would then yield isopropenyl-1,4-benzoquinone which is known[239] to cyclize in at least 60% yield to **235** under similar irradiation conditions.

O, CHO, CMe_2, O — hν → CO, C—Me, CH_2, H, O — $-CO_2$ →

Me, C=CH_2, O — (a) → O, Me, C=CH_2, O — hν → **235**

(47)

(a) Tautomerism, followed by oxidation by ground-state quinone

Irradiation of *N*-chloroacetyl-3,4-dimethoxyphenylethylamine (**236**)[240] in aqueous ethanol with a high-pressure lamp yields 7,8- (**237**), and 8,9-dimethoxy-1,2,4,5-tetrahydro-3*H*-3-benzazepin-2-one (**238**) which, upon hydrolysis and lactonization yields 4-(β-aminoethyl)-7-hydroxy-2(3*H*)-benzofuranone (**239**) (Eq. 48).

MeO, MeO, NH, $COCH_2Cl$ **236** — hν → MeO, MeO, O, NH + **237** (27% yield)

HBr (48%) ↓

OMe, MeO, O, NH **238** (6% yield)

$CH_2CH_2NH_2$, O, O, OH **239**

(48)

An addition reaction has been reported to take place between acetylacetone and maleic anhydride to form, at first, double molecules by substitution addition and then, in a second stage, acetylacetone adds to the double molecules.[241] Treatment of the adduct **240** with sulfuric acid brings out loss of the elements of water with the formation of the benzofuranone (**241**) which, on reflux with hydroiodic acid, yields the carboxy derivative (**242**). The latter is easily decarboxylated to afford **243** (Eq. 49).

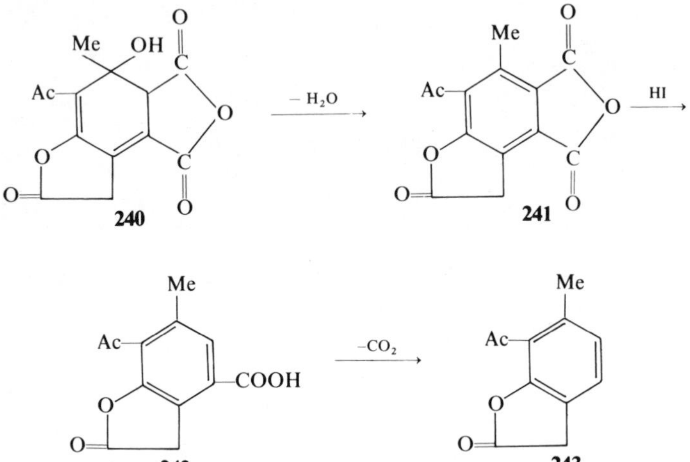

(49)

3-Hydroxy-2(3*H*)-benzofuranone (**244**), which is partially hydrolyzed on dissolution in water to give 2-hydroxymandelic acid, has been prepared (Eq. 50).[242]

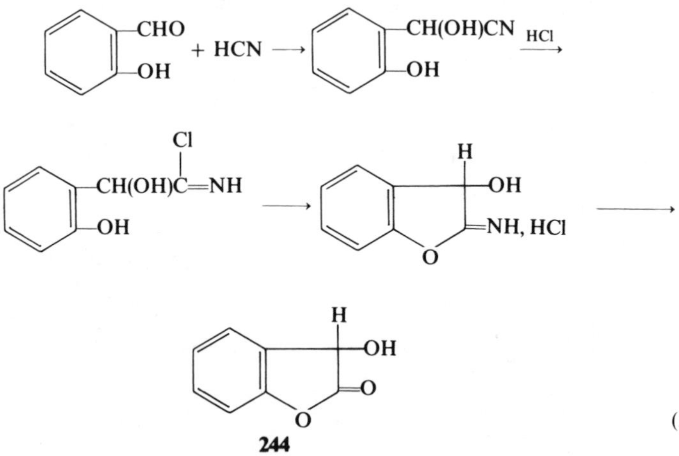

(50)

Synthesis of 3-ethyl-3-(β-dimethylaminoethyl)-7-methoxy-2(3*H*)-benzofuranone (**245**) has been achieved (Eq. 51).[194]

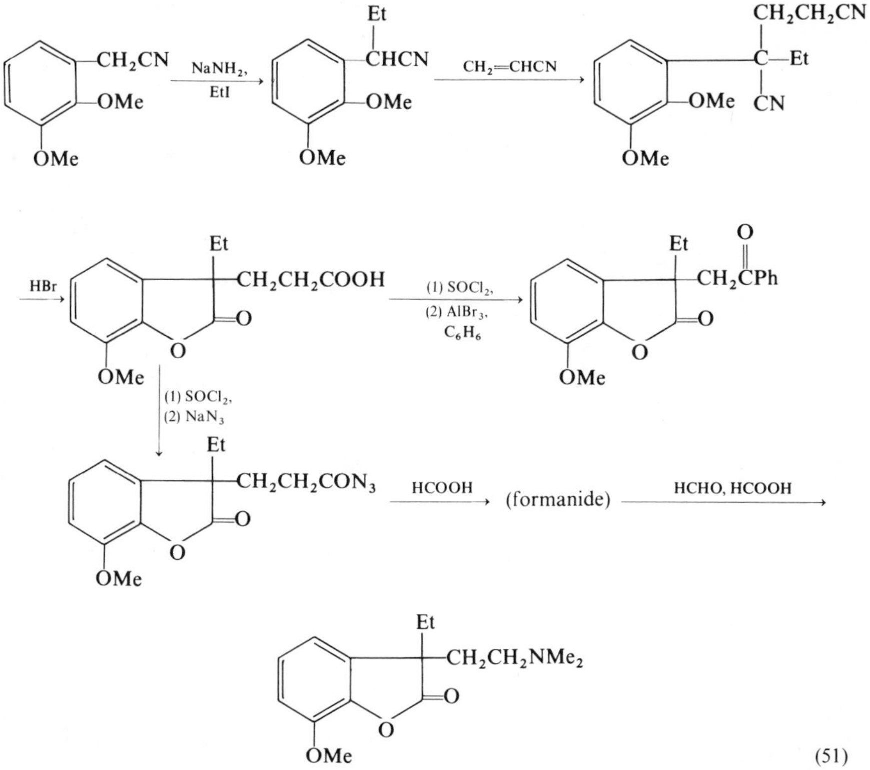

(51)

The α-methylene-γ-lactone function is a common grouping in various naturally occurring sesquiterpenoid lactones. Synthesis of this grouping has been recently achieved by Marshall and Cohen,[243,244] for example (±)-alantolactone[245] and (±)-telekin.[246] Compound **253** was obtained from formylactone (**247**), derived from *cis*-perhydro-2(3*H*)-benzofuranone (**246**) after the sequence of reactions shown in Eq. 52. Reduction of **247** with sodium borohydride in methanol gave an oily α-hydroxymethyl-lactone (**248**); its tosylate was refluxed in pyridine to yield 3-methylene-*cis*-perhydro-2(3*H*)-benzofuranone (**249**).[243] 3-Methylene-*trans*-perhydro-2(3*H*)-benzofuranone (**253**) has been obtained in a similar manner (**250**→**253**). Compound **252** can also be obtained upon catalytic hydrogenation of **251** in presence of 10% palladium–charcoal or Adams catalyst in ethanol.

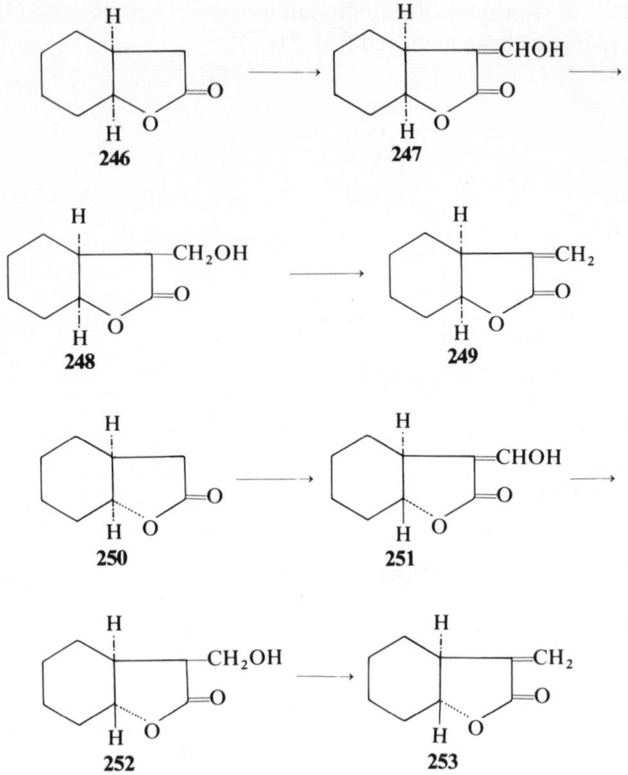

246 → 247 (=CHOH) → 248 (CH$_2$OH) → 249 (=CH$_2$)

250 → 251 (=CHOH) → 252 (CH$_2$OH) → 253 (=CH$_2$) (52)

Hexahydro-2(3*H*)-benzofuranone (**254**) has been prepared as outlined below:[247-250]

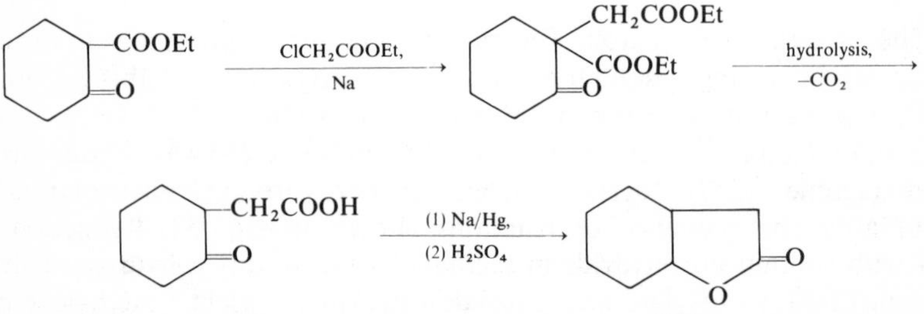

$$\text{2-oxocyclohexanecarboxylic acid ethyl ester (COOEt)} \xrightarrow[\text{Na}]{\text{ClCH}_2\text{COOEt,}} \text{(CH}_2\text{COOEt)(COOEt)cyclohexanone} \xrightarrow[-\text{CO}_2]{\text{hydrolysis,}}$$

$$\text{(CH}_2\text{COOH)cyclohexanone} \xrightarrow[(2)\ \text{H}_2\text{SO}_4]{(1)\ \text{Na/Hg,}} \textbf{254}$$

Condensation of cyclohexanol with hydroxycarboxylic acids yields, for example, 2-hydroxy-1,3-cyclohexane diacetic acid (**255**) and its lactone (**256**).[251]

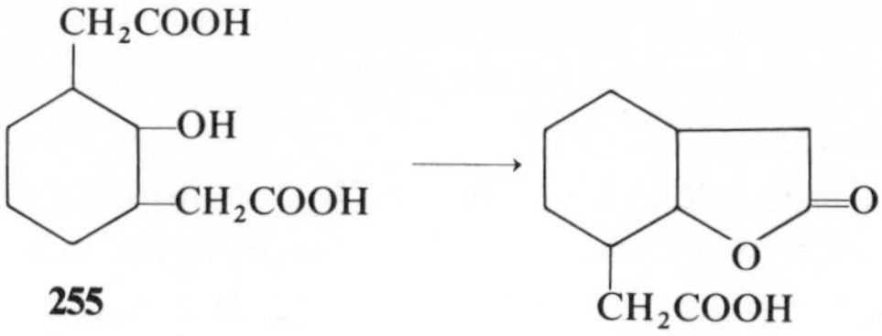

CH_2COOH OH CH_2COOH → O O CH_2COOH

255 **256**

Addition of dimethylketene to *N*-(2,6-xylyl)benzylideneamine *N*-oxide[252] yields 7α-(benzylideneamino)-3α,7α-dihydro-3,3,3α,7-tetramethyl-2(3*H*)-benzofuranone (**257**) by a sigmatropic migration of an enolate ion.

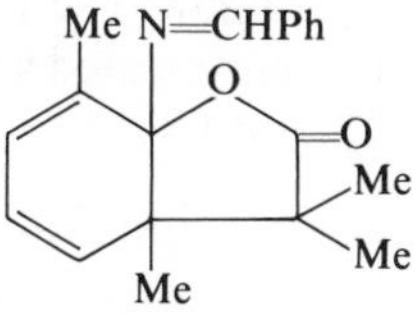

Me N=CHPh O O Me Me Me

257

Bromination of **257** yielded 7α-(benzylideneamino)-4,5-dibromo-3α,4,5,7α-tetrahydro-3,3,3α,7-tetramethyl-2(3*H*)-benzofuranone.

B. Miscellaneous Reactions

Reactions of 2(3*H*)-benzofuranones, in general, are consistent with their lactone character. Thus the lactone ring is readily opened by reagents which customarily accomplish such lactone cleavage with the formation of derivatives of *o*-hydroxyphenylacetic acid.[199] Treatment of 3-phenyl-2(3*H*)-benzofuranone (**258**) either with sodium[253,254] or sodium hydride[255] in dimethylformamide–benzene mixture gave the sodium derivative **259** which, when treated with alkylating agents, affords **260**[253-255] (Eq. 53). Alkylation

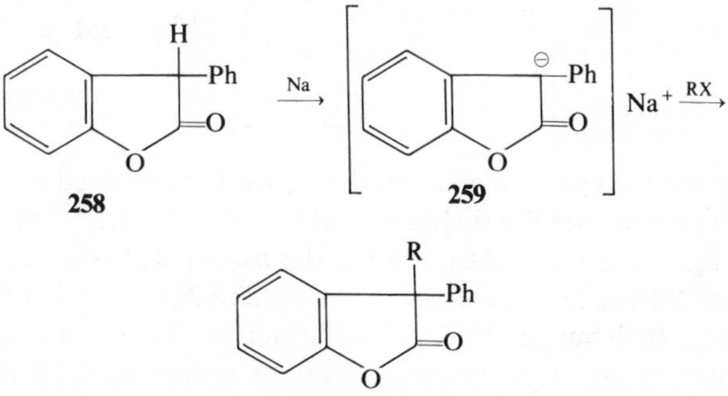

H Ph O O → Na [⊖ Ph O O] Na^+ → RX

258 **259**

R Ph O O

260 (53)

of **259** with β-diethylaminoethyl chloride [256] and $X(CH_2)_nX$ (X = Cl or Br, n = 1,2, or 3)[257] leads to derivatives of benzofuranones, which are useful as antispasmodics.

The nucleophilic attack by alkoxide and hydroxide ion on 3-(ω-haloalkyl)-3-phenyl-2(3*H*)-benzofuranones has been thoroughly studied by Zaugg et al.[258] Thus when 3-halomethyl-3-phenyl-2(3*H*)-benzofuranone (**261**) was treated with sodium methoxide or ethoxide in the appropriate alcohol, intramolecular displacement of the halogen atom took place exclusively. The only product isolated resulted from initial attack of alkoxide ion on the carbonyl atom. The phenoxide ion, which then formed by cleavage of the lactone ring, became the nucleophile and displaced the halogen intramolecularly to give the product **262**. The bromomethyl **261** formed the carboxylic acid **263** in high yield when treated with a slight excess of aqueous sodium hydroxide solution (5%; 100°C).

CH$_2$Br, Ph, O, O — **261** → COOH, Ph, O — **263**

258 → **261**

258 → COOR, Ph, O — **262**

The methoxide ion induces rearrangement of **261** to the corresponding ester (**262**, R = Me) with extraordinary rapidity.[258]

261 $\xrightarrow{^{-}OMe}$ [O$^-$, CH$_2$Br, Ph, COOMe] $\xrightarrow{-(Br^-)}$ **262** R = Me

The 5-membered oxygen is favored over the 6-membered oxygen cyclization in this system, and the displacement is stereospecific.[259] One diastereoisomer of the dibromide, **264a**, affords the *trans*-ester **265a**, and **264b** gives the *cis*-ester **265b** as an oil, which is converted to **266** upon distillation. With benzylamine, **265b** but not **265a** yields the lactone **267**. It seems clear that the general mechanism of these halides can be represented by the sequence shown in Eq. 54.

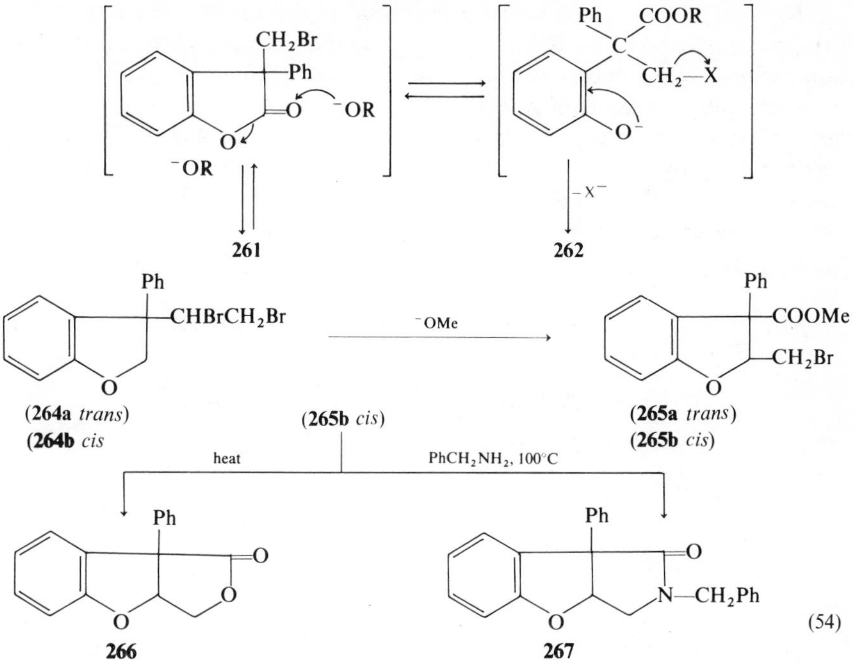

$$\left[\text{261-intermediate}\right] \rightleftharpoons \left[\text{262-intermediate}\right]$$

CH₂Br, Ph, ⁻OR, ⁻OR → 261; Ph, COOR, C, CH₂—X, O⁻ → −X⁻ → 262

(264a *trans*) (264b *cis*) —⁻OMe→ (265a *trans*) (265b *cis*)

(265b *cis*) —heat→ 266; —PhCH₂NH₂, 100°C→ 267 (54)

The reaction of 3-(ω-haloalkyl)-3-phenyl-2(3*H*)-benzofuranones (**268**, n = 0 or 1) with ammonia and/or primary amines[260] may be illustrated as shown in **268** to **271**.

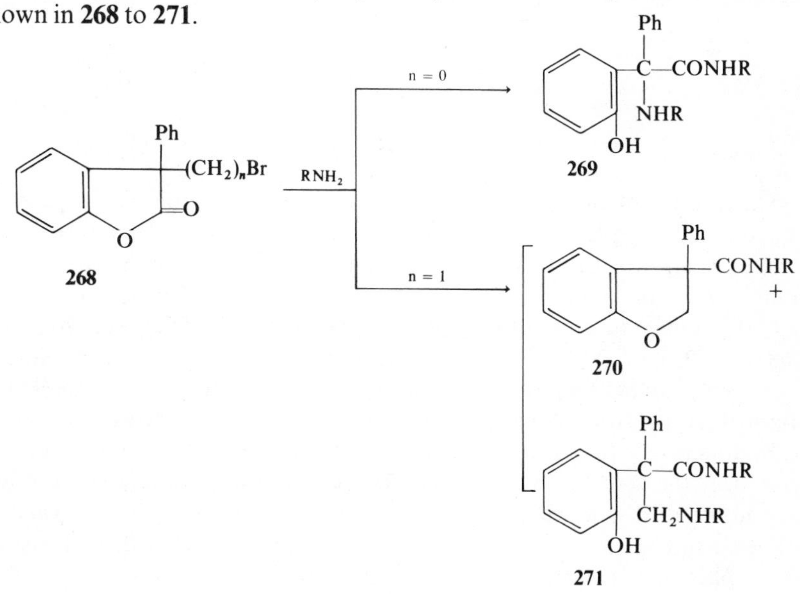

268 —RNH₂, n = 0→ 269; n = 1 → 270 + 271

The unexpected displacement of a neopentyl-type bromine atom by primary amines, that is, **268** ($n = 1$) → **271** under mildly exothermic conditions, is strong indication that, as is clear in the formation of **270**, an intramolecular mechanism is involved. The observation that the seemingly "direct" displacement of halogen in the production of **271** is invariably accompanied by aminolysis of the lactone ring of **268** suggests that the intermediate involved in the 4-membered cyclic imidate **A**. The mechanism can be represented as in **271**.

268 n = 1 —RNH_2→ [$\overset{\oplus}{H_2NR}$, $O^{\ominus}$, CH_2Br, Ph] —$-HBr$→

271 ——→ **271**

OH NR C C O C Ph

H_2 :NHR H

A

The reaction of **268** with secondary amines has been reported.[259] Depending on the length of the haloalkyl side-chain, the amine used, the temperature, and the solvent, any one or several of three products was found. With morpholine, for example, the bromomethyl homolog (**268**, $n = 1$), under all conditions gave the rearranged product **272**; however **268** (n = O, 3, or 4) formed only the product **273**. In excess of morpholine at room temperature the bromethyl homolog **268** ($n = 2$) gave exclusively the trapped tetrahedron intermediate **274**; but at raised temperature (95–100°C) as in dimethylformamide or dimethyl sulfoxide solution at room temperature, only the displacement product **273** ($n = 2$) was obtained. With other secondary amines more basic than morpholine, **268** ($n = 2$) gave varying amounts of the rearranged amide (corresponding to **272**) at the expense of the trapped tetrahedral intermediate, for example, **274**.

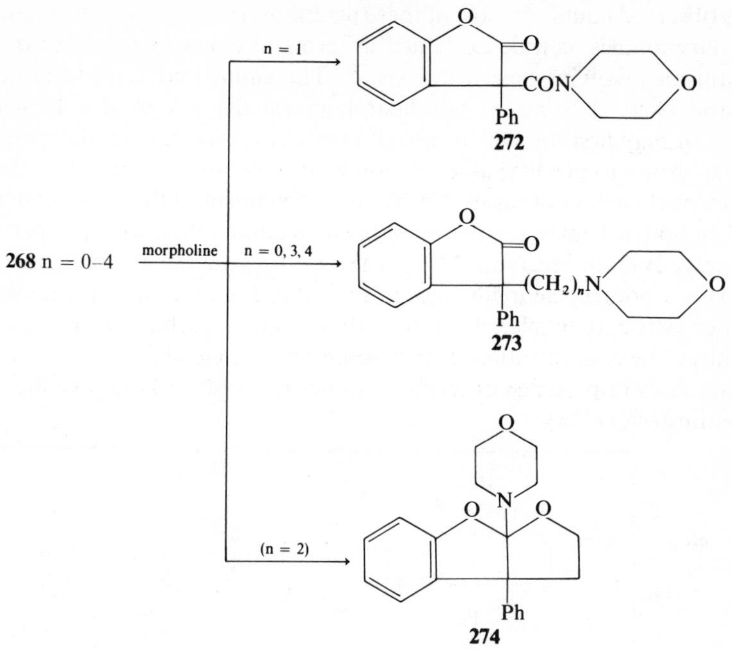

268 n = 0–4 —morpholine→ n = 1 → 272; n = 0, 3, 4 → 273; (n = 2) → 274

Sodium derivatives of tertiary aminoalkanols react with **275** to give the corresponding basic ester (**276**). Compound **275** reacts, in the same way, with tertiary aminoalkanols either alone or in the presence of triethylamine; moreover, it does not react with β-diethylaminoethyl mercaptan under amine-catalyzed conditions, but is attacked by sodium derivative of the mercaptan[261] to yield **277**, containing a small amount of **278**. These results suggest that in the amine-catalyzed reaction of **275**, the tertiary amine functions as a base to remove a proton from a tetrahedral intermediate.

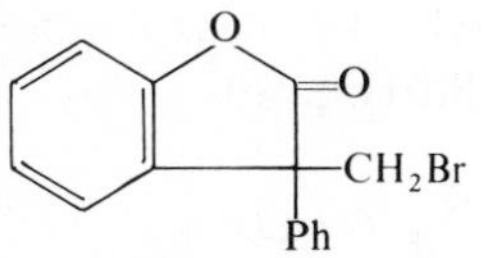

275 CH_2Br, Ph

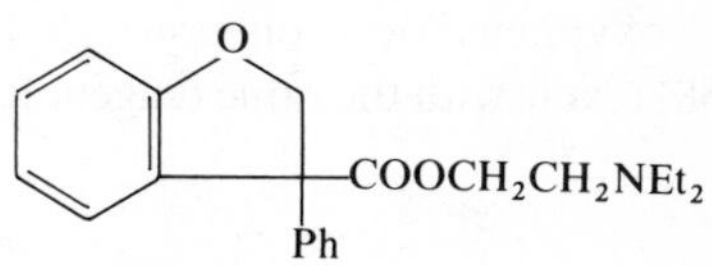

276 $COOCH_2CH_2NEt_2$, Ph

277 $COSCH_2CH_2NEt_2$, Ph

278 $CH_2SCH_2CH_2NEt_2$, Ph

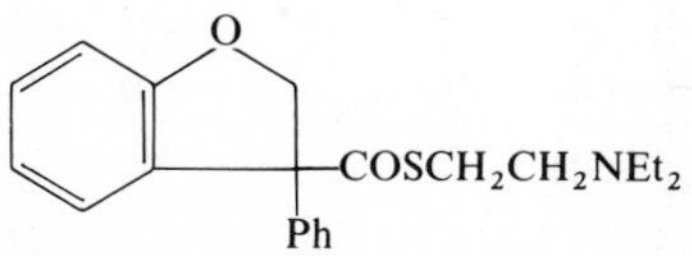

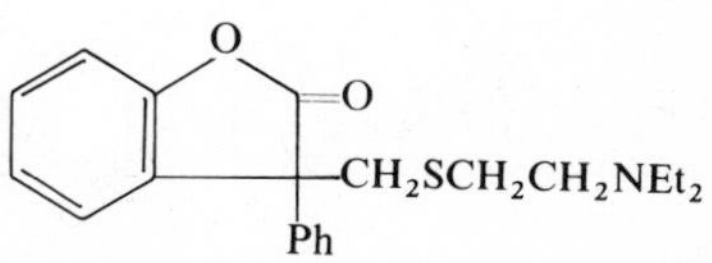

The observed course of most of these reactions, carried out in the presence of tertiary amines, can be explained in terms of either one or both of the mechanisms involving basic catalysts.[262] The amine may serve to abstract a proton (step 2) from the tetrahedral intermediate **A** to give **B**, which cyclizes, it may first (or synchronously) remove a proton from the carbinol hydroxy group to produce alkoxide ion which next (or concertedly) attacks the carbonyl carbon atom of **275** by the mechanism outlined by Zaugg et al.[258] In both schemes, the pre-equilibrium leading to the common penultimate stage **B** favor reactants **275** in one case because the attacking reagent (ROH) is a poor nucleophile and, in the other because the reagent (RO^-), although strongly nucleophilic toward carbonyl carbon, is necessarily present in low concentration. Either route serves to explain why **275** reacts with alcohols in presence of tertiary amines (procedure 1) to give the corresponding ester (**276**).

275 $\underset{(1)}{\overset{ROH}{\rightleftharpoons}}$ [H^+, O—R, O^-, CH_2Br, Ph $\rightleftharpoons$ OR, OH, CH_2Br, Ph] **A**

$\underset{(2)}{\overset{R_3N}{\rightleftharpoons}}$ [O^-, CH_2Br, C—COOR, Ph] **B** $\xrightarrow{(3)}$ $R_3\overset{+}{N}HBr^-$ **276**

Treatment of 2(3*H*)-benzofuranone with phenylmagnesium bromide brought about the lactone ring-opening with the formation of ω-(*O*-hydroxyphenyl)acetophenone.[263] However, 3-phenyl-2(3*H*)-benzofuranone (**258**) reacts with the same reagent to give **279**, (R = Ph) (Eq. 55).

$\xrightarrow[H^+]{PhMgBr,}$ CH_2COPh, OH

Ph, OH, O, R

279 (55)

3-Phenyl-2(3*H*)-benzofuranones, which are alkylated in the 4 position, behave like 2(3*H*)-benzofuranones with two substituents in the 3 position;[264] they do not give the reactions characteristic for active hydrogen atoms and are easily converted into the corresponding 2,3-dihydrobenzofuranols of type **279**.

Ethane formation from the corresponding methane in sunlight in the presence of air has been reported by Schönberg and Mustafa.[265,266] Thus when **258** was insolated in benzene solution in presence of air, 2,2′-diketo-3,3′-diphenyl-3,3′-dibenzofuranyl (**280**) was obtained.[265] It is possible that during exposure to sunlight, removal of one hydrogen atom takes place, but the free radical does not form peroxide,[266] rather it undergoes dimerization to form **280**. This behavior may be because the corresponding free radical is to be regarded as a resonance hybrid (**281**), and because the latter form contains univalent oxygen; such a radical is known to be not readily affected by atmospheric oxygen.[267,268] The reaction finds analogy with the photo oxidation of 2,2′-diphenylthioindigo-white (**283**) by irradiation of 2-phenyl-3-keto-2,3-dihydrothionaphthen (**282**) under the same conditions.[269,270]

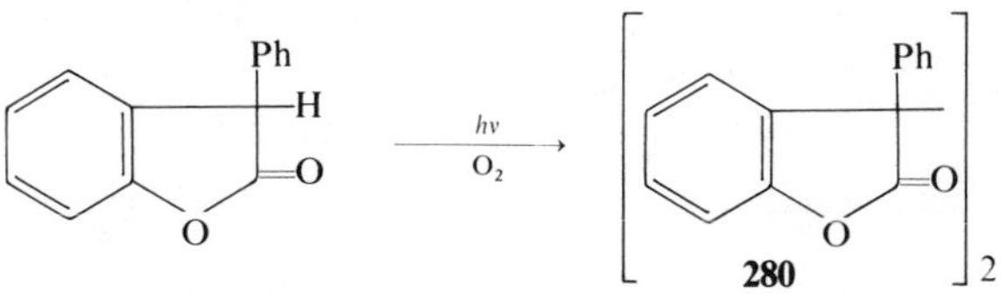

Ph H O O $\xrightarrow[O_2]{h\nu}$ [Ph O O]$_2$ **280**

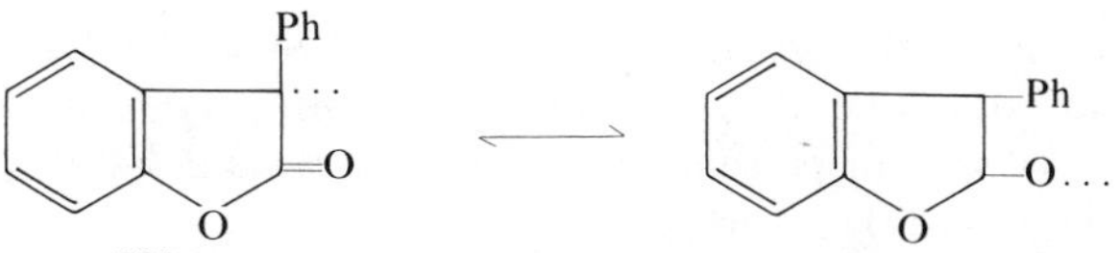

Ph O O **281a** ⟷ Ph O... O **281b**

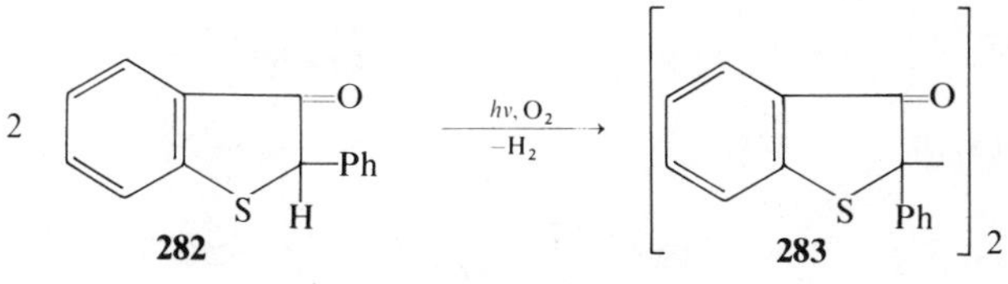

2 O Ph S H **282** $\xrightarrow[-H_2]{h\nu, O_2}$ [O S Ph]$_2$ **283**

Thermal disproportionation of **284** to **285** has been observed.[271] The formation of **285** is believed to be effected by the free radical disproportionation. Reductive fission of the ethane linkage in **284** has been observed by Mustafa,[272,273] upon treatment of **284** with Grignard reagents, yielding **285** (Eq. 56).

(56)

285

2(3*H*)-Benzofuranone in methanol, upon irradiation, gave the ether **286**. The intermediate **287** was trapped as the ortho lactone (**288**) with $CH_2=C(OMe)_2$.[274]

hν, MeOH

CH_2OMe

OH

173

286

CH_2

$(OMe)_2$

287

288

Oxidation of 6-methyl-2(3*H*)-benzofuranone-4-carboxylic acid (**289**) with alkaline potassium permanganate yields 4-methyl-6-hydroxyphthalic acid (**290**); however, in presence of sodium carbonate 4,4′-dicarboxy-6,6′-dimethylisoindigo (**291**) is formed.[275]

COOH

NaOH

HO COOH

COOH

Me

$KMnO_4$

Me

290

Me

COOH

289

Na_2CO_3

Me

291

3-Acyl-2(3*H*)-benzofuranones (**292**) are rearranged easily to the related pyrazolones.[234,276] The acyl derivatives have been prepared by action of *N,N'*-diarylamidines on the appropriate benzofuranone in alcohol, but they are labile in nature; in all cases they undergo rearrangement to pyrazolones either by action of hydrogen chloride or in some cases by heat only. In the case of 3-benzoyl-2(3*H*)-benzofuranone, its normal product with phenylhydrazine, is the pyrazolone **293** (R = Ph).

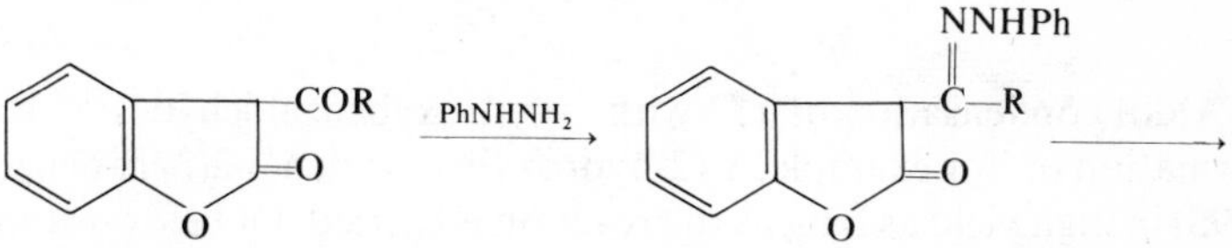

COR PhNHNH$_2$ NNHPh C—R O O O O

292

Ph O=C—N N CH—C OH R

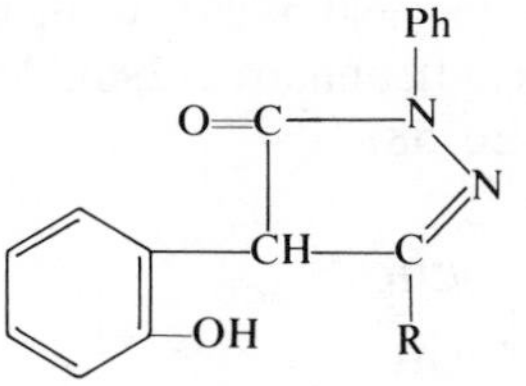

293

Acid-catalyzed rearrangement of **292**[234] could, as well, be applied to anils. 3-Formyl-2(3*H*)-benzofuranone anil (**294**), for example, on boiling with hydrochloric acid underwent hydrolysis, decarboxylation, and rearrangement to benzofuran (Eq. 57).[276]

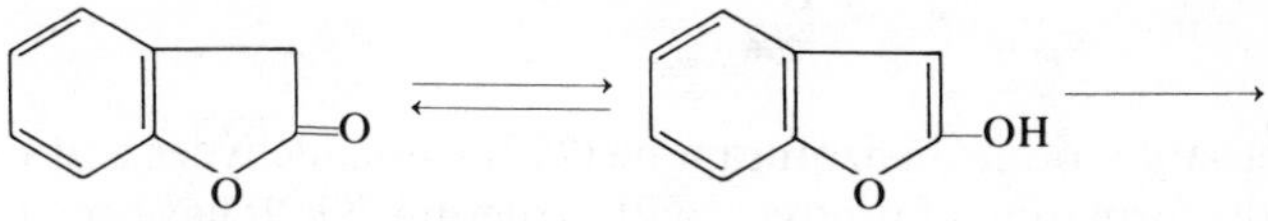

O O OH O

173

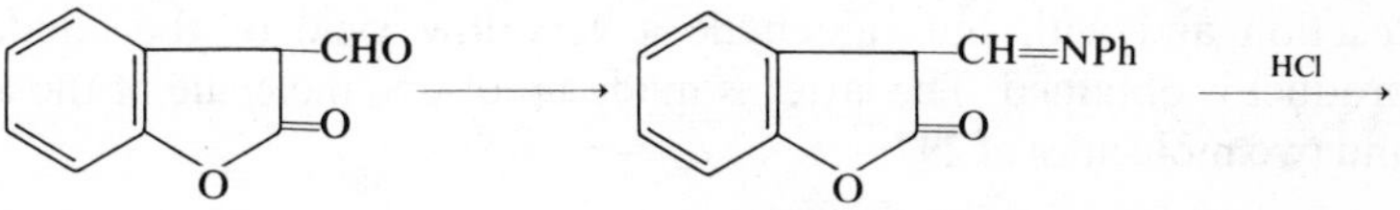

CHO O O CH=NPh O O HCl

294

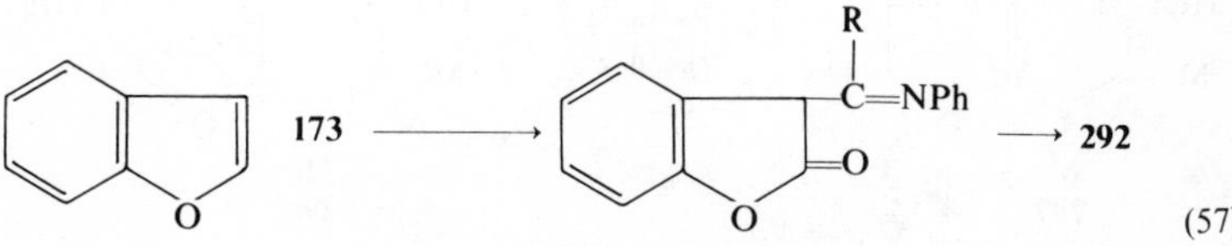

O 173 R C=NPh O O → 292

(57)

That aromatic aldehydes condense with the α-methylene group of 2(3*H*)-benzofuranone in presence of an organic base is well established.[277-279] 3-Benzylidene-2(3*H*)-benzofuranone, however, can also be prepared by an indirect route.[280,281]

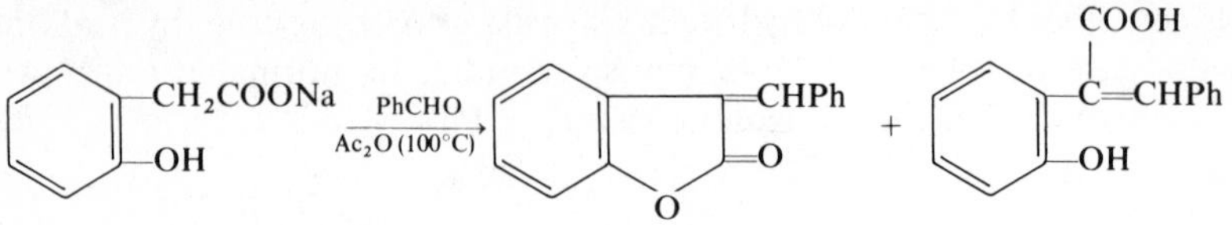

Aldol condensation of **173** with 2-hydroxybenzaldehydes[282] leads to the formation of, for example, 3-(2-hydroxybenzylidene)-2(3*H*)-benzofuranone (**295**) in high yield as long as the reaction is carried out below room temperature.[283] The rise in temperature during condensation of **173** with 2-hydroxybenzaldehyde diminished the yield of **295**, while giving rise to an additional product; 3-(2-hydroxyphenyl)coumarin (**296**). At still higher temperatures, **296** was the sole product (Eq. 58).

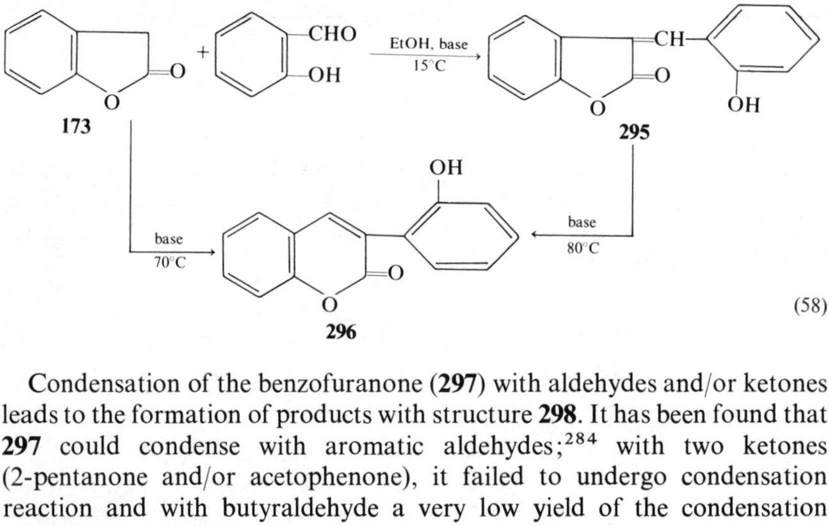

(58)

Condensation of the benzofuranone (**297**) with aldehydes and/or ketones leads to the formation of products with structure **298**. It has been found that **297** could condense with aromatic aldehydes;[284] with two ketones (2-pentanone and/or acetophenone), it failed to undergo condensation reaction and with butyraldehyde a very low yield of the condensation product is obtained. The latter is made up of one molecule of the aldehyde and two molecules of **297**.

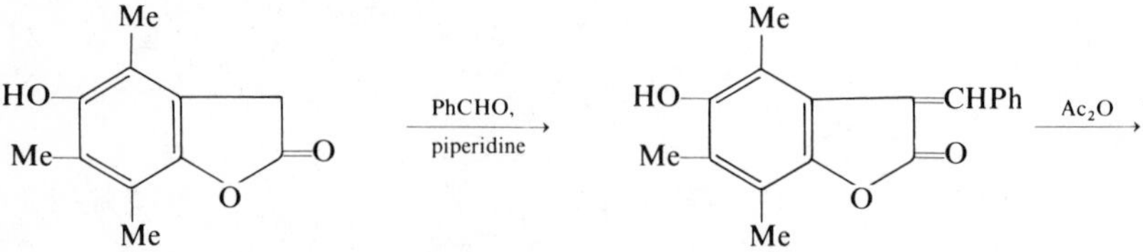

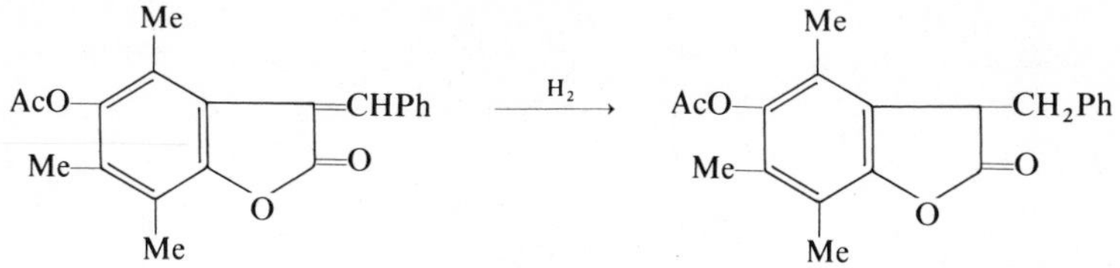

Nitration of 2(3*H*)-benzofuranone (**173**) results in the formation of 5-nitro-2(3*H*)-benzofuranone (**299**) (Eq. 59).[285]

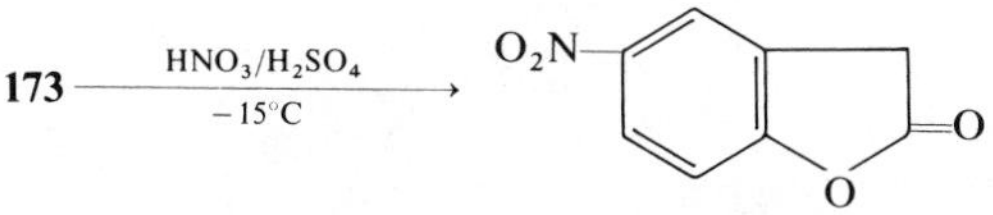

(59)

Chlorination of 6-hydroxy-3,3-diphenyl-2(3*H*)-benzofuranone (**300**) affords highly chlorinated products (**301**) in which the benzenoid system of double bonds has been disrupted.[287] Reduction of **301** yields 6-hydroxy-5-chloro-3,3-diphenyl-2(3*H*)-benzofuranone (**302**). However bromination of **258** yields the 3-bromo derivative (**303**).[286]

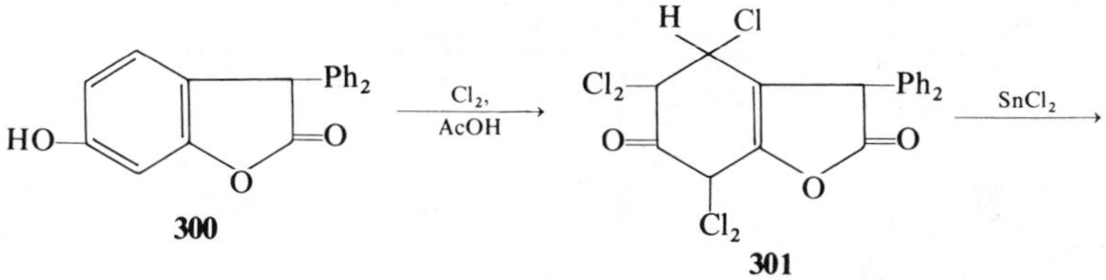

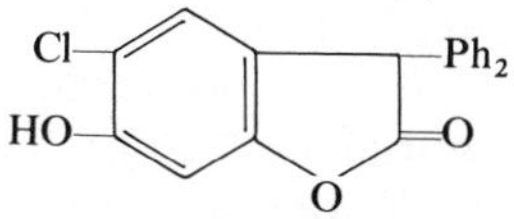

3-Phenyl-2(3*H*)-benzofuranone (**258**) reacts with trialkyl (or aryl) mercapto carbonium salt to yield quinodimethan, the structure of which is either **304** or **305** according to the free para positions.[288]

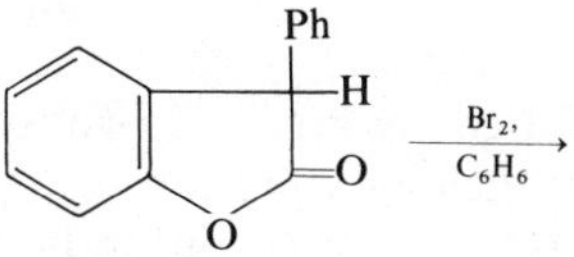

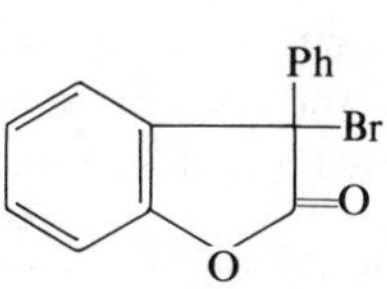

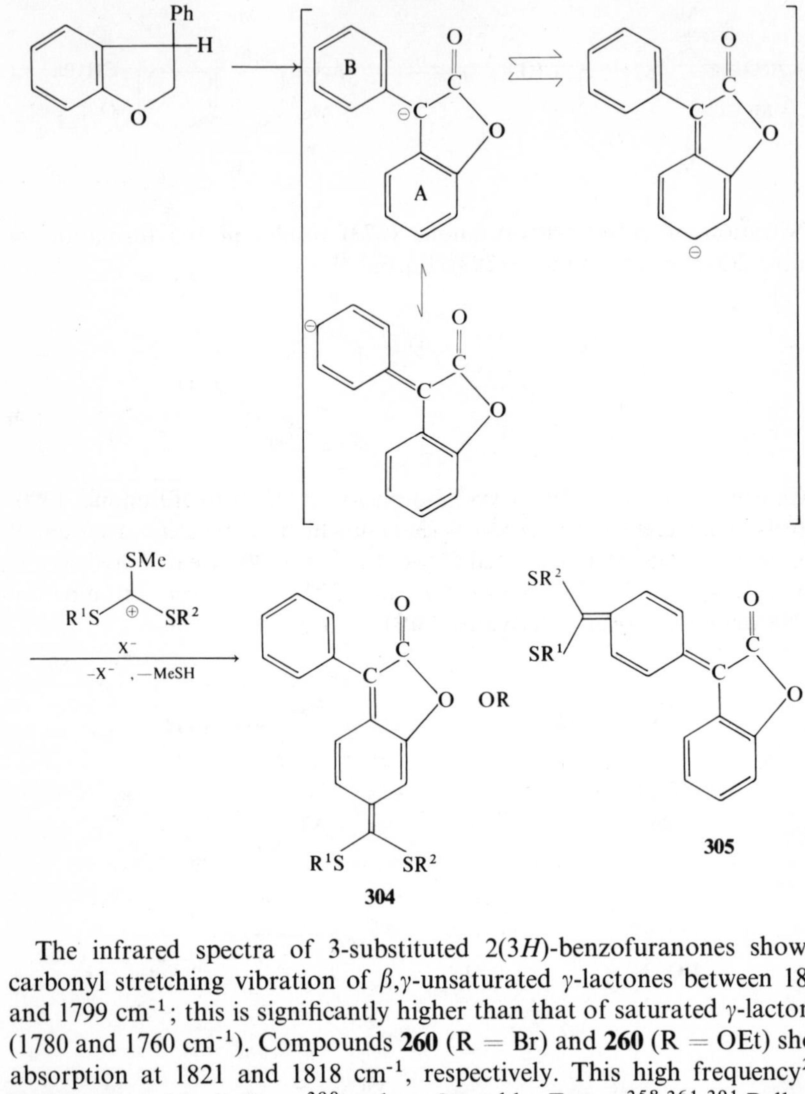

304

305

The infrared spectra of 3-substituted 2(3*H*)-benzofuranones show a carbonyl stretching vibration of β,γ-unsaturated γ-lactones between 1805 and 1799 cm^{-1}; this is significantly higher than that of saturated γ-lactones (1780 and 1760 cm^{-1}). Compounds **260** (R = Br) and **260** (R = OEt) show absorption at 1821 and 1818 cm^{-1}, respectively. This high frequency[289] has been noted by Bellamy[290] and confirmed by Zaugg.[258,261,291] Bellamy has suggested that in vinyl esters of the type CO—O—C=C, normal mesomerism due to the forms such as $-\overset{O^-}{C}=\overset{+}{O}R$ is suppressed. The frequency then is determined solely by the high electronegativity of the oxygen atom. The enhanced high frequency shift observed is due in each case to the strongly electron-withdrawing substituent attached to the carbon atom alpha to the carbonyl group. The benzofuranone nucleus appears to be further characterized by the low-frequency shift of the C—O stretching bond.

The ultraviolet spectra of a number of 3-phenyl-3-substituted 2(3*H*)-benzofuranones (**260**) are reported.[291]

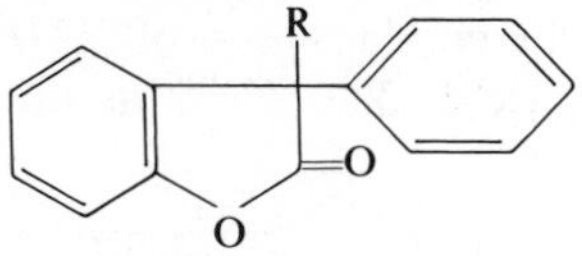

260

4. 2,3-Dihydrobenzo Furandiones

A. Preparation

Orcinol, treated with cyanogen and hydrogen chloride, produces 4-methyl-6-hydroxy-2,3-dihydrobenzofurandione (**306**).[292]

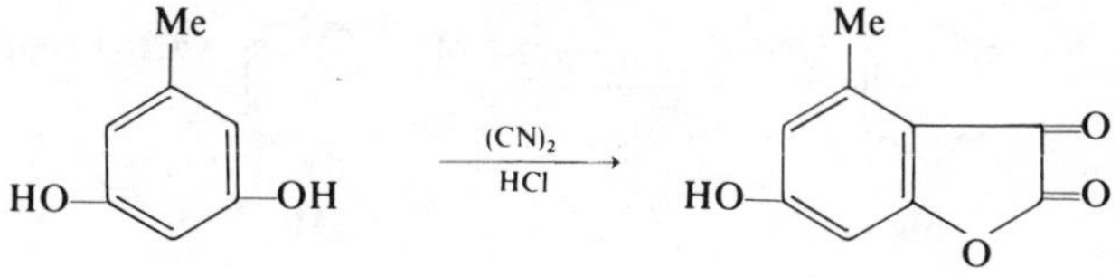

$$\xrightarrow[\text{HCl}]{(\text{CN})_2}$$

306

Aluminum chloride brings about the transformation of aryloxalyl chlorides to dihydrobenzofurandiones; yields of the corresponding substituted diones are higher the more stable they are. This stability is increased by substituents in the position meta to the heterocyclic oxygen and is weakened by ortho and para substituents.[293]

When 2-nitrobenzofuran is treated with sodium ethoxide at room temperature, a reaction analogous to that displayed by mono nitro naphthalenes

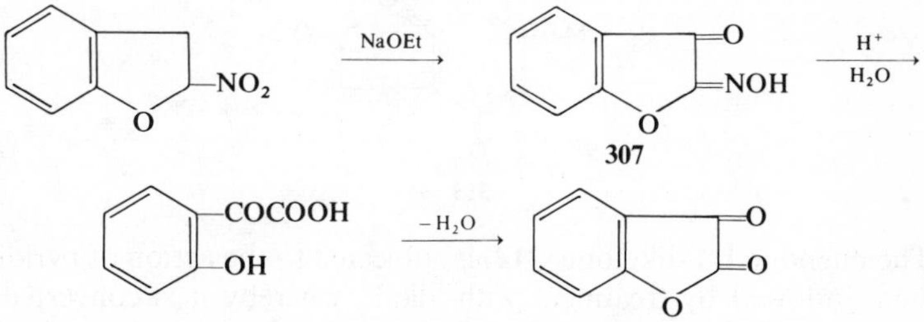

$$\xrightarrow{\text{NaOEt}} \quad \mathbf{307} \quad \xrightarrow[H_2O]{H^+}$$

$$\xrightarrow{-H_2O} \quad \mathbf{308} \qquad (60)$$

and several other nitro compounds occur.[294] Hydrolysis of **307** gives the dione **308**,[295] presumably by way of the intermediate *o*-hydroxyphenylglyoxylic acid. This reaction has been successfully applied to the preparation of 6,7-dimethoxy-,[296] 6-ethyl-,[297] and 4-methyl-7-isopropyl-2,3-dihydrobenzofurandione[25] (Eq. 60). In a similar manner hydrolysis of the azomethine (**309**), obtained by condensation of 3(2*H*)-benzofuranone (**1**) with *p*-nirosodimethylaniline, yields **308**,[298,299] as outlined below.

$Me_2NC_6H_4NO$-p, aq. NaOH → **309** (NMe_2) → H^+ → **308**

The dione **313**, is obtained either by nitrosation of 4,6,7-trimethyl-3(2*H*)-benzofuranone (**311**) to yield the isonitroso compound **312**, followed by hydrolysis (Eq. 61), or directly by the interaction of 2,3,5-trimethylphenol and oxalyl chloride in presence of aluminum chloride in benzene, in contrast to the literature reports.[293,300,301]

ONa → $ClCH_2COOH$ → OCH_2COOH → H_2SO_4 →

311 → HONO → **312** (NOH) → H^+ →

313 (61)

The phenolic 1,4-diketone (**314**) is subjected to the action of pyridine and iodine, followed by treatment with alkali, whereby it is converted to the dione (**313**).[302] The exact mechanism by means of which **313** is formed from **314** is not clear, but transformation of **314** into **313** by this "modified haloform reaction" establishes the presence of the grouping —$COCH_2$— in **314**.

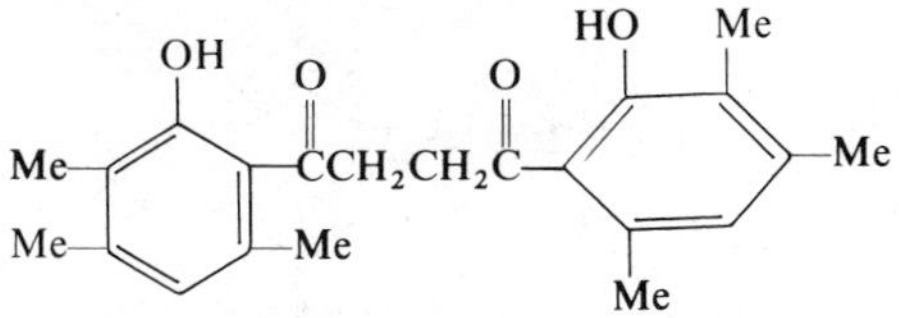

314

Clemmensen reduction of **313** resulted in the formation of 4,6,7-trimethyl-2(3*H*)-benzofuranone (**315**) and an unidentified product. Nitration of **313** leads to the formation of **316**. Coupling **313** with diazotized sulfanilic acid, the resulting azo dye is reduced by action of stannous chloride to the corresponding amino compound and followed by oxidation of the latter with ferric chloride which led to the formation of the dione (**317**). The latter can also be obtained by interaction of trimethylhydroquinone with oxalyl chloride (Eq. 62).

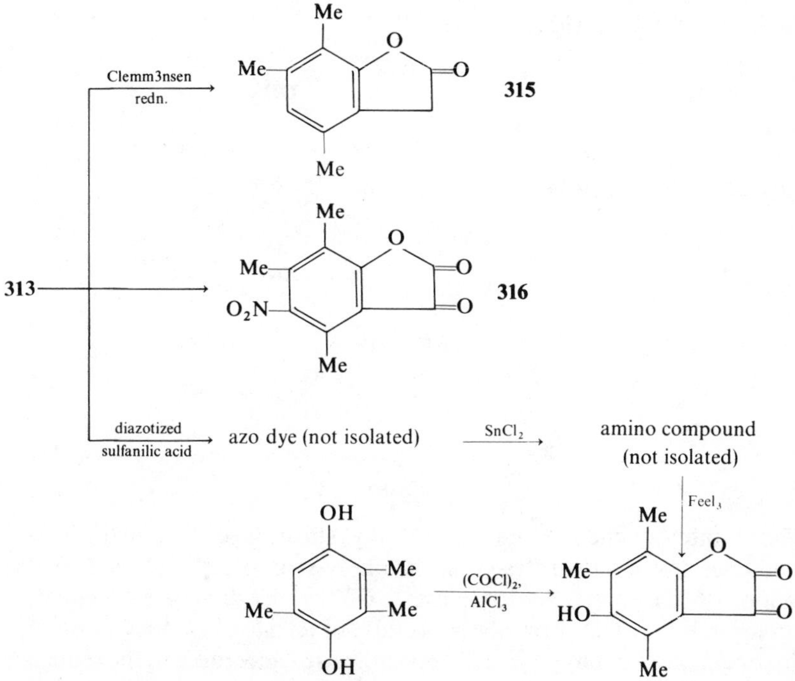

313 —Clemm3nsen redn.→ **315**

313 → **316**

313 —diazotized sulfanilic acid→ azo dye (not isolated) —$SnCl_2$→ amino compound (not isolated) —FeCl$_3$→ **317**

Trimethylhydroquinone —$(COCl)_2$, $AlCl_3$→ **317** (62)

The phenol 1,4-diketone (**318**) differed greatly from its isomer (**314**) in its behavior toward pyridine and iodine, and yielded **319**.[302]

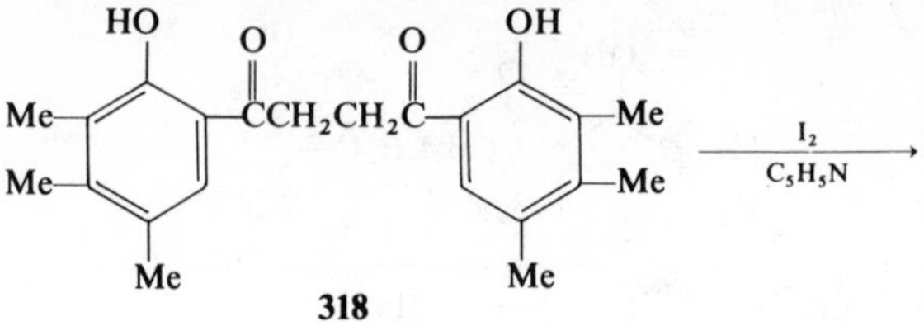

2,4-Dihydroxy-3-ethyl-5-methylglyoxylic acid (**320**), obtained by the action of ethyl oxalyl chloride on 2-ethyl-4-methylresorcinol, is readily lactonized to the dione (**321**) (Eq. 63).[303]

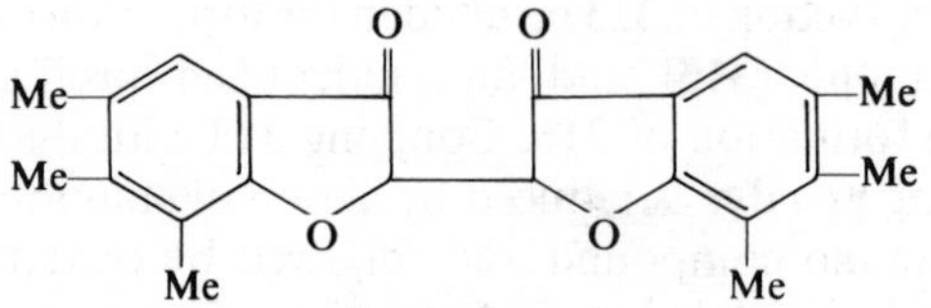

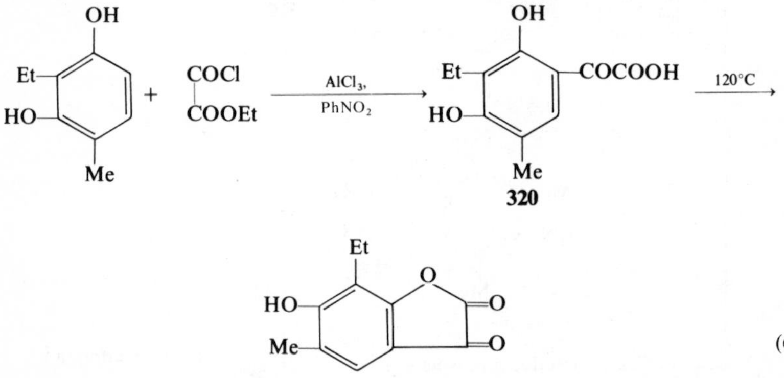

(63)

The stability of the lactone ring in alkyl-substituted 2,3-dihydrobenzofurandiones toward hydrolysis and alcoholysis is greatly affected by the position of the alkyl substituent.[304] 6-Ethyl-4,7-dimethyl-2,3-dihydrobenzofurandione (**322**), a pyrolysis product of tetrahydrophysalin obtained by hydrogenation of physalin **A**,[305] is synthesized according to the sequence of reactions shown in Eq. 64.

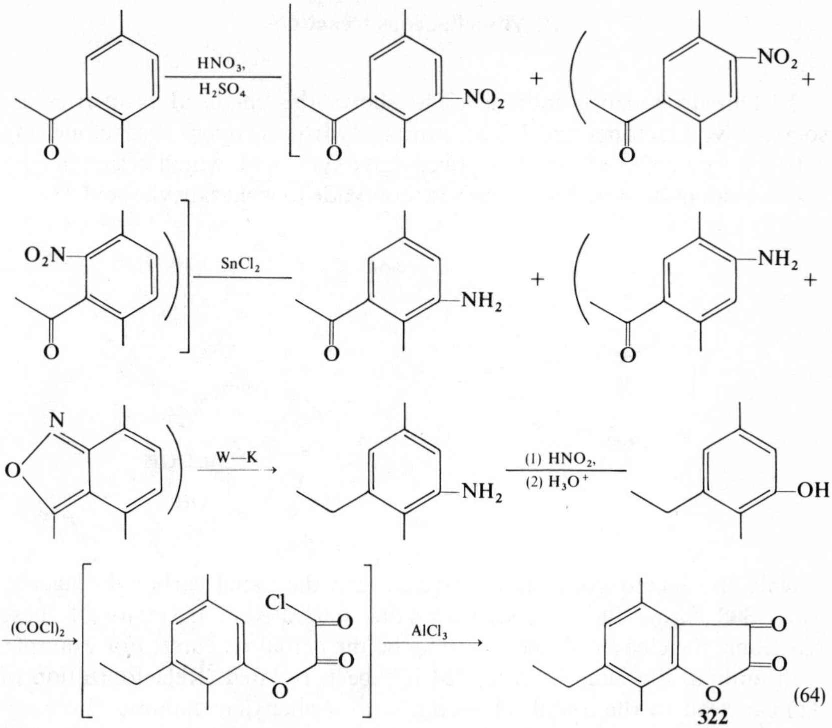

HNO₃, H₂SO₄ ; SnCl₂ ; W—K ; (1) HNO₂, (2) H₃O⁺ ; (COCl)₂ ; AlCl₃

322 (64)

From ultraviolet spectral studies, it has been shown that methyl substitution at the position adjacent to the lactone ring contributes to the stability for steric reasons.

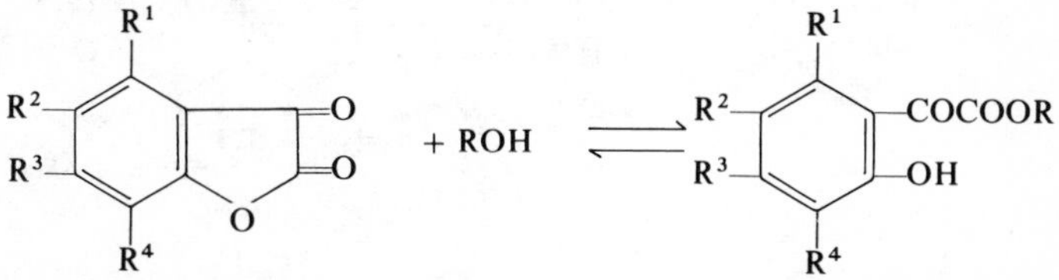

Condensation of substituted cyclohexanone with diethyl oxalate yields the dione (**323**).[306,307]

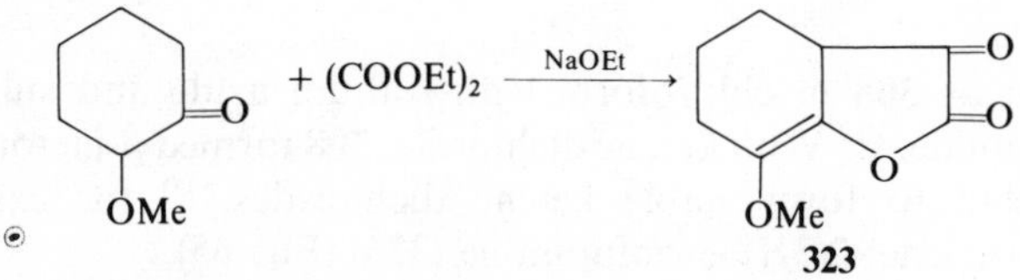

323

B. Miscellaneous Reactions

2,3-Dihydrobenzofurandione (**308**) shows the chemical properties associated with lactones and 1,2-diketones. Hydrolysis opens the lactone ring with the formation of *o*-hydroxyphenylglyoxylic acid, which acting like an α-keto acid, undergoes loss of carbon monoxide to yield salicylic acid.[52]

NaOH
COCOOH
OH
heat
308
H_2SO_4
COOH
OH

Only the 3-keto group is reactive toward the usual carbonyl reagents. Thus **308** forms the 3-phenylhydrazone derivative.[1] In many of these reactions, the cleaved dione ring may be the actual reactant. For example, with aniline, the Schiff's base, **324** has been isolated. Relactonization of **324** can lead to the 3-anil. However, with *o*-phenylenediamine, the usual quinoxaline formation (**325**) characteristic of 1,2-diketones occurs. The lactone ring is simultaneously opened (Eq. 64a).

NPh
308 $\xrightarrow[H_2O]{PhNH_2,}$ CCOOH
OH
NPh
O
324

308 + NH₂ NH₂ ⟶ OH N N

Treatment of **308** in chloroform with sodium azide and sulfuric acid gives salicylamide.[309] With ketene dichloride, **308** formed β-lactones, which decarboxylated to form stable ketene dichlorides,[310] for example, 3-(dichloromethylene)-2(3*H*)benzofuranone (**326**) (Eq: 65).

308 $\xrightarrow[H_2SO_4\ (0°C)]{NaN_3}$ [o-C_6H_4(CN)(OH)] ⟶ [o-C_6H_4($CONH_2$)(OH)]

308 + $Cl_2C{=}C{=}O$ ⟶ [spiro β-lactone adduct] $\xrightarrow{-CO_2}$ **326**

(65)

Irradiation of 2,3-dihydrobenzofurandiones **308** and **327** in presence of certain nucleophiles (water, phenols, and/or carboxylic acids) effected photodecarbonylation by *n*—*π** excitation in benzene solution to produce a high yield of derivatives of salicylic acid, **329a** and **329b**, respectively. A keto-ketene intermediate, **328**, is suggested as the result of decarbonylation.[311]

327 $\xrightarrow{h\nu}$ [**328**] $\xrightarrow{R^1OH}$ [R-substituted o-C_6H_3($COOR^1$)(OH)]

329a R = H, R^1 = COPh
329b R = Me, R^1 = Ph

Methylation of **330** with dry diazomethane has been reported to yield a monomethylated product, which is believed to have structure like **332**.[312] Cram[313] has shown from ultraviolet spectral study of **330** and **332** that the monomethylated product possesses the benzenoid structure **331** rather than the hemiquinone structure **332**.[313] However methylation of **330** with methyl sulfate affords, besides **331**, compound **333**. Furthermore, it has been shown that 2,3-dihydrobenzofurandione (**308**) reacts with diphenyldiazomethane to yield the epoxide, **334**,[314] which undergoes transformation[315] as illustrated in Eq. 66.

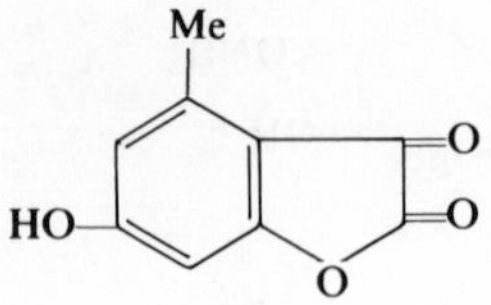

330

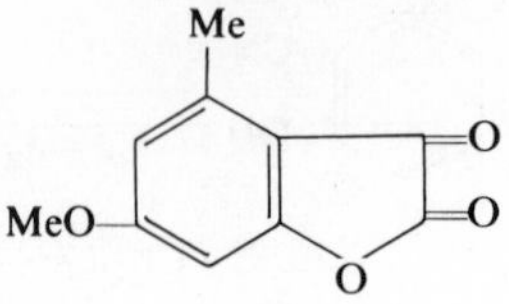

331

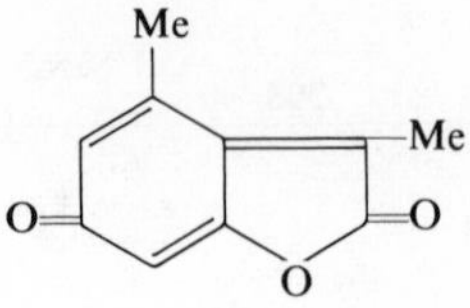

332

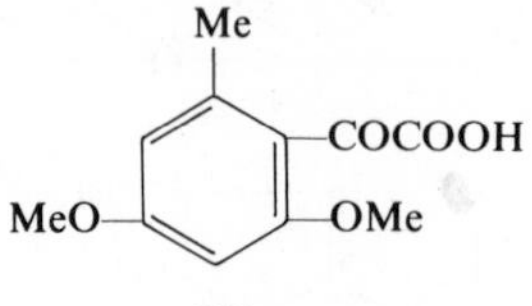

333

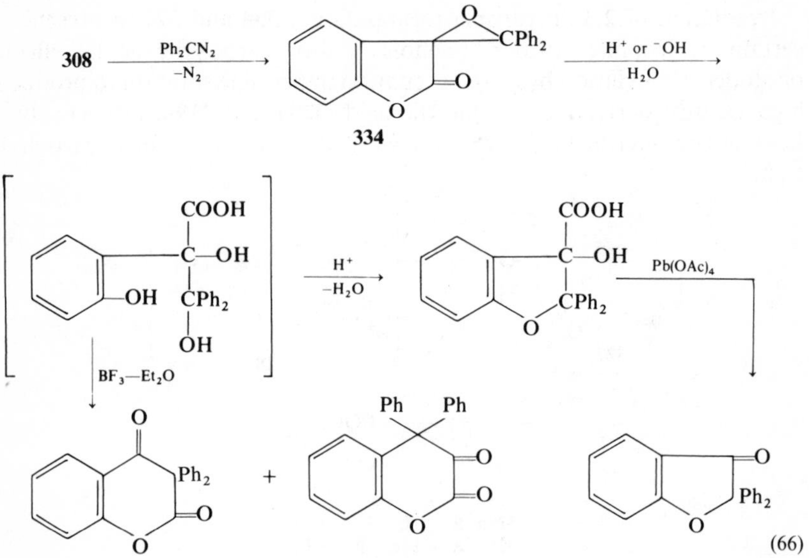

334

(66)

4,6-Dimethyl-2,3-dihydrobenzofurandione (**335**) and 4,6-dimethyl-3(2*H*)-benzofuranone (**336**) in acetic acid containing hydrochloric acid, condense together to yield **337** which, when dissolved in sulfuric acid, furnished **336** and 4,6,4′,6′-tetramethyloxindirubin (**338**) (Eq. 67).[316] In a similar manner, 4,6-dimethyl-2,3-dihydrobenzofurandione-2-anil (**339**) adds **336** in boiling toluene to give 2-anilino-3,3′-dioxo-r,6,4′,6′-tetramethyl-[2,2′-di-(2,3-dihydrobenzofuran)] (**340**),[298] which, upon treatment with sulfuric acid, affords 4,6,4′,6′-tetramethyloxindigo (**341**). Reduction of the latter produces oxindigo-white (**342**) (Eq. 68).

335 + 2 336 —AcOH/HCl→ 337 —H_2SO_4→ 338 (67)

339 + 336 → 340 ⇌ (+ $PhNH_2$ / $-PhNH_2$) 341 —Zn dust, AcOH→

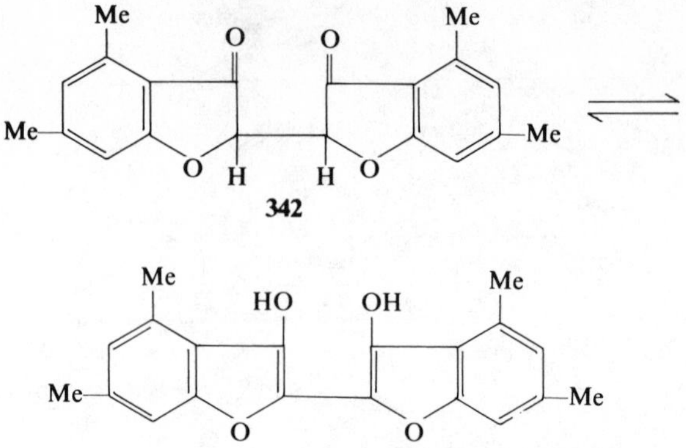

342

(68)

Condensation of **308** with *o*-methoxyphenacyl bromide in alcoholic sodium ethoxide solution gives the ester **343**.[317,318] An obvious variation is the condensation of the dione **308** with *o*-carbmethoxyphenacyl bromide in alcoholic sodium methoxide solution to yield **344**. Similarly, condensation of **308** with *sym*-dichloroacetone in alcoholic sodium ethoxide solution furnishes the keto ester **345** (R = Et), which gives the acid **335** (R = H) on hydrolysis. 6-Methoxy-2,3-dihydrobenzofurandione reacts with *o*-bromophenacyl bromide in the presence of sodium methoxide to yield the benzofuran derivative **346**.[319,320]

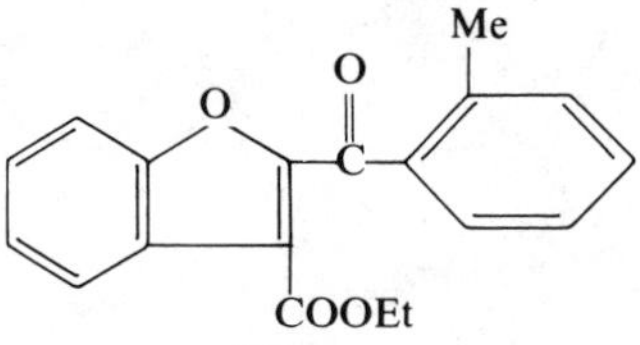

343 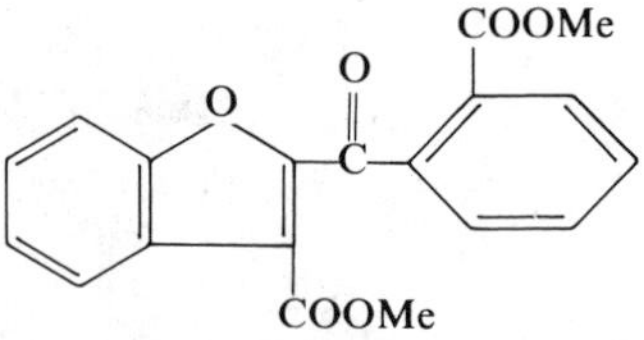 **344**

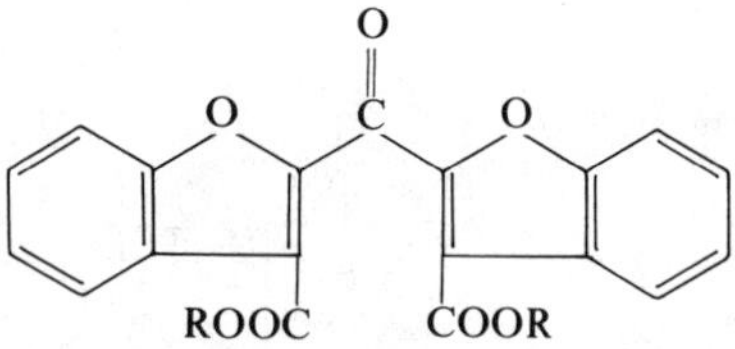

345

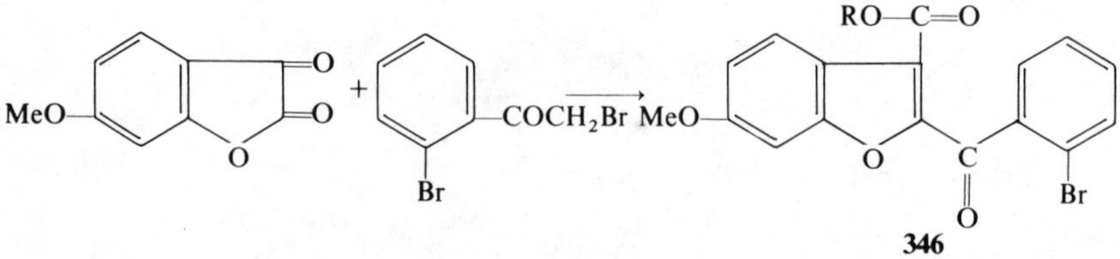

346

Interaction of **308** with ethyl bromoacetate results in opening of the dione ring to give the keto ester **347**, followed by reconstitution to form **348**.[317,321]

$$\mathbf{308} \xrightarrow[\text{NaOEt, EtOH}]{\text{BrCH}_2\text{COOEt,}} \mathbf{347} \longrightarrow \mathbf{348}$$

347: $C_6H_4(O{-}CH_2COOR)(C(=O){-}COOR)$ **348**: benzofuran-2,3-$(COOR)_2$

References

1. K. v. Auwers and W. Müller, *Chem. Ber.*, **50,** 1149 (1917).
2. K. v. Auwers and H. Schütte, *Chem. Ber.*, **52,** 77 (1919).
3. K. v. Auwers and E. Auffenberg, *Chem. Ber.*, **52,** 92 (1919).
4. K. v. Auwers, Th. Bakr, G. Wagner, and C. Wiegand, *Chem. Ber.*, **61B,** 408 (1928).
5. D. C. Schroeder, P. C. Corcoran, C. A. Holden, and M. C. Mulligan, *J. Org. Chem.*, **27,** 586 (1962).
6. R. v. Stoermer and F. Bartsch, *Chem. Ber.*, **33,** 3177 (1900).
7. P. Friedländer and J. Neudörfer, *Chem. Ber.*, **30,** 1081 (1897).
8. A. Blom and J. Tambor, *Chem. Ber.*, **38,** 3589 (1905).
9. C. J. Schoot and K. H. Klaasens, *Rec. Trav. Chim. Pays-Bas,* **75,** 190 (1956).
10. D. Stafanye and W. I. Howard, *J. Org. Chem.*, **20,** 813 (1955).
11. J. A. Carbon and L. S. Fosdick, *J. Am. Chem. Soc.*, **78,** 1504 (1956).
12. K. Fries and W. Pfaffendorf, *Chem. Ber.*, **43,** 212 (1910).
13. R. v. Stoermer and P. Atenstädt, *Chem. Ber.*, **35,** 3562 (1902).
14. M. L. Kalinowski and L. W. Kalinowski, *J. Am. Chem. Soc.*, **70,** 1970 (1948).
15. P. Pfeiffer and E. Enders, *Chem. Ber.*, **84,** 247 (1951).
16. P. Friedländer, *Chem. Ber.*, **32,** 1867 (1899).
17. K. Fries, A. Hasselbach, and L. Schroeder, *Justus Liebigs Ann. Chem.*, **405,** 370 (1914).
18. W. L. F. Armargo, *Aust. J. Chem.*, **13,** 95 (1960).
19. L. Higginbotham and H. Stephen, *J. Chem. Soc.*, **117,** 1534 (1920).
20. K. Fries and G. Finck, *Chem. Ber.*, **41,** 4271 (1908).
21. K. v. Auwers and K. Müller, *Chem. Ber.*, **41,** 4233 (1908).
22. K. v. Auwers, *Chem. Ber.*, **49,** 809 (1916).
23. M. H. Palmer, and G. J. McVie, *J. Chem. Soc., B,* **(1968)**, 745.
24. M. H. Palmer and N. N. Scollick, *J. Chem. Soc., C,* **(1968)**, 2833.
25. E. Mameli, *Gazz. Chim. Ital.*, **52,** 322 (1922); *Chem. Abstr.*, **16,** 2688 (1922).
26. O. Dann, G. Volz, and O. Huber, *Justus Liebigs Ann. Chem.*, **587,** 16 (1954).
27. R. Kuhn and H. R. Hensel, *Chem. Ber.*, **84,** 557 (1951).
28. J. N. Chatterjea, B. K. Banerjee, and H. C. Jha, *J. Indian Chem. Soc.*, **44,** 911 (1967); *Chem. Abstr.*, **69,** 18973 (1968).
29. F. Dallacker and W. Korb, *Justis Liebigs Ann. Chem.*, **694**, 98 (1966); *Chem. Abstr.*, **65**, 12153 (1966).
30. A. K. Bose and P. Yates, *J. Am. Chem. Soc.*, **74,** 4703 (1952).
31. E. E. Marshall, J. A. Kuck, and R. C. Elderfield, *J. Org. Chem.*, 7, 444 (1942).
32. N. Palit and J. N. Chatterjea, *Sci. Cult.*, **17,** 345 (1952); *Chem. Abstr.*, **47,** 2133 (1953).
33. A. Schönberg, K. Praefeke, and J. Kohtz, *Chem. Ber.*, **99,** 3077 (1966).
34. S. Akaboshi and T. Kutsuma, *Yakugaku Zasshi,* **88**(8), 1020 (1968); *Chem. Abstr.*, **70,** 11526 (1969).
35. T. L. Burkoth and E. F. Ullman, *Tetrahedron Lett.*, **(1970)**, 145.

36. C. S. Foote, M. T. Wuesthoff, S. Whexler, I. G. Burstain, R. Denny, G. O. Schenck, and K-H. Schulte-Elte, *Tetrahedron*, **23**, 2583 (1967).
37. G. A. Caplin, W. D. Ollis, and I. O. Sutherland, *J. Chem. Soc., C*. (**1968**), 2302.
38. G. Fodor and O. Kovàcs, *J. Am. Chem. Soc.*, **71**, 1045 (1949).
39. W. Logemann, G. Cavagna, and G. Tosolini, *Chem. Ber.*, **96**, 2248 (1963).
40. R. Howe, B. S. Rao, and H. Heyneker, *J. Chem. Soc., C*, (**1967**), 2510.
41. K. Fries and K. Saftien, *Chem. Ber.*, **59**, 1246 (1926).
42. V. K. Mahesh and R. K. Kela, *J, Indian Chem. Soc.*, **46**, 89 (1969); *Chem. Abstr.*, **70**, 114904 (1969).
43. P. D. Bartlett and E. N. Trachtenberg, *J. Am. Chem. Soc.*, **80**, 5808 (1958).
44. K. P. Barr, F. M. Dean, and H. D. Locksley, *J. Chem. Soc.*, (**1959**), 2424.
45. G. Baddeley, S. M. Makar, and M. G. Ivinson, *J. Chem. Soc.*, (**1953**), 3969.
46. N. Kornblum, T. M. Davies, G. W. Earl, G. S. Greene, N. L. Holy, R. C. Kerber, J. W. Manthey, M. T. Musser, and D. H. Snow, *J. Am. Chem. Soc.*, **89**, 725 (1967).
47. N. Kornblum, T. M. Davies, G. W. Earl, G. S. Greene, N. L. Holy, R. C. Kerber, J. W. Manthey, M. T. Musser, and D. H. Snow, *J. Am. Chem. Soc.*, **89'** 5714 (1967).
48. N. Kornblum and W. Stuchal, *J. Am. Chem. Soc.*, **92**, 1804 (1970).
49. N. L. Holy and J. D. Marcum, *Angew. Chem.*, **10**, 116 (1971).
50. G. A. Russel and W. C. Danen, *J. Am. Chem. Soc.*, **88**, 5663 (1966).
51. N. Kornblum, *Trans. N. Y. Acad. Sci.*, **29**, 1 (1966); *Chem. Abstr.*, **67**, 63455 (1967).
52. P. Friedländer and J. Neudörfer, *Chem. Ber.*, **30**, 1077 (1897).
53. G. Wittig, *Justus Liebigs Ann. Chem.*, **446**, 155 (1926).
54. K. v. Auwers, *Chem. Ber.*, **49**, 820 (1916).
55. K. v. Auwers and W. Müller, *Chem. Ber.*, **50**, 1161 (1917).
56. R. Royer and E. Bisagni, *Bull. Soc. Chim. Fr.*, (**1963**), 1746; *Chem. Abstr.*, **59**, 15265 (1963).
57. M. Descamps, F. Binon, and J. van der Elst, *Bull. Soc. Chim. Belges*, **73**(5–6), 459 (1964); *Chem. Abstr.*, **61**, 10671 (1964).
58. S. R. Cawley and S. G. P. Plant, *J. Chem. Soc.*, (**1938**), 1214.
59. J. W. Conforth, G. K. Hughes,F. Lions, and R. H. Harrzdence, *J. Proc. Roy. Soc. N.S. Wales*, **71**, 486 (1938); *Chem. Abstr.*, **33**, 588 (1939).
60. G. Villere and V. Grinsteins, *Ltvijas Univ. Kim-Fak. Zinatniskie Raksti*, **22**(6), 129 (1958); *Chem. Abstr.*, **53**, 17091 (1959).
61. E. Mameli, *Gazz. Chim. Ital.*, **52**(II), 184 (1922); *Chem. Abstr.*, **17**, 1643 (1923).
62. E. Mameli, *Atti. Congr. Naz. Chim. Pura Appli.* (**1923**), 436; *Chem. Abstr.*, **18**, 3063 (1924).
63. E. Mameli, *Gazz. Chim. Ital.*, **56**, 759 (1926); *Chem. Abstr.*, **21**, 1268 (1927).
64. R. v. Stoermer and E. Barthelmes, *Chem. Ber.*, **48**, 62 (1915).
65. A. Fabrycy, *Tetrahedron Lett.*, (**1962**), 175; *Chem. Abstr.*, **57**, 9774 (1962).
66. J. A. Barltrop, *J. Chem. Soc.*, (**1946**), 958.
67. L. Farkas, L. Pallos, and M. Nogradi, *Magy. Kem. Folyoirat*, **71**(6), 270 (1965); *Chem. Abstr.*, **63**, 9901 (1965).
68. W. Feuerstein and St. v. Kostanecki, *Chem. Ber.*, **31**, 1757 (1898).
69. L. T. Bogolyubskaya and M. A. Al'porovich, *Zh. Obshch. Khim.*, **34**(9), 3119 (1964); *Chem. Abstr.*, **61**, 16055 (1964).
70. K. Fries and G. Finck, *Chem. Ber.*, **41**, 4271 (1908).
71. K. Fries and W. Pfaffendorf, *Chem. Ber.*, **45**, 154 (1912).
72. R. v. Stoermer, *Justus Liebigs Ann. Chem.*, **313**, 79 (1900).
73. K. Fries and W. Pfaffendorf, *Chem. Ber.*, **44**, 114 (1911).
74. K. v. Auwers and R. Apitz, *Justus Liebigs Ann. Chem.*, **381**, 265 (1911);
75. K. v. Auwers and P. Pohl, *Justus Liebigs Ann. Chem.*, **405**, 243 (1914).
76. B. Holmberg, *Chem. Ber.*, **43**, 2192 (1910).
77. K. v. Auwers, K. Müller, and R. Apitz, *Chem. Ber.*, **43**, 2192 (1910).

78. F. M. Dean and K. Manunapichu, *J. Chem. Soc.*, **(1957)**, 3112.
79. J. Algar and J. P. Flynn, *Proc. Roy. Irish Acad.*, **42B**, 1 (1933); *Chem. Abstr.*, **29**, 161 (1935).
80. T. Oyamada, *J. Chem. Soc. Japan*, **55**, 1256 (1934); *Chem. Abstr.*, **29**, 4357 (1935).
81. T. Oyamada, *Bull. Chem. Soc. Japan*, **10**, 182 (1935); *Chem. Abstr.*, **29**, 5112 (1935).
82. M. Murakami and T. Irie, *Proc. Imp. Acad. Tokyo*, **(1935)**, 229. *Chem. Abstr.*, **29**, 6598 (1935).
83. R. Bognar and J. Stefanovsky, *Tetrahedron*, **18**, 143 (1962).
84. T. A. Geissman and D. K. Fukushima, *J. Am. Chem. Soc.*, **70**, 1686 (1948).
85. N. Narasimhachari and T. R. Seshadri, *Proc. Indian Acad. Sci.*, **30A**, 216 (1969).
86. T. H. Simpson and W. B. Whalley, *J. Chem. Soc.*, **(1955)**, 166.
87. N. Anand, R. N. Iyer, and K. Venkataraman, *Proc. Indian Acad. Sci.*, **29A**, 203 (1949); *Chem. Abstr.*, **44**, 1494 (1950).
88. T. S. Wheeler, *Rec. Chem. Progr. Kresge-Hooker Sci. Lib.*, **18**, 133 (1957); *Chem. Abstr.*, **52**, 376 (1958).
89. G. E. Gowan, E. M. Philbin, and T. S. Wheeler, "Chemistry of Vegetable Tannins," *Soc. Leather Trades' Chemists*, **(1956)**, 133.
90. E. M. Philbin, J. Swirski, and T. S. Wheeler, *Chem. Ind. (London)*, **(1956)**, 1018; *Chem. Abstr.*, **51**, 5015 (1957).
91. B. Cummins, D. M. X. Donnelly, E. M. Philbin, J. Swirski, T. S. Wheeler, and R. K. Wilson, *Chem. Ind. (London)*, **(1960)**, 348.
92. D. M. X. Donnelly, J. F. K. Eades, E. M. Philbin, and T. S. Wheeler, *Chem. Ind. (London)*, **(1961)**, 1453.
93. B. Cummins, D. M. X. Donnelly, J. F. Eades, H. Fletcher, F. O'Cinnèide, E. M. Philbin, J. Swirski, T. S. Wheeler, and R. K. Wilson, *Tetrahedron*, **19**, 499 (1963); *Chem. Abstr.*, **59**, 3875 (1963).
94. V. B. Mahesh and T. R. Seshadri, *Proc. Indian Acad. Sci.*, **41B**, 210 (1955).
95. E. J. Corey, E. M. Philbin, and T. S. Wheeler, *Tetrahedron Lett.*, **(1961)**, 429.
96. J. Gripenberg, *Acta Chem. Scand.*, **7**, 1323 (1953).
97. T. Kubota, *J. Chem. Soc. Japan*, **73**, 571 (1952); *Chem. Abstr.*, **48**, 2058 (1954).
98. C. Enebäck and J. Gripenberg, *Acta Chem. Scand.*, **11**, 866 (1957).
99. J. Chopin and M. I. Mouillant, *C. R. Acad. Sci. Paris*, **254**, 3699 (1962).
100. D. M. X. Donnelly, T. P. Lavia, D. P. Melody, and E. M. Philbin, *Chem. Commun.*, **(1965)**, 460.
101. N. Narasimhachari, S, Narayanswami, and T. R. Seshadri, *Proc. Indian Acad. Sci.*, **37A**, 104 (1953); *Chem. Abstr.*, **48**, 7606 (1954).
102. Y. Kimura and M. Hoshi, *Proc. Imp. Acad. Tokyo*, **12**, 285 (1936); *Chem. Abstr.*, **31**, 1807 (1937).
103. F. M. Dean and V. Podimuang, *J. Chem. Soc.*, **(1965)**, 3978.
104. C. H. Hassall and A. T. Scott, "Recent Developments in the Chemistry of Natural Phenolic Products," W. D. Ollis, Ed., Pergamon, London, 1961, p. 119.
105. J. R. Lewis, *Chem. Ind. (London)*, **(1962)**, 159.
106. Y. Kimura, *J. Pharm. Soc. Japan*, **58**, 415 (1938); *Chem. Abstr.*, **32**, 6649 (1938).
107. G. W. K. Cavill, F. M. Dean, A. McGookin, B. M. Marshall, and A. Robertson, *J. Chem. Soc.*, **(1954)**, 4573.
108. Y. Kimura, *J. Pharm. Soc. Japan*, **57**, 147 (1937); *Chem. Abstr.*, **33**, 531 (1939).
109. A. C. Jain, V. K. Rohatgi, and T. R. Seshadri, *Indian J. Chem. Soc.*, **5**(2), 68 (1967); *Chem. Abstr.*, **67**, 82014 (1967).
110. D. M. X. Donnelly, D. F. Melody, and E. M. Philbin, *Tetrahedron Lett.*, **(1967)**, 1023; *Chem. Abstr.*, **67**, 43635 (1967).
111. K. R. Chandorkar, D. M. Phatak, and A. B. Kulkarni, *J. Sci. Ind. Res. (India)*, **21B**, 24 (1962); *Chem. Abstr.*, **57**, 9791 (1962).

112. J. W. Clark-Lewis and R. W. Jemison, *Aust. J. Chem.*, **21,** 815 (1968); *Chem. Abstr.*, **68,** 95457 (1968).
113. J. Chopin, P. Durual, and M. Chadenson, *Compt, Rend.*, **259**(9), 1638 (1964); *Chem. Abstr.*, **61,** 14618 (1964).
114. J. Chopin and R. Durand, *Compt. Rend.*, **256**(24), 5151 (1963); *Chem. Abstr.*, **59,** 8681 (1963).
115. A. Bellino and P. Venturella, *Atti. Accad. Sci. Lett. Arti Palermo,* **21,** 17 (1962); *Chem. Abstr.*, **59,** 552 (1963).
116. A. Chopin, P. Durual, and M. Chadenson, *Bull. Soc. Chim. Fr.*, **(1965)**, 3572; *Chem. Abstr.*, **64,** 9672 (1966).
117. C. Eneback, *Soc. Sci. Fennica, Commebtationes Phys. Math.*, **28**(10), 95 (1963); *Chem. Abstr.*, **60,** 7972 (1964).
118. D. Molho, J. Chopin, and M. Chadenson, *Bull Soc. Chim. Fr.*, **(1959)**, 453; *Chem. Abstr.*, **53,** 20043 (1959).
119. W. D. Ollis, K. L. Ormand, and I. O. Sutherland, *J. Chem. Soc., C.* (**1970**), 119.
120. F. E. King, T. J. King, and A. J. Warwick, *J. Chem. Soc.*, (**1952**), 1920.
121. H. G. C. King and T. White, *Proc. Chem. Soc.*, (**1957**), 341.
122. H. G. C. King and T. White, *J. Soc. Leather Trades' Chemists,* **41,** 368 (1957).
123. K. S. Kirby and T. White, *Biochem. J.*, **60,** 582 (1955); *Chem. Abstr.*, **49,** 15272 (1955).
124. D. Molho, J. Coillard, and C. Mentzner, *Bull. Soc. Chim. Fr.*, (**1954**), 1397; *Chem. Abstr.*, **50**, 325 (1956).
125. G. Lindstedt, *Acta Chem. Scand.*, **4,** 772 (1950); *Chem. Abstr.*, **45,** 2483 (1951).
126. T. Kubota, Y. Naya, and N. Ichikawa, *Nippon Kagaku Zasshi,* **77,** 648 (1956); *Chem. Abstr.*, **52,** 375 (1958).
127. H. G. C. King, T. White, and (in part) R. B. Highes, *J. Chem. Soc.*, (**1961**), 3234.
128. J. Jouanneteau, G. Zwingelstein, and C. Mentzer, *Compt. Rend.*, **239,** 1514 (1954).
129. I. Z. Saiyed, D. R. Nadkani, and T. S. Wheeler, *J. Chem. Soc.*, (**1937**), 1737.
130. J. Chopin, C. Katamna, and J. Jouanneteau, *Compt. Rend.*, **258**(21), 5231 (1964); *Chem. Abstr.*, **61,** 4303 (1964).
131. J. C. Pew, *J. Am. Chem. Soc.*, **70**, 3031 (1948).
132. T. A. Geissman and H. Lischner, *J. Am. Chem. Soc.*, **74,** 3001 (1952).
133. L. H. Hergert, P. Coad, and A. V. Logan, *J. Org. Chem.*, **21**, 304 (1956).
134. C. Enebäck and J. Gripenberg, *J. Org. Chem.*, **22**, 220 (1957).
135. T. Oyamada, *Justus Liebigs Ann. Chem.*, **538,** 44 (1939).
136. M. Kotake and T. Kubota, *Justus Liebigs Ann. Chem.*, **544,** 253 (1940).
137. J. Gripenberg and B. Juselius, *Acta Chem. Scand.*, **8,** 734 (1954); *Chem. Abstr.*, **49,** 10260 (1955).
138. Y. Kimura, *J. Pharm. Soc. Japan,* **57,** 160 (1937),
139. K. J. Balakrishna, N. P. Rao, and T. R. Seshadri, *Proc. Indian Acad. Sci.*, **29A**, 394 (1949).
140. S. K. Grover, N. V. Gupta, A. C. Jain, and T. R. Seshadri, *J. Sci. Ind. Res. (India),* **19B,** 258 (1960); *Chem. Abstr.*, **55,** 3566 (1961).
141. E. H. Charlesworth, J. J. Chavan, and R. Robinson, *J. Chem. Soc.*, (**1933**), 372.
142. Th. Zincke and K. Eismayer, *Chem. Ber.*, **51,** 751 (1918).
143. J. J. Drumm, M. M. MacMahon, and H. Ryan, *Proc. Irish. Acad.*, **36B,** 149 (1924).
144. K. Freudenberg, H. Fikentscher, M. Harder, and A. Schmidt, *Justus Liebigs Ann. Chem.*, **444,** 135 (1925).
145. M. Suzuki, *Nippon Kagaku Zasshi,* **82,** 883 (1961); *Chem. Abstr.*, **57,** 11148 (1962).
146. A. Sonn, *Chem. Ber.*, **50**, 1262 (1917).
147. D. M. Fitzgerald, J. F. O'Sullivan, E. M. Philbin, and T. S. Wheeler, *J. Chem. Soc.*, (**1955**), 860.
148. R. L. Shriner and R. E. Damschroeder, *J. Am. Chem. Soc.*, **60,** 894 (1938).
149. J.B.D.MacKenzie,A.Robertson,A.Bushra, and R.Towers, *J.Chem.Soc.*,(**1949**), 2057.

150. G. H. Jones, J. B. D. MacKenzie, A. Robertson, and W. B. Whalley, *J. Chem. Soc.*, (**1949**), 562.
151. H. Obara and J-I. Onodera, *Bull. Chem. Soc. Japan,* **41,** 2800 (1968).
152. T. A. Geissman and A. Armen, *J. Am. Chem. Soc.,* **77,** 1623 (1955).
153. J. R. Merchant and D. V. Rege, *Tetrahedron Lett.,* (**1969**), 3589.
154. E. M. Philbin, W. I. A. O'Sullivan, and T. S. Wheeler, *J. Chem. Soc.,* (**1954**), 4174.
155. S. D. Limaye, H. K. Pendse, K. R. Chandorkar, and G. V. Bhide, *Rasäyanam,* **2,** 97 (1956); *Chem. Abstr.,* **51,** 5064 (1957).
156. H. K. Pendse and S. D. Limaye, *Rasäyanam,* **2,** 107 (1956); *Chem. Abstr.,* **51,** 5061 (1957).
157. H. K. Pendse and K. S. Moghe, *Rasayanam,* **2,** 114 (1956); *Chem. Abstr.,* **51,** 5062 (1957).
158. H. K. Pendse and N. D. Patwardhan, *Rasäyanam,* **2,** 117 (1956); *Chem. Abstr.,* **51,** 5062 (1957).
159. H. K. Pendse, *Rasäyanam,* **2,** 121 (1956); *Chem. Abstr.,* **51,** 5062 (1957).
160. W. E. Fitzmaurice, W. I. O'Sullivan, E. M. Philbin, T. S. Wheeler, and T. A. Geissman. *Chem. Ind. (London),* (**1955**), 652.
161. R. Bryant and D. L. Haslam, *J. Chem. Soc.,* (**1965**), 2361.
162. R. Bryant, *J. Chem. Soc.,* (**1965**), 5140.
163. S. K. Grover, A. C. Jain, S. K. Methur, and T. R. Seshadri, *Indian J. Chem.,* **1**(9), 382 (1963); *Chem. Abstr.,* **60,** 4094 (1964).
164. J. Arima and T. Okamoto, *J. Chem. Soc. Japan,* **50,** 344 (1929); *Chem. Abstr.,* **26,** 139 (1932).
165. G. Caporale and A. M. Bareggi, *Gazz. Chim. Ital.,* **98,** 444 (1968); *Chem. Abstr.,* **70,** 57701 (1969).
166. D. K. Chatterjea and K. Sen, *Indian J. Chem.,* **9,** 31 (1971).
167. W. K. Slater and H. Stephen, *J. Chem. Soc.,* (**1920**), 309.
168. J. Murai, *Sci. Repts. Saitama Univ.,* **1A,** 129 (1954); *Chem. Abstr.,* **50,** 981 (1956).
169. C. Katamna, *Bull. Soc. Chim. Fr.,* (**1970**), 2309.
170. J. Chopin and C. Katamna, Fr. Pat. 1,502,727 (1967); *Chem. Abstr.,* **69,** 96447 (1968).
171. A. Kogure and T. Kubota, *J. Inst. Polytech. Osaka City Univ., Ser. C,* **2,** 70 (1952); *Chem. Abstr.,* **47,** 10526 (1953).
172. J. W. Daly, J. P. Benigni, R. L. Minnis, Y. Yanaoka, and B. Witkop, *Biochemistry,* **4**(11), 2513 (1965).
173. D. M. X. Donnelly, R. M. Lawless, and R. K. W. Wilson, *Chem. Ind. (London),* (**1961**), 1906; *Chem. Abstr.,* **57,** 3343 (1962).
174. W. J. Horton and E. G. Paul, *J. Org. Chem.,* **24,** 2000 (1959).
175. K. J. Balakrishna, T. R. Seshadri, and G. Viswanath, *Proc. Indian Acad. Sci.,* **33A,** 233 (1951); *Chem. Abstr.,* **46,** 6637 (1952).
176. G. Schenck, H. Huke, and K. Goerlitzer, *Tetrahedron Lett.,* (**1968**), 2375; *Chem. Abstr.,* **69,** 59014 (1968).
177. T. A. Geissman and E. Hinreiner, *J. Am. Chem. Soc.,* **73,** 782 (1951).
178. A. Schönberg and A. Mustafa, *J. Chem. Soc.,* (**1946**), 746.
179. V. C. Farmer, N. F. Hayes, and R. H. Thomson, *J. Chem. Soc.,* (**1956**), 3600.
180. T. P. C. Mulholland and G. Ward, *J. Chem. Soc.,* (**1953**), 1642.
181. A. C. Jain, V. K. Rohatige, and T. R. Seshadri, *Tetrahedron,* **23,** 2499 (1967).
182. W. Logemann, G. Cavagne, and G. Tosolini, *Chem. Ber.,* **96,** 1680 (1963).
183. R. L. Shriner and J. Anderson, *J. Am. Chem. Soc.,* **60,** 1418 (1938).
184. W. I. O'Sullivan and C. R. Hauser, *J. Org. Chem.,* **25,** 839 (1960).
185. R. Bryant and D. L. Haslam, *J. Chem. Soc., C,* (**1967**), 1345.
186. E. Knoevenagel, *Chem. Ber.,* **31,** 2585 (1898).
187. R. L. Shriner and M. Witte, *J. Am. Chem. Soc.,* **63,** 1108 (1941).
188. J. Murai, *Sci. Repts. Saitama Univ.,* **2A,** 59 (1955); *Chem. Abstr.,* **50,** 12014 (1956).

189. S. Kumar, L. Ram, and J. N. Ray, *J. Indian Chem. Soc.*, **23,** 365 (1946); *Chem. Abstr.*, **41,** 6556 (1947).
190. H. G. C. King and T. White, *J. Chem. Soc.*, (**1961**), 3539.
191. A. Baeyer and P. Fritsch, *Chem. Ber.*, **17,** 973 (1884).
192. T. Kariyone and S. Imai, *J. Pharm. Soc. Japan,* **55,** 679 (1935).
193. E. Bernatek and E. Berner, *Acta. Chem. Scand.*, **3,** 1117 (1949); *Chem. Abstr.*, **44,** 4897 (1950).
194. E. C. Horning and R. U. Schock, *J. Am. Chem. Soc.*, **70,** 2441 (1948).
195. O. Kromatka, *Chem. Ber.*, **65,** 123 (1932).
196. O. Aubert, A. Augdahl, and E. Berner, *Acta. Scand.*, **6,** 433 (1952); *Chem. Abstr.*, **47,** 3846 (1953).
197. A. Kamal, A. Robertson, and T. Tittensor, *J. Chem. Soc.*, (**1950**), 3375.
198. J. H. Burckhalter and J. R. Campbell, *J. Org. Chem.*, **26,** 4232 (1961).
199. A. Bistrzycki and J. Flatau, *Chem. Ber.*, **28,** 989 (1895).
200. A. Bistrzycki and F. v. Weber, *Chem. Ber.*, **43,** 2496 (1910).
201. H. v. Liebig, *Chem. Ber.*, **41,** 1644 (1908).
202. A. Bistrzycki and H. Simonis, *Chem. Ber.*, **31,** 2812 (1898).
203. B. I. Arventi, *Ann. Sci. Univ. Jassy,* **24,** 72 (1938); *Chem. Abstr.*, **33,** 1693 (1939).
204. B. I. Arventi, *Ann. Sci. Univ. Jassy,* **24,** 103 (1938); *Chem. Abstr.*, **33,** 1694 (1939).
205. A. W. Weston and W. B. Brownell, *J. Am. Chem. Soc.*, **74,** 653 (1952).
206. F. Wessely and E. Schinzel, *Monatsh. Chem.*, **84,** 425 (1953); *Chem. Abstr.*, **48,** 12021 (1954).
207. D. Molho and J. Coillard, *Bull. Soc. Chim. Fr.*, (**1956**), 78; *Chem. Abstr.*, **50,** 15482 (1956).
208. W. Borsche, C. Walter, and J. Niemann, *Chem. Ber.*, **62,** 1360 (1929).
209. J. T. Marsh and H. Stephen, *J. Chem. Soc.*, **127,** 1633 (1925).
210. G. J. Durant, G. M. Smith, and R. G. W. Spickett, Brit. Pat. 1,097,596 (1968); *Chem. Abstr.*, **69,** 27232 (1968).
211. D. Farge, M. N. Messer, and C. Moutonnier, Ger. Pat. 1,963,824 (1970); *Chem. Abstr.*, **73,** 77035 (1970).
212. H. H. Wasserman, T. C. Liu, and E. R. Wasserman, *J. Am. Chem. Soc.*, **75**, 2056 (1953).
213. S. Yukawa, *J. Pharm. Soc. Japan,* **48,** 816 (1928); *Chem. Abstr.*, **23**, 832 (1929).
214. D. L. Turner, *J. Am. Chem. Soc.*, **73**, 1284 (1951).
215. L. I. Smith and C. W. MacMullen, *J. Am. Chem. Soc.*, **58,** 629 (1936).
216. L. I. Smith and W. W. Pichard, *J. Org. Chem.*, **4,** 342 (1939).
217. L. I. Smith and J. Nichols, *J. Am. Chem. Soc.*, **65,** 1739 (1943).
218. L. I. Smith and P. F. Wiley, *J. Am. Chem. Soc.*, **68,** 894 (1946).
219. L. I. Smith and E. W. Kaiser, *J. Am. Chem. Soc.*, **62,** 133 (1940).
220. L. I. Smith and J. A. King, *J. Am. Chem. Soc.*, **65,** 441 (1943).
221. L. I. Smith and G. A. Boyack, *J. Am. Chem. Soc.*, **70,** 2690 (1948).
222. L. I. Smith and W. J. Dale, *J. Org. Chem.*, **15,** 832 (1950).
223. L. I. Smith and R. B. Carlin, *J. Am. Chem. Soc.*, **64,** 524 (1942).
224. F. Bergel, A. Jacob, A. R. Todd, and T. S. Work, *J. Chem. Soc.*, (**1938**), 1375.
225. M. Murakami and S. Senoh, *Mem. Inst., Sci. Ind. Res. Osaka Univ.*, **12,** 167 (1955); *Chem. Abstr.*, **50,** 15521 (1956).
226. E. Cerutti, *Ann. Sci. Univ. Besancon Chim.*, **31** (1), 79 (1963); *Chem. Abstr.*, **61'** 14651 (1964).
227. S. M. McElvin and E. L. Engelhaedt, *J. Am. Chem. Soc.*, **66,** 1077 (1944).
228. S. M. McElvin and H. Cohen, *J. Am. Chem. Soc.*, **64,** 260 (1942).
229. L. Canonica and A. Fiecchi, *Atti. Accad. Nazl. Lincei Rend. Classe Sci. Fics. Nat. Nat.*, **17,** 385 (1954); *Chem. Abstr.*, **50,** 249 (1956).
230. P. A. Berke and W. E. Rosen, U.S. Pat. 3,471,537 (1969); *Chem. Abstr.*, **72,** 12386 (1970).
231. W. Feuerstein and A. Brass, *Chem. Ber.*, **37,** 821 (1904).

232. W. Baker and R. Banks, *J. Chem. Soc.*, (**1939**), 279.
233. J. N. Chatterjea, *J. Indian Chem. Soc.*, **33**, 175 (1956).
234. J. N. Chatterjea, *J. Indian Chem. Soc.*, **34**, 299 (1957).
235. T. R. Seshadri, *Sci. Proc. Roy Dublin Soc.*, **27**, 77 (1956).
236. D. Molho, *Compt. Rend.*, **248**, 1535 (1959); *Chem. Abstr.*, **53**, 17994 (1959).
237. J. Chopin, M. Chadenson, and P. Durual, *Compt. Rend.*, (**1964**), 6178; *Chem. Abstr.*, **61**, 6983 (1964).
238. J. M. Bruce and D. Creed, *J. Chem. Soc., C*, (**1970**), 649.
239. J. M. Bruce and P. Knowles, *J. Chem. Soc., C*, (**1966**), 1627.
240. O. Yonemitsu, Y. Okuno, Y. Kanacka, and B. Witkop, *J. Am. Chem. Soc.*, **92**, 5686 (1970).
241. E. Berner, *J. Chem. Soc.*, (**1946**), 1052.
242. K. Landenberg, K. Folkers, and R. T. Major, *J. Am. Chem. Soc.*, **58**, 1292 (1936).
243. J. A. Marshall and N. Cohen, *Tetrahedron Lett.*, (**1964**), 1997.
244. J. A. Marshall, N. Cohen, and K. R. Aeznson, *J. Org. Chem.*, **30**, 762 (1965).
245. J. A. Marshall and N. Cohen, *J. Am. Chem. Soc.*, **87**, 2773 (1965).
246. J. A. Marshall and A. R. Hochsteller, *Tetrahedron Lett.*, (**1966**), 55.
247. O. Wallach, *Justus Liebigs Ann. Chem.*, **353**, 284 (1907).
248. S. Coffer, *Rec. Trav. Chim. Pays-Bas*, **42**, 387 (1923).
249. J. v. Braun and W. Munch, *Justus Liebigs Ann. Chem.*, **465**, 52 (1928).
250. R. Ghosh, *J. Indian Chem. Soc.*, **12**, 601 (1935).
251. T. Haga, *Nippon Kagaku Zasshi*, **84**, 193 (1963); *Chem. Abstr.*, **60**, 5349 (1964).
252. D. P. Stokes and G. A. Taylor, *J. Chem. Soc., C*, (**1971**), 2334.
253. A. Löwenbein, H. Simonis, H. Lang, and W. Jacobus, *Chem. Ber.*, **57**, 2040 (1924).
254. A. Löwenbein and H. Simonis, *Chem. Ber.*, **60**, 1851 (1927).
255. H. E. Zaugg, D. A. Dunnigan, R. J. Michaels, Jr., R. J. Swett, T. S. Wang, A. H. Sommers, and R. W. DeNet, *J. Org. Chem.*, **26**, 644 (1961).
256. A. W. Weston and M. A. Spielman, U.S. Pat. 2,513,698 (1950); *Chem. Abstr.*, **44**, 8956 (1950).
257. H. E. Zaugg, R. W. DeNet, and R. J. Michaels, Belg. Pat. 618,528 (1962); *Chem. Abstr.*, **59**, 6372 (1963).
258. H. E. Zaugg, R. W. DeNet, and R. J. Michaels, *J. Org. Chem.*, **26**, 4821 (1961).
259. H. E. Zaugg and R. W. DeNet, *J. Org. Chem.*, **35**, 3567 (1970).
260. H. E. Zaugg, R. W. DeNet, and R. J. Michaels, *J. Org. Chem.*, **28**, 1795 (1963).
261. H. E. Zaugg, R. W. DeNet, and R. J. Michaels, *J. Org. Chem.*, **26**, 4828 (1961).
262. M. L. Bender, *Chem. Rev.*, **60**, 53 (1960).
263. A Spetz, *Acta Chem. Scand.*, **10**, 1422 (1956).
264. B. I. Arventi, *Ann. Sci. Univ. Jassy, Sect. 1*, **27**, 461 (1941); *Chem. Abstr.*, **38**, 2338 (1944).
265. A. Schönberg and A. Mustafa, *J. Chem. Soc.*, (**1945**), 657.
266. A. Schönberg and A. Mustafa, *Chem. Rev.*, **40**, 181 (1947).
267. S. Goldschmidt and W. Schmidt, *Chem. Ber.*, **55**, 3202 (1922).
268. A. Löwenbein, H. Simonis, H. Lang, and W. Jacobus, *Chem. Ber.*, **57**, 2047 (1924).
269. A. Mustafa, "Advances in Photochemistry," W. A. Noyes, Jr., G. S. Hammond, and J. N. Pitts, Jr., Eds., Vol. 2, Interscience, New York, 1964, p. 63.
270. A. Mustafa, A. H. E. Harhash, A. K. E. Mansour, and M. A. E. Omran, *J. Am. Chem. Soc.*, **78**, 4306 (1956).
271. A. Schönberg and A. Mustafa, *J. Chem. Soc.*, (**1949**), 889.
272. A. Mustafa and A. M. Islam, *J. Chem. Soc.*, (**1951**), 1616.
273. A. Mustafa, *J. Chem. Soc.*, (**1949**), 1662.
274. O. L. Chapman and C. L. Mclntosh, *J. Chem. Soc., D*, **(1971)**, 383.
275. R. C. Elderfield and V. B. Meyer, "Heterocyclic Compounds," R. C. Elderfield, Ed., Wiley, New York, 1950, p. 39.

276. J. N. Chatterjea, *J. Indian Chem. Soc.*, **36**, 69 (1959).
277. J. Thiele, R. Tischbein, and E. Lossow, *Justus Liebigs Ann. Chem.*, **319**, 180 (1910).
278. W. F. v. Oettingen, *J. Am. Chem. Soc.*, **52**, 2024 (1930).
279. C. Erlenmeyer and H. Zimmer, *J. Heterocycl. Chem.*, **1**, 205 (1964).
280. S. Czaplicki, St. v. Kostanecki, and V. Lampe, *Chem. Ber.*, **42**, 827 (1909).
281. W. Mosimann and J. Tambor, *Chem. Ber.*, **49**, 1261 (1916).
282. E. Baltazzi and E. A. Davis, *Chem. Ind. (London)*, (**1962**), 1653.
283. R. Walter, H. Zimmer, and T. C. Purcell, *J. Org. Chem.*, **31**, 3854 (1966).
284. L. E. Smith and R. N. Hurd, *J. Org. Chem.*, **22**, 588 (1957).
285. P. Tobias, J. H. Heidema, K. W. Lo, E. T. Kaiser, and F. J. Kèzdy, *J. Am. Chem. Soc.*, **91**, 202 (1969).
286. A. Bistrzycki and J. Flatau, *Chem. Ber.*, **30**, 124 (1897).
287. K. Fries and J. Kohlhass, *Justus Liebigs Ann. Chem.*, **389**, 284 (1912).
288. R. Gompper, H. U. Wagner, and E. Kutter, *Chem. Ber.*, **101**, 4123 (1968).
289. W. H. Washburn, *Appl. Spectroscop.*, **18**(2), 61 (1964); *Chem. Abstr.*, **61**, 180 (1964).
290. L. J. Bellamy, "The Infrared Spectra of Complex Molecules," 2nd. ed., Methuen, London, 1958, pp. 182, 186.
291. H. E. Zaugg, R. W. DeNet, R. J. Michaels, W. H. Washburn, and F. E. Chadde, *J. Org. Chem.*, **26**, 4753 (1961).
292. P. Karrer and J. Ferla, *Helv. Chim. Acta*, **4**, 203 (1921).
293. R. Stollè and E. Knebel, *Chem. Ber.*, **54**, 1213 (1921).
294. R. v. Stoermer, *Chem. Ber.*, **45**, 163 (1912).
295. R. v. Stoermer and B. Khalbert, *Chem. Ber.*, **35**, 1640 (1902).
296. J. N. Chatterjea, S. N. Gupta, and V. N. Mehrotra, *J. Indian Chem. Soc.*, **42**, 205 (1965); *Chem. Abstr.*, **63**, 9899 (1965).
297. J. N. Chatterjea, *Experientia*, **9**, 256 (1953); *Chem. Abstr.*, **48**, 11390 (1954).
298. K. Fries, *Justus Liebigs Ann. Chem.*, **442**, 254 (1925).
299. K. Fries and A. Hasselbach, *Chem. Ber.*, **44**, 124 (1911).
300. H. Staudinger, *Chem. Ber.*, **41**, 3558 (1908).
301. R. Bergdoll, M. Luther, A. Auerhahn, and W. Wacker, *J. Prakt. Chem.*, **128**, 1 (1930).
302. L. I. Smith and R. R. Holmes, *J. Am. Chem. Soc.*, **73**, 4294 (1951).
303. R. D. Sprenger, P. M. Ruoff, and A. H. Frazer, *J. Am. Chem. Soc.*, **72**, 2874 (1950).
304. T. Matsuura, M. Kawai, and K. Butsugan, *Bull. Soc. Chem. Japan*, **43**, 3891 (1970).
305. T. Matsuura, M. Kawai, R. Nakashima, and Y. Butsugan, *Tetrahedron Lett.*, (**1969**), 1083.
306. A. Kötz, K. Blendermann, and J. Meyer, *Chem. Ber.*, **45**, 3702 (1912).
307. A. Kötz and J. Meyer, *J. Prakt. Chem.*, **88**, 266 (1913).
308. K. Fries, *Chem. Ber.*, **42**, 234 (1909).
309. G. Caronna and S. Palazzo, *Gazz. Chim. Ital.*, **98**, 911 (1968); *Chem. Abstr.*, **70**, 37585 (1969).
310. D. Borrmann and R. Wegler, *Chem. Ber.*, **102**, 64 (1969).
311. W. M. Horspool and G. D. Kandelwal, *Chem. Commun.*, (**1970**), 257.
312. I. H. Hunsberger and E. D. Amstutz, *J. Am. Chem. Soc.*, **70**, 671 (1948).
313. D. J. Cram, *J. Am. Chem. Soc.*, **72**, 1028 (1950).
314. A. Schönberg and K. Junghans, *Chem. Ber.*, **97**, 2539 (1964).
315. A. Schönberg and K. Junghans, *Chem. Ber.*, **99**, 531 (1966).
316. R. Stollè and H. Stamm, *J. Prakt. Chem.*, **114**, 242 (1926).
317. J. N. Chatterjea, *J. Indian Chem. Soc.*, **31**, 101 (1954).
318. J. N. Chatterjea, *J. Indian Chem. Soc.*, **32**, 265 (1955).
319. M. F. Sartori, *J. Org. Chem.*, **26**, 3152 (1961).
320. M. F. Sartori, U. S. Pat. 2,995,578 (1961); *Chem. Abstr.*, **56**, 11757 (1962).
321. V. Titoff, H. Mümmer, and T. Reichstein, *Helv. Chim. Acta*, **20**, 883 (1937).

CHAPTER VI

Naturally Occurring Benzofurans

1. Benzofurans

The benzofuran nucleus is a common one in natural products and appears in many forms. Few such products are simple in the structural sense, and many benzofuran derivatives cannot be discussed here. No doubt lignin is largely composed of benzofuran residues but, again, this difficult and largely specialized matter exceeds the scope of this book. The same is true with sesquiterpene acids which are largely composed of hydrobenzofuran residues; the simple ones are discussed only briefly.

A. 5-Methoxybenzofuran

5-Methoxybenzofuran (**1**) is the simplest of the naturally occurring benzofurans. It was discovered as a result of fungal contamination (*Stereum subpileatum* Berk. et Curt.) of oak beer barrels, which led to the beer having a strong, persistent, distasteful scent. The scent is characteristic of the benzofuran, though there is a contribution from cinnamaldehyde, which is also produced by the fungus. The compound has been synthesized by Tanaka's method.[1]

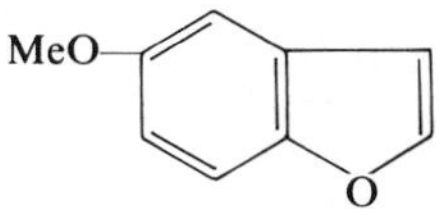

MeO

O

1

B. Furoventalene

Furoventalene (**2**), a new liquid C_{15}-benzofuran possessing an isoprenoid but nonfarensyl skeleton, has been isolated from the steam volatile of the seafan, *Gorgonia ventalina*.[2] The nature of the substituents in the benzofuran skeleton could be deduced from its nuclear magnetic resonance (NMR) spectra. Synthesis of **2** has been achieved.[3] Direct coupling of monoallylic bromide or tosylate (**3**) with the organometallic derivative of 6-bromo-3-methylbenzofuran (**4**), obtained by cyclodehydration of *m*-bromophenoxyacetone with polyphosphoric acid (PPA), failed to give quantities of **2**. The Grignard reagent **4** was condensed with levulinaldehyde ethylene ketal, followed by hydrogenolysis of the resulting benzyl alcohol (**5**), which occurred with concurrent ketal hydrolysis, forming the ketone **6**. Condensation of this ketone with methylmagnesium bromide and subsequent dehydration of the tertiary alcohol (**7**) led to a mixture from which the desired 3-methyl-6-(4′-methylpent-3′-enyl)benzofuran (**2**) was readily isolated (Eq. 1).

Br, Me, O; Mg →; BrMg, Me, O, **4**; levulinaldehyde, ethylene ketal →

O, O, OH, Me, O, **5**; Pd/C, H_2 →

O, Me, O, **6**; MeMgI →

Me, OH, Me, O, **7**; $POCl_3$, C_5H_5N →

$Me_2C{=}CH$, H_2, C, C, H_2, Me, O, **2**

Me, Me, C=CH, H_2, C, C, X, H_2 (1)

3 X = Br or Tos

C. Euparin

Euparin, $C_{13}H_{12}O_3$, **8,** the yellow constituent from the roots of *Eupatorium purpureum* (gravel root),[5-7] has been shown to be a phenolic ketone; it forms derivatives with carbonyl reagents, a methyl ether, and bluish-green ferric reaction occurs.[4] Hydrogenation of euparin established the presence of two ethylenic linkages, yielding tetrahydroeuparin which, like euparin, formed the carbonyl and phenolic derivatives. The presence of a *C*-acetyl group, which is in the ortho position to the hydroxyl group is shown by the fact that *O*-methyltetrahydroeuparin oxime is transformed by Beckmann's procedure to an acetanilide (**9**), giving on hydrolysis the corresponding amine, which on acetylation reformed the anilide. On oxidation with potassium permanganate, the methyl ether yielded 2-hydroxy-4-methoxy-5-acetylbenzoic acid, and on hydrolysis, 2-hydroxy--4-methoxy-5-acetylbenzaldehyde (**10**) is obtained.

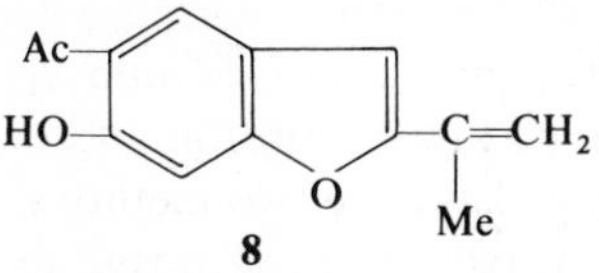

Ac HO O C=CH$_2$ Me **8** AcNH HO O CHMe$_2$ **9** Ac HO CHO OH **10**

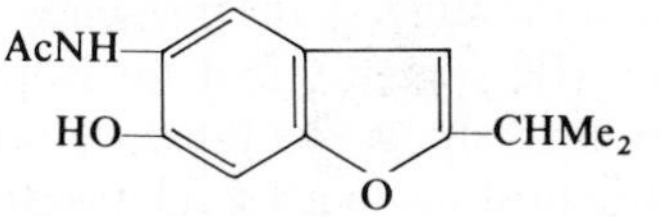

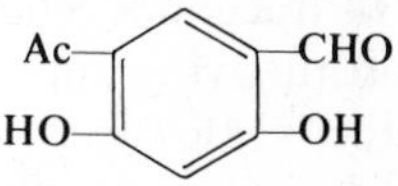

Synthesis of euparin (5-acetyl-6-hydroxy-2-isopropenylbenzofuran) has been achieved (Eq. 2).[8,9]

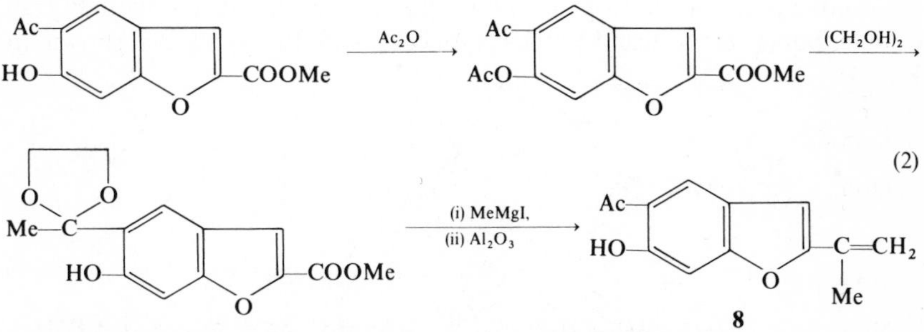

Ac HO O COOMe $\xrightarrow{Ac_2O}$ Ac AcO O COOMe $\xrightarrow{(CH_2OH)_2}$

O O Me—C HO O COOMe $\xrightarrow[\text{(ii) } Al_2O_3]{\text{(i) MeMgI,}}$ Ac HO O C=CH$_2$ Me **8** (2)

Elix[10] has recently reported the synthesis of euparin via the selective attack of $CH_2{=}PPh_3$ on the appropriate 2,5-diacetylbenzofuran.

Further evidence of euparin structure (**8**) is the observation that the interaction of resorcinol with α-chloroisovaleryl chloride gives a ketone from which internal etherification produces the benzofuranone **11.** Conversion to the oxime derivative followed by reduction does not result in the formation of the corresponding amine because ammonia is lost just as the corresponding alcohol would split out water, yielding the benzofuran **12.**

Reduction of the latter to the dihydrobenzofuran, followed by Hoesch reaction with acetonitrile, afforded the formation of tetrahydroeuparin.

11 12

D. 5,6-Dimethoxy-2-isopropenylbenzofuran

Benzene extract of the roots of *Ligularia stenocephala* Matsum et Koidz. (Compositae) afforded 5,6-dimethoxy-2-isopropenylbenzofuran (**13**), $C_{13}H_{14}O_3$.[11] It is unstable in air and acids, and develops a blue color during TLC when 2*N* ceric sulfate is sprayed. On hydrogenation over 5% Pd/C, **13** absorbed two moles of hydrogen to give the tetrahydro derivative **14** (racemic). These data suggest the presence of a benzofuran ring, two methoxyl groups (NMR spectra), and an isopropenyl group. Moreover, the NMR spectrum of **13** shows the presence of one furan β-proton and of two isolated aromatic protons in a 1,4 relationship. Thus the two methoxyl groups and the isopropenyl group must be attached to either positions 2, 5, or 6 of the benzofuran ring. Proton-decoupling experiments afford evidence which leads to the location of the isopropyl group in **14**. Condensation of **13** with maleic anhydride furnished an adduct, which indicates that the double bond of the isopropenyl group in **13** is conjugated with a furan double bond. Structure **13** finds support from the synthesis shown in Eq. 3.

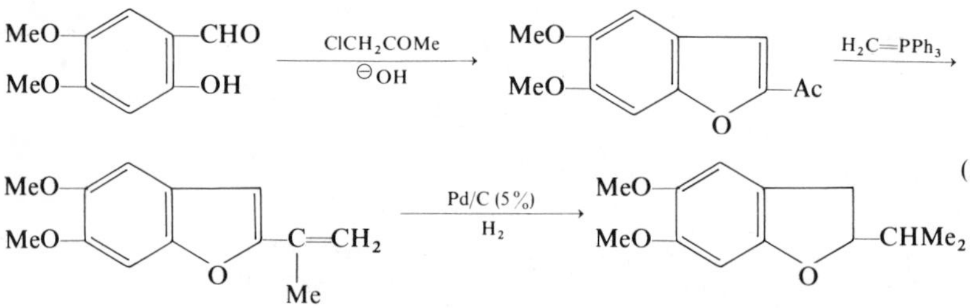

(3)

13 14

E. Ageratone and Dihydroageratone

Ageratone (2-acetoxyisopropenyl-5-acetyl-6-hydroxybenzofuran, **15**) and dihydroageratone (2-acetoxyisopropenyl-5-acetyl-6-hydroxy-2,3-dihydrobenzofuran, **16**) were isolated from the roots of *Ageratum houstonianum* Mill.[12] The genera *Ageratum* and *Eupatorium*, which are closely related botanically, have also been shown to be chemically related.[13–15] Their structural relationship (**15** and **16**) to euparin (**8**) and hydroxytremetone (**17**) further confirms the chemical relationship between the two genera.

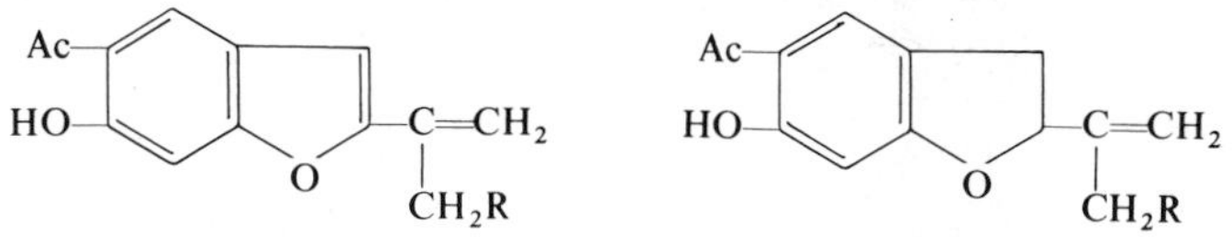

15 R = OAc
8 R = H

16 R = OAc
17 R = H

The infrared spectrum indicates the presence of an acetate ester ($\nu = 1735\ cm^{-1}$) and a hydrogen-bonded aromatic carbonyl group ($\nu = 1632\ cm^{-1}$) which is confirmed by the NMR spectrum. Moreover, the NMR spectrum reveals the presence of two aromatic protons, two well-separated methylene protons, and the furanoid proton. These data are in good agreement with the NMR spectrum of **15.** The structure of dihydroageratone (**16**) also finds support from the spectral data.

F. Pongamol

Pongamol, $C_{18}H_{14}O_4$, is a yellow diketone, isolated from the seed oil (karanjia oil) of *Pongamia glabra*, and closely related to the furoflavone (**22**). It contains one methoxyl group, though it gives a ferric test which is indicative of a phenolic or enolic hydroxyl group. It does not undergo a methylation or benzylation reaction under ordinary conditions. Oxidation with potassium permanganate led to the formation of benzoic acid.[16] Fission of pongamol with ethanolic potassium hydroxide gave benzoic acid and a mixture of karanjic acid (**21**) and *O*-methylkaranjic acid (**20**). However methylation gave pure **20.** The neutral fraction consisted of a mixture of acetophenone and the ketone 5-acetyl-4-methoxybenzofuran (**19**). These results lead to the conclusion that pongamol is benzoyl-*O*-methylkaranjoylmethane (**18**).[17]

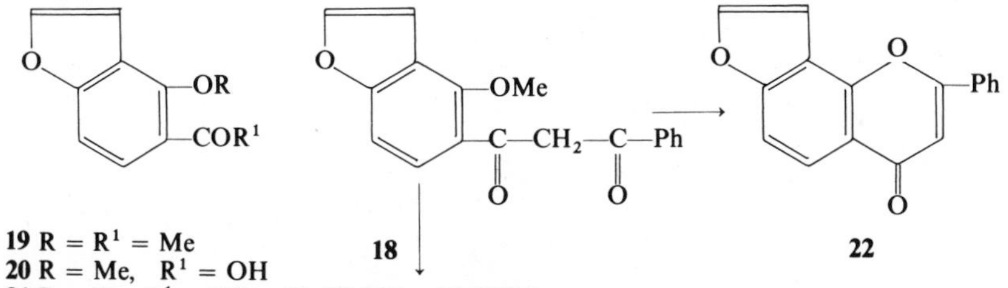

19 R = R[1] = Me
20 R = Me, R[1] = OH
21 R = H, R[1] = OH

18 → PhCOOH + PhCOMe

22

Pongamol (**18**) has been synthesized by condensation of methyl 4-methoxybenzofuran-5-carboxylate with acetophenone, followed by its isolation as the copper complex, which is easily decomposed to give pongamol.[18] Pongamol is identified with lanceolatin C isolated from *Tephrosia lanceolata*.

G. Tremetone, Dehydrotremetone, and Hydroxytremetone

Tremetone (**23**), the principal levorotatory constituent of "tremetol," the crude toxin of white snakeroot (*Eupatorium urticaefolium*), is responsible for trembles in cattle and milk sickness in humans.[19-21] Separation of tremetol into a sterol fraction and a ketone fraction and further separation of some of these fractions into several pure components has been accomplished.[22] The compounds, isolated from the ketone fraction, are toxic to goldfish and have been investigated from a structural viewpoint by degradative techniques with the findings that the principal toxic ketone is tremetone (**23**); the two other minor ketones are dehydrotremetone (**24**) and hydroxytremetone (**17**).[22-25] The related compound toxol, **26**, was isolated by Zalkow and Burke[26] from the rayless goldenrod, *Aplopappus heterophyllus*.

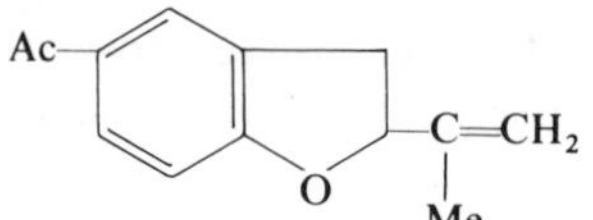

Ac, O, $C{=}CH_2$, Me
23

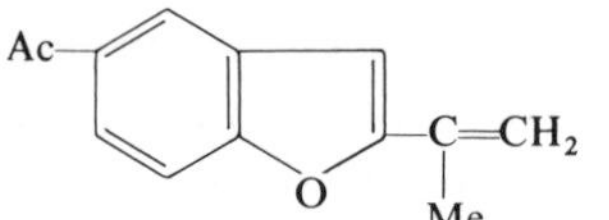

Ac, O, $C{=}CH_2$, Me
24

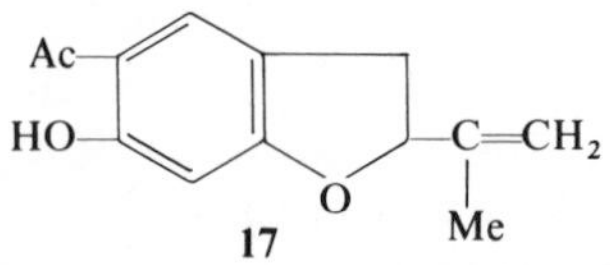

Ac, HO, O, $C{=}CH_2$, Me
17

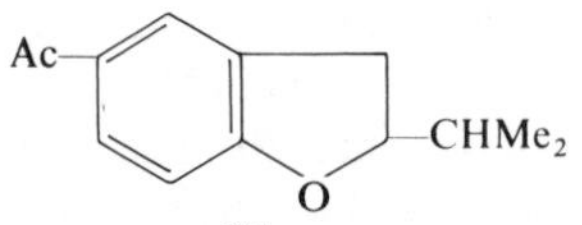

Ac, O, $CHMe_2$
25

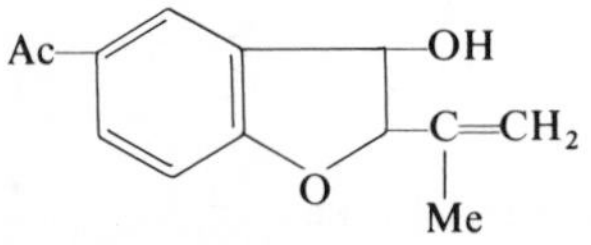

Ac, OH, O, $C{=}CH_2$, Me
26

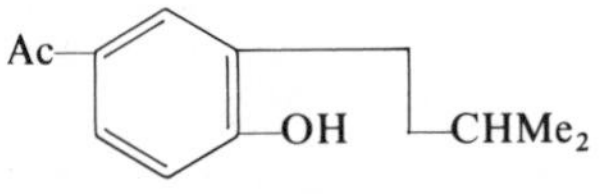

Ac, OH, $CHMe_2$
27

Synthesis of racemic dihydrotremetone (**25**) has been achieved,[25] as outlined in Eq. 4.

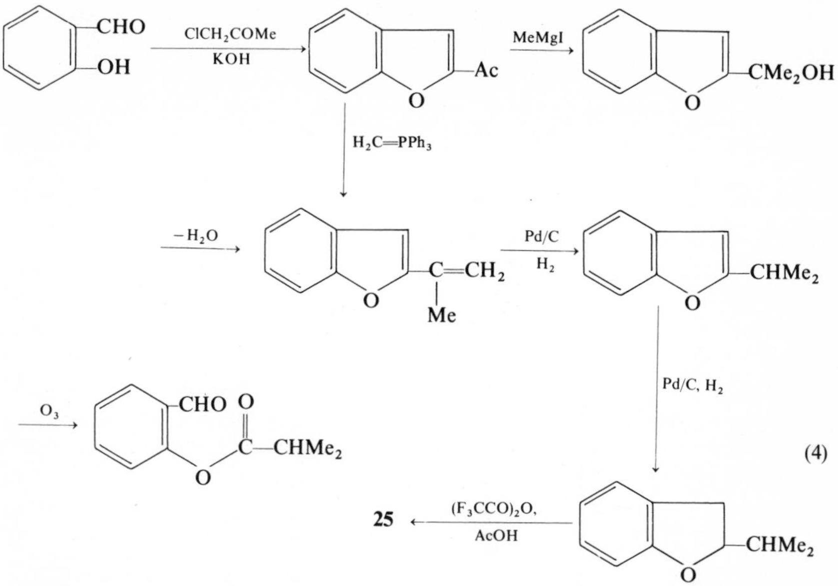

(4)

Hydrogenation of tremetone (**23**) yielded a simple reduction product, dihydrotremetone (**25**), as well as the phenolic ketone hydrogenolysis product, 2-isoamyl-4-acetylphenol (**27**).

Tremetone, the most abundant ketone, has been studied structurally by investigation of its degradation products (Chart 1).[22]

Synthesis of racemic tremetone has been achieved (Eq. 5).[27,28]

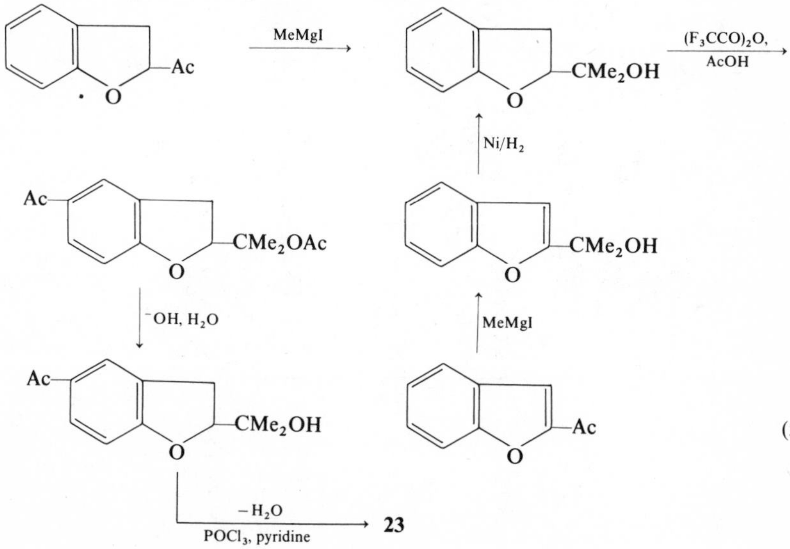

(5)

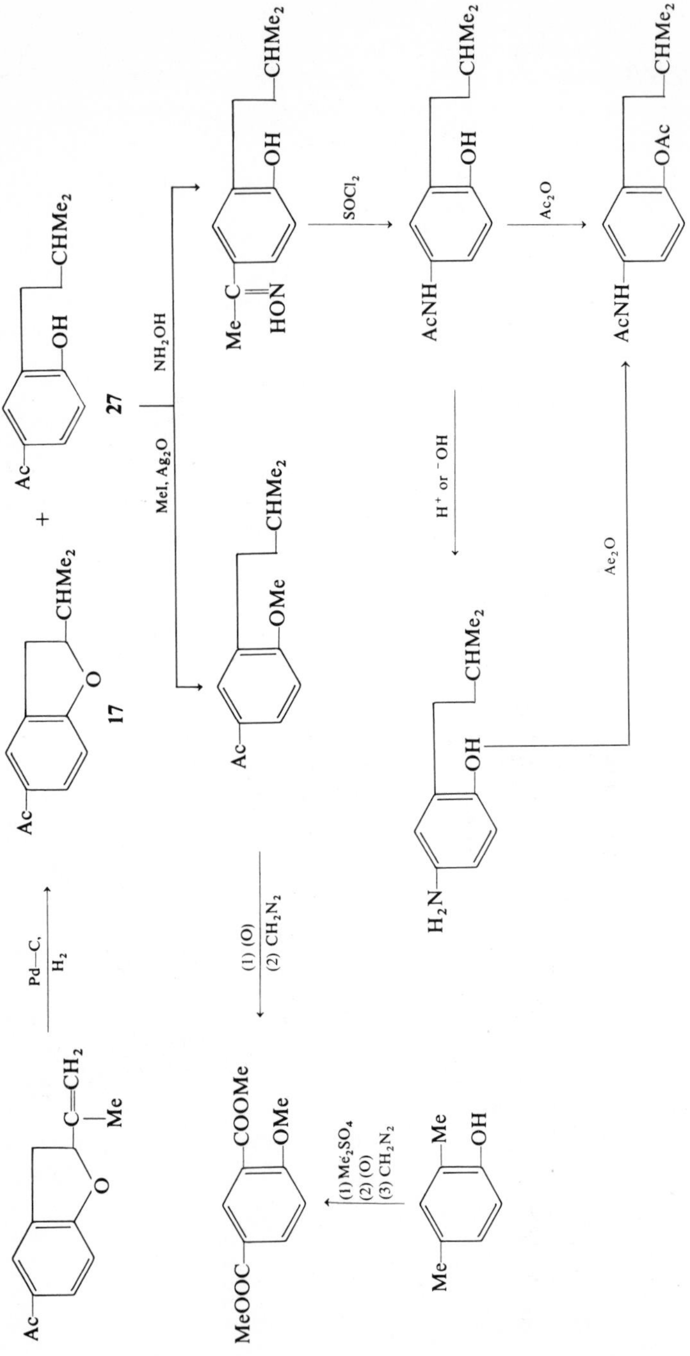

Pd—C, H_2

17

27

NH_2OH

MeI, Ag_2O

$SOCl_2$

Ac_2O

H^+ or ^-OH

Ac_2O

(1) (O)
(2) CH_2N_2

(1) Me_2SO_4
(2) (O)
(3) CH_2N_2

Chart 1.

The molecular formulas of dehydrotremetone, $C_{13}H_{12}O_2$ (**24**), and its oxime, as well as thier two *C*-methyl groups and optical inactivity suggest this ketone is the completely unsaturated analog of tremetone (**23**), 2-isopropenyl-5-acetyl-2,3-dihydrobenzofuran.[24] As with tremetone, ozonization of **24** afforded benzaldehyde; catalytic hydrogenation resulted in the uptake of two moles of hydrogen and formation of dihydrotremetone (**25**), structurally identical with the product isolated by hydrogenation of tremetone. Dehydrotremetone (**24**) has been synthesized by Horton et al.[8-10] Acetylation of methyl benzofuran-2-carboxylate with acetyl chloride in presence of aluminum chloride gave methyl 5-acetylbenzofuran-2-carboxylate (**28**); its ethylene ketal, **29**, was treated with methylmagnesium iodide to yield the tertiary alcohol, **30**. The latter was both cleaved and dehydrated with *p*-toluenesulfonic acid to dehydrotremetone (**24**) (Eq. 6); this is identical to the natural product isolated from the rayless goldenrod (*Aplopappus heterophyllus*)[29] and from snakeroot (*Eupatorium urticaefolium*).[22]

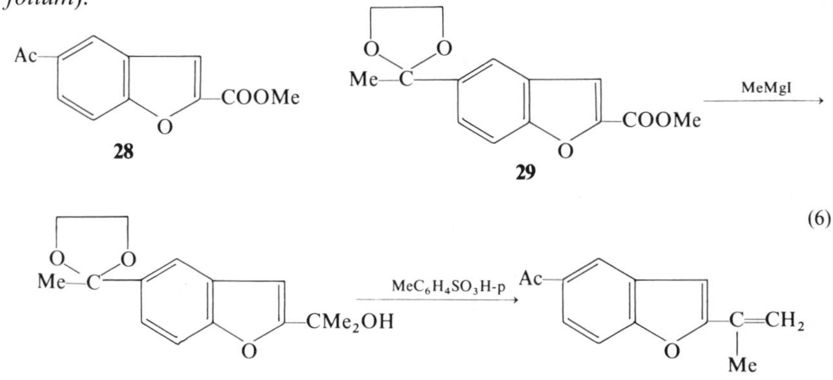

(6)

Hydroxytremetone (**17**)[22] contained two C-Me groups. Its molecular formula, $C_{13}H_{14}O_3$, was suggestive of an oxygenated derivative of tremetone (**23**), a hydroxylated derivative that gives a positive ferric chloride test, a positive iodoform test in methanol, and that forms the *O*-acetyl derivative.[24] Catalytic hydrogenation of **17** proceeded with hydrogenolysis, affording a sample of 4-isoamyl-6-acetylresorcinol (**27**). These derivatives establish the structure of hydroxytremetone as the 6-hydroxy analog of tremetone. For the synthesis of hydroxytremetone, the sequence outlined in Chart 2 has been developed.[30] 2-(α-Hydroxyisopropyl)-6-hydroxy-2,3-dihydrobenzofuran (**31c**) was dehydrated with phosphorus tribromide and pyridine under the same conditions that had been applied for dehydration of α-hydroxyisopropyl-2,3-dihydrobenzofuran to an isopropenyl derivative in the synthesis of tremetone (**23**).[27] It yielded an oily product which has been found to be a mixture of the desired isopropenyl-2,3-dihydrobenzofuran (**32a**) and isopropylbenzofuran (**33a**) by NMR spectral data. The mixture was treated with trifluoroacetic anhydride and acetic acid to give an oily product from which a small quantity of hydroxytremetone (**17**) has been separated. Hydrolysis of the latter with dilute methanolic potassium hydroxide solution yielded (±)-**17**. The crude oily residue, obtained from the mother liquor of the acetylation product, has been confirmed to be isohydroxytremetone (**34a**) by its infrared spectrum which is identical to that synthesized by dehydration of the dihydrobenzofuran (**35b**) with phosphorus pentoxide. Further support for structure **34a** is the conversion of **31c** to isopropylbenzofuran (**33a**). Hydrolysis of the *O*-acetyl group to the hydroxy compound, **33a,** whose structure has been established synthetically,[4] followed by acetylation with trifluoroacetic anhydride yielded the same material (**32a**). This *O*-acetate has been readily saponified to isohydroxytremetone (**34b**) (Chart 2).

Another approach for the synthesis of racemic hydroxytremetone has been reported (Eq. 7).[31]

$PhCH_2O$–(dihydrobenzofuran)–Ac $\xrightarrow{MeMgI}$ $PhCH_2O$–(dihydrobenzofuran)–CMe_2OH $\xrightarrow[H_2]{\text{Raney Ni}}$

HO–(dihydrobenzofuran)–CMe_2OH $\xrightarrow[C_5H_5N]{Ac_2O,}$ AcO–(dihydrobenzofuran)–CMe_2OAc $\xrightarrow[AcOH]{(F_3CCO)_2O,}$

Ac, AcO–(dihydrobenzofuran)–CMe_2OAc $\xrightarrow[(2)\ Ac_2O.C_5H_5N]{(1)\ \text{saporification}}$

Ac, AcO–(dihydrobenzofuran)–CMe_2OH $\xrightarrow[(2)\ \text{saponification}]{(1)\ SOCl_2,\ C_5H_5N}$ Ac, HO–(dihydrobenzofuran)–C(Me)=CH_2 (7)

Zalkow et al.[29] have established the presence of both dehydrotremetone (**24**) and toxol ((—)-2-isopropenyl-3-hydroxy-5-acetyl-2,3-dihydrobenzofuran) (**26**) in crude rayless goldenrod tremetol. The absolute configuration of toxol (**26**) has been established by its conversion through ozonization,

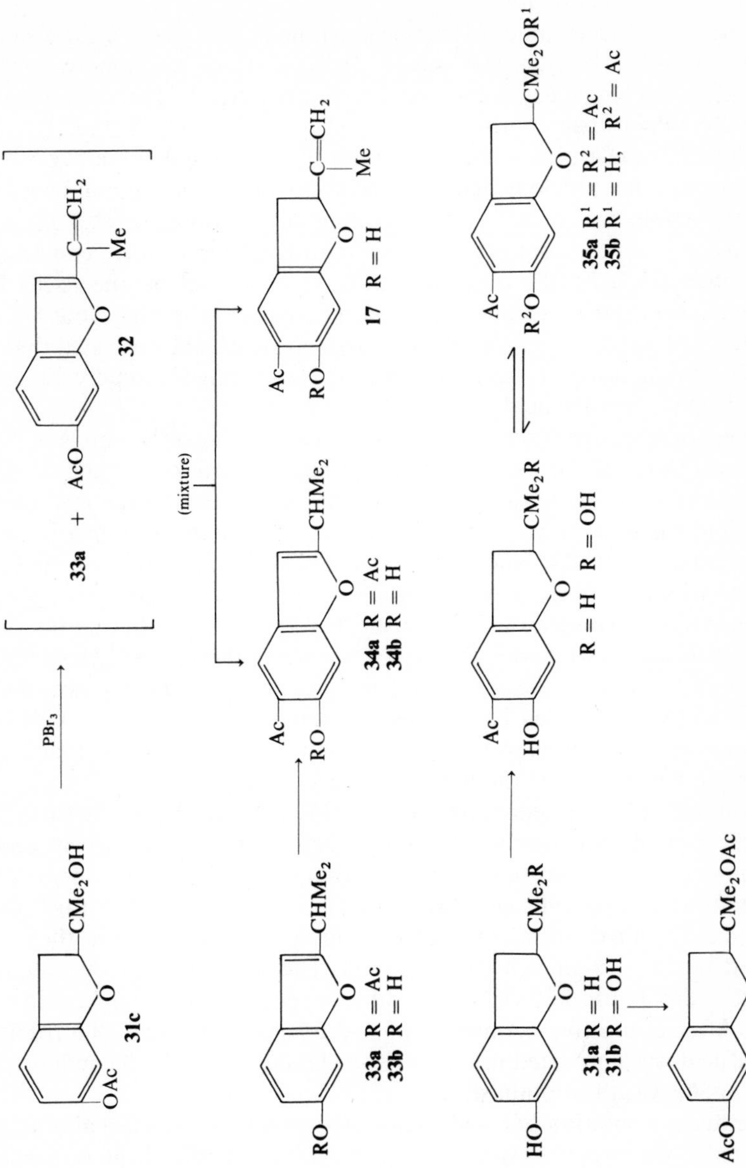

31c $\xrightarrow{PBr_3}$ [33a + 32] $\xrightarrow{}$ (mixture) 34a/34b + 17

33a R = Ac
33b R = H

34a R = Ac
34b R = H

17 R = H

31a R = H
31b R = OH

R = H R = OH

35a $R^1 = R^2 = Ac$
35b $R^1 = H$, $R^2 = Ac$

Chart 2.

hypoiodite degradation, and esterification into methyl (+)-tartarate of known absolute configuration.[26]

Natural (—)-tremetone (**23**) has been synthesized from (+)-2,4-dihydrobenzofuran-2-carboxylic acid (Chart 3) by series of reactions which did not affect the single asymmetric center in the latter.[28] The conversion of (+)-2,3-dihydrobenzofuran-2-carboxylic acid (**36**) into methyl D-(+)-malate (**37**) indicates that the former has the absolute configuration **36** in contrast to the previous tentative configurational prediction based on plant physiological tests[32] and optical rotatory dispersion.[33] Tremetone accordingly has the configuration **23**.[34] Confirmation of these conclusions was undertaken by the direct degradation of (—)-tremetone itself. Preliminary ozonization as applied to **36**, proved to be inapplicable but the sequence of steps, 23–38–39–40–37 (Chart 3), again demonstrated that the single asymmetric center in (—)-tremetone was configurationally related to that of D-(+)-malic acid.

The configurational relationship of the two asymmetric centers in toxol (**26**) with those of (+)-tartaric acid has been reported.[26] In addition, the configuration of the C_2 asymmetric center in **26** (bearing the isopropenyl function) has been directly related to the corresponding asymmetric center in (—)-tremetone (**23**), and directly relates the configuration of the latter ketone with (+)-tartaric acid as well.

Since the assignment of C_2 in tubaic acid is unaffected during the series of transformations (45–46–47–48–43) (Chart 3), the single C_2 asymmetric center in tremetone, the C_5, center in rotenone, and the C_2 asymmetric center in toxol (**26**) may be assigned the same R configuration. This configurational identity suggests the possibility that tremetone, toxol, and rotenone might have a common biosynthetic precursor.

Synthesis of (—)-hydroxytremetone (17)[35] from (R)-6-benzyloxy-2,3-dihydrobenzofuran-2-carboxylic acid (**50a**) eventually led to successful assignment of its absolute configuration. Treatment of 6-benzyloxybenzofuran-2-carboxylic acid (**49**) with 2% sodium amalgam furnished the racemic 2,3-dihydrobenzofurancarboxylic acid (**50a**); this was then converted into (—)-α-phenylethylamine salt. Decomposition of the salt with dilute hydrochloric acid yielded the dihydrobenzofurancarboxylic acid (—)-**50a**. After exhaustive ozonolysis of (—)-**50a** in acetic acid, the resulting malic acid was converted into (—)-(S)-malamide (**51**). The S configuration of (—)-**50a** was thus established.

Reflux of a solution of (—)-**50a** in methanol with *p*-toluene-sulfonic acid resulted in the formation of the methyl ester of (+)-**50b**. Grignard reaction with methylmagnesium bromide modified the side chain of (+)-**50b**, yielding the tertiary alcohols (—)-**52a** and (—)-**52b**. Acetylation of (—)-**52b** with acetic anhydride and stannic chloride afforded (—)-**52c**, whose side chain could be converted into the requisite isopropenyl type by heating at

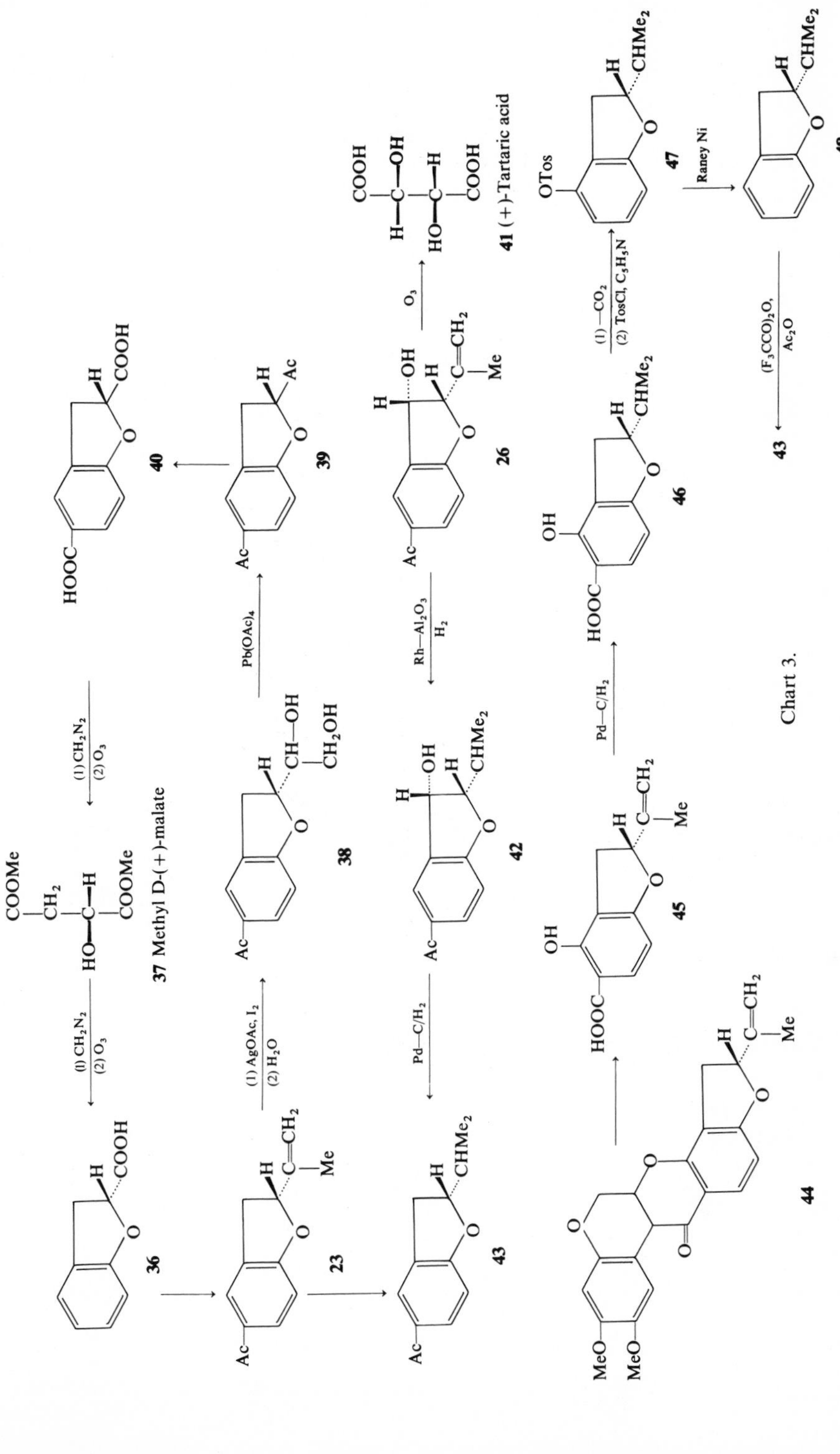

37 Methyl D-(+)-malate

41 (+)-Tartaric acid

36 → (1) CH_2N_2, (2) O_3 → 37

37 → (1) CH_2N_2, (2) O_3 → 40

23 → (1) AgOAc, I_2, (2) H_2O → 38 → $Pb(OAc)_4$ → 39 → 40

43 ← Pd—C/H_2 ← 42 ← Rh—Al_2O_3, H_2 ← 26 → O_3 → 41

44 → 45 → Pd—C/H_2 → 46 → (1) —CO_2, (2) TosCl, C_5H_5N → 47 → Raney Ni → 48 → $(F_3CCO)_2O$, Ac_2O → 43

Chart 3.

PhCH$_2$O— —COOH

49

(−)—(S)—(**51**)

(−)—(S)— **50a** R = H
(−)—(S)— **50b** R = Me

(+)—(R)— **50a** R = H
(+)—(R)— **50b** R = Me

(−)—(R)— **52a** $R^1 = CH_2Ph$, $R^2 = R^3 = H$
(−)—(R)— **52b** $R^1 = R^2 = R^3 = H$
(−)—(R)— **52c** $R^1 = R^2 = R^3 = Ac$

(−)—(R)—**17**

52c

330°C for 75 sec. After hydrolysis with 5% potassium hydroxide, the product was purified by chromatography on silica gel to give (−)-hydroxytremetone (**17**).

These findings established the R configuration of (−)-hydroxytremetone (**17**), showing that this family of compounds, (−)-tremetone (**23**), (−)-toxol (**26**), and (−)-hydroxytremetone isolated from "tremetol," all have the same configuration pertinent to their biogenetic origins.

H. 2-(6-Hydroxy-2-methoxy-3,4-methylenedioxyphenyl)benzofuran

The isolation from baker's yeast of a new crystalline compound, **53**, which acts as an antioxidant and prevents hemorrhagic liver necrosis in rats on diets which would normally induce this condition and also prevents haemolysis of red cells in vitamin E-deficient rats, has been reported.[36-38] The compound, $C_{16}H_{12}O_5$, acquires blue fluorescence and shows the presence of one methoxyl group and a free phenolic hydroxyl group, which is readily methylated by the action of diazomethane. The presence of a methylenedioxy grouping is revealed by production of formaldehyde with sulfuric acid in the Eegrieve test, as well as by its color with chromotropic acid. Oxidation of the compound with sodium chromate in glacial acetic acid resulted in the formation of benzofuran-2-carboxylic acid.[37,38] Degradation of the methyl derivative, **54**, through its oxidation with potassium permanganate, and with lead tetraacetate to the benzoic acid (**55**) is used to establish the position of the oxygen substituents in the unelucidated portion of the molecule. Thus oxidation with permanganate or chromate did not destroy the benzofuran portion of **54;** instead the highly oxygenated benzene ring was attacked and benzofuran-2-carboxylic acid and salicylic acid were obtained. Osmium tetroxide reacted slowly and specifically at the 2- and 3-positions in **54** to yield presumably the glycol (**56**). The latter has not been isolated, since rearrangement in presence of alkali has led to the yellow 2,6-dimethoxy-3,4-methylenedioxy-2′-hydroxybenzoin (**57**).

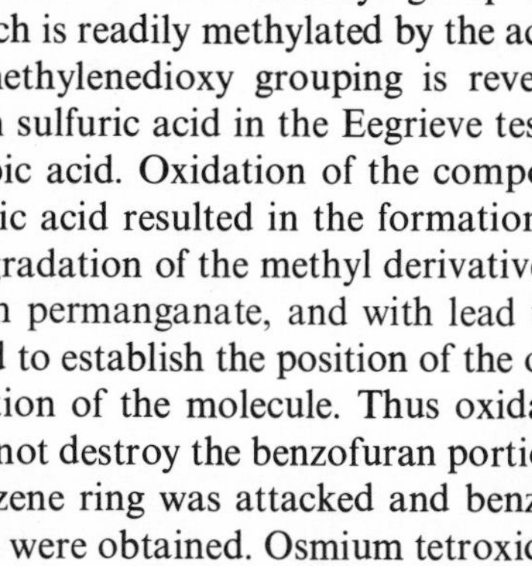

OMe O—CH₂ ... O ... OH

53

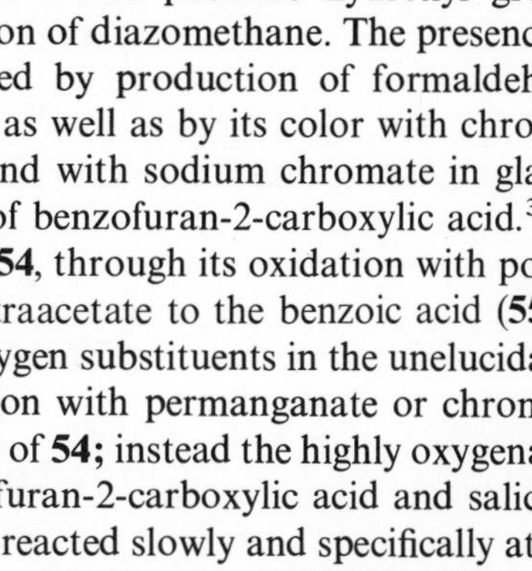

OMe O—CH₂ ... O ... OMe

54

O—CH₂ MeO— —O HOOC— OMe

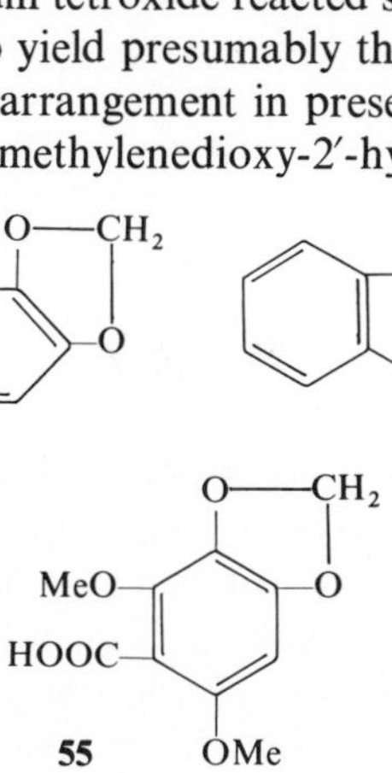

55

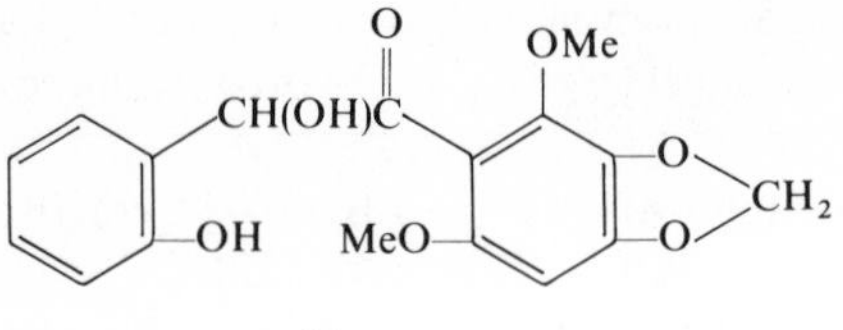

57

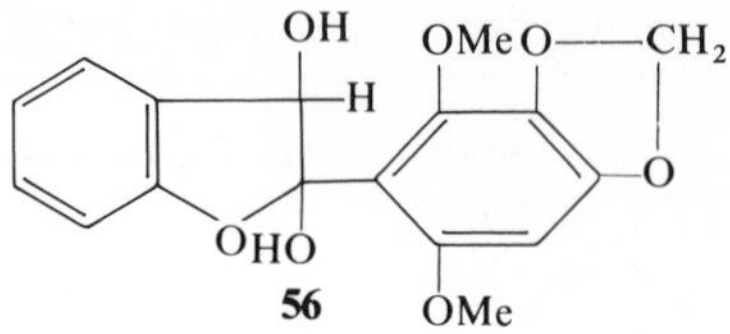

56

Synthesis of **53**, making use of the Hoesch reaction, has been achieved (Eq. 8).[38]

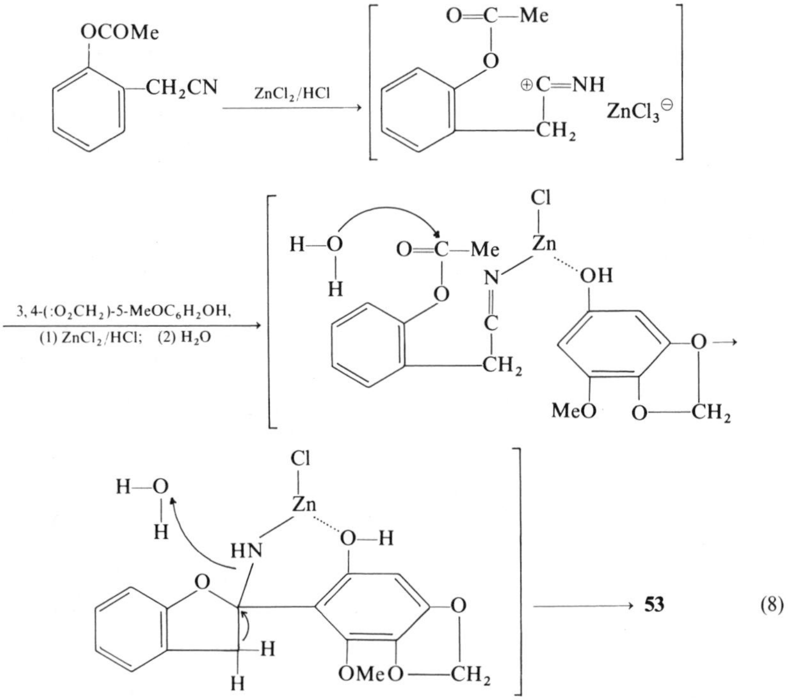

$ZnCl_2/HCl$

$3,4\text{-}(:O_2CH_2)\text{-}5\text{-}MeOC_6H_2OH$, (1) $ZnCl_2/HCl$; (2) H_2O

→ **53** (8)

I. Pterofuran

Pterofuran (**58**), isolated from *Pterocarpus indicus* heartwood,[39] is a 2-arylbenzofuran. It forms monoacetate and dimethyl ether (**59**); ozonolysis of **58** followed by decomposition of the ozonide with zinc and treatment of the crude product obtained by hydrolysis, with methyl sulfate, gave the aldehyde (**61**). The dimethyl ether (**59**) has been obtained via the copper-catalyzed reaction of ω-diazo-2,3,4-trimethoxyacetophenone with *m*-methoxyphenol,[40] together with a small amount of the other isomer, 4-methoxy-2-(2′,3′,4′-trimethoxyphenyl) benzofuran (**60**).

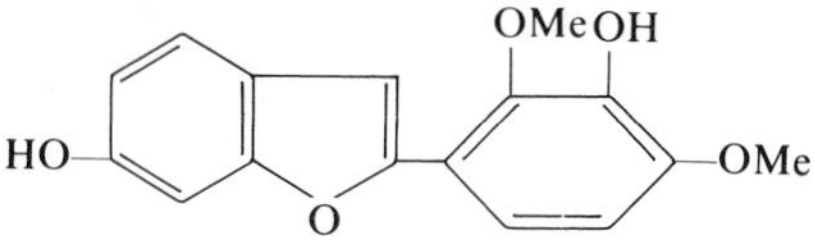

58 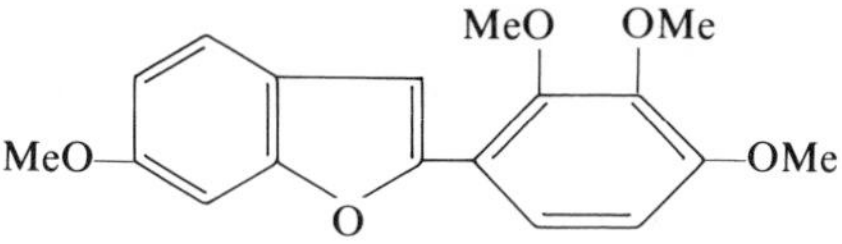 **59**

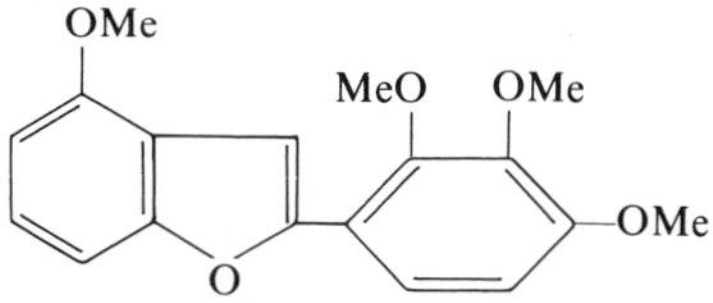

60 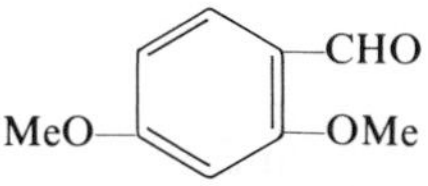 **61**

J. Eupomatene

The bark of *Eupomatia laurina* R. Br. yielded the alkaloid liriodenine, **62**,[41,42] and four new substances. The other four compounds have the molecular formulas: $C_{20}H_{18}O_4$ (m.p. 154–156°C), $C_{25}H_{32}O_8$ (m.p. 170–171°C), $C_{22}H_{30}O_6$ (m.p. 215–218°C), and $C_{24}H_{26}O_6$ (amorphous, m.p. 152–154°C). Eupomatene, the first member of these compounds, has been shown to be 7-methoxy-3-methyl-2-(3′,4′-methylenedioxyphenyl)-5-*trans*-propenylbenzofuran (**63**).[43] The infrared spectrum showed the absence of hydroxyl and carbonyl groups, but absorption at 960 cm^{-1} indicated the presence of a trans-disubstituted double bond, and the high intensity of ultraviolet absorption is consistent with the idea this function is a part of an external chromophore. The NMR spectrum is definitive and indicates of structure **63.** Ozonolysis of eupomatene yielded acetaldehyde, and catalytic hydrogenation afforded dihydroeupomatene (**64**). Ozonolysis of the latter produced the ester **65,** which on hydrolysis, gave the phenol **66.**

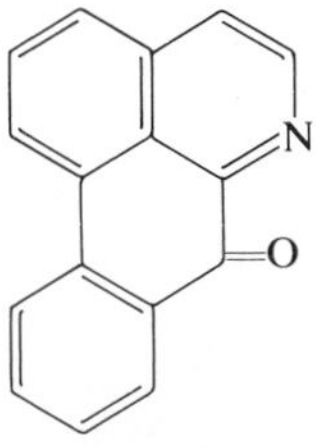

62 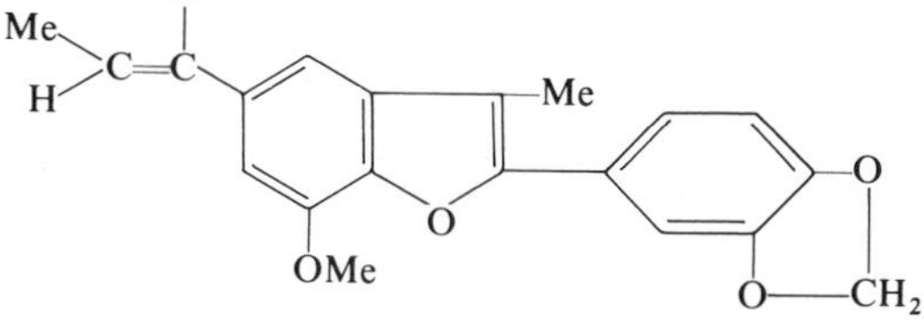 **63**

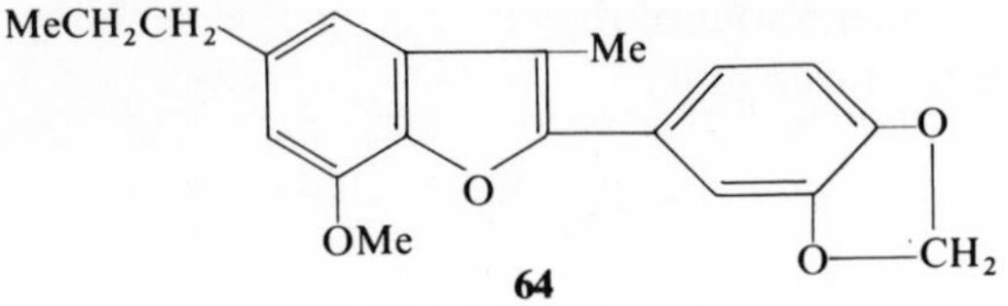

MeCH$_2$CH$_2$ Me O OMe O O CH$_2$

64

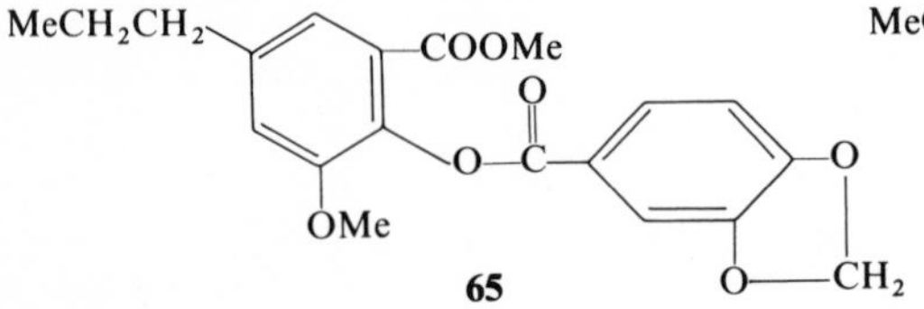

MeCH$_2$CH$_2$ COOMe O O C O OMe O CH$_2$

65

MeCH$_2$CH$_2$ Ac OH OMe

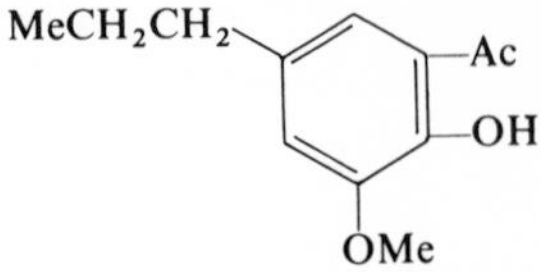

66

Eupomatene and dihydroeupomatene have been synthesized[43] as outlined by the sequence of reactions in Eq. 9.

McCredie, Ritchie, and Taylor[43] have suggested that egonol (**67**), compound **68,** and eupomatene (**63**) (also magnolol, **69,**[44] and futoenone, **70**[45]) be classified as lignan (the more complex substances, hordatines A (**71**) and B (**72**), isolated from barley are of the eupomatene type and recognized as lignans[46]).

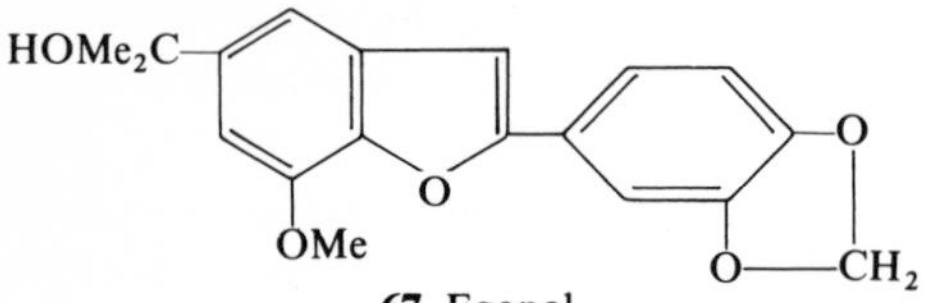

HOMe$_2$C O O OMe O CH$_2$

67 Egonol

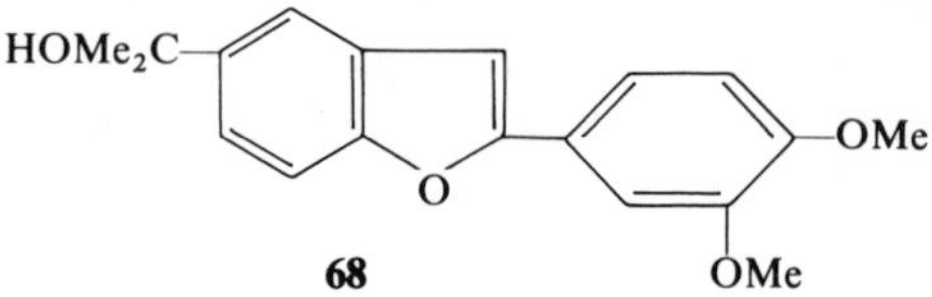

HOMe$_2$C O OMe OMe

68

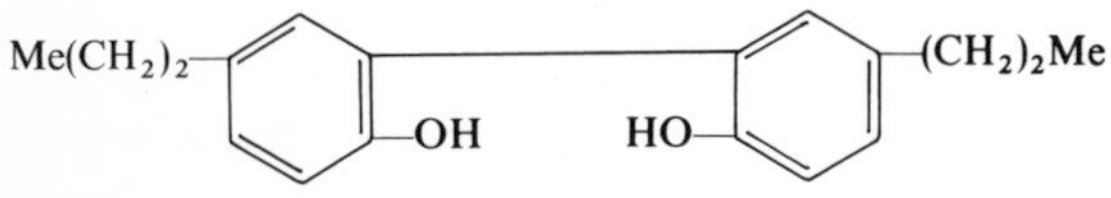

Me(CH$_2$)$_2$ (CH$_2$)$_2$Me OH HO

69 Magnolol

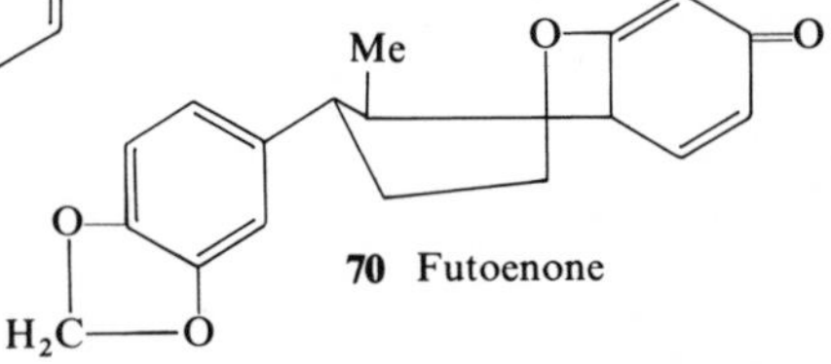

Me O O O H$_2$C O

70 Futoenone

O OR^1 R CONH(CH$_2$)$_4$NHC(:NH)NH$_2$ H HC=C CONH(CH$_2$)$_4$NHC(:NH)NH$_2$

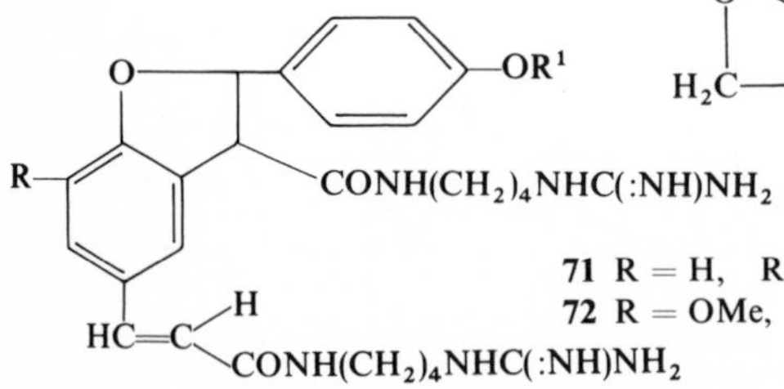

71 R = H, R^1 = D-glucopyranosyl (Hordatine A)
72 R = OMe, R^1 = D-glucopyranosyl (Hordatine B)

It is also proposed that the term "lignan" be extended to cover all natural products of low molecular weight that arise primarily from the oxidative coupling of *p*-hydroxyphenylpropane units.

K. Egonol

Egonol, $C_{19}H_{18}O_5$, was first discovered by Okada[47] in the seed oil of "egonoki" (*Styrax japonicus* Sieb. et Zucc., a plant common to Japan). Other sources are *S. formanosus* Matsum. and *S. obassia* Sieb. et Zucc. According to Kawai,[48] egonol (**67**) is not ketonic, but is a primary alcohol, since it forms an acetate and is also esterified by phthalic anhydride. A methoxyl group is present, though there is no double bond subject to ready hydrogenation; ozonolysis of the acetate gives an aldehydic ester (**73**), hydrolysis of which affords piperonylic acid and styraxinolaldehyde (**74**)—a strong argument for the presence of a benzofuran system. The constitution of styraxinolaldehyde follows from the green ferric reaction, together with the oxidation of the acetate by peracetic acid to styraxinolic acid (**74;** COOH for CHO), which after complete methylation and hydrolysis of the ester grouping, can also be oxidized to isohemipinic acid (**75**)[49]. The nature of the aliphatic side chain follows from the fact that the compound is a primary alcohol devoid of optical activity.[50] To obtain egonol, the aldehyde **74** is etherified with ethyl α-chlorophenylacetate, a simultaneous internal aldol condensation that gives two stereoisomeric esters (**76**). After hydrolysis to the acid and pyrolysis in quinoline, carbon dioxide and water are both lost, and egonol results.[51]

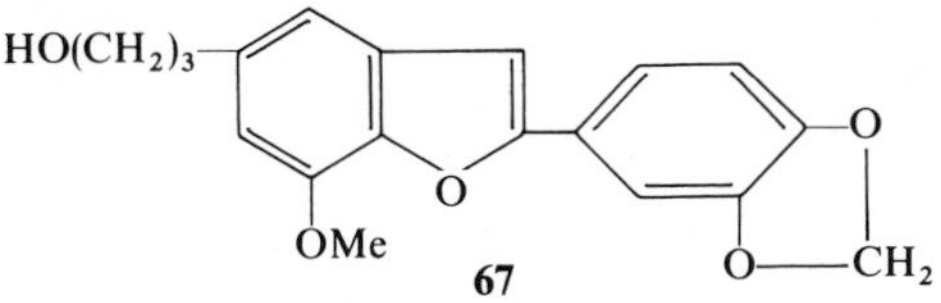

HO(CH₂)₃– ; O ; OMe ; O ; CH₂

67

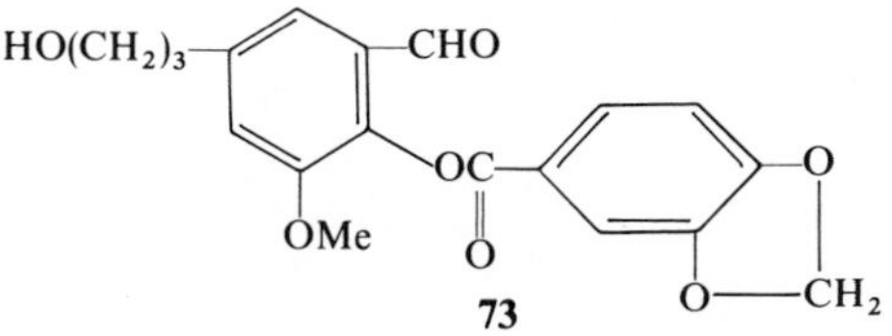

HO(CH₂)₃– ; CHO ; OC ; O ; OMe ; O ; CH₂

73

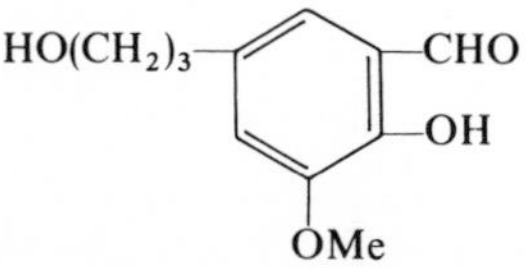

HO(CH₂)₃– ; CHO ; OH ; OMe

74

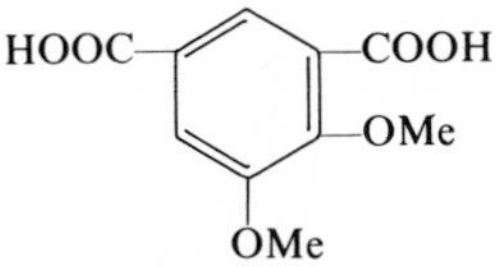

HOOC– ; COOH ; OMe ; OMe

75

$$\mathbf{74} + \mathrm{ArCH(Cl)COOEt} \longrightarrow$$

H_2C (methylenedioxy) Ac + $Me(CH_2)_2$ CHO OH MeO **A** → $Me(CH_2)_2$ MeO O^+ O CH_2 O $\xrightarrow{MeOH,\ H_2O_2}$

$H_2C{=}CHCH_2$ COOMe MeO O O CH_2 O (minor) + $H_2C{=}CHCH_2$ CH_2COOMe MeO O C=O O CH_2 O (major) →

(minor) $\xrightarrow{LiAlH_4}$ $H_2C{=}CHCH_2$ CH_2OH MeO O O CH_2 O $\xrightarrow{MnO_2}$ $H_2C{=}CHCH_2$ CHO MeO O O CH_2 O →

$H_2C{=}CHCH_2$ ... OMe ... S—CH$_2$—CH$_2$—S—CH ... O—CH$_2$—O $\xrightarrow{\text{Raney Ni}}$

$H_2C{=}CHCH_2$ ... MeO ... Me ... O—CH$_2$—O
(Isoeupomatene)
$\xrightarrow[\text{Me}_2\text{SO}]{\text{K-t-butoxide}}$
Me(H)C=CH ... OMe ... Me ... O—CH$_2$—O
(Eupomatene) 63

$H_2C(O)_2C_6H_3COCHN_2$ $\xrightarrow{BF_3/MeOH}$ $H_2C(O)_2C_6H_3COCH_2OMe$ $\xrightarrow{A}$

$Me(CH_2)_2$ ... MeO ... O^+ ... OMe ... O—CH$_2$—O $\xrightarrow[\text{MeOH}]{H_2O_2,}$ $Me(CH_2)_2$ ... MeO ... COOMe ... O—CH$_2$—O $\longrightarrow$ **64** (9)

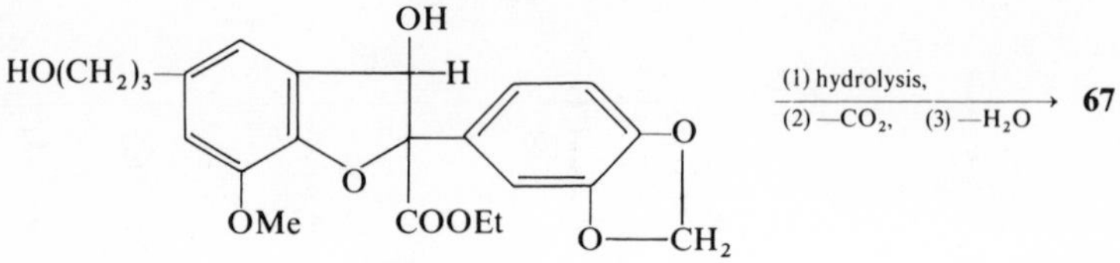

(1) hydrolysis, (2) $-CO_2$, (3) $-H_2O$ → 67 (10)

76

Solutions of egonol in sulfuric acid are orange-red; in chloroform containing antimony trichloride, violet; but in acetic acid with hydrogen peroxide a deep carmine-red color is produced, which forms the basis for spectroscopic estimations of the compound. It is believed that a quinone (**77**) is formed which, in acid medium, might well form salts of type **78**.[52-54] Furthermore, it has been shown that the "egonol reaction" is given only by 2-phenylbenzofurans methoxylated at the 5 or 7 positions; the presence of hydroxyl group at the 6 position does result in a deep color, but the reaction appears to be of a different type.

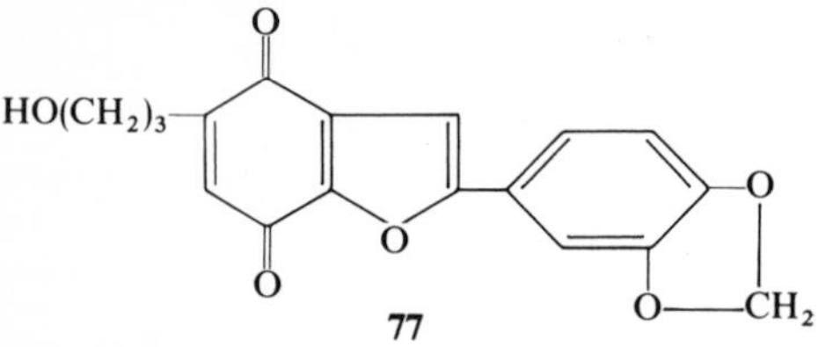

77 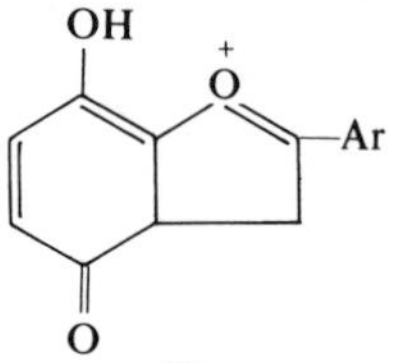 **78**

Egonol (**67**) is partially synthesized by hydrolysis, decarboxylation, and hydroboration of the already known benzofuran ester (**80a**).[56] The latter is prepared[43] from the flavylium salt (**79**); hydrolyzed to the acid (**80b**), and decarboxylated to the benzofuran (**80c**).[56] Markinikoff hydration of the side-chain olefinic double bond was effected by hydroboration with oxidative work-up (Eq. 10).

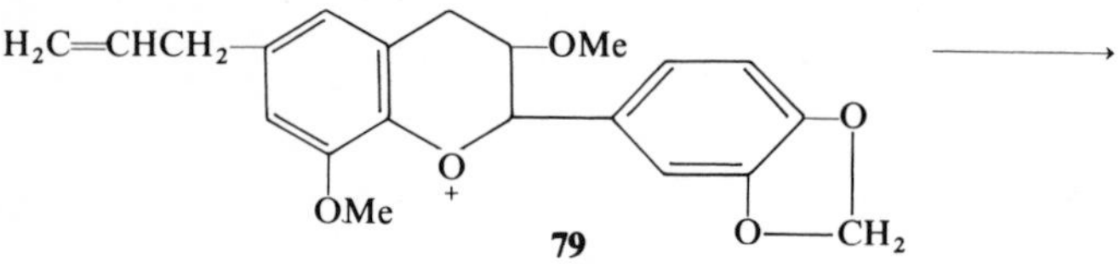

79

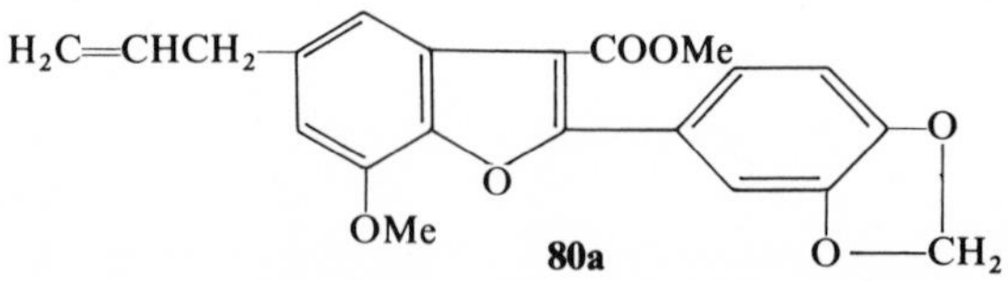

80a

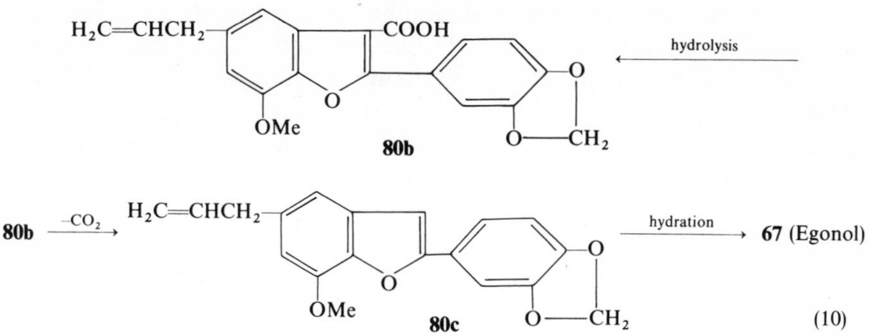

$H_2C{=}CHCH_2$ COOH OMe O O O CH_2 **80b** ← hydrolysis

80b $\xrightarrow{-CO_2}$ $H_2C{=}CHCH_2$ OMe O O O CH_2 **80c** $\xrightarrow{\text{hydration}}$ **67** (Egonol) (10)

Egonol glycoside is isolated from the fruits of 'twain-egonoki'' (''Urazine-Egonoki,'' *Styrax suberifolium* Hook et Arn.) and, on hydrolysis, gives egonol (1 mole) and glucose (2 moles).[55] It shows orange-yellow halochromism in sulfuric acid, but gives a deeper carmine-red egonol reaction than egonol itself; this test illustrates that the sugar residue cannot be attached to C-4.[53] Kawai and Sugimoto,[55] noting the rarity of the 2-phenylbenzofuran system among the natural products, recognized an affinity between egonol and the lignans, the former being based on C_6C_3 and C_6C_2 units, the latter on two C_6H_3 units. The resemblance extends to the oxygenation pattern and even to the biological activites.

The NMR spectra of egonol (**67**), 3-nitroegonol (**81**), egonoic acid (**82**), and egonyl chloroacetate confirmed the structure of egonol, and the content of egonol in oils may be estimated from the intensity of specified bands in the spectra.[57]

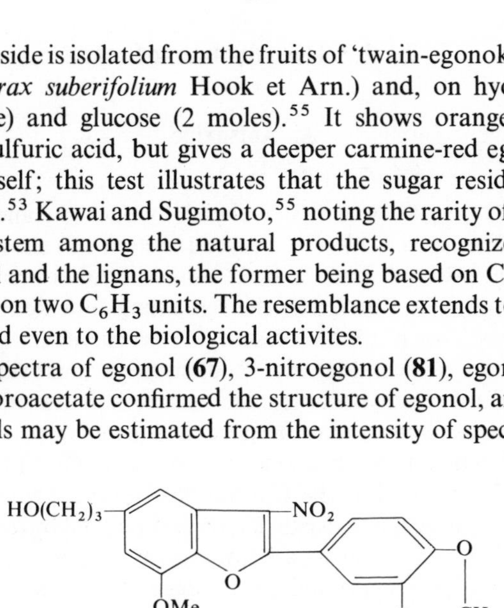

$HO(CH_2)_3$ NO_2 OMe O O O CH_2 **81**

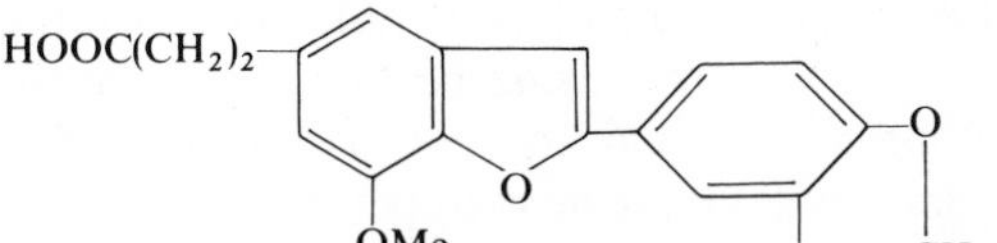

$HOOC(CH_2)_2$ OMe O O O CH_2 **82**

L. 2-(3′,4′-Dimethoxyphenyl)-5-(3″-hydroxypropyl)-7-methoxybenzofuran

The glycosidic fraction of *Styrax officinalis* L. afforded, after hydrolysis and chromatography, besides egonol, a new benzofuran derivative, $C_{20}H_{22}O_5$, which is formulated as **83;** giving a monoacetate.[58] Oxidation of **83** with hydrogen peroxide, followed by basic hydrolysis, gave styraxinolic acid (**74**) and veratric acid (**84**). These results indicate that **74** was related to egonol and that it differed only in the substitution pattern in the phenyl ring.

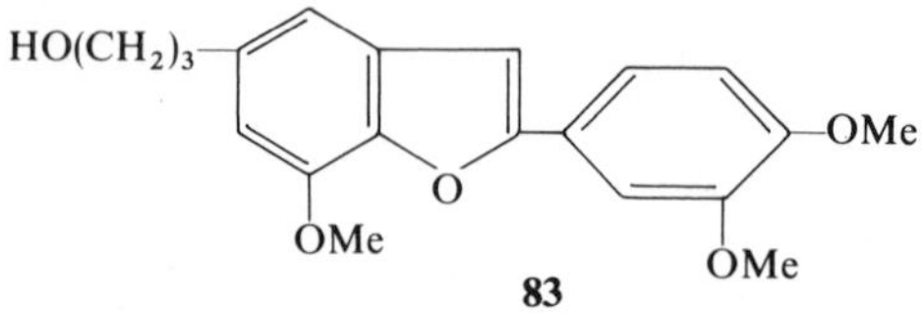

$HO(CH_2)_3$ OMe O OMe OMe **83** COOH OMe OMe **84**

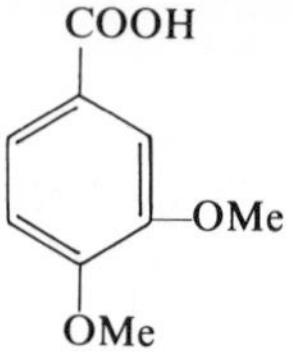

2. 2,3-Dihydrobenzofurans

A. Anisoxide

Anisoxide, $C_{14}H_{18}O$, was first isolated by Jackson and Short[59] from star anise oil (*Illicium verum*). The oxygen function was shown to be present in a cyclic ether, probably not 1:2 or 1:3, since it did not react with phenylmagnesium bromide at 100°C. Of the remaining four double-bond equivalents, one was shown to be present in an ethylidene grouping since ozonolysis produced acetaldehyde. Reduction of anisoxide with sodium and ethanol afforded dihydroanisoxide with loss of the ethylidene group. Energetic catalytic hydrogenation gave a saturated perhydroanisoxide, $C_{14}H_{26}O$, indicating that anisoxide was bicyclic. One further *C*-methyl group, probably attached to the C atom bearing the ethereal oxygen, was shown to be present by Kuhn-Roth determination, and by fusion of the anisoxide followed by oxidation to a ketone giving a positive iodoform test.

The ultraviolet spectrum of anisoxide suggested a close relationship with anethole, which was confirmed by the resemblance of the spectrum of dihydroanisoxide to that of dihydroanethole. The infrared spectrum also suggested a 1:3:4-trisubstituted benzene ring. This, together with Jackson and Short's findings,[59] favored the partial formula **85** for anisoxide.[60] Oxidative degradation of anisoxide with permanganate in pyridine resulted in the formation of the glycol **87**, and the acids **88, 92,** and **93,** respectively; this accounts for all given evidence. In view of the unusual rearrangement of the isoprenoid side chain, structure **86** was confirmed by synthesis. 2-Hydroxy-5-methylbutyrophenone (**94**) was converted by the action of methylmagnesium bromide into the alcohol **95.** This was dehydrated and cyclized by perchloric acid in acetic acid medium directly to the dihydrobenzofuran (**96**). Oxidation with potassium permanganate, followed by dehydration gave the acid **93** which was converted to the acid **88**. Furthermore, **88** was converted into the corresponding ethyl ketone, **91**, by interaction of the acid chloride with diazoethane, followed by reduction with hydroiodic acid and with zinc and acetic acid. Reduction to the alcohol **90** was effected by the action of potassium borohydride, but dehydration of the latter to anisoxide presented some difficulty because of the ready polymerization of anisoxide itself; however, hot aqueous ethanolic hydrochloric acid gave anisoxide in 20% yield.

The unusual method of attachment of the isoprenoid residue of anisoxide has been studied. It appears that isoprenoid residues are in general attached to the aromatic ring at position 1 or 3 of $\underset{1}{C}-\underset{2}{C}-\underset{3}{C}\langle$, corresponding to the electrophilic carbons of a postulated carbonium ion (**98**). The attachment at position 2 observed in anisoxide suggests either a new type of biogenetic mechanism or, more probably, a rearrangement of the carbonium ion (**97**) at some stage in the biogenesis.[61] Anisoxide is also peculiar in being optically inactive although containing an asymmetric center; this may be related either to the relatively drastic method of working up the oil, which caused racemization, or else it is possible that star anise oil really contains a readily cyclized unsaturated precursor, for example **99**, which lacks asymmetry (Chart 4).

B. Remirol

Remirol, $C_{14}H_{16}O_4$ (**100a**), is isolated from the rhizomes of *Remirea maritima* (Cyperceae), a small perennial which is common on the seashores of northern Australia and is a cosmopolitan tropical species.[62] Hydrogenation gave dihydroremirol, $C_{14}H_{18}O_4$, and methylation with methyl sulfate (but not with diazomethane) gave the methyl product **100d,** indicating the presence of an olefin and hydrogen-bonded phenol. Remirol is not methylated upon treatment with diazomethane. The NMR spectrum in conjunction with chemical shift values, suggests that remirol has structures **100a, 100b** or **100c.** Support for **100a** comes from a positive Gibbs test, indicative of a proton para to a free phenol. The ultraviolet spectrum and mass spectrum are in accord with this structure.

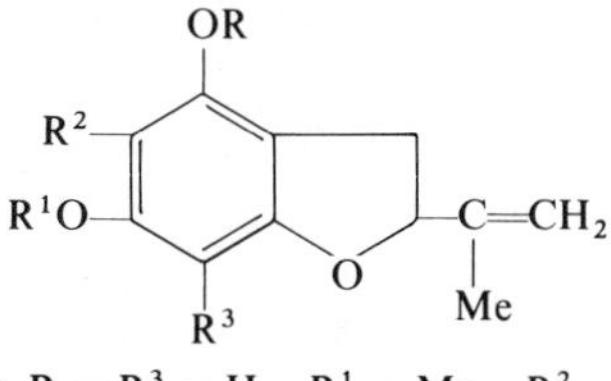

100a $R = R^3 = H$, $R^1 = Me$, $R^2 = Ac$
100b $R^1 = R^3 = H$, $R = Me$, $R^2 = Ac$
100c $R = Me$, $R^1 = R^2 = H$, $R^3 = Ac$
100d $R = R^1 = Me$, $R^3 = H$, $R^2 = Ac$

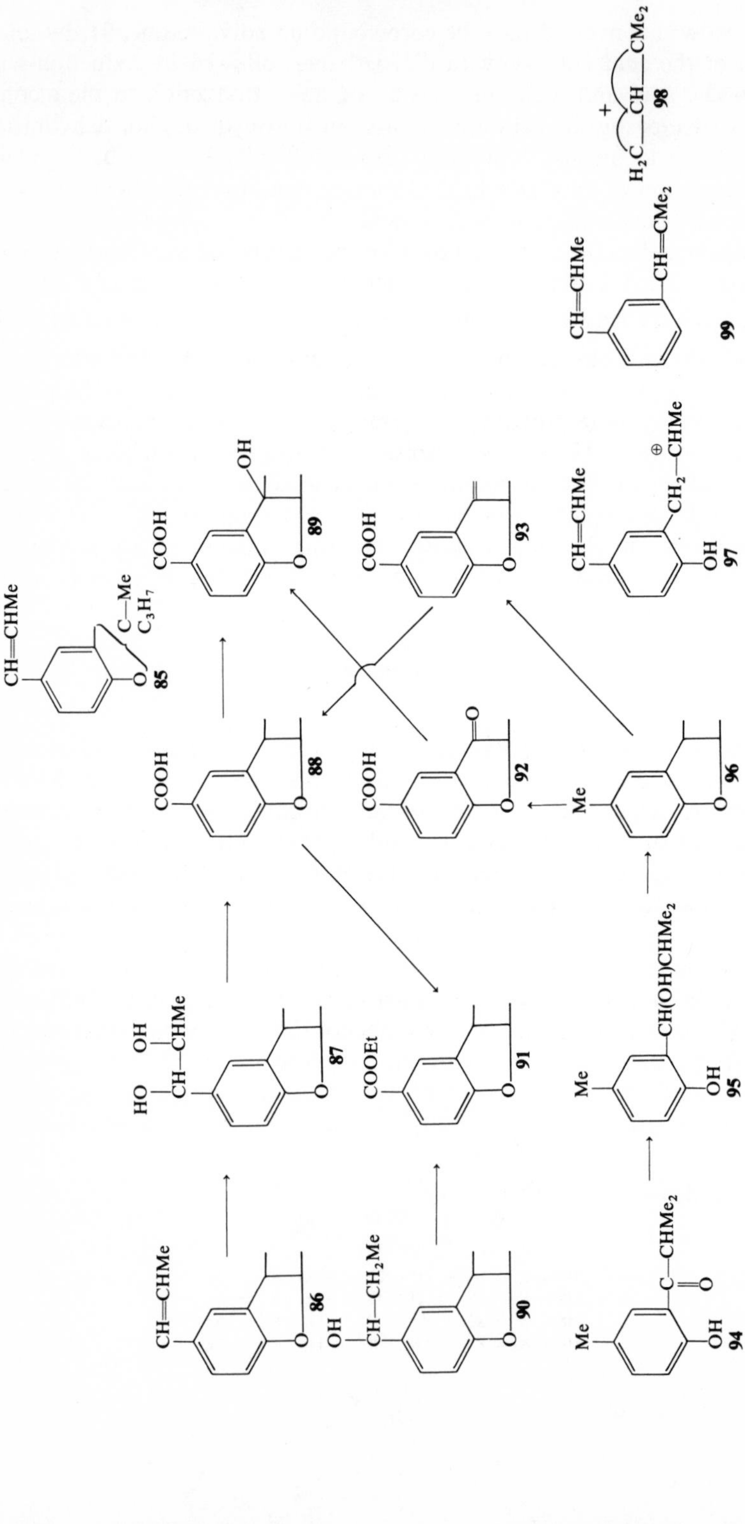

Chart 4.

C. Obtusafuran

Obtusafuran (**101**, $R^1 = H$, $R^2 = Me$), isolated from the heartwood of *D. obtusa* Lecomte (sym. *D. retusa*), was converted by hydrogenolysis over palladium/carbon in acetic acid, followed by air oxidation into yellow quinone (**102**). Its configuration was confirmed by synthesis of its racemate.[63]

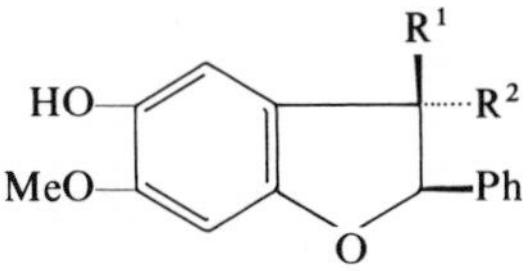

101 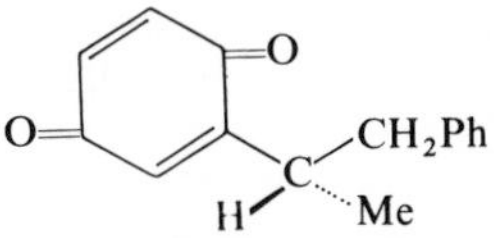 **102**

The stereochemistry of **101,** ($R^1 = H$, $R^2 = Me$), the structure of which has been deduced from the make-up of **102**, was determined by comparison with the synthetic racemic *cis*-2,3-dihydrobenzofuran derivative (**101**, $R^1 = Me$, $R^2 = H$) which was prepared by the controlled hydrogenation of the benzofuran reaction product of 2,5-dihydroxyanisole and 1-bromo-1-phenyl-2-propanone. Ozonolysis of **102** prepared from **101,** followed by decomposition of the ozonide with hydrogen peroxide, gave (-)-(S)-methylsuccinic acid, showing that **102** had the S configuration, and **101** had the (2R,3R) absolute configuration. Biogenetic formation of obtusafuran has been determined.[64]

D. Melanoxin

Melanoxin, a new dihydrobenzofuran, has been isolated from *Dalbergia melanoxylon* Guil. and Perr. (Leguminoseae). Spectroscopic methods established structure **103** for melanoxin and confirmation was obtained by synthesis of *O*-diethyldehydromelanoxin l**104**, R = Et).[65] Ozonolysis of (−)-melanoxin yielded (−)-2(S),3(R)-3-methylsuccinic acid (**105**), while hydrogenolysis and subsequent ozonolysis gave (+)-(R)-methylsuccinic acid (**106**).These results show (−)-melanoxin to have a 2(S),3(S) configuration. The known (S)-4′-hydroxy-4-methoxydalbergı (**107**), malamin (**108**), and 2,5-dimethoxy-*p*-quinone were also isolated from *Dalbergia melanoxylon*.

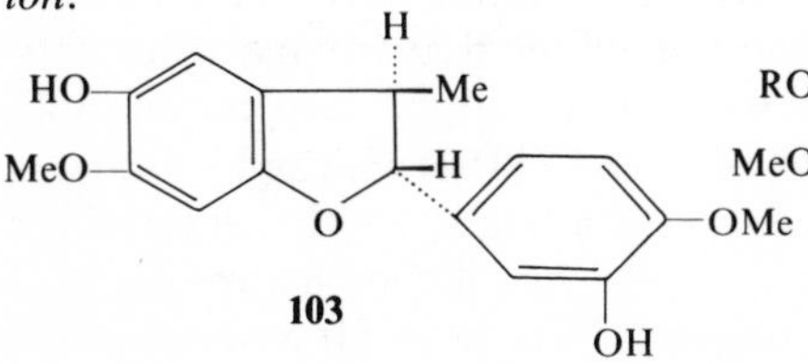

103 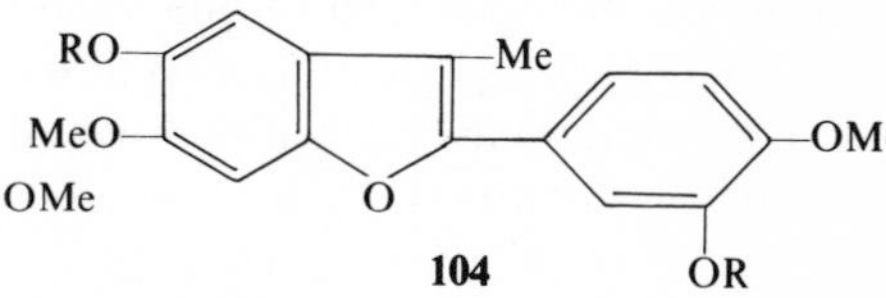 **104**

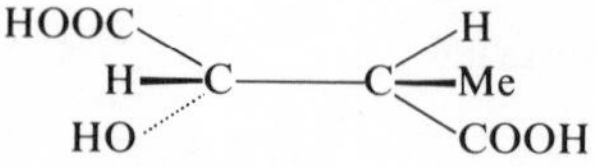

5,6-Dimethoxy-2-(3,4-dimethoxyphenyl)-3-methylbenzofuran (**104,** R = Me) was synthesized by the interaction of sodium salt of 3,4-dimethoxyphenol with 1-bromo-1-(3,4-dimethoxyphenyl) prop-2-one and was found to be identical with *O*-dimethyldehydromelanoxin. Replacement of the second hydroxyl group at position 3′ was confirmed by identity of *O*-diethyldehydromelanoxin with 5-ethoxy-2-(3-ethoxy-4-methoxyphenyl)-6-methoxy-3-methylbenzofuran (**104,** R = Et). The biogenetic formation of melanoxin has been recently discussed.[64]

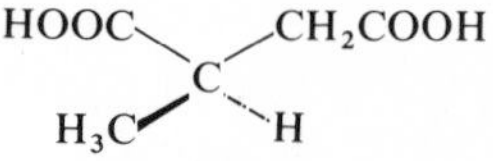

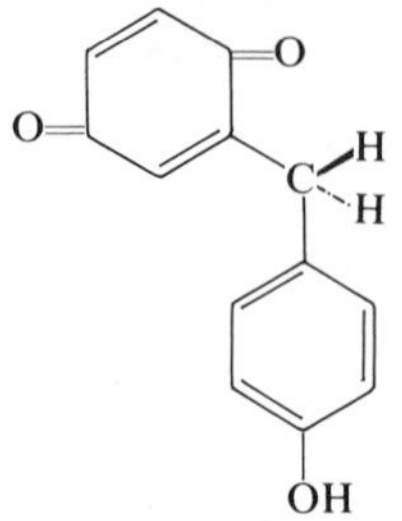

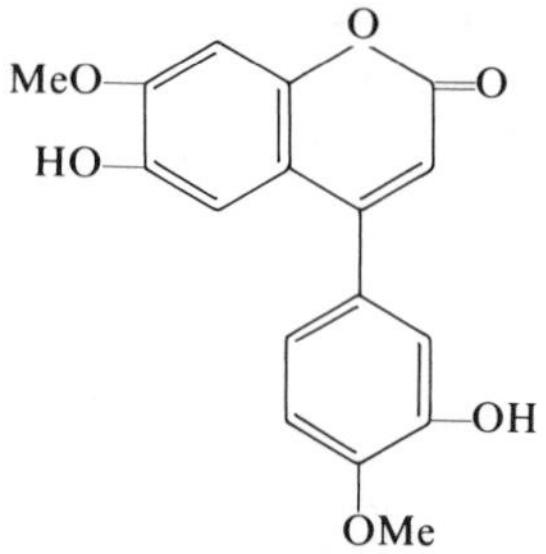

E. Hordatines A and B

One of the fungal factors, isolated from barley coleoptile seedlings, consists essentially of the glucosides of two closely related compounds, hordatine A (**109a**) and hordatine B (**109b**).[66] The isolation of naturally occurring antifungal hordatines is effected by mild methanolysis[68] which liberated the hordatines. A combination of ion exchange chromatography and countercurrent distribution of the hot aqueous extract gave hordatine A and hordatine B. The ultraviolet spectra in neutral and alkaline solutions of the hordatines and of their dihydro derivatives, **110a** and **110b**, respectively, obtained by hydrogenation, showed that they contained a free phenol group not conjugated to the benzofuranyl chromphore. The dihydro derivatives, obtained also by methanolysis of the dihydroglucosides **110c** and **110d,** were partially resolved by countercurrent distribution into fractions containing chiefly dihydrohordatine A (**110a**) with small amounts of dihydrohordatine B (**110b**). Methylation of a fraction with methylsulfate in presence of potassium carbonate, followed by alkaline hydrolysis gave **111** (R = H) and **111** (R = OMe). Compound **111** (R = H) gave on fusion

112 (R = H); upon treatment with dilute base the latter compound is reopened. Ozonolysis of **111**, (R = H) resulted in the formation of salicyclic acid. Evidence for structures **111** (R = H) and **112** (R = H) was found from chemical and spectral data, and independent synthesis.[68]

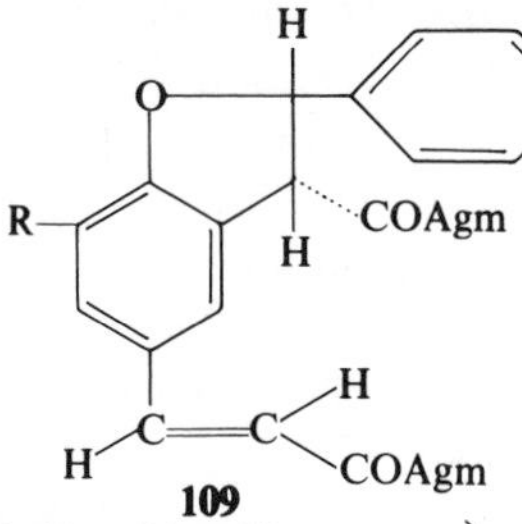

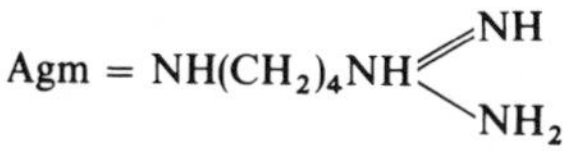

109

109a R = R^1 = H
109b R^1 = H, R = OMe } (or enantiomers)
109c R = H, R^1 = α-D-glucopyranosyl
109d R = OMe, R^1 = α-D-glucopyranosyl

Agm = $NH(CH_2)_4NH$C(=NH)NH_2

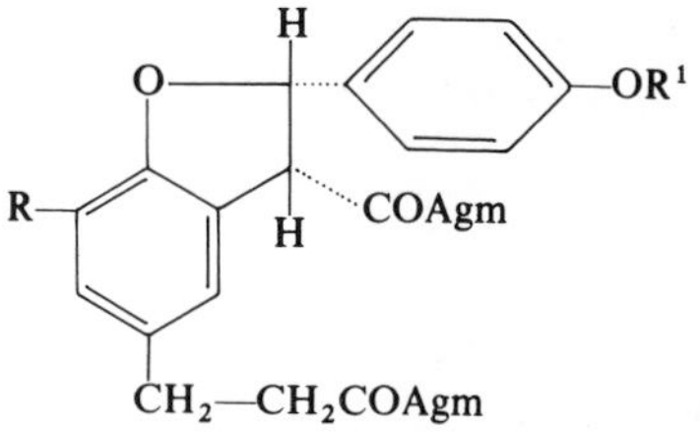

110

110a R = R^1 = H (or enantiomers)
110b R = OMe, R^1 = H (or enantiomers)
110c R = H, R^1 = α-D-glucopyranosyl
110d R = OMe, R^1 = α-D-glucopyranosyl

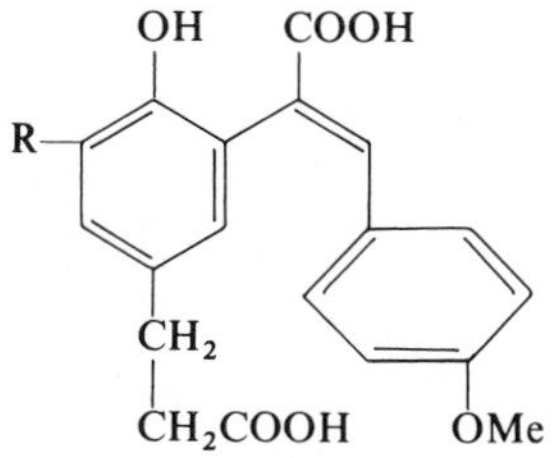

111 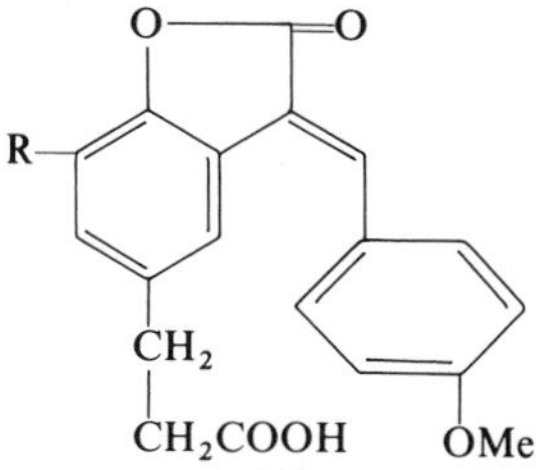 **112**

Addition of allyl bromide to methyl phloretate gave methyl β-4-allyloxyphenylpropionate, which upon Claisen rearrangement at 210°C followed by acetylation of the reaction product, ozonolysis, alkali hydrolysis, and chromatography, gave β-(3-carboxymethyl-4-hydroxyphenyl) propionic acid. The latter acid condensed with anisaldehyde in presence of pyridine to yield **112** (R = H); **111** (R = OMe) heated to fusion yielded **112** (R = OMe). Structure **111** (R = OMe) was confirmed from the spectral data and from the fact that it could be derived from ferulic acid, which was most probable on biogenetic grounds. Synthesis of **111** (R = OMe) from dihydroferulate along the route described above was accomplished. Nuclear magnetic resonance spectra of hordatine A and B glucoside diacetate in deuterium oxide and of dihydrohordatine A and B glucoside diacetate support structures **109c, 109, 110,** respectively.

The hordatine glucosides can then be represented as a mixture consisting predominantly of **109c** and smaller amounts of **109d**; the double bond is tentatively assigned as *cis* configuration by analogy with dehydroconiferyl alcohol. The assigned structures are consistent with all available evidence.

Racemic **109a** is obtained by addition of hydrogen peroxide to solution of benzofuranylagmatine acetate in water at 30–35°C containing 1% solution of horseradish peroxidase in ammonium phosphate. The agmatine acetate was obtained by condensation of methyl *trans-p*-coumarate with agmatine hydrochloride and separated as the acetate.[67]

3. 2(3*H*)-Benzofuranones

A. Xylerythrin

Xylerythrin, $C_{26}H_{16}O_5$, is one of few pigments of fungal origin isolated from decaying wood attacked by the fungus *Peniophora sanguinea* Bres. (*Corticuim sanguinaum* Fr.). It has been shown[69] to be accompanied by 5-*O*-methylxylerythrin; it gives a dimethyl ether, $C_{28}H_{20}O_5$, and a diacetate, $C_{30}H_{20}O_7$. Reductive acetylation of xylerythrin gives a dihydro triacetate. An X-ray analysis[71] of xylerythrin bisbromoacetate led to structure **113**. Structure **114** is thus proposed for xylerythrin. The electronic spectrum of xylerythrin dimethyl ether is very similar to that of **114**, showing that they have the same chromophore.[72] The correct structure for xylerythrin is confirmed from the spectral studies as **114**; 5-hydroxy-3-(*p*-hydroxyphenyl)-4,

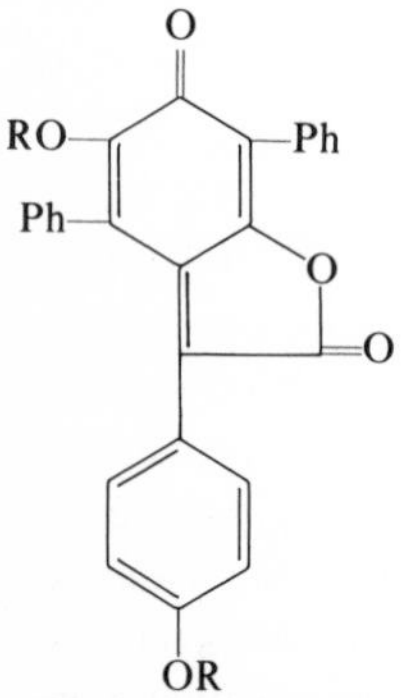

113 R = $COCH_2Br$
115 R = Ac
116 R = Me

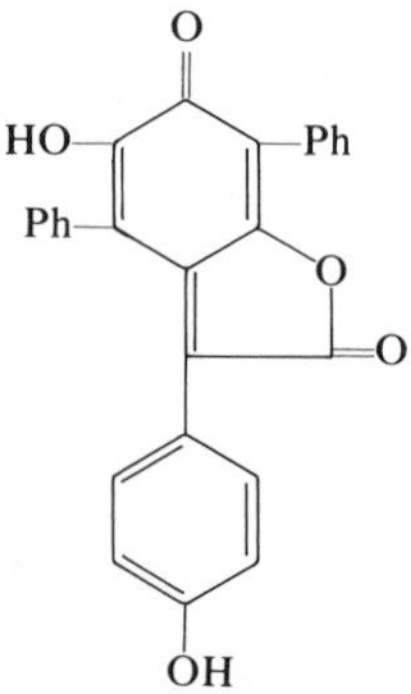

114

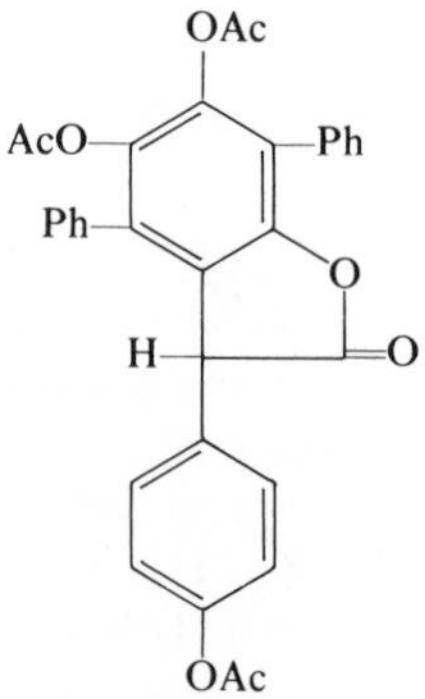

118

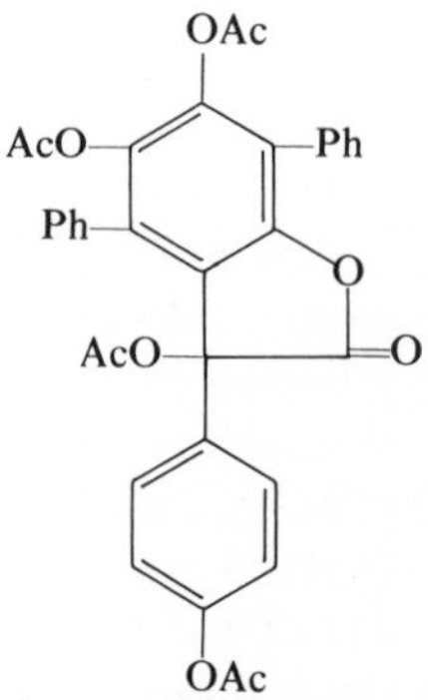

117

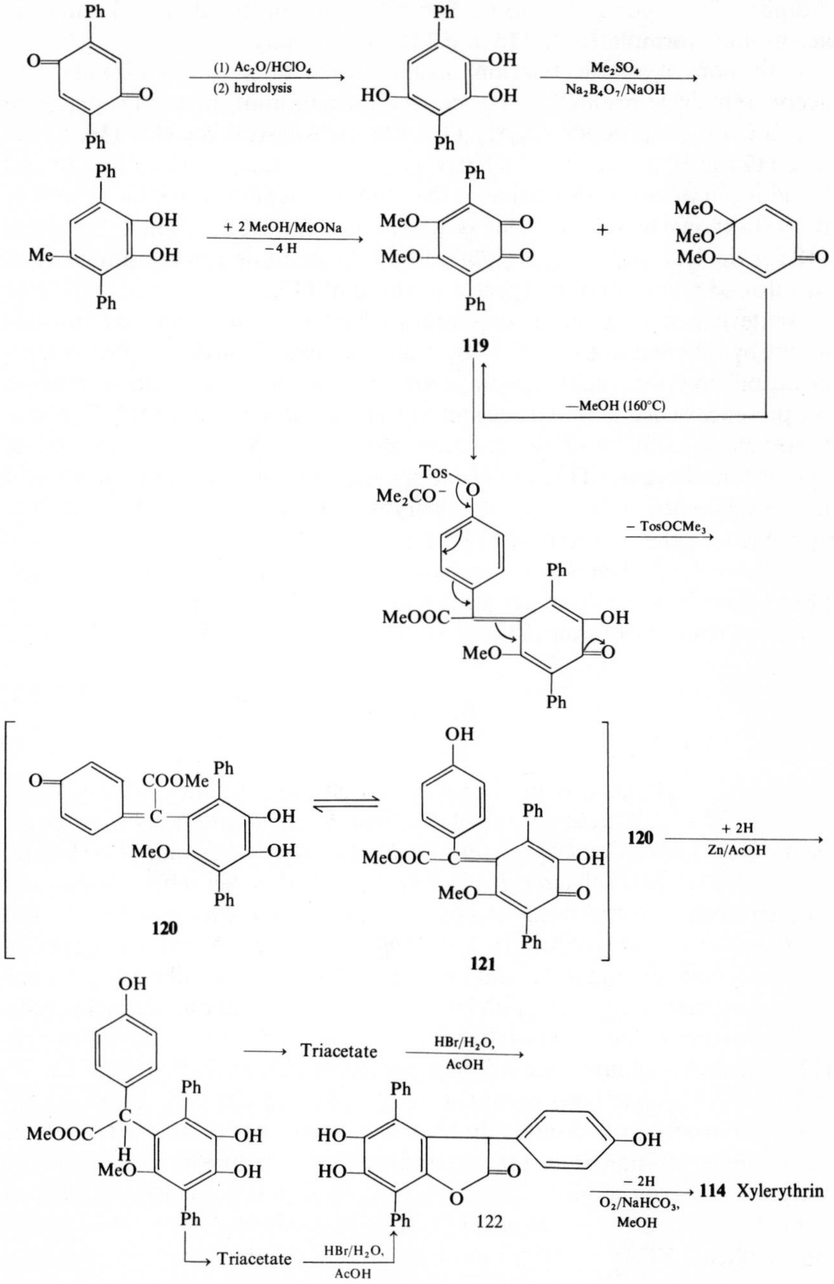

(1) $Ac_2O/HClO_4$
(2) hydrolysis

Me_2SO_4
$Na_2B_4O_7/NaOH$

+ 2 MeOH/MeONa
− 4 H

119

—MeOH (160°C)

− $TosOCMe_3$

120

121

120 + 2H, Zn/AcOH →

→ Triacetate $\xrightarrow{HBr/H_2O,\ AcOH}$

→ Triacetate $\xrightarrow{HBr/H_2O,\ AcOH}$

122

− 2H, $O_2/NaHCO_3$, MeOH → **114** Xylerythrin

Chart 5.

7-diphenylbenzofuran-2,6-dione. The diacetate and the dimethyl ether are accordingly formulated as **115** and **116**, respectively.[70]

Extension of the acetylation reaction time of xylerythrin (**114**) with acetic anhydride resulted in, besides the formation of the orange diacetate **115,** a colorless product, $C_{34}H_{26}O_{10}$. On hydrolysis this yields **114.** Structure **117** can be derived for the latter colorless product, which is thus formed by addition of acetic anhydride to the quinone methide group as a result of the well-known tendency of this system to undergo 1,6 addition.[73] Structure **118** for the leucoacetate is supported both by analysis and the similarity of its infrared and ultraviolet spectra to those of **117**.

Achievement of xylerythrin synthesis has been independently brought about by different authors [70,74] by condensation of *p*-hydroxyphenylacetic acid and polyporic acid (2,5-dihydroxy-3,6-diphenyl-1,4-benzoquinone) in the presence of acetic anhydride and sodium acetate as a catalyst.[70] Chromatographic separation of the reaction mixture yielded a small amount of xylerythrin diacetate **115;** the crude reaction mixture is submitted to acid hydrolysis when a fair yield of xylerythrin (**114**) is obtained.[70] Another approach for the synthesis of **114**[74] has been effected via interaction of 4,5-dimethoxy-3,6-diphenyl-*o*-benzoquinone (**119**) with *p*-toluenesulfonylphenylacetic acid methyl ester via 2-oxo-3-(*p*-hydroxyphenyl)-4,7-diphenyl-5,6-dihydroxybenzofuran (**122**),[74] as outlined in Chart 5.

B. Calycin

Calycin, $C_{18}H_{10}O_5$, is produced by purely fungal cultures from *Sticta coronata* Muell, Arg.; and was isolated from *Sticta aurata* Ach., *S. coronata* Ach., *S. colensoi, Candelaria concolor* Dicks., *Gyalolechia aurella* Hoffm., *Lepraria candelaris* Schaerer, *L. chlorina* Ach., *L. flava* Schreber, *Gasparrima medians* Ngl., *Physia medians* Ngl., *Chryssothrix nolitangere* Mont., *and Callospsima vitellinum* Ehrh. In *Sporopodium leprieurri* Mont. var. citrinum (Zahlbr.) R. Sant. and in *S. xantholeucum* (Müll. Arg.) Zahlbr. the presence of pulvinic acid derivatives, pulvinic dilactone (**123**) and pulvinic acid (**124**), could be demonstrated. The latter species was also found to contain calycin (**125**). These compounds are not uncommon in lichens.[75,76]

Calycin (**125**) has been obtained, upon reflux of **126** with hydrobromic (48%). It reacts with diazomethane to yield the methyl ether; its spectral data[77] make revision of the previously described structure necessary (**127**).[78] Calycin seems not to give a ferric reaction, though it is phenolic. Although demethylation of the dilactone (**127,** R = Me) affords calycin, this cannot be the phenol **127** (R = H), as it is much too acidic and remethylation does not regenerate the original dilactone. Hence calycin must be the isomer **125,** formed from **127** by ring opening and reclosure. Maass[79] has shown

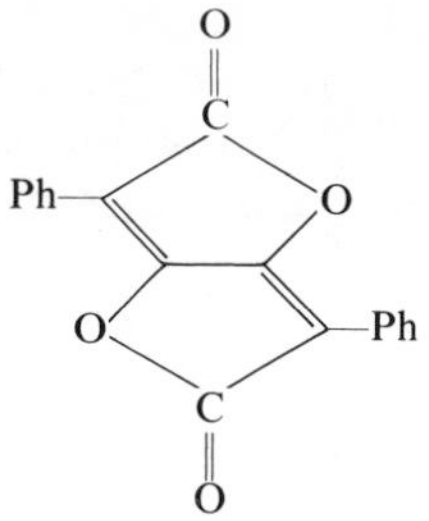

123 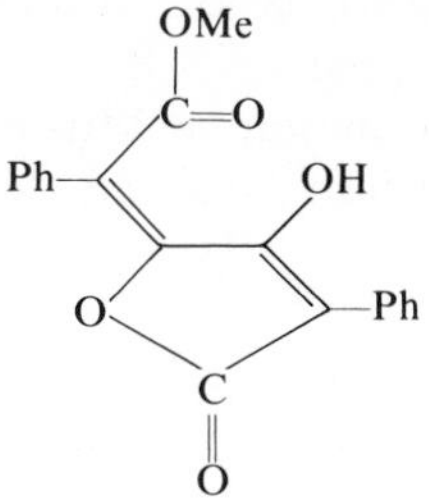 **124**

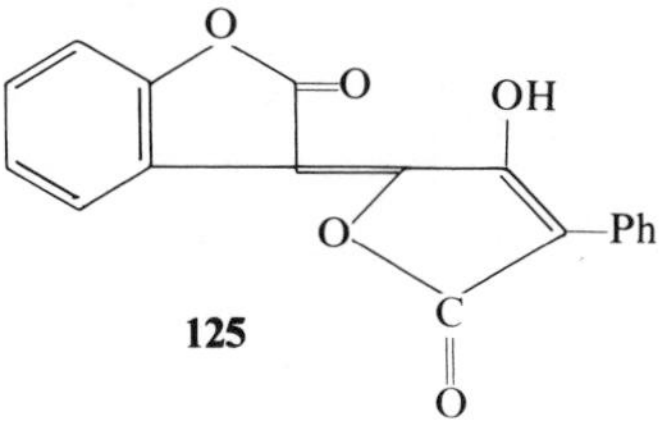

125 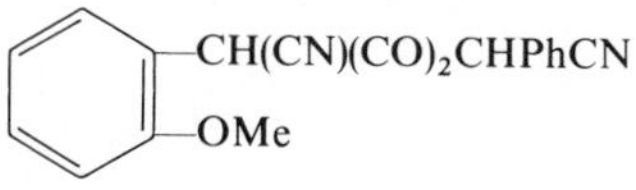 **126**

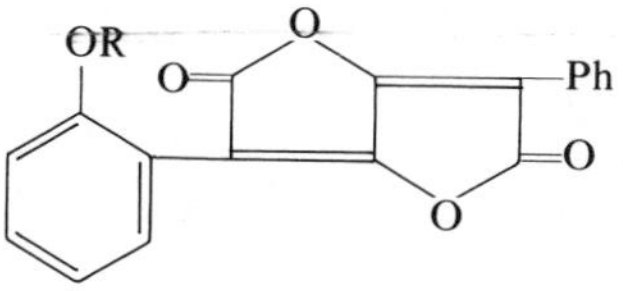

127

that ^{14}C-labeled pulvinic acid, and ethyl pulvinate were precursors of calycin, but not of pulvinic acid dilactone normally present in *Pseudocyhallaria crocata.* Some of the labeled compounds became bound in acetone-insoluble materials before ^{14}C began to accumulate in **125**. The latter **125** probably arise by hydroxylation of a bound form of pulvinic acid.

4. 3(2*H*)-Benzofuranones

A. Maesopsin

The 2-benzyl-2-hydroxy-3-(2*H*)-benzofuranones are now augmented as a group of maesopsin, $C_{15}H_{12}O_6$, a compound isolated from the heartwood of the African timber musizi (*Mazsopsis eminii*, Rhamnaceae) and identified as 2-benzyl-2,4,6,4′-tetrohydroxy-3(2*H*)-benzofuranone (**128**).[80] Fusion of maesopsin with a mixture of sodium and potassium hydroxide resulted in the formation of a yellow crystalline product, $C_{15}H_{10}O_5$ (**129**) which, on methylation, formed two methyl ethers; their structures find support from their NMR and mass spectra.[84] Moreover, the spectral data afford confirmation of the original work.[80-81] Maesopsin is apparently transformed into intermediate **130** and thence into other products.[80]

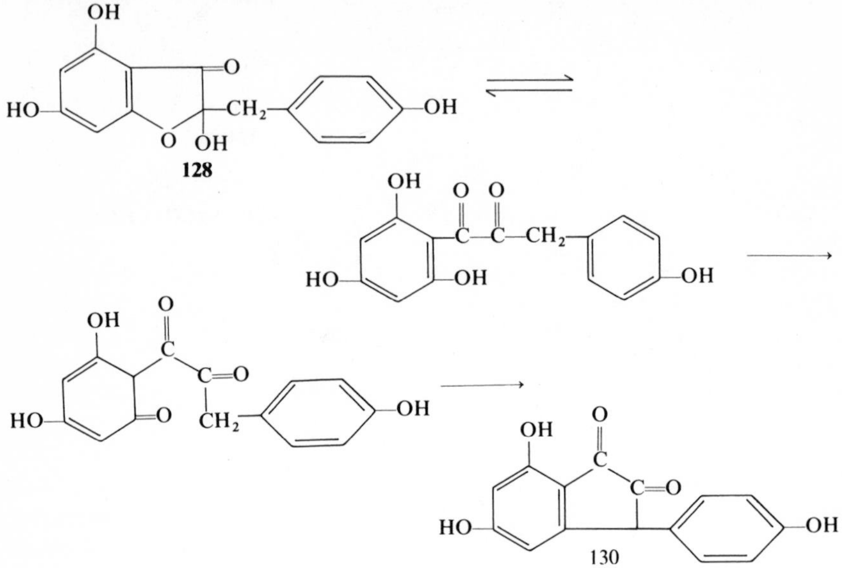

128

130

Synthesis of maesopsin **128** has been achieved Eq. 11.[82]

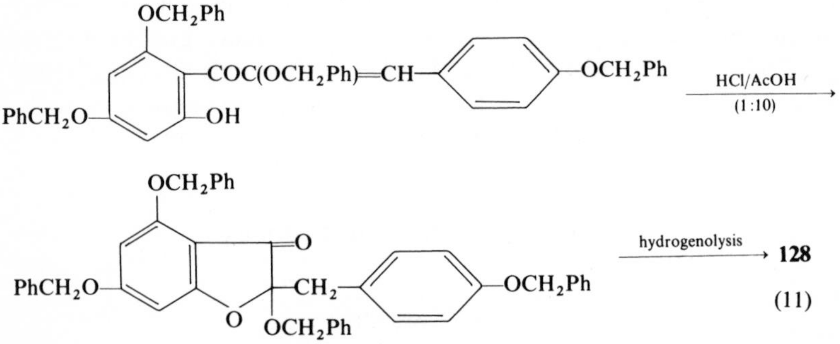

HCl/AcOH (1:10)

hydrogenolysis → **128** (11)

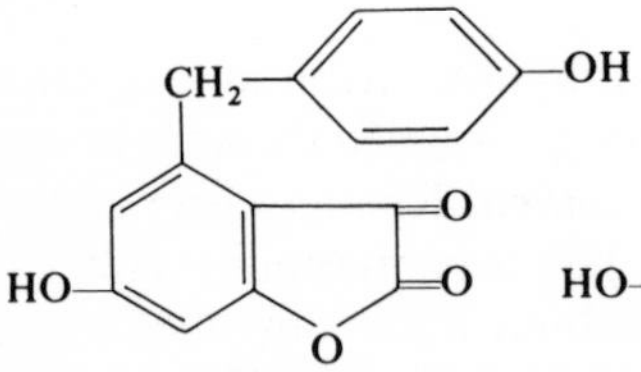

129

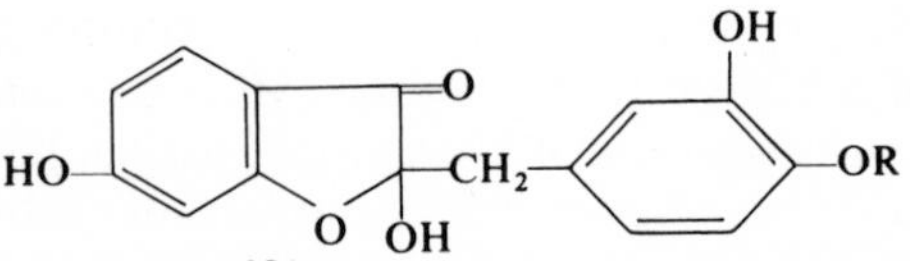

131

Other compounds in this group are found in tannin extracts from *Schinopsis balansae* and *S. lorentzii*,[83] and include the benzofuranone **131** (R = H) and **131** (R = Me). The latter is obtained when isovanillin is allowed to react with ω-resacetophenone in alkaline medium.[84]

B. Alphitonin

Alphitonin, $C_{15}H_{12}O_7$, was extracted from the heartwood of the Australian red ash (*Alphitonia excelsa*); it is optically inactive and gives a purple ferric reaction.[85] Alphitonin (**132a**) could not be fully methylated with diazomethane, but gave a pentamethyl ether, $C_{20}H_{22}O_7$ (**132b**), under the action of methyl sulfate and potassium carbonate. It exhibited no hydroxyl band (infrared spectrum), but an intense absorption peak at 1710 cm^{-1}, probably accounted for by a cyclic ether linkage. The carbonyl group is sterically hindered since the pentamethyl ether gave the oxime under vigorous conditions. Oxidation of the pentamethyl ether by potassium permanganate gave veratric acid; however its reduction with sodium in liquid ammonia yielded a 2-benzyl-3(2*H*)-benzofuranone-substituted derivative (**133**). Structure of alphitonin methyl ether (**132b**) has been confirmed by synthesis.[86] Veratraldehyde was condensed with 2-hydroxy-ω-4,6-trimethoxyacetophenone to yield 2′-hydroxy-α,3,4,4′,6′-pentamethoxychalcone which was cyclized by action of alkali to give **132b.** Alphitonin has been obtained by treatment of taxifolin (**134**) (dihydroquercetin) with alkali together with a small amount of quercetin.[82]

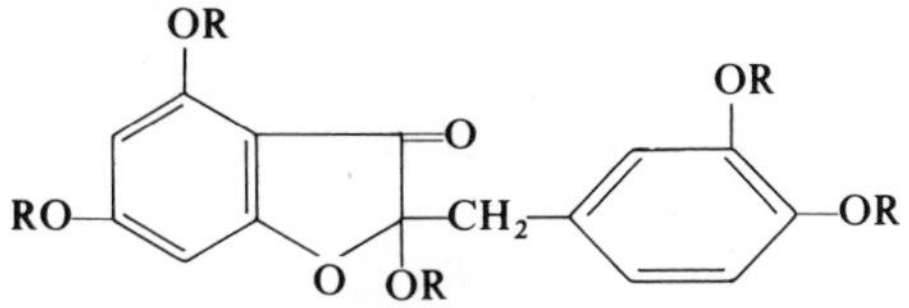

132a R = H
132b R = Me

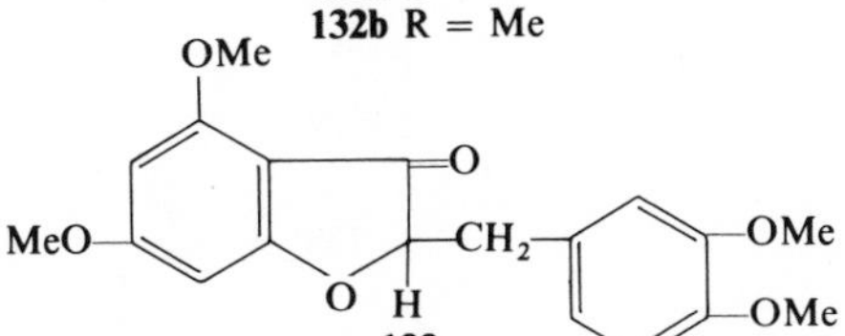

133

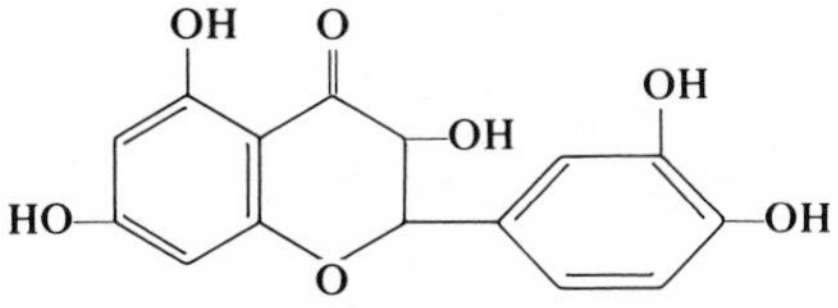

134

C. Aurones

In 1943, Geissman[87] announced the discovery of a new type of plant pigment (anthochlor pigments) from the ray flowers of *Coreopsis grandiflora,* in which the fundamental nucleus is that of benzalbenzofuranone-3-one (2-benzal-3(2*H*)-benzofuranone) (see Chapter V). A convenient name "aurone" was introduced by Bate-Smith and Geissman.[88] The aurones seem to be particularly closely associated with the Coreopsidinae, a subgroup of the Compositae. Aurones are usually orange colored and are isomeric with flavones, not only in structural features and distribution in plants, but also chemically. From the ray flowers of *Coreopsis grandiflora,* Geissman and co-workers[87] have isolated four compounds: one was the known flavone, luteolin (**135**), 8-methoxybutin (**136**), the aglucone "leptosidin" (**137**) and the glucoside "leptosin" (**138**). The orange-yellow pigment "aureusidin" (**139**) was first isolated in the heptaacetate form of its glucoside, "aureusin", from a yellow variety of the garden snapdragon[89-91] and other flowers, for example, *Oxalis cernua.*

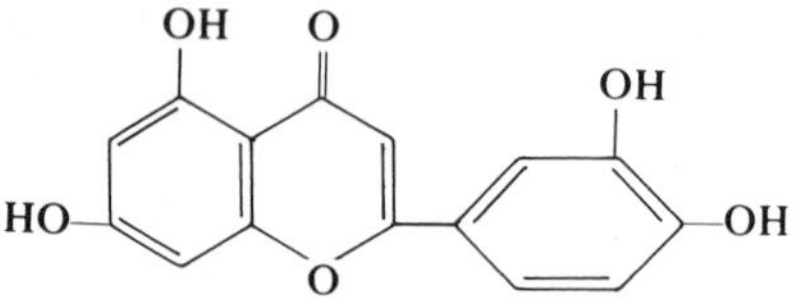

OH O OH HO OH O **135**

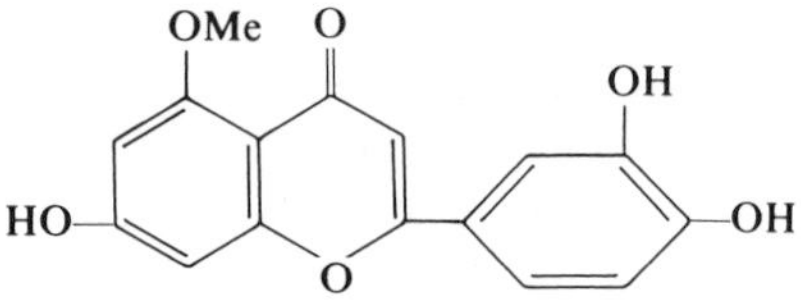

OMe O OH HO OH O **136**

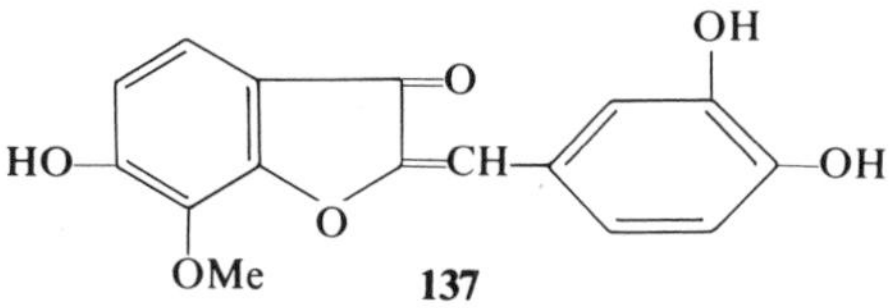

OH O HO CH OH O OMe **137**

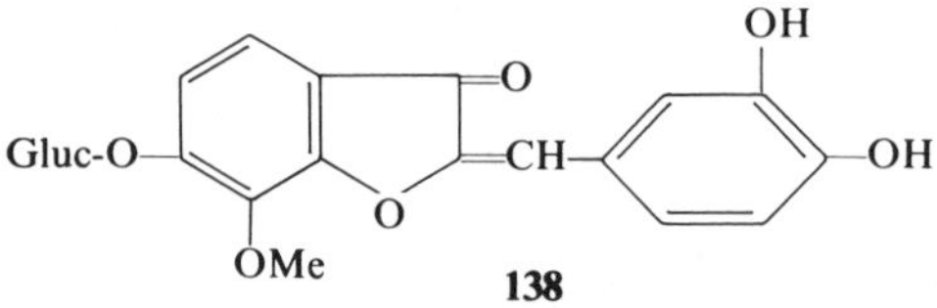

OH O Gluc-O CH OH O OMe **138**

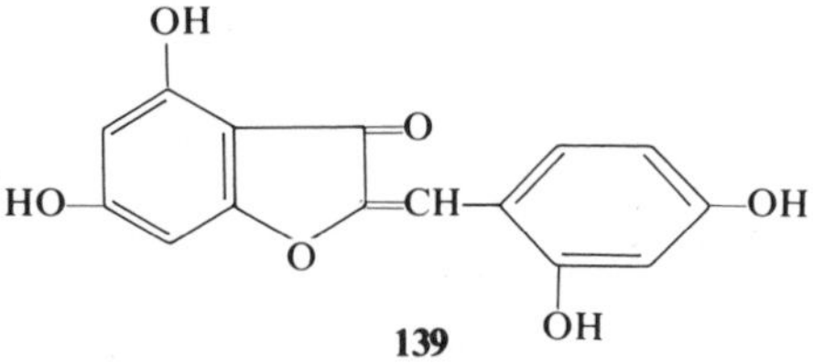

OH O HO CH OH O OH **139**

The coexistence of an aurone glycoside and a chalcone glycoside of corresponding structure was first observed in the ray flowers of sulfureus, *Coreopsis lanceolata* and *C. saxicola*;[92] coreopsin (**140**) and sulfurein (**142**) occur in the former, and lanceolin and leptosin (**138**) in the *Coreopsis* species. Such a coexistence of these two kinds of related pigments has also been shown in *Coreopsis martima, C. gigantia*,[93,94] and *C. tincoria*,[95] thus identifying the coloring constituents as butein (**141**), coreopsin (**140**), sulfurein (**142**), marein (**143**), maritimein (**144**) and luteolin-7-glucoside (**144**). Furthermore, it has been shown that the chalcone pigments, when stored, slowly isomerize to the related flavanones and oxidize to the corresponding aurones.

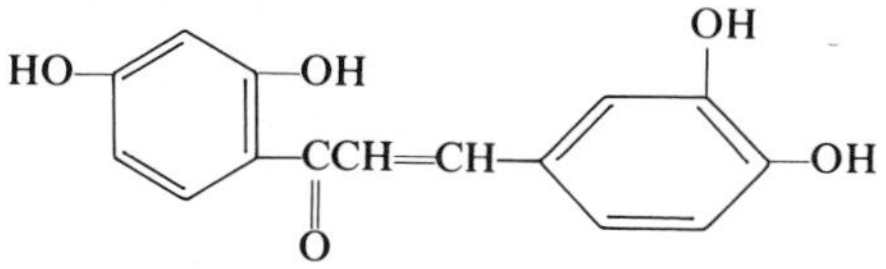

141

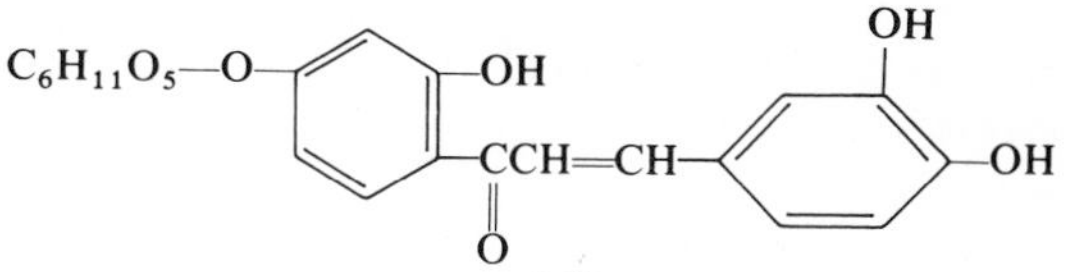

140

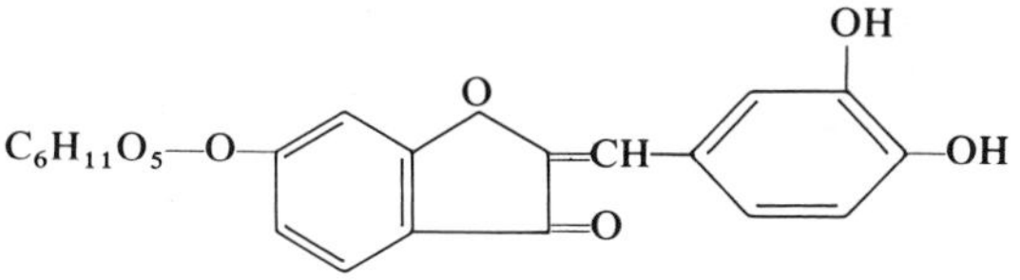

142

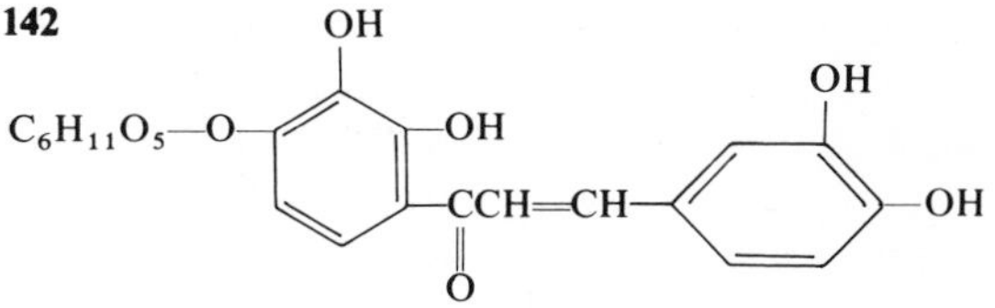

143

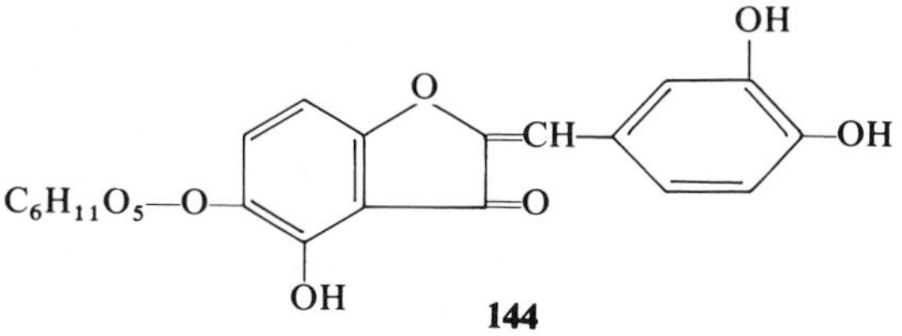

144

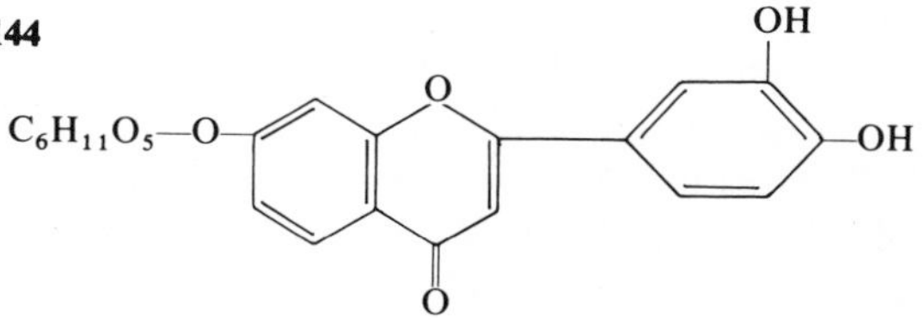

145

It seems likely, moreover, that there is a close biogenetic relation between a chalcone and an aurone of the corresponding structure; but this cannot be generalized. The aurones found in *Antirrhinum* or *Oxalis* have phloroglucinol-type structures and no corresponding flavanones or chalcones have been found to coexist with them. However the phloroglucinol-type flavanones have been found alone with no evidence for the coexistence for a chalcone or an aurone of the corresponding structure.[96,97]

These pigments lend themselves to study on a very small scale, because their ultraviolet spectra are characteristic, especially in conjunction with the techniques of paper chromatography. The naturally occurring chalcones and aurones and their glycosides show an intense absorption band in the region of 360–420 mμ and the absorption spectra have been found to readily differentiate the two types of pigments.[91,98] The differentiation is best with the acetates, either of the aglycons or the glycosides. Acetylation of the phenolic hydroxy groups substantially nullifies their effects upon the ultraviolet absorption, a polyacetoxychalcone will have a spectrum similar to that of benzalacetophenone, and polyacetoxyaurone will show a resemblance to that of benzalfuranone. With chalcone itself and with the acetyl derivatives of polyhydroxychalcones and their glycosides, the longer wavelength band has one maximum. With the corresponding compounds of the benzalfuranone series, two maxima are observed.[99]

a. Aureusin and Aureusidin

The yellow-orange pigment aureusidin, 2-(3,4-dihydroxybenzylidene)-4,6-dihydroxy-3(2*H*)-benzofuranone (**146a**), was first isolated in the hepta-acetate form of its glucoside, aureusin [2-(3,4-dihydroxybenzylidene)-6-(β-D-glucopyranosyloxy)-4-hydroxy-3(2*H*)-benzofuranone], aureusidin-6-glucoside, **146b**, from a yellow variety of the garden snapdragon. *Antirrhinum majus*,[99] and *Oxalis cernua L.*[89,90] contain aureusin (**146b**) and, in larger amounts, another aureusidin glucoside, cernuoside (**147b**). Since both the latter pigments are monoglucosides of aureusidin, the establishment of the point of attachment of the two pigments would determine their complete structures. Particular use was made of the fact that the spectra of aurones, bearing a 4-hydroxy group, are shifted by the addition of aluminum chloride.[100] Aureusin (**146b**) and cernuoside (**147b**), both of which are glucosides of 3′,4′,6-trihydroxy-aurone (aureusidin), have been shown to be the 6- and 4-glucoside, respectively.[101]

Aureusidin (**146a**) has been synthesized by the condensation of 4,6-dihydroxy-3(2*H*)-benzofuranone with 3,4-dihydroxybenzaldehyde in the presence of acetic anhydride to yield the tetraacetate which, on hydrolysis, yielded the yellow **146a**.[102]

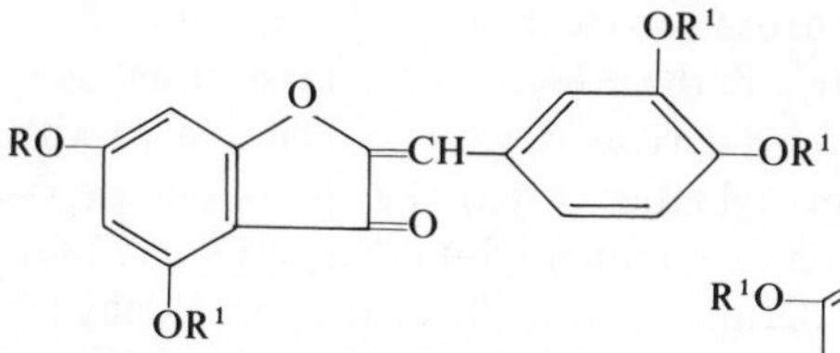

146a R = R^1 = H (aureusidin)
146b R = glucoyl, R^1 = H (aureusin)

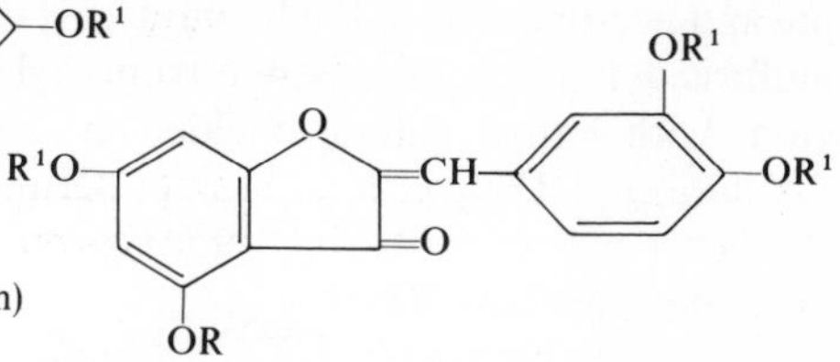

147a R = R^1 = H (cernuine)
147b R = glucosyl, R^1 = H (cernuoside)

Isolation of an *O*-rhamnopyranosyl (1 → 6)-glucopyranoside of aureusidin (3′,4′,4,6-tetrahydroxyaurone, **148**), from a basic alcoholic extract of lemon peel[103] suggested that **148** might be formed during oxidation of eriocitrin (**149**), a 7-rutinoside of eriodictyol (3′,4′,5,7-tetrahydroxyflavanone), liberally present in lemon peel.[104]

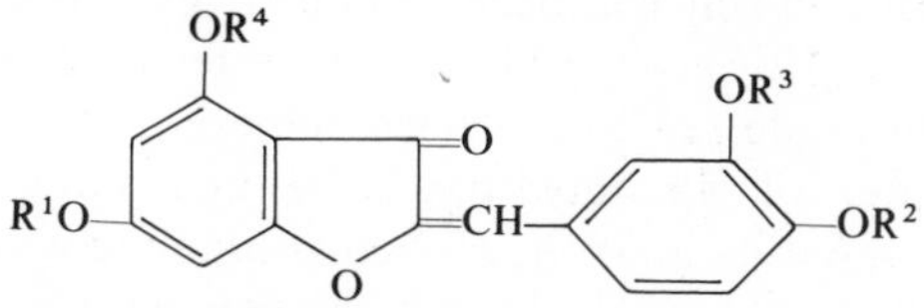

148 R^1 = rhamnoglucosyl, R^2 = R^4 = H, R^3 = H

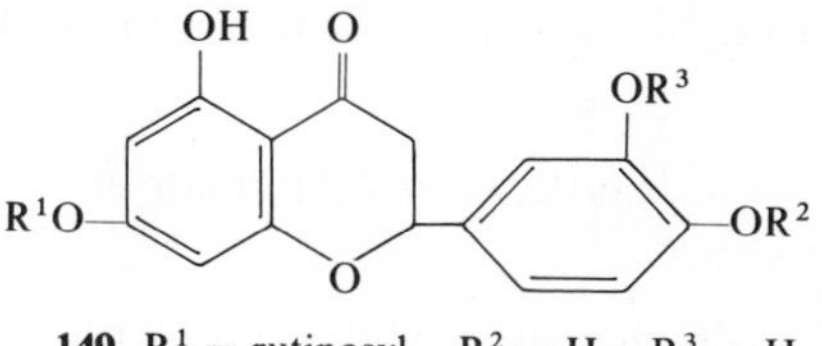

149 R^1 = rutinosyl, R^2 = H, R^3 = H

b. Cernuoside

Cernuoside [2-(3,4-dihydroxybenzylidene)-4-(β-D-glucopyranosyloxy)-6-hydroxy-3(2*H*)-benzofuranone, **147b**, aureusidin-4-glucoside, $C_{21}H_{20}O_{11}\cdot H_2O$], was isolated from the ray flowers of *Oxalis cernua*,[89,90] and flowers of *Petrocosmea kerii.*[105] The ultraviolet spectrum of **147b** was not altered by the addition of aluminum chloride, so there is no free 4-hydroxyl group. That **147b** contains a free 4′-hydroxyl group is indicated by the shift induced by alkali; there is evidence that in aurones, as in flavones, conjugation of the carbonyl group with 4′-hydroxyl group dominates all other conjugation effects. For example, the alkaline shift for 6-hydroxy-4′-methoxyaurone is only 18 mμ, while that for 4,6′-dihydroxyaurone is 66 mμ.[101] Cernuoside is readily methylated with methyl sulfate in the presence of potassium carbonate and acetone to give the heptamethyl derivative which, when treated with sulfuric acid (3%), gave 3′,4′,6-trimethyl ether (**150a**). The latter on ethylation with ethyl sulfate yielded 3′,4′,6-trimethyl-4-ethyl ether (**150b**). Oxidation of **150b** with potassium permanganate afforded veratraldehyde; the latter was also obtained upon ozonolysis of tetramethyl ether **150c** in cold ethyl acetate.[106]

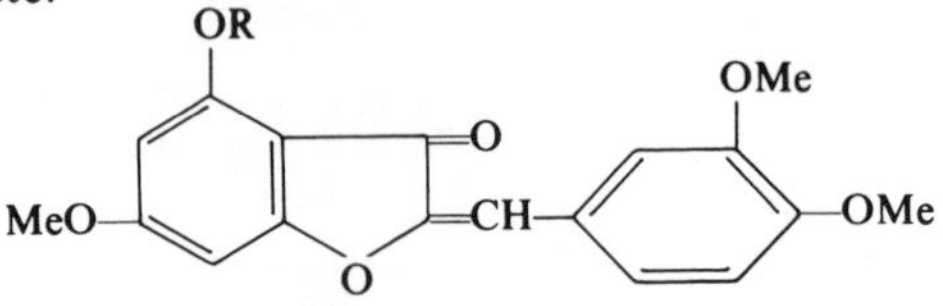

OR, OMe, O, MeO, CH, OMe, O

150a R = H
150b R = Et
150c R = Me

The aglycone, termed "cernuine" [2(3,4-dihydroxybenzal)-4,6-dihydroxy-3(2*H*)-benzofuranone, **147a**] has been obtained upon treatment of **147b** with sulfuric acid (3%); it gives a black–brown ferric chloride test.

Synthesis of cernuoside (**147b**) has been achieved.[102] Condensation of the 4-(β-D-glucoside) of 4,6-dihydroxy-3(2*H*)-benzofuranone, obtained upon treatment of 4,6-dihydroxy-3(2*H*)-benzofuranone with acetobromoglucose with 3,4-dihydroxybenzaldehyde in the presence of acetic anhydride, followed by hydrolysis of the reaction product, gave cernuoside [4,6-dihydroxy-2-(3′,4′-dihydroxybenzal)-3(2*H*)-benzofuranone-4-(D-glucoside)].

c. *Leptosin and Leptosidin*

Leptosin [2-(3,4-dihydroxybenzylidene)-6-(β-D-glucopyranosyloxy)-7-methoxy-3(2*H*)-benzofuranone, leptosidin 6-glucoside, **138**], the first example of the naturally occurring aurone glycosides, and aglycone leptosidin (**137**) were isolated from the ray flowers of *Coreopsis grandiflora* Nutt.[87] The assigned structure was based upon the synthesis of leptosidin trimethyl ether[87] and the color reactions. Interpretation of these reactions led to the assignment of the location of the methoxyl and glucosidoxyl groups. The total synthesis of **138** and **137** has been achieved[107] as outlined in Eq. 12.

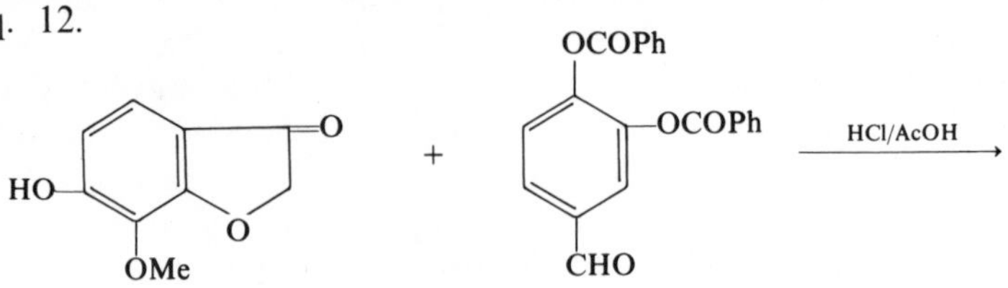

O, HO, O, OMe + OCOPh, OCOPh, CHO $\xrightarrow{\text{HCl/AcOH}}$

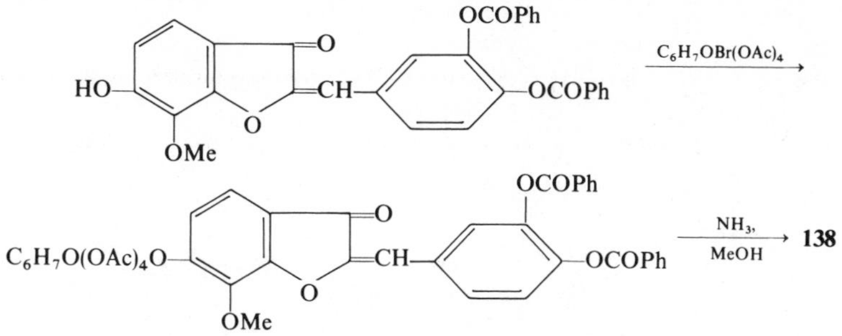

$\xrightarrow{C_6H_7OBr(OAc)_4}$

$\xrightarrow[\text{MeOH}]{NH_3,}$ **138** (12)

In a similar manner, condensation of the 6-hydroxy-7-methoxy-3(2*H*)-benzofuranone-6-tetra-*O*-acetyl-*β*-D-glucopyranosyl derivative with protocatechualdehyde in acetic anhydride gave 2-(3,4-diacetoxybenzal) derivative which was saponified to furnish **138**.[108]

Partial synthesis of an aurone glycoside from the corresponding chalcone glycoside has been accomplished by atmospheric oxidation in alkaline solution.[92] Thus leptosin **138** can be prepared from the corresponding chalcone glycoside, coreopsin **140**. Leptosin has also been synthesized by Puri and Seshadri.[109]

d. Sulfuretin and Sulfurein

Sulfuretin [2-(3,4-dihydroxybenzylidene)-6-hydroxy-3(2*H*)-benzofuranone, **151**] and sulfurein [sulfuretin-6-glucoside, 2-(3,4-dihydroxybenzylidene)-6-(*β*-D-glucopyranosyloxy-3(2*H*)-benzofuranone, **142**] have been isolated from the ray flowers of *Cosmos sulfureus*,[92] *Dahlia variabilis*,[110] and from heartwood constituents of *Rhus javanica*.[111] The flower of petals of *Viguiera multiflora* Nutt. and *Baeria chrysostoma* F. and M. (Compositae), neither of which is a member of the subtribe Coreopsidinae, have been found to be pigmented with chalcones and aurones. The presence of these anthochlor pigments (the designation "anthochlor" has been applied to the polyhydroxychalcones and aurones typified by the widely distributed compounds butein, **141,** and sulfuretin, **151,** in plants not in the subtribe in which they were formerly grouped, seems to be characteristic and offers further evidence concerning their biosynthetic relationships.[112] The formulation of this reaction suggests possible implications in the oxidation of other 3,4-dihydroxylated flavonoid compounds.

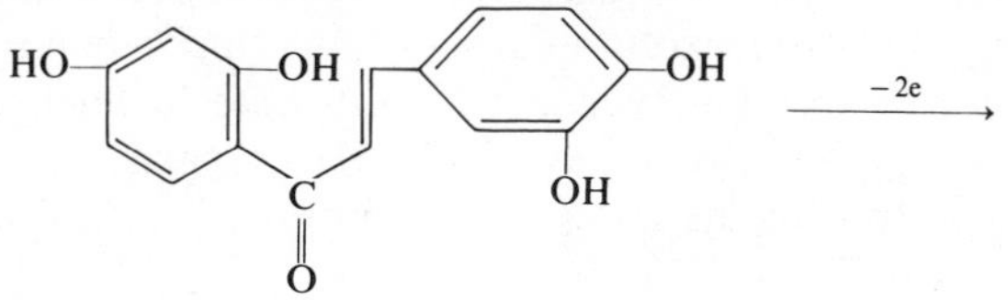

$\xrightarrow{-2e}$

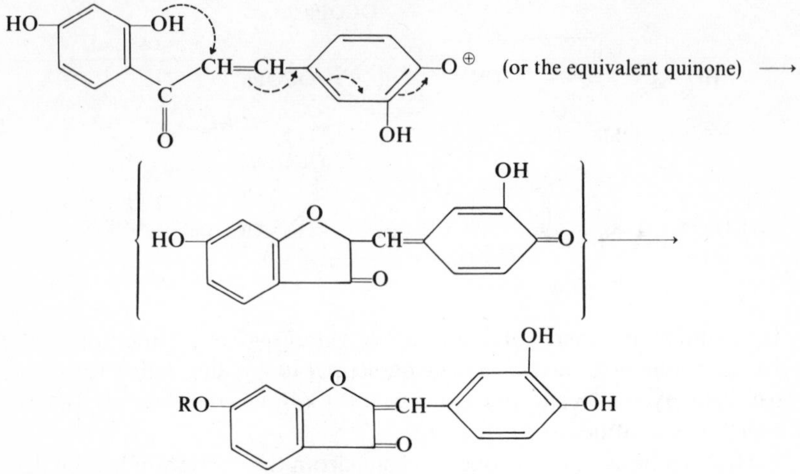

(or the equivalent quinone) ⟶

151 R = H
142 R = $C_6H_{11}O_5$

Sulfurein (**142**) can be partially synthesized from the corresponding chalcone glycoside, coreopsin **140,** by atmospheric oxidation in alkaline medium.[92] Zemplen et al.[113] have prepared (**142**) when 6-hydroxy-3(2*H*)-benzofuran-one glucoside condensed with protocatechuic aldehyde. The properties of the product thus obtained, believed to be identical with sulfurein, differ from those of the natural glucoside. Farkas et al.[114] have synthesized **142**; their product showed all the characteristics ascribed to the natural compound.

e. Palasitrin

Palasitrin [sulfuretin-3′,6-diglucoside, 6-(β-D-glucopyranosyloxy)-2,(3-(β-D-glucopyranosyloxy)-4-hydroxybenzylidene)-3(2*H*)-benzofuranone, **152**], is isolated from the flowers of *Butea frondosa.*[109,115] Methylation followed by hydrolysis affords 6,3′-dihydroxy-4′-methoxyaurone; this is identical to the synthetic material.[107] The related chalcone (3,4′-isobutrin, 3,4′-butrin diglucoside) is found in association with palasitrin,[109,115] and may be converted to it by addition of bromine to the acetate, followed by action of alkali.[107,109]

Direct synthesis of palasitrin has been brought about by condensation of 6-hydroxy-2,3-dihydrobenzofuran-3-one-3-β-D-glucoside-6-tetraacetate[113] in the presence of acetic anhydride and protocatechualdehyde-β-D-glucoside-3-pentaacetate; the nonaacetyl derivative of palasitrin has thus been obtained. Hydrolysis of the latter gave a product identical to the natural one.[116]

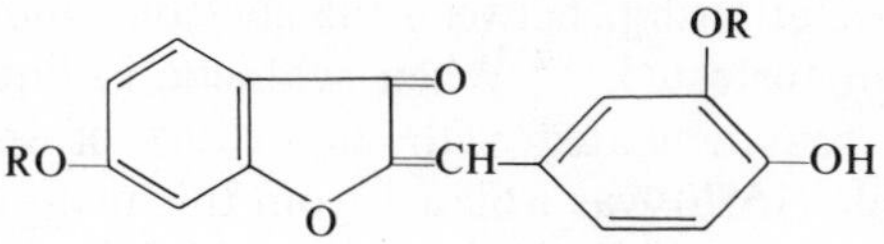

152 R = $C_6H_{11}O_5$

Hydrolysis of **152** yielded 2 moles of glucose and an aglycon (palasetin) identified as **153** by comparison with a synthetic sample obtained by condensation of 6-hydroxy-3(2*H*)-benzofuranone with protocatechualdehyde.

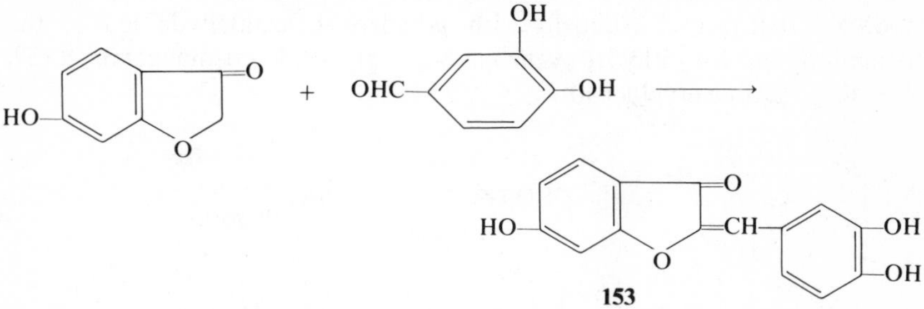

153

Although sulfuretin (**151**) and palasetin (**153**) have been assigned the same structure, their identity does not appear to have been established by a direct comparison, and there are differences in melting points for the two cited compounds and their triacetyl derivatives.

f. Hispidol

Hispidol [6-hydroxy-2-(*p*-hydroxybenzylidene)-3(2*H*)-benzofuranone], was isolated from soybean (*Soya hispida*) seedlings with its flavonoid constituents,[117] for example a chalcone, isoliquiritigenin (**154a**).[119,120]

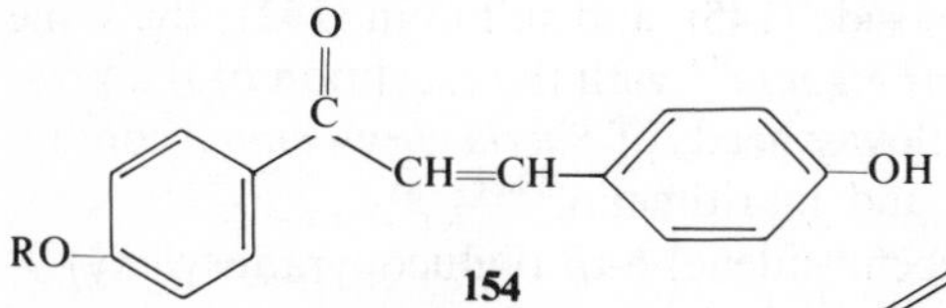

154

154a R = H
154b R = glucosil

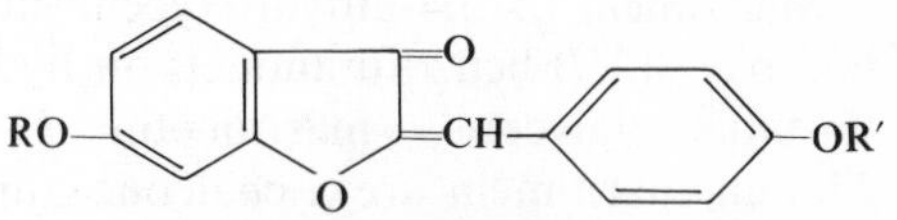

155

155a R = R′ = H
155b R = glucosil, R′ = H
155c A = H, R′ = glocosil
155d R = R′ = glocosil

The biogenetic relationship between the chalcone and aurone constituents had been demonstrated.[117] When synthetic isoliquiritigenin-4′-β-D-glucoside (**154b**) was incubated with an extract of soybean seedlings, hispidol-6-glucoside (**155b**) was isolated from this material. This relationship was further supported by hydrolysis of **154b** to hispidol (**155a**).[118]

Synthesis of hispidol has been effected by condensation of 6-hydroxy-3(2*H*)-benzofuranone (**156**, R = H) and *p*-hydroxybenzaldehyde; subsequent deacetylation of the reaction product (6,4′-diacetoxyaurone) gave rise to hispidol that was identical in every respect with the natural product.[119] Similarly, condensation of 6-hydroxy-3(2*H*)-benzofuranone-β-D-glucoside (**156**, R = tetraacetyl glucosil) with *p*-hydroxybenzaldehyde led to the formation of 6,4′-dihydroxyaurone-6-β-D-glucoside pentaacetate (**157**, R = R^1 = tetraacetyglucosil).

RO–(benzofuranone)=O **156** + R^1O–C_6H_4–CHO $\xrightarrow[\text{(2) hydrolysis}]{\text{(1) } Ac_2O;}$ RO–(aurone)=CH–C_6H_4–OR^1 **157**

Catalytic saponification afforded pure hispidol-6-β-D-glucoside (**155b**).

g. Maritimein and Maritimetin.

Investigation of the constituents of the flowers of *Coreopsis maritima* illustrated the presence of butein (**141**), coreopsin (**140**), marein (**143**), maritimein (**144**), luteolin 7-glucoside (**145**), and sulfurein (**142**); the same was found to be true of *Coreopsis giganta*[93] with the exception of the presence of pigment **145.** The whole flower heads of *Baeria chrysostoma* contain marein, sulfuretin, maritimein, and maritimetin.[112]

Maritimein [2-(3,4-dihydroxybenzylidene)-6-(β-D-glucopyranosyloxy)-7-hydroxy-3(2*H*)-benzofuranone], on hydrolysis, gave glucose and a rather unstable aglycone, maritimetin (3′,4′,6,7-tetrahydroxyaurone, **158**). Marein–maritimein are a chalcone–aurone pair, interrelated in the same way as lanceolin–leptosin,[92,] and coreopsin and sulfurein. Since marein (**143**), a glucoside of **159** (R = H), is converted by aerial oxidation to maritimein (**144**), the latter must have the corresponding aurone structure and must be a glucoside of **158** (R = H).

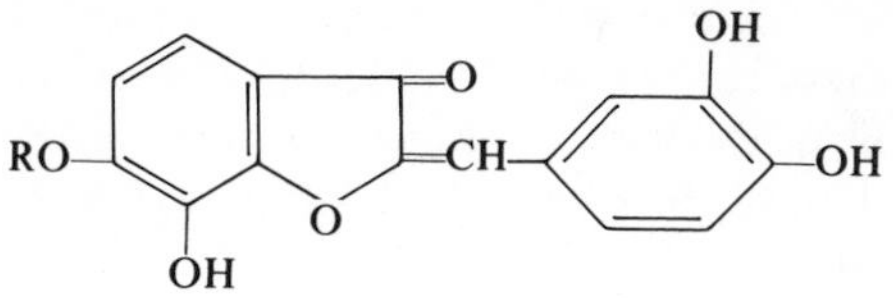

158 R = H (maritimetin)
144 R = $C_6H_{11}O_5$ (maritimein)

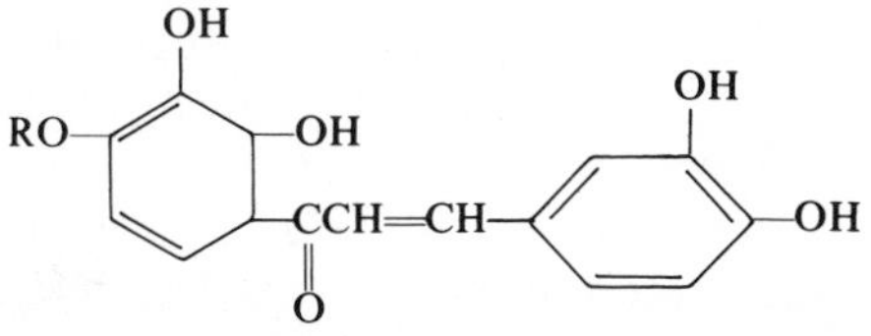

159

This fact was confirmed in different ways. Careful acid hydrolysis of **144** gave the aglycone which is identical in all respects with the synthetic 6,7,3′,4′-tetrahydroxyaurone (**158,** R = H), obtained from 6,7-dihydroxy-3(2*H*)-benzofuranone and protocatechualdehyde.[94,108] The position of the sugar molecule in **144** was determined by complete methylation, yielding a compound presumably having structure **160** which, on acid hydrolysis gave 6-hydroxy-7,3′,4′-trimethoxyaurone (**161**).[94,95]

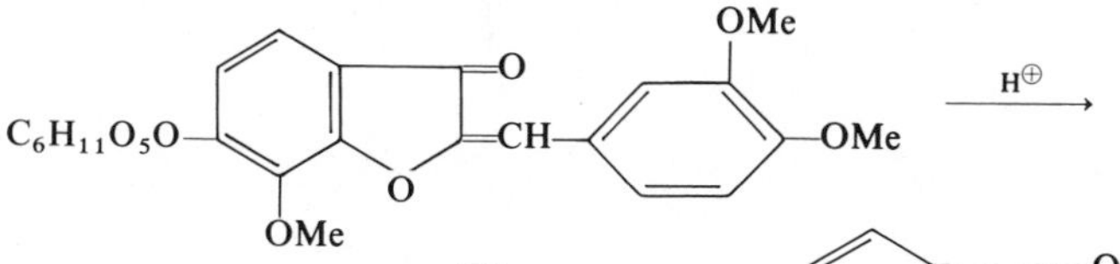

160

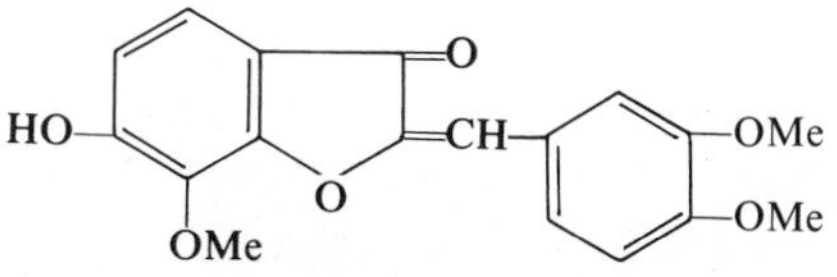

161

h. Rengasin

Examination of the ethereal extract of the ground heartwoods of rengas (*Melanorrhea* Sp.), a timber-producing tree of Malaya, led to the isolation of a sticky orange solid and yellow crystals. The water-insoluble part of the former gave rengasin, $C_{15}H_9O_5 \cdot OMe$. The formation of a trimethyl- and triethyl ether and a triacetate indicated three phenolic hydroxyl groups; the color reactions (ferric reaction—green; alkali—reddish-purple solution) suggested it to be, after King et al.[121] 2-benzylidene-6-methoxy-4,3′,4′-trihydroxy-3(2*H*)-benzofuranone. It gives a brown precipitate with chloropentammine cobalt, which indicated that the oxygen substituents of the benzylidene group were unmethylated. As methoxylation next to a carbonyl group in natural products is rarer than methoxylation in other positions, it was possible that rengasin is the 6-methoxy compound (**162**, R = R′ = H, R″ = Me). This hypothesis was confirmed when a synthetic specimen of the ethyl ether (**162,** R = R′ = Et, R″ = Me) proved to be identical with the ethyl ether of the natural product.

Synthesis of tri-*O*-ethylrengasin has been accomplished from 3′,4′,6-triethoxy-2-hydroxy-4-methoxychalcone (**163**), obtained by base-catalyzed condensation of 3,4-dimethoxybenzaldehyde and 6-ethoxy-2-hydroxy-4-methoxyacetophenone upon treatment with alkaline hydrogen peroxide. Later, Farkas et al.[122] synthesized the four possible monomethyl ethers of **162** (R = R′ = R″ = H) and, contrary to structure **162** (R = R′ = H, R″ = Me) proposed by King et al.,[121] Farkas proposed the 4-methyl ether for rengasin **162** (R = R″ = H, R′ = Me).

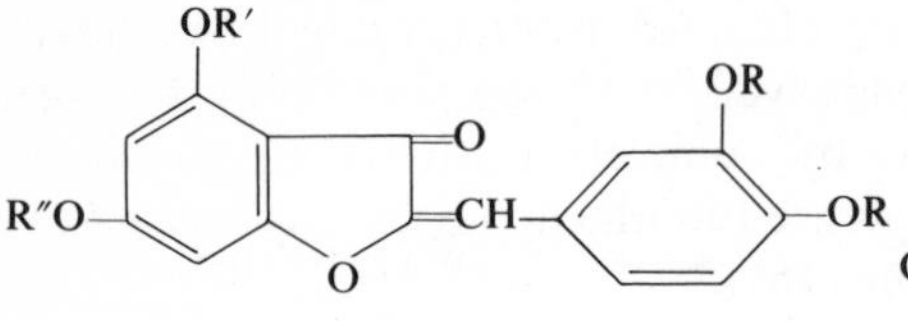

OR′ OR O R″O CH OR O

162

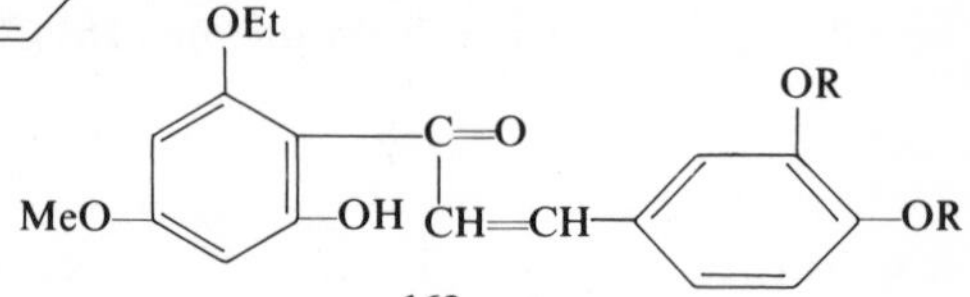

OEt OR C=O MeO OH CH=CH OR

163

i. Bracteatin and Bractein

Harborne[123] has shown that the aurone glycoside, isolated from *Antirrhinum majus*, together with aureusin (aureusidin-6-glucoside, **146b**)[96] gives, on acid hydrolysis or enzymic hydrolysis, glucose and an aglycone having one or more hydroxyl groups in its structure than aureusidin (**146a**). From study of its spectral properties, the most likely structure for this aglycone appeared to be 4,6,3′,4′-5′-pentahydroxyaurone, bracteatin [4,6-dihydroxy-2-(3,4,5-trihydroxybenzylidene)-3(2*H*)-benzofuranone, **164**], a compound isolated as the 4-glucoside, bractein [4-(β-D-glucopyranosyloxy)-6-hydroxy-2-(3,4,5-trihydroxybenzylidene)-3(2*H*)-benzofuranone, **165**], from flower plants of *Helichrysum bracteatum*.[124] A direct comparison of the aglycone and bracteatin showed them to be identical. It is hydrolyzed by the action of β-glucosidase as rapidly as aureusin, and is therefore presumably a simple glucoside; its spectral properties indicate that the glucose residue is attached either to the 4-hydroxyl group or to any of the β rings by hydrolysis; and its R_f values are related to those of the isomeric 4-glucoside (bractein) in the same way as the R_f of the 6- and 4-glucosides of aureusidin.

Condensation of 4,β-D-glucoside of 4,6-dihydroxy-3(2*H*)-benzofuranone with 3,4,5-trihydroxybenzaldehyde and subsequent hydrolysis of the reaction product gave the 4-β-D-glucoside of 4,6,3′,4′,5′-pentahydroxyaurone which is identical with bracetein (**165**).

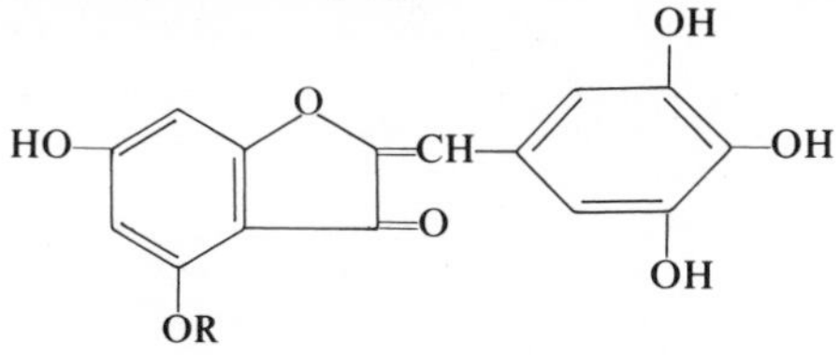

OH

HO— O —CH— —OH

=O

OH

OR

164 R = H
165 R = $C_6H_{11}O_5$

D. Synthetic Aurone Analogs

During the last decade interest has been centered on synthesizing a large number of aurone analogs, making use of the methods discussed in Chapter V, namely: (a) oxidation of 2-hydroxychalcones with alkaline hydrogen peroxide (AFO reaction)[125-129] and/or manganic acetate in acetic acid medium;[130] (b) cyclization of 2′-hydroxy-α,β-dibromodihydro-chalcones with alkali;[131-150] and (c) alkaline rearrangement of 3-hydroxyflavanones.[151-156]

The physical properties of the naturally occurring aurones, their synthetic analogs [2-arylidene-3(2*H*)-benzofuranones], and 3-arylidene-2(3*H*)-benzofuranones are listed in Tables IIa, IIb, IIc, IId, respectively.

E. Chemical Properties

Aurones undergo ring expansion during their rearrangement into the corresponding flavones and flavanones. Chalcone dibromides (**166**) give 2-(*p*-alkoxybenzylidene)-3(2*H*)-benzofuranones (**168**) with hot alcoholic postassium hydroxide.[157] However, in absence of alcohol 4′-alkoxyflavones (**167**) are obtained.[140] Hutchins and Wheeler[138] have shown that **166** also yielded **167** upon treatment with alcoholic potassium cyanide. Furthermore, it has been found that [158] the arylidene-3(2*H*)-benzofuranones (**168**) are intermediates in the cyanide conversion of **166** into **167**. The earliest example of this type of ring enlargement was pointed out by Auwers et al.[159-160] More recently it has been shown that aurones undergo oxidative ring expansion when treated with alkaline hydrogen peroxide (Chart 1).[161]

Aurone epoxides undergo ring expansion upon treatment with alkali to

TABLE Ia. NATURALLY OCCURRING BENZOFURANS

No.	Name and substituents (synonyms, formula)	Natural source	Appearance	$(\alpha)_D$	Ref.
	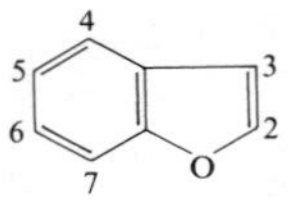				
1	5-OMe (**1**)	*Sterum subpileatum* Berk. et Curt.			1
2	3-Me,6-(4′methylpent-3′-enyl) (**2**) furoventalene,	*Gorgonia ventalina* (seafan)		+ 10°	2
3	2-Isopropenyl, 5-Ac, 6-OH (euparin, **8**)	*Eupatorium purpureum* (gravel roots)	Pale yellow		4,5,6,7
4	2-Isopropenyl, 5,6-di-OMe (**13**)	*Ligularia stenocephala* et Koidz (Compositae)		± 0°	11
5	2-Acetoxyisopropenyl, 5-Ac, 6-OH (ageratone, **15**)	*Ageratum houstonianum* Mill. (roots)			12
6	4-OMe, 5-$COCH_2COPh$ (pongamol, **18**)	*Pongamia glabra*; identical with lanceolatin from *Tephrosia lanceolata*	Yellow		16,17,87, 173, 174
7	2-Isopropenyl, 5-Ac (dehydrotremetone, **24**)	*Eupatorium urticaefolium* (white snakeroot)		± 0°	22,23
	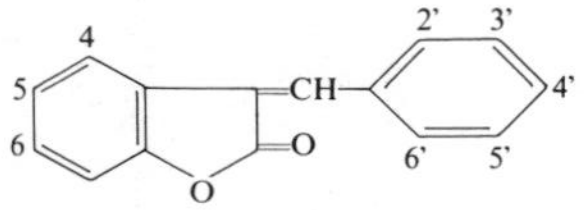				
8	2′-OMe, 6′-OH, 3′,4′(:O_2H_2) (**53**)	Baker's yeast			37
9	6,3′-Di-OH, 2′,4′-(:O_2CH_2) (pterofuran, **58**)	*Pterocarpus indicus* (heartwood)			39,69
10	3-Me, 5-Propenyl, 7-OMe, 3′,4′-(:O_2CH_2) (eupomatene, **63**)	*Eupomatia laurina* R. Br. (bark)			41–43
11	5-γ-Hydroxypropyl, 7-OMe, 3′,4′-(:O_2CH_2) (egeno, **67**)	*Styrax japonicus* Sieb. et Zucc. (seed-oil); *S. formanesus* Matsum. *S. Obassia* Sieb. et Zucc.			47,53,54
	(egonol glycoside)	*Styrax officinalis* L.; S. *suberifolium* Hook et Arn. (fruits)		− 29.8°	55

12	5-γ-Hydroxypropy1–7,3′,4′-tri-OMe (**83**)	*Styrax oficinalis* L. (hydrolysis of the glycosidic fraction)	Colorless	± 0°	58
	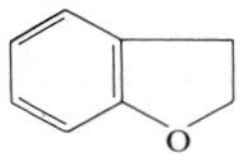				
13	2,3,3-Tri-Me, 5-propenyl (anisoxide, **86**)	*Illicum verum* (star anise oil)	Colorless oil	± 0°	59,60
14	2-Acetoxyisopropenyl, 5-OMe, 6-OH (dihydroageratone, **16**)	*Ageratum houstonianum* Mill. (roots)			12
15	2-Isopropenyl, 5-Ac (tremetone, **23**)	*Eupatorium urticaefolium* (white snakeroot)		− 59.6°	22,23,30
16	2-Isopropenyl, 5-Ac, 6-OH (hydroxytremetone, **17**)	*Eupatorium urticaefolium* (white snakeroot)		− 46.4 -; − 50.7°	22,23,30
17	2-Isopropenly, 4-OH, 5-Ac, 6-OMe (remirol, **100a**)	*Remirea maritima* (rhizomes) (Cyperaceae)	Pale yellow needles	+ 66.5°	62
18	3-Me, 5-OH, 6-OMe (obtusafuran, **101**, $R^1 = R^2 =$ Me)	*D. obtusa* Lecomte (sym. *D. Retusa*) (heartwood)		+ 47°	63
19	5,3′-Di-OH, 6,4′-di-OMe (melanoxin, **103**)	*Dalbergia melanoxyls* Guil. and Perr. (Leguminoseae)		− 99°	65
20	3-COAgm, 5-CH═CH(COAgm), 4′-OH (Hordatine A, **109a**)	Barley coleoptiles seedlings (fungal factors)		+ 69°	66–68
	(Hordatine A glycoside, **109c**)	Barley coleoptiles seedlings			66–68
21	3-COAgm, 5-HC═CH(COAgm), 7-OMe, 4′-OH (Hordatine B, **109b**)			+ 54°	66–68
	(Hordatine B glycoside)	Barley coleoptiles seedlings			66–68

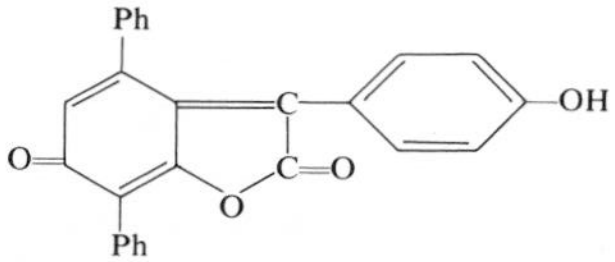

TABLE Ia. NATURALLY OCCURRING BENZOFURANS (continued)

No.	Name and substituents (synonyms, formula)	Natural source	Appearance	$(\alpha)_D$	Ref.
22	5-OH (Xylerythrin, **113**)	*Decaying* wood attacked by fungus (*Peniophora sanguinea* Bres. (*Corticuin sanguinaum* Fr.)	Blackish with green luster		69
23	5-OMe (5-methylerthrin, **116**)	Decaying wood attacked by fungus	Blackish		70
	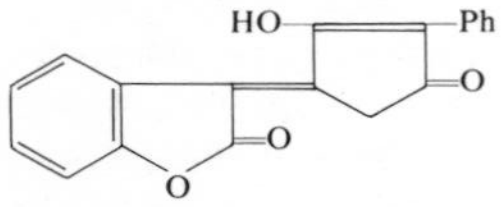				
24	Unsubstituted (calycin, **125**)	*Sticta aurata* Ach.; *S. ceronata* Ach.; *S. colensei*; *Candelaria concolor* Dicks.; *Gyalolechia aurella* Hoffm.; *Lepraria candelaris* Schaerer, *L. chorina* Ach.; *L. flava* Schreber; *Gasparrima medians* Nyl.; *Physcia medians* Byl.; *Chrysothrix nolitangera* Ment.; *Callopsima vitellinum Ehrh*; also produced by purely fungal cultures from *Sticta coronata* Muell. Arg.			75,76
	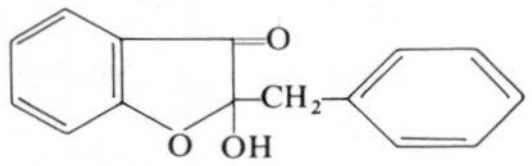				
25	4,6,4′-Tri-OH (maesopsi, **128**)	*Maesopsis eminii* Rhamnaceae (heartwood) (African timber "musizi")	Colorless	± 0°	81,84
26	4,6,3′,4′-Tetra-OH (alphitonin, **132a**)	*Alphitonia excelsa* (heartwood of Australian red ash)	Colorless	+ 44.5°	82,86

TABLE Ib. NATURALLY OCCURRING BENZOFURANS

Benzofuran (No., see Table Ia)	Melting point (°C)	Solvent for crystallization[a]	Ultraviolet spectrum [max mμ, log E]	Remarks	Ref.
1	34	A	221(3.76), 247(3.99), 293(3.59), 301(3.55)		1
2			249 (E, 16000), 282 (E, 4200), 287 (E, 6200)		2
3	118.5–121	A		Green color ($FeCl_3$); 2,4–DNP (K), m.p. 252°; oxime (C + J), m.p. 147–148°; acetate (A), m.p. 80°; *O*-methyl ether (C + J), m.p. 76–77°; tetrahydroeuparin (A), m.p. 57°; oxime (A), m.p. 139°; acetate (A), m.p. 96–97°; 2,4–DNP (B), m.p. 240–241°	4–7
4	72.5–73		324s(4.13), 314(4.21), 287(4.00), 278(4.00), 215(4.18)		11
5	122–124		240 (E, 4700), 325 (E, 10,000)		12
6				Blood red ($FeCl_3$); yellow (H_2SO_4)	47,53,54 58
7	87.5	A	252 (E, 39,000), 280 (E, 19,000), 292 (E, 15,500)	Oxime (B + A), m.p. 131–132°	22,23
8	118	A	302(2.20); blue fluorescence	acetate (L), m.p. 112–113°; methyl ether (A,D), m.p. 117–118.5°; ethyl ether (A), m.p. 112°	37
9	208–208.5	B – C	219.5(4.47), 235.5(4.22), 285.5(4.21), 294.5(4.26), 304.5(4.39), 317(4.58), 332(4.56)	Blue (Gibbs reaction); golden-yellow (H_2SO_4); dimethyl ether (A), m.p. 86–87°; diacetate (C), m.p. 133.5–146.5	39,69

TABLE Ib. NATURALLY OCCURRING BENZOFURANS

Benzofuran (No., see Table Ia)	Melting point (°C)	Solvent for crystal-lization[a]	Ultraviolet spectrum [max $m\mu$, log E]	Remarks	Ref.
10	154–156	D	235(4.42), 266(4.44), 312(4.38)	Dihydroeupomatene (C), m.p. 99°; 4-bromoacetyl deriv. (M), m.p. 124.5–125°; 3-nitro deriv. (N), m.p. 160°; 4-nitro deriv. (N), m.p. 139°; 4-bromo deriv. (B), m.p. 164–165°; red (egonol reaction); orange-red (H_2SO_4)	41–43
11	118	C or E			47,53,54, 176
12	120–122	F	217 (E, 24,000), 313 (E, 20,200), 300–305(infl.)	monoacetate (F), m.p. 90–91°	58
13	41; b.p. 140/11 mm.			Dihydroanisoxide, b.p. 120–122/10 mm; perhydroanisoxide, b.p. 120–122/10 mm	18,59,60
14			222 (E, 16,400), 237 (E, 20,300), 278 (E, 18,700), 325 (E, 12,800)		12
15	oil		227 (E, 11,950), 280 (E, 12,600), 285 (E, 12,300)	Red color (H_2SO_4); positive-iodoform test; semicarbazone (C), m.p. 222°; 2,4-DNP, m.p. 183.8–184.2°	22
16	70–71		236 (E, 38,700), 280 (E, 29,100), 326 (E, 19,300); (in 0.1N NaOH) 248 (E, 38,800), 280 (E, 20,300); 359 (E, 20,000)	Purple black ($FeCl_3$); acetate (A), m.p. 89–89.5°; positive-iodoform test	22, 23,30
17	76.5–77	G	294(4.26), 239(3.98), 216(4.22)	Dihydroremirol, m.p. 61.5–62°, $(\alpha)_D^{25}$ + 61°; positive Gibbs test	62
18	110–113				63

19	107–108	A	207(4.53), 233(infl.)(4.05), 291(3.73), 305(3.80); dehydromelanoxin, 213(4.47), 287(4.04), 325(4.28)	O-diacetate (C), m.p. 111–113°, $(a)_D^{22}$ − 15.5° O-dimethyl ether (F), m.p. 74–75°, $(\alpha)_D^{22}$ + 41°; dehydromelanoxin, pale yellow, m.p. 171–172°	65
20				Dipicrate, m.p. 127–128°	66–68
21				Dipicrate, m.p. 132–135°	66–68
22	265–266	H/B	255(4.37), 360(3.94), 450(4.18); diacetate, 242(4.31), 360(4.14), 400(infl.) (4.09); dimethyl ether, 245(4.38), 358(4.01), 443(4.21)	Diacetate (H/A), m.p. 228–230°; dimethyl Ether (N/J), m.p. 213–215°	69
23	250–256	I	246(4.37), 327(4.03), 448(4.24); acetate, 246(4.33), 384(4.18)	Acetate, m.p. 208–210°	70
24	249–249.5		208(E, 19,800), 237(E, 15,700), 430(E, 24,700)		75,76
25	218–220 (dec.)	D	211(4.38), 290(4.28); tetramethyl ether, 213(4.42), 293(4.28)	Purple color ($FeCl_3$); red color with vanillin–hydrochloric acid; tetramethyl ether, m.p. 130–131°(F)	81,84
26	225–226	J	290(4.31)	Purple ($FeCl_3$); dibromopentamethyl ether (F), m.p. 193–194°; pentamethyl ether oxime (K/F), m.p. 224°	86,82

[a] A, petroleum ether; B, benzene; C, ethyl alcohol; D, ether; E, *n*-butyl alcohol; H, chloroform; I, carbon tetrachloride; J, water.

TABLE IIa. NATURALLY OCCURRING AURONES

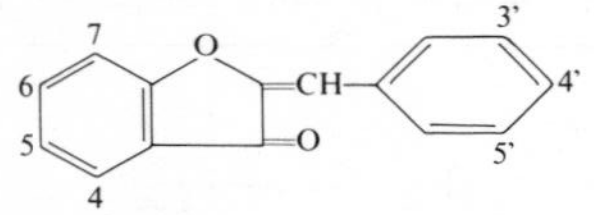

No.	Name and substituents (synonyms, formula)	Natural source	Solvent for crystal-lization[a]	Melting point (°C)	Appearance	Remarks	Ref.
1.	4,6,3′,4′-Tetra-OH (aureusidin, cernuin, **146a**)	*Antirrhinum*; *Oxalis cernua*	A	270 ±, 295 ± (rate of heating)	Deep yellow	Orange-red (10% NaOH), orange-red (H_2SO_4), brown ($FeCl_3$)	91,101
	(Tetraacetate)		B or C − D	184–185	Yellow tint	Slow orange-red color (10% NaOH), orange-red (H_2SO_4)	91,101
	(Tetramethyl ether)		E (75%)	173–174	Greenish-yellow	Dark red (H_2SO_4)	91,101
2.	4,3′,4′-Tri-OH, 6-OGl. (aureusin, Aureusidin-6-glucoside, **146b**)	*Antirrhinum*; *Oxalis cernua*					89,90,99
	(Heptaacetate)		F	264.5–265.5		Rose red (10% NaOH),	91,101
				260–260.5 (dec.)		Vivid red (H_2SO_4)	

3. 6,3′,4′-Tri-OH, 4-OGl. (Cernuoside, **147b**)	*Oxalis cernua* (flowers); *Pterocosmea kerii*	E (30%)	262.3; $(\alpha)_D - 28°$	Yellow		102,105
(Heptaacetate)		B	169–170; $(\alpha)_D - 28°$	Pale yellow		102
4. 6,3′,4′-Tri-OH, 7-OMe (Leptosidin, **137**)	*Corepsis grandiflora Nutt*		254; 236–240; 250–254 (dec.)	Yellow	Red (10% NaOH)	87,107
(Triacetate)		E	164–166; 165.5–167	Pale yellow		87
5. 3′,4′-Di-OH,6-OGl.,7-OMe (Leptosin, **138**)	*Coreopsis grandiflora Nutt (ray flowers); C. Lanceolata*	B or E	230–235; 218–224	Golden yellow	Purple-reddish color on standing (10% NaOH); red (H_2SO_4); brown ($FeCl_3$)	92
(Hexaacetate)		G; F − D	235–236 (dec.)	Pale yellow		92,107
6. Sulfuretin (6,3′,4′-Tri-OH, **151**)	*Bidens* (leaves); *Cosmos sulfureus*; *C. bipennatus*; *Dahlia variabilis*; *Rhus javaerica* (heartwood)	E	280–285 (dec.); above 300		Purple (10% NaOH); NaOH); purple (H_2SO_4)	92,110,111
(Triacetate)		B	191–194	Pale yellow		92
7. Sulfurein (3′,4′-Di-OH,6-OGl, **142**)	*Cosmos sulfureus*; *Cosmos maritima*; *C. gigantea*; *Dahlia variabilis*; *Viguiera multiflora*		200 (dec.)			92,110
8. 4′-OH,6,3′-Di-OGl. (Palasitrin, **152**)	*Butea frondosa*	H	210 (dec.)	Yellow	Purple (10% NaOH); deep red (H_2SO_4); olive-green ($FeCl_3$)	109

TABLE IIa. NATURALLY OCCURRING AURONES (continued)

No.	Name and substituents (synonyms, formula)	Natural source	Solvent for crystallization[a]	Melting point (°C)	Appearance	Remarks	Ref.
9.	6,4′-Di-OH (Hispidol, **155a**)	Soya hispida (seedlings)	D	286–288	Yellow		119,120
	(Diacetate)		B − I	162–163	Pale yellow		119
10.	4′-OH, 6-OGl. (Hispidol-6-glucoside, **155b**)	Soya hispida (seedlings)	B + A	211–212	Yellow		117,119
11.	(Pentaacetate)		B − I	189; $(a)_D - 27°$	Pale yellow		119
12.	6,7,3′,4′-Tetra-OH (Maritimetin, **158**)	*Coreopsis maritima* (flowers); *C. gigantea*; *Baeria chrysostoma*	B				93,112
13.	7,3′,4′-Tri-OH, 6-OGl. (Maritimein, **144**)						93,112
14.	7,3′,4′-Tri-OH,5-OMe (Rengasin, **162**)	heartwood of *Rengas*; *Melanorrhea* spp.	B + A	220 (dec.)	Orange-yellow		121
15.	(Triacetate)		B	209–210	Cream		121
16.	(Trimethyl ether)		E	169–170	Lemon yellow		121
17.	4,6,3′,4′,5′-Penta-OH (Bracteatin, **164**)		B	350 (dec.)	Orange needles		121,123, 158, 175, 122
18.	6,3′,4′,5′-Tetra-OH,4-Gl. (Bractein, **165**)	*Helichrysum bracteatum*; *Antirrhinum majus*	B	244–245	Yellow		121, 122, 123, 158, 175

[a] A, petroleum ether; B, benzene; C, ethyl alcohol; D, ether; E, *n*-butyl alcohol; H, chloroform; I, carbon tetrachloride; J, water.

TABLE IIb. ULTRAVIOLET ABSORPTION DATA OF NATURALLY OCCURRING AURONES

Compound	EOH(95%) [max mμ, log E]
Aureusidin (**146a**)	398.5(4.44), 288(3.66), 268(3.90), 263(3.89)
Tetraacetate	374.5(4.23), 346.5(4.01), 317(4.29), 279(3.93), 251(4.12), 237.5(4.05)
Tetramethyleether	397(4.48), 288(3.71), 254(4.03), 243(4.01)
Aureusin (**146b**) Heptaacetate	368(4.32), 353(4.28), 326(4.37), 276(3.88), 244(4.11), 239(4.11)
Cernuoside (**147b**)	267, 405; not affected with $AlCl_3$; in EtONa–EtOH, 450
Leptosidin (**137**)	405.5(4.45), 292(3.67), 272(3.92), 266(3.91), 257(3.95), 251(3.93)
Triacetate	379(4.20), 350(4.00), 319(4.29), 276(3.89)
Leptosin (**138**)	411(4.09), 352.5(3.83), 328.5(3.95), 293.5(3.90), 276.5(3.92), 266(3.91), 257(2.89)
Hexaacetate	374(4.27), 354(4.22), 325.5(4.35), 278(3.82);
	375(4.30), 328(4.35), 241(4.07)
Hispidol (**155a**)	235(4.02), 254(4.14), 388(4.48); in EtONa–EtOH,454
Hispidol-6-β-D-glucoside	226(3.91), 254(4.11), 390(4.44)
Maritimetin (**158**)	270, 355(infl.), 413; in EtONa–EtOH (dec.)
Maritimein (**144**)	242, 274, 330, 419; in EtONa–EtOH, 505
Rengasin (**162**)	254 (E, 9,900), 403 (E, 40,000)
Trimethyl ether	263 (E, 10,000), 396 (E, 30,000)
Bractein (**165**)	407, 260
Octaacetate	375, 310, 246

[a] A, water; B, methyl alcohol; C, carbon tetracholoride; D, petroleum ether; E, ethyl alcohol; F, ethyl acetate; G, dioxan; H, *n*-butyl alcohol; I, acetic acid; J, chloroform.

TABLE IIc. SYNTHETIC 2-ARYLIDENE-3(2*H*)-BENZOFURANONES

Substituents	Melting point (°C)	Solvent for crystal-lization[a]	Appearance	Ultraviolet spectrum [λ max, mμ, (logE)]	Remarks	Ref.
(*a*) Unsubstituted	107–108	A	Small needles	251 (4.10), 316.5 (4.27), 379 (4.06)		120,177, 182
(*b*) Monosubstituted						
5-Me			Red-yellow needles			159
6-Me	153	B				179
7-Me	111.5–112.5	C	Pale yellow needles			180
5-Cl	169; 166–167[177]	B or B + K	Pale yellow needles			180
2′-OH	256–260 (dec.)	D	Orange-yellow needles	250 (4.07), 270 (4.11), 317 (4.05), 402 (4.27); in EtONa-EtOH, 367 (4.01), 499 (4.29); carmine red (NaOH)		120,182
3′-OH	218–220	B	Yellow needles	252 (4.03), 268 (4.12), 316 (4.29); in EtONa-EtOH, 330 (4.17); yellow (NaOH)		120,182
4-OH	141–143	E	Yellow needles	225(infl.) (4.14), 307 (4.26), 389 (4.25); in EtONa-EtOH, 443 (4.27); yellow (NaOH)		120
4′-OH	275–279 190–191[120]	B or B + K	Yellow needles	260 (4.32), 346 (4.07), 405 (4.47); in EtONa–EtOH, 358(3.70), 487 (4.65); orange (NaOH)		120,182

6-OH	261–262	B		229 (4.10), 257(infl.) (3.92), 344 (4.43); in EtONa–EtOH, 402 (4.39)		120,183
2′-OMe	214–216 (subl.)	B	Yellow needles			136,182
3′-OMe	190–191	B	Needles			182
4-OMe	149; 214–215[185]	E	Pale yellow needles	265 (4.02), 307 (3.77), 345 (3.80)[185]; 225 (4.12), 261 (3.86), 308 (4.25), 387 (4.33)[120]		120,185
4′ OMe	169–170	B	Pale yellow needles			182
(c) Disubstituted						
5-Me,2′-NO_2	154	D	Orange-red			187
5-Me,3′-NO_2	205; 202[188]		Yellow			186,188
5-Me,4′-OH	270		Yellow			186
5-Me,2′-OMe	152	F	Yellow			189
5-Br,6-OMe	251	B				138
5-NO_2,6-OMe	229–230	D			red (H_2SO_4)	190
7-Br,4-OMe	202–203					142
6,2′-Di-OH	309–310 (dec.)			265 (3.93), 375 (4.25)		158
6,3′-Di-OH	278			265 (4.03), 340 (4.35)		158
4,4′-Di-OH	237–238	D + K	Yellow-orange needles	230 (4.14), 253 (4.07), 327 (4.09), 408 (4.51); in EtONa–EtOH, 385 (4.16), 473 (4.66); orange (NaOH), purple ($FeCl_3$)		120
4,6-Di-OH	297–298					183
5,6-Di-OH	245	D + K	Orange-red plates	230(infl.) (3.93), 273 (3.97), 347 (4.49); red (NaOH) in EtONa-EtOH, 382 (4.42), 505 (3.95)		120
6,4′-Di-OH	288 (dec.)	D + K	Yellow needles	234 (4.01), 254 (4.09), 388 (4.44); in EtONa–EtONH, 454 (4.65)		120
2′-OH,3′-OMe	229–230	D	Yellow needles			182
2′-OH,5′-OMe	285–290	D	Orange-yellow			182
3′-OH,4′-OMe	199; 180[120]	B or D	Yellow needles; orange plates	257 (4.08), 276 (4.12), 325 (3.90), 409 (4.32); in EtONa-EtOH, 393 (4.21), 463 (4.07)		120,182

TABLE IIc. SYNTHETIC 2-ARYLIDENE-3(2*H*)-BENZOFURANONES (continued)

Substituents	Melting point (°C)	Solvent for crystallization[a]	Appearance	Ultraviolet spectrum [λ max, mμ, (logE)]	Remarks	Ref.
4′-OH,3′-OMe	238–240; 195–196[120]	G or D + K	Orange-yellow; deep yellow needles	262i (4.07), 275 (4.12), 327 (3.79), 413 413 (4.38); in EtONa–EtOH, 347 (3.79), 510 (4.56); carmine-red (NaOH)		120,182
6-OH,4′-OMe	259	D + K	Pale yellow needles	245 (4.21), 370 (4.47), 387 (4.48); in EtONa–EtOH, 405 (4.60); yellow (NaOH)		120
6-OH,7-OMe	160–162	D + K	Yellow needles	228 (4.16), 315i (4.16), 361 (4.34); in EtONa–EtOH, 375i (4.32), 422 (4.47); yellow (NaOH)		120
2′,3′-Di-OMe	166–168	B	Yellow needles			182
2′,4′-Di-OMe	211–212	D	Orange-yellow agglomerates			182
2′,5′-Di-OMe	219–221	H	Yellow needles			182
3′,4′-Di-OMe	173–174	B	Greenish-yellow			182
4,4′-Di-OMe	173–175			258 (4.01), 332 (3.99), 408 (4.30)		185
5,6-Di-OMe	168	B	Lemon Yellow needles	230 (4.01), 263 (4.02), 344 (4.51)		120
6,4′-Di-OMe						191
(d) Trisubstituted						
5,2′,4′-Tri-Me	157		Yellow			186
4,6-Di-Me,3′-OH	235		Yellow			186
4,6-Di-Me,4′-OH	273		Yellow			186
5-Me,3′,4′-Di-OH	174–176	F	Yellow			161
5-Me,3′,4′-(:O_2CH_2)	208–210	I–E	Yellow needles			161
5-NO_2,6-OMe,2′-OH	214–215	D			red (H_2SO_4)	190
7-NO_2,3′-OMe,4-OH	225–226	D/B			Acetate (m.p. 173–175°); blood red (H_2SO_4); brown ($FeCl_3$)	192

7-NO_2,4′-OMe,4-OH	254	D/B			acetate (m.p. 191–192°); purple (H_2SO_4); deep brown ($FeCl_3$)	192
7-Br,4,6-Di-OMe	258–259					142
5-NO_2,6,2′-Di-OMe	149–150	D			Red (H_2SO_4)	190
5-NO_2,6,4′-Di-OMe	250–251	D			Red (H_2SO_4)	190
2′-NO_2,5,6-Di-OMe	196–196.5	D	Orange-yellow flakes			193
3′-NO_2,5,6-Di-OMe	220	D	Yellow silky needles			193
2′-NH_2,5,6-Di-OMe	211–212	D	Orange-red needles			193
3′-NH_2,5,6-Di-OMe	205.5	B	Lemon yellow needles			193
4-OH,5,7-Di-OMe	183–184		Yellow			183
4-OH,6,7-Di-OMe	234–235		Yellow		Dark brown ($FeCl_3$)	195
6-OH,3′,4′-Di-OMe	220			256 (3.82), 386 (4.24)		158
4,6-Di-OH,2′-OMe	255–260		Yellow plates			161
6,4″Di-OH,3′-OMe	269–270			255 (4.03), 393 (4.50)		158
4,6-Di-OMe, 4′-OCH_2Ph	197	H	Yellow needles	224i (4.17), 248 (4.11), 342i (4.28), 390 (4.42)		197
4,6-Di-OAc,2′-OMe						194
4,6,4′-Tri-OH	295–300 (dec.)	D + K	Yellow needles	225 (4.21), 245i (4.09), 392 (4.27); in EtONa–EtOH, 352 (4.30), 445 (4.59); deep yellow (NaOH)		120
5,6,4′-Tri-OH	320 (dec.)	D + K	Yellow needles	258 (4.05), 271i (3.89), 381 (4.55); in EtONa–EtOH, 435 (4.61); orange (NaOH)		120
4,3′,4′-Tri-OH	310 (dec.)	D + K	Yellow needles	256 (3.91), 274 (4.02), 310 (3.92), 416 (4.47); in EtONa-EtOH, 360, 510 (unstable); crimson (NaOH)		120
4,5,6-Tri-OMe	143–144			252 (3.89), 363 (3.98)		185
4,6,7-Tri-OMe	185–186			315 (4.23), 388 (4.24)		185
4,3′,4′-Tri-OMe	193–194			270 (3.72), 310 (3.63), 412 (4.16)		185
4,2′,3′-Tri-OMe	157			260 (3.95), 312 (4.25), 393 (4.38)		185

TABLE IIc. SYNTHETIC 2-ARYLIDENE-3(2*H*)-BENZOFURANONES (continued)

Substituents	Melting point (°C)	Solvent for crystallization[a]	Appearance	Ultraviolet spectrum [λ max, mμ, (logE)]	Remarks	Ref.
3′,4′,5′-Tri-OMe	215–216	D	Cream needles			182
(e) Tetrasubstituted						
5,2′,4′,6′-Tetra-Me	157		Yellow			186
4,6-Di-Me, 3′,4′-di-OMe	165		Yellow			186
2′,3′-Di-Cl, 4,6-di-OMe	90–92	B				196
2′,4′-Di-Cl, 4,6-di-OMe	235	B + K				196
3′-Br,5-NO_2, 6,4′ di-OMe	179–180	D			Red (H_2SO_4)	190
5′-Br,5-NO_2, 6,2′-di-OMe	167–168	D			Red (H_2SO_4)	190
4,6-Di-OH,3′, 4′-di-OMe	255–260	D	Yellow needles			161
4-OH,3′,4′-(:O_2CH_2), 7-NO_2	278	D/B			Purple (H_2SO_4); deep brown ($FeCl_3$)	192
4′-OH,6,3′-Di-OMe, 5-NO_2	209–210	D			Red (H_2SO_4)	190
6-OH,3′,4′-(:O_2CH_2), 5-NO_2	157		Yellow needles			146
6-OH,3′-OMe, 4′ OCH_2Ph,5-NO_2	232	B/D				146
6,3′,4′-Tri-OMe, 5-NO_2	275–276	D				148
6-OMe,3′,4′-(:O_2CH_2), 5-NO_2)	209–210	D				148
4,6,4′-Tri-OMe,7-Br	243–245					142

4,6,4′-Tri-OH, 3′-OMe	271–273			401 (4.47)		158
4,6,3′-Tri-OH, 4′-OMe	284			265 (3.92), 395 (4.49)		158
4,6,7-Tri-OMe, 2′-OH	245					198
6,7,4′-Tri-OMe,4-OH 4-OH	218	B	Yellow			195
4,6,7-Tri-OMe, 4′-OH	290–296 (dec.);[198] 276–278	F or J − E	Yellow needles	335, 405;[195] 333, 409[198]		195,198
5,6,3′,4′-Tetra-OH	298 (dec.)	J/E	Orange powder	266 (4.10), 395 (4.53); red (NaOH)		120
6,7,3′,4′-Tetra-OH	292 (dec.)	K	Orange needles	252 (4.01), 268 (3.95), 355i (4.19), 415 (4.48); purple (NaOH)		120
6,3′,4′,5′-Tetra-OH	310 (dec.)				Tetraacetate (F), m.p. 213–216° (dec.)	200
4,5,6,4′-Tetra-OMe	188–189;			283 (4.01), 402 (4.40)		185
	147–148	B	Yellow needles		Red (H_2SO_4)	195,199
4,6,7,2′-Tetra-OMe	198–200	F				195
4,6,7,4′-Tetra-OMe	199–210			332 (4.18), 408 (4.41)		185
4,6,2′,3′-Tetra-OMe	165–167			265 (3.82), 375 (4.50)		185
4,6,3′,4′-Tetra-OMe	171.5–172				Crimson magenta (H_2SO_4)	155
(f) Pentasubstituted						
5-NO_2,3′,5′-Di-Br, 2′-OH,6-OMe	195–196	D			Red (H_2SO_4)	190
5-NO_2,5′-Br,4′-OH, 6,3′-Di-OMe	195–196	D			Red (H_2SO_4)	190
4,6,4′-Tri-OH,3′, 5′-Di-OMe	260			260 (3.96), 399 (4.43)		158
4,6,2′,4′-Tetra-OMe, 5-Br	above 300	H/I	Pale yellow plates			144
6,5,3′,4′,5′-Penta-OH	360 (dec.)	F + K			Pentaacetate, m.p. 235–255° (dec.)	200
6,7,3′,4′,5′-Penta-OH	360 (dec.)	F + K			Pentaacetate, m.p. 214–217°	200
4,6,2′,3′-Tetra-OMe, 5-OH	175–176			270 (3.87), 416 (4.33)		185

TABLE IIc. SYNTHETIC 2-ARYLIDENE-3(2*H*)-BENZOFURANONES (continued)

Substituents	Melting point (°C)	Solvent for crystal-lization[a]	Appearance	Ultraviolet spectrum [λ max, mμ, (logE)]	Remarks	Ref.
4,6,7,2′-Penta-OMe	187	B	Yellow			195
4,6,7,2′,6′-Penta-OMe	188	B	Yellow			195
4,5,6-Tri-OMe,3′,4′-(:O_2CH_2)	195–196			254 (3.62), 399 (3.99)		185
4,6,7-Tri-OMe,3′,4′-(:O_2CH_2)	242–243			274 (3.86), 320 (4.00), 412 (4.40)		185
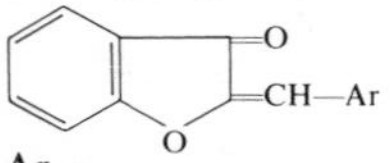 Ar =						
2′-C_5H_4N[b]	128–129	B + K	Small needles			177
3′-C_5H_4N	121–122	C/E	Needles			177
4′-C_5H_4N	169–170	F	Long needles			177
5-Cl,2′-C_5H_4N	162–163	B	Small needles			177
5-Cl,3′-C_5H_4H	192–194	F	Long needles			177
5-Cl,4′-C_5H_4N	205–207 (dec.)	F	Long needles			177
5-Me,5′-$O_2NC_4H_3O$[c]	178		Yellowish-red needles			178
5-Cl,5′-$O_2NC_4H_3O$	228		Yellow needles			178
6-OH,5′-$O_2NC_4H_3O$	282 (dec.)		Yellow powder			178
6-OMe, 5′-$O_2NC_4H_3O$	239 (dec.)		Yellow needles			178
6-OAc, $O_2NC_4H_3O$	172–174		Yellow needles			178
5-Cl,6-OAc, 5′-$O_2NC_4H_3O$	228 (dec.)		Yellow needles			178

[a] A, hexane; B, ethyl alcohol; C, benzene; D, acetic acid; E, petroleum ether; F, methyl alcohol; G, dioxan; H, acetone; I, chloroform; J, ethyl acetate; K, water.

[b] Pyridyl.

[c] 5′-Nitrofuryl.

Substituents	Melting point (°C)	Solvent for crystal-lization[a]	Appearance	Ultraviolet spectrum [λmax(mμ)]	Remarks	Ref.
(a) Unsubstituted	167–168;[201] 169.5–170.5[202]					201, 202
(b) Monosubstituted						202
4-OMe	167					202
6-OMe	129			268, 387		184
5′-Cl	227–229					201
5′-Br	190					201
5′-NO_2	233–234					201
(c) Disubstituted						
3′,4′-Di-OMe	173–174					202
6,4′-Di-OMe	171			233, 255, 402		184
3′,5′-Di-Cl	207					201
3′,5′-Di-Br	202					201
(a) Trisubstituted						
3′,4′,5′-Tri-OMe	159–160					202
6,3′,4′-Tri-OMe	183–184					202
(e) Tetrasubstituted						
4,6,7-Tri-OMe, 5-OH	182.5–184.7	A	Yellow needles			201
4,6′,3′,4′-Tetra-OMe	173.5–174				Green color with H_2SO_4 changing to brown	155
6,3′,4′,5′-Tetra-OMe	149					154
4,6,7-Tri-OMe,5-OAc	198.5–199.5	B				201
(f) Pentasubstituted						
4,6,7-Tri-Me,4′–OMe,5-OH	170.5–173	A				201
4,6,7-Tri-Me,4′-OMe,5-OAc	222–223	C				201

[a] A, alcohol, B, ethyl acetate; C, acetone.

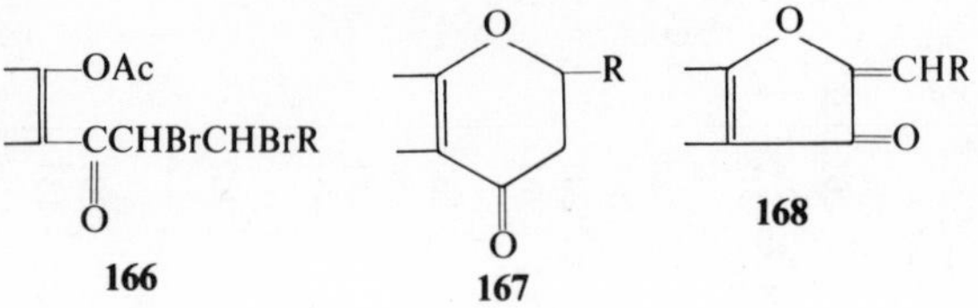

166 167 168

yield favonols (**169**); however, under the action of trifluoroboron-ethereate and/or concentrated sulfuric acid, they yield 4-hydroxy-3-phenylcoumarins (benzotetronic acids) (**170**) (Eq. 13).[162]

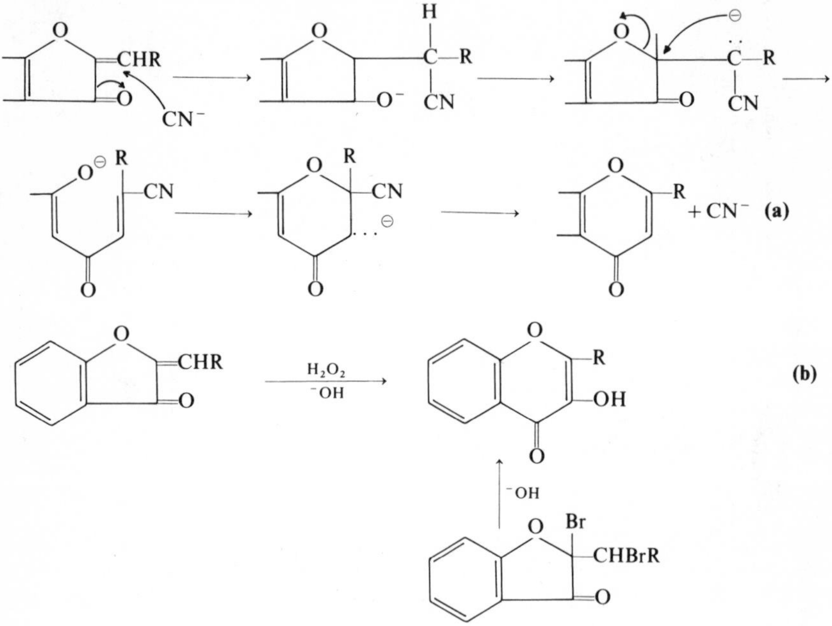

(a)

H_2O_2 / ^-OH (b)

^-OH

Chart 1.

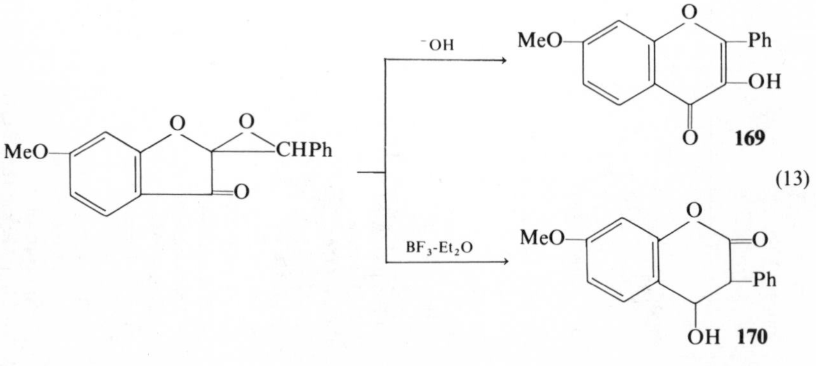

^-OH

BF_3-Et_2O

169

(13)

170

Rearrangement of aurone epoxides into 4-hydroxycoumarins, for example **170**, involves benzyl migration[162] and is analogous to the formation of 2-phenylcyclohexan-1,3-dione from 2-benzylidenecyclopentanone epoxide[163] and the rearrangement of the epoxide **171** into 2,3-diphenylchroman-2,4-dione (**172**).[164]

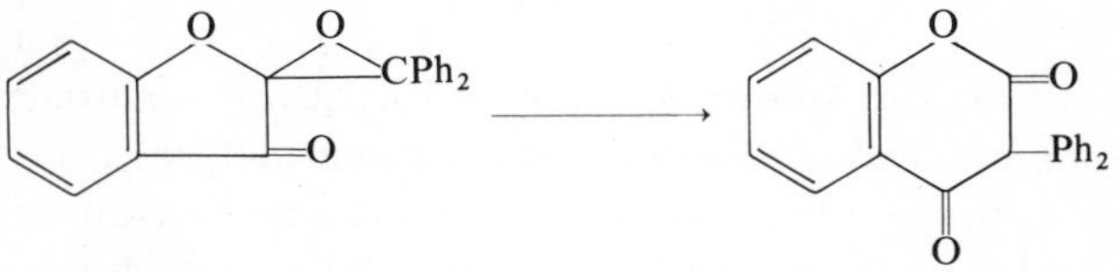

171 → **172**

Examination of the reactions with 2-*p*-methoxybenzylidene-3(2*H*)-benzofuranone (**173**) indicates that those reactions exhibited by the keto-ethylene group in chalcones are unaffected by the cyclic linkage. Thus one of the bromine atoms is replaceable by an alkoxyl group on treatment with alcohol (**174**). Furthermore, it undergoes Michael addition with cyclohexanone to yield **175a**, and with desoxybenzoin (**175b**) condenses with ethyl acetoacetate to give **176.**[165,166]

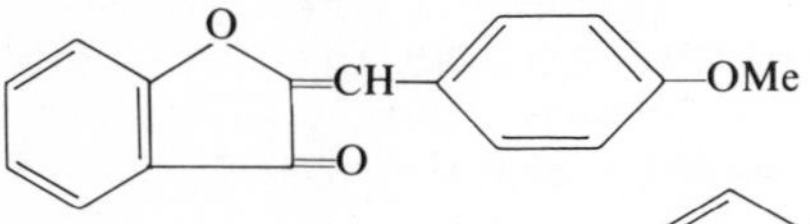

173

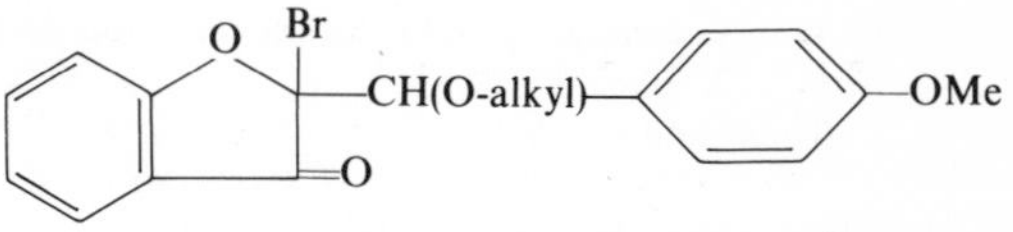

174

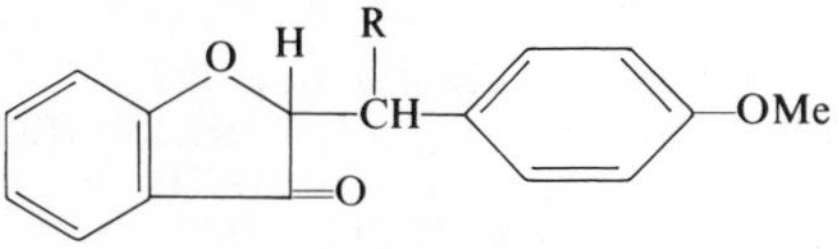

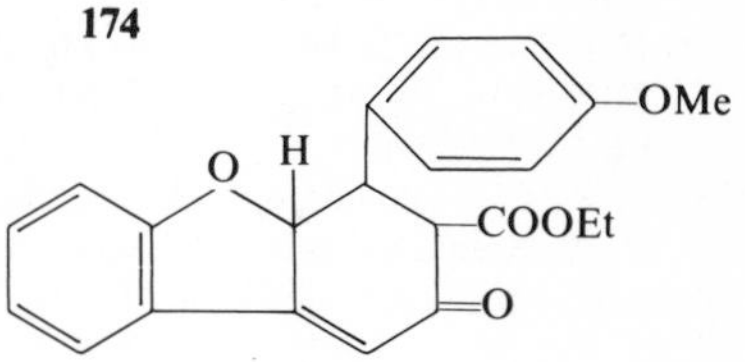

175a R = $CO(CH_2)_4CH$

175b R = PhCOCHPh

176

In the presence of palladium-black, certain hydrogen donors, such as tetralin, can reduce 2-benzylidene-3(2*H*)-benzofuranone which is not hydroxylated in the phenyl group on C-2 and contains a free hydroxyl group. Reduction proceeds with rupture of the heterocycle ring and the formation of a dihydrochalcone.[167,168] Reductive acetylation of 4,6,3′,4′-tetramethoxyaurone gave 1,2-bis(3-acetoxy-4,6-dimethoxybenzofuran-2-yl)-1,2-bis(3,4-dimethoxyphenyl) ethane.[169]

Nuclear methylation with methyl iodide and methanolic potassium hydroxide of 4,6-dihydroxyaurone and its 4′-methoxy and 3′,4′-dimethoxy derivatives yields not only the expected C-5 methyl-4,6-di-*O*-methyl derivatives, but also compounds having a *gem*-dimethyl group in the 5 position. These structures are supported by NMR spectra. C-5 methyl compounds have also been obtained by an unambiguous synthesis.[170]

Hydroxylation of 4,6-dimethoxyaurones with alkaline persulfate takes place smoothly in the 7 position to yield 4,6,7-trihydroxyaurones. However, condensation of 6-hydroxyaurones with hexamine affords a mixture of two aldehydes: 5-formyl-6-hydroxy- and 7-formyl-6-hydroxyaurones.[171] This behavior is different from that of flavonoids; Dakin's oxidation of 7-formyl-6-hydroxyaurone gives 6,7-dihydroxyaurone.

The carbonyl group in 2-benzylidene-3(2*H*)-benzofuranone gives the expected 3-keto-oxime; sodium borohydride effected its reduction to 2-benzofuranylphenylcarbinol.[172–174]

References

1. J. H. Birkinshaw, P. Chaplen, and W. P. K. Findlay, *Biochem. J.*, **66,** 188 (1957).
2. A. J. Weinheimer and H. Washecheck, *Tetrahedron Lett.*, **1969,** 3315.
3. A. J. Weinheimer, F. J. Schmitz, and L. S. Ciereszko, "Drugs Sea Trans. Symp.," H. D. Freudenthal, Ed. Marine Technol. Soc., Washington, D. C., 1967, p. 135; *Chem. Abstr.*, **72,** 70562 (1970).
4. B. Kamthong and A. Robertson, *J. Chem. Soc.*, **1939,** 925.
5. Z. Jerzmanowska, *Polska Akad. Umiej, Prace Kom. Farm. Diss. Pharm.*, **3,** 165 (1951); *Chem. Abstr.*, **48,** 5848 (1954).
6. F. v. Gizychi, *Süddeut. Apoth.-Ztg.*, **90,** 503 (1950); *Chem. Abstr.*, **44,** 9118 (1950).
7. T. Nakaoki, N. Morita, and S. Nishino, *Yakugaku Zasshi*, **78,** 557 (1958); *Chem. Abstr.*, **52,** 13190 (1958).
8. P. K. Ramachandran, T. Cheng and W. J. Horton, *Tetrahedron Lett.*, **1963,** 907.
9. P. K. Ramachandran, T. Cheng, and W. J. Horton, *J. Org. Chem.*, **28,** 2744 (1963).
10. J. A. Elix, *Austr. J. Chem.*, **24,** 93 (1971); *Chem. Abstr.*, **74,** 42235 (1971).
11. T. Murae, Y. Tanahashi, and T. Takahashi, *Tetrahedron*, **24,** 2177 (1968).
12. T. Anthonsen and S. Chantharasakul, *Acta Chem. Scand.*, **24,** 721 (1970).
13. A. R. Albertsen, *Acta Chem. Scand.*, **9,** 1725 (1955).
14. T. Anthonsen, *Acta Chem. Scand.*, **23,** 2605 (1969).
15. T. R. Kasturi and T. Manithomas, *Tetrahedron Lett.*, **1967,** 2573.
16. S. Rangaswami and T. R. Seshadri, *Proc. Indian Acad. Sci.*, **15A,** 417 (1942); *Chem. Abstr.*, **36,** 7025 (1942).
17. S. Naryanaswamy, S. Rangaswami, and T. R. Seshadri, *J. Chem. Soc.*, **1954,** 1871.
18. S. K. Mukerjee and T. R. Seshadri, *J. Chem. Soc.*, **1955,** 2048.
19. J. F. Couch, *J. Am. Chem. Soc.*, **51,** 3617 (1929).
20. J. F. Couch, *J. Agr. Res.*, **35,** 547 (1927).
21. J. F. Couch, *J. Am. Med. Assoc.*, **91,** 234 (1928).
22. W. A. Bonner, J. I. DeGraw, Jr., D. M. Bowen, and V. R. Shah, *Tetrahedron*, **17,** 417 (1961).
23. W. A. Bonner and J. I. DeGraw, Jr., *Tetrahedron*, **18,** 1295 (1962).

24. J. I. DeGraw, Jr., and W. A. Bonner, *J. Org. Chem.*, **27,** 3917 (1962).
25. J. I. DeGraw, Jr., and W. A. Bonner, *Tetrahedron,* **18,** 1311 (1962).
26. L. H. Zalkow and N. Burke, *Chem. Ind. (London)*, **1963,** 292.
27. J. I. DeGraw, Jr., D. M. Bowen, and W. A. Bonner, *Tetrahedron,* **19,** 19 (1963).
28. D. M. Bowen, J. I. DeGraw, Jr., V. R. Shah, and W. A. Bonner, *J. Med. Chem.*, **6,** 315 (1963); *Chem. Abstr.*, **58,** 13881 (1963).
29. L. H. Zalkow, N. Burke, G. Cabat, and E. A. Grula, *J. Med. Pharm. Chem.*, **5,** 1342 (1962); *Chem. Abstr.*, **58,** 2598 (1963).
30. J. Oda, H. Fukami, and M. Nakajima, *Agr. Biol. Chem. (Tokyo)*, **30,** 59 (1966); *Chem. Abstr.*, **64,** 15818 (1966).
31. W. W. Epstein, O. Gerike, and W. J. Horton, *Tetrahedron Lett.*, **1965,** 3991.
32. A. Fredga and C. V. de Castro Sarmiento, *Arkiv. Kemi,* **7,** 387 (1954).
33. B. Sjöberg, *Arkiv. Kemi,* **15,** 481 (1960); *Chem. Abstr.*, **54,** 24579 (1960).
34. W. A. Bonner, N. I. Burke, W. E. Fleck, R. K. Hill, J. A. Joule, B. Sjöberg, and J. H. Zalkow, *Tetrahedron,* **20,** 1419 (1964).
35. I. Harada, Y. Hirose, and M. Nakazaki, *Tetrahedron Lett.*, **1968,** 5463.
36. M. Forbes, F. Zilliken, G. Roberts, and P. György, *J. Am. Chem. Soc.*, **80,** 385 (1958).
37. M. A. P. Meisinger, F. A. Kuehl, Jr., E. L. Rickes, N. G. Brink, K. Folkers, M. Forbes, F. Zilliken, and P. György, *J. Am. Chem. Soc.*, **81,** 4979 (1959).
38. A. F. Wagner, E. Walton, A. N. Wilson, J. O. Rodin, F. W. Holly, N. G. Brink, and K. Folkers, *J. Am. Chem. Soc.*, **81,** 4983 (1959).
39. R. G. Cooke and I. D. Rae, *Austr. J. Chem.*, **17,** 379 (1964); *Chem. Abstr.*, **60,** 14462 (1964).
40. R. G. Cooke and R. M. McQuilkin, *Austr. J. Chem.*, **22,** 2395 (1969); *Chem. Abstr.*, **72,** 2488 (1970).
41. M. A. Buchanan and E. E. Dickey, *J. Org. Chem.*, **25,** 1389 (1960).
42. I. R. C. Bick and G. K. Douglas, *Tetrahedron Lett.*, **1964,** 1629.
43. R. S. McCredie, E. Ritchie, and W. C. Taylor, *Austr. J. Chem.*, **22,** 1011 (1969); *Chem. Abstr.*, **71,** 61097 (1969).
44. A. I. Scott, *Quart. Rev.*, **19,** 1 (1965).
45. A. Ogiso, M. Kurabayashi, H. Mishima, and M. C. Woods, *Tetrahedron Lett.*, **1968,** 2003.
46. A. Stoessl, *Tetrahedron Lett.*, **1966,** 2287, 2849.
47. H. Okada, *J. Pharm. Soc. Japan,* **1915,** 657.
48. S. Kawai and K. Yamagami, *Chem. Ber.*, **71,** 2438 (1938).
49. S. Kawai and F. Yoshimura, *Chem. Ber.*, **71,** 2415 (1938).
50. S. Kawai and N. Sugiyama, *Chem. Ber.*, **72,** 367 (1939).
51. S. Kawai, T. Nakamura, and N. Sugiyama, *Chem. Ber.*, **72,** 1146 (1939).
52. S. Kawai, K. Sugimoto, and N. Sugiyama, *Chem. Ber.*, **72,** 953 (1939).
53. S. Kawai, T. Nakamura, and M. Yoshida, *Chem. Ber.*, **73,** 581 (1940).
54. S. Kawai, N. Sugiyama, T. Nakamura, and K. Kosmatsu, *Chem. Ber.*, **73;** 586 (1940).
55. S. Kawai and K. Sugimoto, *Chem. Ber.*, **73,** 774 (1940).
56. E. Ritchie and W. C. Taylor, *Austr. J. Chem.*, **22,** 1329 (**1969**); *Chem. Abstr.*, **71,** 61094 (1969).
57. C. Y. Hopkins, D. F. Ewing, and M. J. Chisholm, *Can. J. Chem.*, **45,** 1425 (1967); *Chem. Abstr.*, **67,** 64144 (1967).
58. R. Segal, I. Milo-Goldzweig, S. Sokoloff, and D. V. Zaitscheck, *J. Chem. Soc., C,* **1967,** 2402.
59. R. W. Jackson and W. F. Short, *J. Chem. Soc.*, **1937,** 513.
60. D. H. R. Barton, A. Bhati, P. De Mayo, and G. A. Morrison, *J. Chem. Soc.*, **1958,** 4393.
61. J. R. Price and R. Robinson, *J. Chem. Soc.*, **1939,** 1522; **1939,** 1493.

62. R. D. Allan, R. L. Correll, and R. J. Wells, *Tetrahedron Lett.*, **1969,** 4673; *Chem. Abstr.*, **72,** 78814 (1970).
63. M. Gregon, W. D. Ollis, B. T. Redman, I. O. Sutherland and H. H. Dietvichs, *Chem. Commun.*, **1968,** 1394; *Chem. Abstr.*, **70,** 19849 (1969).
64. S. K. Mukerjee, T. Saroja, and T. R. Seshadri, *Indian J. Chem.*, **8,** 21 (1970).
65. B. J. Donnelly, D. M. X. Donnelly, A. M. O'Sullivan, and J. P. Prendergast, *Tetrahedron*, **25,** 4409 (1969).
66. K. Koshimizu, E. Y. Spencer, and A. Stoessl, *Can. J. Botany*, **41,** 744 (1963); *Chem. Abstr.*, **59,** 6721 (1963).
67. A. Stoessl, U.S. Pat. 3,475,459 (1969); *Chem. Abstr.*, **73,** 14665 (1970).
68. A. Stoessl, *Tetrahedron Lett.*, **1966,** 2287; *Chem. Abstr.*, **65,** 2195 (1966).
69. J. Gripenberg, *Acta Chem. Scand.*, **19,** 2242 (1965); *Chem. Abstr.*, **64,** 16198 (1966).
70. J. Gripenberg and J. Martikkala, *Acta Chem. Scand.*, **23,** 2583 (1969).
71. S. Abrahamsson and M. Innes, *Acta Chem. Scand.*, **19,** 2246 (1965); *Chem. Abstr.*, **64,** 16198 (1966).
72. S. Abrahamsson and M. Innes, *Acta Cryst.*, **21,** 948 (1966).
73. A. B. Turner, *Quart. Rev.*, **18,** 347 (1964).
74. H. W. Wanzlick and U. Jahnke, *Chem. Ber.*, **101,** 3753 (1948).
75. C. Culberson, "Chemical and Botanical Guide to Lichens Products," Univ. of N. Carolina Press; Chapel Hill, 1969.
76. J. Santesson, *Acta Chem. Scand.*, **24,** 371 (1970).
77. B. Abermark, *Acta Chem. Scand.*, **15,** 1695 (1961); *Chem. Abstr.*, **57,** 2137 (1962).
78. M. Asamo and Y. Kameda, *Chem. Ber.*, **68,** 1568 (1935).
79. W. S. G. Maass, *Can. J. Biochem.*, **48,** 1241 (1970); *Chem. Abstr.*, **74,** 61623 (1971).
80. N. F. Janes, F. E. King, and J. W. W. Morgan, *Chem. Ind. (London)*, **1961,** 346; *Chem. Abstr.*, **55,** 26136 (1961).
81. N. F. Janes, F. E. King, and J. W. W. Morgan, *J. Chem. Soc.*, **1963,** 1356.
82. J. Chopin and M. Chadenson, *C. R. Acad. Sci. Paris, Ser. C*, **236,** 729 (1966); *Chem. Abstr.*, **66,** 10803 (1967).
83. H. G. C. King, T. White, and R. S. Hughes, *J. Chem. Soc.*, **1961,** 3234.
84. J. H. Bowie and J. W. W. Morgan, *Austr. J. Chem.*, **20,** 117 (1967); *Chem. Abstr.*, **66,** 46004 (1967).
85. J. Read and H. G. Smith, *J. Proc. Roy. Soc., N. S. Wales*, **56,** 253 (1922).
86. A. J. Birch, E. Ritchie, and R. N. Speake, *J. Chem. Soc.*, **1960,** 3593.
87. T. A. Geissman and C. D. Heaton, *J. Am. Chem. Soc.*, **65,** 677 (1943); **66,** 486 (1944).
88. E. C. Bate-Smith and T. A. Geissman, *Nature*, **167,** 688 (1951).
89. R. Lamonica and G. B. Marini-Bettòlo, *Ann. Chem.*, **42,** 496 (1952); *Chem. Abstr.*, **48,** 11719 (1954).
90. A. Ballio, S. Dittrich, and G. B. Marini-Bettòlo, *Gazz. Chim. Ital.*, **83,** 224 (1953); *Chem. Abstr.*, **47,** 12378 (1953); see *Chem. Abstr.*, **48,** 8228 (1954).
91. M. K. Seikel and T. A. Geissman, *J. Am. Chem. Soc.*, **75,** 2277 (1953).
92. M. Shimokoriyama and S. Hattori, *J. Am. Chem. Soc.*, **75,** 1900 (1953).
93. T. A. Geissman, J. B. Harborne, and M. K. Seikel, *J. Am. Chem. Soc.*, **78,** 825 (1956).
94. J. B. Harborne and T. A. Geissman, *J. Am. Chem. Soc.*, **78,** 829 (1956).
95. M. Shimokoriyama, *J. Am. Chem. Soc.*, **79,** 214 (1957).
96. E. C. Jorgensen and T. A. Geissman, *Arch. Biochem. Biophys.*, **54,** 72 (1955); *Chem. Abstr.*, **49,** 4807 (1955); **55,** 389 (1955).
97. T. A. Geissman, E. C. Jorgensen, and B. L. Johnson, *Arch. Biochem. Biophys.*, **49,** 368 (1954); *Chem. Abstr.*, **48,** 8340 (1954).
98. M. K. Seikel and T. A. Geissman, *J. Am. Chem. Soc.*, **72,** 5720 (1950).
99. M. K. Seikel and T. A. Geissman, *J. Am. Chem. Soc.*, **72,** 5725 (1950).
100. J. B. Harborne, *Chem. Ind. (London)*, **1954,** 1142.

101. T. A. Geissman and J. B. Harborne, *J. Am. Chem. Soc.*, **77**, 4622 (1955).
102. L. Farkas, L. Pallòs, and G. Hidasi, *Chem. Ber.*, **94**, 2221 (1961).
103. J. Chopin, G. Dellamonica, and P. Lebreton, *Comlt. Rend.*, **257**, 534 (1963); *Chem. Abstr.*, **59**, 9731 (1963).
104. J. Chopin and G. Dellamonica, *Compt. Rend.*, **260**, 5582 (1965); *Chem. Abstr.*, **64**, 3670 (1966).
105. J. B. Harborne, *Phytochemistry*, **5**, 589 (1966); *Chem. Abstr.*, **65**, 17363 (1966).
106. A. Ballio and G. B. Marini-Bettòlo, *Gazz. Chim. Ital.*, **85**, 1319 (1955); *Chem. Abstr.*, **52**, 1990 (1956).
107. T. A. Geissman and W. Moje, *J. Am. Chem. Soc.*, **73**, 5765 (1951).
108. L. Farkas, L. Pallòs, and M. Nogradi, *Magy. Kem. Folyoirat*, **71**, 272 (1965); *Chem. Abstr.*, **63**, 9901 (1965).
109. B. Puri and T. R. Seshadri, *J. Chem. Soc.*, **1955**, 1589.
110. C. N. Nordström and T. Swain, *Biochem. Biophys.*, **60**, 329 (1956).
111. H. Imamura, H. Kurosu, and T. Takahashi, *Nippon Mokuzai Gakkaishi*, **13**, 295 (1967); *Chem. Abstr.*, **68**, 41224 (1967).
112. M. Shimkoriyama and T. A. Geissman, *J. Org. Chem.*, **25**, 1956 (1960).
113. G. Zemplen, L. Mester, and L. Pallòs, *Acta Chim. Acad. Sci. Hung.*, **12**, 259 (1957); *Chem. Abstr.*, **52**, 11030 (1958).
114. L. Farkas, L. Pallòs, and Z. Paal, *Chem. Ber.*, **92**, 2847 (1959).
115. B. Puri and T. R. Seshadri, *J. Sci. Res. (India), Ser. B*, **12**, 462 (1953); *Chem. Abstr.*, **48**, 4773 (1954).
116. L. Farkas and L. Pallòs, *Chem. Ber.*, **93**, 1272 (1960).
117. E. Wang, Phytochemistry, **5**, 463 (1966).
118. E. Wang, *Chem. Ind. (London)*, **1966**, 598.
119. L. Farkas, E. Berènyl, and L. Pallòs, *Tetrahedron*, **24**, 4213 (1968).
120. T. A. Geissman and J. B. Harborne, *J. Am. Chem. Soc.*, **78**, 832 (1956).
121. F. E. King, T. J. King, and D. W. Rustidge, *J. Chem. Soc.*, **1962**, 1192.
122. L. Farkas, M. Nogradi, and L. Pallòs, *Chem. Ber.*, **97**, 1044 (1964).
123. J. B. Harborne, *Phytochemistry*, **2**, 327 (1963); *Chem. Abstr.*, **60**, 4464 (1964).
124. R. Hänsel, L. Langhammer, and A. G. Albrecht, *Tetrahedron Lett.*, **1962**, 599.
125. J. Algar and J. P. Flynn, *Proc. Roy. Acad.*, **42B**, 1 (1954); *Chem. Abstr.*, **29**, 161 (1935).
126. T. Oyamada, *J. Chem. Soc. Japan*, **55**, 1256 (1934); *Chem. Abstr.*, **29**, 4358 (1935).
127. T. Oyamada, *Bull. Chem. Soc. Japan*, **18**, 182 (1935); *Chem. Abstr.*, **29**, 5112 (1935).
128. T. A. Geissman and D. K. Fukushima, *J. Am. Chem. Soc.*, **70**, 1686 (1948).
129. K. J. Balakrishma, T. R. Seshadri, and G. Viswanath, *Proc. Indian Acad. Sci.*, **30A**, 120 (1949); *Chem. Abstr.*, **44**, 5876 (1950).
130. T. R. Seshadri and N. Narasimhachari, *Proc. Indian Acad. Sci.*, **30A**, 216 (1949); **37A**, 104 (1953).
131. T. Emilewicz and St. v. Kostanecki, *Chem. Ber.*, **31**, 696 (1898).
132. St. v. Kostanecki, *Chem. Ber.*, **31**, 705 (1898).
133. W. Feuerstein and St. v. Kostanecki, *Chem. Ber.*, **31**, 1757 (1898).
134. St. v. Kostanecki and A. Ludwig, *Chem. Ber.*, **31**, 2951 (1898).
135. W. Feuerstein and St. v. Kostanecki, *Chem. Ber.*, **32**, 315 (1899).
136. F. Herstein and St. v. Kostanecki, *Chem. Ber.*, **32**, 118 (1899).
137. St. v. Kostanecki and J. Tambor, *Chem. Ber.*, **32**, 2260 (1899).
138. W. A. Hutchins and T. S. Wheeler, *J. Chem. Soc.*, **1939**, 91.
139. A. M. Warriar, A. P. Khanolkar, W. A. Hutchins, and T. S. Wheeler, *Current Sci.*, **5**, 475 (1937); *Chem. Abstr.*, **31**, 4667 (1937).
140. S. M. Nadakarni, A. M. Warriar, and T. S. Wheeler, *J. Chem. Soc.*, **1938**, 1798.
141. K. v. Auwers and L. Anschütz, *Chem. Ber.*, **54**, 1543 (1921).

142. D. J. Donnelly, J. A. Donnelly, J. J. Murphy, E. M. Philbin, and T. S. Wheeler, *Chem. Commun.*, **1966,** 351; *Chem. Abstr.*, **65,** 5434 (1966).
143. H. P. Vandrewalla and G. V. Jadhav, *Proc. Indian Acad. Sci.*, **28A,** 125 (1948); *Chem. Abstr.*, **44,** 3982 (1950).
144. N. M. Cullinane and D. Philpott, *J. Chem. Soc.*, **1929,** 1761.
145. K. R. Chandorkar, D. M. Phatak, and A. B. Kulkarni, *J. Sci. Ind. Res. (India)*, **21B,** 24 (1962); *Chem. Abstr.*, **57,** 9791 (1962).
146. V. G. Naik and M. G. Marathey, *J. Univ. Poona Sci. Technol.*, **18,** 61 (1960); *Chem. Abstr.*, **55,** 1599 (1961).
147. G. V. Bhide, K. R. Chandorkar, H. K. Pendse, and S. D. Limaye, *Rasayaram*, **2'** 135 (1956); *Chem. Abstr.*, **51,** 5063 (1957).
148. V. G. Kulkarni and G. V. Jadhav, *J. Univ. Bombay*, **23A**(Part 5), 14 (1955); *Chem. Abstr.*, **51,** 11301 (1957).
149. M. G. Marathey, *Sci. and Cult.*, **17,** 86 (1951); *Chem. Abstr.*, **46,** 11188 (1952).
150. R. P. Dodwadmath, *J. Univ. Bombay*, **9**(Part 3), 172 (1940); *Chem. Abstr.*, **35,** 6959 (1941).
151. F. F. Kurth, *Ind. Eng. Chem. (Anal.)*, **45,** 2096 (1953); *Chem. Abstr.*, **48,** 10016 (1954).
152. H. C. Hergert, O. Coad, and A. V. Logan, *J. Org. Chem.*, **21,** 304 (1956).
153. T. Tsukamoto and T. Tominaga, *J. Pharm. Soc. Japan*, **73,** 1172, 1175, 1179 (1953).
154. D. Molho, J. Chopin, and M. Chadenson, *Bull. Soc. Chim. Fr.*, **1959,** 454.
155. C. Eneback and J. Gripenberg, *J. Org. Chem.*, **22,** 220 (1957).
156. H. Kotake, T. Sakan, and T. Kubota, *Chem. Ind. (London)*, **1954,** 1562.
157. T. Emilewicz and St. Kostanecki, *Chem. Ber.*, **32,** 309 (1899).
158. R. Hänsel, L. Langhammer, J. Frenzel, and G. Ranft, *J. Chromatog.*, **11,** 369 (1963).
159. K. v. Auwers and K. Müller, *Chem. Ber.*, **41,** 4233 (1908).
160. K. v. Auwers and P. Pohl, *Justus Liebigs Ann. Chem.*, **405,** 243 (1914).
161. D. M. Fitzgerald, E. M. Philbin, and T. Wheeler, *Chem. Ind. (London)*, **1952,** 130; *Chem. Abstr.*, **47,** 2164 (1953); D. M. Fitzgerald, J. F. O'Sullivan, E. M. Philbin, and T. S. Wheeler, *J. Chem. Soc.*, **1955,** 860.
162. M. Geoghan, W. T. O'Sullivan, and E. M. Philbin, *Tetrahedron*, **22,** 3211 (1966); **22,** 3203 (1966).
163. H. O. House and R. L. Wasson, *J. Am. Chem. Soc.*, **78,** 4394 (1956).
164. A. Schönberg and K. Junghans, *Chem. Ber.*, **99,** 531 (1966).
165. T. B. Panse, R. C. Shah, and T. S. Wheeler, *J. Indian Chem.*, **18,** 453 (1941); *Chem. Abstr.*, **36,** 4507 (1942).
166. S. N. Rao and T. S. Wheeler, *J. Chem. Soc.*, **1939,** 1004.
167. C. Mentzer and J. Massicot, *Bull. Soc. Chim. Fr.*, **1956,** 144; *Chem. Abstr.*, **50,** 13004 (1956).
168. K. Freudenberg, H. Fikentscher, and M. Harder, *Justus Liebigs Ann. Chem.*, **441,** 157 (1925).
169. J. W. Clark-Lewis and R. W. Jemison, *Austr. J. Chem.*, **23,** 89 (1970); *Chem. Abstr.*, **72,** 66724 (1970).
170. A. C. Jain, V. K. Rohatgi, and T. R. Seshadri, *Indian J. Chem.*, **7,** 540 (1969); *Chem. Abstr.*, **71,** 49680 (1969).
171. A. C. Jain, V. K. Rohatgi, and T. R. Seshadri, *Indian J. Chem.*, **7,** 543 (1969); *Chem. Abstr.*, **71,** 49681 (1969).
172. A. Holy and A. Vystrcil, *Collection Czech. Commun.*, **27,** 1861 (1962); *Chem. Abstr.*, **57,** 15040 (1962).
173. S. Rangaswami and B. V. R. Sastry, *Current Sci.*, **24,** 13 (1955).
174. R. Adams, R. C. Morris, D. J. Butterbaugh, and E. C. Kirkpatrick, *J. Am. Chem. Soc.*, **60,** 2191 (1938).
175. L. Farkas and L. Pallòs, *Chem. Ber.*, **98,** 2930 (1965).

176. S. Kawai and N. Sugiyama, *Chem. Ber.*, **91,** 2421 (1938).
177. A. R. Katrizky, P. D. Kennewell, and M. Snarey, *J. Chem. Soc., B*, **1968,** 544.
178. S. Toyoshima, K. Shimada, and K. Kawal, *Yakugaku Zasshi*, **88,** 589 (1968); *Chem. Abstr.*, **69,** 995877 (1968).
179. K. v. Auwers and P. Pohl, *Justus Liebigs Ann. Chem.*, **405,** 291 (1914).
180. K. v. Auwers, *Chem. Ber.*, **49,** 814 (1916).
181. K. Fries, A. Hasselbach, and L. Schroeder, *Justus Liebigs Ann. Chem.*, **405,** 370(1914).
182. A. N. Grinev, I. A. Zaitsev, N. K. Venevtseva, and A. P. Terent'ev. *Zh. Obshch. Khim.*, **28,** 1853 (1958); *Chem. Abstr.*, **53,** 1299 (1959).
183. K. J. Balakrishma, N. P. Rao, and T. R. Seshadri, *Proc. Indian Acad. Sci.*, **29A,** 394 (1949); *Chem. Abstr.*, **44,** 5350 (1950).
184. J. Chopin, M. Chadenson, and P. Durual, *Compt. Rend.*, **258,** 6178 (1964); *Chem. Abstr.*, **61,** 6983 (1964).
185. P. Venturella and A. Bellino, *Ann. Chim.* (*Rome*), **50,** 202 (1960); *Chem. Abstr.*, **54,** 22610 (1960).
186. F. M. Dean and V. Podimuang, *J. Chem. Soc.*, **1965,** 3978.
187. R. Stoermer and E. Barthelmes, *Chem. Ber.*, **48,** 69 (1915).
188. K. v. Auwers, *Chem. Ber.*, **49,** 809 (1916).
189. K. v. Auwers and L. Anschutz, *Chem. Ber.*, **54,** 1556 (1921).
190. V. G. Kulkarni and G. V. Jadhav, *J. Indian Chem. Soc.*, **32,** 97 (1955); *Chem. Abstr.*, **50,** 4135 (1956).
191. T. B. Panse, R. C. Shah, and T. S. Wheeler, *J. Univ. Bombay*, **10,** 84 (1941) ; *Chem. Abstr.*, **36,** 4507 (1942).
192. T. R. Seshadri and P. L. Trivedi, *J. Org. Chem.*, **25,** 841 (1960).
193. D. Price and M. T. Bogert, *J. Am. Chem. Soc.*, **56,** 2442 (1934).
194. J. Kalff and R. Robinson, *J. Chem. Soc.*, **1925,** 1969.
195. B. Cummins, D. M. Philbin, J. F. Eades, H. Fletcher, and R. K. Wilson, *Tetrahedron*, **19,** 499 (1963).
196. M. V. Kadival, J. G. Hinyakaanawar, V. Badigar, and S. Rajagopal, *J. Prakt. Chem.*, **17,** 1 (1962); *Chem. Abstr.*, **57,** 15058 (1962).
197. L. Farkas, A. Gottsegen, and M. Nogradi, *Acta Chim. (Budapest)*, **55,** 311 (1968); *Chem. Abstr.*, **69,** 51938 (1968).
198. P. Venturella and A. Bellino, *Ann. Chim. Italy*, **50,** 1510 (1960).
199. R. L. Shriner, E. J. Matson, and R. E. Danishroder, *J. Am. Chem. Soc.*, **1939,** 2322.
200. G. Schenck, M. Huke, and K. Goerlitzer, *Tetrahedron Lett.*, **1967,** 2063.
201. R. Walter, H. Zimmer, and T. C. Purcell, *J. Org. Chem.*, **3,** 3854 (1966).
202. J. Gripenberg and B. Juselius, *Acta Chem. Scand.*, **8,** 734 (1954); *Chem. Abstr.*, **49,** 10260 (1955).

CHAPTER VII

Naturally Occurring Spirobenzofuranones

1. Spiro-3 (2*H*)-Benzofuranones

A. Griseofulvin

Griseofulvin, $C_{17}H_{17}O_6Cl$(**1**), a colorless natural compound, was isolated from the mycelium of *Penicillium griseofulvum* Dierckx.[1] Subsequently, it was isolated from *P. janczewskii* Zal [= *P. nigricans* (Bainier) Thom.] and its unique biological activity on molds was noted[2-4] by McGowan,[5] who originally called it "curling factor" before the identity of griseofulvin was established.[6,7] The relation between structure and activity is being studied but is still obscure.[8,9] Earlier chemical work has been reviewed by many authors.[10-13] This chapter is concerned only with those aspects that have some connection with the most recent observations on griseofulvin and/or have a particular bearing on its chemistry with reference to its important applications.

a. Nomenclature

The systematic nomenclature[14] is based on the trivial name "grisan" for the tricyclic system **2,** numbered as shown. The absolute configuration of griseofulvin, 2(S),6′(R)-7-chloro-4,6,2′-trimethoxy-6′methylgris-2′-en-3,4′-dione (**1,** R = Cl), has been determined,[15,16] but since correlation of configuration in a series of related compounds is not always evident on the nomenclature system of Cahn, Ingold, and Prelog, the configuration at the two asymmetric centers 2 and 6′ is designated by *d* or *l*. Thus, griseofulvin in prefixed by *d,d*- and the diastereoisomer **3** by *l,d*-, the spiran center is that first mentioned. In an alternative system,[17] the natural isomer is referred to as (+)-griseofulvin and the diastereoisomer (**3**) as epi-(+)-griseofulvin. This system is less cumbersome than the *d,d*-notation but, like that of Cahn et al., is confusing when used to correlate transformation products (e.g. dihydrogriseofulvin has $(a)_D - 20°$).

The nomenclature of griseofulvin, after the *Chemical Abstracts* system,[18] is 6′β-methyl-2′,4,6-trimethoxy-7-chloro-spiro(benzofuran-2(3*H*)-1′-[2]cyclohexene-3,4′-dione; in this chapter the trivial name, "griseofulvin," is retained.

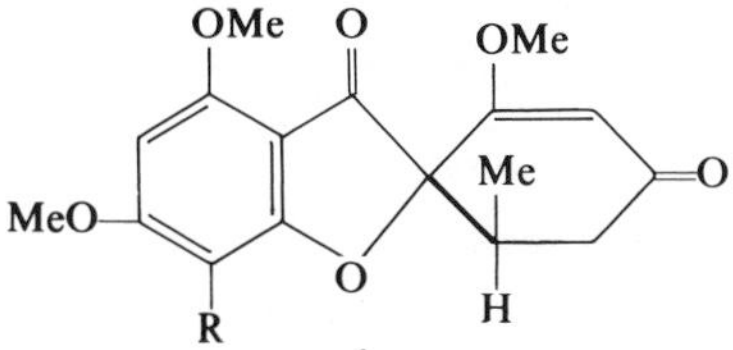

OMe O OMe Me MeO O R H **1**

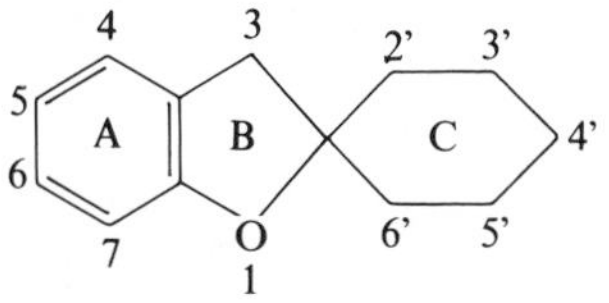

4 3 2' 3' 5 A B C 4' 6 7 O 1 6' 5' **2**

OMe O H Me MeO O Cl OMe **3**

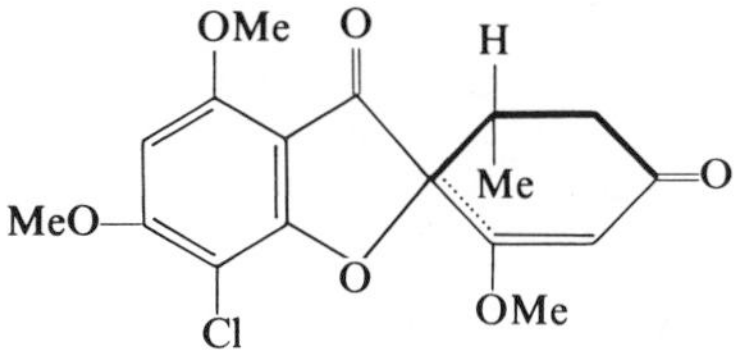

b. Structure

Oxford et al.[1] have shown that griseofulvin contains one reactive double bond, one reactive carbonyl group (in ring C), and three methoxyl groups, one of which is present as enol ether, while the α,β-unsaturated β-alkoxyketonic system reacts with methanolic ammonia to give the enamine, $C_{16}H_{16}O_5NCl.H_2O$, which may be represented as the amine, **4**, or the imine, **5** or **6**.

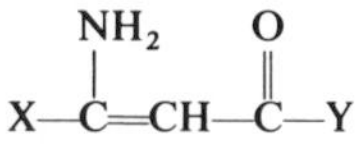

$$\underset{\mathbf{4}}{X-C(NH_2)=CH-C(=O)-Y} \qquad \underset{\mathbf{5}}{X-C(=NH)-CH_2-C(=O)-Y} \rightleftharpoons \underset{\mathbf{6}}{X-C(=NH)-CH=C(OH)-Y}$$

Oxidative degradation of griseofulvin (**1**) and griseofulvic acid (**7**) with potassium permanganate results in the formation of 3-chloro-2-hydroxy-4,6-dimethoxybenzoic acid (**8**) and 7-chloro-2-hydroxy-4,6-dimethoxy-3(2*H*)-benzofuranone-2-β-butyric acid (**9**, R = OH). Alkaline hydrogen peroxide oxidation of griseofulvic acid gave **9** (R = H) from which **9** (R = OH) can be obtained upon treatment with permanganate. Furthermore, yellow mercuric oxide has brought out the oxidation of **7** to chlorohydroxydimethoxymethyldibenzofuran (**10**), which was similarly obtained upon treatment of decarboxygriseofulvic acid (**11**) with air under alkaline conditions. 3-Methoxy-2,5-toluoquinone (**16**) was isolated when chromic oxide in acetic acid medium was used.[19,20]

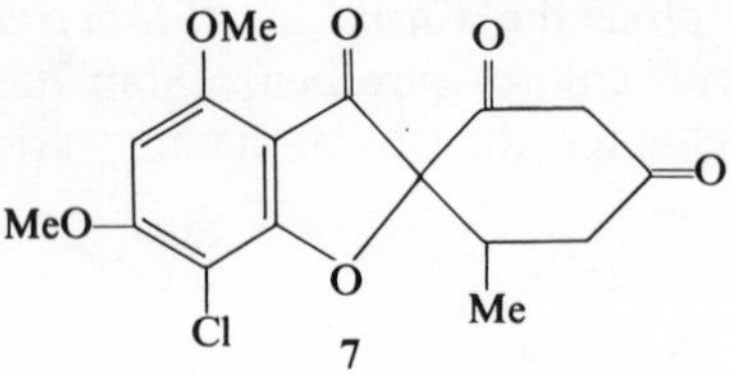

7 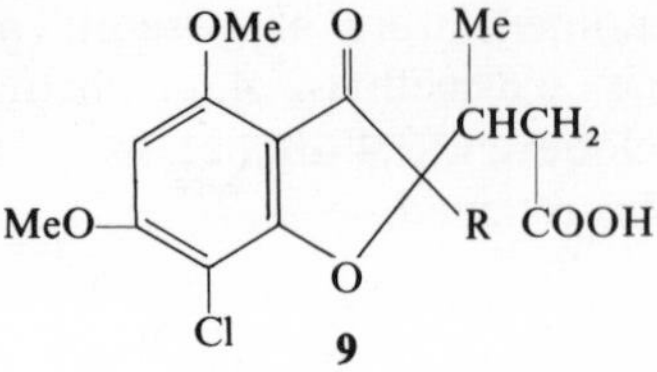 9 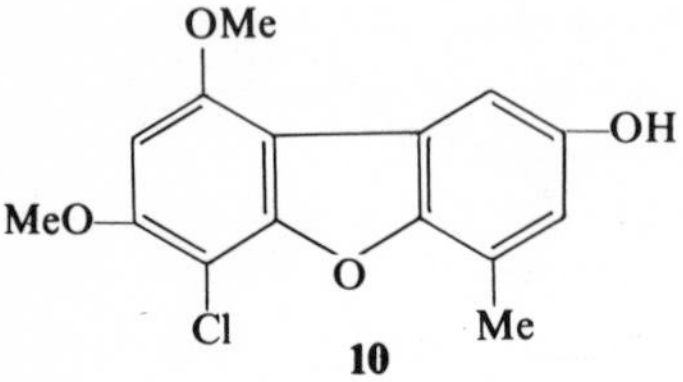 10 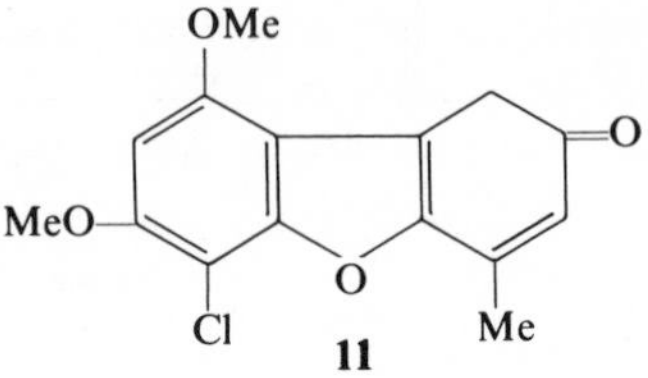 11 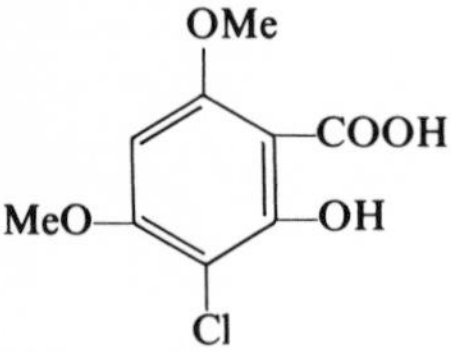 8 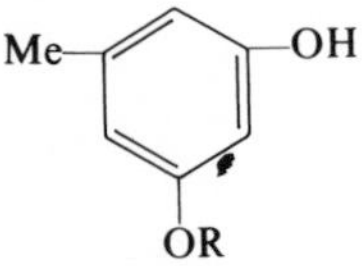 12 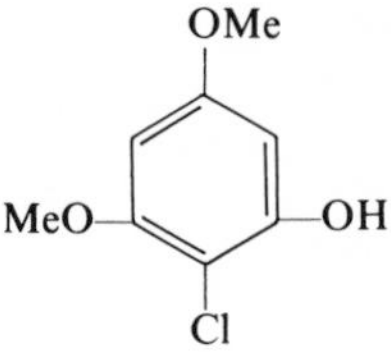 13 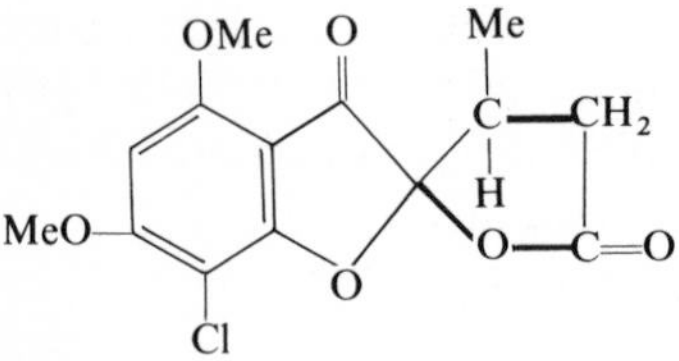 14 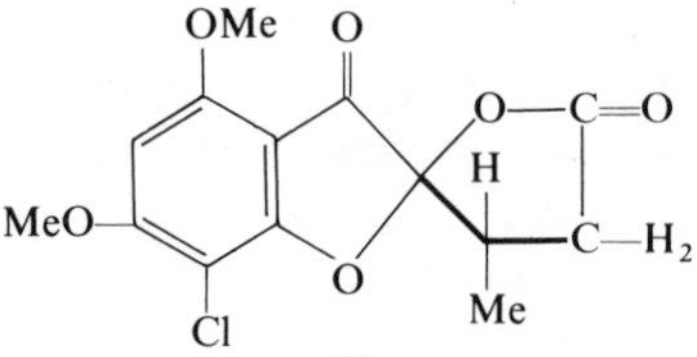 15 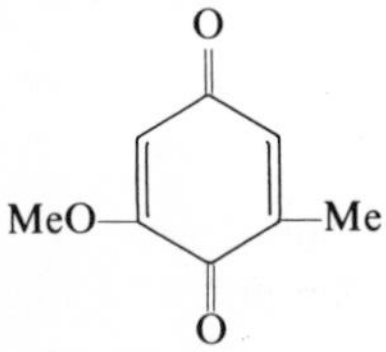 16 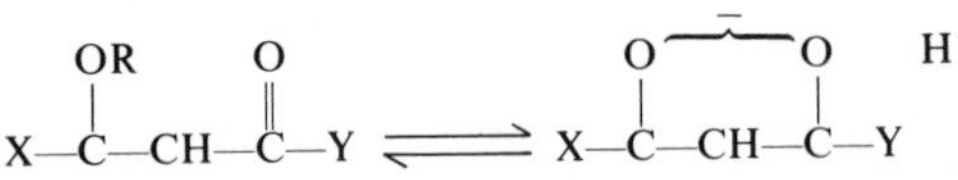 17 18

In permanganate oxidation experiments, **8** is readily decarboxylated to the phenol **13**. The genesis of **9** (R = OH) is by way of the β-ketodibasic acid **9** (R = COOH) followed by spontaneous decarboxylation to **9** (R = H), with subsequent hydroxylation of the reactive α-hydrogen in the dihydrobenzofuranone moiety. The acid **9** (R = OH) furnishes two stereoisomeric lactones, **14** and **15**. One lactone, which is produced by distillation of the acid or by the action of acetic anhydride is optically inactive, while cyclization with sulfuric acid or a mixture of acetic anhydride and pyridine gives the second lactone which is optically active.

The butyric acid **9** (R = OH) contains one *C*-methyl residue, which appears as (+)-methylsuccinic acid when **9** (R = OH) is oxidized with potassium permanganate. The carboxyl group of the dihydrobenzofuranone system is inert towards the usual reagents, but is readily identified by its infrared spectrum. Structure of phenol **13** has been confirmed by synthesis.

The presence of a second 6-membered ring (*C*) in griseofulvin is indicated by oxidation with chromic oxide to 3-methoxy-2,5-toluquinone (**16**) and the formation of orcinol by potassium hydroxide fission.[1] These products could not arise from rings A in **8** and **9** (R = OH). Since the three methoxyl groups in griseofulvin appear in oxidation products **8** and **16**, it is evident that it is not the methyl ester of a carboxylic acid.[14] Furthermore, it has been established that the methoxyl group in the quinone **16** is derived from a methyl enol ether of the type **17**.[21] Moreover, griseofulvic acid (**7**), treated with methanolic diazomethane, generates a mixture of griseofulvin and an isomer with very similar properties, isogriseofulvin (**19**, R = Me, 7-chloro-4,6,4′-trimethoxy-6′methylgris-3′-en-3,2′-dione); thus the tautomeric system **17** in griseofulvin ⇌ **18** in griseofulvic acid is present. The enolic methoxy group in ring C of griseofulvin is readily removed by dilute alcoholic potassium hydroxide solution to give the 1,3-diketone, griseofulvic acid (**7**), which is not a carboxylic acid, and is more accurately named as 7-chloro-4,6-dimethoxy-6′-methylgrisan-3,2′,4′-trione or 6′-methyl-2′-4,6-dimethoxy-7-chloro-spiro(benzofuran-2(3*H*)-1′-cyclohexane)-2′,3,4′-trione.[18]

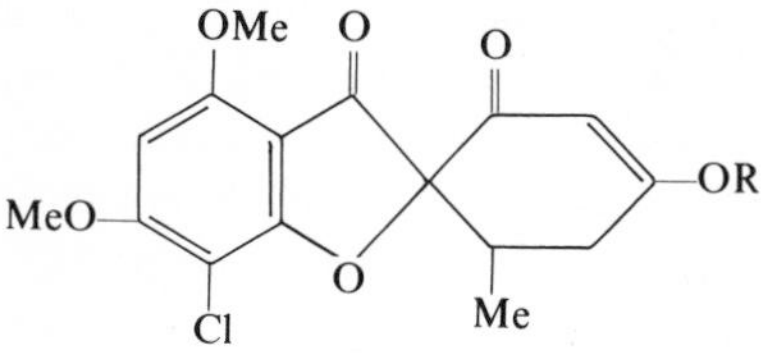

OMe O O MeO OR O Cl Me **19**

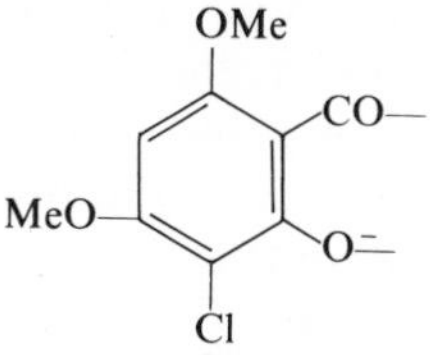

OMe CO— MeO O⁻ Cl **20**

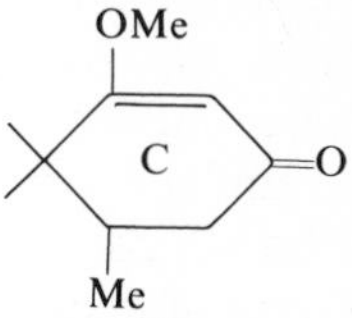

OMe C O Me **21**

All the carbon atoms in griseofulvin are accounted for in oxidation products **19** and **16**. The formation of **8** under hydrolytic conditions provides convincing chemical proof that there is a carbonyl group directly attached to the benzenoid ring (C), as suggested by spectroscopic evidence.[21] The hydroaromatic ring must contain the *C*-methyl group and the olefinic double bond. Moreover, if it is assumed that rearrangement does not occur in the formation of orcinol (**12**, R = H) and its monomethyl ether (**12**, R = Me), two potential hydroxyl groups must be located in the 3,5 positions with respect to the *C*-methyl group. Thus ring C must contain at least the methyl ether group of the system **17** (R = Me), and the most likely skeleton of ring C, therefore, appears to be **20.**

The manner in which the two partial structures **20** and **21** are linked in griseofulvin seems to be unequivocally determined by the formation of the acids **9** (R = H) and **9** (R = OH), obtained from griseofulvic acid **(7)** by alkaline hydrogen peroxide and/or by permanganate oxidation, respectively. Union of partial structures **20** and **21** in a like manner leads to the two spiran structures **22** and **23,** which are isomeric methyl ethers of the tautomeric methyl ethers of the tautomeric enol **7.**[14] It is considered that **22,** (R = Me) is more likely to give rise to **16** upon chromic oxide, oxidation and therefore represents griseofulvin. However isogriseofulvin does not yield a quinone with chromic oxide,[19,22] and therefore is considered to be 7-chloro-4,6-dimethoxy-3(2*H*)-benzofuranone-2-spiro-1′-(4′-methoxy)-6′-methylcyclohex-3′-en-2′-one **(23)**

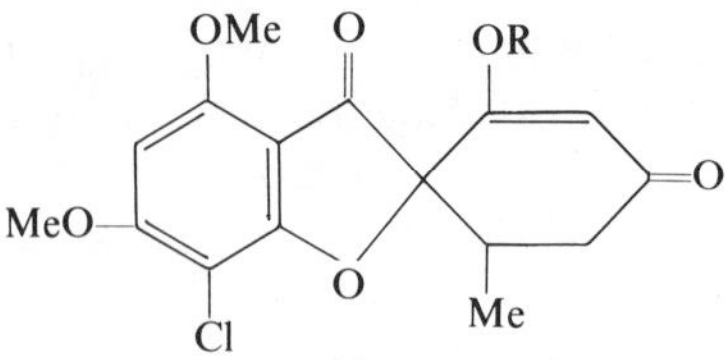

22 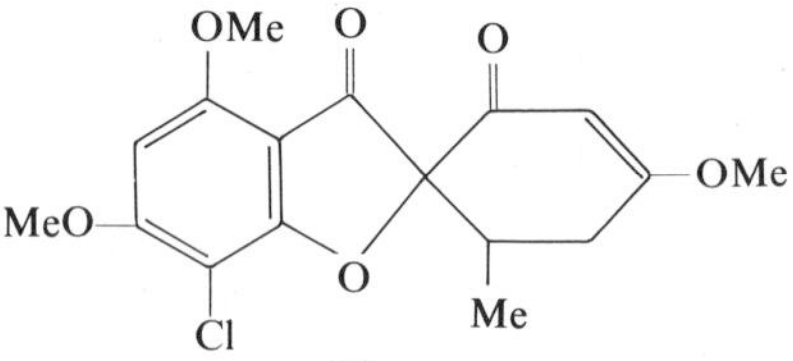 **23**

Alkaline degradation of griseofulvin, which may be regarded as a vinylog of a 1,3-diketone, is complex and extremely dependent on the reaction conditions. Fusion with alkali gives rise to orcinol, while the action of 2*N* sodium methoxide yields **8** together with **12** (R = Me). Under mild conditions, this permits epimerization at the spiran atom, but under more vigorous conditions the intermediate **(24)** suffers fission of bond *b*, so that **8** and **12** (R = Me) are formed. However degradation of griseofulvin with baryta results in the formation of (−)-decarboxygriseofulvic acid **(11)**; the action of aqueous alkali on griseofulvin furnishes, inter alia, decarboxygriseofulvic acid together with norgriseofulvic acid **(25)**. It was the early discovery of this reaction that so misled the first workers in the field about the nature of griseofulvin. It is believed that **26** is presumably an intermediate in this reaction, a view supported by synthetical evidence. The removal of a methoxyl group from ring A at the 6 position of griseofulvin by the action of alkaline reagents[23] is an unusual but characteristic feature of the chemistry of griseofulvin; it is ascribed to the influence of the adjacent chlorine, since dechlorogriseofulvin **(26)** behaves normally. Similar phenomena are observed among the phthalides, but in the absence of a halogen atom.[24]

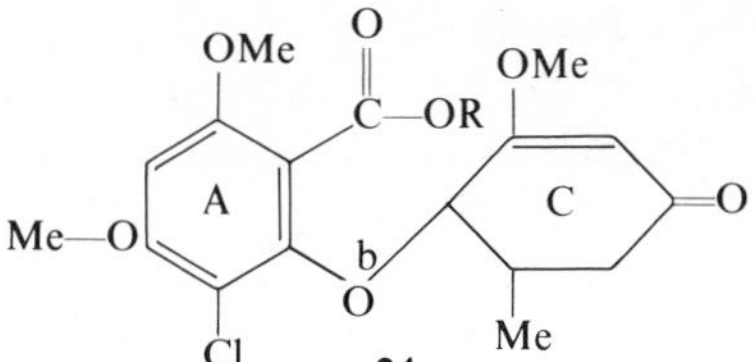

Treatment of griseofulvin with 0.5*N* sodium methoxide avoids degradation, but induces another change; it produces an isomeric product of griseofulvin. An equilibrium is attained with 40% of the latter and 60% of the isomer. The reaction seems to be an inversion of the spiro atom, since oxidation of the isomer still affords (+)-methylsuccinic acid. The structure for the diastereoisomers is **27a** for griseofulvin (the less stable of the two), in which the 6′-methyl group is cis to the relatively bulky benzofuranone carbonyl group. In the isomer, the 6′-methyl group and the (relatively small) oxygen atom at the 1 position are also cis. The mechanism of epimerization is easy to explain, as it is only necessary to suppose that the ring fission leading to **24** is reversible. For convenience, griseofulvin (**27a**) is referred to as the (*d*,*d*-isomer), and the diastereoisomer (**27b**) as the (*l*,*d*-isomer), the spiran carbon atom is the first mentioned.[15,25] Strictly speaking, griseofulvin has the 2(S),6(R′)-configuration, and the diastereoisomer, 2(R),6′(R).

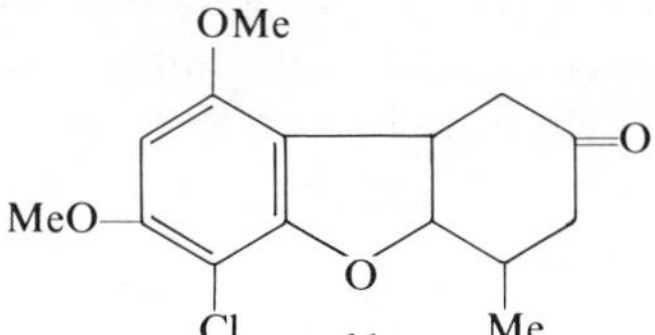

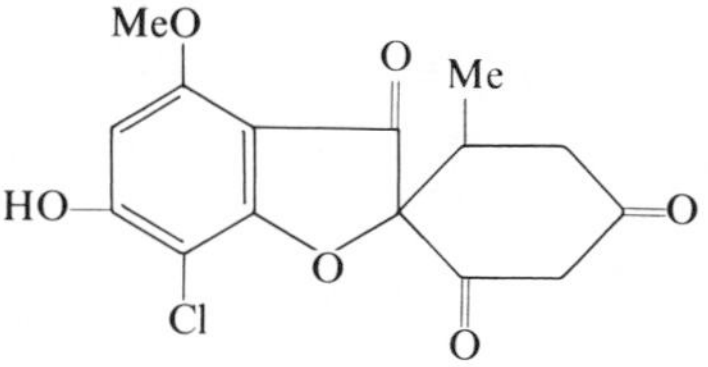

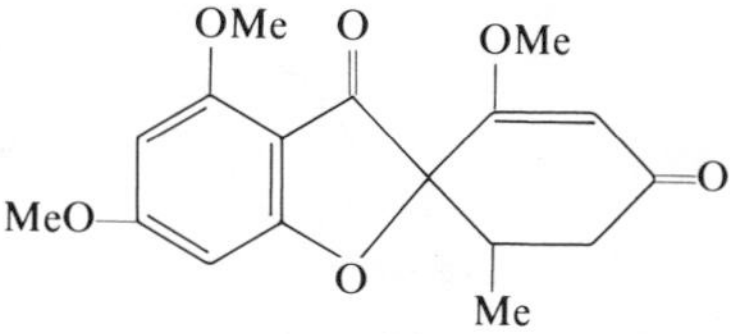

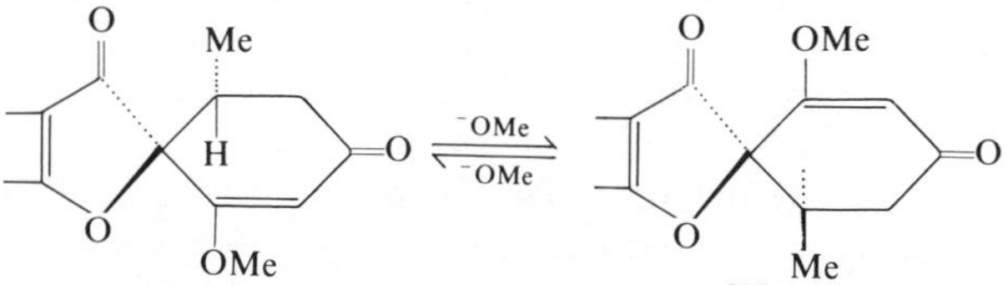

Most of the above work has been confirmed synthetically.[26] 7-Chloro-4,6-dimethoxy-3(2*H*)-benzofuranone (**28**) condensed with 1,5-dibromohexane by means of potassium *t*-butoxide to afford the spiran (**29**) which had been obtained as a minor hydrogenation product of griseofulvin when Adam's catalyst is used.[27,28] Addition of methylvinyl ketone to

2-acetyl-3(2*H*)-benzofuranones yields compounds of the type **30** but, when dehydrated with sulfuric acid, they do not cyclize to the desired 2′-methylgrisan-4′-ones but to the isomeric 4′-methylgrisen-2′-ones[29,30] (**31**).

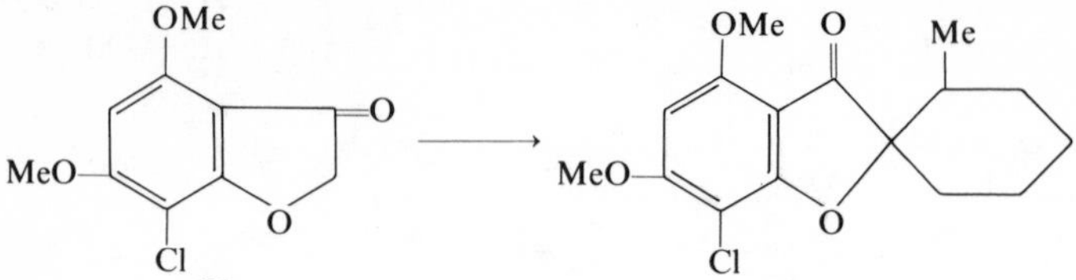

OMe MeO Cl O O **28** → OMe O Me MeO Cl O **29**

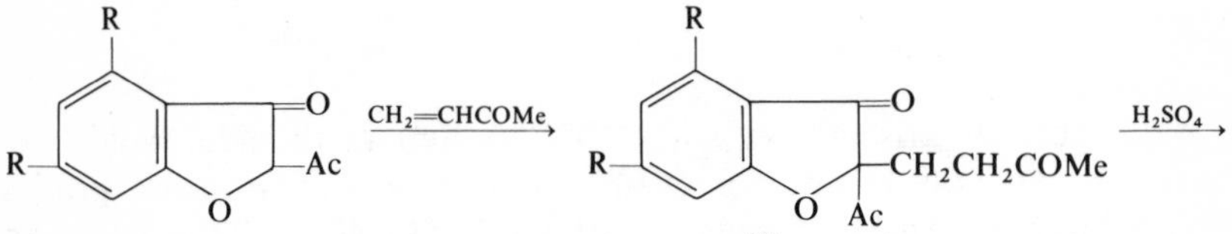

R R O O Ac $\xrightarrow{CH_2{=}CHCOMe}$ R R O O CH_2CH_2COMe Ac **30** $\xrightarrow{H_2SO_4}$

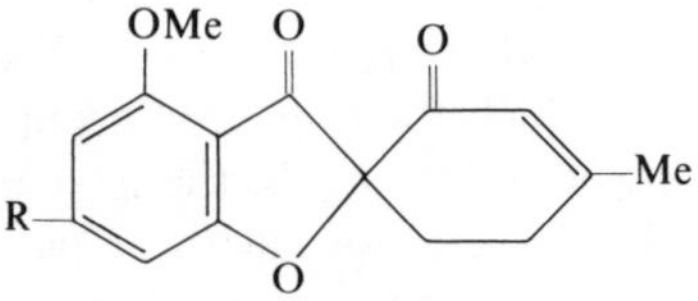

OMe O O R O Me **31**

Hydrogenation of griseofulvin[31,32] proceeds in three ways: (*a*) reduction of the double bond only to give a product which, on further hydrogenation, effects the reduction of the reactive keto group and hydrogenolysis of the chlorine; (*b*) reduction to an alcohol and saturation of the ethylenic linkage; or (*c*) reduction of the keto group to an alcohol, which is hydrogenolyzed to methylene, followed by reduction of the ethylenic linkage. Thus according to (*a*) griseofulvin (**1**) yields 7-chloro-4,6,6′-trimethoxy-2′-methylgrisan-3,4′-dione (dihydrogriseofulvin) (**32**) which undergoes further reduction to 4′-hydroxy-4,6,6′-trimethoxy-2′-methylgrisan-3-one (**34**, R = H). According to (*b*), reduction of the carbonyl to an alcohol group with saturation of the double bond yields 7-chloro-4′-hydroxy-4,6,6′-trimethoxy-2′-methylgrisan-3-one (**34**, R = Cl). By (c), the carbonyl group is reduced, followed by hydrogenolysis to yield 7-chloro-4,6,2′-trimethoxy-6′-methylgris-2′-en-3-one (desoxygriseofulvin, **35**), and then by saturation of the double bond to 7-chloro-4,6,2′-trimethoxy-6′-methylgrisan-3-one (**33**, tetrahydrodesoxygriseofulvin).

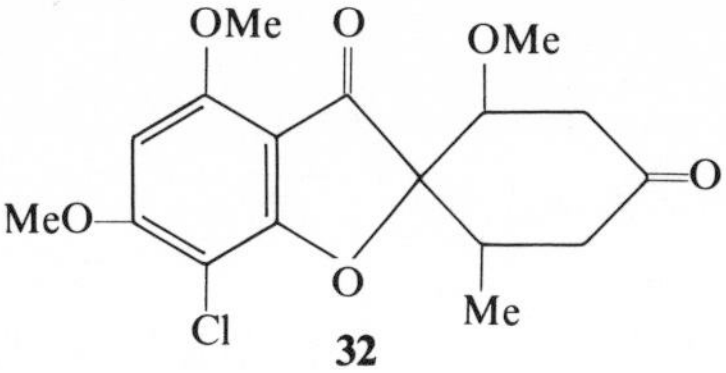

The formation of **35** is of particular interest in that the reduction of the carbonyl group of ring C to a methylene residue precedes saturation of the α,β-double bond. Further hydrogenation of **35** led to the formation of **33**.

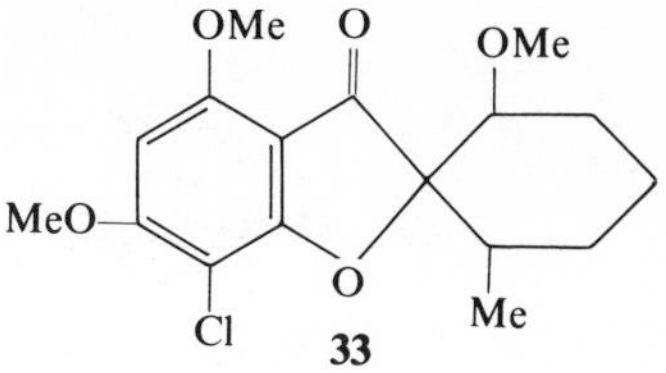

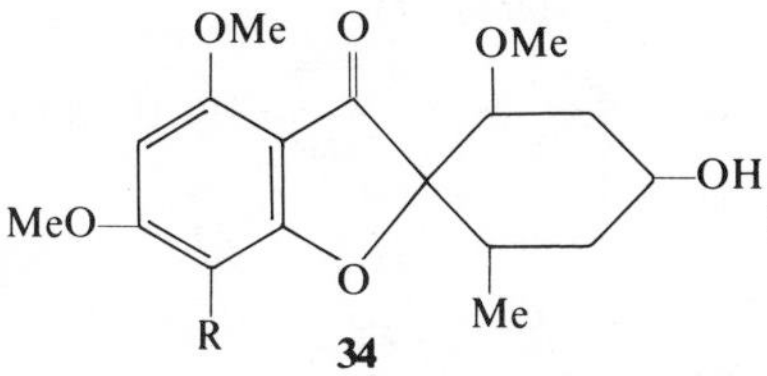

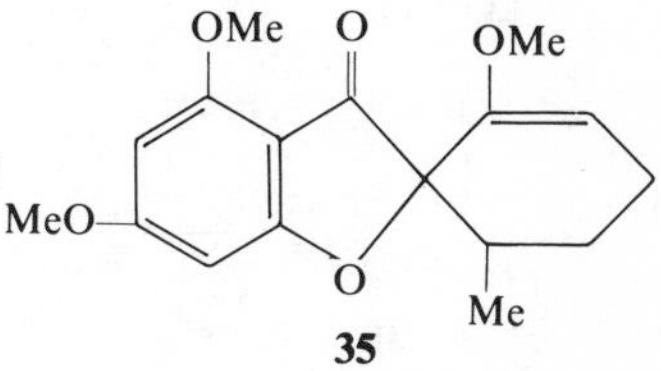

Hydrogenation of 7-chloro-4,6,4′-trimethoxy-6′-methylgris-3′-en-3,2′-dione (isogriseofulvin, **23**) in presence of palladium—charcoal gives as a main product 7-chloro-6′-hydroxy-4,6,4′-trimethoxy-2′-methylgrisan-3-one (**36**). Oxidation of the latter yields the ketone 7-chloro-4,6,4′-trimethoxy-6′-methylgrisan-3,2′-dione (dihydroisogriseofulvin, **37**).[31]

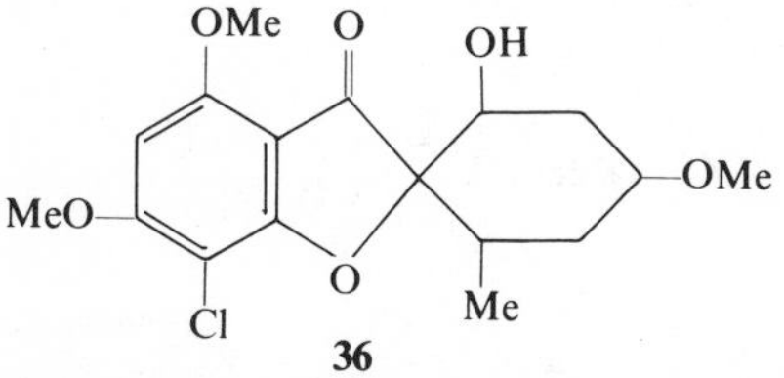

The detailed chemistry of the reduction products of griseofulvin (**1**) and isogriseofulvin (**23**) has been reported;[32] their reactions are consistent with the proposed formulas. A comparative examination of reactions of **32** and **37** enables an unequivocal differentiation to be made between structures **1** and **23**.[32]

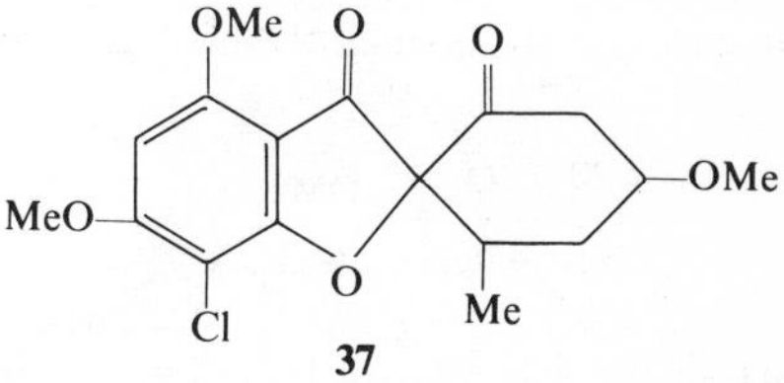

The β-methoxyketonic systems in **32** and **37** readily lose a molecule of methanol in the presence of sulfuric acid to yield α,β-unsaturated ketones **38** and **40**, respectively, and in accordance with its structure, degradation of **38** with alkali gives **8**. Hydrogenation of **38** affords the alkali-stable dihydro compound **39**, while **40** furnishes **41** which, in accordance with its formulation as a β-diketone, is readily converted by alkali into (−)-dihydrobenzofuranone acid (**42**) (Eq. 1).

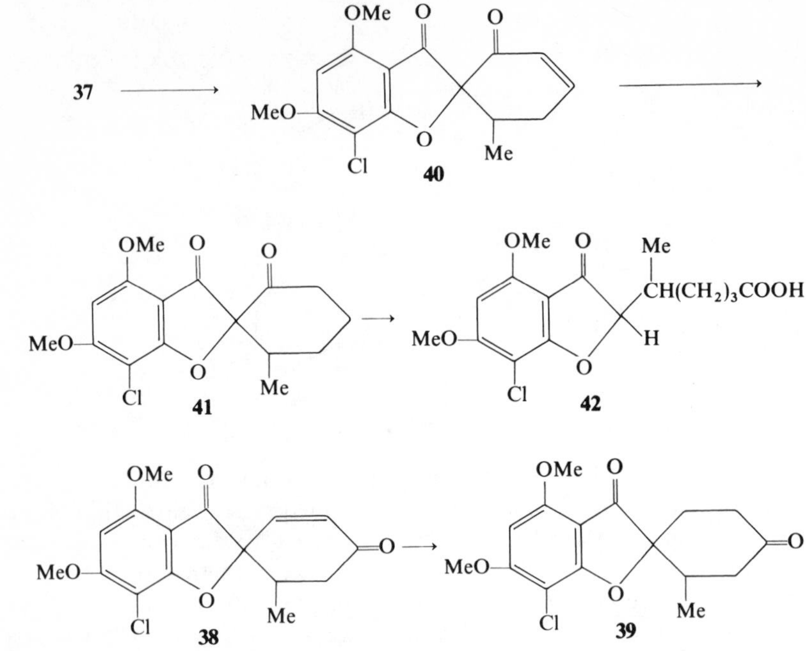

37 → 40 →

41 → 42

38 → 39 (1)

Reduction of **1** and of **23** with sodium borohydride gives **43** and **44**, respectively.[33] They can be converted to **45**, which has also been obtained by catalytic hydrogenation of griseofulvinic acid.[35]

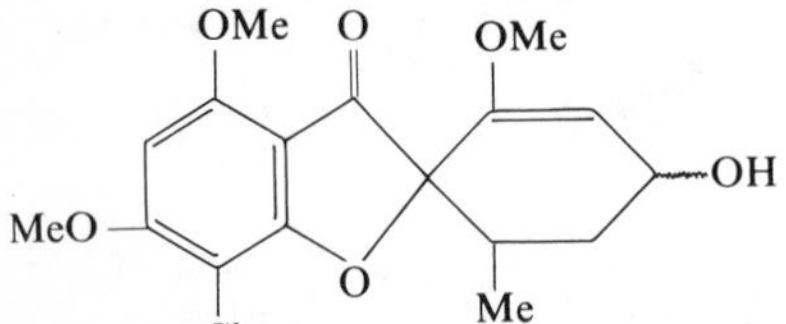

43 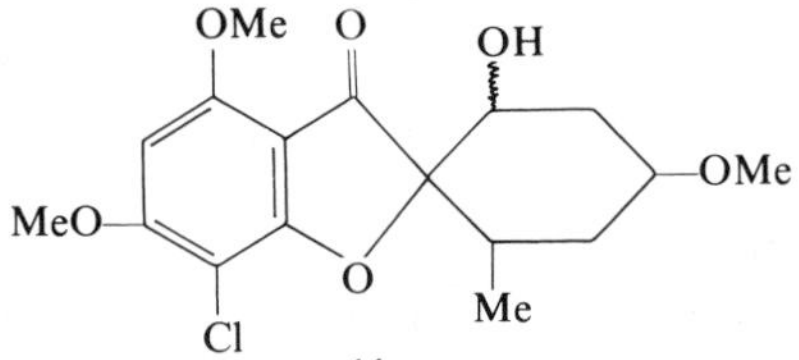 **44**

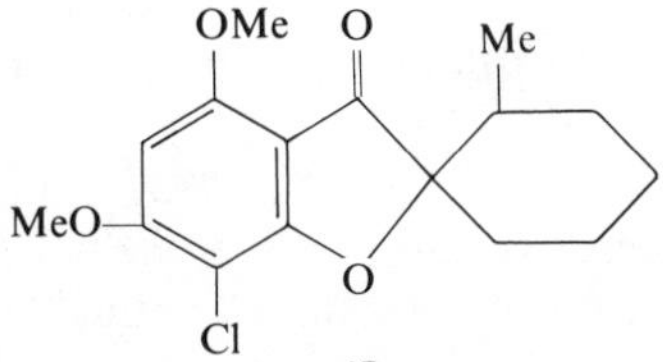

45

The chemistry of griseofulvol (**43**) and isogriseofulvol (**44**) has been studied.[34]

Epoxidation of griseofulvin and its 5′-bromo derivative has recently been investigated.[35] They are converted into their corresponding epoxy derivatives in high yield upon treatment with hydrogen peroxide and base. Epoxidation of griseofulvin and its 5-methylsulfonyl derivative is also achieved with benzoyl peroxide–methoxide in good yield, but with poor conversion. A tentative stereochemical structure is assigned to the product, 2′,3′-epoxy-7-chloro-2′,4,6-trimethoxy-6′β-methylspiro(benzofuran-2(3*H*)-1′-(2)-cyclohexane)-3,4′-dione (**46**). 5′-Diazogriseofulvin (**47**),[36] prepared by allowing the readily available 5-formylgriseofulvin to react with tosyl azide in the presence of diethylamine, is a potential intermediate for the preparation of a variety of 5′-substituted griseofulvins.

c. *Synthesis of Griseofulvin and Its Analogs*

The literature includes different syntheses of griseofulvin (**1**) and its analogs by Brossi et al.,[17,37-39] Taub, et al.,[40-43] Stork and Tomasz,[44,45] and Scott et al.[46,47]

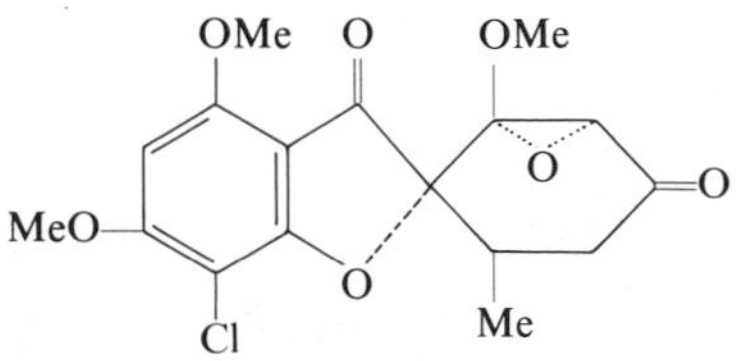

OMe O OMe MeO O Cl Me O O

46

OMe O OMe MeO O Me R O

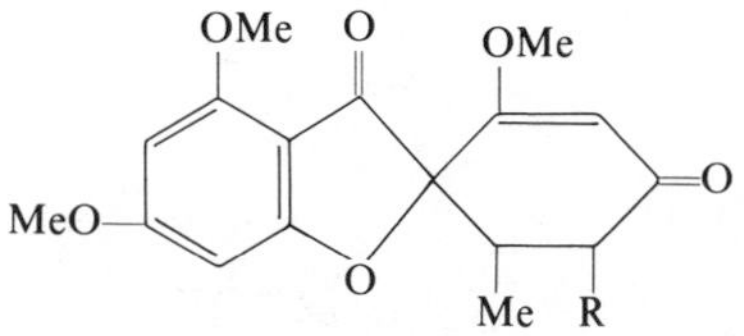

47 R = N_2 R = CHO R = H

The synthesis of the orally active antifungal antibiotic griseofulvin presents interesting structural and stereochemical problems which are well illustrated by the previous syntheses of this compound: both epimers at the 2′ carbon are obtained (±), while the intermediate of the β-diketone (griseofulvic acid, **7**) leads to the two possible enol ethers with diazomethane.[37] An interesting solution to these difficulties, based on a biogenetically likely pathway, has been recorded.[40,46]

The double Michael addition of methoxyethynylpropenyl ketone to 7-chloro-4,6-dimethoxy-3(2*H*)-benzofuranone in presence of *t*-butoxide in diethylene glycol dimethyl ether at room temperature gives *d,l*-griseofulvin, **48** (Eq. 2). It is of interest that the synthesis appears to produce **48** free from its epimer *d,l*-epigriseofulvin (**49**); this synthesis is stereospecific. It should be noted that the epimer actually obtained is known to be the less stable one. Equilibration of (+)-griseofulvin has been shown to produce an equilibrium mixture containing 60% of the epimer in which the carbonyl group of the benzofuranone ring is trans to the neighboring methyl group.[15] Formation of *d,l*-griseofulvin (**48**) must, therefore, be under kinetic control. It is suggested that the observed stereospecifity is the result of better overlap of the electron donor system of the enolate ion with the unsaturated ketone acceptor in the transition state **50** which leads to *d,l*-griseofulvin, than in the transition **51** which leads to *d,l*-epigriseofulvin (**49**).

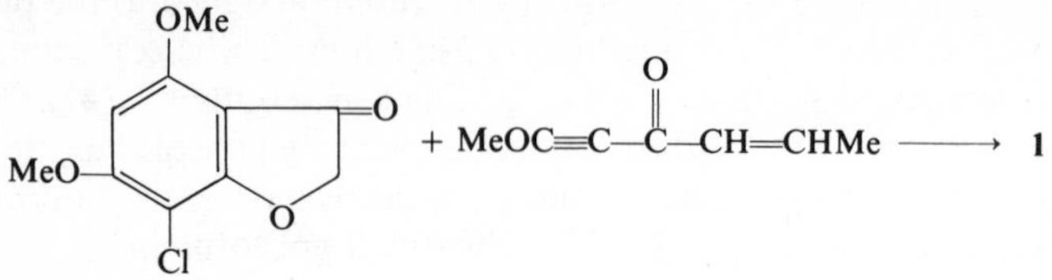

+ MeOC≡C—C(=O)—CH=CHMe ⟶ **1** (2)

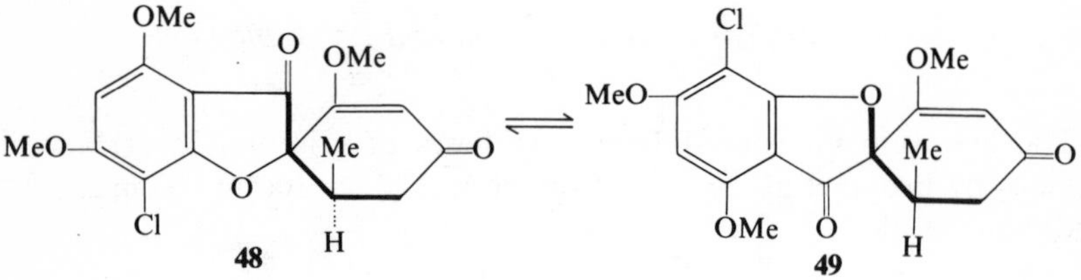

48 ⇌ **49**

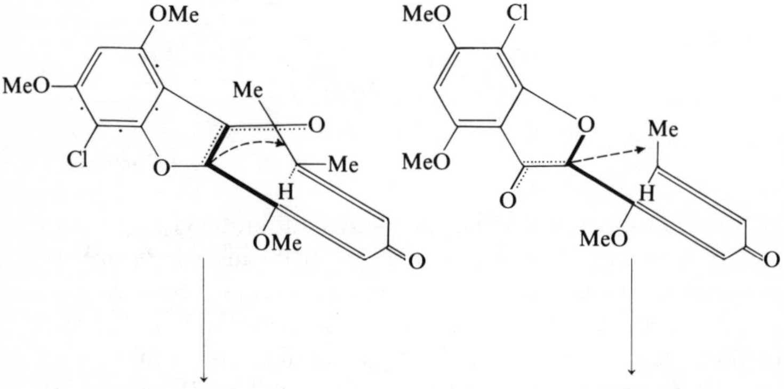

50 *d,l*-Griseofulvin

d,l-Epigriseofulvin

The application of current concepts of biosynthesis to griseofulvin has been reviewed.[12,48-51] Synthesis of griseofulvin after Brossi et al.[17] is illustrated in Eq. 3.

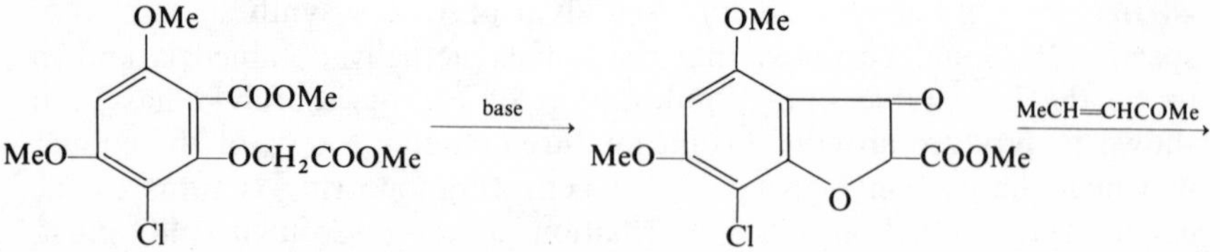

⟶ (base) ⟶ (MeCH=CHCOMe)

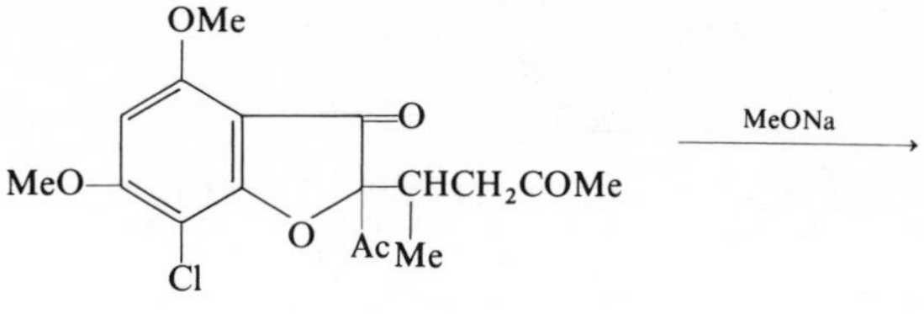

(Two stereoisomeric racemates)

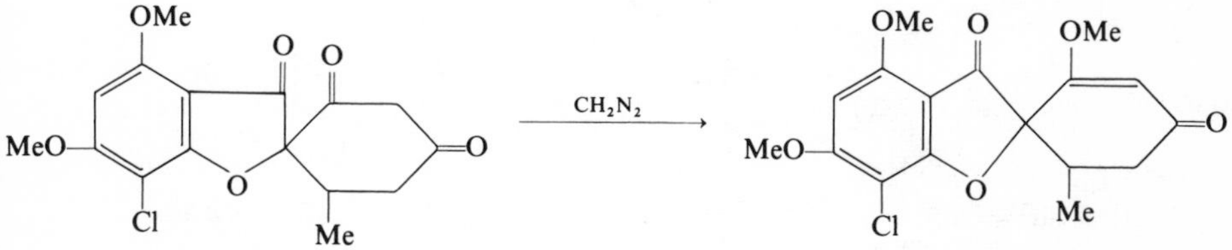

MeONa

(*l,d*) + (*d,l*)

CH_2N_2

(+ Related isocompound)

(*d,d*)- and (*l,l*)-griseofulvin $\xrightarrow{H_3O^+}$ (*d,d*)- and (*l,l*)-griseofulvic acid $\xrightarrow{\text{brucine}}$ (*d,d*)-griseofulvic acid $\xrightarrow{CH_2N_2}$ (*d,d*)-griseofulvin + isogriseofulvin)

(3)

Another approach for the synthesis of griseofulvin was inspired by the theory that, in vivo, such spiro compounds would probably be formed by oxidative cyclization of suitable hydroxybenzophenones.
The benzophenone derivative was obtained by means of a Friedel-Craft's reaction[47,50,52] and/or by Fries photochemical rearrangement.[40] Oxidation of the benzophenone with potassium ferricyanide[37] or with selenium dioxide[40] furnished (±)-dehydrogriseofulvin; this could be easily recognized as the (−)-compound that may be obtained by dehydrogenation of griseofulvin with selenium dioxide. Hydrogenation of dehydrogriseofulvin over rhodium—selenium catalyst resulted in the formation of racemic griseofulvin. However the last steps in this synthesis are hydrolysis to racemic griseofulvin acid and resolution by the quinone methosalt and remethylation (Eq. 4).[47,50]

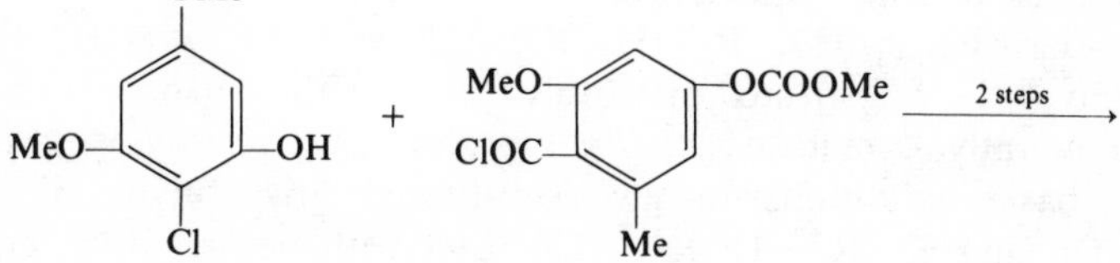

2 steps

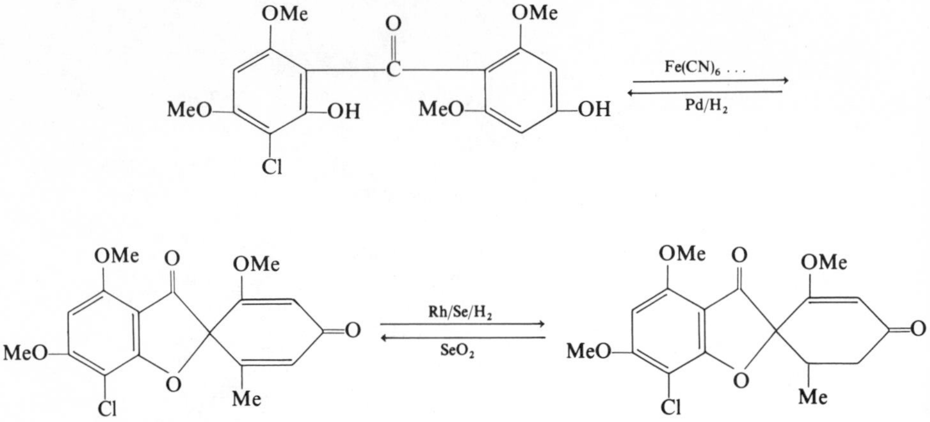

(Dehydrogriseofulvin) (Griseofulvin)

(4)

Nucleophilic reagents react with spiro compounds containing a functional group, which could be displaced leading to substituted griseofulvins. Ether (**52**, $R^1 = O$-alkyl) homologs are readily obtained upon treatment of griseofulvic acid (**7**) with diazoalkanes,[53] with alcohols in presence of acids,[53] and/or with alkyl halides in presence of potassium carbonate or silver oxide.[54] All these methods have the disadvantage that some of the 4′-ether is produced at the same time. Moreover, griseofulvic acid and its 3′-substituted derivatives are not particularly suitable starting materials for the preparation of *S*-alkyl compounds.

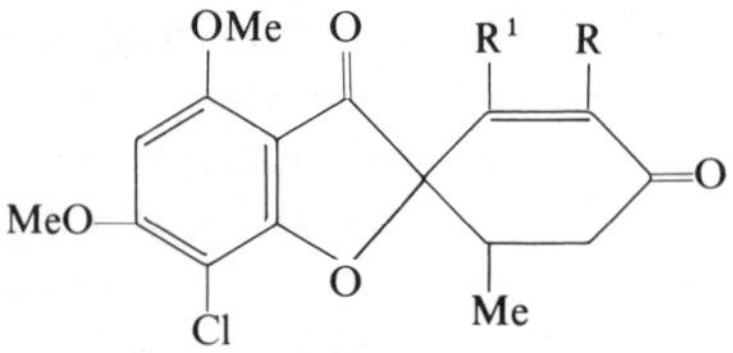

52 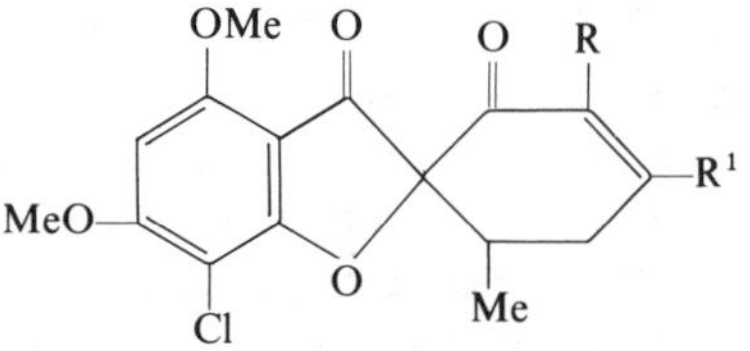 **53**

Griseofulvic acid (**7**) is transformed by the action of phosphoryl chloride in presence of water and lithium chloride into a mixture of 2′- and 4′-chloro-derivatives, (**52**, R = H, R^1 = Cl and **53**, R = H, R^1 = Cl), respectively. 3′-Substituted griseofulvic acid, with the same reagent, gave predominantly 4′-chlorides.[55] 2′-Chlorides react readily with methanol under basic conditions to give substituted griseofulvin. **52** (R = H, R^1 = Cl) and **53** (R = H, R^1 = Cl) react with benzenethiol, containing triethylamine, to yield the 2′- and 4′-*S*-phenylthio compounds; in a similar manner, the corresponding amines are obtained when the appropriate amines are used.

Synthesis of chloro analogs of griseofulvin (**54**) has been achieved, (Eq. 5).[41,43]

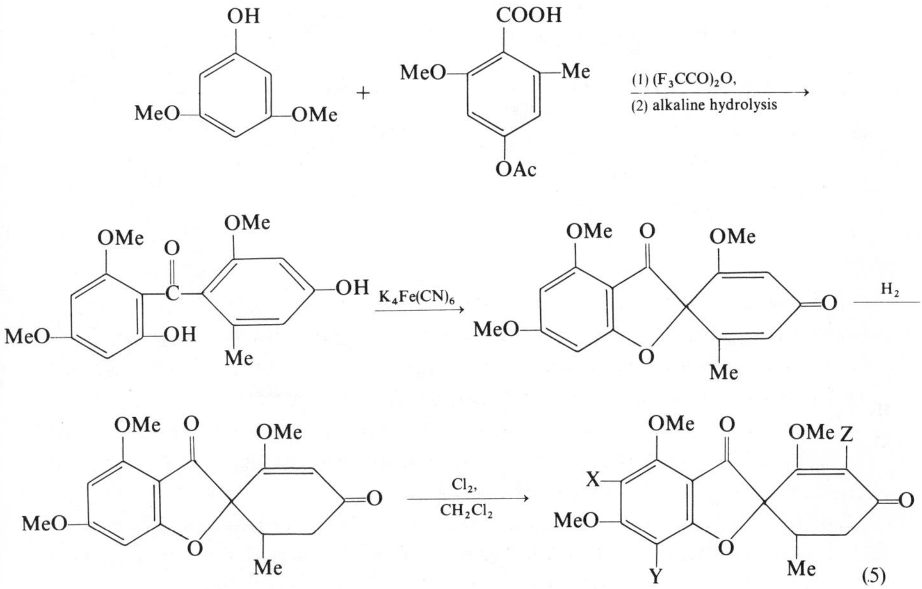

(1) $(F_3CCO)_2O$, (2) alkaline hydrolysis

$K_4Fe(CN)_6$

H_2

Cl_2, CH_2Cl_2

(5)

54a X = Z = H, Y = Cl
54b X = Cl, Y = Z = H

Lead tetraacetate has been found to effect the oxidative ring closure of **55** with the formation of **56**.[56]

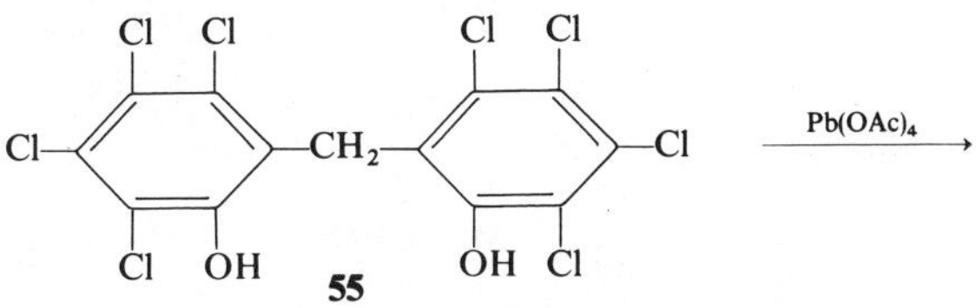

$Pb(OAc)_4$

55

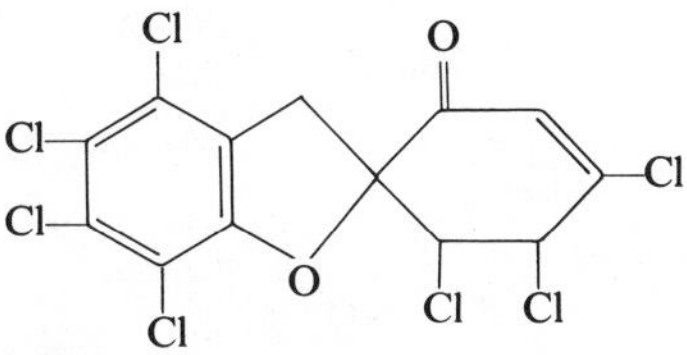

56

Griseofulvin in acetic acid medium does not react with bromine alone, but in the presence of mercuric acetate it yields the 5,3′-dibromo derivative (**57**); similar treatment of griseofulvic acid (**7**) affords the tribromo derivative (**58**).[57]

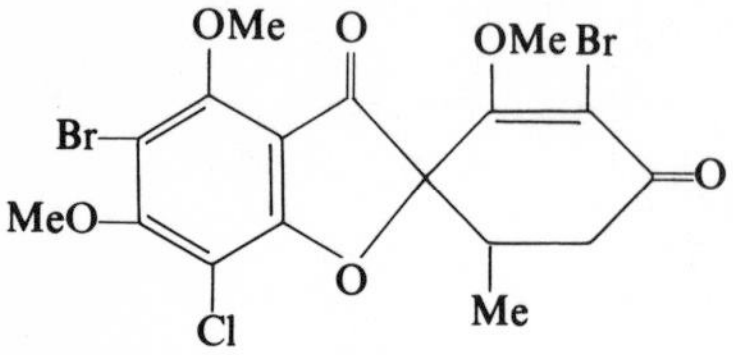

57

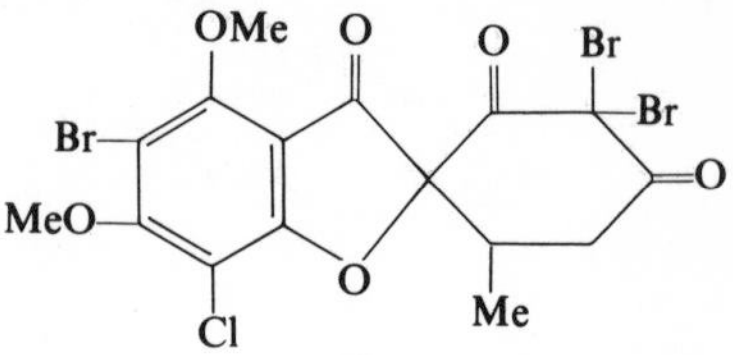

58

7-Bromo-4,6,2′-trimethoxy-6′-methylgris-2′-en-3,4′-dione, the bromo analog of griseofulvin, has been isolated from the culture filtrates of both *Penicillium griseofulvum* Dierckx and *nigricans* (Bainier) Thom., grown on a synthetic medium containing potassium bromide.[16]

The antifungally active 5′-hydroxygriseofulvin (**61**) (less active than the very potent griseofulvin), has been obtained by the microbiological oxidation of griseofulvin but the poor yield of this conversion and the tedious procedure involved made this route rather unattractive. 5′-Formylgriseofulvin (**59**), as its sodium salt in methanol, reacts with benzoyl peroxide to yield 5′-hydroxydehydrogriseofulvin (**60**).[58] Structure **60** finds support from the spectral data, its reduction to the benzophenone (**62**), and its preparation from 5′-hydroxygriseofulvin (**61**),[59,60] by oxidation with bismuth trioxide.

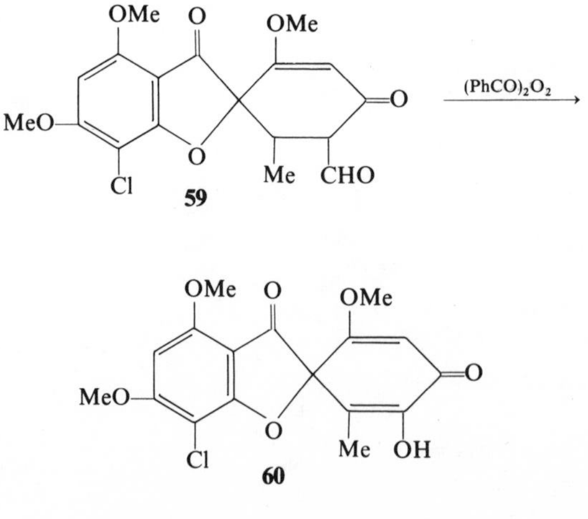

59

$(PhCO)_2O_2$

60

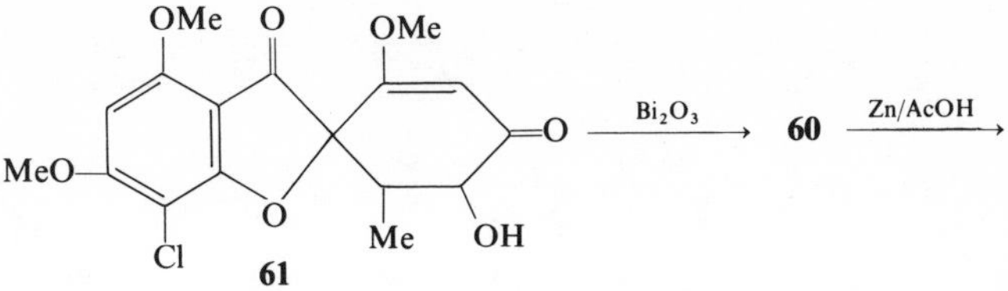

OMe O OMe

MeO — O — $\xrightarrow{Bi_2O_3}$ **60** $\xrightarrow{Zn/AcOH}$

O

Cl Me OH

61

O

OMe ‖ OMe

C

MeO OH

OH

Cl Me OH

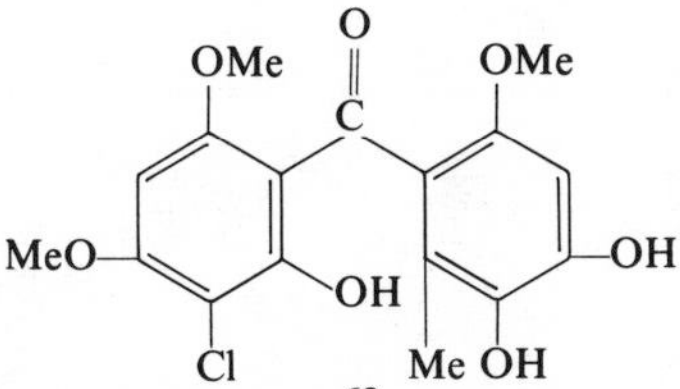

62

Transformation of dehydrogriseofulvin (**63**) into (±)-griseofulvin (**48**) and (+)-5′-hydroxygriseofulvin (**61**) has been achieved by the action of *Streptomyces cinereocrocatcus* NRRL 3443.[59]

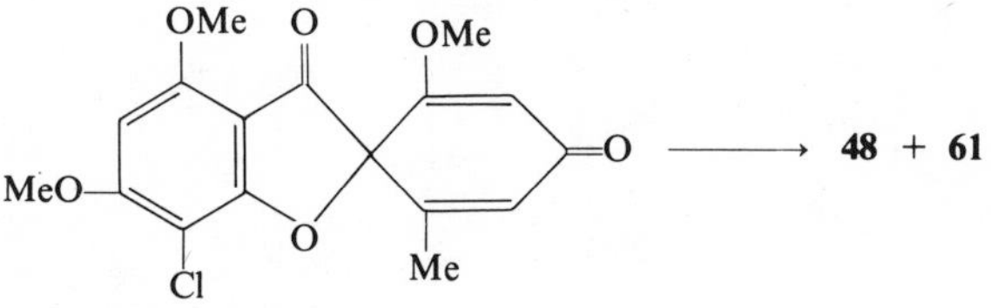

OMe O OMe

MeO — O ⟶ **48** + **61**

O

Cl Me

63

Taub et al.[61,62] have synthesized 6′-demethylgriseofulvin (**67**) as outlined in Eq. 6. Demethylation of griseofulvin has been brought about by the action of fungi, giving rise to various demethylated metabolites: *Microsporium canis* yields 4-demethylgriseofulvin; *Botrytis allii*, 2′-demethylgriseofulvin; and *Coreospora melonis*, 6-demethylgriseofulvin.[63]

d. Dehydrogriseofulvin

Dehydrogriseofulvin, $C_{17}H_{15}O_6Cl$ (**63**), is found in the mother liquors of *Penicillium patulum* together with several other new metabolites of this mold. These were shown to include the benzophenone (**68**), dehydrogriseofulvin(6′-methyl-2′,4,6-trimethoxy-7-chloro-spiro(benzofuran-2(3*H*)-1′-(2,5)cyclohexadiene)-3,4′-dione (**63**), and **69**. In addition, a new benzophenone, griseophenone, $C_{17}H_{17}ClO_5$ (**70**), was isolated; its mass spectrum defined the groups attached to the aromatic ring.

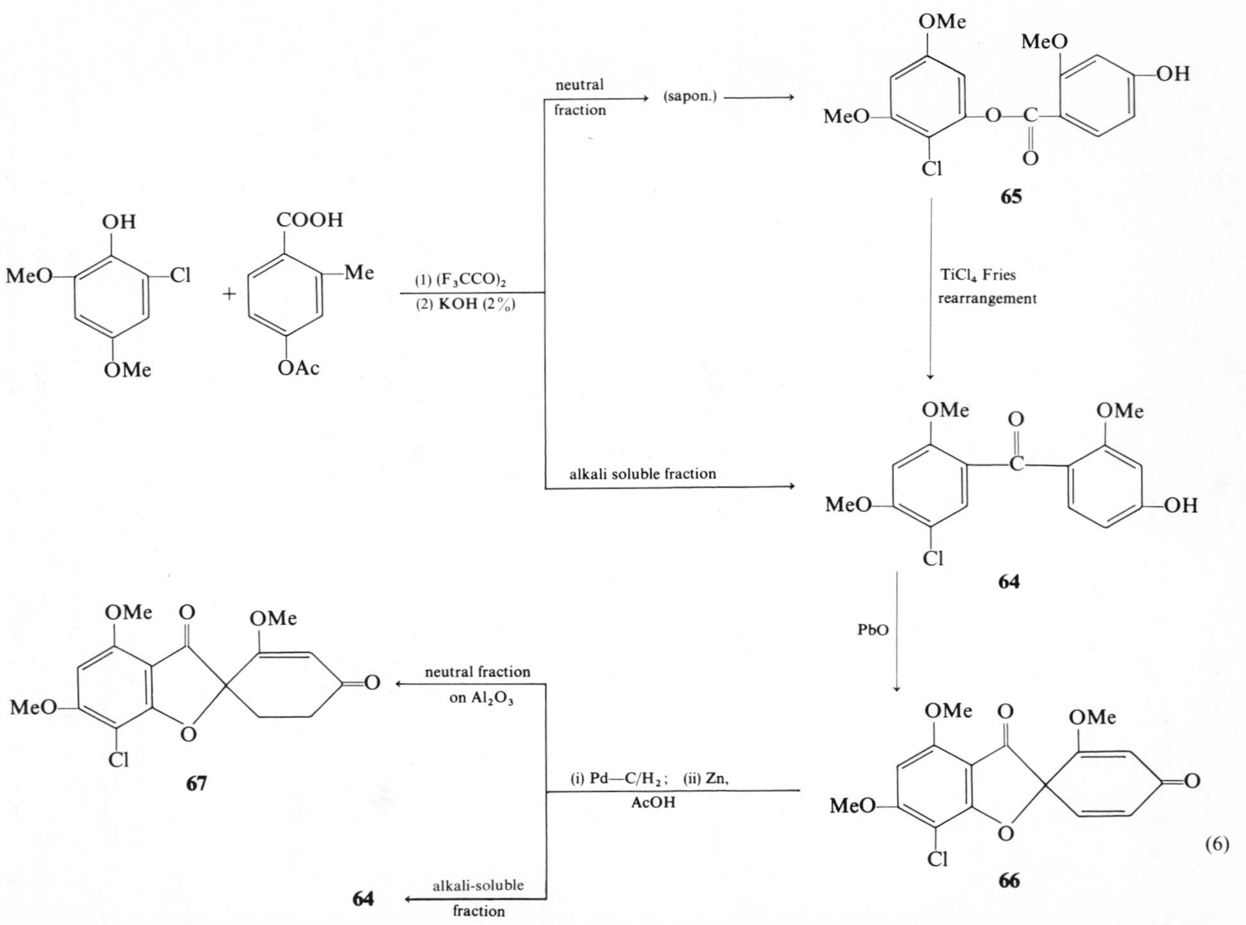
(1) $(F_3CCO)_2$
(2) KOH (2%)
neutral fraction
(sapon.)
alkali soluble fraction
65
$TiCl_4$ Fries rearrangement
64
PbO
66
(6)
(i) Pd—C/H_2; (ii) Zn, AcOH
neutral fraction on Al_2O_3
alkali-soluble fraction
67
64

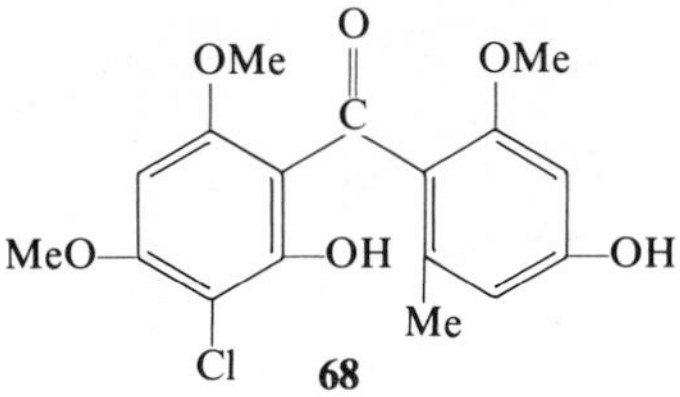

68 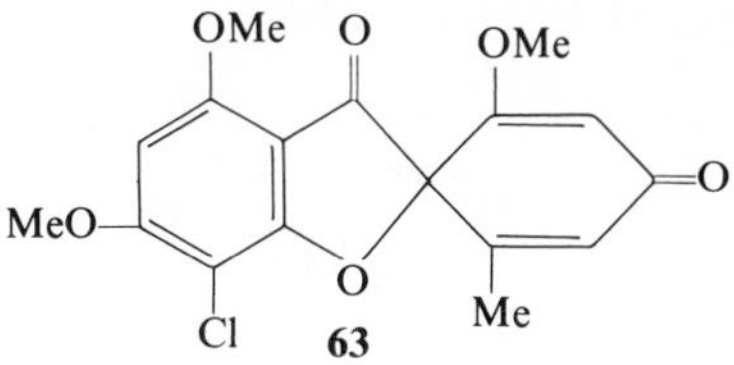 **63**

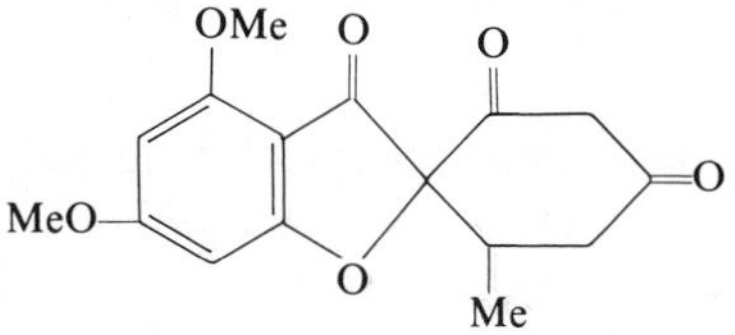

69 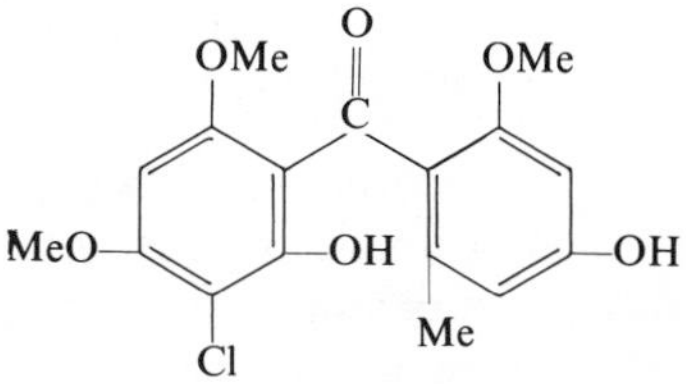 **70**

When dechlorogriseofulvin was allowed to absorb 2 to 6 moles of hydrogen, the dihydro derivative **71** was not isolated; instead the hydrogenolysis product, **72,** and the unsaturated ketone, **73,** were obtained. The latter product must have arisen from the dihydro derivative **71** during work-up, because it was readily obtained by treatment with acid. Hydrogenation of **73** with palladium—charcoal yielded 4,6-dimethoxy-2′-methylgrisan-3,4′-dione (**74**).[30]

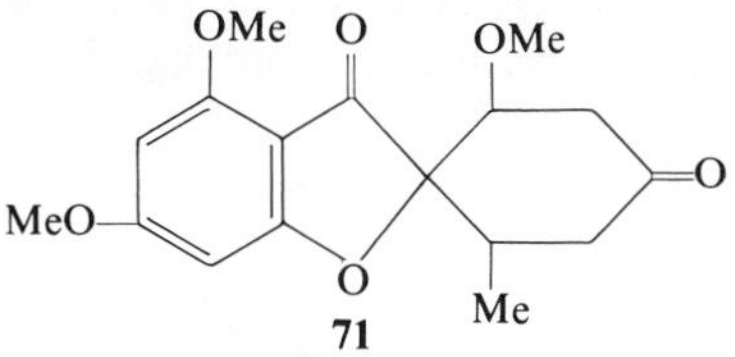

71 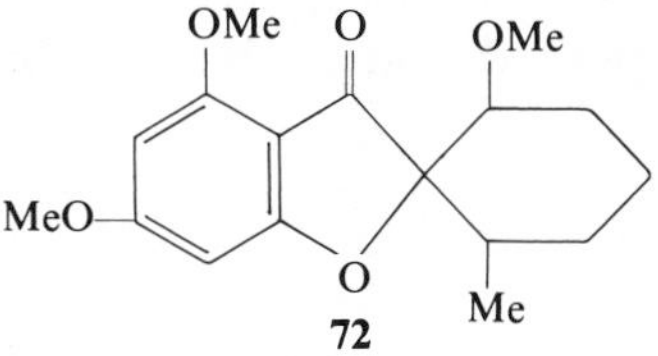 **72**

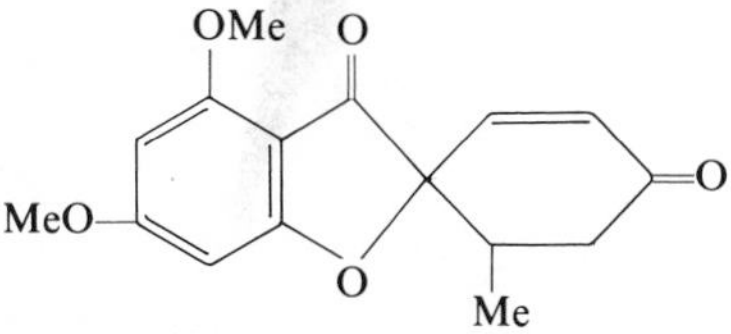

73 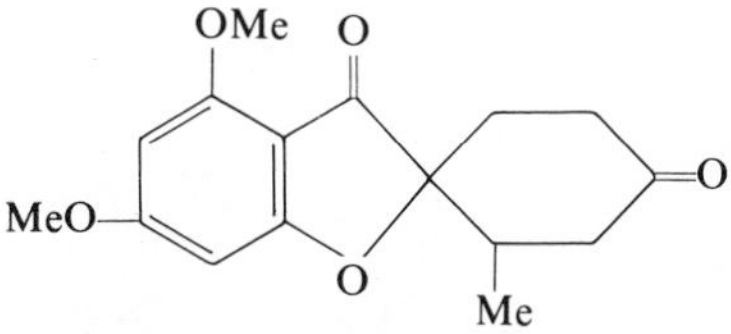 **74**

Application of the oxidative coupling reaction[46,47,52,62] to **75** establishes a route for the preparation of the sulfur analog of griseofulvin (**76**).[60]

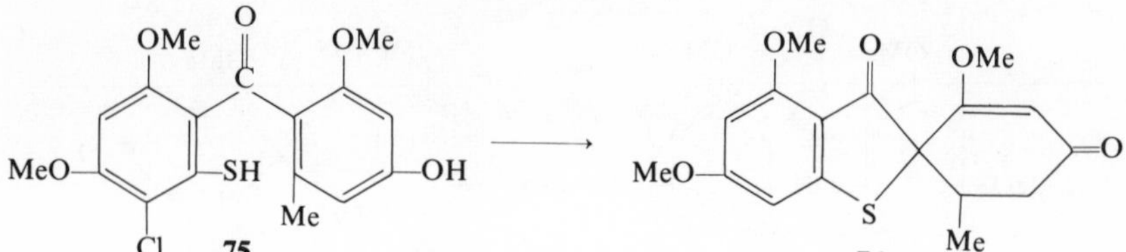

e. Dechlorogriseofulvin

Dechlorogriseofulvin, $C_{17}H_{18}O_6$ (**26**), was isolated together with griseofulvin from the culture filtrate of *Penicillium griseofulvim* Dierckx and certain strains of *janczewskii* Zal.[64] It was shown to be the dechloro analog of griseofulvin via catalytic reduction to 4′-hydroxy-4,6,6′-trimethoxy-2′-methylgrisan-3-one (**77**),[14] which has been obtained by reductive dechlorination of griseofulvin.[31] Like griseofulvin, **26** reacts with semicarbazide in pyridine and with methanolic ammonia solution to give the basic derivatives. Upon hydrolysis of **26** with dilute mineral acid, 1,3-diketone is obtained (**78a** ⇌ **78** ⇌ **78b**). With hydroxylamine, it gives the dioxime **79** rather than **80**. With 0.5*N* sodium hydroxide, **26** yields **78** and 1,2,3,4-tetrahydro-3-keto-5,7-dimethoxy-1-methyldibenzofuran (**81**). Both **26** and **78** are oxidized with alkaline mercuric oxide to 3-hydroxy-5,7-dimethoxy-1-methyldibenzofuran (**82**). Zinc permanganate oxidation of **26** yields the expected 2-hydroxy-4,6-dimethoxybenzoic acid (**83**) and 2-hydroxy-4,6-dimethoxy-3(2*H*)-benzofuranone-2-β-butyric acid (**84**); the latter was isolated only as the lactone and **85**.[65]

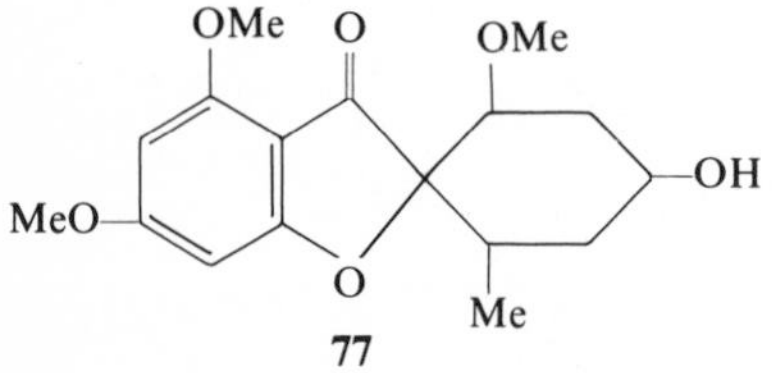

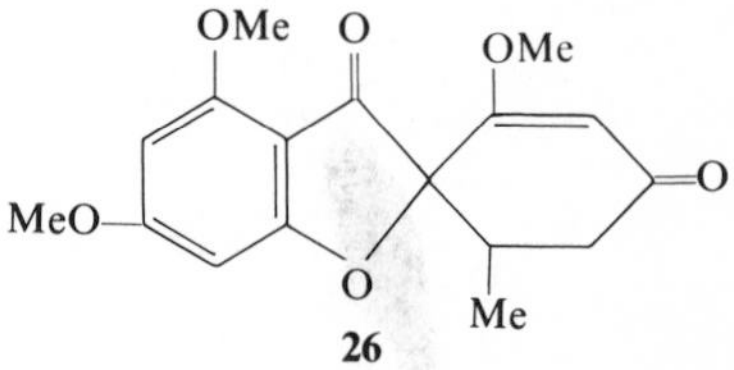

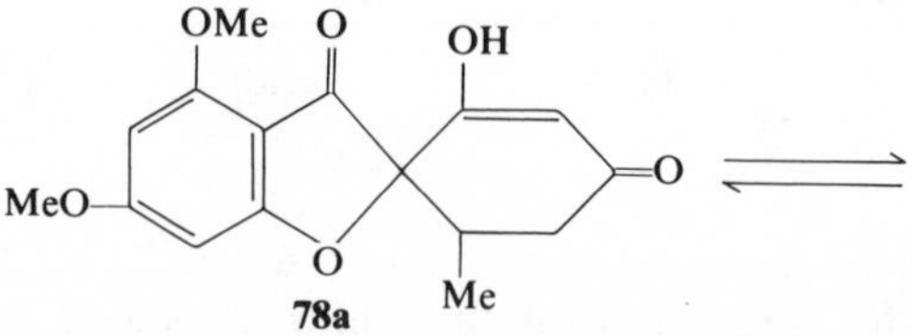

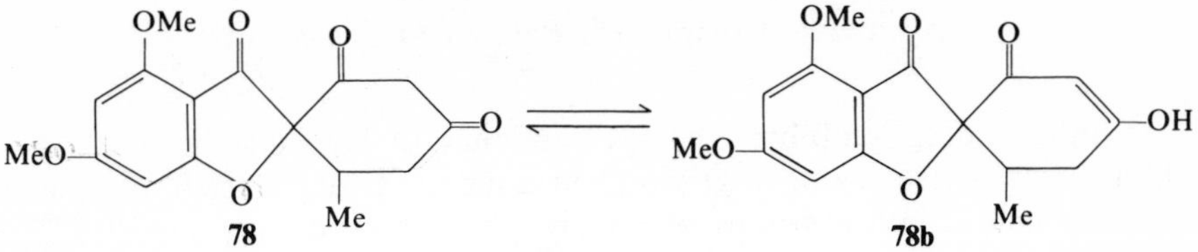

78 ⇌ 78b

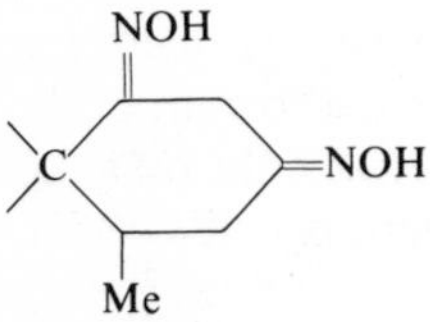

79 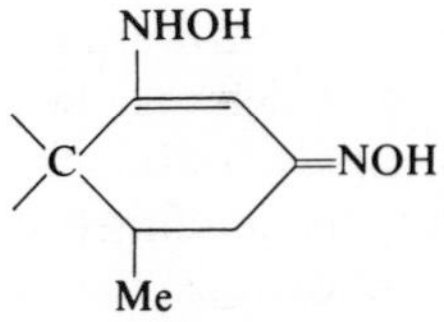 80 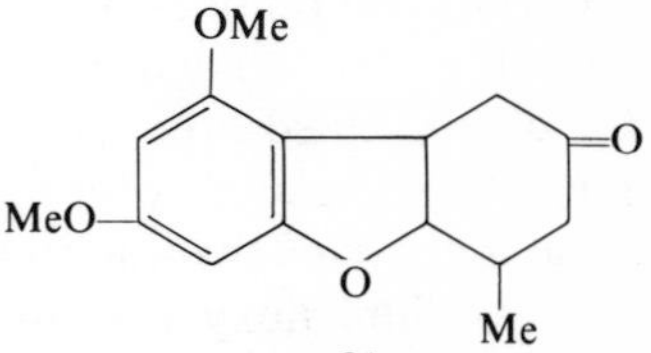 81

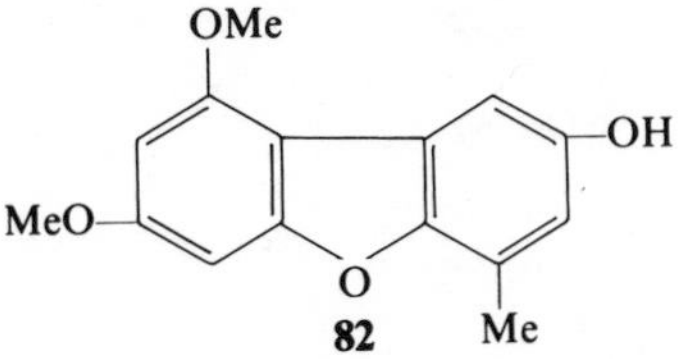

82 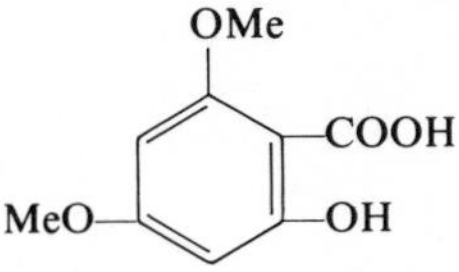 83

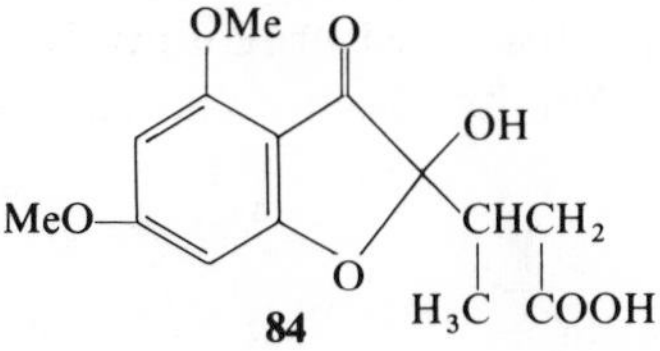

84

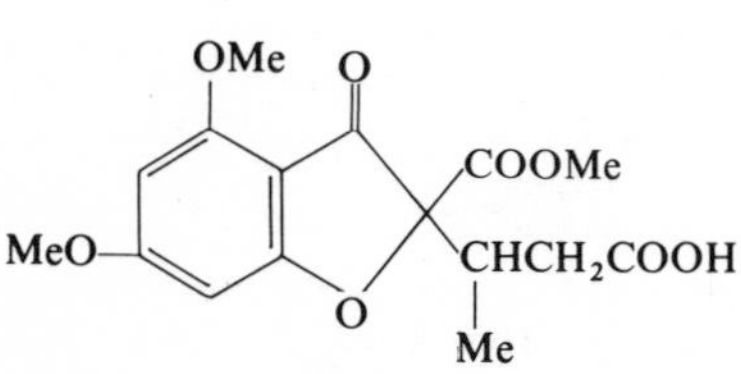

85

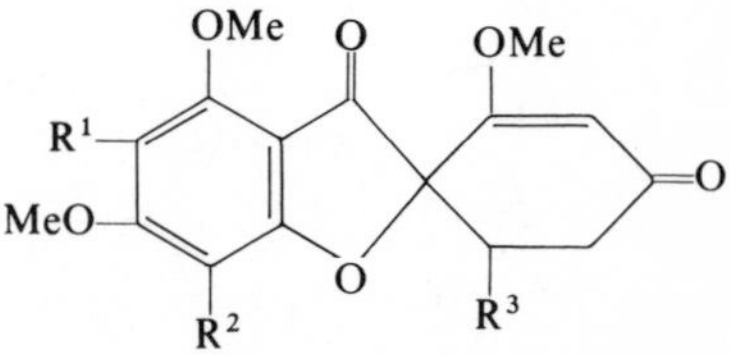

86 R^1 = Cl, R^2 = H, R^3 = Me

Synthesis of (±)-7-dechlorogriseofulvin (**26**) and of (±)-5-chloro-7-dechlorogriseofulvin (**86**) have been achieved.[62] Compound **26** is degraded to **74** and upon catalytic hydrogenation it affords **71**.[30]

f. Spirocyclic Compounds Related to Griseofulvin

Synthesis of griseofulvin analogs continues to be an active field. A biologically active analog of griseofulvin, a methyl group at the 6′-position lacking among other groups of analogs, has been recently reported.[69] Attempts to synthesize spirocyclic systems similar to those of griseofulvin[14] from benzofuranones of type **87** have led to spirocyclic diketones of type **31,** (R = H). Alkylation of 3(2*H*)-benzofuranones generally leads to 2,2-disubstituted products, some of which may be converted to spirocyclic compounds related to griseofulvin. Cyclization is achieved only by acid-catalyzed cyclodehydration leading to 6,4′-dimethyl-4-methoxy-(R = Me), and 4,6-dimethoxy-4′-methyl-3,2′-dioxogris-3′-en (**31,** R = OMe).[29]

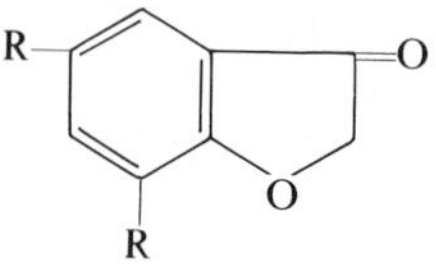

87

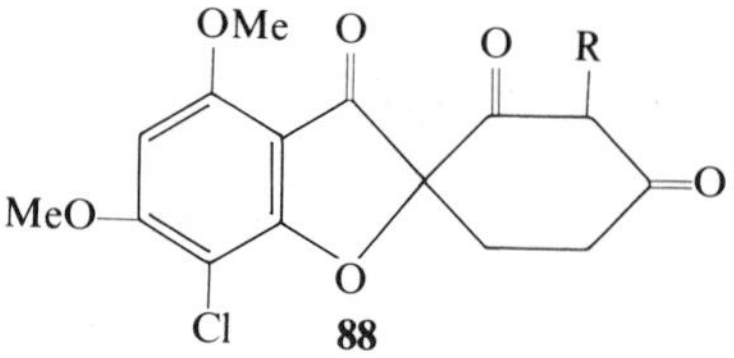

88

Structural modifications of antifungal antibiotic griseofulvin have been extended in the preparation of analogs bearing an alkyl substituent at position 3′ and 3′-alkyl-substituted griseofulvic acid (**88**).[33,66] Synthesis of grisan-3,2′,4′-trione (**89**) has been achieved[67] as outlined in Eq. 7.

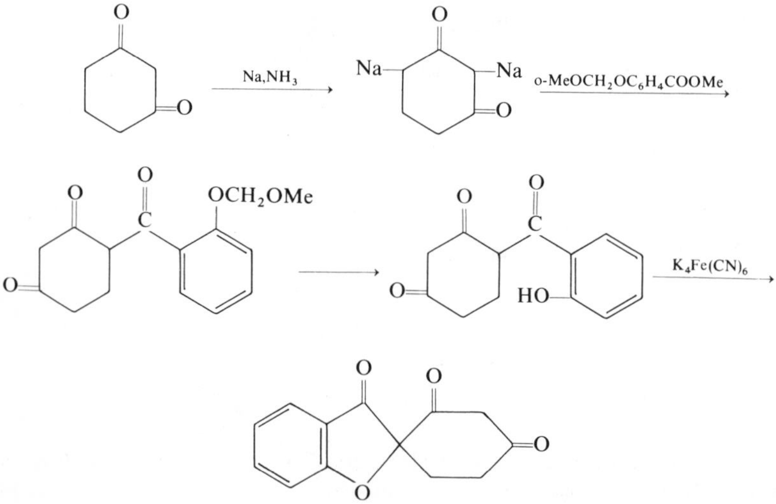

Na,NH$_3$

o-MeOCH$_2$OC$_6$H$_4$COOMe

$K_4Fe(CN)_6$

89

(7)

Dawkins and Mulholland[68] have reported the synthesis of **91** from the diketo ester (**90**). Ozonolysis of **91** provides **92** and formaldehyde.

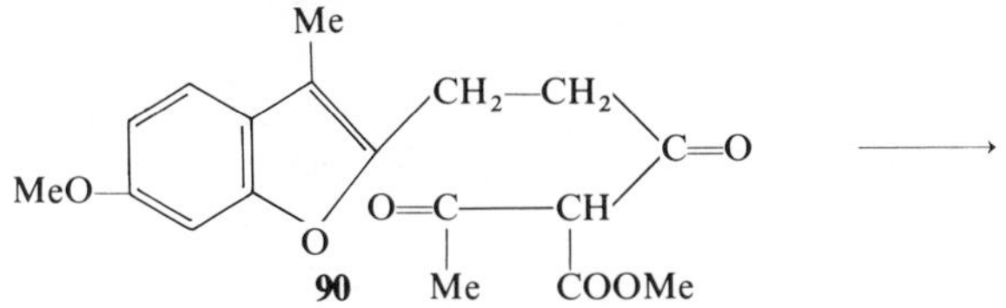

90 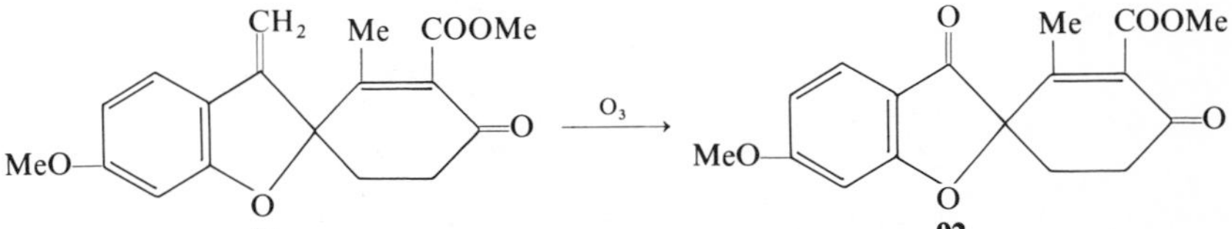 → **91** $\xrightarrow{O_3}$ **92**

Interaction of 2-acetyl-3-ethyl-4,6-dimethylbenzofuran (**93**) with ethyl bromomalonate[71] under Reformatsky conditions afforded ethyl β-(3-ethyl-4,6-dimethyl-2-benzofuranyl)crotonate (**94**, R = OEt), which hydrolyzed to the acid (**94**, R = OH). The acid chloride (**94**, R = Cl) reacted with cadmium methyl furnishing the ketone (**95**) which, when subjected to Mannich reaction, afforded the compound **96**. Treatment of its quaternary salt (**97**) with a base, resulted in the formation of **98**.

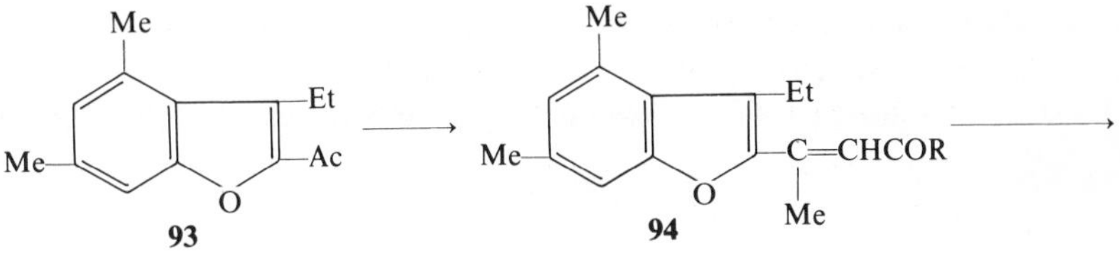

93 → **94** 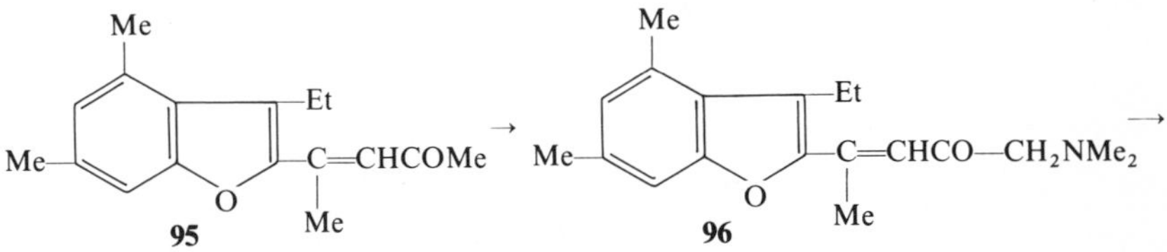 → **95** → **96** 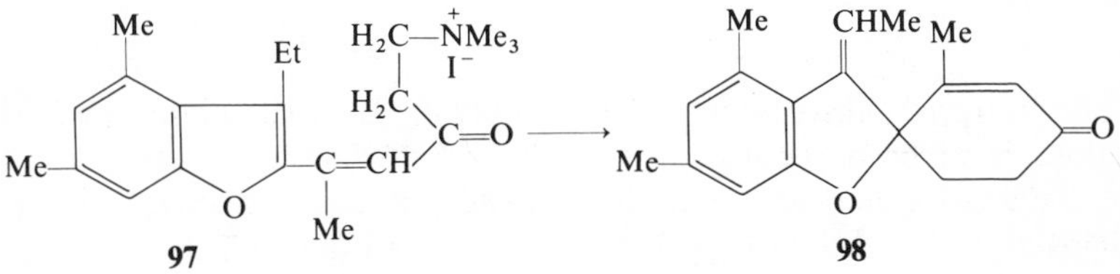 → **97** → **98**

Dean et al.[70] have reported the cyclization of diketobenzofurans (**99**) with sulfuric acid with the formation of **100**.

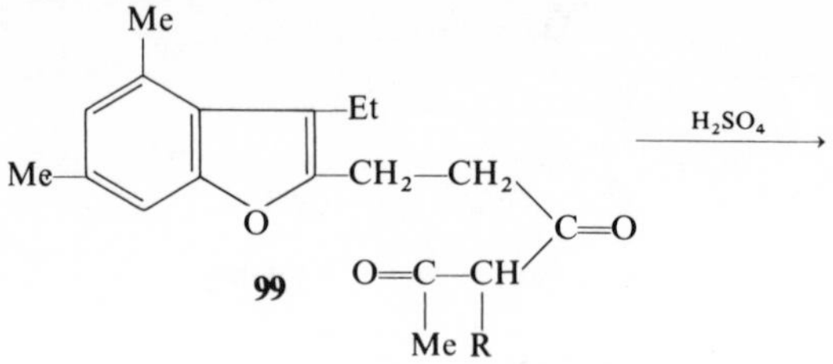

Me Et Me CH_2—CH_2 C=O O O=C—CH Me R H_2SO_4

99

CH_2 Me R =O Me O

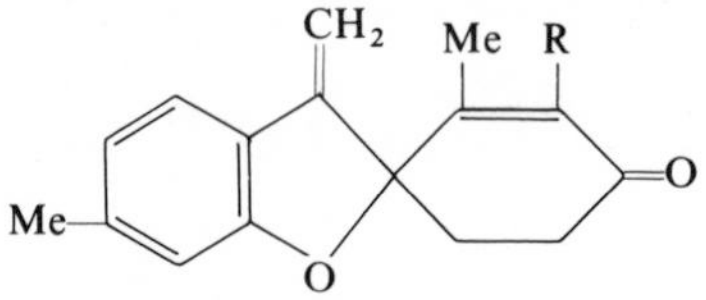

100

Synthesis of dienones of type **101** has been brought about by potassium ferricyanide oxidation of 2-hydroxyphenyl-(4-hydroxyphenyl) methane and 1-(2-hydroxyphenyl)-2-(4-hydroxyphenyl)ethane, respectively.[72,73] Oxidation with 2,3-dichloro-5,6-dicyano-1,4-benzoquinone leads to the formation of less than 0.5% of the dienone; however upon oxidation with active manganese dioxide, 2,4′-dihydroxybenzophenone is obtained (10%). The dienone (**101**) in acetic anhydride and sulfuric acid underwent rearrangement to 2-acetoxyxanthen (**102**), showing the occurrence of aralkyl migration.

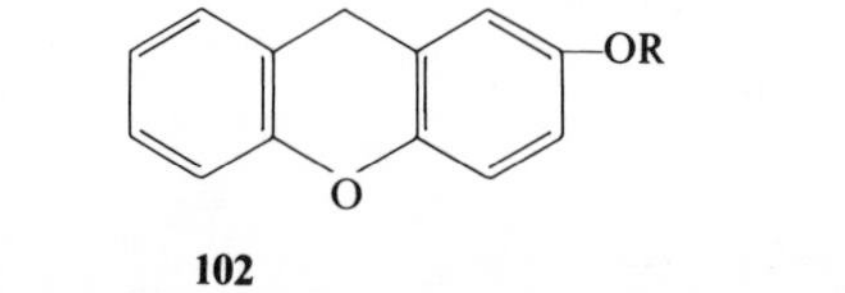

=O $H^{\oplus}$ O

101

OR O

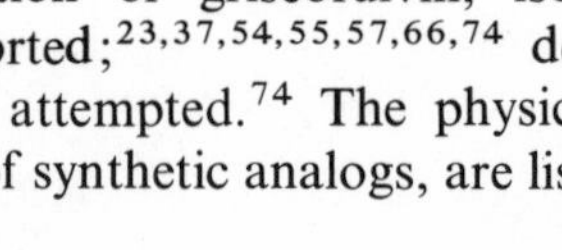

102

The infrared absorption of griseofulvin, isogriseofulvin, and their analogs have been reported;[23,37,54,55,57,66,74] detailed interpretation of this data has not been attempted.[74] The physical properties of griseofulvins as well as those of synthetic analogs, are listed in Table 1.

TABLE I. GRISANS

R	R^1	R^2	R^3	X	Y	Z	Melting point (°C)	$(\alpha)_D^{22}$	Solventifor crystal-lization[a]	Ultraviolet spectrum [λ max., (mμ) (log E)]	Ref.
Cl	OMe	H	OMe	OMe	H	Me1	181	−34°	A	322(3.68), (287(4.28), 234(4.10) (tetra-hydrogriseofulvin)	31
H	OMe	H	OMe	OMe	OH	Me	222–223	−10°	A	314(3.69), 282 (4.31)	31
Cl	OMe	H	OMe	OMe	OH	Me	197–198	−17°	B	323(3.67), 287(4.29), 230(4.11); mono-acetyl-deriv., m.p. 208–209° (A)	31
Cl	OMe	H	OMe	OH	OMe	Me	228–230	−29°	A	322(3.70), 289(4.33), 240(4.07)	31
Cl	OMe	H	OMe	H	OH	Me	160[32] ; 193–195		C;	236(4.20), 286(4.32), 320(3.71); 324(3.71), 287(4.37), 238(4.08); *l,d*) 2(S);2′(R)	30
H	OMe	H	OMe	OMe	OH	Me	222				126
Cl	OMe	H	OMe	H	H	Me	142–146; 149–153	−24° ± 3	B/C	(*l,d*) 2(S);2′(R)	27
Cl	OMe	H	OMe	OH	H	Me		−27° ± 3		(*l,d*) 2(R);6′(R)	30
Cl	OMe	H	OMe	OH	OH	Me	213–215	+43° ±	D	(*l,d*) 2(R);6′(R)	30
Cl	OMe	H	OMe	H	H	H	176–177		A		27
H	H	H	H	H	OH	H	105–106.5 (α-form)		E	249(3.98), 324.5(3.70); 3,5-dinitrobenzeate (m.p. 221–222°F); acetyl deriv. (132–133°C)	28

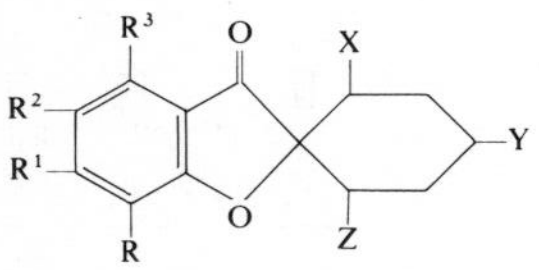

TABLE I (continued)

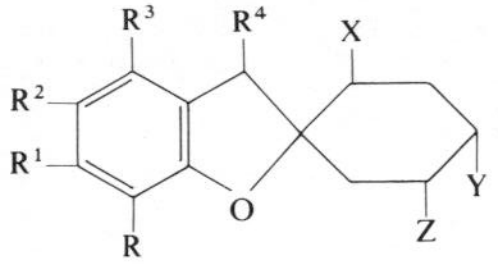

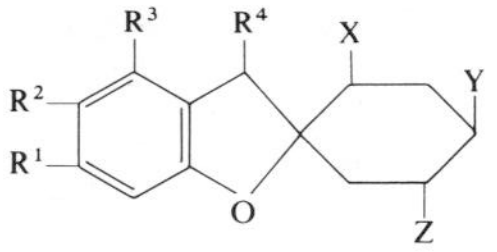

R	R^1	R^2	R^3	R^4	X	Y	Z	Melting point (°C)	$(\alpha)_D^{22}$	Solvent for crystallization[a]	Ultraviolet spectrum [λ max., (mμ) (log E)]	Ref.
H	H	H	H	OH	H	OH	H	175.5–180.5			281(3.49), 288(3.44); (±)-α-form	28
H	H	H	H	OH	H	OH	H	203–205	+73°	B/F	(+)-α-form	28
H	H	H	H	OH	H	OH	H	206–207	−73°	B/F	(−)-α-form	28
H	H	H	H	OH	H	OH	H	171–175		B or G	(±)'β'form; 281(3.51), 287.5(3.46)	28
R	**R^1**	**R^2**	**R^3**	**X^1**	**X^2**	**X^3**	**X^4**					
Cl	Cl	Cl	Cl	Cl	Cl	Cl	Cl	246		D/C		56
Cl	H	Cl	Cl	Cl	H	Cl	Cl					56

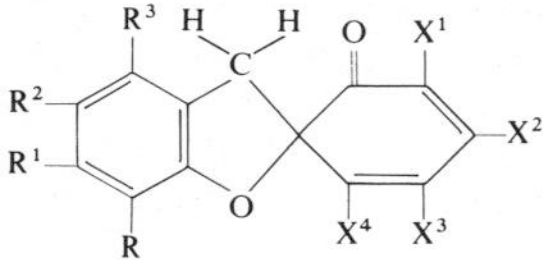

R	R^1	R^2	R^3	X	Y					
Grisan-3,2′-diones		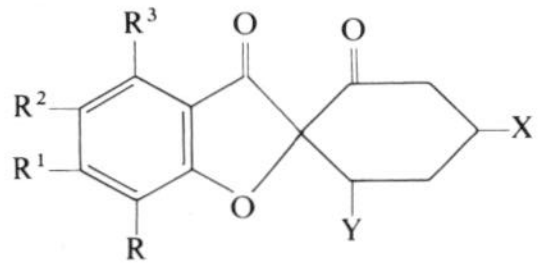								
H	OMe	H	OMe	OMe	Me	177–178		B/C or B	323(3.73), 289(4.20), 242(3.99)	31
Cl	OMe	H	OMe	H	Me	204–205		B/C	326(3.69), 291(4.28), 239(3.98), 216(4.24)	32
Cl	OMe	H	OMe	H	Me	203–204	(+) + 43°			34
Cl	OEt	H	OEt	H	Me	270–272	(+) 309°			34
Grisan-3,4′-diones		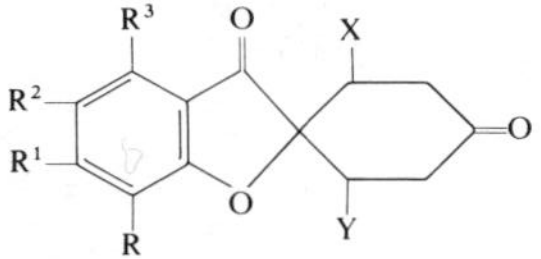								
Cl	OMe	H	OMe	OMe	Me	198	−20°	A	323(3.70), 288(4.31), 234(4.10); oxime [m.p. 255–256° dec.); A] (dihydrogriseo–fulvin)	31
Cl	OMe	H	OMe	H	Me	178–199		B/C	324(3.71), 288(4.37), 238(4.08), 213(4.35)	32
Cl	OMe	H	OMe	OMe	Me	180–182	0° ±3		(*l,d*) 2(R);6′(R)	30
Cl	OMe	H	OMe	H	Me	168–170	−30° ±3	G	(*l,d*) 2(S);2′(R)	30
Cl	OMe	H	OMe	H	Me	211–212		G	(*d,d*) 2(R);2′(R)	30
Cl	OMe	H	OMe	H	Me	173–174	(+) 54°			34
H	H	H	H	H	H	89.5–90.5		E	(2,4-Dinitrophenyl-hydrazone, m.p. 230–231°; 3′-COOMe, m.p. 122.5–124); 250 (4.0), 328(3.70)	28

TABLE I (continued)

						Melting point (°C)	$(\alpha)_D^{22}$	Solvent for crystal-lization[a]	Ultraviolet spectrum [λ max., (mμ) (log E)]	Ref.
Cl	OEt	H	OMe	OMe	Me	215–217	+314°			66
Cl	OPr;n	H	OMe	OMe		187–189	+305°			66
Cl	OPr.iso	H	OMe	OMe	Me	190–191	+310°			66
Cl	OBu;n	H	OMe	OMe	Me	181–183	+299°			66
Cl	Oallyl	H	OMe	OMe	Me	173	+305°			66
Cl	$OC_{16}H_{33}$	H	OMe	OMe	Me		+201°			66
Cl	OCH_2COOMe	H	OMe	OMe	Me	206–208	+271°			66
Cl	OCH_2Ph	H	OMe	OMe	Me	192–194	+260°			66
Cl	OMe	H	OEt	OMe	Me	211–213	+324°			66
Cl	OMe	H	OPr;n	OMe	Me	186–9	+313°			66
Cl	OMe	H	OPr;iso	OMe	Me	124	+295°			66
Cl	OMe	H	OBu;n	OMe	Me	168–169	+300°			66
Cl	OMe	H	Oallyl	OMe	Me	202–204	+308°			66
Cl	OMe	H	OCH_2Ph	OMe	Me	244	+284°			66
Cl	OMe	H	$O(CH_2)_2NMe_2$	OMe	Me	160	+272°			66
Cl	OMe	H	$O(CH_2)_2NEt_2$	OMe	Me	183–186	+262°			66
Cl	OMe	H	$OOH(Me)CH_2NHPh$	OMe	Me	201–202	+279°			66
Cl	OMe	H	$OCH_2CONHPh$	OMe	Me	257–258	+277°			66
Cl	OMe	H	$O(CH_2)NC_4H_8$	OMe	Me	157–159	+258°			
Grisan-3,2′,4′-triones										
H	OMe	H	OMe	H	H	204–206		D		125

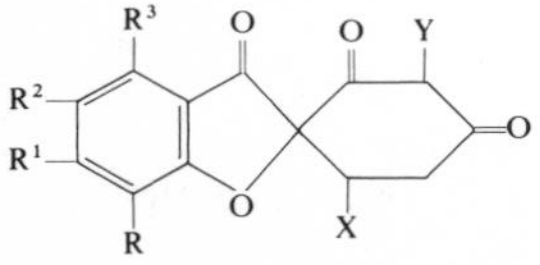

Cl	OMe	H	OMe	H	H	240–241		D		125
Cl	OMe	H	OMe	Me	H	233–235				125
Me	OMe	H	OMe	H	H	250				38,125
H	OMe	H	OMe	Me	H	248–250 (dec.)	+480°	H	(dioxime, m.p. 231–233°)	126
Cl	OMe	H	OMe	Me	H	260–263 (dec.)	+414°	H	(*d,d* isomer)	15
Cl	OEt	H	OEt	Me	H	226–228 (dec.)	+1457°	H	(*d,d* isomer)	15
Cl	OEt	H	OMe	Me	H	234–6 (dec.)			(*d,d*; i.e. 2(S);6′(R)	30
Cl	OMe	H	OMe	Me	H	215–216	+112°	H	(*l,d*; i.e. 2(R);6′(R)	30
Cl	OMe	H	OMe	Me	H	233.5			231, 289, 322; max. (in alc. NaOH) 230; 291, 322 (*d,l*-epigriseofulvin acid)	37
Cl	OMe	H	OMe	Me	H	242–244			255, 271, 289, 326; (in alc. NaOH) 233, 288, 326; *d*-brucine salt of griseofulvic acid monohydrate, m.p. 240.5, $(a)_D^{25}$ 131°; *d*-griseofulvic acid, m.p. 258–260°).	17
H	OMe	Cl	OMe	H	H	212–213				38
Cl	OMe	H	OMe	H	H	250–251				38
Cl	OMe	H	OMe	Me	Me	248–250.5	+404°	J		54
Cl	OMe	H	OMe	Me	Et	217–218	+267°	H	232 (E, 16100), 282 (E, 31,600), 320 (E,6000)	54
Cl	OMe	H	OMe	Me	allyl	162–168	+220°			54
Cl	OMe	H	OMe	Me	*n*-Pr	125–126				54
Cl	OMe	H	OMe	Me	CH_2 Ph	207–209	+192°	H	230 (E, 16,200), 286–287 (E, 31,200), 383 (E, 5700); in 0.1 *N* NaOH, 239.5 (E, 49,100)	54

TABLE I (continued)

						Melting point (°C)	$(\alpha)_D^{22}$	Solvent for crystal-lization[a]	Ultraviolet spectrum [λ max., (mμ) (log E)]	Ref.
Cl	OMe	H	OH	Me	CH_2Ph	143–146	+205°	H + I		54
Cl	OMe	H	OH	Me	H	144	+257°			66
Cl	OPr.n	H	OMe	Me	H	211–214	+246°			66
Cl	OPr.iso	H	OMe	Me	H	221–224	+365°			66
Cl	OBu	H	OMe	Me	H	193	+239°			66
Cl	Oallyl	H	OMe	Me	H	200	+244°			66
Cl	$OC_{16}H_{33}$	H	OMe	Me	H	122–127	+155°			66
R	R^2	R^3	R^1	X	Y					
Cl	H	OMe	OCH_2COOH	Me	H	190	+181°			66
Cl	H	OMe	OMe	Me	Br	248–250 (dec.)	+263°	K		57
Cl	H	OMe	OMe	Me	I	209	+188 -	K		57
Cl	H	OMe	OMe	Me	Cl	286 (dec.)	+331	K		57
Cl	Cl	OMe	OMe	Me	Cl	214.5–216.5	+220°	B		57
Cl	Cl	OMe	OMe	Me	H	243 (dec.)	+272°	L		57
Cl	OMe	OMe	OMe	Me	Br	200–202	+163°			57
Cl	Br	OMe	OMe	Me	H	237 (dec.)	+310°	M + trace of J		57
H	H	OMe	OMe	Me	H	233–234				125
Cl	H	OMe	OMe	H	H	240–241				125
Me	H	OMe	OMe	H	H	250				125
H	H	OMe	OMe	H	H	294–296		D		125
Cl	H	OMe	OMe	Me	Ph	120–125 (dec.); melts at 133–135°	+110°			130

Cl	H	OMe	OMe		Me	o-$C_6H_4NO_2$	157–160 (dec.)	+19.5°			130
Cl	H	OMe	OMe		Me	4-Cl-2-NO_2-C_6H_3	157–185	+62°			130
Cl	H	OMe	OMe		Me	-CH_2Ph	144–53	+122°			130
Cl	H	OMe	OMe		Me	t-Bu	130 (dec.)	+201°			130
Cl	H	OMe	OMe		Me	Me	110–115 (dec.)	+162°			130
Cl	H	OMe	OMe		Me	Et	120–130 (dec.)	+145°			130
Cl	H	OMe	OMe		Me	o-C_6H_4OMe	128–130	+184°			130
Cl	H	OMe	OMe		Me	iso-Pr	113–117	+160°			130
Cl	H	OMe	OMe		Me	2,4-$C_6H_3Me_2$	227–229				130
Cl	H	OMe	OMe		Me	*n*-Pr	103	+103°			130
Cl	H	OMe	OMe		Me	H	219–223	+111°		(*l,d*; i.e. 2(R);6′(R)	15

R	R^1	R^2	R^3	X	Y	Z					
Cl	OMe	H	OMe	Me	Me	Me	128; 152	+30°	A	292 (E, 23,000), 324–328 (E,55000)	54
Cl	OMe	H	OMe	Me	CH_2Ph	CH_2Ph	146–149.5	−159°	A	292i (E, 21,800), 322 (E, 5190)	54
Cl	OMe	Cl	OMe	Me	Cl	Cl	153–155	−3.3°	H		57
Cl	OMe	Br	OMe	Me	Br	Br	98–101 (dec.)	0°	H		57

TABLE I (continued)

					Melting point (°C)	$(\alpha)_D^{22}$	Solvent for crystal-lization[a]	Ultraviolet spectrum [λ max., (mμ) (log E)]	Ref.
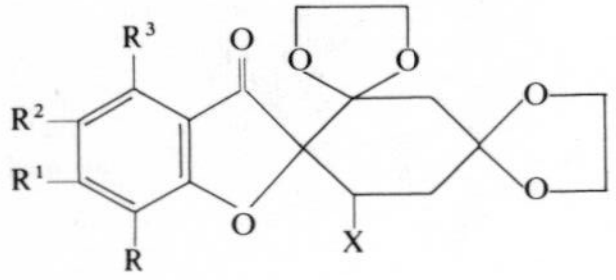 R² R¹ R³ R X									
Cl	OMe	H	OMe	Me	198–199	−98°			66
Cl	OH	H	OMe	Me	271–272	−94°			66
Cl	OMe	H	OH	Me	249	−105°			66
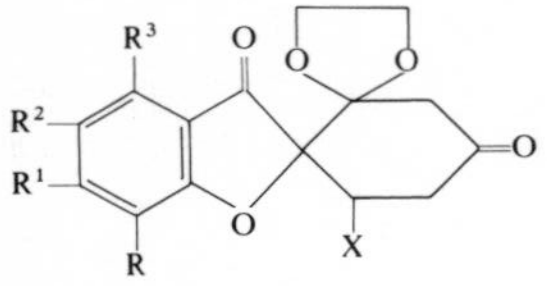 R² R¹ R³ R X									
Cl	OMe	H	OMe	Me	205–207	−147°			66
Cl	OH	H	OMe	Me	218–222°	−133°			66
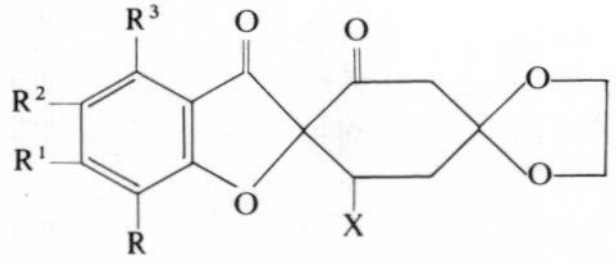 R² R¹ R³ R X									
Cl	OMe	H	OMe	Me	260–263	+54°			66

R	R^1	R^2	R^3	X	Y					
Gris-2′-en-3,4′-diones										
H	OMe	H	OMe	OMe	H	162–163			236 (E, 23,000), 291 (E, 24,000), 322 (E, 4750)	125
Cl	OMe	H	OMe	OMe	H	228–229				125
Cl	OMe	H	OMe	OMe	Me	247–248			(Griseofulvin)	125
Cl	OMe	H	OMe	H	Me	177–178; 174–175	+442° +443°	A	2,4-DNP; m.p. 279–280	32,34
H	OMe	H	OMe	OMe	Me	179–181; 218–219	+390°	B or G	(Dechlorogriseofulvin), 325 (3.62), 286(4.39), 250(4.18); 2,4-DNP, m.p. 264–265 (red cryst. from G/N)	42,126
Br	OMe	H	OMe	OMe	Me	204–205		B or G	325(3.75), 292(4.365°, 255(4.22), 235(4.38); yellow color with HNO_3 (bromogriseofulvin)	16

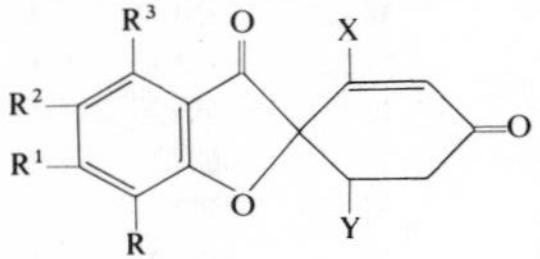

TABLE I (continued)

						Melting point (°C)	$(\alpha)_D^{22}$	Solvent for crystallization[a]	Ultraviolet spectrum [λ max., (mμ) (log E)]	Ref.
Cl	OMe	H	OMe	OMe	Me	214–216	+88°	B	323(3.78), 290(4.41), 252(4.25), 236(4.46) (*l,d*; i.e. 2(R)6′(R) (griseofulvin)	15
Cl	OMe	H	OMe	OMe	Me	240–242	+179° (NaOMe); +202° ($CHCl_3$); +181° (Me_2Co)	B/C; F	(*d,d*) and (*l,d*)	15
Cl	OMe	H	OMe	OMe	Me	219–222	+342 ($CHCl_3$)	A	(*d,d*)	15
Cl	OEt	H	OEt	OEt	Me	173–175	+297°	B/C	323(3.79), 292.5(4.38), 256(4.19), 237(4.33)	15
Cl	OEt	H	OEt	OMe	Me	208–212		A	325(3.87), 292(4.46), 255(4.29), 233(4.48)	15
H	OMe	H	OMe	OMe	Me	184	+39° ± 3°	B/C	(*l,d*; i.e. 2(R);6′(R)	30
Cl	OMe	H	OMe	OPr-*n*	Me	156–157	+271°	B/C		53
Cl	OMe	H	OMe	OBu-*n*	Me	155	+263	G		53
Cl	OMe	H	OMe	OBu-*t*	Me	135–136	+264°	C		53
Cl	OMe	H	OMe	OC_5H_{11}-*n*	Me	154	+254°	G		53
Cl	OMe	H	OMe	OC_7H_{15}-*n*	Me	159–160	+242°	B/C		53
Cl	OMe	H	OMe	OC_6H_{13}-*n*	Me	161	+248°	B/C		53
Cl	OMe	H	OMe	OC_8H_{17}-*n*	Me	134	+237°	B/C		53
Cl	OMe	H	OMe	$OC_{10}H_{21}$-*n*	Me	89–91	+184°	A or C		53
H	OMe	Cl	OMe	OMe	H	182–184				38
Cl	OMe	H	OMe	OMe	H	228–229				38
Me	OMe	H	OMe	OMe	H	208–209; 205	125			38,125
Cl	OMe	H	OMe	OEt	Me	206–207	+319°	A		54

Cl	OMe	H	OMe	OCH_2Ph	Me	204–206	+202°		226 (E, 23,200), 291 (E, 24,000), 322 (E, 4,750)	54
Cl	OMe	allyl	OH	OMe	Me	300	+405°			66
Cl	OMe	allyl	OMe	OMe	Me	132–132.5	+278°			66
Cl	OMe	*n*-Pr	OMe	OMe	Me	113–114	+252°			66
Cl	OMe	H	OMe	Cl	Me	200–201.5	+370°	M		55
Cl	OMe	H	OMe	OEt	Me	203–206	+317°		236(4.35), 253(4.21), 291(4.39), 322 (3.74); [*d,d*; m.p. 200–202 (A)] HCl salt (A), m.p. 230–232°	53,55 111
Cl	OMe	H	OMe	$NHEt_2$	Me	132–134	+421	F/N		55
H	OMe	H	OMe	OMe	Me	218–219			323s (E, 4700), 286 (E, 25,000), 249 (E, 15,200)	41
H	OMe	Cl	OMe	OMe	Me	213–214			331 (E, 5,600), 280 (E, 16,000), 237 (E, 23,600) (5-chlorogriseofulvin)	41,42
Cl	OMe	Cl	OMe	OMe	Me		+278°		345 (E, 4050), 231 (E, 14,500) (5,7-dichlorogriseofulvin)	41
Cl	OMe	H	OMe	OMe	H	227–228			(6′-demethylgriseofulvin)	61,125
H	Me	H	Me	Me	H	98		C	268(4.31), 329(3.71); green cocor with alkaline nitroprusside; 2,4-DNP, m.p. 225° [red needles (A-D)]	128
H	OMe	H	H	Me	H	118			231(4.32), 271(4.22), 319(4.05); 2,4-DNP, m.p. 231° 213(4.83), 286(4.71), 315(4.04)	68,128

TABLE I (continued)

						Melting point (°C)	$(\alpha)_D^{22}$	Solvent for crystal-lization[a]	Ultraviolet spectrum [λ max., (mμ) (log E)]	Ref.
H	OMe	H	OMe	Me	Me	196		G	213(4.83), 286(4.71), 315(4.04)	128
Cl	OMe	H	OMe	$O(CH_2)_2OMe$	Me	208–210		D/C	236(4.43), 253(4.23), 289(4.36), 320(2.76)	111
H	H	Cl	H	OEt	Me	111–112			223(4.38, 248(4.39), 252(4.40), 340(3.69)	111
H	H	H	H	OEt	Me	111–112		D/C	212(4.34), 251(4.47), 329(3.73)	111
Cl	H	Me	H	OEt	Me	126–127		D/C	218(4.41), 255(4.40), 262(4.35), 343(3.70)	111
H	Cl	H	H	OEt	Me	148; 135°	(dimorph.)	C/E	223(4.31), 229(4.31), 257(4.52), 327(3.88)	111
Cl	H	H	H	OEt	Me	157–158		D/C	216(4.38), 254(4.41), 331(3.71)	111
H	H	H	H	Me	Me	170		D	251(4.43), 326(3.73)	111
Cl	H	H	H	Me	Me	165–166		A	216(4.38), 255(4.37), 335(3.68)	111
F	OMe	H	OMe	OMe	Me	209–211		E/F	(fluorogriseofulvin)	42
Cl	OMe	H	OMe	H	Me	247–248				125
Cl	OMe	H	OMe	H	Me	196–198	−148±3	G	(*l,d*; i.e. 2(S);6′(R)	30
Cl	OMe	H	OMe	*O*-allyly	Me	173–137	+294°		234 (E, 24,000), 290–299 (E, 25,000), 324 (E, 6060)	54

Gris-2′-en-3,4′-diones

R	R[1]	R[2]	R[3]	X	Y	Z					
Cl	OMe	H	OMe	OEt	Me	Et	133–139	+286°	A		54
Cl	OMe	H	OMe	OMe	Me	Me	166.5–170.5	+291°		232i(19.3), 289(25.3), 320(5.5) (10^{-3}E)	54
Cl	OMe	H	OMe	OMe	OMe	allyl	140.5–142.5	+264°		231i(20.2), 290(26.1), 318i(6.05) (10^{-3}E)	54
Cl	OMe	H	OMe	OMe	Me	CH_2Ph	148–150	+278°		231i(19.0), 291(25.6), 319(6.3) (10^{-3}E)	54
Cl	OMe	H	OMe	OEt	Me	CH_2Ph	152–153	+271°		231i(19.7), 290(25.4), 316i(5.9) (10^{-3}E)	54
Cl	OMe	H	OMe	Pr-*n*	Me	CH_2Ph	89	+248°		232i(18.0), 291(25.3), 320(5.75) (10^{-3}E)	54
Cl	OMe	H	OMe	Bu-*n*	Me	CH_2Ph		+225°		232i(18.7), 292(25.0), 320(5.5) (10^{-3}E)	54
Cl	OMe	H	OMe	$Br(CH_2)_2$	Me	CH_2Ph		+210°		230i(20.1), 290(25.3), 320(5.7) (10^{-3}E)	54
Cl	OMe	H	OMe	Me	Me	Pr-*n*	115–116	+284°		230i(19.4), 290.5(26.2), 321(6.1) (10^{-3}E)	54
Cl	OMe	H	OMe	Pr-n	Me	Pr-*n*	106.5–107.5	+260°		232i(20.2), 289.5(25.0), 321(5.9) (10^{-3}E)	54
Cl	OMe	H	OMe	OMe	Me	$(CH_2)_2$ COMe	129.5–130.5	+259°	A	290 (E, 25,800), 321 (E, 5650)	54
Cl	OMe	Br	OH	OMe	Me	Br	215 (dec.)	+172°			57
Cl	OMe	H	OMe	OMe	Me	Br	112–116	+249°	G		57
Cl	OMe	Br	OMe	OMe	Me	Br	188–190	+180°	O		57
Cl	OMe	H	OMe	OMe	Me	Br	187–189°	+250°			57
Cl	OMe	H	OMe	OEt	Me	Br	144–146	+243°			57
Cl	OMe	H	OMe	OPr-*n*	Me	Br	160–1	+220°			57
Cl	OMe	H	OMe	OBu-*n*	Me	Br	181–182	+215°			57

TABLE I (continued)

							Melting point (°C)	$(\alpha)_D^{22}$	Solvent for crystallization[a]	Ultraviolet spectrum [λ max., (mμ) (log E)]	Ref.
Cl	OMe	H	OMe	OCH_2Ph	Me	Br	211 (dec.)	+191°			57
Cl	OMe	H	OMe	OMe	Me	I	193.5–195	+235°			57
Cl	OMe	H	OMe	OEt	Me	I	174–176.5	+233°			57
Cl	OMe	H	OMe	OPr-*n*	Me	I	140–141	+216°			57
Cl	OMe	H	OMe	OBu-*n*	Me	I	177–178	+206°			57
Cl	OMe	H	OMe	OCH_2Ph	Me	I	179–181	+138°			57
Cl	OMe	H	OMe	OMe	Me	Cl	188.5–190	+278°			57
Cl	OMe	H	OMe	OEt	Me	Cl	149–152	+241°			57
Cl	OMe	H	OMe	OPr-*n*	Me	Cl	146–148	+235°			57
Cl	OMe	H	OMe	OBU-*n*	Me	Cl	175	+232°			57
Cl	OMe	H	OMe	OCH_2Ph	Me	Cl	209 (dec.)	+190°			57
Cl	OMe	H	OMe	OMe	Me	Cl	169–170	+221°			57
Cl	OMe	H	OMe	OMe	Me	H	90–93				57
Cl	OMe	H	OMe	SMe	Me	H	250–253	+500°	P		55
Cl	OMe	H	OMe	SEt	Me	H	185–187	+475	A		55
Cl	OMe	H	OMe	SPr-*n*	Me	H	166.5–168	+459	G		55
Cl	OMe	H	OMe	SBu-*n*	Me	H	154.5–156.5	+429°	A		55
Cl	OMe	H	OMe	SCH_2CH_2-OH	Me	H	181–183		D		55
Cl	OMe	H	OMe	SCH_2CH_2X[b]	Me	H	204–206	+273°	A		55
Cl	OMe	H	OMe	SY[c]	Me	H	215–216	+350° (H_2O)	HCl salt (A–F)		55
Cl	OMe	H	OMe	SPh	Me	H	229–231; 244–245	+110°	A or P		55
Cl	OMe	H	OMe	$S(CH_2)_2NEt_2$	Me	H	135.5–137.5	+378°	Q		55
Cl	OMe	H	OMe	SCH_2Ph	Me	H	210–212	+336°	D		55
Cl	OMe	H	OMe	S-allyl	Me	H	153–154	+443°	A		55
Cl	OMe	H	OMe	SMe	Me	Br	190–193	+408°	G		55
Cl	OMe	H	OMe	SMe	Me	Cl	192.5–195	+471°	G		55
Cl	OMe	H	OMe	SEt	Me	Cl	184.5–186.5	+385°	A		55

Cl	OMe	H	OMe	SPr-*n*	Me	Cl	135–137	+373°	Q	55
Cl	OMe	H	OMe	SBu-*n*	Me	Cl	103–106.5	+350°	Q	55
Cl	OMe	H	OMe	S-allyl	Me	Cl	133.5–135.5	+342	A	55
Cl	OMe	H	OMe	SCH_2Ph	Me	Cl	136.5–138.5	+202.5°	G	55
Cl	OMe	H	OMe	SMe	Me	CH_2Ph	219–220	+455°	A	55
Cl	OMe	H	OMe	SEt	Me	CH_2Ph	178–180	+465°	A	55
Cl	OMe	H	OMe	SPr-*n*	Me	CH_2Ph	115–116	+451°	Q	55
Cl	OMe	H	OMe	SBu-*n*	Me	CH_2Ph		+390°		55
Cl	OMe	H	OMe	*S*-allyl	Me	CH_2Ph	127–129	+465°	Q	55
Cl	OMe	H	OMe	SCH_2Ph	Me	Pr-n	159–161	+337°	A	55
Cl	OMe	H	OMe	SPr-*n*	Me	Pr-*n*	112–113	+423°	A	55
Cl	OMe	H	OMe	SBu-*n*	Me	Pr-*n*	90–91	+402°		55
Cl	OMe	H	OMe	*S*-allyl	Me	Pr-*n*	109–110°	+425°	A	55
Cl	OMe	H	OMe	Cl	Me	Cl	212.5–215; 221–224.5	+335°	H	55
Cl	OMe	H	OMe	Cl	Me	CH_2Ph	123–125	+279°	A	55
Cl	OMe	H	OMe	Cl	Me	Pr-n	196–198	+307°		55
Cl	OMe	H	OMe	OEt	Me	Cl	260–262	+171°	K	55
Cl	OMe	Cl	OMe	OMe	Me	Cl	167–168	+226°		41
H	Me	Me	H	Me	H	COOMe	171		G	128
H	OMe	H	OMe	Me	Me	COOMe	157		G	128
H	OMe	H	H	Me	H	COOEt	110–111		A	68

Gris-3′-en-3,2′-diones

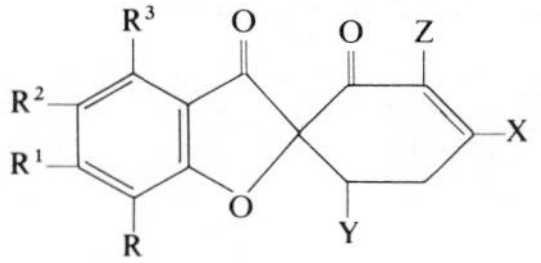

H	OMe	H	OMe	OMe	H	H	128–130			125
Cl	OMe	H	OMe	OMe	H	H	207–208			125

TABLE I (continued)

							Melting point (°C)	$(\alpha)_D^{22}$	Solvent for crystal-lization[a]	Ultraviolet spectrum [λ max., (mμ) (log E)]	Ref.
Me	OMe	H	OMe	OMe	H	H	205				125
Cl	OMe	H	OMe	H	Me	H	273 (dec.)		A	326(3.69), 290(4.27), 228(4.40)	32
H	OMe	H	OMe	OMe	Me	H	177–179		G	323(3.70), 286(4.425), 266(4.24)	126
Cl	OEt	H	OEt	OEt	Me	H	180–181		A	324(3.86), 293(4.33), (264(4.35), 237(4.40)	15
Cl	OEt	H	OEt	OEt	Me	H	174–178		A	(*d,d*)	15
Cl	OEt	H	OEt	OMe	Me	H	218–219		G	320(3.76), 292(4.27), 235(4.31) (*d,d*)	15
Cl	OMe	H	OMe	OPr-*n*	Me	H	179	+196°	G	234(4.32), 263(4.33), 292(4.36), 325(3.74)	53
Cl	OMe	H	OMe	iso-OPr	Me	H	189–190	+208°	G	234(4.28), 265(4.35), 292(4.31), 325(3.70)	53
Cl	OMe	H	OMe	OBu-*n*	Me	H	157.8	+192°	G	234(4.32), 264(4.33), 293(4.32), 325(3.80)	53
Cl	OMe	H	OMe	OBu-iso	Me	H	152–155	+212°	G	236(4.31), 267(4.36), 292(4.32)326(3.75)	53
Cl	OMe	H	OMe	OBu-sec.	Me	H	210–211	+210°	A	235(4.28), 268(4.37), 292(4.32), 335(3.73)	13
Cl	OMe	H	OMe	OC_5H_{11}-*n*	Me	H	143–144	+197°	A	235(4.32), 265(4.32), 292(4.31), 326(3.66)	53
Cl	OMe	H	OMe	OC_6H_{13}-*n*	Me	H	117–8	+195°	A	234(4.31), 264(4.34), 292(4.31), 326(3.71)	53
Cl	OMe	H	OMe	OC_7H_{15}-*n*	Me	H	120–122	+193°	A	235(4.31), 265(4.32), 293(4.30), 326(3.72)	53
Cl	OMe	H	OMe	OC_8H_{17}-*n*	Me	H	129	+179°	A	234(4.28), 265(4.30), 293(4.29), 326(3.72)	53

Cl	OMe	H	OMe	$OC_{10}H_{21}$-*n*	Me	H	84–86	+169°	A or C or E	235(4.34), 265(4.35), 293(4.33), 326(3.76)	53
Cl	OMe	H	OMe	OBu-*n*	Me	Bu-*n*	214–216	+180°	E or G	234(4.24), 285(4.40), 327(3.77)	53
Cl	OMe	H	OMe	OBu-*n*	Me	H	136	+185°	B/C	235(4.33), 265(4.32), 295(4.32), 325(3.72)	53
Cl	OMe	H	OMe	OEt	Me	H	192–194	+218°	A		54
Cl	OMe	H	OMe	OEt	Me	Et	185–186	+280°	A		54
Cl	OMe	H	OMe	OCH_2Ph	Me	H	184		A	234.5 (E, 24,500), 264 (E, 24,600), 291 (E, 23,400), 323 (E, 6100)	54
Cl	OMe	H	OMe	OMe	Me	Me	249–250	+348°		230i(16.6), 281.3(33.2), 320(5.6) (10^{-3}E)	54
Cl	OMe	H	OMe	OMe	Me	allyl	169–170.5	+274°		230i(15.8), 281.5(33.0), 320(5.6) (10^{-3}E)	54
Cl	OMe	H	OMe	OMe	Me	CH_2Ph	150–152	+224°		230i(15.9), 283(31.4), 321(5.3) (10^{-3}E)	54
Cl	OMe	H	OMe	OEt	Me	CH_2Ph	137.5–141.5	+218°		230i(16.7), 282.5(30.2), 321i(5.6) (10^{-3}E)	54
Cl	OMe	H	OMe	OPr-*n*	Me	CH_2Ph	112–113	+222°		230i(16.8), 283(33.1), 322(5.2) (10^{-3}E)	54
Cl	OMe	H	OMe	OBu-*n*	Me	CH_2Ph	152–153.5	+220°		231i(16.7), 283(34.6), 322(5.65)	54
Cl	OMe	H	OMe	*O*-allyl	Me	CH_2Ph	102–103	+205°		230i(16.3), 282.5(31.2), 321(4.9) (10^{-3}E)	54
Cl	OMe	H	OMe	$O(CH_2)_3Br$	Me	CH_2Ph	145–147	+197°		231i(16.8), 282.5(33.4), 322i(5.3) (10^{-3}E)	54
Cl	OMe	H	OMe	OMe	Me	Pr-*n*	206–207	+305°		230i(16.4), 281.5(34.4), 320(5.6) (10^{-3}E)	54
Cl	OMe	H	OMe	OPr-*n*	Me	Pr-*n*	148–149	+282°		231i(15.3), 283(26.1), 320i(5.25) (10^{-3}E)	54

TABLE I (continued)

							Melting point (°C)	$(\alpha)_D^{22}$	Solvent for crystallization[a]	Ultraviolet spectrum [λ max., (mμ) (log E)]	Ref.
Cl	OMe	H	OMe	OMe	Me	$(CH_2)_2COMe$	187.5–190	+286°	A		54
Cl	OEt	H	OMe	OMe	Me	H	202–203	+206°			66
Cl	OPr-*n*	H	OMe	OMe	Me	H	162–163	+198°			66
Cl	OPr-iso	H	OMe	OMe	Me	H	185–187	+196°			66
Cl	OBu-*n*	H	OMe	OMe	Me	H	161–164	+194°			66
Cl	*O*-allyl	H	OMe	OMe	Me	H	182–183	+199°			66
Cl	OCH_2	H	OMe	OMe	OMe	H	215–217	+174°			66
Cl	OMe	H	OMe	OMe	Me	Br	246–249	+264°			57
Cl	OMe	H	OMe	OEt	Me	Br	204–206	+199°			57
Cl	OMe	H	OMe	OPr-*n*	Me	Br	219	+240°			57
Cl	OMe	H	OMe	OBu-*n*	Me	Br	188–190	+225°			57
Cl	OMe	H	OMe	*O*-allyl	Me	Br	193–194	+253°			57
Cl	OMe	H	OMe	OCH_2Ph	Me	Br	221–223	+208°			57
Cl	OMe	H	OMe	OMe	Me	I	240 (dec.)	+192°			57
Cl	OMe	H	OMe	OEt	Me	I	229–230	+194°			57
Cl	OMe	H	OMe	OPr-*n*	Me	I	225 (dec.)	+192°			57
Cl	OMe	H	OMe	OBu-*n*	Me	I	200–201	+176°			57
Cl	OMe	H	OMe	*O*-allyl	Me	I	188–190	186.5°			57
Cl	OMe	H	OMe	OCH_2Ph	Me	I	188–190	+165°			57
Cl	OMe	H	OMe	OMe	Me	Cl	257–209.5	+324°			57
Cl	OMe	H	OMe	OEt	Me	Cl	213–216	+315°			57
Cl	OMe	H	OMe	OPr-*n*	Me	Cl	200–201	+309°			57
Cl	OMe	H	OMe	OBu-*n*	Me	Cl	182–184	+296°			57
Cl	OMe	H	OMe	*O*-allyl	Me	Cl	194–195	+311°			57
Cl	OMe	H	OMe	OCH_2Ph	Me	Cl	245–247 (dec.)	+251°			57
Cl	OMe	Cl	OMe	OMe	Me	Cl	222.5–224.5	+289°			57
Cl	OMe	Cl	OMe	OMe	Me	H	188.5–190.5				57

Cl	OMe	H	OMe	Cl	Me	H	248–250.5	+278°	H		55
Cl	OMe	H	OMe	Cl	Me	Br	213–216	+212°	H		55
Cl	OMe	H	OMe	Cl	Me	Cl	235–237.5	+258°			55
Cl	OMe	H	OMe	Cl	Me	CH_2Ph	170.5–171.5	+158°	O		55
Cl	OMe	H	OMe	NH_2	Me	H	321–322	+359°	G	(isogriseofulvin)amine	55
Cl	OMe	H	OMe	NH_2	Me	Cl	320 (dec.)	+285°	P	(3′-Chloroisogriseofulvin)amine	55
Cl	OMe	H	OMe	NH_2	Me	Br	290–295 (dec.)	+227°	A	(3′-Bromoisogriseofulvamine)	55
Cl	OMe	H	OMe	NH_2	Me	CH_2Ph	183.5	+196°	G	(3′-Benzylisogriseofulvamine)	55
Cl	OMe	H	OMe	NH_2	Me	Pr-*n*	147–150	+225°	G–E	(3′-Propylisogriseofulvamine)	55
Cl	OMe	H	OMe	SH	Me	H	244.5 (dec.)		H	Disulfide, m.p. 289° (dec.)	55
Cl	OMe	H	OMe	SMe	Me	H	229.5–232.5		K		55
CL	OMe	H	OMe	OPh	Me	H	207–208.5	+93.5°	A		55
Cl	OMe	H	OMe	SEt	Me	H	177–179	+236°	P		55
Cl	OMe	H	OMe	SCH_2Ph	Me	H	145–147	+181°	G		55
Cl	OMe	H	OMe	SPh	Me	H	229–231	+110°	P		55
Cl	OMe	H	OMe	$SNHCH_2CHMe_2$	Me	H	97–102		O		55
Cl	OMe	H	OMe	OCH_2CH_2OH	Me	H	138–140	+232°	B–E	Acetate, m.p. 162–163° (A); $(a)_D^{20} = 196°$	55
H	OMe	Cl	OMe	OMe	H	H	163–164				38
Cl	OMe	H	OMe	OMe	H	H	207–208				38
Me	OMe	H	OMe	OMe	H	H	196–197				38

Gris-2′,5′-dien-3,4′-diones

Gris-2', 5'-dien-3, 4'-diones

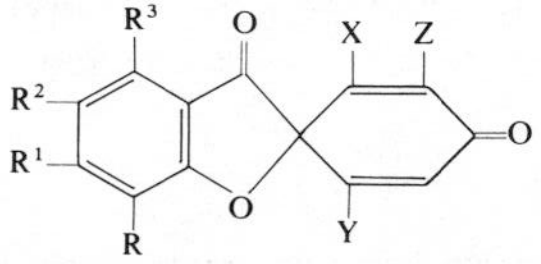

R^3 O X Z

R^2 O

R^1

O Y

R

TABLE I (continued)

							Melting point (°C)	$(\alpha)_D^{22}$	Solvent for crystallization[a]	Ultraviolet spectrum [λ max., (mμ) (log E)]	Ref.
Cl	OMe	H	OMe	OMe	Me	H	291–293; 278–280[37]		D–F	(racemic Dehydro–griseofulvin)	37,42
H	OMe	H	OMe	OMe	Me	H	241–244			288 (E, 32,400), 255 (E, 24,000)	41,42
H	OMe	H	OMe	OMe	Me	H	278–9			232, 92, 326 (*d,l*-dehydrogriseofulvin)	17
H	OMe	Cl	OMe	OMe	Me	H	204–207			230 (E, 5800), 283 (E, 26,000)	41,42
F	OMe	H	OMe	OMe	Me	H	222–5		D–E		42
H	Me	H	OH	COOMe	OMe	H	189–91		B/C	284(4.20); green color with ferric chloride	127
Gris-2′-en-3-ones											
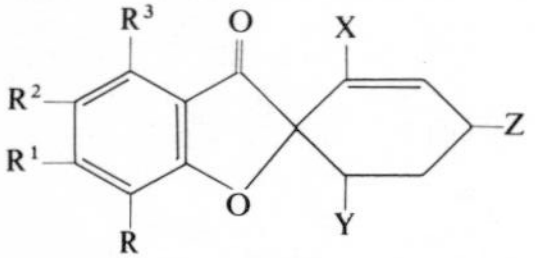											
Cl	OMe	H	OMe	OMe	Me	H	194–195; 189–190	+155°	A	325(371), 288(4.32), 236(4.16) (deoxygriseofulvin)	31,34
Cl	OMe	H	OMe	OMe	Me	OMe	201–203	+185°			34
Cl	OMe	H	OMe	OMe	Me	OAc	205–206	+117°			34
Cl	OMe	H	OMe	OMe	Me	OEt	185–186	+94°			129
Cl	OMe	H	OMe	OMe	Me	OCH_2Ph	190–191	+95°			129
Cl	OMe	H	OMe	OMe	Me	pivalyoxy	192	+81°			129
Cl	OMe	H	OMe	OMe	Me	OCOPh	156–157	+89°			129
Cl	OMe	H	OMe	OMe	Me	undecenyloxy	84–85	+81°			129
Cl	OMe	H	OMe	OMe	Me	p-nitrobenzy	160; 200	+78°			129

R	R^1	R^2	R^3	R^4	X	Y	Z	T				
H	Me	H	Me	H	Me	H	H	=CHMe	106	C/E	2,4-DNP, m.p. 25°	71
H	Me	H	Me	H	Me	H	COOMe	=CHMe	128	G + R		71
3-alkylidenegris-2′-en-4′ones												
H	Me	H	Me	H	Me	COOH	H	=CHMe	174 (hydrate); 192	G + R or C		70
H	Me	H	Me	H	Me	COOMe	H	=CHMe	162		2,4-DNP, orange cryst. (S) m.p. 244°; oxime, m.p. 220°(B)	70
H	Me	H	Me	H	Me	COOEt	H	=CHMe	142	A or C	2,4-DNP, red (A), m.p. 233°; oxime, m.p. 172° (B/C)	70
H	Me	H	Me	H	Me	COMe	H	=CHMe	167	A + R	221(4.51), 229(4.55), 236(4.30), 269(4.21), 309(3.98), 323(4.0)	70
H	Me	H	Me	H	Me	H	H	=CHMe	106	C	228(4.57), 235(4.57), 256(4.25), 266(4.14), 306(4.0), 314(4.0), 317(4.02); 2,4-DNP, orange (D), m.p. 250–252°	70

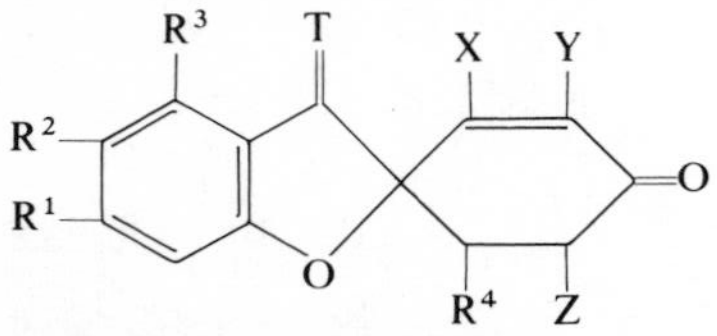

TABLE I (continued)

									Melting point (°C)	$(\alpha)_D^{22}$	Solvent for crystal-lization[a]	Ultraviolet spectrum [λ max., (mμ) (log E)]	Ref.
H	Me	H	Me	H	Me	COOEt	H	$=CH_2$	141		C	2,4-DNP, orange (D) m.p. 233°	70
H	OMe	H	OMe	Me	Me	COOMe	H	$=CH_2$	169		G/C	234(4.51), 241(4.47), 272(4.30), 308(3.89)	128
H	OMe	H	H	H	Me	COOMe	H	$=CH_2$	126		G	216(4.36), 232(4.40), 259(4.18), 315(4.06)	128
H	OMe	H	H	H	Me	COOEt	H	$=CH_2$	97–98		A + R	326(4.05), 316(4.09), 273(4.18), 260(4.43), 233(4.37); 2,4-DNP, m.p. 182° (dec.) (A–H)	128

Racemic spirans related to griseofulvin

R	R^1	R^2	R^3	X	Melting point (°C)	$(\alpha)_D^{22}$	Solvent for crystal-lization[a]	Ultraviolet spectrum [λ max., (mμ) (log E)]	Ref.
H	OMe	H	H	Me	167–168		A, sub-limation	267(4.01), 233(4.12), 208(4.05)	27
H	OMe	H	H	H	192–193		A, sub-limation	331(3.68), 332(3.69), 289(4.30), 238(4.20), 213(4.37)	27

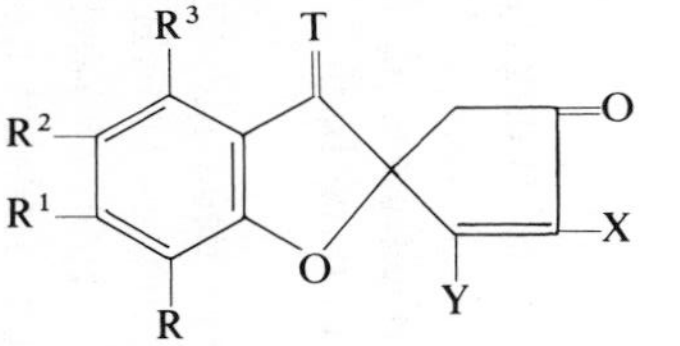

R	R^1	R^2	R^3	T	X	Y				
H	Me	He	Me	=O	COOEt	Me	91–92	C	270(4.34), 330(3.7P)	128,131
H	OMe	H	H	=O	H	Me	193–195	F + R	320(3.92), 273(4.07), 232(4.16), 209(4.40); 2,4-DNP, m.p. 238–240° (D)	68
H	OMe	H	H	=O	COOEt	Me	115–117	C	321(3.98), 275(4.21), 232(4.23); blue color with HNO_3; mono-2,4-DNP, yellow (B/C), m.p. 243° (dec.)	68
H	OMe	H	H	=O	COOH	Me	175–177	A	318(4.05), 273(4.25), 232(4.30), 210(4.45)	68
H	OMe	H	H	$=CH_2$	COOEt	Me	122	C	327(3.98), 263(4.112), 231(4.35), 214(4.31): 2,4-DNP, orange (B/C), m.p. 175–176°	68,131

[a]A, ethyl alcohol; B, benzene; C, petroleum ether; D, ethyl acetate; E, ether; F, acetone; G, methyl alcohol; H, acetic acid; I, water; J, chloroform; K, acetonitrile; L, nitromethane; M, carbon tetrachloride; N, hexane; O, 2-propanol; P, nitroethane; Q, *n*-propyl alcohol; R, aqueous ethyl alcohol; S, methyl acetate. [b]Phthalimide: [c]2-Imidazoyl.

g. Detection and Estimation

The bioassay,[2,6] involving observation of morphological changes in the germ-tubes of *B. allii*, is reliable and widely used but suffers from that degree of imprecision inherent in all $\times$ 2 serial dilution bioassays. It is not specific for griseofulvin, and many derivatives can interfere. Many investigators have preferred to use physical methods, particularly ultraviolet spectrophotometry,[75] by utilizing the strong absorption band in griseofulvin at 291 mμ, and the more sensitive but temperature-dependent spectrofluoremetry,[76] by measuring the emission at 450 mμ. The spectrophotometric method is relatively nonspecific since all relatives containing the chloroacetophenone chromophore absorb strongly near 285 mμ. Depending on the accuracy required, various mathematical procedures have been adopted for correcting the background absorption present when griseofulvin is estimated in fermentation broths[77-79] or plant tissue extracts.[8] The fluorometric method is rather more specific and, coupled with a suitable extraction procedure, has been used extensively for detecting and estimating griseofulvin in biological fluids.

Addition of hydrochloric acid and a fragment of magnesium to an alcoholic solution of griseofulvin develops a yellow, then brownish-yellow color in isoamyl alcohol after aqueous dilution. Moreover, addition of $Na_2S_2O_4$ and sodium hydroxide to the solution of griseofulvin and heating to boiling forms a lemon-yellow color. Both reactions are sensitive up to 100 ml γ. Furthermore, a solution of griseofulvin in sulfuric acid containing a crystal of potassium dichromate develops a rose or orange color, which is visible with 50 γ of griseofulvin.[81]

A relatively simple method for spectrofluorometric estimation is by thin-layer chromatography of serum extracts.[82] An isotope dilution assay involving griseofulvin labeled with ^{35}Cl has also been described.[80]

h. Biosynthesis of Griseofulvin

P. griseofulvum and *P. patulum* give rise to a similar series of metabolic products, and in addition to griseofulvin and 6-methylsalicylic acid (**103**), patulin (**104**) and gentisic acid and a series of simple phenolic acids (which presumably arise from glucose by shikimic acid route) have been isolated from both organisms.[83,84]

Mycelianamide (**105**)[85] and fulvic acid (**106**)[86] have been identified as metabolic products of *P. griseofulvum*. The benzophenones **108** (R = Cl, R^1 = Me; R = Cl, R^1 = H; R = R^1 = H) and **107**, (**109**), (—)-dehydrogriseofulvin (**63**) and the dechlorotrione (**69**) have been isolated from the fermentation broths of *P. patulum*,[50,51] although some of these were obtained by using chloride-deficient media.

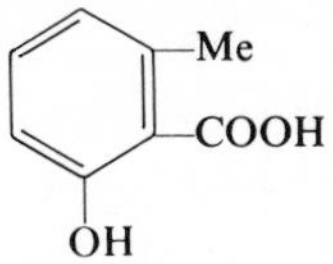

103

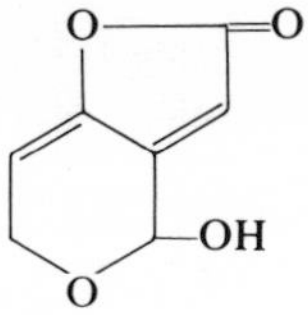

104

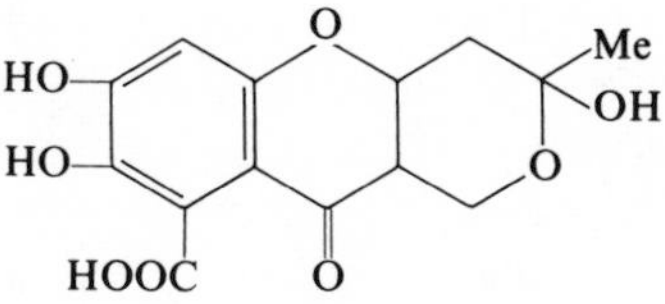

106

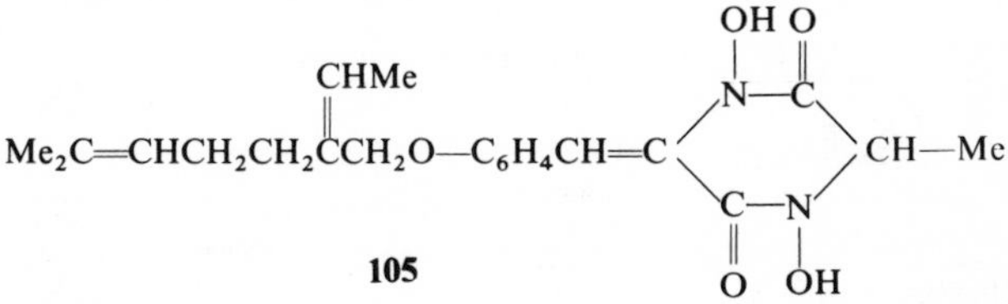

$Me_2C{=}CHCH_2CH_2C({=}CHMe)CH_2O{-}C_6H_4CH{=}C$ …

105

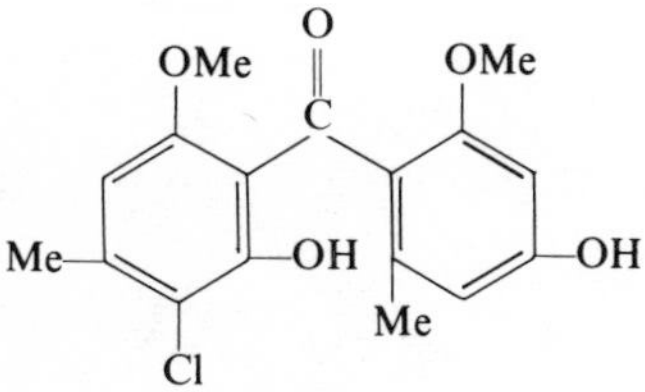

107

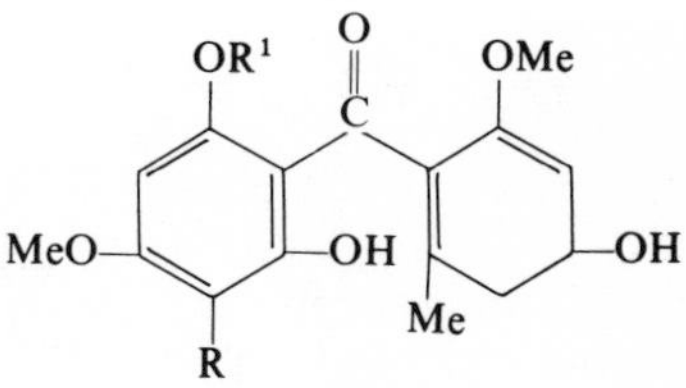

108

The biosynthesis of many fungal metabolites by head-to-tail linkage of "acetate units" is well established and griseofulvin biosynthesis is formed from seven (1-^{14}C) acetate units according to Chart 1; this method of degradation represents one of the most important methods of identifying the labeled atoms in phenolic nuclei.[87] In the Chart 1, the radioactivity of griseofulvin is assumed to represent the expected total of seven labeled atoms, and on this basis the corresponding activities (italicized) of the degradation products closely follow the predicted values. In his review, Rickards[88] refers to the work by Birch and his colleagues who have lately confirmed the earlier results by using (2-^{14}C) acetate.

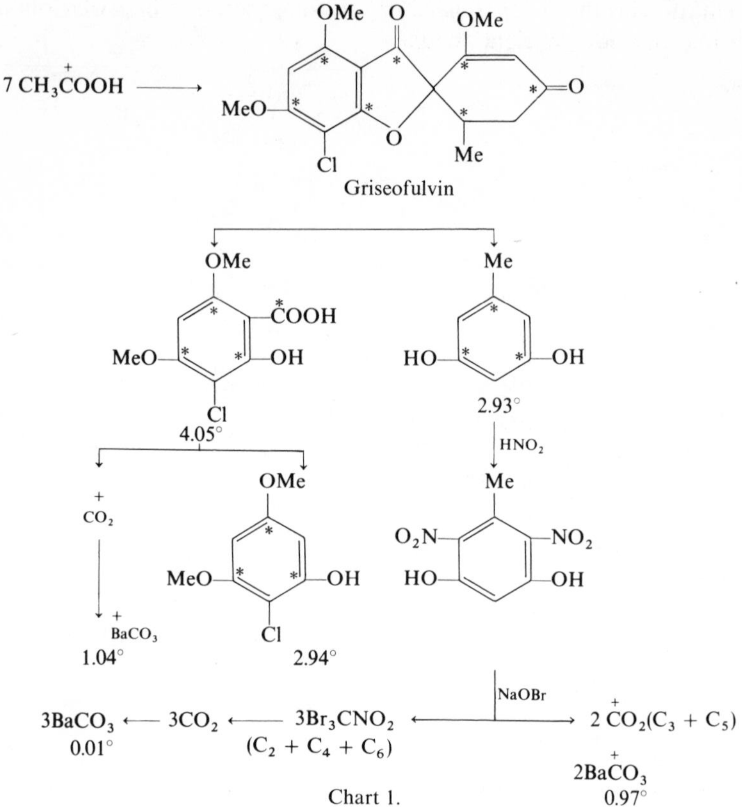

$7\ \overset{+}{C}H_3COOH \longrightarrow$ Griseofulvin

$4.05°$ · $2.93°$ · $\overset{+}{CO_2} \rightarrow \overset{+}{BaCO_3}$ $1.04°$ · $2.94°$ · HNO_2 · NaOBr

$3BaCO_3 \leftarrow 3CO_2 \leftarrow 3Br_3CNO_2\ (C_2 + C_4 + C_6)$; $0.01°$

$2\ \overset{+}{C}O_2(C_3 + C_5)$; $2\overset{+}{B}aCO_3$ $0.97°$

Chart 1.

i. Metabolism

Griseofulvin owes much of its unique therapeutic value to its resistance to many of the common metabolic processes—the chief reaction in mammals is merely demethylation of the 4-methoxyl group[89] A slow breakdown of griseofulvin is observed;[90] demethylation is known to be the first step in the degradative sequence, since phenols, griseofulvic acid **(7)**, and **109** and **110** have been isolated[63] after destruction of griseofulvin by mycelia of *Microsporum canis, Botrytis allii* and *Cercospora melonis*. Amounts of the products did not account for all the griseofulvin which was lost, and further degradation to less easily recognizable products may have occurred. Quantitative liberation of the chloride ion during decomposition of griseofulvin by *Pseudomonas*[91,92] suggested that rupture of the aromatic ring is a feature of microbial degradation.

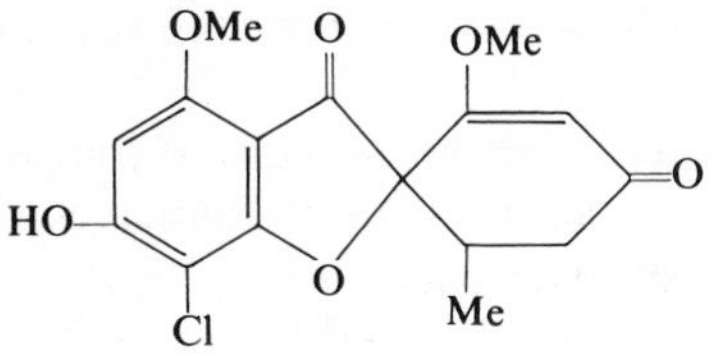

109

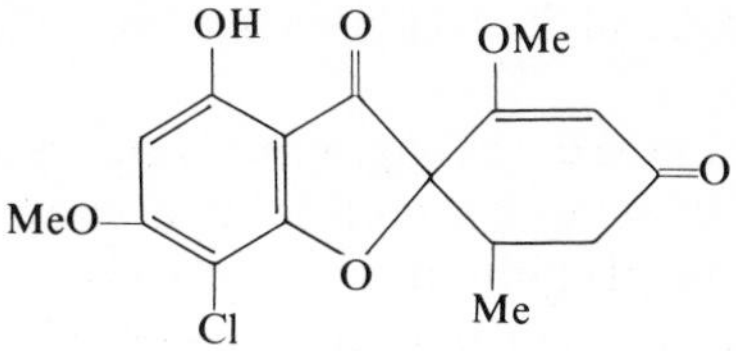

110

The salicylic acid (**8**)[93] and the phenol (**109**)[89] have been isolated from mammalian urine after oral administration of griseofulvin: in man the amount of griseofulvin in the urine was less than 1% of that administered, but large amounts of phenol **109** were present.[89]

Griseofulvin was slowly degraded in higher plants to unidentified acidic and phenolic compounds; the halflife in bean tissue is about 4 days.[94] It was relatively stable in roots, but the isomer, isogriseofulvin (**19**) was rapidly degraded; the salicylic acid (**8**) was a major product.[8]

j. Biological Activities

Although this antibiotic, a product of *P. griseofulvum*, was discovered in 1939, its antifungal activity was not utilized until 1951. At that time the compound was shown to be active against plant pathogenic fungi by systemic as well as topical administration.[95] Of the four stereoisomers, only griseofulvin itself is active.[96] The fluoro analog retains the potency of the original drug,[43] but the bromo- and dechloro analogs are ineffective. Neither the methoxy group on ring A nor the keto group on ring C is required for the potency. Replacement of the methoxy substituent in ring C with either a propoxy or butoxy group increases activity 20 to 50-fold.[96] The molecular site of the antibiotic action apparently is associated with DNA replication. In addition to causing the formation of abnormal fungal cells, the drug inhibits mitosis of animal and plant cells, causes multipolar mitosis, and produces abnormal nuclei.[97]

Griseofulvin is not active against animals and humans when used topically, but was found in 1958 to be quite effective when administered orally.[98-100] It has a curative effect on dermatophytic infections,[98,99] and has found important medical and veterinary applications in vitro against ring worms.[100] Apparently a small amount of griseofulvin can diffuse from the blood to the site of fungal multiplication in the skin and hair. This low concentration sufficiently slows the rate of hyphal penetration so that the outward thrust of keratinized host cells is able to deprive the fungus of access to nutrients.[95] Because of its low toxicity, the antibiotic can be given to patients for a period of many weeks. Unfortunately, griseofulvin has no effect on either candidiasis or the systematic mycoses.[101-103]

Large doses of griseofulvin injected intravenously in rats cause a transient depression of mitosis, and intraperitoneal administration damages the seminal epithilium;[104] but when it is taken orally, toxicity is very low,[101,104] and in man only minor side effects have been observed.

Griseofulvin is also active in vitro against many important fungal pathogens of plants.[3,90] The antibiotic effects of griseofulvin are essentially specifical against fungi; bacteria, including actinomycetes, are unaffected.[3] Griseofulvin was found to control a number of plant diseases attributed to the fungal pathogens of plants.[105,106] It is a systemic antifungal compound and its action is essentially fungistatic.[107,108]

Griseofulvin is of increasing importance as a fungicide or fungistat. Fungicides of suitable activity are only rarely so harmless to the host. Unlike the present compound, which is absorbed through the roots and distributed internally throughout the plants, they have to be applied superficially.[94,107]

Griseofulvin is relatively nonphytotoxic; some minor modifications of root growth[109] are probably associated with an interference with mitosis.[104] It is slowly degraded in plant tissue[75,94] and the protective action against fungal pathogens is therefore transient. 2-Ethoxy-2,3-dihydro-3,3-dimethyl-5-benzofuranylmethanesulfonate reduced turfgrass growth.[110]

k. Structure–Activity Relations

The relation between structure and activity is being studied, but is still obscure.[8,66] The antifungal activity in derivatives of griseofulvin has been limited mainly to grisan derivatives containing alkoxyl or hydroxyl groups on the aromatic ring.[111] Of the racemic spirans, lacking methoxyl groups on ring A, those also lacking a chlorine substituent showed griseofulvin-like activity, but only at high concentration. Increased activity resulted from the introduction of a chlorine atom at position 7, but not at positions 5 or 6, and the 7-chloro analog (**111a**) showed activity comparable with that of (*d,d*)-dechlorogriseofulvin. Introduction of a 5-methyl group (**111b**) reduced the activity.

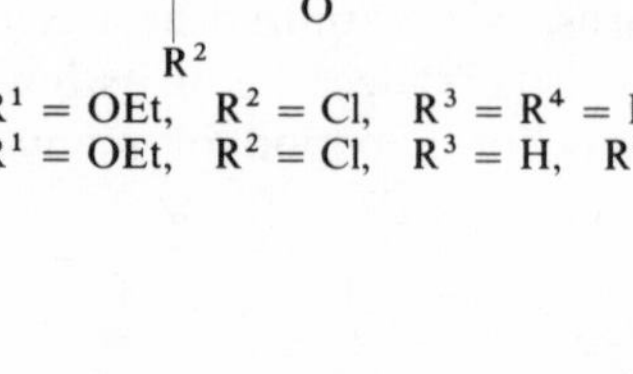

111a R^1 = OEt, R^2 = Cl, R^3 = R^4 = H
111b R^1 = OEt, R^2 = Cl, R^3 = H, R^4 = Me

Activity is only present in neutral molecules; acidic and basic analogs were ineffective;[9] compounds more active than griseofulvin can be obtained by increasing the oil–water position coefficient. Cross, et al.,[112] in a study of a large number of griseofulvin analogs, have found none to be active against a bacterium (*Bacillus subtillis*) or a yeast (*Candida albicans*). On the whole, increased activity was shown to a greater extent against plant pathogens than against dermatophytic fungi. Replacement in the 2′ position, particularly with higher alkoxy groups, and substituents in the 3′ position produced the greatest improvements, that is, the 2′-butoxy-3′-iodo analog was over 500 times as active as griseofulvin in a "hyphal curling" test against *B. allii*. The 2′,3′ analogs showed much less marked improvement in a similar test on *B. cinerea*, or in tests of radical growth on agar with a variety of other fungi. It was suggested that the favorable effect of 3′ substitution on the activity of *B. allii* might result from a hindrance of 2′ dealkylation by the fungus; 2′ dealkylation of griseofulvin to the biologically inactive griseofulvic acid is a known activity of *B. allii*.[63,96]

Stereochemistry has a decisive influence and of the four stereoisomers, only griseofulvin is active.[39] Racemic griseofulvin had half the activity of griseofulvin;[40] the racemic (*l,d*)-diastereoisomer together with all the transformation products with the *l,d* stereochemistry[25] were inactive.[9] The synthetic ketone **112**[68] had half of activity of the *d,d* isomer. On elimination of the asymmetric 6′ center some activity is retained and dehydrogriseofulvin (**63**) and the synthetic griseonone (**113**) showed typical (weak) griseofulvin-like activity.

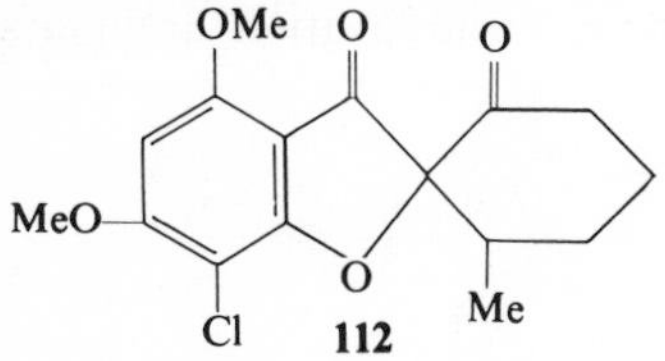

112

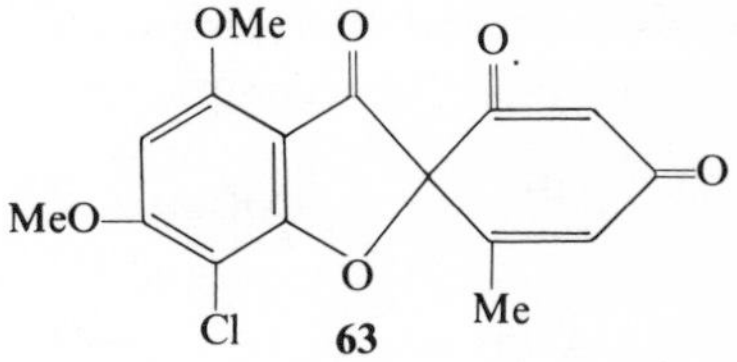

63

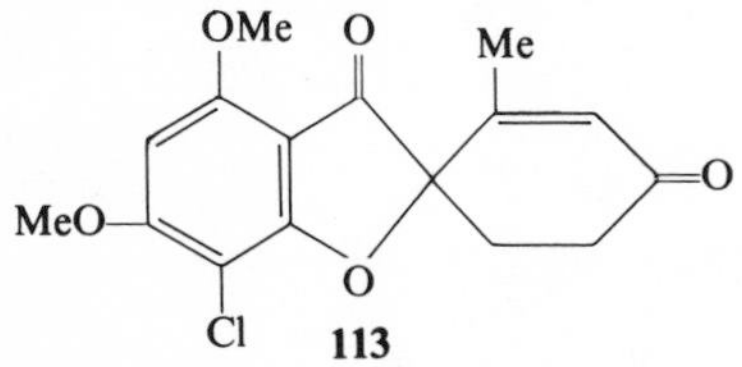

113

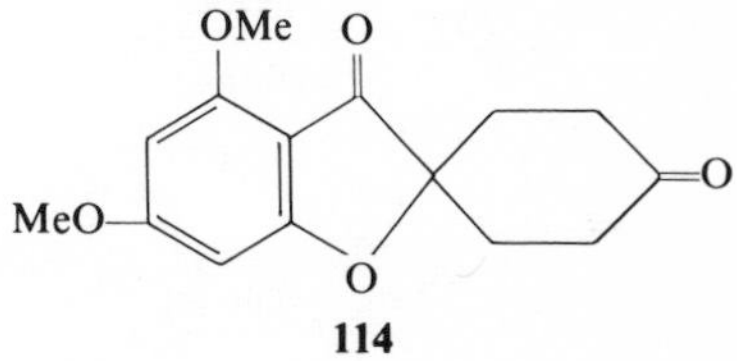

114

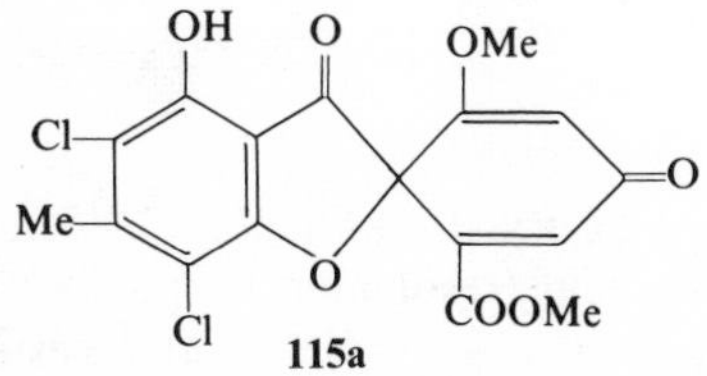

115a

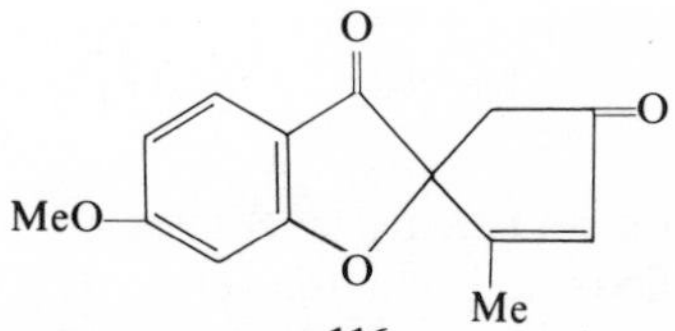

116

Grisan-3,4′-dione (**114**) was inactive. A halogen substituent in the 5 instead of 7 position reduced activity,[38] and although geodin (**115a**), in common with many dienones, showed fungistatic activity, the typical griseofulvin-like effects were lacking both in the natural product and in its methyl ether. A free 5 position on the aromatic ring is therefore advantageous. Activity is also recognized and confirmed in compounds with the grisan nucleus; synthetic compounds with 5-membered rings (C), for example, **116**, were inactive.

B. Geodin and Erdin

Smith and Raistrick[113] have isolated from a strain of *Aspergillus terreus* Thom (strain No. 45 and Ac 100, grown on a Czapck-Dox medium containing glucose as the source of carbon and potassium chloride as the source of chlorine), two chlorine-containing metabolites, geodin, $C_{17}H_{12}O_7Cl_2$ [m.p. 235°, $(a)_{D461}$ + 179° ($CHCl_3$)], and erdin, $C_{16}H_{10}O_7Cl_2$ (m.p. 211°). Erdin is optically inactive. Komatsu[117] has shown that geodin is also produced by *Penicillium paxilli* var. echinulatium.

The chemistry of geodin has been the subject of extensive investigation.[114,115] The close connections between geodin [3′,4′-dioxo-6-methyl-6′-methoxy-4-hydroxy-5,7-dichloro-spiro(benzofuran-2(3*H*)-1′-(2,5)cyclohexadiene)-2′-carboxylic acid methyl ester, **115b**] and griseofulvin (**1**) is marked; ring C, however, is more nearly aromatic in geodin than in griseofulvin and the tendency to complete the aromatization dominates the reactions of this compound.

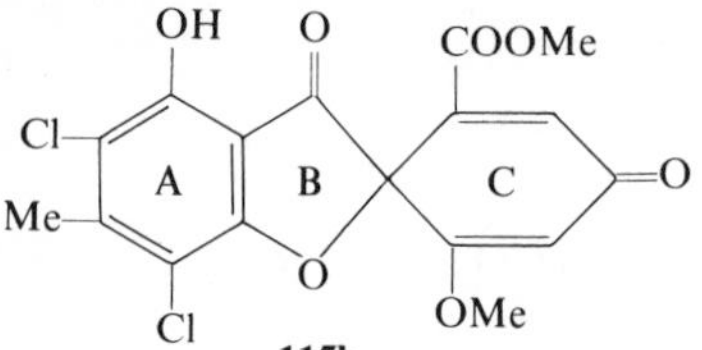

OH O COOMe Cl A B C =O Me O Cl OMe

115b

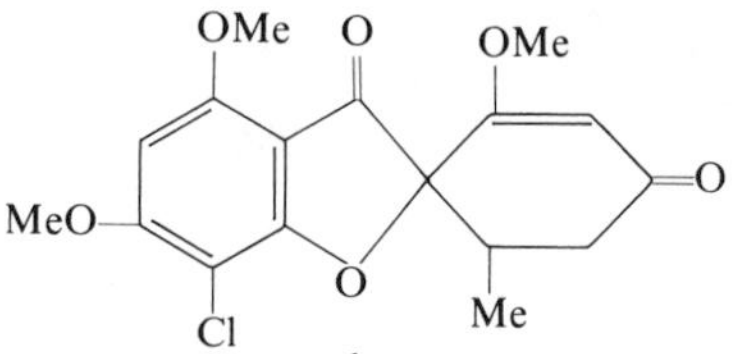

OMe O OMe MeO =O O Cl Me

1

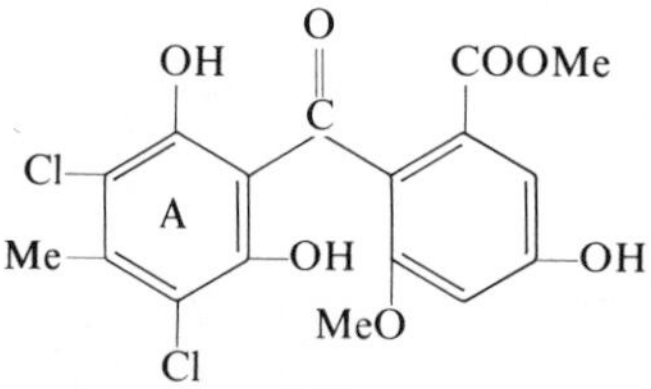

O OH C COOMe Cl A Me OH OH MeO Cl

117

Barton and Scott[118] completed the work of Raistrick et al.,[115,116] and clarified the point that erdin is a racemic acid (resolvable by the use of quinine methohydroxide), and geodin is a (+)-erdin methyl ester.

Both geodin and erdin are readily hydrogenated to give (±)-dihydrogeodin and dihydroerdin, respectively. The close relationship between geodin and erdin is shown by methylation of these two dihydro compounds, with diazomethane, to the same tetra-*O*-methyl ether methyl ester which is readily converted by hydrolysis to the monobasic acid, $C_{15}H_6O_3Cl_2(OMe)_4$. The same reaction sequence with dihydroerdin and dihydrogeodin with diazoethane furnishes the monobasic acid, $C_{14}H_5OCl_2(OMe)(OEt)_3(COOH)$. This acid contains the methoxyl group present in dihydroerdin, from which it follows that dihydrogeodin (**117**) is the methylation product of dihydroerdin.

Dihydrogeodin (**117**), a benzophenone, can be split hydrolytically like other resorcinol ketones so that, when treated with 80% sulfuric acid, it affords **118** and **119**. Fully methylated dihydrogeodin has been

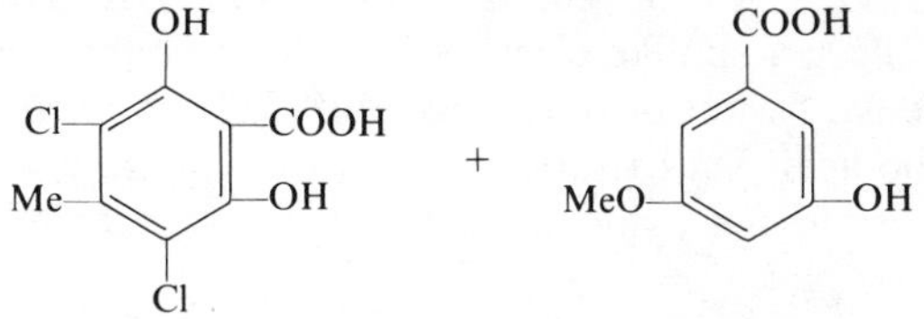

118 + **119**

synthesized[115] by a Friedel-Crafts reaction with fully methylated **119** as one component, and the acid chloride of the dimethyl ether of **118** as the other. This leaves in doubt only the orientation of the C ring in **117**.

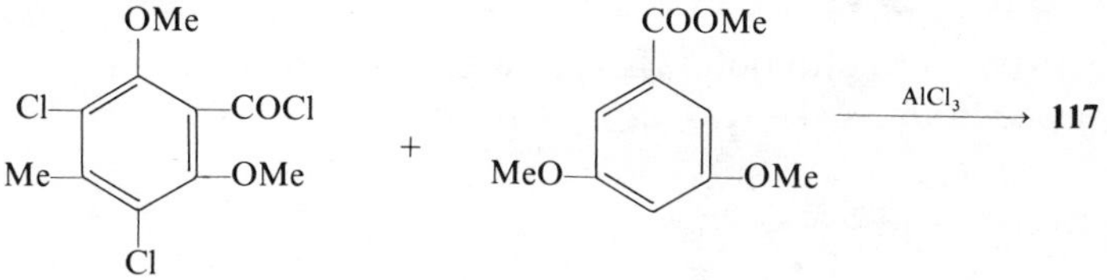

$\xrightarrow{AlCl_3}$ **117**

Alkaline hydrolysis of dihydroerdin (**120**) yields the phthalic acid (**121**), which is devoid of ferric chloride, together with phenol **122**. Evidently, hydrolytic scission by alkaline reagents occurs at a point different from that afforded by acids—a fortunate result that dispells the doubt as to the orientation of the C ring.

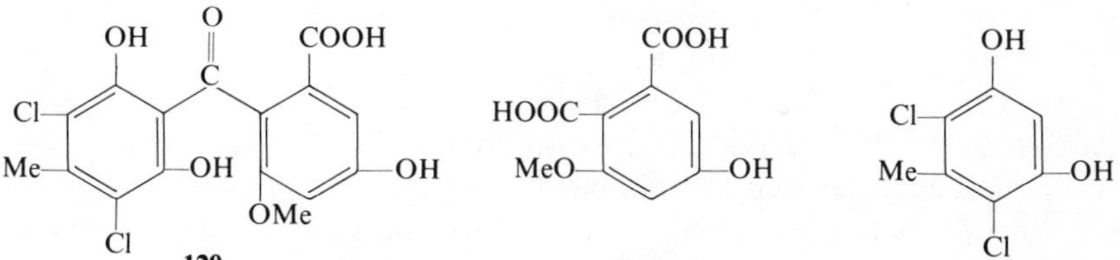

120 **121** **122**

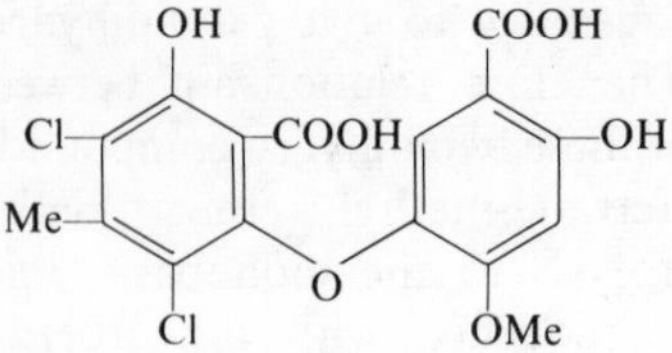

123

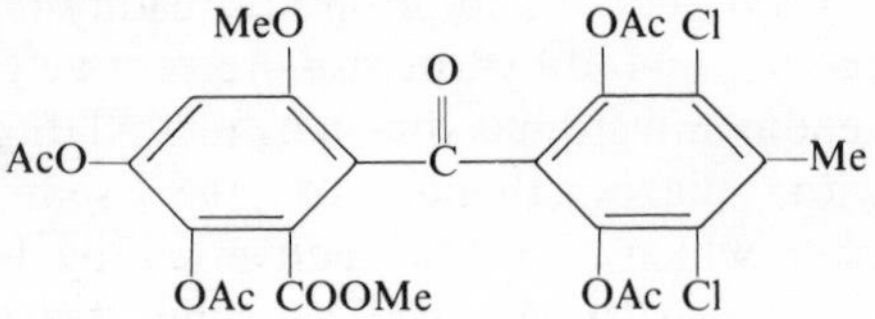

124

The spectroscopic data and the fact that geodin exhibits a muddy ferric reaction favor **116** for its structure. Scott[52] achieved the oxidative cyclization of dihydrogeodin (**117**) to geodin (**115b**) using ferricyanide and later adopted the method for synthesis of griseofulvin.

Geodin is aromatized, upon treatment with sulfuric acid (80%), to geodin hydrate (**123**); it is produced by *β*-diketonic cleavage. Basic hydrolysis does not effect a similar cleavage but gives compounds of type **124** instead. Reductive fission of dihydrogeodin (**117**) and dihydroerdin (**120**) with hydroiodic acid gives rise to 3,5-dihydroxybenzoic acid and oxinol, with carbon dioxide and 2 and 1 molecular proportions of methyl iodide, respectively.

Thermal rearrangement of geodin, erdin, and dihydroerdin at 250° gives 2,6-dichloroorcinol; oxidation of dihydroerdin tri-*O*-methyl ether with potassium permanganate furnishes a dibasic acid, $C_{19}H_{16}O_9Cl_2$, which contains the same number of carbon atoms as its progenitor.

Geodin on acetylation and alkylation, under certain conditions, may also take abnormal courses. The former induces fission of the heterocyclic ring with the formation of **124**; the latter, if effected with diazomethane, involves addition of the reagent to the acrylic ester system (**125**).

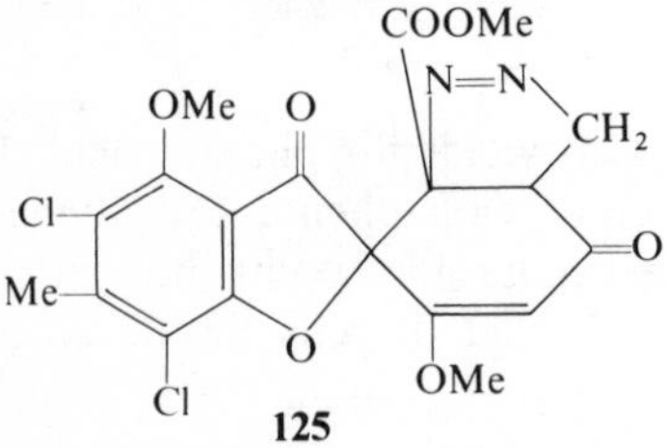

125

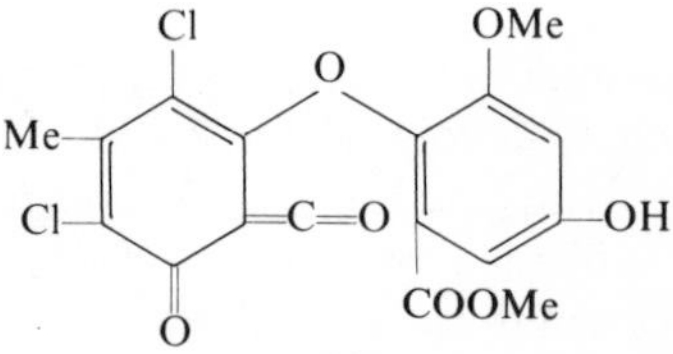

126

Sulochrin [5-hydroxy-2-(3,5-dichloro-4-methyl-Y-resorcyloxy)-*m*-anisic acid, methyl ester, **127**] is converted to dechlorogeodin (**128**) by treatment with alkaline potassium ferricyanide.[119] This is similar to the conversion of dihydrogeodin (**117**) and *O*-methyldihydrogeodin to geodin and *O*-methylgeodin, respectively.[52,119]

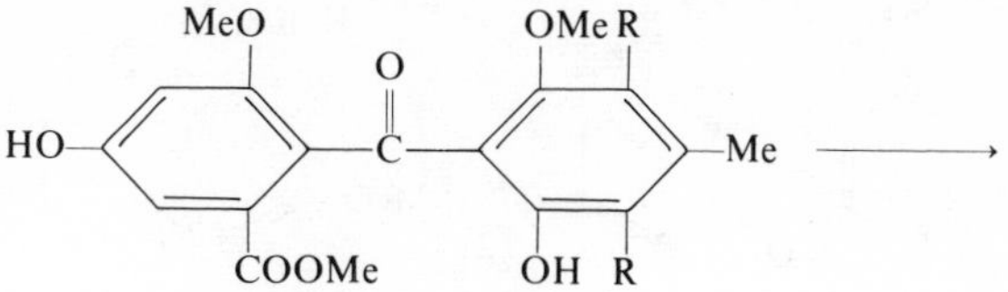

MeO OMe R O HO C Me COOMe OH R

127 Sulochrin R = H

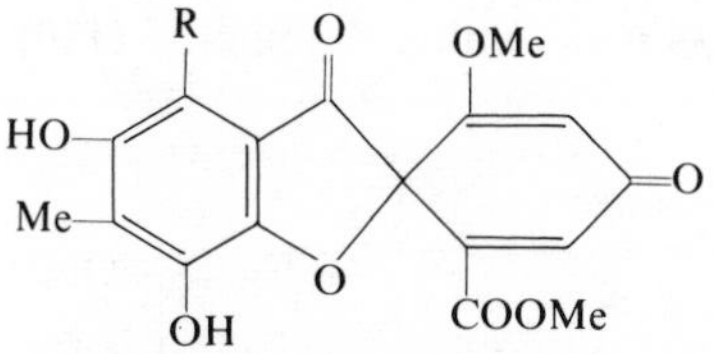

R O OMe HO Me O O OH COOMe

128 Dechlorogeodin (R = H)

Geodin inhibited the growth of gram-positive bacteria in glucose broth at dilutions from 1:1,000 to 1:128,000 (*B. anthracis* NCTC 5444, *Staph. aureus* NCTC 3761 and NCTC 3098; *C. diphteriae* var. gravis, *Staph. aureus* NCTPC 4163, *Str. viridans* NCTC 3166, and *Vibria cholerae*).[120]

2. Spiro-2 (3*H*)-Benzofuranones

A. Picrolichenic Acid

Picrolichenic acid [4,6′-diphenyl-2,4′-dioxo-2′-methoxy-6-hydroxy-spiro-(benzofuran-3(2*H*)-1′-(2,5)cyclohexadiene)-5-carboxylic acid, **129**, $C_{25}H_{30}O_7$, m.p. 180°], a bitter compound elaborated by the crustose lichen *Pertusaria amara* (Ach.) Nyl., has many chemical features that recall griseofulvin. Picrolichenic acid (**129**) affords a phenol (**130**) when heated with hydrobromic acid; zinc chloride dehydrates **130** to the dibenzofuran (**131,** R = H, $R^1 = C_5H_{11}$) and then by methylation and oxidation to the known acid **132**. The C_5 side chain accounts for the oxidation of picrolichenic acid to *m*-caproic acid.

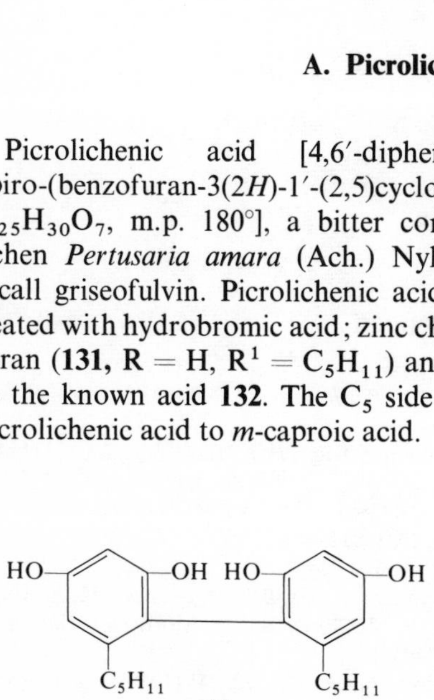

HO OH HO OH C_5H_{11} C_5H_{11}

130

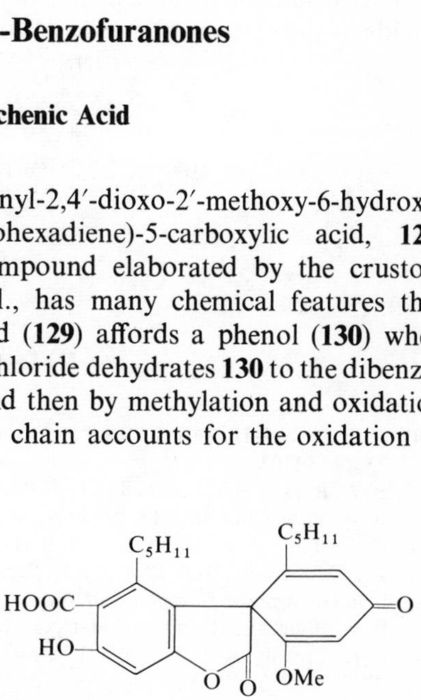

C_5H_{11} C_5H_{11} HOOC O HO O O OMe

129

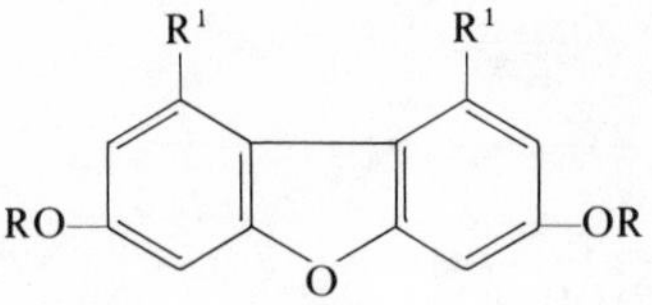

131 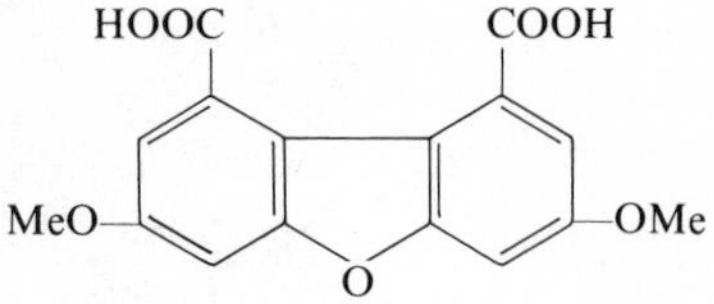 **132**

Compound **129** is decomposed by cold alkali, and evolves carbon dioxide and forms the acid **133** from which the phenyl (**130**) is obtained by hydrobromic acid.[121-123]

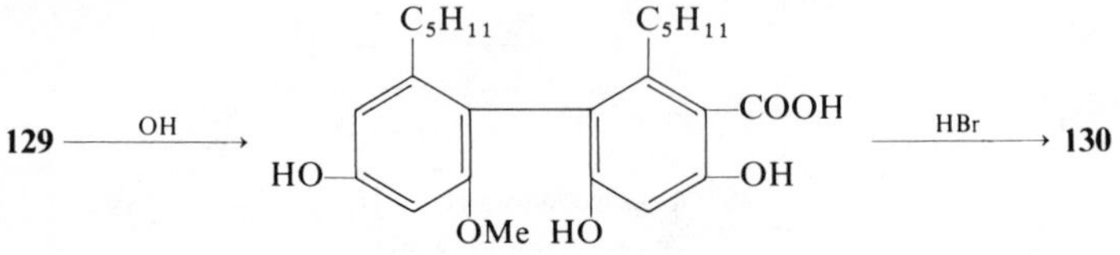

133

The infrared spectrum is evidence for structure **129**. Approaches to the synthesis have so far led to dihydropicrolichenic (**134**), but this acid has not been affected by treatment with alkaline ferricyanide under vigorous conditions. Synthesis of picrolichenic acid (**129**) from the oxidation of the dihydropicrolichenic acid (**134**) has been achieved using manganese dioxide as the oxidant.[124]

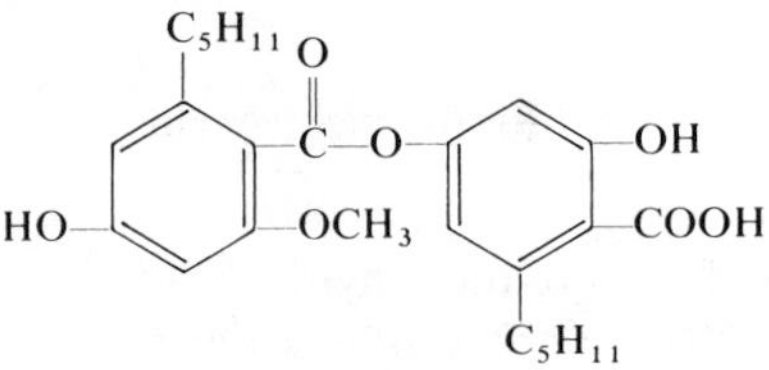

134

References

1. A. E. Oxford, H. Raistrick, and P. Simonart, *Biochem. J.*, **33,** 240 (1939); *Chem. Abstr.*, **33,** 6833 (1939).
2. P. W. Brian, P. J. Curtis, and H. C. G. Hemming, *Trans. Brit. Mycol. Soc.*, **29,** 173 (1946).
3. P. W. Brian, *Ann. Bot.*, **13,** 59 (1949).
4. G. H. Banbury, *J. Expt. Botany*, **3,** 86 (1952); *Chem. Abstr.*, **48,** 3480 (1954).
5. J. C. McGowan, *Trans. Brit. Mycol. Soc.*, **29,** 188 (1946).
6. J. F. Grove and J. C. McGowan, *Nature*, **160,** 574 (1947).
7. P. W. Brian, P. J. Curtis, and H. C. G. Hemming, *Trans. Brit. Mycol. Soc.*, **32,** 30 (1949).
8. S. H. Crowdy, A. P. Green, J. F. Grove, P. McClosky, and A. Morrisen, *Biochem. J.*, **72,** 230 (1959); *Chem. Abstr.*, **53,** 15212 (1959).
9. S. H. Crowdy, J. F. Grove, and P. McClosky, *Biochem. J.*, **72,** 241 (1959); *Chem. Abstr.*, **53,** 15212 (1959).

10. W. B. Whalley, "Progress in Organic Chemistry," Vol. 4, J. W. Cook, Ed., Butterworths, London, 1958, p. 98.
11. F. M. Dean, "Naturally Occurring Oxygen Ring Compounds," Butterworths, London, 1963, p. 153.
12. C. H. Hassall and A. I. Scott, "Recent Developments in the Chemistry of Natural Phenolic Compounds," W. D. Ollis, Ed., Pergamon, London, 1961, p. 122.
13. J. F. Grove, *Quart. Rev. (London)*, **17**(1), 1 (1963).
14. J. F. Grove, J. MacMillan, T. P. C. Mulholland, and M. A. T. Rogers, *J. Chem. Soc.*, **1952,** 3977.
15. J. MacMillan, *J. Chem. Soc.*, **1959,** 1823.
16. J. MacMillan, *J. Chem. Soc.*, **1954,** 2585.
17. A. Brossi, M. Baumann, M. Gerecke, and E. Kyburz, *Helv. Chim. Acta*, **43,** 1444 (1960).
18. Chemical Abstracts, "Index Guide," Vol. 69, July–December, 1968.
19. J. F. Grove, D. Ismay, J. MacMillan, T. P. C. Mulholland, and M. T. Rogers, *J. Chem. Soc.*, **1952,** 3958.
20. J. F. Grove, J. MacMillan, T. P. C. Mulholland, and J. Zealley, *J. Chem. Soc.*, **1952,** 3967.
21. J. F. Grove, J. MacMillan, T. P. C. Mulholland, and M. A. T. Rogers, *J. Chem. Soc.*, **1952,** 3949.
22. J. F. Grove, D. Ismay, J. MacMillan, and M. A. T. Rogers, *Chem. Ind. (London)*, **1951,** 219.
23. L. A. Duncanson, J. F. Grove, J. MacMillan, and T. P. C. Mulholland, *J. Chem. Soc.*, **1957,** 3555.
24. W. R. Logan and G. T. Newbold, *J. Chem. Soc.*, **1956,** 4980.
25. A. W. Dawkins and T. P. C. Mulholland, *J. Chem. Soc.*, **1959,** 1830.
26. J. MacMillan, T. P. C. Mulholland, A. W. Dawkins, and G. Ward, *J. Chem. Soc.*, **1954,** 429.
27. A. W. Dawkins and T. P. C. Mulholland, *J. Chem. Soc.*, **1959,** 2203.
28. P. McCloskey, *J. Chem. Soc.*, **1958,** 4732.
29. F. M. Dean and K. Manunapichu, *J. Chem. Soc.*, **1957,** 3112.
30. J. MacMillan and P. J. Suter, *J. Chem. Soc.*, **1957,** 3124.
31. T. P. C. Mulholland, *J. Chem. Soc.*, **1952,** 3987.
32. T. P. C. Mulholland, *J. Chem. Soc.*, **1952,** 3994.
33. E. Kyburz and A. Brossi, *Angew. Chem.*, **10,** 84 (1961).
34. E. Kyburz, H. Geleick, J. R. Fray, and A. Brossi, *Helv. Chim. Acta*, **43,** 2083 (1960).
35. H. Newman, *J. Org. Chem.*, **35,** 3990 (1970); ibid, **36,** 2375 (1971).
36. Th. L. Fields and R. B. Angier, *J. Med. Chem.*, **13,** 1242 (1970); *Chem. Abstr.*, **74,** 22628 (1971).
37. A. Brossi, M. Baumann, M. Gerecke, and E. Kyburz, *Helv. Chim. Acta*, **43,** 2071 (1960).
38. M. Gerecke, E. Kyburz, C. v. Planta, and A. Brossi, *Helv. Chim. Acta*, **45,** 2241 (1962).
39. A. Brossi, M. Baumann, and F. Burkhardt, *Helv. Chim. Acta*, **45,** 1292 (1962).
40. C. H. Kuo, R. D. Hoffsommer, H. L. Slates, D. Taub, and N. L. Wendler, *Chem. Ind. (London)*, **1961,** 1627.
41. D. Taub, C. H. Kuo, and N. L. Wendler, *Chem. Ind. (London)*, **1962,** 1617; *Chem. Abstr.*, **57,** 15040 (1962).
42. D. Taub, N. L. Wendler, and C. H. Kuo, U.S. Pat. 3,254,113 (1966); *Chem. Abstr.*, **65,** 15326 (1966).
43. D. Taub, C. H. Kuo, and N. L. Wendler, *Chem. Ind. (London)*, **1962,** 557.
44. G. Stork and M. Tomasz, *J. Am. Chem. Soc.*, **84,** 310 (1962).
45. G. Stork and M. Tomasz, *J. Am. Chem. Soc.*, **86,** 471 (1964).
46. A. C. Day, J. Nabney, and A. I. Scott, *J. Chem. Soc.*, **1961,** 4067.
47. A. C. Day, J. Nabney, and A. I. Scott, *Proc. Chem. Soc. London*, **1960,** 284.
48. A. J. Birch, R. W. Rickards, and H. Smith, *Proc. Chem. Soc.*, **1958,** 98.

49. A. J. Birch, "Biosynthetic Relationship of Some Natural Phenolic and Enolic Compounds,"in "Fortschritte der Chemie Organischer Naturstoffe," L. Zechmeister, Ed. Wien-Springer-Verlag, 1951, p. 186.
50. W. J. McMaster, A. I. Scott, and S. Trippett, *J. Chem. Soc.*, **1960,** 4628.
51. A. Rhodes, B. Boothroyd, M. P. McGonagle, and G. A. Somerfield, *Biochem. J.*, **81,** 28 (1961).
52. A. I. Scott, *Proc. Chem. Soc.*, **1958,** 195.
53. L. A. Duncanson, J. F. Grove, and P. W. Jeffs, *J. Chem. Soc.*, **1958,** 2929.
54. G. I. Gregory, P. J. Holton, H. Robinson, and T. Walker, *J. Chem. Soc.*, **1962,** 1269.
55. L. Stephenson, T. Walker, W. K. Warburton, and G. B. Webb, *J. Chem. Soc.*, **1962,** 1282.
56. N. Greenhalgh, T. Leigh, and W. A. Sexton, Brit. Pat. 913,535 (1962); *Chem. Abstr.*, **58,** 12514 (1963).
57. T. Walker, W. K. Warburton, and J. B. Webb, *J. Chem. Soc.*, **1962,** 1777.
58. H. Newman, *J. Heterocyc. Chem.*, **7**(4), 957 (1970).
59. W. W. Andres, W. J. McGahren, and M. P. Kunstmann, *Tetrahedron Lett.*, **1969,** 3777.
60. H. Newman and R. B. Angier, *J. Org. Chem.*, **34,** 1463 (1969).
61. D. Taub and N. L. Wendler, *Angew Chem.*, **74,** 586 (1962); *Chem. Abstr.*, **57,** 11139 (1962).
62. D. Taub, C. H. Kuo, and N. L. Wendler, *J. Org. Chem.*, **28,** 3344 (1963).
63. B. Boothroyd, E. J. Napier, and G. A. Somerfield, *Biochem. J.*, **80,** 34 (1961); *Chem. Abstr.*, **55,** 26102 (1961).
64. J. MacMillan, *Chem. Ind. (London)*, **1951,** 719.
65. J. MacMillan, *J. Chem. Soc.*, **1953,** 1697.
66. V. Arkley, J. Attenburrow, G. I. Gregory, and T. Walker, *J. Chem. Soc.*, **1962,** 1260.
67. T. Tanaka, *Chem. Pharm. Bull. (Tokyo)*, **12,** 214 (1964); *Chem. Abstr.*, **60,** 14459 (1960).
68. A. W. Dawkins and T. P. C. Mulholland, *J. Chem. Soc.*, **1959,** 2211.
69. R. F. Curtis, C. H. Hassall, and D. W. Jones, *J. Chem. Soc.*, **1965,** 6960.
70. F. M. Dean, T. Francis, and K. Manunapichu, *J. Chem. Soc.*, **1958,** 4551.
71. D. S. Deorha and P. Gupta, *Tetrahedron*, **22,** 2015 (1966).
72. A. M. Choudhury, K. Schofield, and R. S. Ward, *J. Chem. Soc., C*, **1970,** 2543.
73. D. H. R. Barton, Y. L. Chow, A. Cox, and G. W. Kirby, *J. Chem. Soc.*, **1965,** 3571.
74. J. E. Page and S. E. Staniforth, *J. Chem. Soc.*, **1962,** 1292.
75. S. H. Crowdy, D. Gardner, J. F. Grove, and D. Pramer, *J. Expt. Bot.*, **6,** 371 (1955); *Chem. Abstr.*, **50,** 9524 (1956).
76. C. Bedford, K. J. Child, and E. G. Tomich, *Nature*, **184,** Suppl. No. 6, 364 (1959); *Chem. Abstr.*, **54,** 15512 (1960).
77. G. C. Ashton and A. P. Brown, *Analyst,* **81,** 220 (1956); *Chem. Abstr.*, **50,** 9683 (1956).
78. G. C. Ashton and J. P. R. Tootill, *Analyst*, **81,** 225 (1956); *Chem. Abstr.*, **50,** 9683 (1956).
79. C. Daly, *Analyst*, **86,** 129 (1961).
80. G. C. Ashton, *Analyst*, **81,** 228 (1956); *Chem. Abstr.*, **50,** 9683 (1956).
81. H. Laubie, *Bull. Soc. Pharm. Bordeaux*, **99,** 4 (1960); *Chem. Abstr.*, **54,** 17793 (1960).
82. H. Kadner, H. Jacobi, and S. Eagel, *Pharmazie*, **26,** 94 (1971); *Chem. Abstr.*, **74,** 138921 (1971).
83. E. W. Bassett and S. W. Tanenbaum, *Experientia*, **14,** 38 (1958); *Chem. Abstr.*, **52,** 12985 (1958).
84. P. Simonart, A. Wiaux, and H. Verachtert, *Bull. Soc. Chim. Biol.*, **41,** 541 (1959); *Chem. Abstr.*, **53,** 18166 (1959).
85. A. J. Birch, R. A. Massy-Westropp, and R. W. Rickards, *J. Chem. Soc.*, **1956,** 3717.
86. F. M. Dean, R. A. Eade, and R. A. Moubasher, *J. Chem. Soc.*, **1957,** 3497.
87. A. J. Birch, R. A. Massy-Westropp, R. W. Rickards, and H. Smith, *J. Chem. Soc.*, **1958,** 360.

88. R. W. Rickards, "Chemistry of Natural Phenolic Compounds," W. D. Ollis, Ed., Pergamon, Oxford, **1961,** p. 1.
89. M. J. Barnes and B. Boothroyd, *Biochem. J.*, **78,** 41 (1961).
90. M. T. J. Abbot and J. F. Grove, *Expt. Cell Res.*, **17,** 105 (1959).
91. J. F. Grove, P. W. Jeffs, and D. W. Rustidge, *J. Chem. Soc.*, **1956,** 1956.
92. J. M. Wright and J. F. Grove, *Ann. Appl. Biol.*, **43,** 288 (1955).
93. S. Tomomatsu and J. Kitamura, *Chem. Pharm. Bull. (Tokyo)*, **8,** 755 (1960); *Chem. Abstr.*, **55,** 14565 (1961).
94. S. H. Crowdy, J. F. Grove, H. Hemming, and K. C. Robinson, *J. Expt. Bot.*, **7,** 42 (1956).
95. A. H. Campbell, "Experimental Chemotherapy", Vol. 3, R. J. Schnitzer and F. Hawking, Eds., Academic, New York, 1964, p. 461.
96. K. J. Bent and R. H. Moore, "Biochemical Studies of Antimicrobial Drugs," B. A. B. Newton and P. E. Reynolds, Eds., *16th Symposium of the Society for General Microbiology*, Cambridge University Press, 1966, p. 82.
97. F. M. Huber, "Antibiotics," Vol. 1, D. Gottlieb and P. D. Shaw, Eds., Springer, Berlin, 1967, p. 181.
98. J. C. Gentles, *Nature*, **182,** 476 (1958).
99. A. Martin, *J. Invest. Derm.*, **32,** 525 (1959).
100. J. C. Gentles and M. J. Barnes, *Arch. Dermatol.*, **81,** 703 (1960).
101. R. Brown, "Experimental Chemotherapy," Vol. 3, R. J. Schnitzer and H. Hawking, Eds., Academic, New York, 1964, p. 418.
102. H. M. Robinson, Jr., *Med. Clin. North. Am.*, **51,** 1181 (1967).
103. E. D. Weinberg, "Medicinal Chemistry," Part I, A. Burger, Ed., Interscience, New York, 1970, p. 606.
104. G. E. Paget and A. L. Walpole, *Nature*, **182,** 1320 (1958); *Chem. Abstr.*, **53,** 9485 (1959).
105. P. W. Brian, J. M. Wright, J. Stubbs, and A. M. Way, *Nature*, **167,** 347 (1951); *Chem. Abstr.*, **45,** 5351 (1951).
106. A. Rhodes, R. Cross, R. McWilliam, J. P. R. Toothill, and A. T. Dunn, *Ann Appl. Biol.*, **45,** 215 (1957); *Chem. Abstr.*, **51,** 15860 (1957).
107. P. W. Brian, *Trans. Brit. Mycol.*, **43,** 1 (1960).
108. A. Rhodes, in "Antibiotics in Agriculture," *Proc. of the University of Nottingham, 9th Easter School in Agriculture Science,* 1962, p. 101.
109. A. Stokes, *Plant Soil*, **5,** 132 (1954); *Chem. Abstr.*, **48,** 8341 (1954).
110. R. E. Engel, K. J. McVeigh, R. M. Schmit, and R. W. Duell, *Proc. Northeast. Weed Sci. Soc.*, **25,** 131 (1971); *Chem. Abstr.*, **74,** 139851 (1971).
111. T. P. C. Mulholland, and (in part) R. I. Honeywood, H. D. Preston, and D. T. Rosevear, *J. Chem. Soc.*, **1965,** 4939.
112. R. Cross, R. McWilliam, and A. Rhodes, *J. Gen. Microbiol.*, **34,** 51 (1964); *Chem. Abstr.*, **61,** 8662 (1964).
113. H. Raistrick and G. Smith, *Biochem. J.*, **30,** 1315 (1936); *Chem. Abstr.*, **30,** 8215 (1936).
114. P. W. Clutterbuck, W. Köerber, and H. Raistrick, *Biochem. J.*, **31,** 1089 (1937); *Chem. Abstr.*, **32,** 124 (1938).
115. C. T. Calam, P. W. Clutterbuck, A. F. Oxford, and H. Raistrick, *Biochem. J.*, **33,** 579 (1939); *Chem. Abstr.*, **33,** 7768 (1939).
116. C. T. Calam, P. W. Cutterbuck, A. E. Oxford, and H. Raistrick, *Biochem. J.*, **41,** 458 (1947); *Chem. Abstr.*, **42,** 4580 (1948).
117. E. Komatsu, *Nippon-Kagaku*, **31,** 349 (1957); *Chem. Abstr.*, **52,** 16473 (1958).
118. D. H. R. Barton and A. I. Scott, *J. Chem. Soc.*, **1958,** 1767.
119. C. H. Hassall and A. I. Scott, "Recent Developments in the Chemistry of Natural Phenolic Compounds," W. D. Ollis, Ed., Pergamon, London, 1961, p. 129.
120. H. W. Florey, E. Chain, N. G. Heatley, M. A. Jennings, A. G. Sanders, E. P. Abraham, and M. E. Florey, "Antibiotics," Vol. 2, Oxford University Press, London, 1949, p. 1520.

121. C. A. Wachtmeister, *Acta Chem. Scand.*, **12,** 147 (1958); *Chem. Abstr.*, **53,** 18935 (1959).
122. H. Erdtman and C. A. Wachtmeister, *Fortsch. A. Stoll*, Basle, Birkhauser, 1957, p. 144.
123. H. Erdtman and C. A. Wachtmeister, *Chem. Ind. (London)*, **1957,** 1042; *Chem. Abstr.*, **52,** 1114 (1958).
124. T. A. Davidson and A. I. Scott, *Proc. Chem. Soc.*, **1960,** 390; *Chem. Abstr.*, **55,** 9336 (1961).
125. A. Brossi, M. Gerecke, and E. Kyburz, Swiss. Pat. 383,994 (1965); *Chem. Abstr.*, **62,** 16193 (1965).
126. J. MacMillan, *J. Chem. Soc.*, **1953,** 1697.
127. R. F. Curtis, C. H. Hassall, D. W. Jones, and T. S. Wheeler, *J. Chem. Soc.*, **1961,** 4838.
128. F. M. Dean and J. C. Knight, *J. Chem. Soc.*, **1962,** 4745.
129. F. Hoffman-La Roche & Co., A.-G., Brit. Pat. 914,502 (1963); *Chem. Abstr.*, **59,** 1592 (1963).
130. J. S. Morley, Brit. Pat. 921,665 (1963); *Chem. Abstr.*, **59,** 7491 (1963).
131. F. M. Dean, P. Halewood, S. Mongkolsuk, A. Robertson, and W. B. Whalley, *J. Chem. Soc.*, **1953,** 1250.

CHAPTER VIII

Less Common and Modified Naturally Occurring Benzofurans

1. Less Common Benzofurans

A. Menthofuran

While some of the benzofurans are plant products, a few are of fungal origin. Menthofuran can be prepared in the laboratory from terpenoid olefins, (for example, pulegone by epoxidation and rearrangement, which might therefore be a possible mode). Menthofuran, evodone, and so on, are considered terpenoid substances like furans, yet they can be considered to be tetrahydrobenzofuran derivatives. Menthofuran, $C_{10}H_{14}O$,(1), is isolated from the flower heads and leaves of peppermint, *Mentha piperita vulgaris* S. Charles[1] regarded menthofuran as an oxide; but Weinhaus et al.,[2-4], on the grounds that vigorous oxidation of **1** gave β-methyladipic acid and vigorous reduction yielded p-menthane, assigned structure **1**. According to Bedoukian [5] it is best characterized as the maleic anhydride adduct.[6] Menthofuran can be hydrogenated on a platinum catalyst to tetrahydromenthofuran (3,6-dimethyloctahydrobenzofuran, **2**); treatment with zinc chloride and hydrogen chloride yielded a dichloride (**3**) which in turn yielded *p*-menthadiene (**4**).[3]

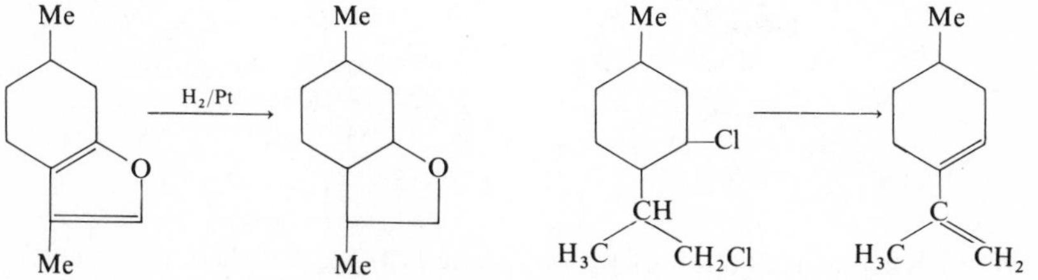

1 **2** **3** **4**

Furthermore, Tagaki and Litsui[13] have shown that direct hydrogenation of menthofuran afforded not only **2**, but also a mixture of (+)-neomenthol and (+)-neoisomenthol with structure **5**. The hydroxyl and isopropyl groups were cis to each other.

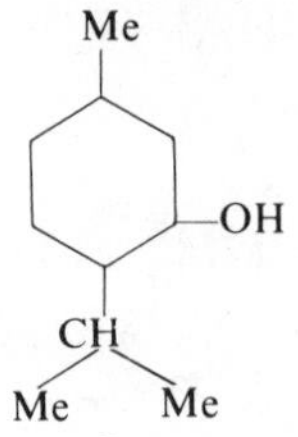

5

The first synthesis of menthofuran, by Triebs,[8-9] involved treating pulegone (**6**) with sulfuric acid in cold acetic acid medium, so as to produce the cyclic enol sulfonate (**7**); when heated, this lost sulfur dioxide giving menthofuran almost quantitatively.

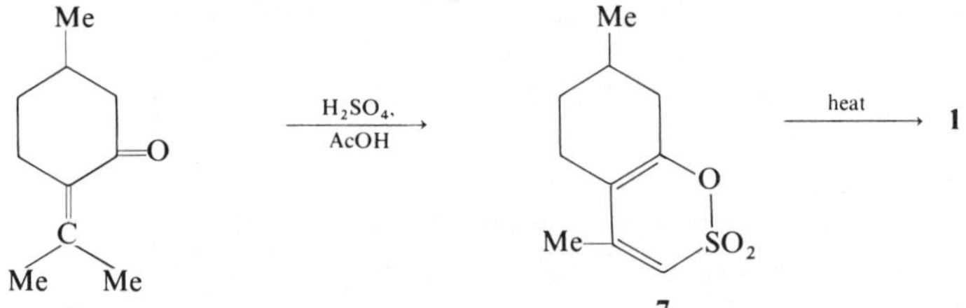

6 —H_2SO_4, AcOH→ **7** —heat→ **1**

The syntheses in which isopulegone (**8**) is converted by perbenzoic acid into the oxide (**9**) and then, by hot dilute acid, into menthofuran (presumably via a ketonic aldehyde), is notable because of its possible implications for biosynthesis.[10]

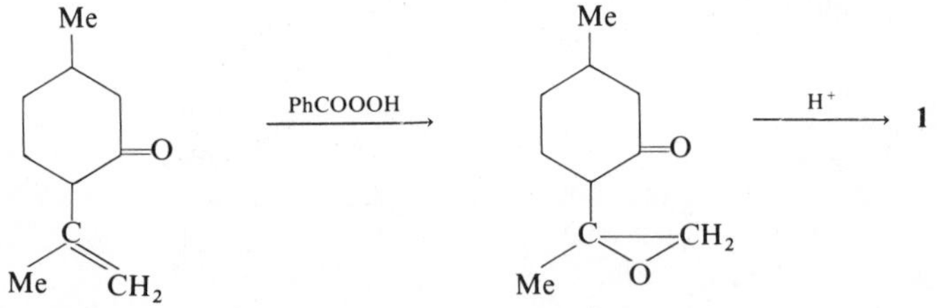

8 —PhCOOOH→ **9** —H^+→ **1**

Wolff-Kishner reduction of (±)-evodone (**10**)[7] established a convenient method for the preparation of menthofuran.

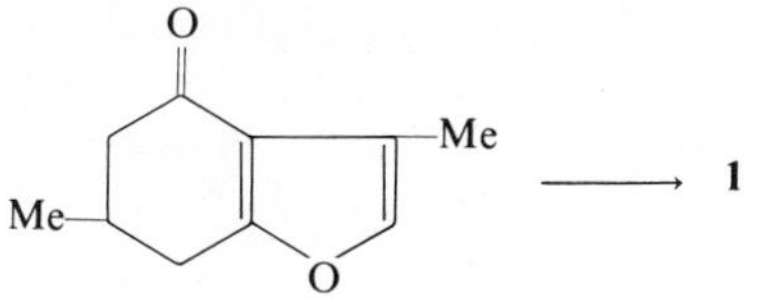

10 → 1

Autooxidation of menthofuran led to the formation of piperic acid (**11**)[4, 11] When kept in air, menthofuran turns yellow, green, and finally brownish, but in acetic acid it turns blue rapidly. Schenck and Foote[12] noted that in methanol a peroxide of type **12** resulted which generated piperic acid (**11**) by ready loss of methanol. Reduced by triphenylphosphine, the peroxide (**12**) gave a new substance. The structure of this substance is not known, but it can also be made by hydrolyzing the ether (**13**) obtained by subjecting menthofuran to Clausen-Kaas reagent, (bromine and methanol).[14]

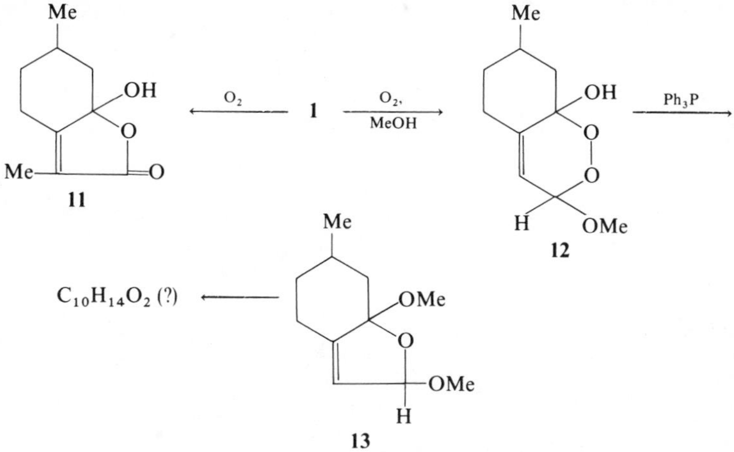

11 ←(O_2)— 1 —(O_2, MeOH)→ 12 —(Ph_3P)→

$C_{10}H_{14}O_2$ (?) ← 13

B. Evodone

Evodone, $C_{10}H_{12}O_2$(14), is isolated from *Evodia hortensis* Forst.[15,16] The relation to menthofuran (**1**) is obvious. In addition to a carbonyl group, the substance contains one unreactive oxygen atom, and as it can be hydrogenated to a ketonic tetrahydro derivative, two double bonds. The ultraviolet band at 266 mμ (log E, 3.60) shows that the system is conjugated. Ozonolysis of evodone generates compound (**15**) which has all the properties, including the fine reaction, of the triketone formed when dihydroorcinol (**16**) is acetylated with acetic anhydride and sodium acetate. No direct identification of the two has been reported, however.

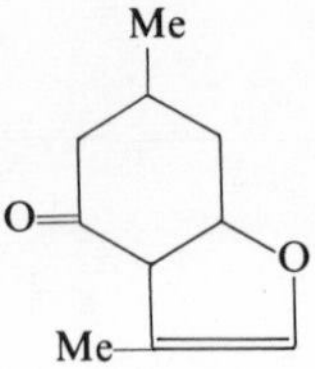

Condensation of 1-methyl-3, 5-cyclohexandione with ethyl α-chloroaceto-acetate resulted in the formation of ethyl 3,6-dimethyl-4-oxo-4,5,6,7-tetra-hydrobenzofuran-2-carboxylate (**17**); saponification of the latter yielded the corresponding acid (**18**). The acid is separable into optical isomers via the cinchonidine salt. Decarboxylation of 1–**18** gave **14**.[17]

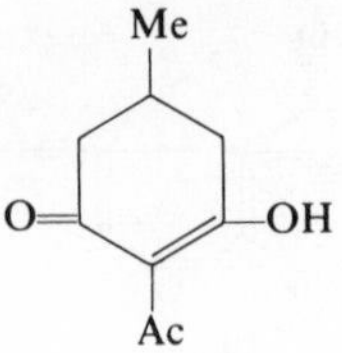

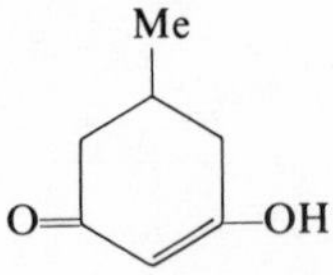

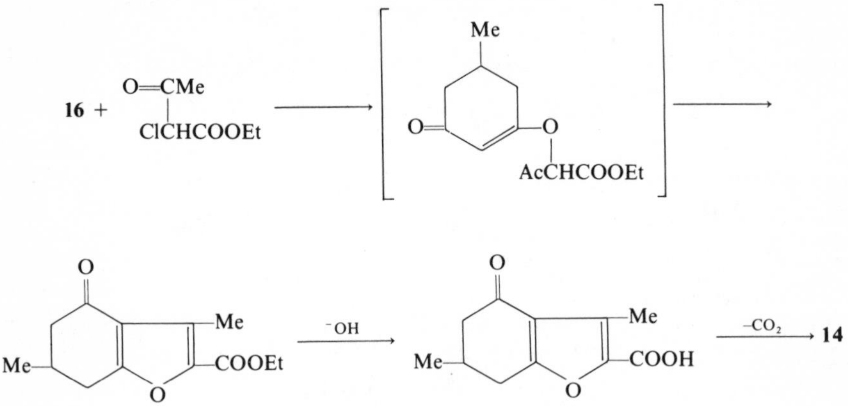

C. Bisabolangelone

Bisabolangelone, $C_{15}H_{12}O_3$, was isolated from freshly ripened *Angelica sylvestris* L. seeds.[18, 19] The infrared spectrum shows α,β-unsaturated ketone and hydroxyl groups. On hydrogenation in ethyl acetate in presence of $Pd/SrCO_3$, compound **19** is obtained. Autooxidation of bisabolangelone (**20**) by exposure to daylight and air gave the main product **21**, $C_9H_{10}O_2$, identified as 2-hydroxy-4-methylacetophenone. Furthermore, the hydroxy acid, $C_6H_{10}O_3$, was identified as 4-methyl-4-hydroxypenten-2-oic acid (**22**). Besides the formation of **21** and **22**, *m*-cresol has been isolated in small amount from the eluates.

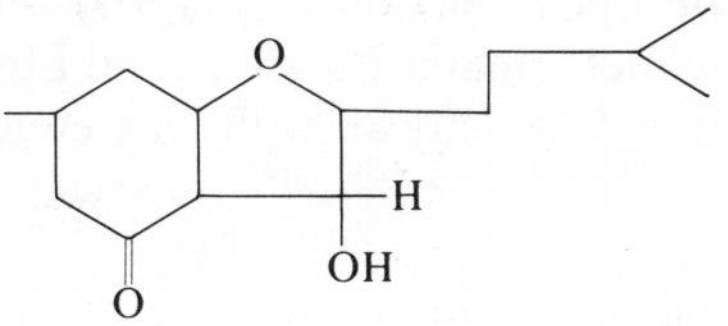

19

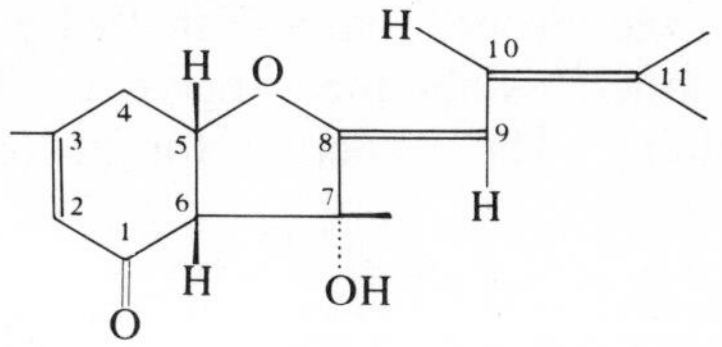

20

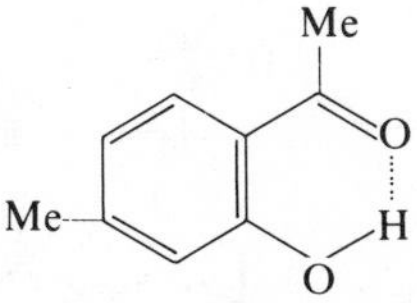

21

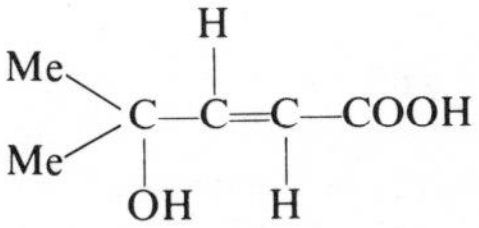

22

Degradation of the hexahydro derivative (**19**) in alkaline medium gave *m*-cresol as the main product and lower amounts of **21** after pyrolysis of **20**. Further information was obtained after ozonolysis of **20**, which afforded acetone, and after oxidation with hydrogen peroxide oxalic acid is also obtained. Detailed study of the NMR spectra of **20** and **19**, using the solvent effect in C_6D_6 and field-swept decoupling experiments, illustrate the relative configuration of bisabolangelone as shown in **20**.[20]

D. Curzerenone, Epicurzerenone, and Isofuranogermacrene

From the rhizome of Zedoary (*Curcuma zedoaria* Roscoe, Zingiberaceae), three new furan-containing sesquiterpenoids designated as curzerenone [6,7-dihydro-5β-isopropenyl-3,6β-dimethyl-6-vinyl-4(5*H*)-benzofuranone, **23**], epicurzerenone [6,7-dihydro-5α-isopropenyl-3,6β-dimethyl-6-vinyl-4 (5*H*)-benzofuranone, **24**], and isofuranogermacrene (curzene, 4,5,6,7-tetrahydro-5-isopropenyl-3,6-dimethyl-6-vinylbenzofuran, **25**) were isolated.[21,22] The formulae are expressed in terms of the enantiomeric pairs.[23]

Curzerenone, (**23**, $C_{15}H_{18}O_2$, $(\alpha)_D \pm 0°$), as revealed by its ultraviolet, infrared, and NMR spectra, contains a carbonyl group involved in a β-furoyl system, an α-hydrogen and a β-methyl on a furan ring, a methylene flanked by quaternary carbons and weakly coupled with the α-hydrogen of the furan, a tertiary methyl, a vinyl on a quaternary carbon, an isopropenyl, and an isolated hydrogen. Partial hydrogenation of curzerenone over Raney nickel in methanol resulted in the saturation of the vinyl group, yielding the dihydro derivative (**26**), which was further hydrogenated with platinum in methanol to the tetrahydro derivative (**27**). Curzerenone must thus be represented as **23** (disregarding the stereochemistry).[24]

When methanolic extract of *Curcuma zedoaria* is subjected to steam distillation and the distillate is chromatographed on silica, a new sesquiterpene, zedoarone, 6-vinyl-3-methyl-5-isopropenyl-6,7-dihydro-4(5*H*)-benzofuranone,[24] is obtained. It gives positive Liebermann-Burchard and Ehrlich reactions. The structure for zedoarone is identical with curzerenone.

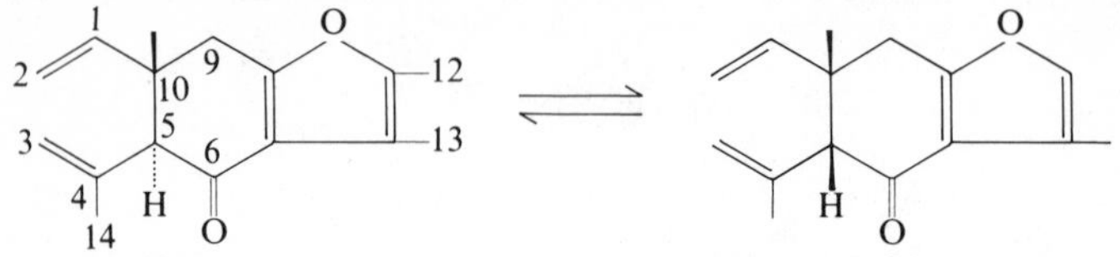

23 ⇌ **24**

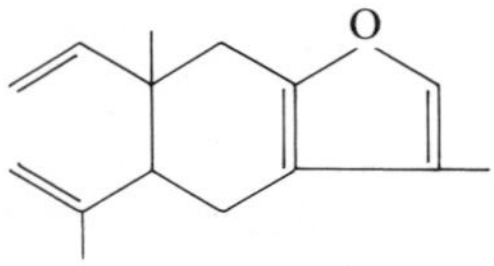

25

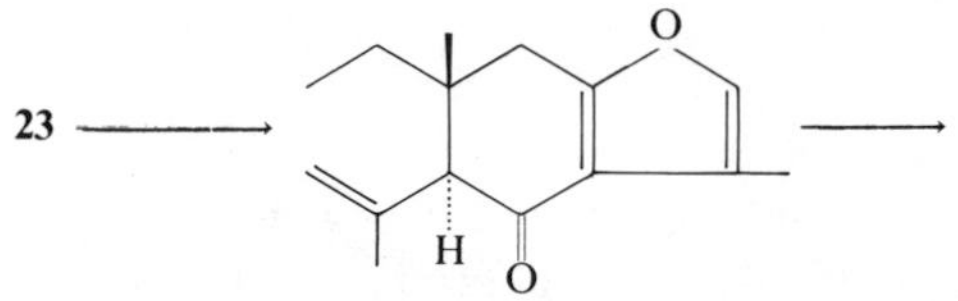

23 ⟶ **26** 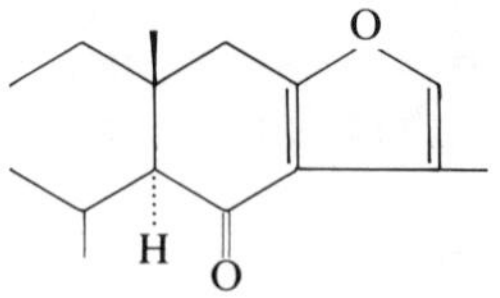 ⟶ **27**

Epicurzerenone ($C_{15}H_{18}O_2$, **24**, $(\alpha)_D \pm 0°$), possesses very similar spectral properties to those of curzerenone. Furthermore, epicurzerenone was also converted into the dihydro derivative **28** and the tetrahydro derivative (**29**). These observations led Hikino et al.[23] to conclude that substances might be epimeric in the position alpha to the carbonyl function, (i.e., C-5). As confirmation, alkali treatment of curzerenone and epicurzerenone led to the same equilibrium mixture of the same substances in both cases.

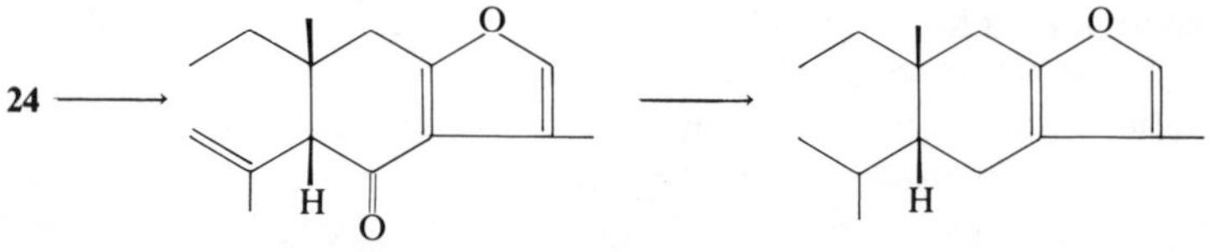

24 ⟶ **28** ⟶ **29**

Since **23** and **24** show no optical activity, both substances are racemic. The relative configurations at C-5 and C-10 are obtained by study of their NMR spectra.[23]

Although epicurzerenone has been obtained from the extract of zedoary, it may be an artifact formed from curzerenone during the process of isolation, rather than a natural product present in the plant.

Isofuranogermacrene (curzerene)[22,23,25] has an NMR spectrum closely resembling that of the curzerenones. Its empirical formula, $C_{15}H_{20}O$, $(\alpha)_D \pm 0°$, contains one less oxygen atom than that of curzerenone. This is accounted for by its ultraviolet and infrared spectra, which exhibit the absence of a conjugated carbonyl system. The NMR spectrum indicates the following elements of structure: an α-hydrogen and a β-methyl on a furan ring, a tertiary methyl, a vinyl on a quaternary carbon, an isopropenyl, and five allylic hydrogens. On oxidation with 2,3-dichloro-5,6-dicyano-p-benzoquinone, isofuranogermacrene yielded the conjugated lactone **30**. This exhibits in its NMR spectrum a signal because of a newly vinyl hydrogen on C-9 as a singlet, demonstrating the quaternary nature of the adjacent carbon (C-10). It is concluded from these facts that isofuranogermacrene must have structure **25**, or else the isopropenyl group at C-5 could be at C-6. The conclusion that the isopropenyl group is at C-5 was confirmed by the finding that pyrolysis of furanodiene (**31**), another constituent of zedoary, resulted in a Cope rearrangement giving isofuranogermacrene (curzerene).[25]

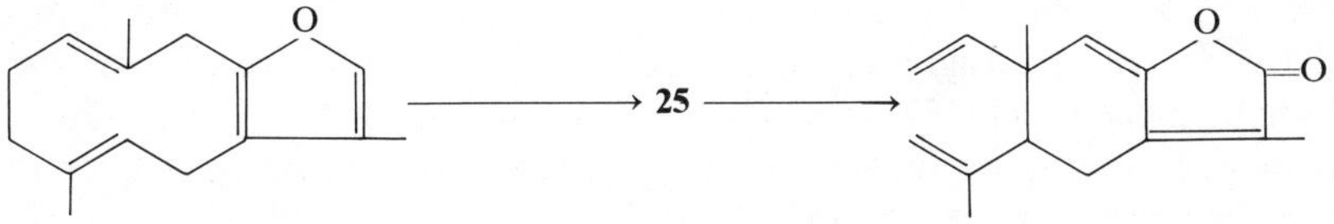

O → **25** → O, O

31 **30**

The mass spectrum of **25** showed the expected molecular peak at *m/e* 216. The peak appears at *m/e* 108 because of a fragmentation of a retro Diels-Alder type reaction, indicating that the isopropenyl group is not situated at C-6, but at C-5.

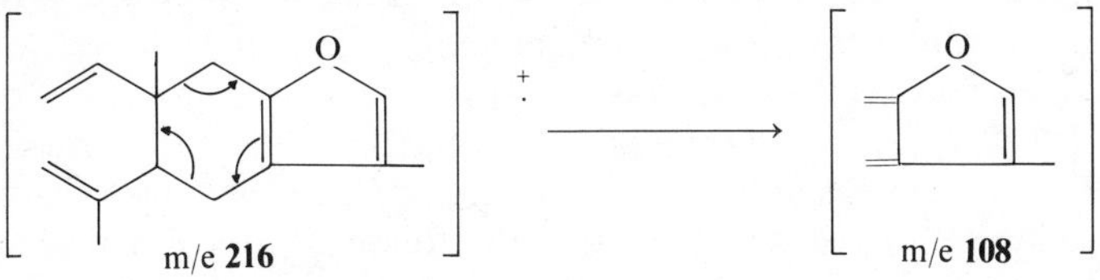

m/e **216** → m/e **108**

It is believed, at present, that **25** is a natural product, since heat sufficient to cause the Cope rearrangement was not used at any stage of the isolation.[25-27]

E. Isosericenin, Sericealactone, and Deoxysericealactone Methyl Esters

From the essential oil that was obtained from the leaves of *Neolitsea sericea* Koidz. (Japanese name, Shrimodamo), by steam distillation, the methyl esters of new elemane-type sesquiterpene acids containing a furan ring, namely, isosericenin (**32**),[28] sericealactone (**33**),[29] and deoxysericealactone (**34**)[29] were isolated.

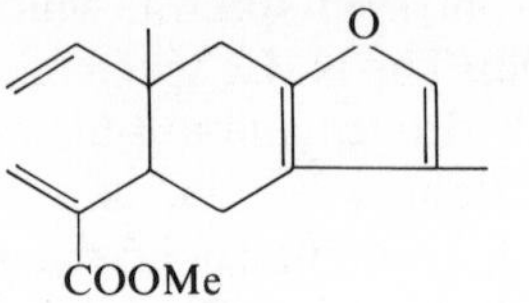

Isosericenin (methyl ester of 1-vinyl-α-methylene-5,9-epoxy-*p*-mentha-4, 8-diene-2-acetic acid, **32**) was isolated as an oily substance, $C_{16}H_{20}O_3$, that is very susceptible to autooxidation in air. The mass spectrum showed that the molecule contains a β-methylfuran ring, vinyl and tertiary methyl groups, together with a terminal double bond attached to a methoxycarbonyl group.[28] It showed a positive color reaction for the Ehrlich and Liebermann-Burchard tests, and produced maleic anhydride adduct; all of which illustrate the presence of a furan system.

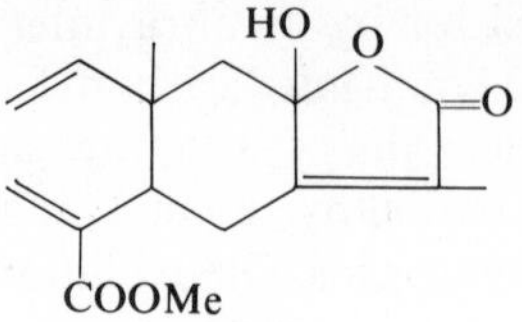

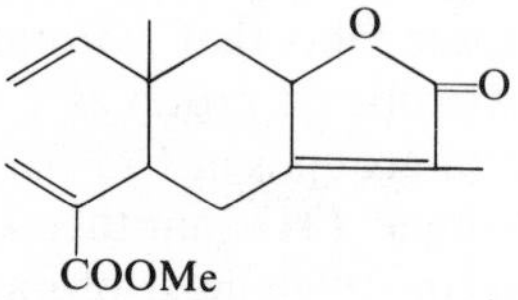

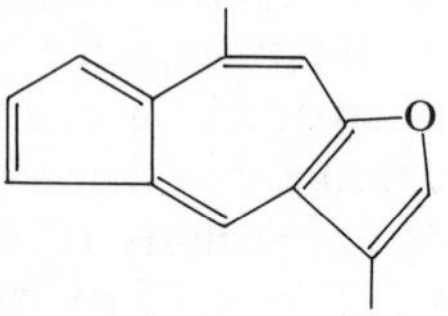

When **32** was hydrogenated over Adams catalyst in acetic acid medium, it absorbed hydrogen equivalent to two molecules to produce tetrahydroisosericenin while leaving the furan intact. Selenium dehydrogenation produced ujacazulene (**35**); thus α-**32** represents isosericenin.

The NMR spectrum of sericealactone (methyl ester of 6-vinyl-2-oxo-α-methylene-3,6-dimethyl-7α-hydroxy-2,4,5,6,7,7α-hexahydro-5-benzofuranacetic acid, **33**, $C_{16}H_{20}O_5$), revealed the following functional groups: an angular methyl group, an ABX-type vinyl group attached to a tertiary carbon atom, methine and methylene groups, a methyl group on a double bond, and a methoxycarbonyl group conjugated with a terminal methylene group. In addition to these groups, sericealactone contained a hydroxy group and α, β-unsaturated γ-lactone system. On the other hand, sericealactone is identical with the autooxidation product of isosericenin in contact with oxygen together with Adams catalyst. Saponification of **33** produced acid **36**; catalytic hydrogenation by using Adams catalyst in acetic acid produced a saturated ester (**37**), and dehydration with thionyl chloride gave an anhydro compound (**38**).

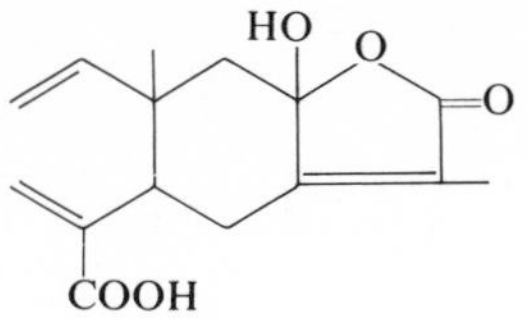

HO O =O COOH 36

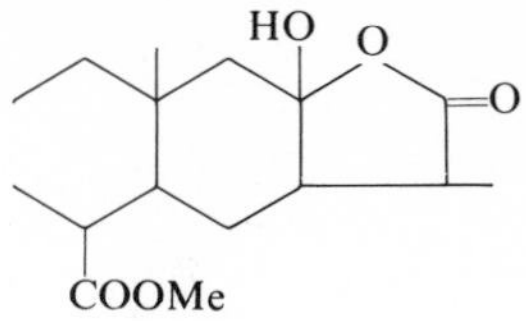

HO O =O COOMe 37

O =O COOMe 38

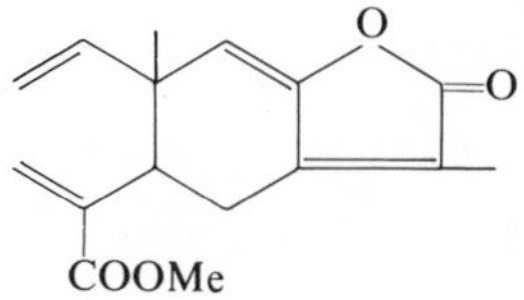

Although the NMR spectrum of deoxysericealactone (methyl ester of 6-vinyl-2-oxo-α-methylene-3,6-dimethyl-2,4,5,6,7,7α-hexahydro-5-benzofuranacetic acid, **34**, $C_{16}H_{20}O_4$, did not have a hydroxyl signal, it could be superposed almost exactly on that of sericealactone. The integral strength of the proton signals in the methylene and methine proton region correspond to 6*H* (in contrast to 5*H*) for sericealactone (**33**). The mass spectrum exhibited the parent ion at *m/e* 276 which is smaller by 16 mass units than the molecular weight of sericealactone. Its structure is represented by formula **34** as the deoxy compound of sericealactone.

F. (±)-Loliolide [(±)-Digiprolactone]

Loliolide [5,6,7,7α-Tetrahydro-6β-hydroxy-4,4,7α,β-trimethyl-2(4*H*)-benzofuranone], is isolated from *Lolium perenne*[30] and has been shown to have structure **39**. This compound proved to be identical with a substance obtained from *Fumaria officinalis*,[31] and from *Digitalis lanata*,[32] as well as with digiprolactone from *Digitalis purpurea* (which was independently found to possess structure **39**.[33-36] The latter has been synthesized from isophorone **40** by two routes.[32] The first utilized the oxidative lactonization of the mono-unsaturated acetoxy ester (**41b**) with selenium dioxide as a key step. The second involved the cyclization of the doubly unsaturated keto ester (**41a**) with concentrated sulfuric acid. The synthetic racemic compound proved to be identical with a sample prepared from natural optically active loliolide.

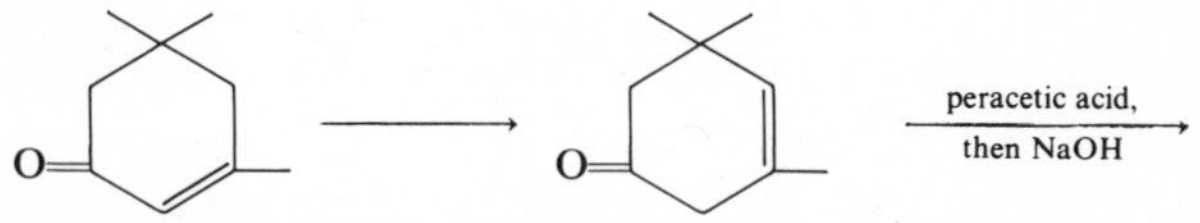

O= **40** ⟶ O= —peracetic acid, then NaOH⟶

-OH $\xrightarrow{MeC_6H_4SO_3H\text{-}p}$ =O $\xrightarrow[(CH_2OH)_2]{MeC_6H_4SO_3H,}$

=O $\xrightarrow[\text{acetylide}]{\text{Lithium ethoxy-}}$ OH C≡COEt $\xrightarrow{\text{dil. } H_2SO_4}$

=CHCOOEt $\xrightarrow{NaBH_4}$ RO— =COOEt $\xrightarrow[AcOH]{SeO_2,}$

41a R = H
41b R = Ac

HO— =O

39

Oxidation of natural loliolide with chromium trioxide has been shown to give rise to the optically active ketone **43**,[30,34,36] which could be reconverted to loliolide by the action of sodium borohydride.[36] Furthermore, the optically active ketone **43** is racemized on treatment with sodium hydroxide and subsequent acidification.[30] The fact that the optically active ketone **43** is racemized with base, presumably via the salt of the *cis*-isomer of the dienone acid (**42b**),[30] suggested the following synthesis:[32]

-OH $\xrightarrow[\text{Jones reagent}]{CrO_3}$ =O $\xrightarrow[MeC_6H_4SO_3H\text{-}p]{(CH_2OH)_2,}$

=O

lithium ethoxy-acetylide →

OH, C≡COEt, O, O

dil. H_2SO_4 →

=CHCOOR, O= →

O=, =O, O

$NaBH_4$ ⇌ **39**

42a R = Et
42b R = H

43

2. Modified Benzofurans

A. Usnic Acid

Usnic acid ($C_{18}H_{16}O_7$, m.p. 203°, $(\alpha)_D$-495° $CHCl_3$) is a yellow pigment first isolated from a lichen, *Ramalina fraxinea,* by Rochleder and Heldt[37] and found in many species of *Cetraria* and *Cladonia*, and also some *Planc-odium, Alectonia,* and *Haematomma.* (+)-Usnic acid is also well known: it occurs in a large number of species of *Usnea*, *Evernia*, *Parmelia*, *Leccanora*, *Nephroma*, and *Cladonia.* The racemate has been isolated from *Cetraria islandica* and *Cladonia silvatica,* but is relatively uncommon. Generally, it has been found in its optically active forms as well as the racemic form in 75 lichen species. It is not a carboxylic acid, but a highly acidic enol. It has been the subject of very extensive chemical investigation.[38-41] Several papers, dealing with the earlier chemical work, have been published.[42-46]

The interrelationship of some of the major degradation products of usnic acid is outlined as follows:

Decarbousnic acid ($C_{17}H_{18}O_6$) + CO_2	Usnoic acid ($C_{18}H_{16}O_7$)	—CO_2 →	Decarbousnol ($C_{17}H_{16}O_5$)		
↑ EtOH, pressure (from Usnic acid)	↑ H^+				
	Usnic acid ($C_{18}H_{16}O_7$)	—CO_2 →	Usnetic acid ($C_{14}H_{14}O_6$) + Me_2CO	—CO_2 →	Usnetol ($C_{18}H_{14}O_4$)
↓ (from Usnic acid)	↓ oxidn.		↓		↓ —AcOH, + H_2O
Usnic acid ethoxylate ($C_{20}H_{22}O_7$)	Usnonic acid ($C_{18}H_{16}O_8$)		Pyrousnic acid ($C_{12}H_{12}O_5$)	—CO_2 →	Usneol ($C_{11}H_{12}O_3$)

Usnic acid undergoes decarboxylation (**44**) to decarbousnic acid (**45**) when heated under pressure in the presence of ethanol. Alkaline hydrolysis of **44** under mild conditions yields acetone and acetic acid together with usnetic acid (**46**) and the ketone (**47**). Under drastic conditions, pyrousnic acid (**48**) is produced. Usnetic acid decarboxylation affords usnetol (**50**) and, on elimination of the acetyl group by alkaline hydrolysis, usneol (**49**) is obtained.[47,48] Synthesis of the dimethyl ether of usneol from 4-hydroxy-2, 6-dimethoxytoluene and 3-chlorobutan-2-one, followed by cyclization of the resulting ether (**52**), confirmed structure **49**.[39] Moreover, ozonolysis of usneol (**49**) afforded an acetate (**51**) of *C*-methylphloroacetophenone, providing enough evidence that usneol and pyrousnic acid are benzofuran derivatives.

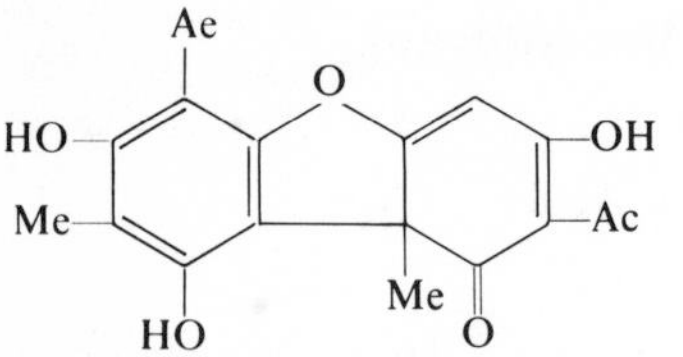

Ae HO Me HO O Me O OH Ac
44

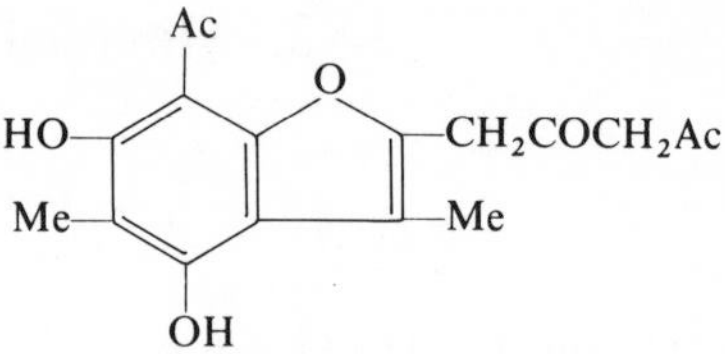

Ac HO Me OH O CH_2COCH_2Ac Me
45

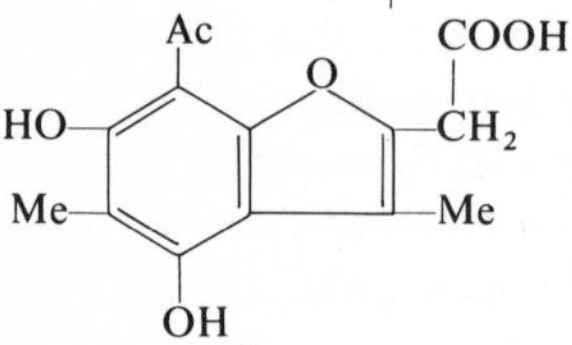

Ac HO Me OH O COOH CH_2 Me
46

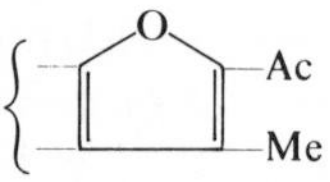

O Ac Me
47

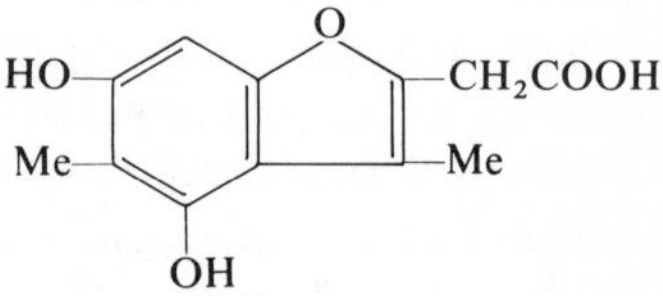

HO Me OH O CH_2COOH Me
48

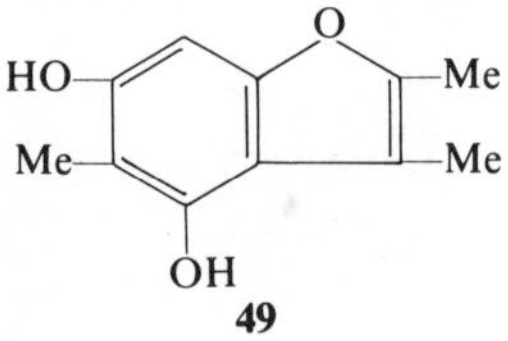

HO Me OH O Me Me
49

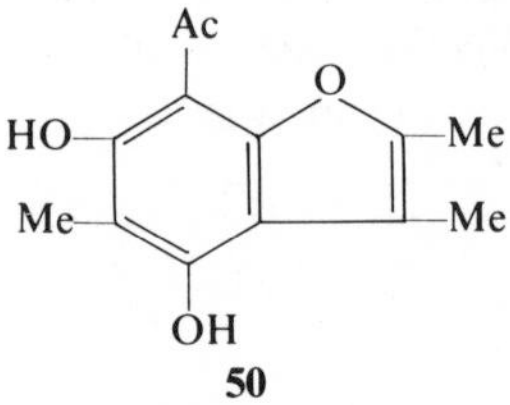

Ac HO Me OH O Me Me
50

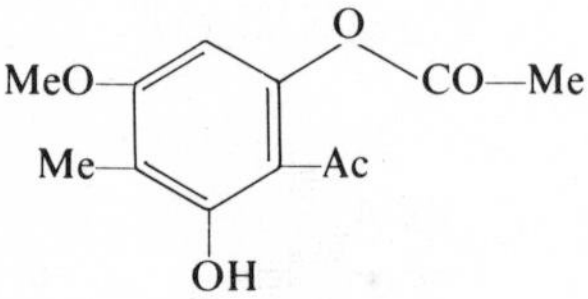

MeO Me OH O CO—Me Ac
51

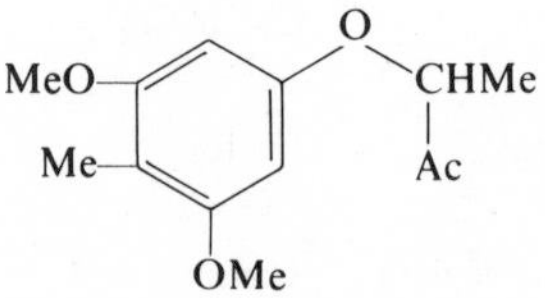

MeO Me OMe O CHMe Ac
52

The position of the carboxyl group in usnetic acid (**46**) was first indicated by Asahina and Yanagita[49,50] who obtained the tricarboxylic acid (**53**) upon oxidation of **46** with hydrogen peroxide. This location was later confirmed through synthesis of the *O*-dimethyl ether of pyrousnic acid[51] by subjecting the benzofuran (**54**) first to Gattermann aldehyde, then to Erlenmeyer azlactone synthesis. Thus pyrousnic acid (**48**) and usnetic acid (**46**) can be formulated as benzofuryl-2-acetic acids.

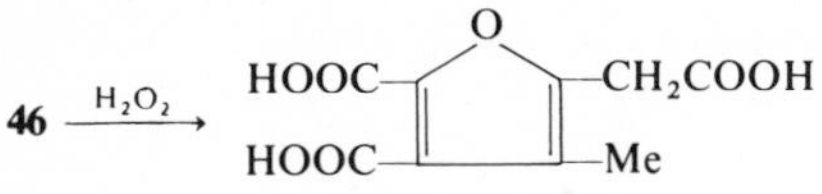

46 $\xrightarrow{H_2O_2}$ **53**

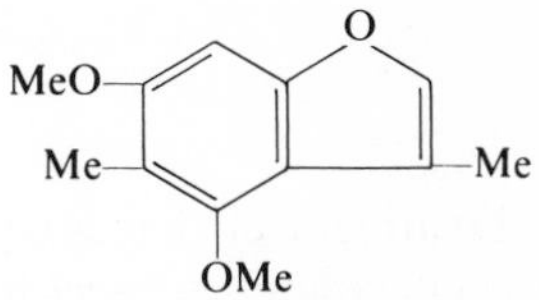

54

The structure of decarbousnic acid (**45**) followed from the nature of the hydrolysis products: **46**, **47** and **48**, as well as from the formation of the tetraacetyl derivative and the characteristic behavior as a β-diketone on treatment with substituted hydrazines.

47 ⟵ **45** ⟶ **46** ⟶ **48**

Diacetoxyusnic acid ethoxylate, formed on ethanolysis of diacetylusnic acid, has been shown to have structure **55**.[40,50,52] Moreover, the observation that 3-methyl-1-phenylpyrazole-4,5-dicarboxylic acid (**56**) may be obtained from usnic acid anhydrophenylhydrazone diacetate (**57**) ozonolysis[53] is further confirmation. Ozonolysis of decarbousnic acid diacetate afforded *C*-methylphlorodiacetophenone diacetate (58);[54] its structure has been confirmed by synthesis.[55]

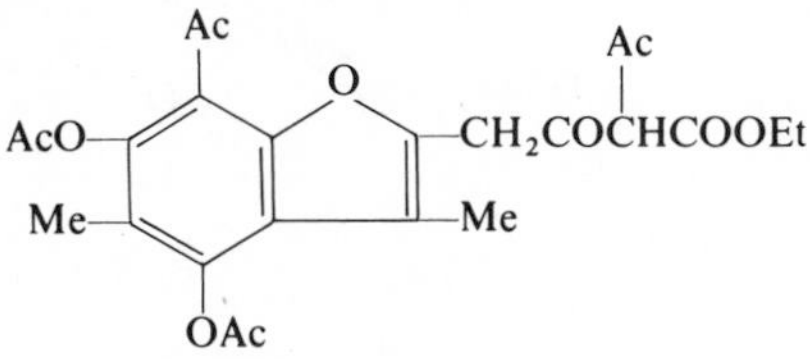

55 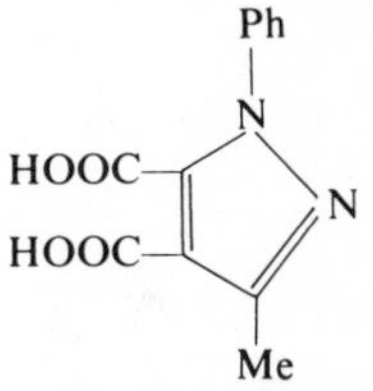 **56**

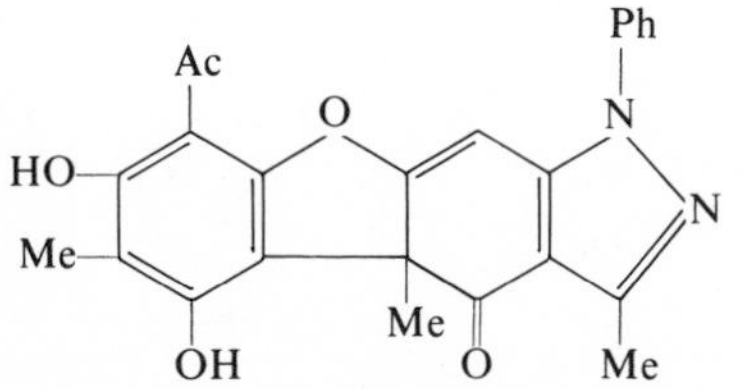

57 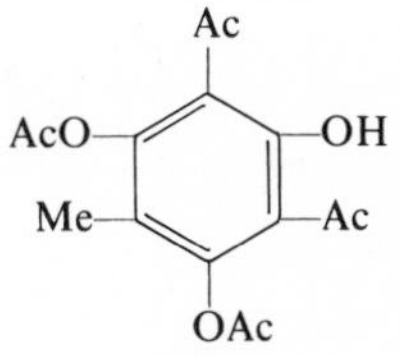 **58**

The position of the double bonds in usnic acid was indicated by ozonolysis of usnic acid diacetate, yielding **59**, whose structure was confirmed by synthesis.[54,56,57]

Usnic acid diacetate $\xrightarrow{O_3}$ **59** + HOOC–CO–CH_2COCH_3

Treatment of diacetoxyusnic acid (usnic acid diacetate) with potassium permanganate in acetone gave 3,7-diacetyl-3,5-dimethyl-4,6-diacetoxy-2(3*H*)-benzofuranone (**60**) which, when heated with saturated methanolic hydrogen chloride, yielded the 6-hydroxy analog of **60** and **61**. The latter compound afforded, when distilled in vacuo, 7-acetyl-4-acetoxy-6-hydroxy-3,5-dimethyl-2(3*H*)-benzofuranone (**62**). Upon deacetylation by the action of sulfuric acid it afforded **63**. However deacetylation of **60** under the same conditions gave **64**.[58]

60 $\xrightarrow{\text{MeOH, HCl (100°C, 2 hr.)}}$ **61** $\xrightarrow{\text{distil. in vacuo (240 C)}}$ **62** $\xrightarrow{KMnO_4}$ **60**

62 $\xrightarrow{H_2SO_4 \text{ (10 min)}}$ **63**

60 $\xrightarrow{\text{conc. } H_2SO_4}$ **64**

All this evidence is accommodated by structure **44** which was first proposed by Robertson.[39,59] However this evidence did not define conclusively the point of attachment in usnic acid of the carbonyl group that gives rise to the carbethoxy grouping in diacetoxyusnic acid ethoxylate (**55**). As a result, various alternative structures were suggested;[38,40,43,49] these have been excluded by a two-stage synthesis that clearly establishes structure **44** for usnic acid,[41,60] including union of two molecules of *C*-methylphloroacetophenone brought about by the action of ferricyanide, followed by dehydration of the product with acetic anhydride, yielding (±)-usnic acid diacetate. A similar oxidation converts *P*-cresol into an usnic acid analog (**65a**), in which the reversible nature of the ring opening to **65b** has been demonstrated.[61,62]

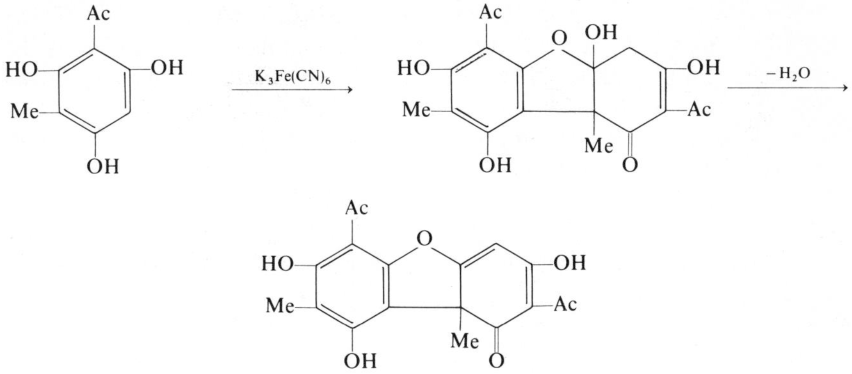

$K_3Fe(CN)_6$; $-H_2O$

44

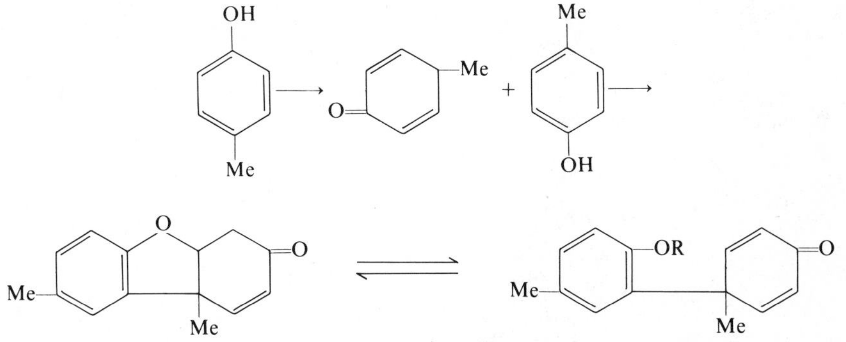

65a **65b**

Usnic acid is biosynthesized in lichens[63] by oxidative coupling of two *C*-methylphloroacetophenone moieties whose methyl group is introduced into the molecule prior to the completion of aromatic ring closure. As for the introduction of the methyl group, this result is in accord with the finding that incorporation of atranorin[64,65] in the C_1 fragment occurs in the course of biosynthesis.

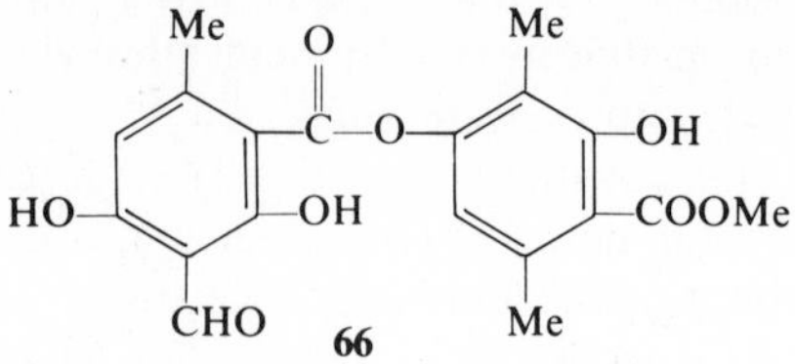

66

Usnic acid is not racemized by bases, but it is racemized when heated in xylene or toluene, or by heating in acetic acid. These unusual circumstances can be explained[66,67] by the dissociation of bond *a* in **67** to an entity of type **68** which, having lost its asymmetric carbon atom, can recyclize to the racemate. Though **68** is written as a biradical (both centers being stabilized by the adjacent double-bond systems), it can equally well be written as an ionic entity or even as a ketone (**69**).[66] Such an intermediate (**69**) would explain the formation of decarbousnic acid (**45**) by action of alcohol on usnic acid under pressure; the failure of usnic acid anhydrophenylhydrazone (**57**) to racemize is explained by the fact that formation of a ketone by homolysis would involve the more difficult disruption of the aromatic pyrazole ring system.

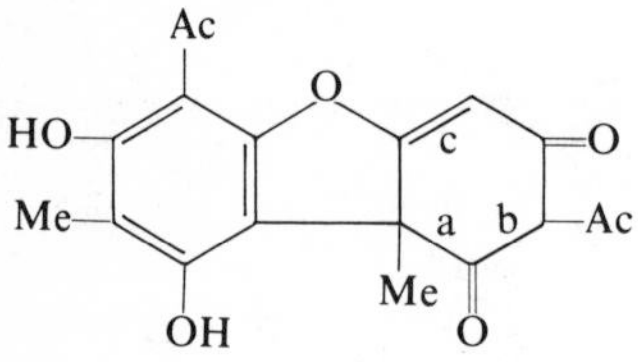

67

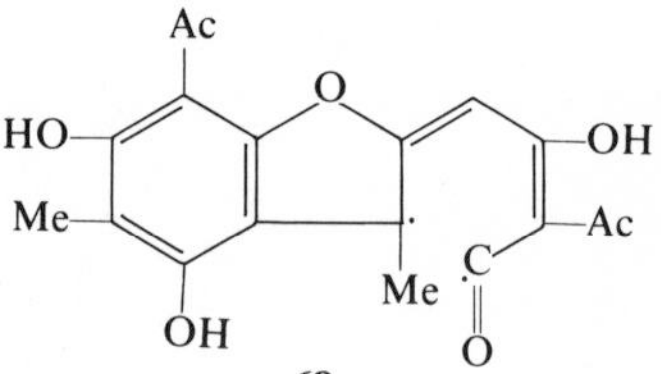

68

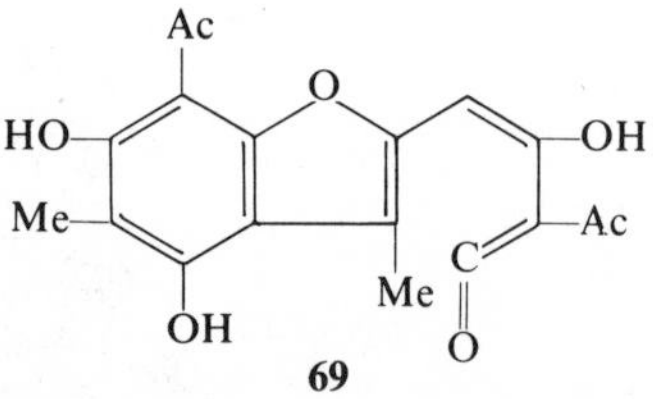

69

Several reactions of usnic acid call for special comment. Usnic acid diacetate is hydrogenated only with difficulty because the double bond is part of an enolic system. When pyrolyzed with calcium chloride, dihydrousnic acid diacetate (**70**) is a split into 2(3*H*)-benzofuranone (**71**).[68,69]

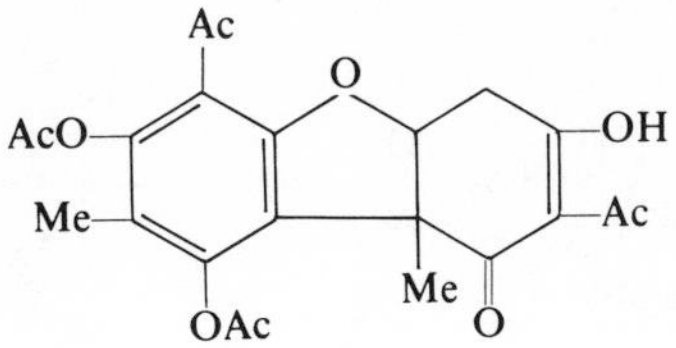

Ac AcO Me OAc O OH Ac Me O **70**

Ac HO Me OH O O Me **71**

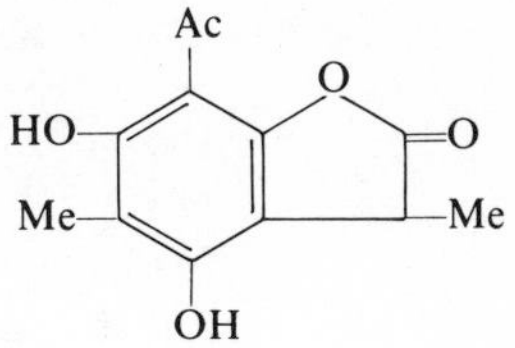

Methylusnic acid (**72**), when refluxed in methanol, yielded methyl methylacetusnate (**73**); the latter, on hydrolysis in presence of alkaline medium followed by decarboxylation, gave methylacetusnetol (**74**).[70]

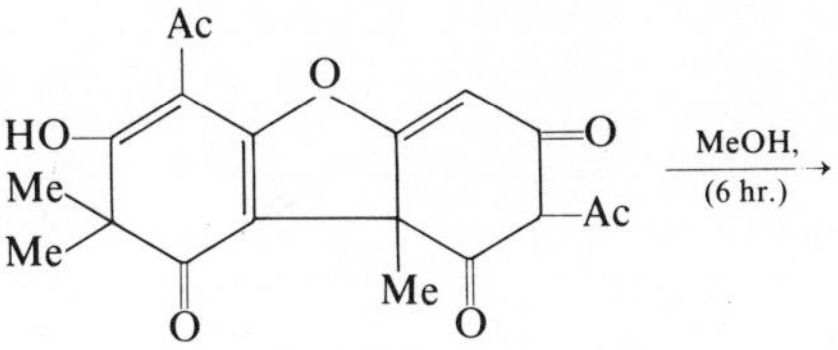

Ac HO Me Me O O O Ac Me O **72** $\xrightarrow[\text{(6 hr.)}]{\text{MeOH,}}$

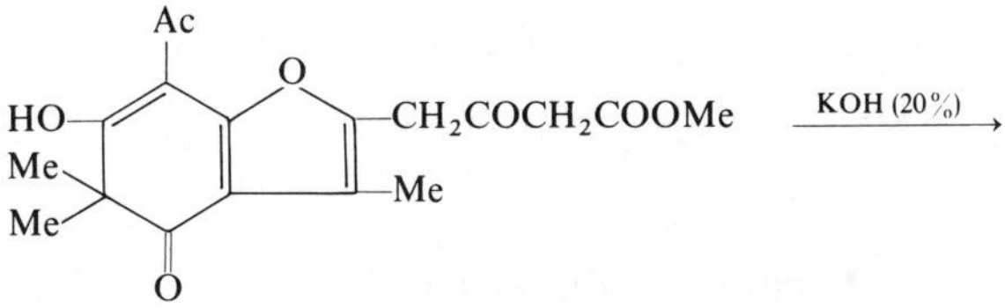

Ac HO Me Me O O CH_2COCH_2COOMe Me **73** $\xrightarrow{\text{KOH (20\%)}}$

Ac HO Me Me O O CH_2COCH_3 Me **74**

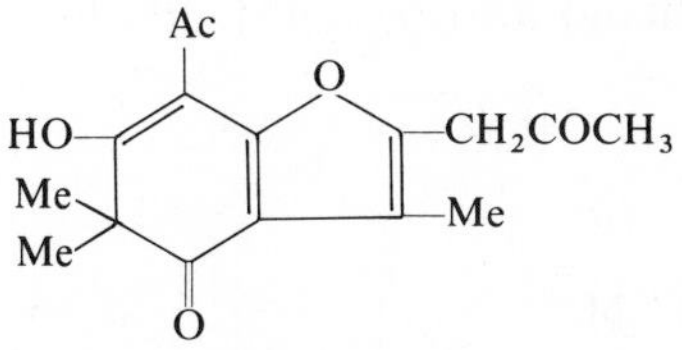

Ozonolysis of anhydromethyldihydrousnic acid acetate (**76**), obtained from methyldihydrousnic acid (**75**), yielded **77a** which gave rise to monomethyl ester **77b**; ozonolysis opened ring A. However hydrolysis of **77a** gave rise to **77c** which was also obtained from **77b** by hydrolysis; **77c** gave a methyl ester and a monoxime. Decarboxylation of **77c** afforded **78** which yielded a bioxime. Ozonolysis of **76**, (R = H) resulted in the formation of **77d**.[71,72]

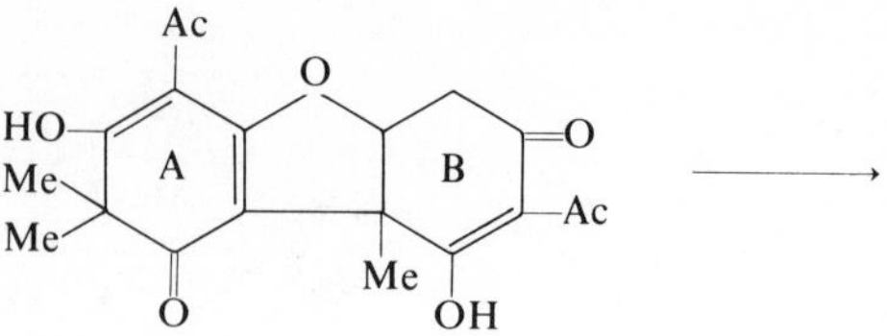

75

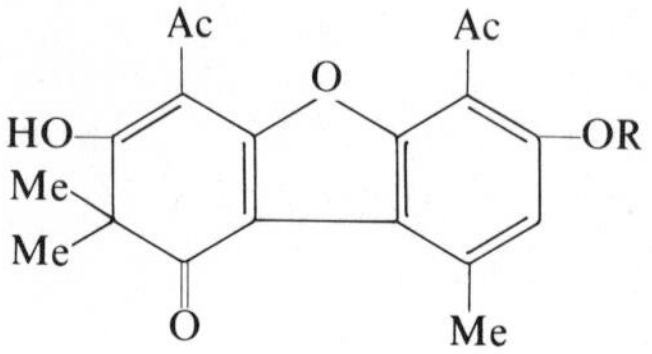

76 R = Ac; or R = H

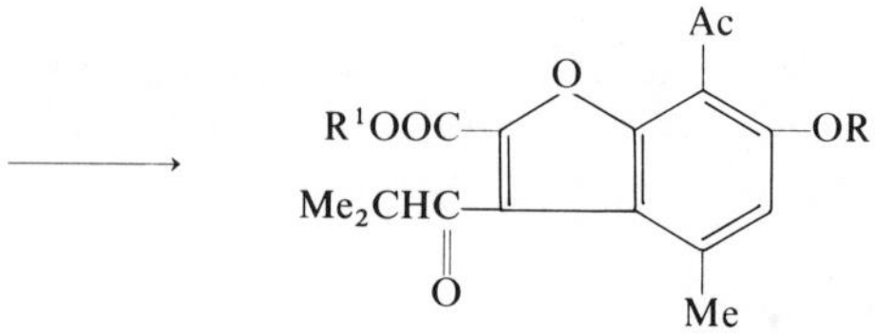

77a R = Ac, R^1 = H
77b R = Ac, R^1 = Me
77c R = R^1 = H
77d R = H, R^1 = Me

77c $\xrightarrow[(260\text{–}270°\text{C})]{-CO_2}$ **78**

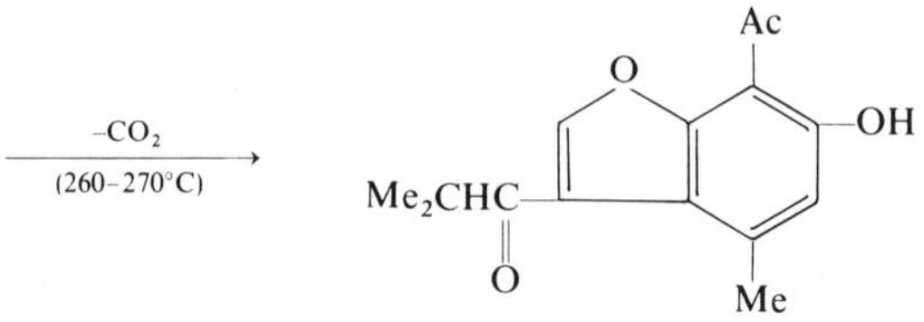

Usnonic acid is formed by the action of potassium permanganate or lead tetraacetate on usnic acid.[50] It is converted back to usnic acid upon treatment with zinc and mineral acid. The evidence available favors structure **79a** or **79b**. It should also be noted that the starting rearrangement

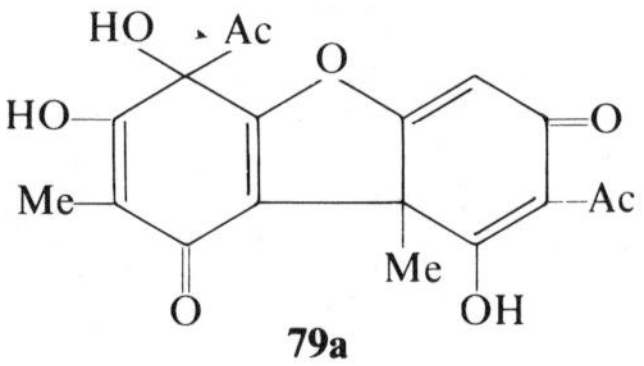

79a

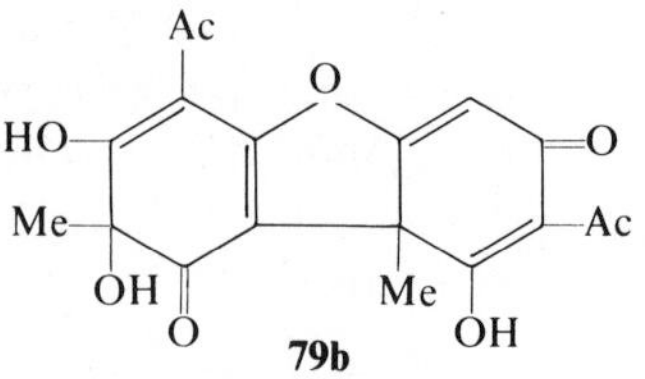

79b

suffered by usnic acid when treated with concentrated sulfuric acid yields an isomer, usnolic acid (**80**), formed presumably by way of **81** which, in dilute acids, would have lost carbon dioxide, yielding decarbousnic acid (**45**). Structure **80**[73-75] has been confirmed by synthesis.[76]

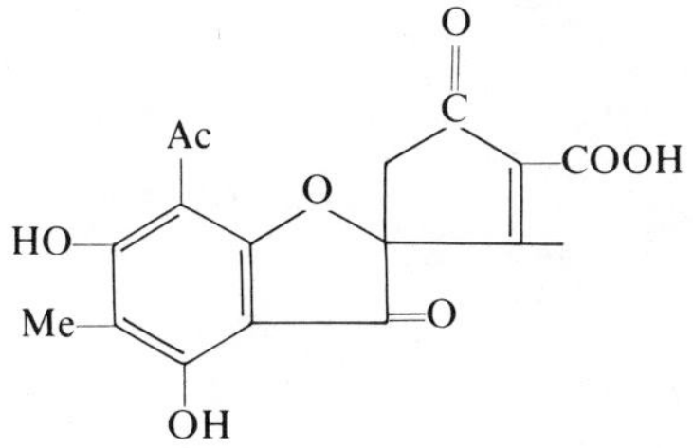

80

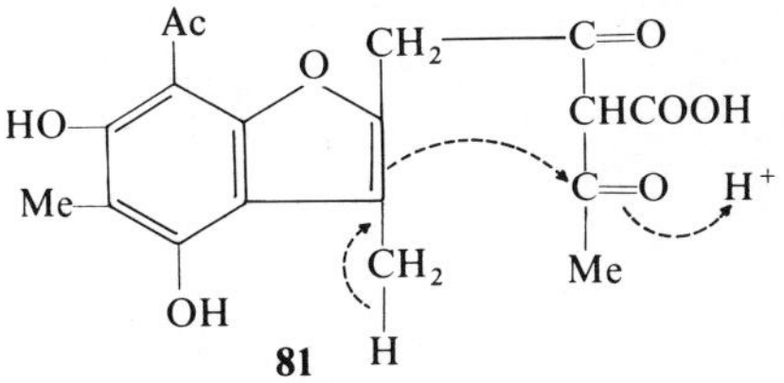

81

It seems evident that the formation of these spirans involves, as an initial step, a process similar to that concerned with the racemization of usnic acid. In this connection, it has been shown that **82** is transformed by the action of sulfuric acid to **83**.[77]

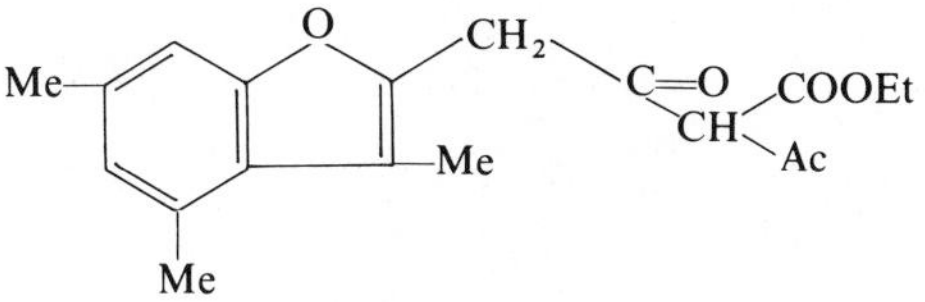

82

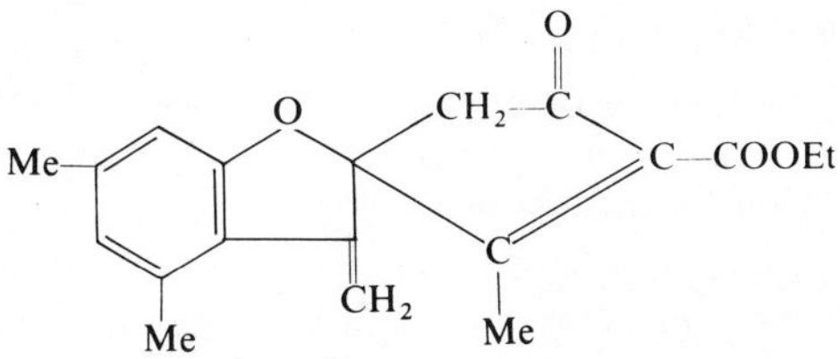

83

A number of usnic acid analogs have been synthesized [78,79] as outlined below:

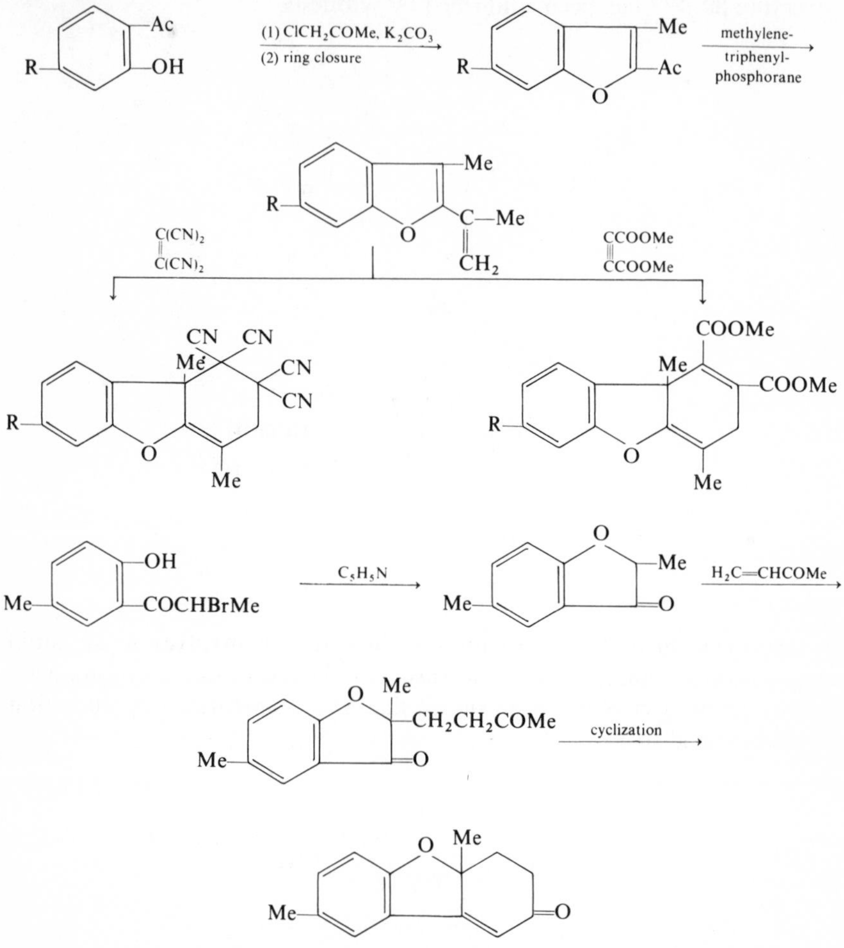

(1) $ClCH_2COMe$, K_2CO_3 (2) ring closure

methylene-triphenylphosphorane

$C(CN)_2=C(CN)_2$

$CCOOMe \equiv CCOOMe$

C_5H_5N

$H_2C{=}CHCOMe$

cyclization

Usnic acid strongly inhibits many gram-positive organisms;[80,81] several reports[82-85] suggest that it works by interfering with oxidative phosphorylation in the nucleus and, in general with functions associated with ribonucleic acids. Usnic acid is not particularly toxic,[86,87] though when injected into dogs, it can induce paralysis (e.g. of the central nervous system). The most interesting activity of the compound, however, is the inhibiting effect[88,89] on *Myobacterium tuberculosis* against which it is potent in conjunction with small quantities of streptomycin.

Amines react easily with usnic acid. The products may have physical properties which make them superior, in some respects, to the parent acid for therapeutic use.[90]

In general usnic acid has no effect on fungi, yeasts, or coxackie virus, *E. coli, Proteus valageoris,* or *Salmonella typhosa.*

B. Isousnic Acid

(+)-Isousnic acid ($C_{18}H_{16}O_7$, m.p. 150–152°C, $(\alpha)_D + 500°$, dioxan), was isolated from the ethereal extracts of *Cladonia mitis* Sandst., as yellow prisms; it shows a close similarity to (+)-usnic acid in its ultraviolet and infrared spectra.[91] The established biosynthetical pathway of usnic (**85a**)[63,60] suggests that there are three other possibilities of ring closure after the first oxidative coupling of the two methylphloroacetophenone units. Acetylation of (+)-isousnic acid afforded (+)-diacetylisousnic acid, $C_{22}H_{20}O_9$, which showed very similar ultraviolet and infrared spectra to those of (+)-diacetylusnic acid. Ozonolysis of diacetylisousnic acid yielded 3,5-dimethyl-4,6-diacetoxy-7-acetyl-2(3*H*)-dihydrobenzofuranone (**86a**). However the ozonolytic product, obtained from (+)-diacetylisousnic acid, from NMR spectral evidence, has structure **86b**. Deacetylation of **86a** and **86b** with concentrated sulfuric acid afforded **87**. This would have resulted from the rearrangement of lactone linkage on deacetylation of **86b**. The NMR spectrum of isousnic acid, as well as the evidence provided, led to the conclusion that isousnic acid, occurring in lichens, should be formulated as **85b**. Catalytic hydrogenation of (+)-isousnic acid in presence of palladium black afforded (+)-isodihydrousnic acid ($C_{18}H_{18}O_7$, m.p. 109–115°C, $(a)_D^{23} + 157°$, dioxan). Isodihydrousnic acid is readily converted into dihydrousnic acid even by recrystallization.

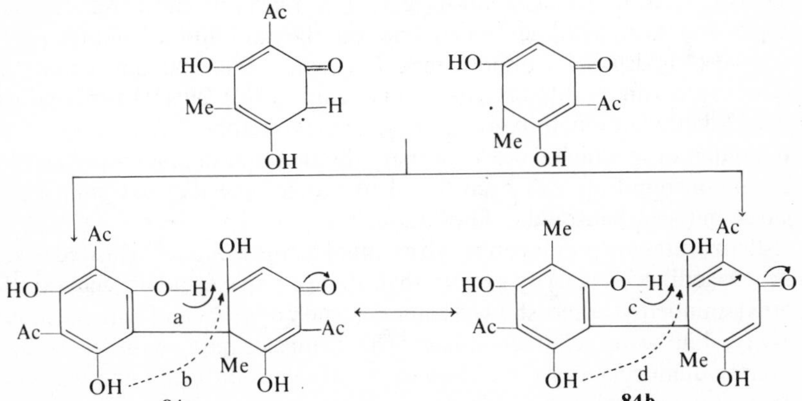

84a **84b**

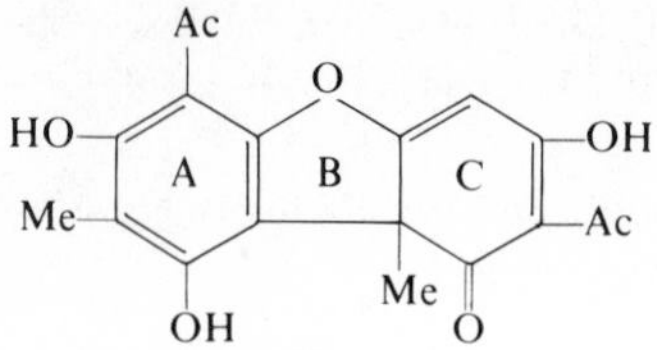

85a

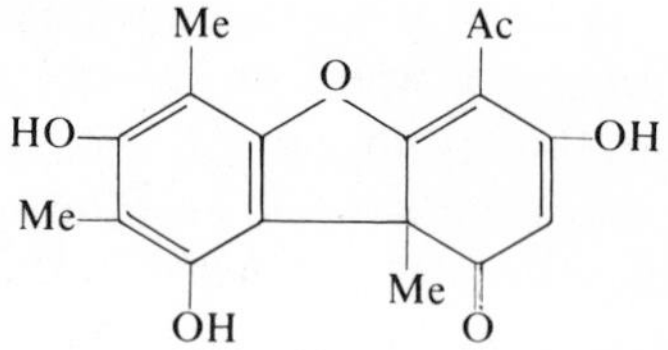

85c

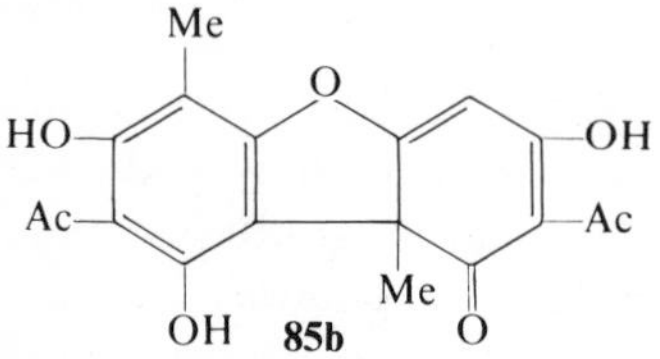

85b

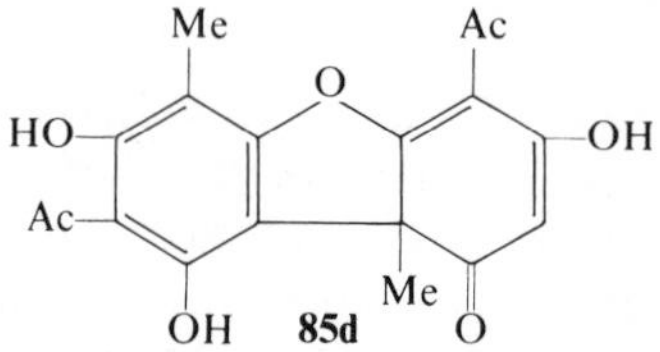

85d

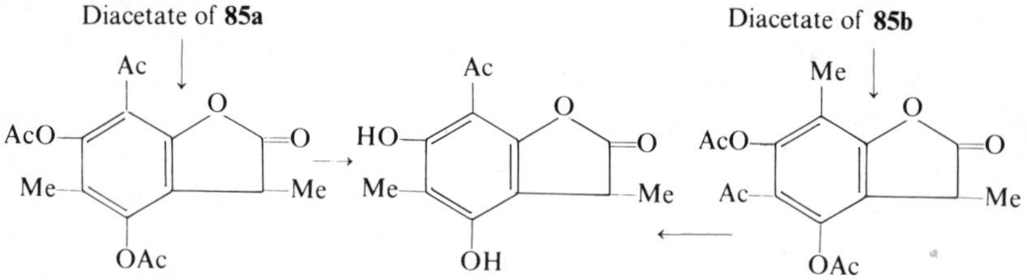

Diacetate of 85a

Diacetate of 85b

86a

87

86b

3. Applications of Benzofurans

Substituted benzofurans show marked pharmacological properties. Thus 2-(4-hydroxybenzoyl)-benzofuran exerts a relaxant effect on the histamine and acetylcholine spasm and on the mobility of guinea-pig;[92] 2-ethyl-3-(4-hydroxy-3,5-diiodobenzoyl)benzofuran is 100 times more effective as a coronary dilator agent than khellin.[93-95] 2-Butyl-3-benzoylfuranyl-4-(2-diethylaminoethoxyl)-3,5-diiodophenyl ketone, "Amidarone," an antianginal drug which causes coronary dilation and depresses myocardial oxygen consumption, has been found to protect anesthetized guinea pigs against induced ventricular fibrillation.[96]

2-Benzofuranyl-(*p*-chlorophenyl) carbinol hemisuccinate[97] showed a coronary vasodilatory activity greater than that of papaverine. In general, the diethylaminoethyl ethers showed some coronary dilatory and antispasmodic activity, but were relatively toxic. The hemisuccinate, with a stronger coronary dilatory activity, was free of in vitro cardiotoxicity and was well-tolerated perenterally.

Petta and Zaccheo[98] have studied the comparative profile of 13428 [2-butyl-3-[3,5-diiodo-4(β-diethylaminoethoxy)]benzofuran] and other anti-anginal agents on cardiac hemodynamics.

A solution of 3-(2-hydroxy-3,5-dichlorophenyl)-5,7-dichlorobenzofuran in aqueous lauryl sulfate showed bactericidal and bacteriostatic activity against *Staph aureus.*[98a,99] However 2-(4-nitrophenyl)benzofuran has shown high activity against *Staphylococci, Pyocyanea,* and *E. Coli* test species of bacteria.[100]

The antimycotic activity of a number of substituted 3-(3,3,3-trichloro-2-hydroxyprop-1-yl)-benzofuran is reported by Zawadowski et al.[101,102]

Diuretic and saluretic activities have been acquired by a number of substituted aminobenzofuran-2-carboxylic acid derivatives (**88**).[103,104] Ethyl-2-(3-benzofuranyl)-cyclopropane carboxylate is useful as an anti-depressive, ataractic, and hypotensive.[105] On the other hand, α-(2-benzofuranyl) acrylic acid has exhibited a marked activity; while its esters and amides are either less active or inactive.[106] Long and Hornby[107] have shown that the principal metabolite of 2-(1-hydroxy-2-isopropylamino-ethyl)-6,7-dimethylbenzofuran, showing adrenosceptive blocking properties, is 6-carboxy-2-(1-hydroxy-2-isopropylaminoethyl)-7-methoxybenzofuran. The adrenergic, cholinergic properties, and central nervous system effects of 5-(*N*-piperidinomethyl) coumarylamide have been studied.[108,109] Furthermore, 4′,5-diamidine-2-phenylbenzofuran (**89**) has shown activity against *Trypanosoma rhodesiense.*[110,111]

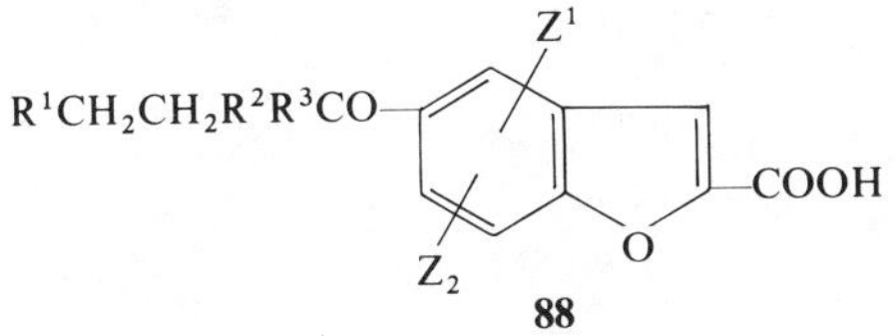

$R^1CH_2CH_2R^2R^3CO$— ; Z^1 ; Z_2 ; —COOH ; O

88

H_2N, H_2N ⊕ C ; ⊖Cl ; O ; C ⊕ NH_2 NH_2 Cl⊖

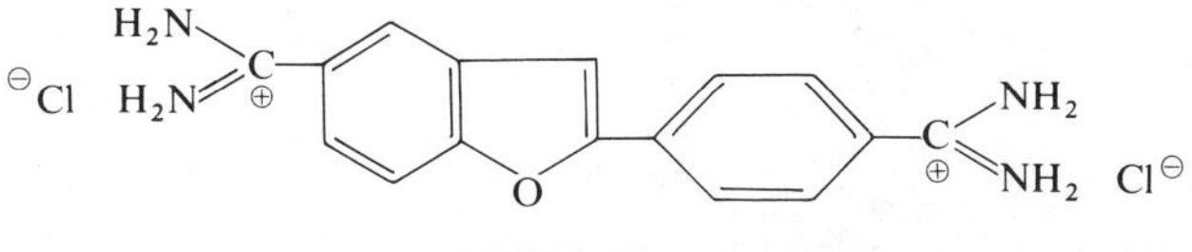

89

N-Alkyl-*N*-benzofuranyl-*N*,*N*-dialkylpolymethylenediamines of the type **90**, having two carbon atoms in the chain, had a greater ganglion-blocking activity than hexonium. Those containing 6 to 7 carbon atoms showed the most ganglion-blocking activity, but they appeared toxic. Diamines with 2 to 3 carbon atoms in the chain are not toxic.[112] 5-Cinnamoylbenzofurans of type **91** have been tested in mice for hypotensive, vasodilating, and spasmolytic activities.[113]

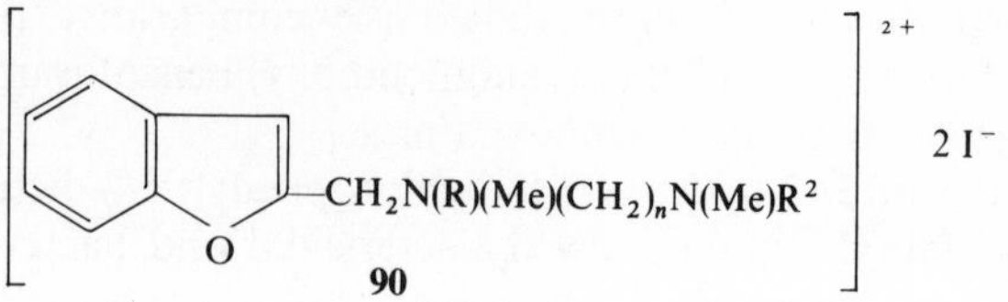

90

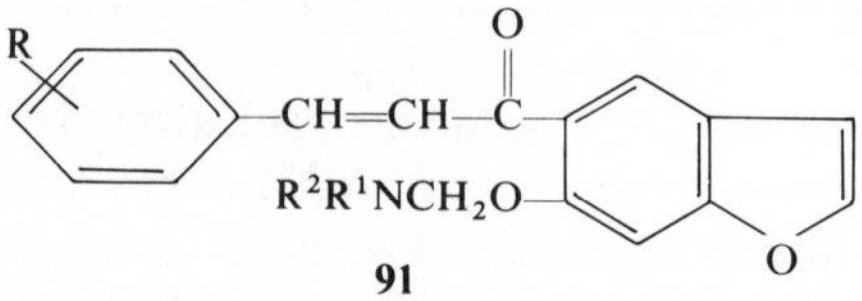

91

2,3-Dihydrobenzofurans exhibit pharmacological properties which have attracted the interest of many investigators. In general,[114] the 2-(3-alkoxy-propylaminomethyl)-2,3-dihydrobenzofuran analogs (**92**) studied with H, Me, or Cl at C-7 on the benzofuran ring showed relatively potent analgesic, spinal reflex-depressing, and adrenergic α-blocking activity in vivo; Me, Cl, or OMe at C-5 decreased these pharmacological activities. The benzofuran ring, having an OEt group as a terminal alkoxy group in the side chain, such as 2-(3-ethoxypropylaminomethyl)-5-methoxy-2,3-dihydrobenzofuran (**92**, X = 5-OMe, R = Et), had more potent pharmacological activity than **92** with OMe or O-iso-Pr in the side-chain.

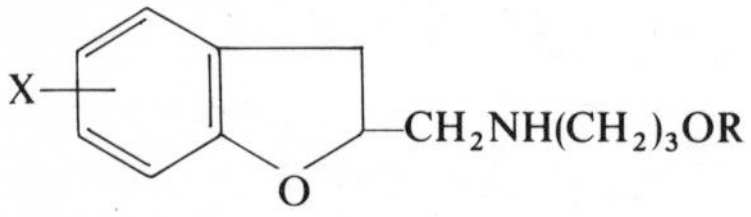

92

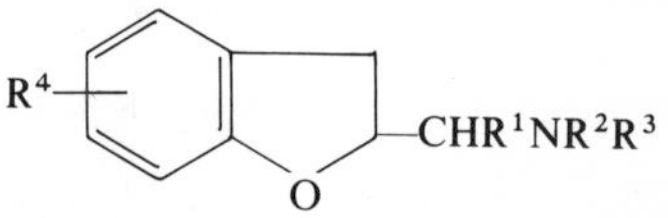

93

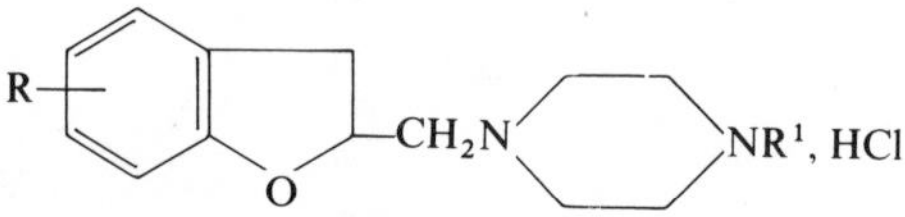

94

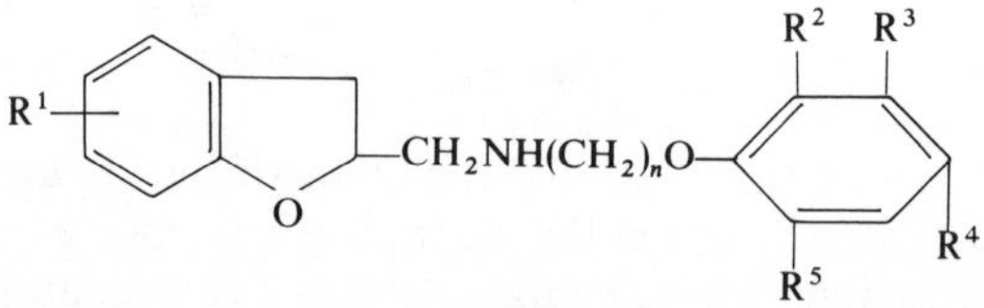

95

Funke and Däniken[115] have studied the activity of a large number of 2-aminoethyl-substituted 2,3-dihydrobenzofurans. 2-Amino-2,3-dihydro-benzofurans of type **93** and **94** are useful as antidepressants and hypotensives.[116-118] 2-(Aryloxyalkylaminoethyl)-2,3-dihydrobenzofurans of type **95** showed hypotensive properties.[119] 2,3-Dihydrobenzofuran derivatives with choleretic activity have been reported.[120-122] It has also been shown that substituted amino-2,3-dihydrobenzofuran-2-carboxylic acids of type **96** possess diuretic and saluretic activities.[104]

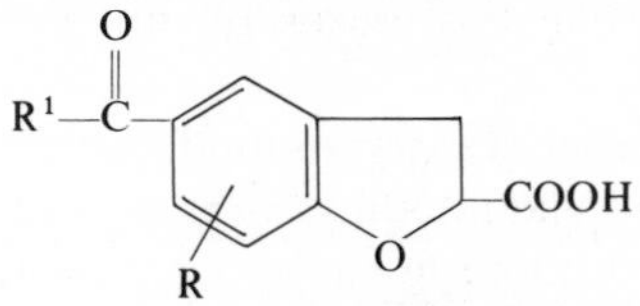

96

Derivatives of basic 3-phenyl-2(3*H*)-benzofuranone of type **97** exhibit a musculotropic action comparable to that of papaverine, and in many cases, a strong anesthetic effect, a property often associated with antispasmodic drugs. The most promising compound, 3-(β-diethylaminoethyl)-3-phenyl-2(3*H*)-benzofuranone, (**98**), possesses about one tenth the activity of atropine against acetylcholine and twice the activity of papaverine against barium chloride spasms of the isolated ileum.[123-125] Amethone is used against bronchial asthma, and at the same time has slight antihistaminic activity.[125]

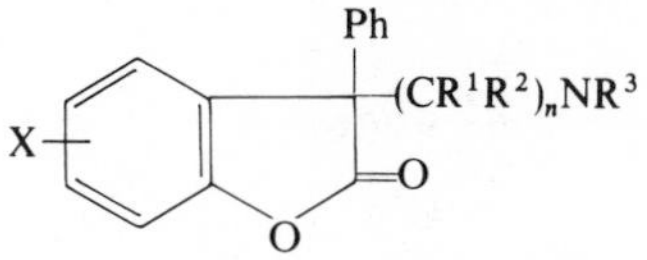

97

98 $X = R^1 = R^2 = H$, $R^3 = Et$, $n = 2$ (Amethone)

Several carboxy derivatives of benzofuran have been shown to be active growth stimulators. 2-methyl-5-hydroxybenzofuran-3-carboxylate, 5-(2-methylbenzofuranoxy) acetate, and 2-phenyl-5-methoxybenzofuran-3-carboxylate are more active than heteroauxin.[126,127] 2-Methyl-3-carboxy-5-hydroxybenzofuran, the sodium salt, its ethyl ether, and the 6-bromo and 6-iodo derivatives, in general, stimulated the rate of germination and seedling growth, rye, and alfalfa, but the pattern response was varied.[128] However, 5-benzofuranyl esters of type **99** are plant growth regulators of value in controlling grasses.[129]

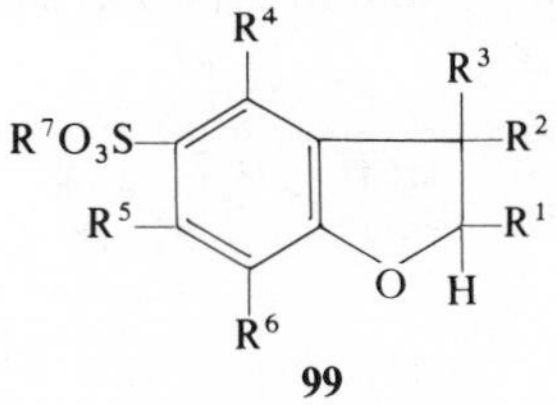

99

Benzofuranyl carbamates exhibit marked insecticidal activity, for example 2,2-dimethyl-2,3-dihydrobenzofuranyl *N*-acetylmethylcarbamate kills house flies (*Musca domestica*), adult vetch aphids (*Megoura vicia*) and adult mosquitos (*Aedes aegypti*).[130] Similarly, 3-chloro-2,3-dihydrobenzofuran-2-thiol *S*(*o*,*o*-dialkylphosphorodithioate) showed insecticidal activity.[131] Benzofuran carbamates are useful for controlling arthropods and nematodes.[132]

As antioxidant preservatives, arylaminodihydrobenzofurans, for example 2,2-dimethyl-5-β-naphthylaminodihydrobenzofuran, are suitable for use in a proportion of 0.1% or more in rubber compositions.[133]

Benzofuran derivatives have found wide application in photographic processes. Thus 2-cycnobenzofuran-5-sulfonic acid esters and/or amides,[134] β-(cyanoacetyl-3-methyl-6-benzofuranyloxy)propionaldehyde ethyl glycol acetal, and/or the corresponding polyvinyl acetal[136,137] are used in color developers in photographic conclusions. The same is true with *N*-benzoyl-α-2-benzofuranylacetamide.[135]

Cyanine and merocyanine dyes, **100** and **101**, are used as photographic sensitizers.[138,139]

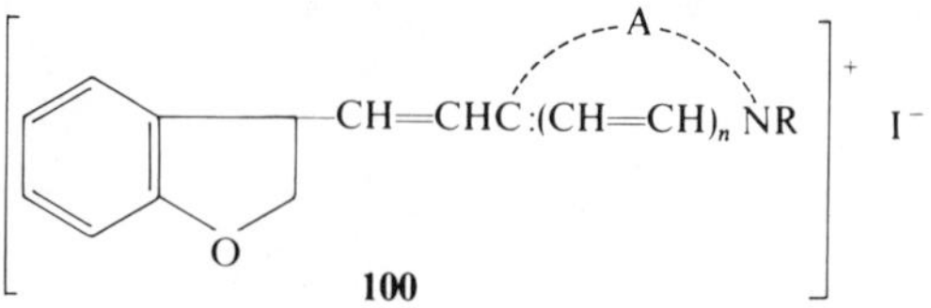

$$\left[\text{(2-benzofuranyl)}\text{—CH=CHC}{:}(\text{CH=CH})_n\,\overset{+}{\text{N}}\text{R} \;(\text{ring A}) \right] \text{I}^-$$

100

Me, Me—, =CHCH=, N, Me, O, O, NEt, S

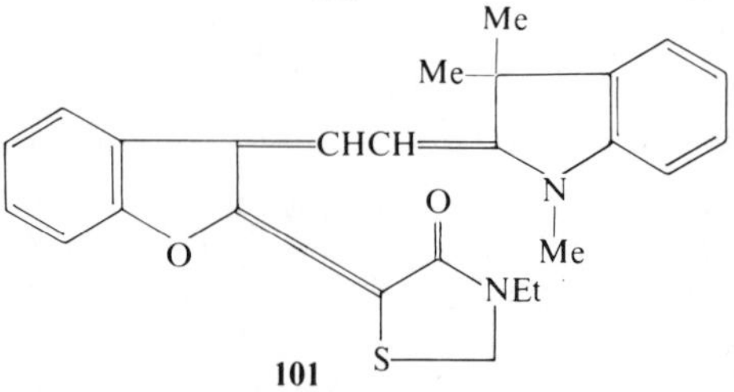

101

Aminobenzofurans are brightening agents for textiles and paper.[140] Among other substituted benzofurans used as brightening agents are 5-carboxy- and 6-carboxy-2,3-dimethylbenzofurans of type **102a** and **102b**, for fluorescent whitening,[141] of type **103a** and **103b**,[142] and of type **104** for cellulose, wool, nylon,[143] and of type **105** and **106**.[144,145]

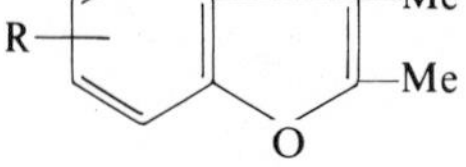

R, Me, Me, O

102a R = 5-COOH
102b R = 6-COOH

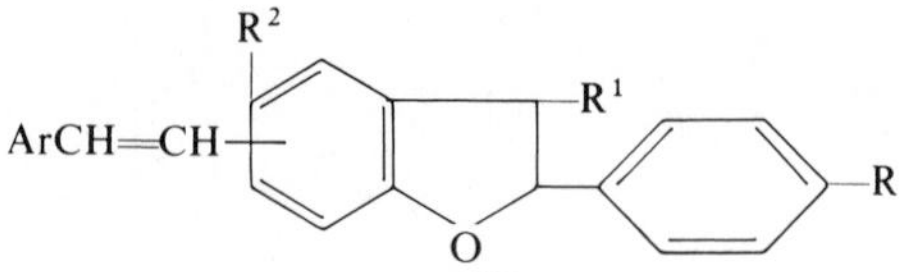

R^2, ArCH=CH—, R^1, O, R

103

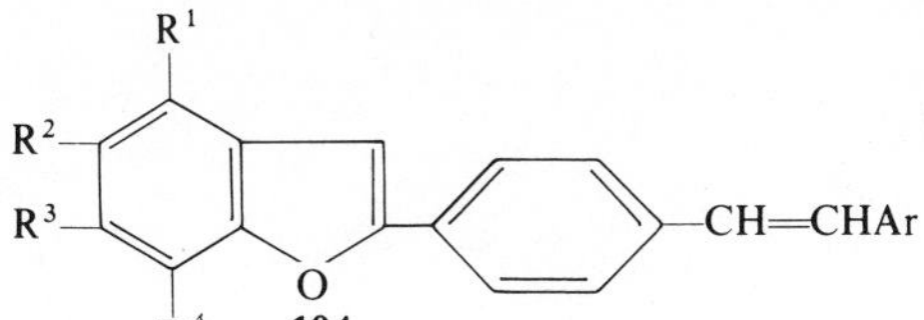

104

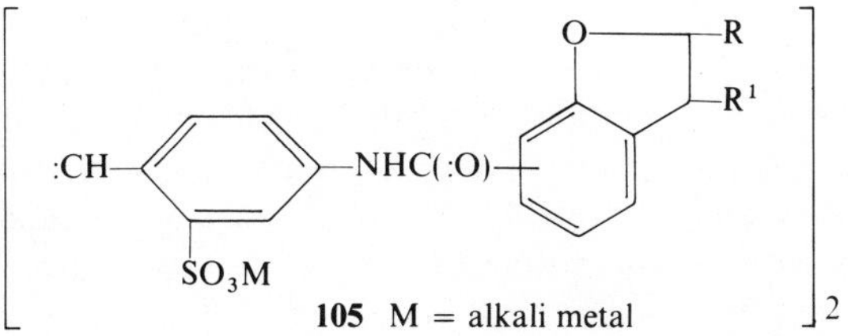

105 M = alkali metal

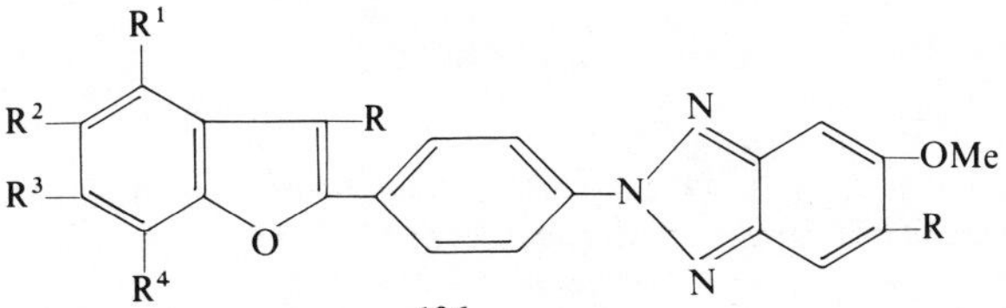

106

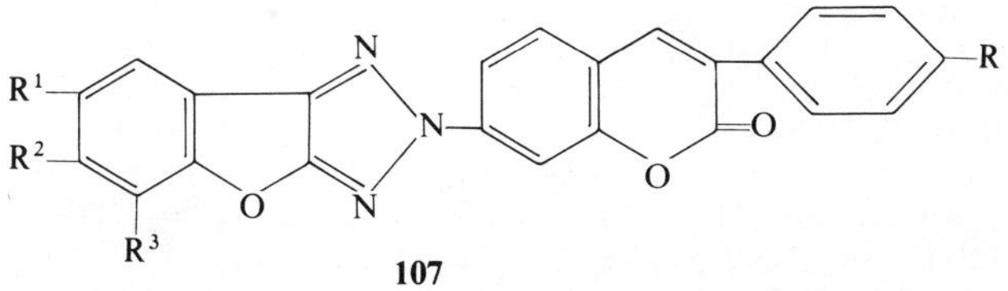

107

Fluorescent whitening agents for poly(acrylonitrile), cotton, polyester, rayon, triacetate, or polypropylene textiles of type **106**, **109**, **110** and **111** have been reported.[146]

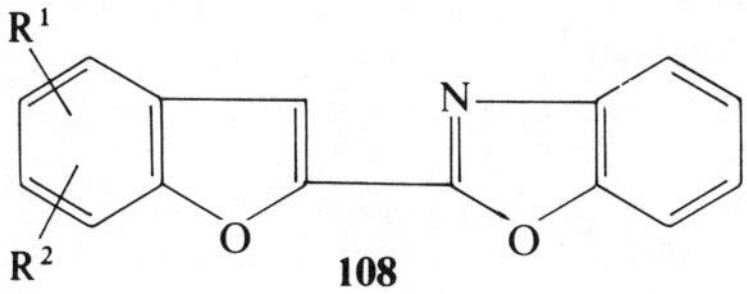

108

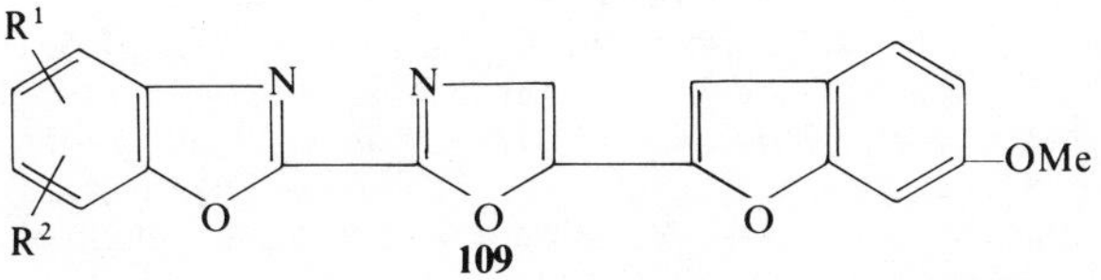

109

$R^1 = R^2 =$ OMe, Cl or H

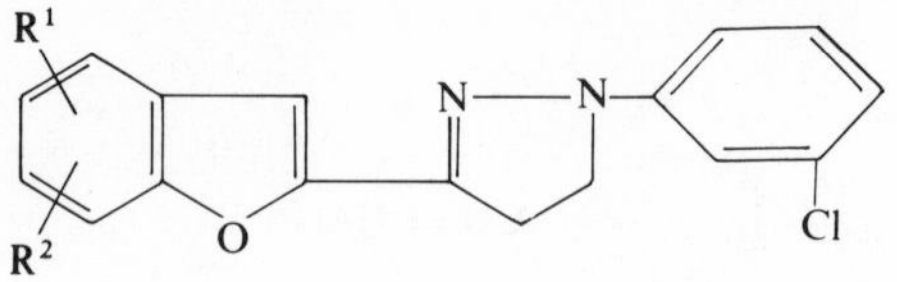

110

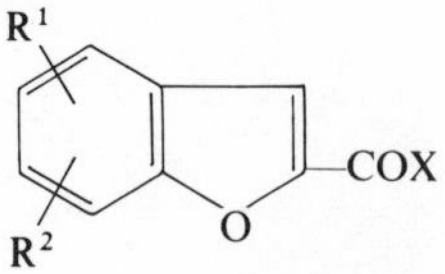

111 $R^1 = R^2 = 6\text{-OMe}$
$X = NHC_6H_4NO_2\text{-o}$

The scintillating properties of some benzofuran derivatives have been investigated.[147] 4-, 5-, 6-, and 7-Methyl-substituted 2,3-diphenylbenzofurans are suitable; for radiation measurements, when using electronic counters with low counting thresholds and because of the small duration of the pulses, they are especially suitable for rapid measurements. 5,6-Dimethyl-2,3-diphenylbenzofuran is the most effective in toluene as a scintillator.[148]

References

1. H. Carles, *Perfum. Mod.,* **22,** 615 (1929); *Chem. Zentr. Blatt,* **1,** 758 (1930).
2. H. Weinhaus and H. Dewein, *Z. Angew. Chem.*, **47,** 415 (1932).
3. H. Weinhaus and H. Dewein, *Chem. Ber.*, **91,** 256 (1958).
4. H. Weinhaus and H. G. Dässler, *Chem. Ber.*, **91** 260 (1958).
5. P. Z. Bedoukian, *J. Am. Chem. Soc.*, **70,** 621 (1948).
6. R. H. Eastman, *J. Am. Chem. Soc.*, **72,** 5313 (1950).
7. H. Stetter and R. Lauterbach, *Chem. Ber.*, **93,** 603 (1960).
8. W. Triebs, *Chem. Ber.*, **70,** 85 (1937).
9. T. Morel and P. E. Verkade, *Rec. Trav. Chim. Pays-Bas*, **67,** 539 (1948).
10. H. Fritel and M. Fètizon, *J. Org. Chem.*, **23,** 481 (1958).
11. R. B. Woodward and R. H. Eastman, *J. Am. Chem. Soc.*, **72,** 399 (1950).
12. G. O. Schenck and C. S. Foote, *Angew. Chem.*, **70,** 505 (1958).
13. W. Tagaki and T. Mütsui, *J. Org. Chem.*, **25,** 301 (1960).
14. C. H. Schmidt, *Angew. Chem.*, **68,** 175 (1956).
15. C. J. van Hulssen, *Ing. Nederland Indie*, **8**(9), VII, 89 (1941); *Chem. Abstr.*, **36,** 4970 (1942).
16. A. J. Birch and R. W. Rickards, *Austr. J. Chem.*, **9,** 241 (1956); *Chem. Abstr.*, **50,** 15502 (1956).
17. H. Stetter and R. Lauterbach, *Angew. Chem.*, **71,** 673 (1959); *Chem. Abstr.*, **54,** 6677 (1960); *Chem. Ber.*, **93** 603 (1960).
18. L. Hörhammer, H. Wagner, and W. Eyrich, *Z. Naturforsch.*, **180,** 639 (1963); *Chem. Abstr.*, **59,** 11187 (1963).
19. Ya. I. Khadzkai and V. E. Sokolova, *Farmakol. i Toksikol.*, **23,** 37 (1960); *Chem. Abstr.*, **55,** 2920 (1961).
20. L. Novotny, Z. Samek, and F. Sörm, *Tetrahedron Lett.*, **1966,** 3541.
21. K. Takeda, M. Ikuta, M. Miyawaki, and K. Tori, *Tetrahedron*, **22,** 1159 (1966).
22. H. Hikino, K. Agatsuma, and T. Takemoto, *Abstract of the 6th Annual Meeting of the Tokoku Branch of the Pharmaceutical Society of Japan*, Sendai, October 1967, p. 6.
23. H. Hikino, K. Agatsuma, and T. Takemoto, *Tetrahedron Lett.*, **1968,** 2855.
24. S. Fukushima, M. Kuroyanagi, Y. Akahori, and Y. Saiki, *Yakugaku Zasshi*, **88**(6), 792 (1968); *Chem. Abstr.*, **69,** 77524 (1968).
25. H. Hikino, K. Agatsuma, and T. Takemoto, *Tetrahedron Lett.*, **1968,** 931.

26. H. Hikino, K. Agatsuma, and T. Takemoto, *Abstract of the 11th Symposium on the Chemistry of Terpenes, Essential Oils, and Aromatics*, Matsuyama, October 1967, p. 168.
27. H. Hikino, K. Agatsuma, C. Konno, and T. Takemoto, *Tetrahedron Lett.*, **1968,** 4417; *Chem. Abstr.*, **69,** 87214 (1968).
28. S. Hayashi, N. Hayashi, and T. Matsuura, *Tetrahedron Lett.*, 1999 **1968,** *Chem. Abstr.*, **69,** 44048 (1968).
29. S. Hayashi, N. Hayashi, and T. Matsuura, *Tetrahedron Lett.*, 2647 **1968,**; *Chem. Abstr.*, **69,** 52332 (1968).
30. R. Hodges and A. L. Porte, *Tetrahedron*, **20,** 1463 (1964).
31. R. H. Manske, *Can. J. Res.*, **B16,** 438 (1938).
32. J. N. Marx and Sondheimer, *Tetrahedron*, Suppl. 8, Part I, 1 (1966).
33. D. Satoh, H. Ishii, Y. Oyama, T. Wada, and T. Okumura, *Chem. Pharm. Bull. Japan*, **4,** 284 (1956).
34. T. Wada and D. Satoh, *Chem. Pharm. Bull. Japan*, **12,** 752 (1964).
35. T. Wada, *Chem. Pharm. Bull. Japan*, **12,** 1117 (1964).
36. T. Wada, *Chem. Pharm. Bull. Japan*, **13,** 43 (1965).
37. F. Rochleder and W. Heldt, *Justus Liebigs Ann. Chem.*, **48,** 11 (1843).
38. C. Schopf and K. Heuck, *Justus Liebigs Ann. Chem.*, **459,** 233 (1927).
39. F. H. Curd and A. Robertson, *J. Chem. Soc.*, **1937,** 894.
40. Y. Asahina and M. Yanagita, *Chem. Ber.*, **72,** 1140 (1939).
41. D. H. R. Barton, A. M. DeFlorin, and O. E. Edwards, *J. Chem. Soc.*, **1956,** 530.
42. F. M. Dean, *Sci. Prog.*, **40,** 635 (1952).
43. Y. Asahina, *Fortschr. Chem. Org. Naturst.*, **1950,** 8.
44. Y. Asahina and S. Shibata, "The Chemistry of Lichen Substances," Tokyo, 1954.
45. C. H. Hassall, "Progress in Organic Chemistry," Vol. 4, J. W. Cook, Ed., Butterworths, London, 1958, p. 127.
46. F. M. Dean, "Naturally Occurring Oxygen Ring Compounds," Butterworths, London, 1963, pp. 148.
47. S. Shibata, Y. Miura, H. Sugimura, and Y. Toyoizumi, *J. Pharm. Soc. Japan*, **68,** 300 (1948); *Chem. Abstr.*, **45,** 6691 (1951).
48. C. A. Wachtmeister, *Acta Chem. Scand.*, **10,** 1404 (1956); *Chem. Abstr.*, **52,** 12836 (1958).
49. Y. Asahina and M. Yanagita, *Chem. Ber.*, **69,** 1646 (1936).
50. Y. Asahina and M. Yanagita, *Chem. Ber.*, **71,** 2260 (1938).
51. H. F. Birch and A. Robertson, *J. Chem. Soc.*, **1938,** 306.
52. S. Shibata, Y. Hizumi, and T. Takahashi, *J. Pharm. Soc. Japan*, **72,** 825 (1952); *Chem. Abstr.*, **48,** 3336 (1954).
53. D. H. R. Barton and T. Bruun, *J. Chem. Soc.*, **1953,** 603.
54. C. Schöpf and F. Ross, *Justus Liebigs Ann. Chem.*, **546,** 1 (1941).
55. F. M. Dean and A. Robertson, *J. Chem. Soc.*, **1953,** 1241.
56. F. M. Dean and A. Robertson, *J. Chem. Soc.*, **1955,** 2166.
57. S. Shibata, J. Shoji, N. Tokutake, and Y. Kaneko, *Chem. Ind. (London)*, **1961,** 320; *Chem. Abstr.*, **55,** 25902 (1961).
58. K. Takahashi and S. Shibata, *J. Pharm. Soc. Japan*, **71,** 1083 (1951); *Chem. Abstr.*, **46,** 5037 (1952).
59. F. H. Curd and A. Robertson, *J. Chem. Soc.*, **1933,** 714; **1933,** 1173.
60. D. H. R. Barton, A. M. DeFlorin, and O. E. Edwards, *Chem. Ind. (London)*, **1955,** 1039.
61. V. Arkley, F. M. Dean, A. Robertson, and P. Sidisunthorn, *J. Chem. Soc.*, **1956,** 2322.
62. R. Pummerer, H. Puttfarcken, and P. Schopflocher, *Chem. Ber.*, **58,** 1808 (1925).
63. H. Taguchi, U. Sankawa, and S. Shibata, *Tetrahedron Lett.*, **1966,** 5211.
64. M. Yamazaki, M. Matsuo, and S. Shibata, *Chem. Pharm. Bull. (Tokyo)*, **13,** 1015 (1965); *Chem. Abstr.*, **63,** 16769 (1965).

65. M. Yamazaki and S. Shibata, *Chem. Pharm. Bull. (Tokyo)*, **14,** 96 (1966); *Chem. Abstr.*, **64,** 20208 (1966).
66. G. Stork, *Chem. Ind. (London)*, **1955,** 915.
67. S. MacKenzie, *J. Am. Chem. Soc.*, **77,** 2214 (1955).
68. Y. Asahina and K. Kin, *Proc. Imp. Acad. Japan*, **20,** 371 (1944); *Chem. Abstr.*, **49,** 999 (1955).
69. S. Shibata, K. Takahashi, and Y. Tanaka, *Pharm. Bull. (Japan)*, **4,** 65 (1956); *Chem. Abstr.*, **51,** 2717 (1957).
70. K. Takahashi, Y. Honda, and S. Miyashita, *Chem. Pharm. Bull. (Tokyo)*, **11,** 1229 (1963); *Chem. Abstr.*, **60,** 5432 (1964).
71. K. Takahashi and S. Miyashita, *Chem. Pharm. Bull. (Tokyo)*, **11,** 209 (1963); *Chem. Abstr.*, **59,** 3855 (1963).
72. K. Takahashi and S. Miyashita, *Chem. Pharm. Bull. (Tokyo)*, **16,** 988 (1968); *Chem. Abstr.*, **69,** 86730 (1968).
73. Y. Asahina and K. Okazaki, *J. Pharm. Soc. Japan*, **63,** 618 (1943).
74. K. Obazaki, *J. Pharm. Soc. Japan*, **63,** 629 (1943).
75. F. M. Dean, P. Halewood, S. Mongkolsuk, A. Robertson, and W. B. Whalley, *J. Chem. Soc.*, **1953,** 1250.
76. F. M. Dean, C. A. Evans, T. Francis, and A. Robertson, *J. Chem. Soc.*, **1957,** 1577.
77. C. S. Gibson, A. R. Penfold, and J. L. Simonsen, *J. Chem. Soc.*, **1930,** 1184.
78. R. S. Cahn, C. S. Gibson, A. R. Penfold, and J. L. Simonsen, *J. Chem. Soc.*, **1931,** 286.
79. D. Diavis and J. A. Elix, *Tetrahedron Lett.*, **1969,** 2901.
80. A. Stoll, A. Brack, and J. Renz, *Experientia,* **3,** 115 (1947); *Chem Abstr.,* **41,** 4535 (1947).
81. J. B. Stark, E. D. Walter, and H. S. Owens, *J. Am. Chem. Soc.*, **72,** 1819 (1950).
82. R. B. Johnson, G. Feldott, and H. A. Lardy, *Arch. Biochem.*, **28,** 317 (1950); *Chem. Abstr.*, **45,** 3458 (1951).
83. A. G. Marshak and J. Fager, *J. Cell. comp. Physiol.*, **35,** 315 (1950).
84. J. Brachet, *Experientia*, **7,** 344 (1951); *Chem. Abstr.*, **46,** 1171 (1952).
85. M. Steinert, *Biochem. Biophys. Acta*, **10,** 427 (1953); *Chem. Abstr.*, **47,** 5565 (1953).
86. O. E. Virtanen and N. Kärki, *Suomen. Kem.*, **29B,** 225 (1956).
87. K. Mikoshiba, *Japan J. Med. Sci., IV. Pharmacol.*, **9,** 77 (1936); *Chem. Abstr.*, 31, 2688 (1937).
88. A. Stoll, A. Brack, and J. Renz, *Z. Schweiz. Path. u Bakt.*, **13,** 729 (1950); *Chem. Abstr.*, **45,** 5234 (1951).
89. V. C. Barry, L. O'Rourke, and D. Twomey, *Nature*, **160,** 800 (1947).
90. A. E. Kortekangas and O. E. Virtanen, *Suomen. Kem.*, **29B,** 2 (1956).
91. S. Shibata and H. Taguchi, *Tetrahedron Lett.*, **1967,** 4867.
92. N. P. Ph. Buu-Hoii, E. Bisagni, R. Royer, and C. Routier, *J. Chem. Soc.*, **1957**, 625.
93. M. Ghelardoni, F. Russo, and V. Pestellini, *Bull. Chim. Farm.*, **109,** 48 (1970).
94. G. Deltour, F. Binon, F. Henaux, and R. Charlier, *Arch. Intern. Pharmacodyn.,* **131, 84** (1961).
95. G. Deltour, J. Brockhuyssen, M. Ghialain, F. Burgeois, and F. Binon, *Arch. Int. Pharmacodyn. Ther.*, **165,** 25 (1966); *Chem. Abstr.*, **66,** 8587 (1966).
96. V. B. Singh and W. E. M. Vaughan, *Brit. J. Pharmacol.*, **39,** 657 (1970); *Chem. Abstr.*, **73,** 97166 (1970).
97. N. Pisanti and G. Volterra, *Farmaco. Ed. Sci.*, **26,** 312 (1971); *Chem. Abstr.*, **75,** 3829 (1971).
98. J. M. Petta and V. J. Zaccheo, *J. Pharm. Exp. Ther.* 176(2), 328 (1971); *Chem. Abstr.*, **74,** 878 (1971).
98.a R. Royer and L. Rene, *Bull. Soc. Chim. Fr.*, **1970,** 3601; *Che. Abstr.*, **74,** 53358 (1971).
99. R. W. Wynn and S. A. Glickman, U.S. Pat. 2,636,885 (1953); *Chem. Abstr.*, **48,** 2779 (1955).

100. P. K. Sharma, K. Mehta, O. P. Gupta, M. M. Mahawar, and S. K. Mukerjee, *J. Pharm. Sci.*, **56,** 1007 (1967); *Chem. Abstr.*, **67,** 79928 (1967).
101. M. Merkel, T. Rokicka, T. Zawadowski, and A. Szuchnik, *Bull. Acad. Poln. Sci. Biol.*, **11,** 449 (1963); *Chem. Abstr.*, **63,** 16325 (1965).
102. Z. Zawadowski, M. Merkel, A. Szuchnik, and J. Swiderski, *Rocz. Chem.*, **36,** 1175 (1962); *Chem. Abstr.*, **59,** 8716 (1963).
103. J. Zeregenyi and E. Habicht, *S. African Pat.* 6804,338 (1969); *Chem. Abstr.*, **71,** 91280 (1969).
104. B. Libis and E. Habicht, Ger. Pat. 1,927,393 (1969); *Chem. Abstr.*, **72,** 43426 (1970).
105. C. Kaiser and Ch. L. Zirkle, U.S. Pat. 3,010,971 (1960); *Chem. Abstr.*, **56,** 15484 (1962).
106. F. Pan and T.-C. Wang, *J. Chinese Chem. Soc., Ser. II*, **8,** 347 (1961); *Chem. Abstr.*, **58,** 13881 (1963).
107. R. F. Long and J. Hornby, *Biochem. J.*, **115,** 60 (1969); *Chem. Abstr.*, **72,** 53390 (1970).
108. F. Chaillet, R. Charlier, A. Christiaens, and G. Deltour, *Arch. Intern. Pharmacodyn. Ther.*, **164,** 451 (1966); *Chem. Abstr.*, **66,** 4279 (1966).
109. F. Chaillet, E. Philippe, and M. J. Dallemagne, *Arch. Intern. Pharmacodyn. Ther.*, 164, 466 (1966); *Chem. Abstr.*, **66,** 8754 (1966).
110. O. Dann and G. Bergen, *Justus Liebigs Ann. Chem.*, **749,** 68 (1971).
111. O. Dann, Fr. Pat. 1,586,113 (1970); *Chem. Abstr.*, **73,** 109674 (1970).
112. A. L. Madzhoyan and V. M. Avakyan, *Izv. Akad. Nauk. Arm. SSR, Biol. Nauki*, **16,** 3 (1963); *Chem. Abstr.*, **60,** 7342 (1964).
113. C. Feuran, G. Raynand, J. Eberle, and B. Pourrias, Ger. Pat. 1,933,178 (1970); *Chem. Abstr.*, **72,** 100486 (1970).
114. T. OhGoh, N. Hirose, N. Hashimoto, A. Kitahara, and K. Miyao, *Jap. J. Pharmacol.*, **21,** 119 (1971); *Chem. Abstr.*, **75,** 47095 (1971).
115. A. Funke and K. V. Daniken, *Bull. Soc. Chim. Fr.*, **1953** 457.
116. G. L. Willey and A. M. Roe, U.S. Pat. 3,513,239 (1970); *Chem. Abstr.*, **73,** 35212 (1970).
117. A. A. Stolyarchuk, M. M. Tarasyuk, K. S. Shadurskii, A. N. Grinev, I. I. Mukhanova, and F. A. Trofinov, *Farmakol. Tsect. Kholinolitkov Drugikh Nierotropyykh. Sredstv.*, **1969,** 109; *Chem. Abstr.*, **73,** 118926 (1970).
118. C. F. Huekner and L. H. Werner, U.S. Pat. 3,459,860 (1969); *Chem. Abstr.*, **71,** 81410 (1969).
119. P. N. Green and M. Shapiro, *S. African Pat.* 6,707,290 (1967): *Chem. Abstr.*, **72,** 100489 (1970).
120. P. N. Geraldi, A. Fojanesi, W. Logeman, G. P. Tosoline, E. Dradi, and M. Bergamaschi, *Arzneim. Fortsch.*, **20,** 676 (1970).
121. F. Lauria, P. N. Geraldi, and W. Logeman, U. S. Pat. 3,452,085 (1969); *Chem. Abstr.*, **71,** 70480 (1969).
122. M. Pesson and M. Jeanine, Ger. Pat. 1,932,933 (1970); *Chem. Abstr.*, **72,** 902 64 (1970).
123. R. K. Richards, G. M. Everett, and K. E. Kuster, *J. Pharmocol.*, **38,** 387 (1945).
124. A. W. Weston and Wm. Brownell, *J. Am. Chem. Soc.*, **74,** 653 (1952).
125. A. W. Weston, *J. Am. Chem. Soc.*, **69** 980 (1947).
126. N. A. Bazilevskya, K. K. Bragina, A. N. Grinev, and A. P. Terent'ev, *Vestik. Moskov. Univ. Ser. VI*, **15,** 33 (1960); *Chem. Abstr.*, **55,** 1998 (1961).
127. A. N. Grinev, N. K. Venevtseva, and A. P. Terent'ev, *Zh. Obshch. Khim.*, **28,** 1850 (1958); *Chem. Abstr.*, **53,** 1296 (1959).
128. C. A. Giza and S. M. Siegel, *Physiol. Plantarum*, **16,** 52 (1963); *Chem. Abstr.*, **60,** 8265 (1964).
129. P. S. Gates and D. T. Saggers, Ger. Pat. 1,926,139 (1969); *Chem. Abstr.*, **72,** 100487 (1970).
130. P. J. Brooker, J. Gillon, and G. T. Newbold, S. African Pat. 6,704,114 (1968); *Chem. Abstr.*, **71,** 91281 (1969).

131. G. A. Buntin, U.S. Pat. 2,749,272 (1956); *Chem. Abstr.*, **50,** 14173 (1956).
132. W. G. Scharpf, U.S. Pat. 3,564,605 (1964); *Chem. Abstr.*, **75,** 20175 (1971).
133. C. F. Gibbs, U.S. Pat. 2,342,135 (1944); *Chem. Abstr.*, **38,** 4834 (1944).
134. H. D. Porter and A. Weissberger, U.S. Pat. 2,350,127 (1944); *Chem. Abstr.*, **39,** 1364 (1945).
135. A. Weissberger and H. D. Porter, U.S. Pat. 2,439,352 (1948); *Chem. Abstr.*, **42,** 4082 (1948).
136. P. M. Mader, U.S. Pat. 2,865,747 (1958); *Chem. Abstr.*, **53,** 4990 (1959).
137. E. I. du Pont de Nemours and Co., Brit. Pat. 701,237 (1953); *Chem. Abstr.*, **48,** 8099 (1954).
138. G. E. Ficken and J. D. Kendall, Brit. Pat. 966,865 (1964); *Chem. Abstr.*, **62,** 7915 (1965).
139. R. H. Sprague, U.S. Pat. 2,706,193 (1955); *Chem. Abstr.*, **49,** 9420 (1955).
140. T. Th. Bakers, *J. Soc. Dyers Colourists*, **69,** 109 (1953); *Chem. Abstr.*, **47,** 6144 (1953).
141. Y. Kawase and M. Takashima, Japan Pat. 6,924,897 (1969); *Chem. Abstr.*, **72,** 43425 (1970).
142. Ciba, Ltd., Fr. Pat. 1,562,477 (1969); *Chem. Abstr.*, **73,** 110913 (1970).
143. M. L. Hoefle and R. W. Wynn, U.S. Pat. 2,674,604 (1954); *Chem. Abstr.*, **48,** 8553 (1954).
144. G. Kabas, H. Schlaepfer, Ger. Pat. 2,031,819 (1969); *Chem. Abstr.*, **75,** 22479 (1971).
145. R. Kirchmayer, Ger. Pat. 2,045,795 (1971); *Chem. Abstr.*, **75,** 22480 (1971).
146. H. Schlaepfer, and G. Kabas, Ger. Pat. 2,031,774 (1971); *Chem. Abstr.*, **74,** 113265 (1971).
147. M. Tibu, I. Viscrian, B. Arventiev, H. Offenberg, and T. Nicolaescu, *Anal. Stiint. Univ. "A. I. Cuza," Isai. Sect*, **19** (i), 261 (1963); *Chem. Abstr.*, **60,** 2433 (1964).
148. M. Tibu and B. Ariventiev, *Anal. Stiint. Univ. "A. I. Cuza," Isai. Sect.*, **18**(1), 261 (1962); *Chem. Abstr.*, **60,** 14091 (1964).

AUTHOR INDEX

This author index is designed to enable the reader to locate an author's name and work with the aid of the reference numbers appearing in the text. Page numbers are printed in normal type, followed by the reference numbers in parentheses.
If reference is made to the work of the same author in different chapters, the above arrangement is repeated separately for each chapter.

SUBJECT INDEX

Compounds which are referred to only in tables and which can be found by examining the tables mentioned under general headings covering the types of compounds are not mentioned individually in the index.